Intercalation Chemistry

This is a volume in the
Materials Science and Technology series.
Editors: *A. S. Nowick and G. G. Libowitz*

A complete list of the books in the series appears at the end of the volume.

Intercalation Chemistry

Edited by

M. Stanley Whittingham

Technology Department
Exxon Research and Engineering Company
Florham Park, New Jersey

Allan J. Jacobson

Corporate Research
Exxon Research and Engineering Company
Linden, New Jersey

1982

ACADEMIC PRESS

A Subsidiary of Harcourt Brace Jovanovich, Publishers

New York London

Paris San Diego San Francisco São Paulo Sydney Tokyo Toronto

ACADEMIC PRESS, INC.
111 Fifth Avenue, New York, New York 10003

United Kingdom Edition published by
ACADEMIC PRESS, INC. (LONDON) LTD.
24/28 Oval Road, London NW1 7DX

Library of Congress Cataloging in Publication Data
Main entry under title:

Intercalation chemistry.

Includes index.
1. Clathrate compounds. I. Whittingham, Michael Stanley, Date. II. Jacobson, Allan J.
QD474.I56 541.2'2 81-20576
ISBN 0-12-747380-7 AACR2

PRINTED IN THE UNITED STATES OF AMERICA

82 83 84 85 9 8 7 6 5 4 3 2 1

Contents

6 The Intercalation Chemistry of β-Alumina

B. C. TOFIELD

7 Organic and Organometallic Intercalation Compounds of the Transition Metal Dichalcogenides

ALLAN J. JACOBSON

8 Intercalation Chemistry of Metal Phosphorus Trichalcogenides

JACK W. JOHNSON

12 Intercalation Chemistry of Metal Chalcogenohalides

THOMAS R. HALBERT

13 Hydrogen-Containing Materials

K. H. J. BUSCHOW AND H. H. VAN MAL

14 Intercalation in Biological Systems

W. DAVID WILSON AND ROBERT L. JONES

15 Reactions in Crystallographic Shear Structures

J. S. ANDERSON

16 Oxide Insertion Compounds

P. G. DICKENS AND M. F. PYE

17 Lithium Intercalation Compounds of Vanadium Chalcogenides

D. W. MURPHY

18 Physical Properties of Intercalation Compounds of Transition Metal Dichalcogenides

A. H. THOMPSON AND F. J. DI SALVO

List of Contributors

Numbers in parentheses indicate the pages on which the authors' contributions begin.

G. Alberti (147), Department of Chemistry, University of Perugia, 06100 Perugia, Italy

J. S. Anderson[1] (503), University College of Wales, Aberystwyth, Dyfed, Wales

Neil Bartlett (19), Department of Chemistry, Materials and Molecular Research Division, Lawrence Berkeley Laboratory, University of California, Berkeley, California 94720

K. H. J. Buschow (405), Philips Research Laboratories, Eindhoven, The Netherlands

John D. Corbett (361), Ames Laboratory-DOE, Department of Chemistry, Iowa State University, Ames, Iowa 50011

U. Costantino (147), Department of Chemistry, University of Perugia, 06100 Perugia, Italy

E. G. Derouane (101), Laboratoire de Catalyse, Facultés Universitaires de Namur, B-5000-Namur, Belgium

F. J. Di Salvo (573), Bell Laboratories, Murray Hill, New Jersey 07974

P. G. Dickens (539), Inorganic Chemistry Laboratory, University of Oxford, Oxford, OX1 3QR, England

Thomas R. Halbert (375), Corporate Research, Exxon Research and Engineering Company, Linden, New Jersey 07036

Tjipke Hibma (285), Brown Boveri Research Center, CH-5405 Baden-Dättwil, Switzerland

Allan J. Jacobson (229), Corporate Research, Exxon Research and Engineering Company, Linden, New Jersey 07036

Jack W. Johnson (267), Corporate Research, Exxon Research and Engineering Company, Linden, New Jersey 07036

Robert L. Jones (445), Department of Chemistry, Laboratory for Microbial and Biochemical Sciences, Georgia State University, Atlanta, Georgia 30303

[1] Present address: Research School of Chemistry, Australian National University, Canberra, A.C.T. 2600, Australia

B. W. McQuillan[2] (19), Materials and Molecular Research Division, Lawrence Berkeley Laboratory, University of California, Berkeley, California 94720
D. W. Murphy (563), Bell Laboratories, Murray Hill, New Jersey 07974
M. F. Pye (539), Inorganic Chemistry Laboratory, University of Oxford, Oxford, OX1 3QR, England
R. Schöllhorn (315), Anorganisch-Chemisches Institut der Universität, 4400 Münster, West Germany
J. M. Thomas (55), Department of Physical Chemistry, University of Cambridge, Lensfield Road, Cambridge CB2 1EP, England
A. H. Thompson[3] (573), Corporate Research, Exxon Research and Engineering Company, Linden, New Jersey 07036
B. C. Tofield (181), Materials Development Division, A. E. R. E. Harwell, Didcot, Oxon, OX11 ORA, England
M. Stanley Whittingham (1), Exxon Research and Engineering Company, Technology Department, Florham Park, New Jersey 07932
W. David Wilson (445), Department of Chemistry, Laboratory for Microbial and Biochemical Sciences, Georgia State University, Atlanta, Georgia 30303
H. H. van Mal[4] (405), Philips Research Laboratories, Eindhoven, The Netherlands

[2] Present address: Instituto Technologico de Tijuana, Tijuana, BC Mexico.
[3] Present address: Exxon Production Research Company, Houston, Texas 77001
[4] Present address: University of Technology, Eindhoven, 560 MB Eindhoven, The Netherlands.

Preface

The study of reactions between guest molecules or ions and solid host lattices, which retain the major features of their structures, has continued since 1841 when Schauffautl first reported the intercalation of graphite by sulfate ions. However, it was not until the 1960s that interest in intercalation chemistry began to grow significantly and to extend into many scientific disciplines. Although the term intercalation has been used widely to describe reactions of guests with layered host lattices, many other reactions have characteristics in common. We decided to draw together in this volume a number of these related areas in which host lattices maintain some essential structural features during the reaction. The volume therefore aims to introduce the specialist reader to the breadth of intercalation chemistry and the newcomer to the diverse research opportunities and challenges available in synthetic and reaction chemistry and also in the controlled modification of physical properties.

After an introductory paper, the next chapter describes the intercalation chemistry of graphite. Graphite is perhaps the simplest host lattice structure but shows varied chemistry due to its ability to react with both oxidants and reductants and to incorporate neutral molecules. Its chemistry is further complicated by staging, which is more prevalent than in any other layered system. The second group of materials described are complex oxides with both two(clays and acid phosphates)- and three(zeolites)-dimensional structures. These systems are currently of considerable interest, because of technological applications in heterogeneous catalysis, as sorbents and inorganic ion exchangers. Their chemistry is dominated primarily by the intercalation of neutral molecules and by ion exchange rather than by the redox chemistry observed in both graphite and the layered chalcogenides. The β-aluminas (Chapter 6) have been mostly studied for applications as solid ion conductors in electrochemical cells, but much of their chemistry is analogous to that of the clays, though limited to small molecules and ions by the fixed interlayer separation imposed by the bridging oxygen.

Four chapters describe the intercalation chemistry of layered chalcogenides and halides with simple and hydrated cations and organic and organometallic ions. The reactions are mainly characterized by reduction of the host lattice; unlike graphite, no host oxidations accompanied by anion insertions have been observed, though attempts have been made to intercalate electron-accepting molecules, such as TCNQ, into the filled d-band group-VI dichalcogenides, for example, MoS_2. Once formed, these intercalation compounds show ion-exchange behavior that is similar to clays. In the redox systems, however, the ion-exchange capacity is set by the degree of host reduction, whereas in clay and zeolite chemistry, it is determined by cation substitution, e.g., Si for Al, in zeolites.

The next two chapters discuss areas that are not normally considered by the inorganic chemist. The first of these is concerned with the chemistry, thermodynamics, and applications of intermetallic compounds that incorporate hydrogen. The reader will observe many similarities and differences in behavior when compared with lithium intercalation in the dichalcogenides and oxides. Chapter 14 discusses intercalation in the context of biological systems and shows how the intercalation model was developed for the interaction of molecules with DNA. Crystallographic shear structures (Chapter 15) are not usually considered as related to intercalation compounds. However, there are important similarities, particularly in the way in which the structure imposes constraints on reactions. Reduction reactions, for example, leave the main part of the crystal lattice unperturbed and are accommodated by rearrangements of small numbers of atoms on specific sites. In the following two chapters, intercalation reactions of oxides and chalcogenides of vanadium, molybdenum, and tungsten are described. The final chapter touches on the physical properties of some intercalation compounds of the dichalcogenides.

In collecting these contributions, we have not attempted to cover all the aspects of intercalation chemistry, nor to describe, in any detail, their technological applications. Rather, we have concentrated on the chemistry and structural principles of a wide range of systems. It is our hope thereby to stimulate wider interactions among researchers in the various materials science disciplines.

M. Stanley Whittingham
Allan J. Jacobson

1

Intercalation Chemistry: An Introduction

M. STANLEY WHITTINGHAM

1. INTRODUCTION

The term *intercalation* literally refers to the act of inserting into the calendar some extra interval of time, such as February 29 in a leap year. In chemistry, it describes the reversible insertion of guest species into a lamellar host structure with maintenance of the structural features of the host. The term can equally be applied to one- and three-dimensional solids and to inorganic materials and biochemical phenomena such as the complexation of planar molecules by spirals of DNA.

The essential feature of the intercalation reaction, and that which makes its study so exciting and profitable, is that the guest and host experience some degree, along a spectrum from subtle to extreme, of perturbation in their geometric, chemical, electronic, and optical properties. There is considerable latitude available to the worker for controlling many of the parameters in order to tailor the behavior desired. Thus, hosts span the range of electronic conductivity from insulators (such as MoO_3, zeolites, and clays) through semiconductors (graphite and transition metal dichalcogenides) to metals ($LaNi_5$). This conductivity behavior can change markedly on intercalation depending on the degree of electron transfer between guest and host. Thus, WO_3, on being intercalated by hydrogen or the alkalis, goes from insulating

INTERCALATION CHEMISTRY

ISBN 0-12-747380-7

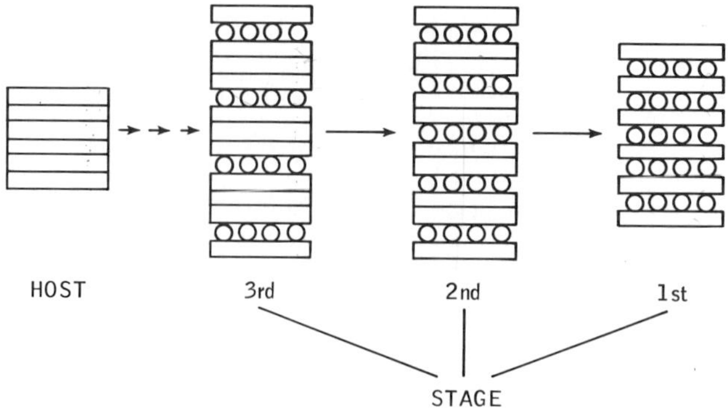

Fig. 1. Staging behavior during intercalation (*1*).

to semiconducting to metallic with concomitant changes in optical properties. The bonding between guest and host also varies broadly from essentially van der Waals dipolar (pseudophysisorption) to metallic and ionic. The guest species can often be easily exchanged for others; ion exchange is one pervasive phenomenon (and utility) for many of these compounds.

Both kinetics and thermodynamics play a critical role in the formation of these compounds. During reaction, a new phase must be nucleated followed by diffusion of the guest species into the host lattice, which normally expands. This expansion imposes an energy penalty, which can be minimized by the phenomenon of staging. This term refers to the situation in layer compounds in which, on a regular basis, certain interlayer regions are filled, whereas others are totally vacant, as shown in Fig. 1 (*1*). When odd-numbered stages are found, it appears that in going from one staging to another an entire layer of guest must exit and reintercalate. Although this phenomenon has been observed only for layered compounds, the formation of the shear structures (Magneli phases) on reduction of transition metal oxides is a related phenomenon, e.g., $WO_3 \rightarrow W_{20}O_{58}$, which is one of a series given by W_nO_{3n-2}.

II. TRANSITION METAL DICHALCOGENIDES

For some compounds the above picture is complicated by the existence of more than one "full"-layer configuration. Thus, as shown in Fig. 2, the long-chain amines form complexes with tantalum disulfide in which the amines form a bilayer (*2*) and then an intermeshed single layer (*4*) as the amine partial pressure is decreased. At still lower partial pressures, a second-stage compound is formed. The behavior of the bilayer is reminiscent of

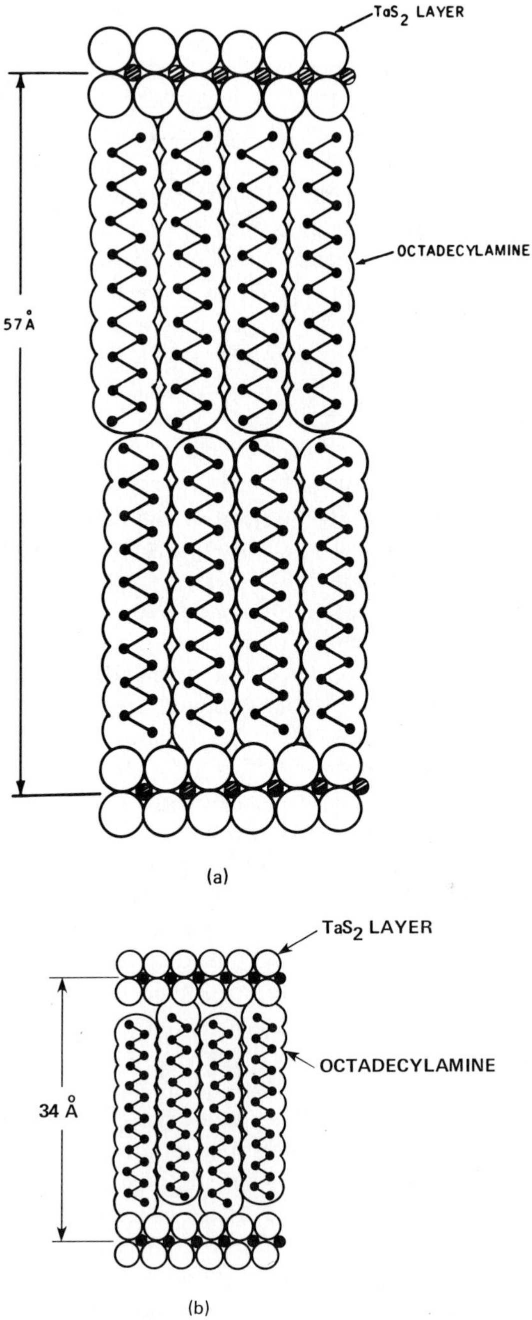

Fig. 2. Structure of *n*-alkylamines: (a) bilayer (*2*) [reproduced with permission from Gamble *et al.*, *Science* **174,** 493–497 (1971); copyright 1971 by the American Association for the Advancement of Science] and (b) interleaved amines [see ref. (*3*)].

soap–water solutions; these layers also show melting behavior just like that of the pure amines (*4*).

There is no conclusive evidence yet that any Lewis base intercalate exists in which there is partial occupancy in any occupied layer. This is probably dictated by the guest–guest van der Waals forces, which make a substantial contribution to the total intercalation energy. When the intercalation energy is much greater, such as in the ionic complexes of the dichalcogenides, e.g., $Na_x^+TiS_2^{x-}$, partial occupancy is quite usual. Thus, all values of x from zero to unity are found for Li_xTiS_2; that is, a single phase is found from TiS_2 to $LiTiS_2$ with smoothly varying structural and thermodynamic parameters (*5*). Lithium is, however, a special case because its small size is the same as that of the octahedral cavity in the van der Waals layer of the disulfides. Thus, here there is no need to expand the crystal lattice to insert the first lithium. In contrast, for sodium and other larger alkali ions, energy must be expended to generate the larger sites required. This energy requirement dictates a finite two-phase region at low x values: If the energy to open the layers is 4 kcal per mole alkali and the intercalation energy 40 kcal per mole alkali, then the two-phase region would extend up to at least $x = 0.1$. When staging occurs, the overall x value can be lower. For K_xTiS_2, three phases are found; a first stage, $0.28 < x \leq 1$; a second stage, $0.14 < x < 0.16$; and a fourth stage, $0.06 < x < 0.08$. Although the small lithium ion must reside in the octahedral sites, the larger cations can occupy trigonal prismatic sites in which the sulfur planes have shifted to lie directly above one another. Potassium takes up such a configuration. Sodium is intermediate in size and is found in both sites—trigonal prismatic, $\sim 0.38 < x < \sim 0.75$ and octahedral, $\sim 0.79 < x \leq 1$, for Na_xTiS_2.

Another aspect of these alkali intercalates is their ability to incorporate polar molecules, which solvate the alkali metals just as in liquid solutions. Thus, TaS_2 gives two compounds exemplified by $K_{1/3}(H_2O)_{2/3}TaS_2$ for the larger cations and $Li_{1/3}(H_2O)_2TaS_2$ for the smaller cations on treatment with the aqueous hydroxide; the alkali ions are as mobile as in a liquid solution (*5*). Many organic compounds, such as ethers, amines, and dimethyl sulfoxide, can take the place of the water and in many cases can be readily removed on evacuation and/or heating. Ion exchange is readily accomplished in these solvated complexes by immersion in aqueous salt solutions. The simple unsolvated complexes can also be ion-exchanged without incorporation of polar solvents by appropriate choice of ion-exchange media; thus, for example, Na_xTiS_2 exchanges sodium for lithium when immersed in solutions of lithium salts in dioxolane (*6*). Figure 3 (*5*) shows schematically the possible reactions for both the solvates and the alkali intercalates themselves.

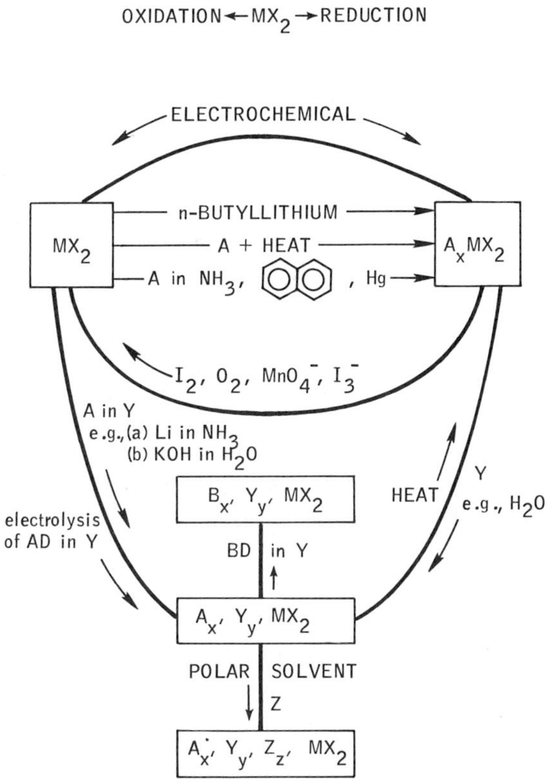

Fig. 3. Schematic of the reactions of dichalcogenides. [Reproduced with permission from Whittingham, *Prog. Solid State Chem.* **12,** 41 (1978). Copyright 1978, Pergamon Press, Ltd.]

Protons can also be readily intercalated into 2H—TaS_2 by electrolysis of aqueous acids (*5*). A considerable amount of water is also taken up between the sulfide layers so that they can be separated by up to 50 Å. The degree of incorporation of water is determined by the water vapor pressure; thus, on placing $H_x(H_2O)_yTaS_2$ on an X-ray slide in air, the disappearance and appearance of a series of new phases is readily observed until all the water is lost, leaving H_xTaS_2. Addition of water causes reexpansion of the lattice. The TaS_2 can even be dispersed in solution by adding a surfactant to the highly swelled product.

Some rather large organometallic molecules can be intercalated into the dichalcogenides. The sandwich metallocenes (*7*), such as cobaltocene, intercalate by ionization to give a double-sandwich compound, $[Co(C_5H_5)_2]_{1/4}TaS_2$, with the orientation shown in Fig. 4. The X-ray

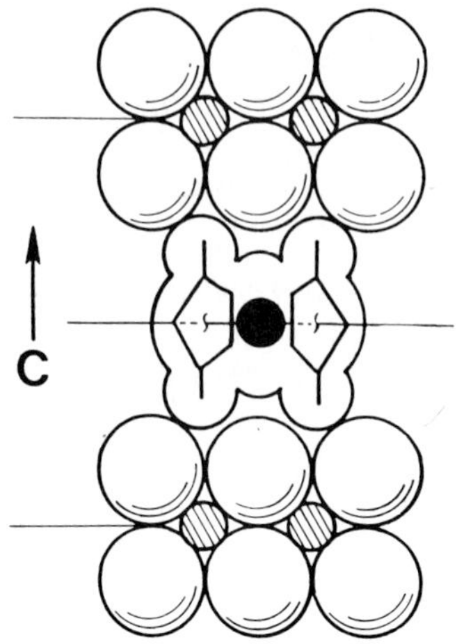

Fig. 4. Double-sandwich compounds between metallocene and tantalum disulfide. [Reproduced with permission from Dines, *Science* **188,** 1210–1211 (1975). Copyright 1975 by the American Association for the Advancement of Science.]

powder spectra (*8*) are remarkably sharp, indicating that the sulfide layers are locked together strongly by the metallocene ions; this is surprising considering the size of the ions, ~6 Å in diameter. At room temperature these ions diffuse readily in the van der Waals gap, but at very low temperatures motion is restricted to rotation of the cyclopentadienyl rings. These rotate even at liquid-helium temperatures, in contrast to the pure guest molecule. A range of other metallocenes, with different metals and ring sizes, can also be intercalated, provided that they not only have a sufficiently low ionization potential, below about 6.5 eV, but also form stable cations. For other hosts the ionization potential limit can be different; thus, ferrocene intercalates FeOCl but not TaS_2; these complexes are discussed in detail in Chapters 12 and 7.

Mixed-guest intercalates can be readily formed. Thus, $(C_5H_5Co)_{0.25}TaS_2$ reacts with 1.25 lithium both chemically and electrochemically; presumably the sulfide reacts with 1 lithium, the cobaltocinium ion being oxidized and also reacting with 1 lithium (*33*). An ordered mixed intercalate can be synthesized by reacting a second-stage compound with a guest molecule; $(NH_3)_{0.5}(\text{pyridine})_{0.25}TaS_2$ was prepared in this way (*34*). Many other atomic and molecular species can be intercalated, including morpholine (*33*) and methyl isocynanide (*4*) but not the cyanide; these gave 3.4 and 3.7 Å expansions, respectively, in TaS_2. The only totally organic species is the tropylium ion (*33*), which expanded the lattice just a little over 3 Å, so that its molecular plane must be parallel to the sulfide layers. Surprisingly, no electron acceptors have been found to intercalate. One might have expected some of the Group VIb or VIIb compounds, such as MoS_2 or ReS_2, to intercalate electron acceptors, but there are no reported examples.

III. GRAPHITE

In contrast to the dichalcogenides, graphite (see Chapter 2) can be intercalated by both electron donors and acceptors. Graphite also differs from the dichalcogenides in a number of other ways. The alkali metal compounds do not exhibit wide ranges of nonstoichiometry but rather have definite compositions, e.g., C_8K, $C_{10}K$, and $C_{12n}K$, and these compounds have formation energies more typical of metallic alloys than of the saltlike $Li^+TiS_2^-$. An increase in electronic conductivity on intercalation is consistent with electron donation from the alkali metal to the graphite conduction band. Similarly, the strong colors, from brasslike for the first stage through copper red, steel blue, blue-green, and bluish black to the black of graphite itself may also be associated with the change of electron concentration. These alkali intercalates can take up hydrogen, presumably as the hydride; the golden first-stage KC_8 is converted to a blue second-stage compound of composition $KH_{0.67}C_8$ (*9*). This ready reaction with hydrogen may be related to the report that ammonia is formed on passing nitrogen and hydrogen over the potassium intercalate (*10*); whether the potassium acts catalytically or as a chemical reagent has not been conclusively determined. The alkali-intercalated dichalcogenides do not react with hydrogen, presumably because the alkali metal activity is too low. The two hosts are similar in their ability to form solvated complexes, as discussed earlier in relation to the sulfides.

Graphite forms compounds with electron acceptors such as bromine and, in so doing, loses electrons from the valence band, and the conductivity becomes *p* type. With bromine the limiting composition C_8Br corresponds to a second-stage compound with the bromine atoms arranged in chains (*11*). The bond length of the bromine molecule, 2.27 Å, is close to that of hexagonal carbon groups (2.56 Å) in the sheets, and this may explain why chlorine and iodine do not intercalate but iodine monochloride does (2.30 Å). This also suggests that the molecules do not dissociate on intercalation. Many metal fluorides, such as TiF_4, SbF_5, UF_6, and XeF_6, are also readily taken up into the graphite structure. It is becoming apparent (*12*) that this is not true intercalation, since permanent oxidation of the graphite lattice by the fluorine can take place, giving some C—F bonds as in CF.

IV. TRANSITION METAL OXIDES

A number of transition metal oxides can also be intercalated by alkali metals and other electron donors. The best known of these are the metal tungsten bronzes, M_xWO_3. These are nonstoichiometric, highly colored compounds that exhibit a number of crystalline structures. In the case of sodium, where $0 \le x \le 1$, the highest members are bronze-colored; hence, their common name. The actual colors, listed below, are a direct result of the donation of electrons to the WO_3 matrix by the sodium atoms, as discussed above for graphite. These colors are independent of the electron donor, as expected for a simple Drude model. This coloration behavior is being used in displays in which hydrogen is intercalated electrochemically into the lattice, giving H_xWO_3 with a concomitant sharp color change. On removal of the hydrogen, the WO_3 electrode reverts back to its essentially colorless state (*13*). These changes also readily occur by addition of zinc and aqueous HCl to WO_3.

WO_3 —— 0.2 —— 0.4 —— 0.6 —— 0.8 —— $NaWO_3$

Green, Dark gray, Dark blue, Royal blue, Purple, Brick red, Orange, Yellow

The WO_3 itself is monoclinic and becomes more symmetric as the electron concentration increases, attaining cubic symmetry for x above about 0.4. Tungsten oxide is an interesting example of an intercalation host because it can be obtained in a number of crystalline modifications. Thus, the frequently opposing forces of thermodynamics and kinetics can be studied. Both the empty monoclinic and hexagonal tunnel structures (Fig. 5) can react with

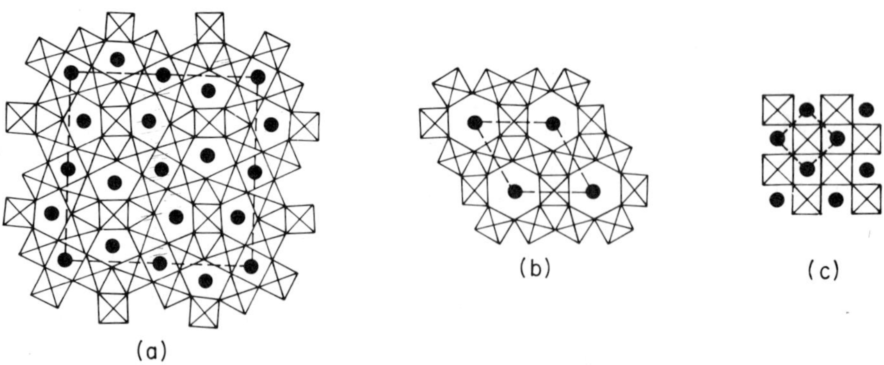

Fig. 5. The (a) tetragonal, (b) hexagonal, and (c) cubic structures of the tungsten bronzes (*32*). ⊠, WO_6 octahedra; ●, metal ions.

two lithiums per tungsten, whereas the partially occupied $Na_{0.6}WO_3$ does not react at all (*14*). Sodium and potassium also intercalate the empty hexagonal structure (*15, 16*). However, if the tunnels are partially filled by larger cations such as potassium, then the lithium ions can intercalate only the trigonal sites, giving a compound of formula $Li_{0.67}K_{0.33}WO_3$. In contrast, $(NH_4)_xWO_3$ is intercalated by two lithiums, presumably because the ammonium ion can be readily reduced. Just as for the ammonium compound, the tunnel compound $W_{18}O_{49}$ also reduces to a formal oxidation state of W^{4+} and must have multiple occupancy of the large tunnel sites.

The reduction of tungsten trioxide by hydrogen occurs by the removal of oxygen atoms along planes at regular intervals throughout the structure, as shown in Fig. 6 (*17*) for the formation of $W_{20}O_{58}$. This is a deintercalation reaction that occurs without change in the remainder of the structure and, as for most intercalation reactions, without any change in the measured surface area, 0.88–0.89 m^2/gm, after reduction (*18*). In contrast, when the structure is destroyed, such as in the reduction of $W_{20}O_{58}$ to β-tungsten, "W_3O," a marked increase in area to 8 m^2/gm is noticed (*18*). These shear structures are discussed in detail in Chapter 15. The tungsten bronzes themselves (*19*) are reduced by the abstraction of oxygen from the surface of the crystallites with tungsten formation there, and the sodium migrates into the center of the particle, leading to a core of ever-increasing sodium content. When x in the core reaches unity, further oxygen abstraction occurs, with the formation of sodium tungstate Na_2WO_4. Whereas with alkali-free tungsten oxide, β-tungsten is the final reduction product below 600°C, when an alkali metal is present, α-tungsten (the pure metal) is formed. This is important for the formation of tungsten for light filaments. Presumably it is the mobility of the sodium that "catalyzes" the formation of the α-tungsten; it also suppresses the increase in surface area. This mobility of the metal atoms in the bronzes

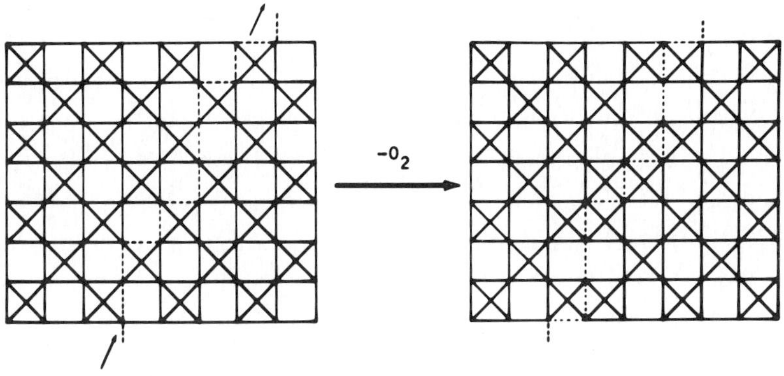

Fig. 6. Shear plane formation in perovskite-type lattice, e.g., in WO_3. ⊠, WO_6 octahedra.

has also allowed them to be used as electrodes that are reversible to both ions and electrons for the measurement of ionic conductivity of materials such as β-aluminas (*20*). Other intercalation compounds can be similarly used.

The tungsten bronzes are the only intercalation compounds that have been systematically studied as heterogeneous catalysts, even though the major hydrotreating catalysts used today in petroleum refining are cobalt- or nickel-promoted molybdenum/tungsten disulfides. The reaction studied for the bronzes was the recombination of oxygen atoms to give oxygen molecules at ambient temperature and 30 torr pressure. The choice of such a simple reaction rules out possible side reactions, so that changes in the measured activity can be associated solely with the catalyst and not with the reaction. The activity, given as the recombination coefficient γ, is plotted as a function of x in Fig. 7 (*21*). The initial drop in activity from $x = 0$ to $x = 0.25$ can be associated with an increase in the n-type nature of the solid. That is, as

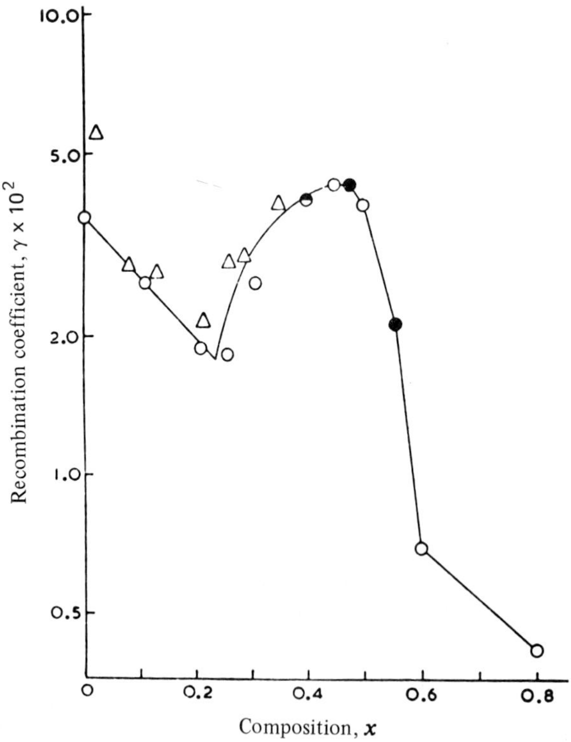

Fig. 7. Recombination coefficient of oxygen atoms on tungsten bronzes, M_xWO_3. Key: ○, Na; △, Li; ●, K (*21*).

the concentration of electrons increases, so does their energy as the low-lying states are being filled. However, the rate-determining step of the reaction is believed to be the reaction between an absorbed O^- ion with a gas-phase, or physisorbed, O atom. This step involves the donation of an electron back to the lattice, and hence the higher the lowest vacant levels the greater is the necessary energy expenditure. Thus, the activity drops for x from 0 to 0.25, but at this concentration Na_xWO_3 becomes a metallic conductor so that the electron changes occurring on reaction can be delocalized throughout the solid; this is believed to be the cause of the sharp increase in activity. The subsequent drop can again be associated with the generally increasing electron concentration of the bulk solid. Thus, for the bronzes, electronic factors appear to be critical in determining catalytic activity. In support of this hypothesis, Li_xWO_3 and K_xWO_3 have corresponding activities even though they have different structures than Na_xWO_3.

V. ALUMINUM OXIDE HOSTS

Aluminum oxide host materials, which are typified by many alumina silicate minerals and the β-aluminas, differ from the preceding hosts in that they generally do not contain reducible or oxidizable cations, so that intercalation can involve only neutral species unless ion exchange takes place.

In the simple silicates such as talc, $Mg_3(OH)_2(Si_4O_{14})$, the structure is built up by corner sharing of three corners of the SiO_4 tetrahedra, giving infinite sandwiches of two sheets of tetrahedra with the bases on the outside of the sandwich, as shown in Fig. 8b (*22*). The sandwiches are only weakly bound together by van der Waals forces, so that ready cleavage occurs along the basal planes. This property leads to the use of talc as a lubricant (french chalk). In the related mineral muscovite, $K[Al_2(OH)_2(Si_3AlO_{10})]^-$, the potassium ion compensates for the introduction of the Al into the Si sheet matrix; the other two aluminums take the position of the magnesium as sheet binders (Fig. 8c). The potassium now holds the sandwiches together and can be readily exchanged for other cations such as sodium (paragonite) and calcium (margenite). The SiO_4 tetrahedra can also share corners to give three-dimensional structures. Among these are the very interesting zeolites, which have been extensively investigated because of their catalytic and separations properties; they are discussed in detail in Chapter 4.

The micas themselves are anhydrous and are not intercalated by other polar molecules; but if the charge per unit area in the van der Waals gap is lowered, then structures are found that contain a considerable amount of

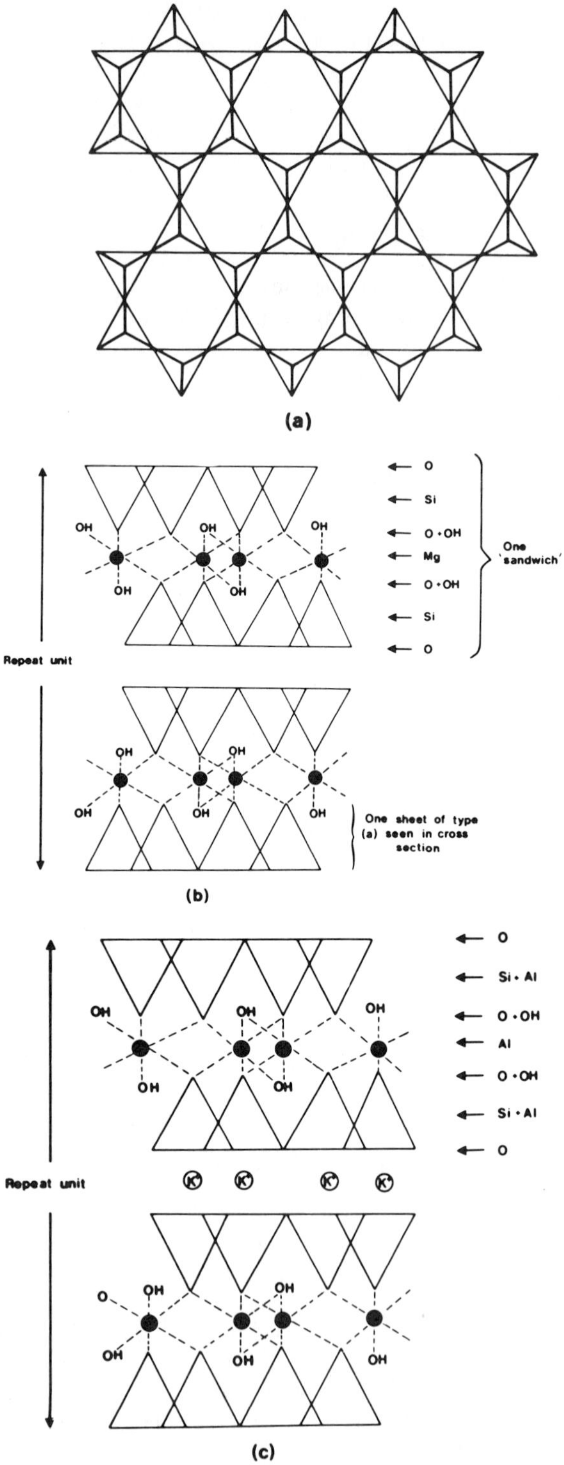

Fig. 8. Comparison of the structures of (b) talc and (c) muscovite. Part a shows arrangement of tetrahedra in basal plane (*22*).

water in this layer. An example is montmorillonite, in which the van der Waals charge density has been reduced to one-third of that of muscovite by replacing some of the Al by Mg:

$$Na_{1/3}(H_2O)_{3.5}[Mg_{1/3}Al_{5/3}(OH)_2(Si_4O_{10})]^{0.33-}$$

The actual cations present determine the degree of hydration. Thus, whereas sodium bentonite (montmorillonite) swells readily in water and is used as a thickener in oil-well drilling muds, the calcium form does not, due to the much higher electrostatic forces. Generally, these clays readily exchange ions in the van der Waals layer and can intercalate many other polar molecules into the structure. In this respect they resemble such other inorganic oxides as dititanates (*23*) and zirconium phosphate (*24–26*). As an example, when primary amines are intercalated into montmorillonite, smectite, or zirconium phosphate, they give expanded products containing ammonium ions quaternized by protons of the solid. When neutral organic compounds are intercalated, a series of compounds is frequently found. For example, the strontium montmorillonite gives three compounds with γ-butyrolactone having interlamellar spacings of 23.1, 18.3, and 13.2 Å; the corresponding compositions are 2.4, 1.6, and 0.8 γ-butyrolactones per Si_4O_{10} unit (*27*). Their stabilities are presumably related to the partial pressure of the γ-butyrolactone just as for the previously discussed case of amines in TaS_2.

The clays discussed above have cations in the van der Waals layer. Other layer compounds, such as the β-aluminas, have alkali cations and Al—O—Al bridging groups that hold the sandwiches together. These compounds have the nominal formula $MAl_{11}O_{17}$, where M is a monovalent cation. In practice, there is an excess of about 15% of the cations, which are charge-compensated by aluminum vacancies and excess oxygen. The packing in the intersandwich layer is very loose, there being sufficient room for three times the nominal M concentration. This leads to a high conductivity and ready exchangeability of the normal resident sodium ions for almost any other singly charged cation, including Ga^+ and NO^+ (*28*). Unlike most of the other layered compounds discussed in this chapter, the lattice cannot readily change dimensions to accommodate cations of different size due to the Al—O—Al bridges. Thus, the mobility of cations larger than sodium is seriously impeded as expected, but the smaller lithium ion also has a much lower conductivity, as shown in Fig. 9 (*29*). This is thought to be related to the lithium ions residing in off-center sites, which cause a deepening of the potential energy wells and a reduction in the binding between the γ-Al_2O_3 blocks (sandwiches). The latter causes an increase in the lattice dimension perpendicular to the sandwiches, and the former an increase in the activation energy of diffusion. The sodium form is under extensive development as the electrolyte for the high-temperature (350°C) sodium–sulfur battery.

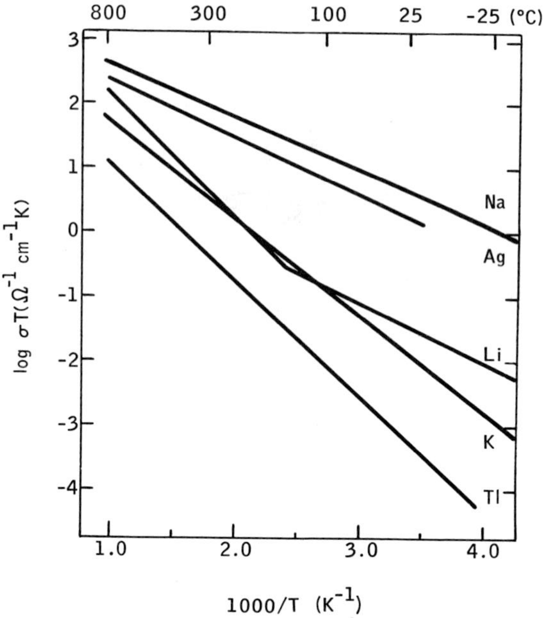

Fig. 9. Ionic conductivities of substituted β-aluminas (*39*).

VI. ENERGY OF INTERCALATION

The thermodynamics of intercalation reactions have been extensively investigated because of the interest in this class of materials as components of energy storage devices. For example, the following reactions are of commercial interest:

$$\begin{aligned} &Li + Al \rightarrow LiAl && \Delta G = -0.3\ \text{mV (battery anode)} \\ &Li + TiS_2 \rightarrow LiTiS_2 && \Delta G \approx -2\ \text{V (battery cathode)} \\ &3H_2 + LaNi_5 \rightarrow LaNi_5H_6 && \Delta G = 0.1\ \text{V (hydrogen storage)} \end{aligned}$$

As is clear, the energy of intercalation varies by an order of magnitude from the saltlike chalcogenides to the metal alloys. Some of these data are shown in Fig. 10. The very different compositional and energy behavior of alkali intercalation into titanium disulfide and graphite has already been discussed. There is even a wide variation in the compositional variation of the energy of formation of the lithium intercalates of the dichalcogenides, as indicated by Fig. 11. However, the total free energies of intercalation are remarkably close, all falling within 10% of 200 kJ/mole (*5*).

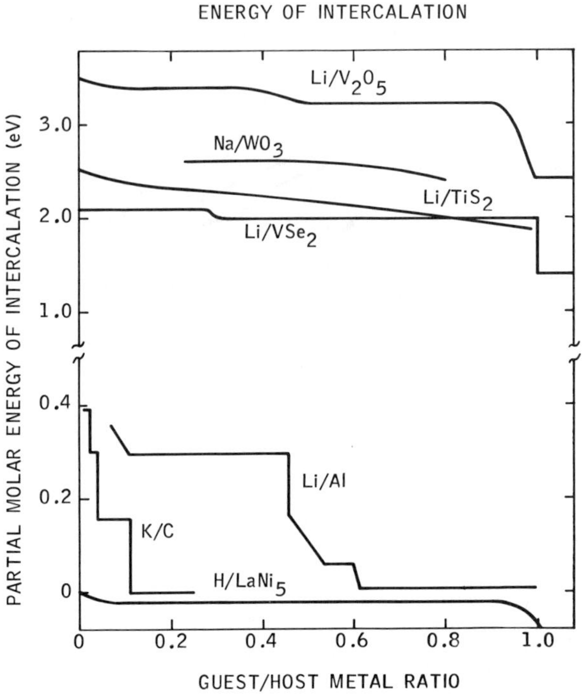

Fig. 10. Energy of intercalation reactions. All data at 25°C except graphite (275°C) and aluminum (423°C).

Titanium disulfide has been extensively studied (*5*, *30*) as the cathode for lithium batteries because it has many positive attributes.

1. A high free energy of reaction
2. A small free-energy change during reaction
3. Little structural change on reaction
4. High reversibility of the chemical reaction
5. A good electronic conductivity
6. High diffusivity of lithium ions in the structure
7. No solubility in common electrolytes

The high free energy of reaction leads to a theoretical energy density (based on weight of electrodes and no current load) of 480, and 455 W hr/kg at a load of 10 mA/cm^2. The intercalation and deintercalation of lithium has been accomplished more than 600 times, as shown in Fig. 12 (*31*), indicating that true intercalation reactions can form the basis of secondary batteries. Such

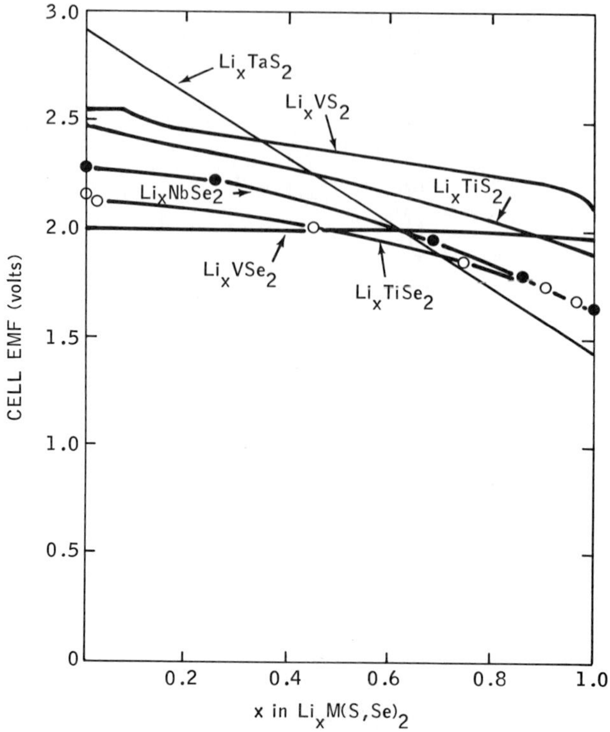

Fig. 11. Energies of lithium intercalation into the dichalcogenides. [Reproduced with permission from Whittingham, *Prog. Solid State Chem.* **12,** 41 (1978). Copyright 1978, Pergamon Press, Ltd.]

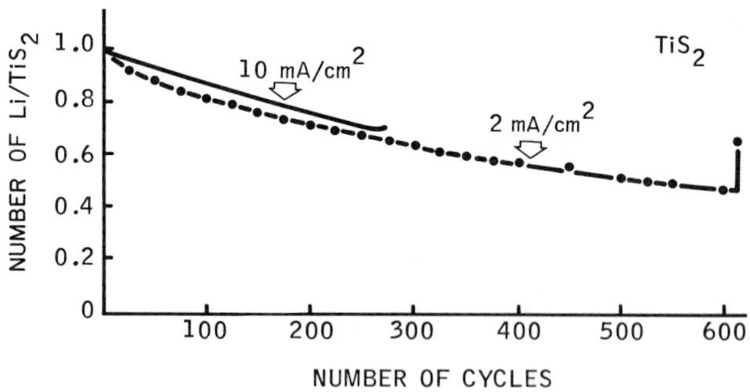

Fig. 12. Extended cycling of Li/TiS_2 cells at 2 and 10 mA/cm^2 discharge rate. Electrolyte is 2.5 *M* $LiClO_4$ in dioxolane. [Reproduced with permission from Whittingham and Panella, *Mater. Res. Bull.* **16,** 37 (1981). Copyright 1981, Pergamon Press, Ltd.]

TiS_2 cells have been manufactured as small button cells for watches. However, before large lithium batteries can be made commercially available, safe high-rate electrolytes have to be found. Other electrochemical cells involving intercalation reactions are the electrochromic displays in which tungsten trioxide is colored by the intercalation of hydrogen or lithium. Such cells have been switched several thousand times and may replace liquid-crystal displays for some applications.

VII. SUMMARY

This survey has touched on a few areas of intercalation chemistry in order to serve as a brief introduction to the chapters that follow. It is not meant to be comprehensive, and no attempt has been made to introduce all the contributions. This area of science is still in its infancy, and its interdisciplinary nature combined with a unique opportunity for molecular engineering is likely to create an even larger interest in the future.

REFERENCES

1. M. S. Whittingham and M. B. Dines, *Surv. Prog. Chem.* **9,** 55 (1980).
2. F. R. Gamble, J. H. Osiecki, M. Cais, R. Pisharody, F. J. DiSalvo, and T. H. Geballe, *Science* **174,** 493 (1971).
3. M. S. Whittingham and R. R. Chianelli, *J. Chem. Educ.* **57,** 569 (1980).
4. M. B. Dines, *Inorg. Chem.* **17,** 763 (1978).
5. M. S. Whittingham, *Prog. Solid State Chem.* **12,** 41 (1978).
6. A. J. Jacobson, unpublished observations (1978).
7. M. B. Dines, *Science* **188,** 1210 (1975).
8. R. P. Clement, W. B. Davies, K. A. Ford, M. L. H. Green, and A. J. Jacobson, *Inorg. Chem.* **17,** 2754 (1978).
9. P. Lagrange, A. Metrot, and A. Herold, *C. R. Hebd. Seances Acad. Sci. Ser. C* **278,** 701 (1974).
10. M. Sudo, M. Ichikawa, M. Soma, T. Orisha, and K. Tamaru, *J. Phys. Chem.* **73,** 1174 (1969).
11. W. T. Ecles and J. A. Turnbull, *Proc. R. Soc. London, Ser. A* **283,** 179 (1965).
12. L. B. Ebert and H. Selig, *Mater. Sci. Eng.* **31,** 177 (1977).
13. S. K. Deb, *Appl. Opt., Suppl.* p. 192 (1969).
14. K. H. Cheng and M. S. Whittingham, *Solid State Ionics* **1,** 151 (1980).
15. K. H. Cheng, A. J. Jacobson, and M. S. Whittingham, *Am. Chem. Soc. Meet. INOR 151, Las Vegas, Nev., 1980.*
16. M. Figlarz and B. Gerand, *Proc. Int. Symp. React. Solids, 9th, Cracow* p. 660 (1980).
17. P. Gado, *Acta Phys. Hung.* **18,** 111 (1965).
18. M. S. Whittingham and P. G. Dickens, *React. Solids, Proc. Int. Symp., 7th, Bristol, Engl.* p. 640 (1972).
19. M. S. Whittingham, Ph.D. Thesis, Oxford Univ., 1967.
20. M. S. Whittingham and R. A. Huggins, *J. Chem. Phys.* **54,** 414 (1971).

21. P. G. Dickens and M. S. Whittingham, *Trans. Faraday Soc.* **61,** 1226 (1965).
22. J. M. Adams, "Inorganic Solids." Wiley, New York, 1974.
23. A. Weiss and A. Weiss, *Angew Chem.* **72,** 413 (1960).
24. S. Yamanaka, *Inorg. Chem.* **15,** 2811 (1976).
25. D. Behrendt, K. Beneke, and G. Lagaly, *Angew. Chem., Int. Ed. Engl.* **15,** 544 (1976).
26. S. Alluli, C. Ferragina, A. LaGinestra, M. A. Massucei, and N. Tomassini, *J. C. S. Dalton* p. 2115 (1976).
27. J. M. Thomas, J. M. Adams, S. H. Graham, and D. T. B. Tennakoon, *Adv. Chem. Ser. No.* 163, p. 298 (1977).
28. J. T. Kummer, *Prog. Solid State Chem.* **7,** 141 (1972).
29. M. S. Whittingham and R. A. Huggins, *React. Solids, Proc. Int. Symp., 7th, Bristol, Engl.* p. 125 (1972).
30. M. S. Whittingham, *Science* **192,** 1126 (1976).
31. M. S. Whittingham and J. A. Panella, *Mater. Res. Bull.* **16,** 37 (1981).
32. P. G. Dickens and M. S. Whittingham, *Q. Rev. Chem. Soc.* **22,** 30 (1968).
33. M. S. Whittingham and M. B. Dines, unpublished observations (1975).
34. M. B. Dines, *J.C.S. Chem. Commun.* p. 220 (1978).

2
Graphite Chemistry

NEIL BARTLETT

and

B. W. McQUILLAN

I. INTRODUCTION

The first intercalated compound, graphite sulfate, was described by Schauffaütl in 1841 (*1*). Intercalation of graphite involves the penetration of guest species between the carbon layers with consequent expansion of the

INTERCALATION CHEMISTRY

ISBN 0-12-747380-7

graphite in the *c*-axis direction. Intercalation usually proceeds in stages (Fig. 1). The region between adjacent carbon layers, which may be occupied by guest species, constitutes a gallery. The known intercalating reagents are numerous. They range from strong oxidants to powerful reductants (*2*). The metallic electrical conductivity of many of the intercalated graphites, as well as the possibility of their use in storing and transporting electricity, has increased interest in them. These compounds also pose fundamental questions, which advances in synthetic techniques, physical measurements, and theory have made timely. There have been several recent comprehensive reviews devoted to graphite intercalation compounds, of which those by Hérold (*3*), Ebert (*4*), and Selig and Ebert (*5*) deal thoroughly with structural and chemical aspects of the compounds. A review by Fischer (*6*) is written from the perspective of a physicist. There have also been two recent conferences (*7*, *8a*,*b*) devoted to intercalation compounds of graphite, the conference proceedings of which include much of the current work in the field. In this chapter we do not provide an exhaustive survey. Instead, we have selected certain systems that have been studied extensively. Our aim is to indicate principles that may be of value for future research directions in this field.

Many of the known intercalants of graphite can be represented as either oxidants or reductants of graphite. Two of the sections in this chapter therefore deal with species that can be placed in either of these categories.

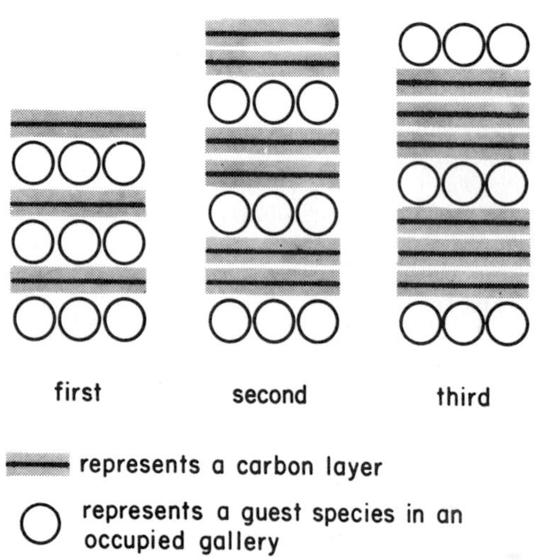

Fig. 1. Staging in graphite intercalation compounds.

At the outset, however, it is appropriate to consider the features of graphite itself that are pertinent to intercalation and then some general features of graphite intercalation compounds.

II. GRAPHITE

A. Nature of the Material

The degree of perfection of the graphite used in the preparation of intercalation compounds and even the size of the graphite crystallites have an important influence on synthesis. Hooley (*9a*, *9b*) documented this influence and demonstrated that graphite specimens of a given quality that are thin in the *c*-axis direction undergo intercalation with a lower threshold activity of intercalant than do thicker specimens. Also, for oxidizing intercalation, those carbons that are highly imperfect (as a consequence of much cross-linking of the layers) intercalate less readily than those that are more perfect. On the other hand, for reducing intercalants, such as the alkali metals, experimental evidence appears to indicate that interlayer cross-linking aids intercalation. This has been attributed to lowering of the Fermi level by the cross-linking (*3*).

It must be remembered that, when intercalation occurs, some of the defects of the parent carbon may carry over into the intercalation compounds (*3*). Such defects can have a gross impact on physical properties such as electron transport. Ubbelohde discussed such effects on charge scattering (*10a*, *10b*) and the ordering of intercalants. In carbons in which there are many linkages between carbon layers (such linkages are referred to by Ubbelohde as "rogue bonds"), intercalation of bulky species can proceed only if such bonds are broken. (Indeed, Ubbelohde suggested that reversible intercalation could provide for increasing the perfection of graphite by removing rogue bonds in this way.) Carbons in which much rogue bonding occurs are usually obtained by the pyrolysis of solid organic compounds and include glassy carbons and polybinylidene cokes. However, because these are often highly porous, physical absorption can readily occur in them, and this may occasionally be confused with intercalation. The pyrolytic graphites made by the pyrolysis of liquid organic compounds (*11*) can also be rich in defects, unless they have been well annealed at ~3000°C. The highest-quality graphites, and the most satisfactory for intercalation studies, are obtained by the pyrolysis of small-molecule hydrocarbons at ~3000°C (*12*). High-temperature annealing again improves the degrees of crystallite perfection.

B. Bonding in Graphite

A fragment of the infinite two-dimensional network of sp^2 hybridized carbon atoms of a perfect graphite is shown in Fig. 2a. The bonding of the carbon atoms of the planar σ-bonded layer (*ab* plane) is enhanced by the π bonding derived from the combination of the p_z orbitals provided by each carbon atom. Since each pair of π electrons in the graphite carbon layer contributes to the bonding of three carbon–carbon linkages, whereas in benzene such an electron pair enhances the bonding of only two carbon–carbon linkages, the bonding in the graphite layers is not as strong as that in benzene. The in-plane bonding in a layer of graphite is, however, stronger than the carbon–carbon bonding in diamond. Indeed, the carbon layers are physically and chemically robust and appear to remain intact even when reduced or oxidized. The addition of electrons to the conduction band (antibonding π orbitals) of the removal of electrons from the valence band (π orbitals) decreases the bonding in the layers. Small changes in the dimensions of the *ab* plane, therefore, do occur in association with intercalation (*3*). In all of the high-quality graphites (Fig. 2b), the effective thickness of a carbon layer is always close to 3.35 Å. Since the layers are merely van der Waals bonded, this distance represents the effective van der Waals thickness of such sheets. Electron oxidation or reduction of the carbon layer does not appear to have much effect on this thickness.

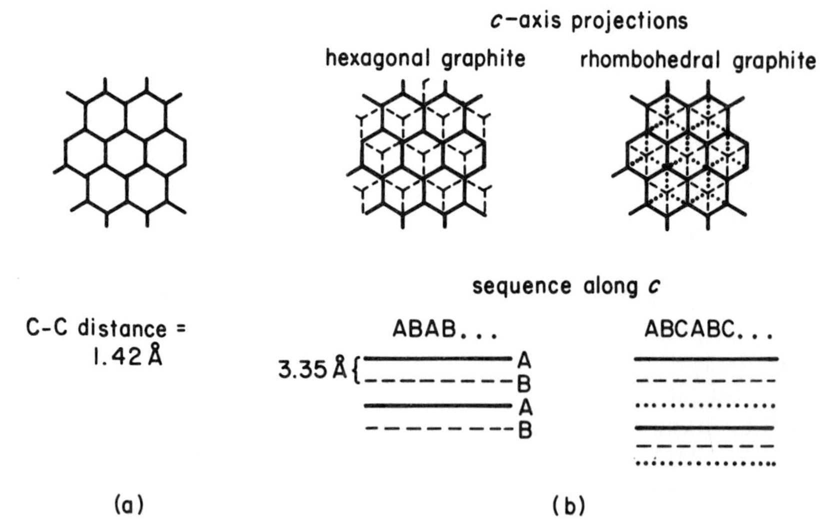

Fig. 2. (a) Fragment of a graphite layer; (b) layer–layer registry in different graphites.

C. Band Structure of Graphite

Since, in most instances, the intercalation of graphite does not lead to gross changes in the carbon layers, it is useful, at least as a first step, to use the band structure of graphite as a basis for understanding the electronic properties of the intercalated graphites. The first detailed discussion of the bands generated by the combination of the p_z orbitals of the carbon layers was that of Wallace (*13*) in 1947. There have been a number of significant refinements since then which are important in accounting for optical, magnetic, and electrical properties (*14a–15*). As far as intercalation chemistry is concerned, at least at the outset, it is sufficient to treat the band structure of an isolated carbon layer, i.e., to deal simply with the two-dimensional network, for which the unit cell and Bravais lattice are shown in Fig. 3a. It must, however, be remembered that the three-dimensional pristine graphite situation is more complex than the one we are about to discuss. A

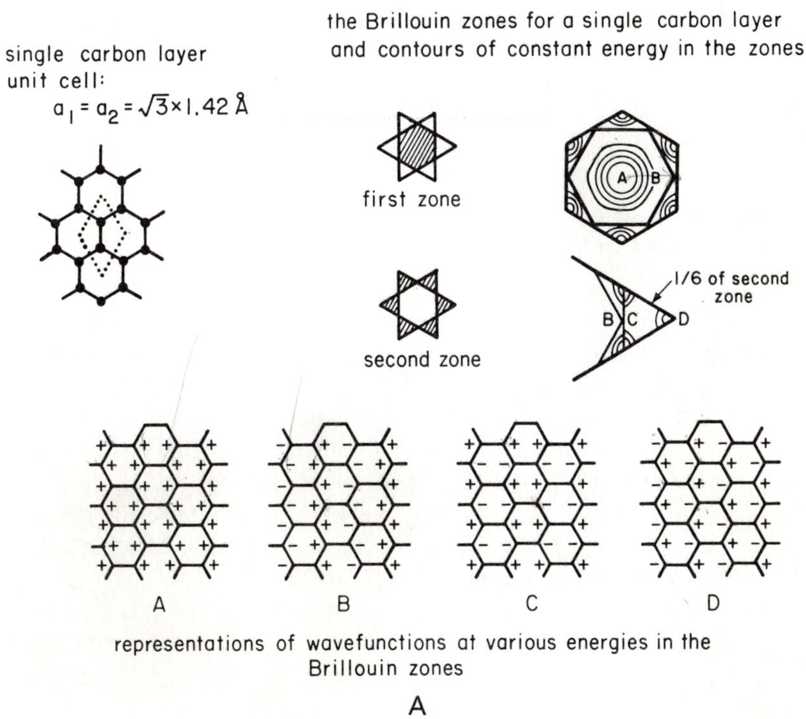

Fig. 3A. The unit cell and first and second Brillouin zones for a single graphite layer. The first Brillouin zone is a hexagon whose sides are distant, $1/(3a)^{1/2}$, from its center. Only at each point of this hexagon is there a zero energy gap with the second Brillouin zone.

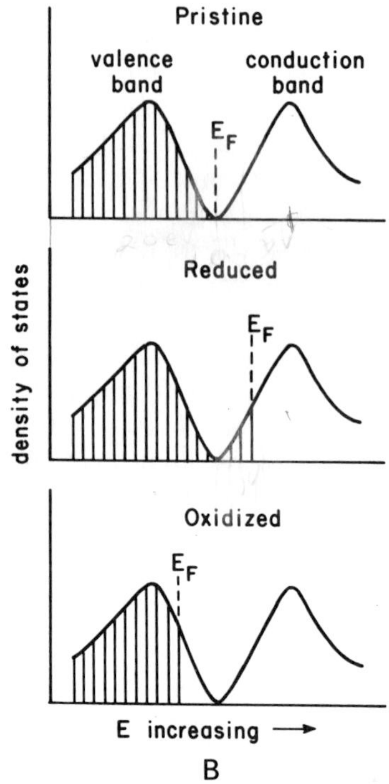

Fig. 3B. Density of states diagrams (for 0 K) of a single graphite carbon layer, in pristine, reduced, and oxidized condition.

major difference is that, in the three-dimensional case, the layer–layer interactions, although merely van der Waals, do result in overlap of the lowest band (the valence band) and the band next in energy (the conduction band). It is this overlap* of approximately 0.04 eV that produces the semi-metal character of graphite. The overlap means that the conduction band contains electrons and the valence band has holes, each with a density of $\sim 10^{-4}$ carrier per carbon atom. Both carriers contribute to the electrical conductivity. The effect of oxidation or reduction on the band occupancy of the isolated carbon layer is shown in Fig. 3B.

For the single layer, as shown in Fig. 3A, the Brillouin zone is a hexagon. The combination of the p_z atomic wavefunctions gives the two π bands (the valence band and the conduction band), which are degenerate only at

* The overlap of the valence and conduction bands for three-dimensional graphite is represented in a three-dimensional Brillouin zone picture by a complex Fermi surface for holes and electrons that has a very small extent in the *xy* plane, its hole or electron character alternating in the k_z direction. See ref. *15*.

the six Brillouin zone corners (*13*). There is no overlap of these bands, nor is there a band gap. At 0 K the valence band is filled and the conduction band is empty; therefore, the Fermi level is at the point of contact of the bands (see Fig. 3B). The width of each of the π bands (*15*) is about 20 eV, and since in the three-dimensional graphite case the energy of interaction of the layers is only about 0.5 eV, the overall character of the π bands is not changed much in the formation of the three-dimensional structure, although the previously mentioned slight overlap, 0.04 eV, is, of course, of profound importance.

The two-dimensional model for the band structure may be a satisfactory approximation for *first-stage* graphite compounds when the intercalants are large closed-shell species, which provide for no effective overlap of orbitals with the p_z orbitals of the carbon layers. It may therefore be valid for the graphite salts in which stable anions such as NO_3^-, ClO_4^-, HSO_4^-, and AsF_6^- are intercalated between layers which bear net positive charge. Certainly the measure of the anisotropy of the electrical conductivity given by σ_{ab}/σ_c indicates (*16*) that such salts are more two-dimensional in conductivity than is pristine graphite itself, where experimental values for σ_{ab}/σ_c range from 10^2 to 10^5 depending on the quality of the graphite.

In the cases of graphite compounds in which the intercalants possess vacant or partially filled orbitals, their overlap with the p_z orbitals of the carbon layers results in chemical bonding. This must raise questions concerning the validity of the two-dimensional band structure as an approximation for such systems. It is therefore unlikely that such a simple model can be applied to the graphite compounds of the alkali and alkaline earth metals. The enhancement of *c*-axis conductivity in these compounds, relative to pristine graphite (they approach three-dimensional metallic behavior), points to such an overlap of valence orbitals of the intercalant and the p_z orbitals of carbon.

III. SOME GENERAL ASPECTS OF GRAPHITE INTERCALATION COMPOUNDS

A. Structural Features

The characterization of the guest species in graphite intercalation compounds is a major experimental challenge. The metallic nature of most of the intercalation compounds means that structural studies based on longer-wavelength electromagnetic radiation are limited, because it ordinarily does not penetrate beyond the surface layers. Thus, infrared spectroscopy usually samples the surface, whereas with the shorter wavelength available in Raman spectroscopy, depths of up to ~1000 Å of the sample can be excited (*1ϵ*). Similar difficulties may arise in NMR spectroscopy, although the information from the surface region is often valuable (*18*). Highly penetrating, hard X rays can measure bulk structural features, and, in the neighborhood of an absorption edge, diagnostically valuable chemical shifts (*20*, *21*) and extended X-ray absorption fine structure (EXAFS) data can be obtained.

It is often difficult to obtain well-ordered single crystals of graphite intercalation compounds; therefore, high-quality structures are rare. Structural information from (00*l*) data is relatively common, however, since structural ordering with respect to the *c* axis is easy to maintain if the graphite used in the preparation was a single crystal or a specimen of highly oriented pyrolytic graphite.

Many of the graphite intercalation compounds have well-ordered structures at the microdomain level, but highly disordered structures can also arise. In stages higher than the first, adjacent carbon layers have half the carbons eclipsed as in graphite itself (Fig. 2b). Their relative disposition can be conveniently represented as *AB* (or *ABC*). Those graphite layers that contain an occupied gallery are commonly eclipsed. If an intercalated layer is represented by a vertical stroke, a shorthand representation for a first-stage compound, with the carbon layers eclipsing in the *c*-axis direction, is $A|A|A$ . . . . A second-stage compound might be $AB|BA|AB$, etc., and a third stage $ABA|ABA|ABA$ . . . .

B. Energetics of Intercalation

Since the energy of binding of carbon layer to carbon layer in graphite (*21*) is approximately 1.5 kcal/mole, the separation of the layers on intercalation involves work. The formation of a compound also commonly involves a decrease in entropy, which can be sizable for gaseous species becoming guests in a highly ordered arrangement. The interaction between guest and host must therefore meet these enthalpy and entropy burdens if the compound is to be thermodynamically stable. In many compounds, this energy comes from the coulomb interaction consequent upon the guest species either oxidizing or reducing the host (*22*). Graphite is relatively easy to oxidize or reduce. For the graphite compounds derived from the heaviest alkali metals (Section IV,A) and for those containing hexafluoro anions (Section V,C), thermodynamic factors have been quite well defined. Thermodynamic factors appear not to be the only obstacle to synthesis, however (*9a*, *9b*).

C. Mechanisms of Intercalation

Studies related to the mechanism of intercalation have been carried out on very few systems, that of bromine intercalation being the best documented. Hooley (*9a*, *9b*) was the first to demonstrate that Br_2 intercalation is determined by basal plane adsorption—no intercalation of bromine occurs if the basal planes are covered with wax. He also showed that the regions nearest the basal planes open first, the stage obtained depending on the partial pressure of the intercalant. The first stage is produced at the highest activity.

A low-energy electron diffraction study (*23*) indicates that a monolayer coverage of the (00*l*) surface with bromine precedes gallery opening. An X-ray absorption study (*24*) indicates that molecular bromine lies flat on the surface at pressures corresponding to near monolayer coverage. The bromines lie 2.53 Å above the carbon plane with a Br–Br distance of 2.31 Å. Polarized EXAFS measurements (*25*) on dilute intercalated samples show two different Br–Br distances at 2.34 and 2.53 Å. The latter distance is consistent with the Br_3^- ion.

Young postulated (*26*) an electronic effect to explain the action of adsorption of Br_2 on the (00*l*) plane in opening a gallery as far away as 700 Å from that surface. Filling of that gallery causes others to open both closer to and farther from the basal plane. Hence, gallery opening progresses from the basal plane.

The domain model first proposed (*27*) by Daumas and Hérold, which is represented in Fig. 4a, has received experimental support from direct imaging studies of the graphite/$FeCl_3$ system (*28*). The model provides for an understanding of how the changes in stage (e.g., third to second) can

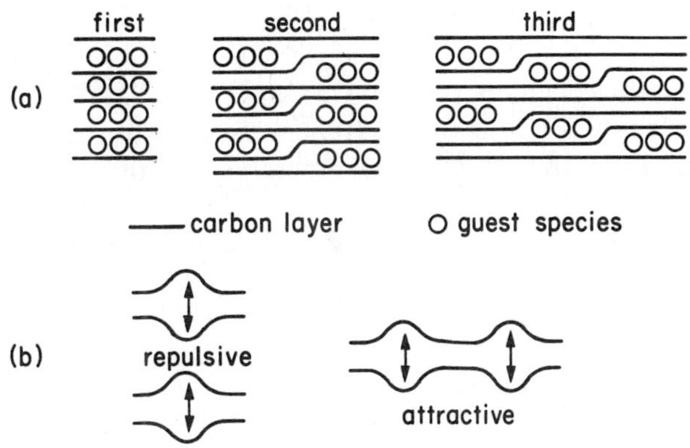

Fig. 4. The domain model for graphite staging. (a) Staging; (b) interaction of elastic dipoles.

occur with such facility. With this model, and from a consideration of the effects of elastic coherency strains, Safran *et al.* (*29*) have provided an explanation for staging. In their analysis, each guest is represented by a couple of forces with zero moment, which expand the gallery from the original height of 3.35 Å. These elastic dipoles are illustrated in Fig. 4b. When two such dipoles are situated as in the left-hand side of Fig. 4b they repel one another, whereas those in the position on the right-hand side attract. Thus, guest islands are formed, which in their stacking along the c axis show the typical staging periodicity (Fig. 4a). These strains drive a mixed-stage or randomly staged crystal to pure-stage ordering.

It seems quite probable that all intercalation into graphite involves adsorption of guest molecules, initially on the basal plane $(00l)$, accompanied by electron transfer between the guest and graphite, with subsequent steps being much the same as described for bromine.

D. Electronic Properties

Most studies of electronic properties of graphite intercalation compounds have been made to further our understanding of their highly anisotropic metallic conductivity. In spite of its oversimplifications, the Drude Model† provides for the relationship between the specific conductivity σ and the number of carriers n. The relationship is

$$\sigma \frac{\text{length}}{\Omega \times \text{cross-sectional area}} = \frac{ne^2\tau}{m^*}$$

Here, τ is an averaged scattering or relaxation time, and m^* is the effective mass of the predominant charge carrier, defined by

$$\frac{1}{m^*} = \frac{1}{\hbar^2}\frac{\partial^2 E}{\partial k^2}$$

where the E versus k contour of the band structure is known. Thus, one has the variables n, τ, m^*, and the cross-sectional area of the current flow to measure in order to interpret the experimental conductivities of these intercalates.

Intercalation affects both m^* and n. Knowledge of the number and charge of the intercalated species gives the number of electrons withdrawn from or added to the graphite. From the knowledge of the structural unit cell, the

† The theory and experimental details can be found in standard solid-state physics texts. Three good texts are: C. Kittel, "Introduction to Solid State Physics," 5th Ed., Wiley, New York, 1979; N. W. Ashcroft and N. D. Mermin, "Solid State Physics," Holt, New York, 1976; M. H. B. Stiddard, "The Elementary Language of Solid State Physics," Academic Press, New York, 1975.

Brillouin zone can be derived. With such information band theory calculations, can, in principle, yield n and m^*. The dependence of the electron energy E on the vector k is illustrated for the two-dimensional Brillouin zones of graphite in Fig. 3A. As we have seen, the effective mass of the charge carrier m^* is defined in terms of those quantities. The only unknown in this simple Drude model is τ, which can only be treated as an empirical factor (the value of which may be expected to change upon intercalation). Thus, in virtually all cases, conductivity measurements have not been interpretable in terms of microscopic paramcters.

Comparisons of the published data of several groups on various compounds leads to different values of these parameters, even for nominally equivalent material. Many measurements have been made using different qualities of graphite, which makes comparisons of σ difficult due to the unknown impact of defects on τ. Most investigators minimize, but cannot eliminate, this effect by measuring each piece before and after intercalation to obtain the relative change in σ. A further difficulty occurs when different techniques for measuring σ are used. Most early workers used four-probe bridge measurements *(30)*. Zeller *et al.* *(31)* showed that a contactless radio frequency (rf) method eliminates the consequences of attack by the reactants on the electrodes and the effects of nonuniform current flow. This rf method seems to be the preferred means for measuring the large in-plane conductivities of many oxidized graphites. The contactless measurements also indicate that the earlier σ_{ab} values from four-probe techniques may have been undervalued (32).

The conventional σ values for intercalation compounds involve, of course, the cross-sectional area across which the carrier flux passes. For σ_{ab}, this involves the thickness of the sample along the c axis. However, this cross-sectional area grows as the c axis expands and the unit area has fewer carbon layers. Most investigators renormalize their σ_{ab} for various stages so as to always maintain a unit cross-sectional area. Thus, their units are $\Omega^{-1}\ cm^{-1}$. However, if the carriers are located primarily in each graphite layer (as appears to be the case for intercalation compounds of oxidized graphite), it is useful to compare the σ_{ab} normalized to a constant number of carbon layers. This can be obtained conveniently by multiplying the σ_{ab} value by the relative change in thickness t/t_0 as measured along the c axis.

IV. INTERCALATION WITH GRAPHITE REDUCTION

A. Binary Systems

The compounds of graphite intercalated by the alkali metals are among the most studied of graphite materials. By comparison, the alkaline earth, lanthanide, and transition metal compounds have been little investigated. To our knowledge, only one review has been devoted solely to the alkali metal intercalants (*33*), although they have been thoroughly discussed in other major reviews (*34*).

The synthetic methods used in the preparation of the alkali metal intercalates distinguish two classes. The heavier metals K, Rb, and Cs react readily with graphite at 200°C or less when a standard two-bulb technique is used. Lithium reacts only at higher temperatures or pressures, and lithium carbide is a common impurity in the product. The sodium compound is difficult to make directly and is relatively poorly characterized. The distinction between the two groups (K, Rb, Cs and Li, Na) has generated speculation* that the ionization potentials of the gaseous atoms provide the basis for distinction between the two classes. The ionization potentials (in electronvolts) of K (4.34), Rb (4.18), and Cs (3.89) lie below the electron affinity of graphite, which has been estimated (*38*) to be 4.6 eV, whereas those of Li (5.39) and Na (5.14) lie above the affinity of graphite.

The conditions of synthesis of the alkaline earth (*36*) (Ca, Ba, Sr) and lanthanide (Eu, Yb, Sm, Tm) (*37a*) intercalates parallel the conditions for lithium intercalation. Temperatures of 500°–600°C for several weeks are required to reach first stage. The first ionization energies of all these metals lie above the electron affinity of graphite.

The distinction between the two metal classes M (K, Rb, Cs) and M′ (Li, Na, Ca, Ba, Sr, Eu, Yb, Tm, Sm) in synthetic conditions also carries over to their composition and staging. The first-stage C_8M and higher-stage $C_{12n}M$ compositions contrast with the first-stage C_6M' (except Na). The second-stage lithium compound appears (*3*) to be of variable composition: $C_{12}Li$ to $C_{18}Li$. No pure higher-stage materials C_xM' have been obtained.

Table I gives the structural parameters for these compounds. The experimental values for the unit-cell *c* dimensions yield carbon layer–guest–carbon layer distances I_c that are compatible with cationic guests. Indeed, the effective guest diameter obtained by subtracting a carbon layer thickness of 3.35 Å from I_c is often less than the diameter of the guest as a cation. This may simply be a consequence of the guest species nestling in the centers of hexagons of the carbon layers. The heavy alkali metals do not form a simple one-gallery-height *c* spacing in the C_8M composition. The unit cell within a gallery is shown in Fig. 5. This unit cell allows for any of four sites to be occupied. In low-temperature forms of C_8K and C_8Rb, all four sites are occupied sequentially in consecutive layers, such that the *c* spacing is four times the interlayer distance. This *c*-axis ordering is denoted by $A\alpha A\beta A\gamma A\delta$,

* Hérold compares the sum of the sublimation energy and the first ionization energy (*3*). Such a sum has more thermodynamic significance than the first ionization potentials alone.

TABLE I

Gallery Height Data for Graphite Compounds of the More Electropositive Metals

Compound	Stage	Color	Occupied gallery height I_c (Å)	$I_c - 3.35$[a] (Å)	Metal ion[b] diameter (Å)
C_8K	1	Gold	5.41	2.06	3.02
$C_{24}K$	2	Blue	8.77	2.07	3.02
$C_{36}K$	3	Blue	12.12	2.07	3.02
$C_{48}K$	4	Blue	15.49	2.09	3.02
C_8Rb	1	Gold	5.65	2.30	3.32
$C_{24}Rb$	2	Blue	8.97	2.27	3.32
C_8Cs	1	Gold	5.94	2.59	3.62
$C_{24}Cs$	2	Blue	9.25	2.55	3.62
C_6Li	1	Brass yellow	3.73	0.38	1.80
C_6Ca	1	Pale yellow	4.55	1.20	2.28
C_6Sr	1	Yellow	4.95	1.60	2.64
C_6Ba	1	Yellow	5.25	1.90	3.00
C_6Eu	1	Yellow	4.85	1.50	Eu^{2+} 2.61 Eu^{3+} 2.17
C_6Sm	1	Yellow	4.72	1.37	Sm^{2+} 2.72 Sm^{3+} 2.20
C_6Tm	1	Yellow	4.62	1.27	Tm^{2+} 2.34 Tm^{3+} 2.04
C_6Yb	1	Gold	4.57	1.22	Yb^{2+} 2.36 Yb^{3+} 2.02

[a] The metal species nestle in the dimples of the carbon layers; hence, $I_c - 3.35$ Å should be less than the species diameter.

[b] Shannon (*37b*).

where the $A \ldots A \ldots A$ represents eclipsing graphite layers. In C_8Cs, however, it is claimed that only three of these sites are occupied, the layering sequence being $A\alpha A\beta A\gamma$. In all C_8M compounds, the *c*-axis ordering changes as a function of temperature, and at high temperatures ordering from one gallery to the next is lost. In the C_6Li compound, the unit cell provides three equivalent sites (Fig. 5). Apparently, the Li occupies one site throughout the structure, and the *c* dimension equals the interlayer spacing. Parry and Nixon (*38*) examined higher-stage $C_{12n}M$ compounds and found that graphite layers eclipse one another across a metal gallery, whereas they are staggered, as in graphite, when the carbon layers are adjacent.

The alkaline earth and lanthanide intercalates are just beginning to be characterized by X-ray diffraction studies. Some structural information on the C_6M compounds (M = Ca, Sr, Ba, Sm, Eu, and Yb) has been obtained (*36*, *37a*). It seems that the metal-layer sequences may be $A\alpha A\alpha A\alpha$

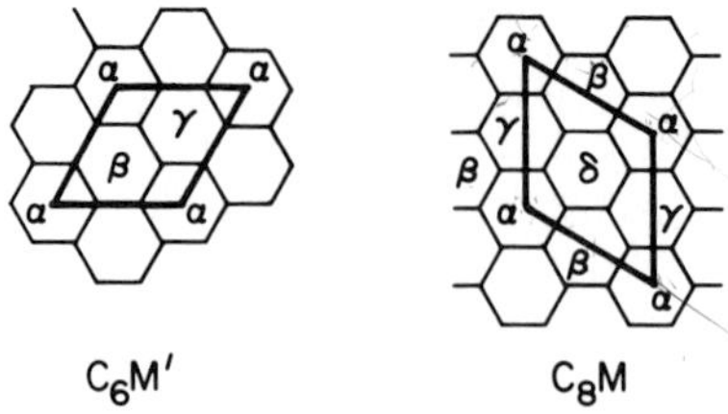

Fig. 5. Unit cells for ordered C_6M' and C_8M intercalation compounds. With guest species in sites α in one gallery, those in succeeding galleries may or may not be in sites β, γ, or δ.

(as for Ca) or $AA\alpha A\beta A\alpha A\beta$ . . . (as for Sr, Ba). It can be seen from Table I that the Eu appears to have an effective size closer to that of Sr than Ca, whereas Yb is like Ca. These two lanthanides have the most stable M^{2+} oxidation states among the lanthanides. Magnetic susceptibility measurements on these compounds have not been made to confirm that the metals are M^{2+}. In all these cases, however, it is doubtful that the metal has been completely oxidized by the graphite. Metal–metal bonding may well occur between the guests in these compounds.

B. Ternary Systems

Knowledge of ternary systems has come chiefly from the work of Hérold (*39*) and co-workers. Intercalation of mixed metals from molten alloys includes Ba–Na and pairwise alloys of Na, K, Rb, and Cs. In the sodium alloys, the presence of Na within the gallery is only temporary. Upon prolonged contact with the alloy, the Na is eventually replaced by other metal (*40*). This behavior parallels the apparent instability of C_xNa compounds. In the phase diagrams for the binary alloy intercalates, the interplanar spacing remains similar to that for the compounds of the pure components near the metal-rich extremes (*41*). Over a particular composition range, the interplanar spacing changes by several tenths of an angstrom, and in this same range the magnetic susceptibility shows an anomalous spike. It may be that some ordering or clustering of the different alkali guest species is involved in this behavior. Structural studies that focus on the question of guest arrangement should clarify the situation.

In the graphite Na–Ba system, at a composition of C_8Na_2Ba, the gallery height and composition imply (*42*) that a single gallery contains a layer of Ba atoms sandwiched between layers of Na atoms, this triple layer being bounded by the graphite layers. Analogous compounds involving Hg sandwiched by K or Rb (first-stage C_4KHg and second-stage C_8KHg) are also known (*43*). Hérold (*3*) has assigned formal charges $C^-K^+Hg^-K^+C^-$

to the metal layers (although metal–metal bonding would eliminate the presence of isolated Hg^-). Perhaps the Hg atoms sandwiched between the alkali cations are somewhat positive. Positively charged mercury would be smaller than neutral or negatively charged mercury.

Another class of ternary compounds includes the graphite–alkali–hydrogen compounds. A first-stage C_8K, when treated with H_2, forms a second-stage material of composition $C_8KH_{0.66}$. In the latter, a gallery contains a K–H–K layer sequence (*44*). The action of additional alkali metals on this product produces a new "first stage" in which the added metals enter the empty galleries. Within these galleries, the newly added metal forms the usual C_8M type of gallery. Lithium removes some of the hydrogen as LiH, without intercalating, whereas Na does not react. Higher-stage $C_{12n}K$ can also be "hydrogenated" to form compounds with K–H–K layers. For these sandwiched hydrogen compounds, it seems likely that the hydrogen is hydridic.

A third class of reduced ternary intercalation compounds is derived from solutions (NH_3, tetrahydrofuran, aromatic compounds) of metals (alkali, alkaline earth, lanthanide, aluminum) (*39*). In these compounds, the solvent molecules as well as the strongly reducing metals are intercalated. The inter-layer spacings appear to be determined primarily by the dimensions of the intercalated solvent molecule. An example is the compound in which both benzene and sodium are intercalated together (*45*). Two models have been proposed. In one, the benzene ring lies adjacent to a graphite layer and parallel with it. The sodium ions are sandwiched between the benzene and the other graphite layer. In the second model, the plane of the ring of the benzene molecule is set perpendicular to the graphite layers. The latter arrangement is a satisfactory one if, as seems reasonable, the benzene rings participate in bearing the negative charge. Thus, the negatively charged benzene rings will screen the sodium ions from one another. Some tilting of the rings from the perpendicular would still provide for this, however, and could give better packing.

The recent discoveries of the mercury and hydrogen ternary compounds with alkali graphites are surprising and pose important questions concerning their formulation and bonding. It seems likely that many more such novel combinations with graphite will be found.

C. Electronic Properties of Reduced Graphite Intercalates

The electronic properties of the alkali metal intercalates have received much attention. The sizable diamagnetic susceptibility of the graphite is lost, and a small Pauli paramagnetism arises (*46a, 46b*). Electrical conductivity increases 10-fold in the *ab* plane and approximately 200-fold along the

c axis at the first stage (*24, 47*). The a-axis conductivity levels off (*48*) in the region of the fourth to eighth stage and apparently decreases slightly for the heavy alkali metals. However, correcting the conductivity for the expansion in the c-axis direction, which accompanies intercalation, shows that the conductivity *per plane* is nearly constant (Table II). The c-axis conductivity increases monotonically up to saturation.

From Raman studies of the carbon-layer vibrations, Solin (*14*) has provided persuasive evidence that the charge transferred to the carbon, from the heavy alkali metal guests, is confined largely to the layers containing the metal atom layer.

One of the most significant measurements has proved to be that of the Hall coefficient of these materials (*49*). In the reduced graphite, for example, $C_{36}K$, the R_H is negative in sign, indicating that electrons are the carriers. The negative sign also is clear electronic evidence that the graphite is reduced. There is some evidence that R_H for C_8K changes sign at very low temperatures (below 77 K) (*50*). This evidence gives some support to the speculation that C_8K may actually be a semimetal with both electron and hole carriers. However, further work is needed to confirm this proposal.

TABLE II

Basal Plane Conductivities for Some Alkali Metal–Graphite Compounds (*48*)

Compound stage (n)	Alkali	t/t_0[a]	σ_0 ($10^4\ \Omega^{-1}\ cm^{-1}$)	$\sigma_0 t/t_0$ ($10^4\ \Omega^{-1}\ cm^{-1}$)[b]
1	C_8K	1.62	11	18
			9	15
2	$C_{24}K$	1.31	12	15
3	$C_{36}K$	1.21	17	20
1	C_8Rb	1.69	8	14
2	$C_{24}Rb$	1.34	—	—
3	$C_{36}Rb$	1.23	15	18
4	$C_{48}Rb$	1.17	19	22
1	C_8Cs	1.77	11	20
			8	15
2	$C_{24}Cs$	1.39	12	16
3	$C_{36}Cs$	1.26	—	—
4	$C_{48}Cs$	1.19	14	17

[a] t/t_0 represents ideal c-axis expansion from the formula

$$t/t_0 = [3.35(n - 1) + I_c]/3.35n.$$

[b] σ_0 normalized to account for thickness t_0 having the same number of carbon layers as thickness t.

In order to measure the density of states in first-stage compounds, magnetic susceptibility (*51–53*) and heat capacity experiments (*52, 53*) have been performed. The value for density of states derived from heat capacity measurements differs by a factor of 2 from that obtained from magnetic susceptibility data. The heat capacity data appear to be in better agreement with band calculations (*54*). Electron–phonon coupling has been invoked to explain the departures from theoretical expectations (*52*). A strong coupling of the electrons with the lattice clearly occurs below 1 K, when these first-stage compounds become superconducting.

In summary, the electronic properties of C_xM compounds are consistent with reduction of the graphite. In the first-stage material, there appears to be some anomalous electronic and spectroscopic behavior which merits further investigation.

V. INTERCALATION WITH GRAPHITE OXIDATION

A. Graphite Salts from Strong Mineral Acids

Preparative and structural work by Rudorff and his group (*55, 56*) laid the basis for the recent studies of graphite salts containing strong-acid anions. However, the impetus for current activity in the field sprang from the discovery of the high *ab*-plane conductivities of these salts (*57, 58*), which came from the work of Ubbelohde and co-workers, following their successful production of large pieces of highly oriented graphite bearing relatively few defects.

The salts can be prepared by chemical oxidation (e.g., by nitric acid), but anodic electrochemical oxidation has also been used. The latter has made it possible to apply accurate coulometry (*57, 58*) to the synthesis of the salts containing HSO_4^-. Since such oxidations in mineral acids also result in the neutral mineral acid molecules being intercalated, coulometry enables one to determine the fraction of the guests that are charged. From the mineral acid preparations, the typical formulations are $C_{24}^+X^-2HX$ ($X = ClO_4^-$ or HSO_4^-) and, for nitric acid, $C_{24}^+NO_3^-3HNO_3$. There appears to be no reason why any strong-acid anion (except, perhaps, very bulky ones) should not be intercalated into graphite. If the neutral acid is present, it will also be intercalated.

Data from some mineral acid salts are given in Table III. They show that the occupied gallery height I_c in the salts is much the same for perchlorate, bisulfate, and some of the nitrates. Indeed, the value of ~7.9 Å and the stoichiometries are compatible with a rather close-packed double layer of

TABLE III

Intercalated Gallery Spacings for Some Strong-Acid Salts

First-stage formula[a]	I_c (Å)	Reference
$C_8H_pSO_4$	7.98	*56*
C_7SO_3F	7.73	*62*
$C_7H_tSO_3F$	8.01	*62*
$C_8H_tClO_4$	7.85	*56*
$C_{4.3}H_xNO_3$	7.8	*56*

[a] $p \simeq 1 + 2/3$; $t \simeq 2/3$; $x \simeq 5/6$.

oxygen ligands. For the bisulfate and perchlorate, this implies a close packing of the "tetrahedra" according to one or more of the models shown in Fig. 6a. It is also clear that, if planar nitrate groups are present in the galleries, they cannot be parallel to the carbon layers in those salts for which $I_c \approx 7.8$ Å, unless double-stacked as in Fig. 6b(iii). Touzain (*59*) used (00*l*) powder data to support the model represented in Fig. 6b(ii) He concluded that the stoichiometry should be $C_{4.3n}H_xNO_3$ ($x = 5/6$) in these materials. The newly described (*60*) nitrates having $I_c = 6.55$ Å. obtained from the $I_c = 7.8$ Å material by desorption of intercalant, appear to obey the staging formula $C_{8n}H_xNO_3$ ($x < 1$ but not defined). This gallery height is consistent with planar NO_3 species lying parallel to the carbon layers. Figure 6b(i) shows a possible ordered arrangement. The conversion of $I_c = 7.8$ Å to $I_c = 6.55$ Å material must involve loss of HNO_3. As these neutral molecules (which serve as dielectric spacers for the NO_3^- ions) leave the galleries, the repulsive interactions of the anions should favor the structure in Fig. 6b(i). Another factor favoring a small I_c when there is a large anion fraction of guests is the increased coulomb attraction between guests and carbon layers.

The $C_{8n}X$ formulation is the same as that observed (*19*, *61*) in the close-packed, first-stage, hexafluoride-intercalated C_8MF_6 (M = As, Os, Ir), in which, of course, two such layers of relatively close-packed guest ligands (F and O are similar in size) are required in place of the single layers seen here. The X-ray diffraction studies of Rudorff (*56*) indicated that the guest disposition in one gallery is sensed by guests in neighboring galleries. Thus, the bisulfate has the structure shown in Fig. 7, and this and other ordered arrangements have been observed in other salts.

From coulometric studies (*57*, *58*) of bisulfate, perchlorate, and nitrate intercalation done in concert with Hall coefficient measurements, Ubbelohde and co-workers demonstrated the *p* type of conductivity of these materials.

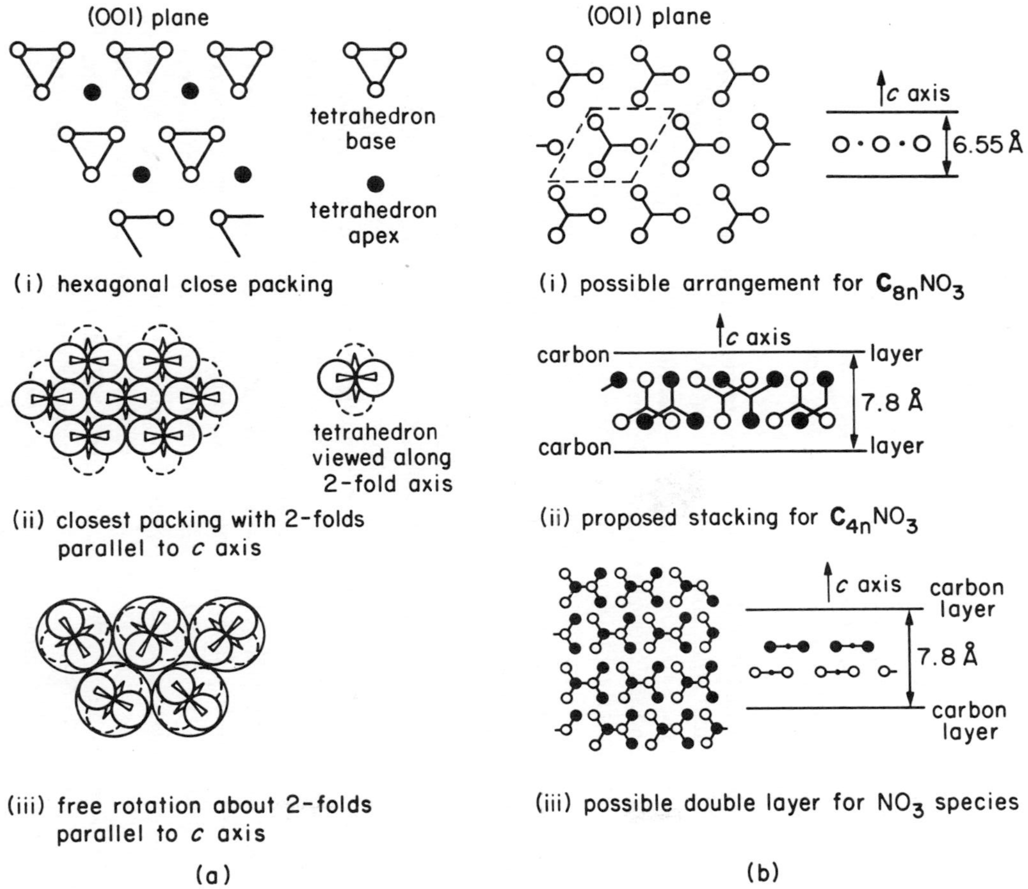

Fig. 6. Alternative packing arrangements for (a) tetrahedral species and (b) planar NO_3 species.

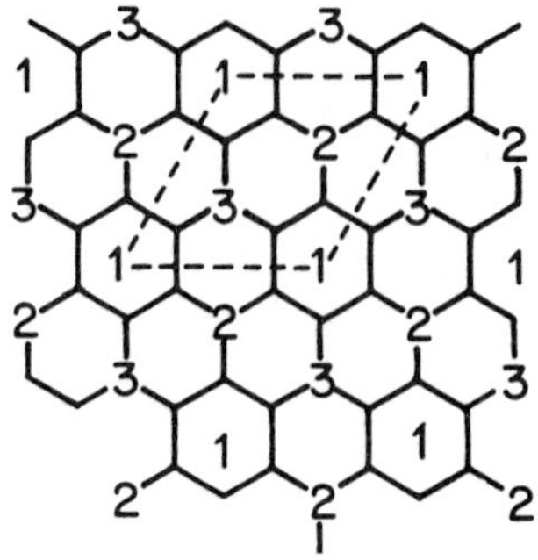

Fig. 7. Projection into (00*l*) plane of a $C_{24}^{+}HSO_4^{-}2H_2SO_4$ structural modification. The full *c*-axis sequence of guest layers is 1, 2, 3, 1, 3, 2, 1. [After Rudorff (*56*).]

They also showed that the salts are very good *ab*-plane conductors, having specific conductivities comparable with or higher than that of aluminum. The conductivity in the *c*-axis direction is usually much lower than in pristine graphite, although the values clearly depend on short circuiting by defects that link the carbon layers. Indeed, as Ubbelohde has discussed (*10a, 10b*), defects are an important source of scattering of the charge carriers. All other things being equal, an increase in the pervection of the graphite salts increases the *ab*-plane conductivity and decreases that in the *c*-axis direction. An approximately tenfold increase in σ_{ab} was noted (*10a, 10b*) for graphite nitrate cooled from room temperature to 63 K. This was attributed to a decrease in the carrier scattering associated with a "freezing-in" of the guest species in their lattice sites.

Much higher positive charges in the carbon layers can be obtained by avoiding the use of mineral acids in the synthesis. Thus, peroxydisulfuryl difluoride, $S_2O_6F_2$, oxidizes graphite spontaneously (*62*) to produce a material of approximate formulation C_7SO_3F. This is an *ab*-plane conductor that is inferior to related material, $C_7(SO_3F)_{1/3}(HSO_3F)_{2/3}$, made by treating graphite with an appropriate mixture of $S_2O_6F_2$ and HSO_3F. It is not clear why this is so. Comparison of X-ray diffraction data from the two materials made from the same parent carbon shows that the acid-containing compound is highly crystalline and that C_7SO_3F is not. In the acid salts, proton mobility (*63*) must result in haphazard and fleeting charging of the guest species. This and the relatively low positive charge per carbon layer may be important to the preservation of undistorted hexagonal carbon layers. Low conductivity and poor crystallinity of C_7SO_3F could, on the other hand, signify a distortion of the carbon layers. Low conductivity and poor crystallinity have also been observed (*61*) in $C^{+}MF_6^{-}$ salts (see Section V,C.)

TABLE IV

Intercalated Gallery Spacings for Some Compounds with Oxides

Reagent	Graphite compound formula	Stages	I_c (Å)	Reference
SO_3	$C_{5n}SO_3$	1, 2, 3	7.96–7.90	*3*
Cl_2O_7	$C_{12n}Cl_2O_7$	1, 2, 3	7.92–7.80	*3*
N_2O_5	$C_{8n}N_2O_5$	2, 3, 4	7.8	*60*

B. Intercalation of Oxides

Claims for the intercalation of a number of oxides into graphite have been advanced (*3, 5*), but definitive identification of the guest species is generally lacking. The similarity of the I_c values for the N_2O_5, Cl_2O_7, and SO_3 derivatives (Table IV) to the I_c values for the acid anion relatives in Table III is striking. Since the familiar forms of the oxides themselves have geometries related to the anions, it is, of course, possible that much of the intercalant, in all cases, is merely the neutral species. It does seem likely, however, that at least some electron oxidation of the graphite and guest anion formation has occurred, to provide the driving force for the intercalation. This is plausible for the N_2O_5 and Cl_2O_7, for which oxidation by the following reactions* might occur:

$$N_2O_5(g) + e^- \text{ (from C)} \rightarrow NO_2(g) + NO_3^-(g) \qquad \Delta H^0 \simeq -76 \text{ kcal/mole}$$

$$Cl_2O_7(g) + e^- \text{ (from C)} \rightarrow (ClO_3)(g) + ClO_4^-(g) \qquad \Delta H^0 \simeq -159 \text{ kcal/mole}$$

The case for SO_3 oxidation of the graphite is less compelling, but the possibility of the intercalation being initiated by some polymeric anion formation is great. A general equation for such a reaction is

$$(x + 3)SO_3 + 2e^- \text{ (from C)} \rightarrow O_3S \cdot O \cdot (SO_3)_x \cdot O \cdot SO_3^{2-} + SO_2$$

Some high-oxidation-state transition metal oxides (for example, CrO_3) have also been reported to intercalate. Electron oxidation of the graphite seems to be involved (*64*).

* A heat of formation for NO_3^- of -81 kcal/mole [T. W. Waddington, *Adv. Inorgan. Chem. Radiochem.* **1**, 157 (1959)] and standard H_f in kilocalories per mole [*NBS Tech. Note* (*U.S.*) No. 270-3 (1969)] for $N_2O_5(g) = 3$; $NO_2(g) = 8$; $NO(g) = 22$; $Cl_2O_7(g) = 65$; $ClO_3(g) = 34$. We estimate $H_f(ClO_4^-)$ to be -131 kcal/mole [from $\Delta H_f(NO^+ClO_4^-) = -37$ kcal/mole].

C. Intercalation of Hexafluoro Anions, Fluorine, and Arsenic Pentafluoride

High carbon-layer charges have been produced with the intercalation of hexafluoro anion species such as MF_6^- (M = P, As, Sb, Os, Ir) and $M'F_6^{2-}$ (M′ = Ge, Sn, Pt), the limiting formulations being, respectively, $C_8^+MF_6^-$ and $C_{12}^{2+}M'F_6^{2-}$. All these salts are blue. Studies involving the third transition series hexafluorides have demonstrated (*61*) that a hexafluoride must have an electron affinity of at least 108 kcal/mole if it is to intercalate into pristine graphite spontaneously and oxidatively.

Data for the third transition series hexafluorides and relatives are given in Table V. Since the hexafluorides are very similar in size (*65*), the coulomb energy for any salt $C_x^+MF_6^-$ will be essentially the same for all M. In a Born–Haber cycle comparison of the salts $C_x^+MF_6^-$, the only term affecting the enthalpy of formation that changes much from one salt to another, is the electron affinity of MF_6. The inability of WF_6 and ReF_6 to oxidize and intercalate graphite can be attributed to the lower electron affinities of these molecules. (ReF_7 does intercalate, and WF_6 and ReF_6 each intercalate in the presence of fluorine to yield MF_7^- salts.) Magnetic susceptibility measurements have shown that the limiting composition formulations of the osmium, iridium, and platinum salts are, respectively, $C_8^+OsF_6^-$, $C_8^+IrF_6^-$, and $C_{12}^{2+}PtF_6^{2-}$.

The essentially close packing of the hexafluoro species in the galleries of the C_8MF_6 cases is required by the unit cell, which is illustrated in Fig. 8a. It is probable that the less close-packed situation pertaining in $C_{12}PtF_6$ (see Fig. 8b) is a consequence of the combination of the greater work function required to produce C_{12}^+ (as compared with C_8^+) and the considerable repulsive interaction of the PtF_6^{2-} ions in a close-packed configuration. Note (Fig. 8c) that each species has *six* nearest neighbors in a C_8MF_6 type of close-packed configuration but only *three* in an ordered $C_{12}MF_6$ arrangement.

The first-stage materials are obtained at ~20°C using the hexafluoride at vapor pressures close to 100 torr. Higher stages may be obtained by using lower pressures or temperatures. The first-stage composition ranges from C_{12} to C_8MF_6. The staging formula for the highest stage is, however, $C_{12n}MF_6$. This is probably a consequence of the large repulsive interactions that the close-packed $C_{8n}MF_6$ formulation requires (Fig. 8b and c). It appears, then, that the energy expended in opening another gallery is less than would be lost in adopting a $C_{8n}MF_6$ packing rather than a $C_{12n}MF_6$ packing. In the former, each anion would have six nearest neighbors, whereas in $C_{12n}MF_6$ there are three nearest neighbors only.

The first-stage osmium and iridium compounds do not conduct electricity (*74*) in the *ab* plane much better than does graphite itself. This is also the

TABLE V

Electron Affinities *A* (kcal/mole) of Third Transition Series Hexafluorides and Their Graphite Salt Limiting Compositions

WF_6	ReF_6	OsF_6	IrF_6	PtF_6
81[a]	107[b]	135[b]	161[b]	188[b]
—	—	$C_8{}^+OsF_6{}^-$[c,d]	$C_8{}^+IrF_6{}^-$[c]	$C_{12}{}^{2+}PtF_6{}^{2-}$[c]
	MoF_6	UF_6		
	117[e]	113[b]		
	C_8MoF_6[f]	C_9UF_6[g]		

[a] George and Beauchamp (*70*).

[b] The oxidative chemistry of the third transition series hexafluorides had established a smooth increase in oxidizing capability from W to Pt, and Bartlett (*65*) had estimated that the electron affinity increases by ~1 eV for each unit increase in atomic number across the series. There has, however, been much uncertainty about the absolute values of the electron affinities $A(MF_6)$; some of this springs from uncertainties in evaluations of lattice energies for salts, such as $NO^+MF_6{}^-$ and $O_2{}^+PtF_6{}^-$. Here, for the first time, we have tied $A(MF_6)$ to $A(UF_6)$. Two independent measurements for $A(UF_6)$ give 113 (*66*) and 117 (*67*) kcal/mole. The former value derives from ion-cyclotron resonance studies, from which studies a rather precise value is also available for $A(WF_6)$; hence, we take it as our fixed point for $A(UF_6)$. Calorimetric studies (*68*) give $\Delta H^0[NO(g) + UF_6(g) \rightarrow NO^+UF_6{}^-(c)] = -52$ kcal/mole. Since the first ionization potential $I(NO)$ is 213 cal/mole and $A(UF_6) = 113$ kcal/mole, it follows that the lattice energy $U[NO^+UF_6{}^-(c)]$ is -152 kcal/mole. A similar calorimetric study (*69*) gives $\Delta H^0[O_2(g) + PtF_6(g) \rightarrow O_2{}^+PtF_6{}^-(c)] = -60$ kcal/mole. Therefore, because $U[O_2{}^+PtF_6{}^-(c)]$ must be slightly more exothermic than $U[NO^+UF_6{}^-(c)]$, since $PtF_6{}^-$ is smaller than $UF_6{}^-$, the electron affinity of PtF_6 must be slightly greater than $I(O_2) + U[NO^+UF_6{}^-(c)] - \Delta H^0[O_2(g) + PtF_6(g) \rightarrow O_2{}^+PtF_6{}^-(c)] = (278 - 152 + 60)$ kcal/mole. Accordingly, we set the electron affinity of PtF_6 at 188 kcal/mole. Since $A(WF_6)$, derived from ion-cyclotron resonance studies (*70*), is in harmony with less precise evaluations (*65*) from other experiments and is also compatible with the known oxidative chemistry of WF_6, we accept this value as our other fixed point for the third transition series. The estimates of $A(MF_6)$ for PtF_6, OsF_6, and IrF_6 were derived, assuming a smooth increase with atomic number of M, between $A(WF_6) = 81$ and $A(PtF_6) = 188$ kcal/mole. Although the lattice energy $U[O_2{}^+PtF_6(c)] = -154$ kcal/mole is higher than that derived from a Kapustinskii lattice energy, the $A(MF_6)$ values based on the higher lattice energy are in excellent agreement with the known oxidative chemistry.

[c] Bartlett *et al.* (*61*).

[d] Bartlett *et al.* (*19*).

[e] Annis and Datz (*71*).

[f] Opalovskii *et al.* (*72*).

[g] Binenboym *et al.* (*73*).

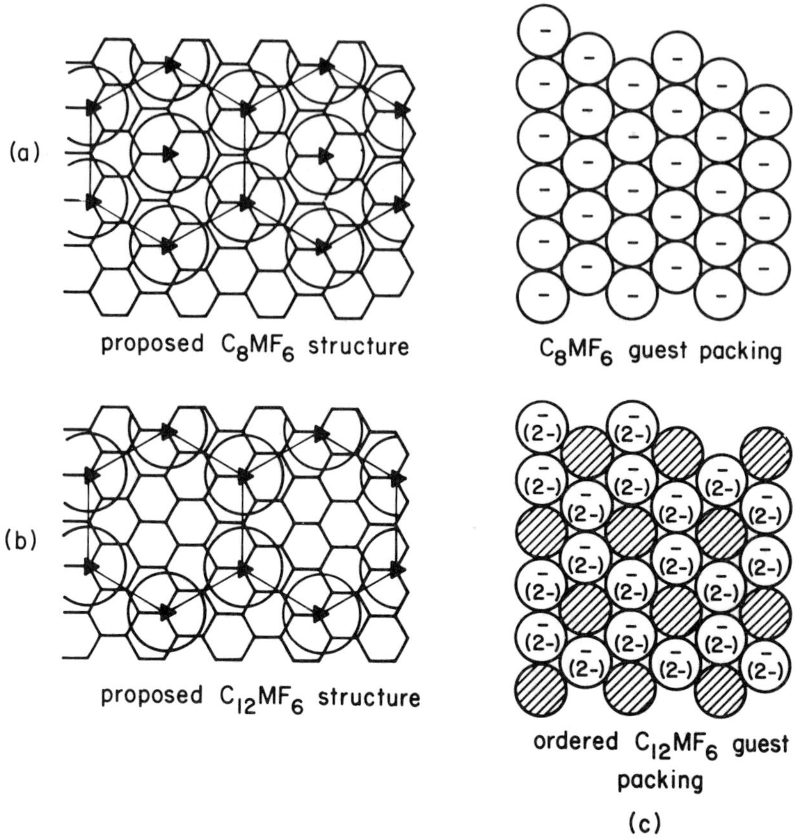

Fig. 8. Ordered structural arrangements for (a) C_8MF_6, (b) $C_{12}MF_6$, and (c) appropriate close packing of guest models for (a) and (b).

case for the platinum compound, even in the second stage. On the other hand, for the second and higher stages derived from OsF_6 or IrF_6, the σ_{ab} values are approximately an order of magnitude greater than for graphite itself. Bartlett *et al.* (*61*) pointed out that the lower conductivity in the more highly charged graphites may be a consequence of distortion of the carbon layers. They noted eventual loss of crystallinity in $C_8{}^+MF_6{}^-$ (M = As, Os, Ir). When freshly prepared and kept at low temperatures, these salts exhibit the diffraction characteristics required for the cell given in Fig. 8a, but at room temperature the crystallinity is lost within a few hours. The salts C_8MF_6 (M = P or As) can be made (*75, 76*) by treating graphite with a mixture of the pentafluoride and fluorine until there is no further consumption of either gas. The arsenic salt conductivity is comparable to that of graphite, in contrast with the graphite–arsenic pentafluoride relative C_8AsF_5, which is comparable to aluminum metal (*74, 77, 78*).

It is not possible to intercalate BF_3, PF_5, or WF_6 alone into pristine graphite. Evidently, as Lewis acids, their interaction with graphite π clouds cannot provide for the energy of the gallery opening. The data in Table VI show, however, that the fluoride ion affinities of each of these molecules is large. This enthalpy evidently drives the intercalation of the ions derived from them.

Munch and Selig (*84*) and McCarron and Bartlett (*76*) independently noted that the consumption of fluorine by graphite in the presence of arsenic pentafluoride is double that required to form AsF_6^-. Bartlett and co-workers (*74, 76, 88*) showed that, for stages $n \geq 2$, the limiting composition

TABLE VI

Enthalpy Changes for the Formation of Some Fluoro Anions

$\Delta H[MF_x(g) + \frac{1}{2}F_2(g) \rightarrow MF_{x+1}^-(g)]^a$

MF_x	ΔH (kcal/mole)	References
BF_3	−155	*81*
PF_5	−158	*82, 83*
AsF_5	−170	*19, 82*
SbF_5	−193	*82* [c]
WF_6	−141	[b]

[a] In all calculations, $\Delta H_f[F^-(g)] = -64$ kcal. There are two reports of the value of $\Delta H[BF_3(g) + F^-(g) \rightarrow BF_4^-(g)]$. Altshuller (*79*) gives a value of −71 kcal, whereas Bills and Cotton (*80*) give −91 kcal. We prefer the −91 kcal value because it is compatible with the marginal stability of $O_2^+BF_4^-(c)$. All other values of fluoride ion affinities are based on the −91 kcal value for BF_4^-. These fluoride ion affinities also depend on the lattice energy evaluations, which are probably no more precise than ±10 kcal/mole.

[b] The value reported for $\Delta H_f[WF_7^-(g)]$ by George and Beauchamp (*70*) has been lowered to -552 ± 10 kcal/mole based on the use of −91 kcal/mole rather than −71 kcal/mole for the fluoride ion affinity of $BF_3(g)$.

[c] This value is for liquid SbF_5 converted to $SbF_6^-(g)$.

of these fluorine-rich systems is $C_{12n}AsF_6 \cdot F$. The fluorine cannot be removed under vacuum, but it can be titrated with AsF_3 to form an AsF_6^- salt,

$$3C_{4y}AsF_6 \cdot F + AsF_3 \rightarrow 4C_{3y}AsF_6$$

and with an equimolar quantity of AsF_3 there is complete conversion to a material that appears to be identical to that prepared directly from graphite and AsF_5. The interaction is evidently

$$2C_{3y}AsF_6 + AsF_3 \rightarrow 3C_{2y}AsF_5$$

The extra fluorine for the $C_{12n}AsF_6 \cdot F$ occupies only those galleries that contain AsF_6^-. It seems likely that the fluorine, as F^-, is located in the vacancies available in the $C_{12}AsF_6$ gallery. Since each of these vacancies (see Fig. 8b and c) can accommodate an AsF_6^- (as in C_8AsF_6), the requirement that each accommodate two F^- is easily met. Of course, this means that the positive charge in the carbon layers in these fluorine-rich compounds must be high, namely, $C_{12n}^{2+}AsF_6^- \cdot F^-$.

Thompson *et al.* (*74*) found that the $C_{12n}AsF_6 \cdot F$ materials are poor *ab*-plane conductors and that their reduction by AsF_3 leads to conductivity enhancement. This low conductivity of $C_{12n}^{2+}AsF_6^- \cdot F^-$ parallels the low conductivity noted for the first-stage $C_8^+MF_6$ salts and those derived from PtF_6^{2-}.

These discoveries are pertinent to the long-standing problem of the non-intercalation of elemental fluorine in pristine graphite (*3*, *9a*, *9b*). It is clear that fluorine is unable (being unaided) to open a gallery, and this continues to be the case for $C_{12}AsF_6$, which takes up fluorine ($C_{12}AsF_6 + \frac{1}{2}F_2 \rightarrow C_{12}AsF_6 \cdot F$) only within the occupied gallery. Perhaps [to follow a suggestion of Hooley (*9a*, *9b*)], F^- ions formed at the opening to a gallery become fixed at the entrance and cannot move forward into the gallery. Such localization of F^- is compatible with their small size (radius 1.33 Å) and high charge density.* Once they are close to a somewhat positively charged carbon atom, the progression to another carbon (at least 1.42 Å distant) requires appreciable energy.

It may be that, with the AsF_6^- in attendance, the progression into the occupied galleries is aided by the fleeting formation of such species as AsF_7^{2-} or even AsF_7^-. These species, with their overcrowded coordination spheres, would slough off a ligand readily, but the one discarded need not be the one taken up. By such a ligand exchange, the F^- could progress rapidly into (and

* The difference between a fluoride ion attracted strongly to a positively charged graphite layer, for example, $C_x^+F^-$, and a covalently attached F ligand is that the former bonding does not move C atoms out of the plane, whereas a covalent bond would result in a roughly tetrahedral coordination for the carbon atom.

out of) the gallery. However, there is currently no evidence to give firm support to such conjectures.

The materials that have aroused perhaps the most interest and controversy are those derived from graphite and AsF_5. Selig and co-workers (*89*) were the first to report such a material, but it was the report of conductivities comparable with that of copper (*77*) which attracted a great deal of attention. More recent work indicates that the conductivities are lower than first assessed, but values comparable with those for aluminum metal are established (*74*, *78*).

Most of the controversy has centered on the nature of the guests. Bartlett *et al.* (*19*, *20*) have contended, on the basis of X-ray absorption edge studies, that arsenic pentafluoride oxidizes graphite according to

$$3AsF_5 + 2e^- \rightarrow 2AsF_6^- + AsF_3$$

and that the conversion to AsF_6^- and AsF_3 is virtually complete. They have also shown that there are substantial fractions of arsenic trifluoride present in the volatile compounds removed from the intercalated graphite (*76*). Moreover, they have shown (*74*) that, when AsF_3 is added to a $C_{12n}AsF_6$ salt ($n \geq 2$), there is no detectable change in stage and the conductivity remains unchanged. The same holds for removal of AsF_3. These observation are in accord with a $C_{12n}AsF_6$ structure (see Fig. 8b) into the vacancies of which can be placed neutral AsF_3 molecules to yield a material $C_{12n}AsF_6 \cdot \frac{1}{2}AsF_3$ (that is, $C_{8n}AsF_5$). A persuasive feature is that the staging formula for graphite/AsF_5 (*90*) is $C_{8n}AsF_5$ (that is, $C_{12n}AsF_6 \cdot \frac{1}{2}AsF_3$), whereas for the AsF_6^- salts it is $C_{12n}AsF_6$ (*76*).

The thermodynamic data in Table VII, when taken in conjunction with those in Table V, indicate that AsF_5 (and SbF_5) should be capable of oxidizing graphite, since the electron affinity per equivalent of electron oxidation of the graphite is comparable with that of OsF_6. On the other hand, PF_5 has a lower electron affinity (for the reaction $3PF_5 + 2e^- \rightarrow 2PF_6^- + PF_3$) than does WF_6. It is therefore not surprising that PF_5 alone will not oxidize graphite spontaneously.

In spite of this evidence, the early views that the guest species were largely molecular AsF_5 (or AsF_5 bonded to the graphite π system) have persisted. The major problems appear to be chemical. (Unfortunately, many physical measurements have been made on materials that have been ill-defined chemically.)

Although AsF_6^- and AsF_3 are the dominant guest species, they are in equilibrium with AsF_5. The latter becomes a more important species as the charge builds up in the carbon layers (which it does, dramatically, in passing from second to first stage). Thus, we have

$$G + 3AsF_5 \rightleftarrows G^{2+}2AsF_6^- + AsF_3$$

Indeed, if a large excess of arsenic pentafluoride is used in the synthesis, as AsF_3 leaves the material it can be replaced by AsF_5. Hence, unless the preparative conditions are clearly specified, it is unlikely that a pure graphite/AsF_5 compound will be obtained. The ready loss of AsF_3 means that it is possible to make a graphite/AsF_6^- salt from graphite and AsF_5 simply by repeating a cycle of operations. Thus, C_8AsF_5 under vacuum passes to $C_{\sim 14}AsF_{5-6}$. The latter is treated with more AsF_5 to a composition C_8AsF_{5-6}, again pumped down, and again treated with AsF_5. After five such cycles, the composition stands (after evacuation) close to the first-stage composition $C_{12}AsF_6$ (*88*).

Although there is no detectable difference between the *ab*-plane conductivities of second and higher stages of $C_{8n}AsF_5$ and $C_{12n}AsF_6$ (*74*), there are large differences between the first-stage materials. The compound C_8AsF_5 is a good conductor (about 12 times better than graphite), whereas $C_{12}AsF_6$ [like $C_{12}MF_6$ (M = Os, Ir)] has a conductivity close to that of graphite itself. There is no evidence of loss of crystallinity in C_8AsF_5 as there is for C_8AsF_6, and it appears that the graphite layers have their hexagonal symmetry preserved in the former.

A somewhat haphazard arrangement of the neutral and charged species (AsF_6^-, AsF_3, and some AsF_5) in the galleries of C_8AsF_5 results in a lower tendency toward charge localization at specific sites in the carbon layer than in $C_8^+AsF_6^-$. Thus, the delocalization π electron system of graphite may be less perturbed in C_8AsF_5 than in $C_8^+AsF_6^-$, in which a guest on each

TABLE VII

Enthalpy Changes for Electron Oxidation by Group V Pentafluorides[a]

	P	As	Sb
(1) $\Delta H^0[2MF_5(g) + F_2 + 2e^- \rightarrow 2MF_6^-(g)]$[b]	≤ -316	≤ -340	-386
(2) $\Delta H^0[MF_5(g) \rightarrow MF_3(g) + F_2(g)]$	152[c]	98[d]	90[e]
(3) $\Delta H^0[3MF_5(g) + 2e^- \rightarrow 2MF_6^-(g) + MF_3(g)]$	≤ -164	≤ -242	-296
Electron affinity [according to Eq. (3)]	≥ 82	≥ 121	≥ 148

[a] Values in kilocalories per mole; values for P and As species are for the gas phase, whereas for Sb Eq. (1) represents conversion of polymeric liquid SbF_5 and Eq. (2) the conversion of liquid SbF_5 to solid SbF_3. Since the heat of vaporization and the enthalpy of formation of the gaseous monomer $SbF_5(g)$ are both endothermic, the enthalpy change for Eq. (3) in the Sb case must be more exothermic than given here.

[b] See Table VI.

[c] Rudzitis *et al.* (*84*).

[d] Woolf (*85*).

[e] O'Hare and Hubbard (*86*).

side of a carbon layer cooperates to localize positive charge. Such cooperative charge localization could lead to structural change in the carbon layers.

Unfortunately, most of the studies on graphite/AsF_5 have been made on materials that must now be regarded as suspect. When close attention is given to the chemical problems, the points of dispute may well disappear.

D. Intercalation of Halogens and Halides Except Fluorides

Hérold has thoroughly discussed the intercalation of halogens and halides into graphite (*3*) as well as much of the structural work; therefore, these aspects are not treated here.

From the enthalpies of formation of the halide ion (Table VIII), electronegativities, and lattice energetics, it is determined that the lightest of the halogens are the most favorable for the formation of salts $C_x^+X^-$ ($X =$ halogen). With the possible exception of F^- in salts such as $C_{12n}AsF_6 \cdot F$ ($n \geq 2$), there is no evidence that discrete halide ions ever occur in intercalation compounds of graphite. Chlorine forms compounds (*95*) that are stable below room temperature. Bromine is the only halogen that spontaneously generates a graphite compound at ordinary temperatures and pressures. Iodine does not intercalate, yet ICl and IBr do. Hooley (*9a*, *9b*) has

Table VIII

Some Enthalpies of Formation (kcal/mole) for Halide and Complex Halide Ions[a]

$\Delta H_f^0[X_2(g) + e^- \rightarrow X^-(g)]$ ($X = F, Cl, Br, I$):	F, −64	Cl, −57	Br, −58	I, −56
$\Delta H^0[\frac{1}{2}M_2Cl_6(g) + \frac{1}{2}Cl_2 + e^- \rightarrow MCl_4^-(g)]$ ($M = Al, Ga$):	Al, −128[b]			Ga, −133[b]
Reported limiting compositions of graphite compounds:	$C_9AlCl_{3.5}$[c]			$C_9GaCl_{3.5}$[d]
$\Delta H[MBr_3(g) + \frac{1}{2}Br_2 + e^- \rightarrow FeBr_4^-(g)]$	−133[b]			−128[b]
$\Delta H[FeCl_3(g) + \frac{1}{2}Cl_2 + e^- \rightarrow FeCl_4^-(g)]$			F, −161[e]	
$\Delta H[FeCl_3(g) \rightarrow FeCl_2(c) + \frac{1}{2}Cl_2]$			F, 22[e]	
$\Delta H[2FeCl_3(g) \rightarrow FeCl_2(c) + FeCl_4^-(g)]$			−139	

[a] NBS (*94*).
[b] Jenkins *et al.* (*98*).
[c] Bach and Ubbelohde (*99*).
[d] Bach and Ubbelohde (*100*).
[e] Calculated from the enthalpies of reaction of KCl(c) or NaCl(c) with FeCl (c) to make $KFeCl_4$(c) and $NaFeCl_4$(c) (*103*), the lattice energy of NaCl (*104*), and a Kapustinskii radius of 3.58 Å (*105*). If the radius is reduced to 3.18 Å, the lattice energy is increased and the chloride ion affinity of $FeCl_3$(g) is decreased by 13 kcal/mole.

pointed to the close compatibility of the internuclear distances in those molecules that do intercalate (Br_2, 2.27 Å; ICl, 2.40 Å, IBr, 2.49 Å) with the center-to-center distance of adjacent hexagons in graphite of 2.42 Å. Perhaps this preserves the electron delocalization of the carbon layers by spreading any positive charge, associated with electron removal by the bromine, more or less evenly over the hexagons of carbon atoms.

There is another feature of these Br_2, ICl, and IBr molecules, as distinct from F_2 and Cl_2, that may contribute to their ease of intercalation into graphite: their readiness to bond to related species to form extended halogen-bridged systems (*96*). Thus, Br_3^-, Br_5^-, and even more extended species may have a role in bromine intercalation. Similarly, ICl_2^-, IBr_2^-, or $I_2Cl_3^-$ and $I_2Br_3^-$ may contribute to the intercalation of ICl and IBr. In all of these simple halogen and interhalogen intercalation compounds, the physical evidence points clearly to a small charge transfer per halogen atom. It is clear that the transfer of one electron brings about the intercalation of many halogen atoms. The bridging ability of the heavier halogens may well be crucial to this intercalating capability. But this capability alone cannot be sufficient since iodine forms such extended systems (for example, I_3^-, I_5^-, I_7^-) with ease and with appreciable energy benefit. It seems that, in comparison with the other halogens, because of its larger size (and resultant poor lattice energy), low electronegativity, and poorer energy of anion formation, iodine is unable to meet the energy requirements for opening a gallery in graphite.

Although a large number of halides have been reported to intercalate into graphite, many of these are high-oxidation-state metal halides that are recognized as strong Lewis acids (for example, $SbCl_5$, $NbCl_5$). Except for species such as $AuCl_4^-$, which prefer a square planar geometry (*97*), the I_c values (for example, 9.5 Å for chlorides) indicate that the gallery contains a double layer of halogen ligands.

In a number of cases arguments have been advanced that the guest species have much the same geometry as in the pure guest precursor (*3*). If this is so, the reason for intercalation is not evident. It seems likely that, in most if not all of such halide intercalations, anion formation may be a necessary energy-providing step. The anion may be a very large, halogen-bridged species bearing small charge per halogen atom.

The clearest indication that anions can be important comes from the intercalation of the Al, Ga, and In chlorides and bromides. These halides intercalate into graphite only in the presence of excess halogen. The data for the Al and Ga compounds in Table VIII show that the formation of the simple MX_4^-(g) species is, in each case, highly energetically favorable for larger species M_nX_{3n+1}. The energy liberated per mole of electrons taken up in forming the anion is probably comparable to that in the MX_4^- case.

Of course, the larger the anion, the smaller is the lattice energy, but then the number of electrons to be removed from the carbon layers for each mole of MX_3 is also less.

It will be realized that species such as MX_3 (M = Al, Ga; X = Cl, Br) can provide halogen for M_nX_{3n+1} production only by reduction to a lower halide. Such reduction is energetically unfavorable for these halides. This may not be the case for most of the transition metal halides in higher oxidation states. Ferric chloride may be such a case.

In spite of the considerable attention given to the graphite/$FeCl_3$ system, what the guest species are is not yet certain. It is clear from Mössbauer studies (*100*) that the iron is all Fe(III). Electron excitation studies (*101*) have also shown that one electron is transferred from the carbon layers for approximately ten iron atoms intercalated. This implies that there is one extra chlorine (chloride) for every ten $FeCl_3$ molecules intercalated. Analytical data are scarce, and there is no explicit claim for a first-stage $C_xFe_{10}Cl_{31}$ stoichiometry as far as the present authors are aware. The thermodynamic data in Table VIII do lend support to the hypothesis that reduction of some $FeCl_3$ to $FeCl_2$ could provide the halogen for the intercalation of a complex Fe_nCl_{3n+1} guest. The $FeCl_2$ would presumably remain external to the graphite. Comparable redox schemes are also plausible for many of the other metal halides that are reported to intercalate spontaneously into graphite.

Stumpp and co-workers succeeded in preparing a large number of graphite intercalation compounds from transport of those halides by forming complexes with $AlCl_3$ (*105*). Schaeffer gave a comprehensive survey (*104*) of the thermodynamic features that form the basis of this vapor transport. Stumpp noted that, in many instances, intercalation of the metal halide occurs only in the presence of free halogen. Again, it seems that at least some guest anion formation provides the energetic basis for the intercalation.

VI. INTERCALATION WITHOUT APPARENT OXIDATION OR REDUCTION OF THE GRAPHITE

In many cases the assumption has been made that the guests within graphite galleries are uncharged. This is unlikely to be so in the majority of cases, but there are a number of cases in which spontaneous intercalation of the graphite does not appear to involve oxidation or reduction.

A group of compounds in this category are the noble gas fluorides. Although they are powerful oxidizers, it appears to be unlikely that they electron-oxidize graphite. Thus, krypton difluoride intercalates spontaneously and appears to have an enhanced stability when intercalated. The graphite/KrF_2 material is also a less extreme fluorinating and oxidizing agent than KrF_2 itself. Yet reduced species such as KrF_2^- cannot be formed. Nor are such species as KrF_3^- very probable. It appears that KrF_2 must be present as the molecular species.

A related group of compounds that appear to behave similarly to the noble gas fluorides are halogen fluorides. These that intercalate spontaneously include BrF_3, IF_5, and IOF_5. Most of the work on these fluorides and the noble gas fluorides was initiated by Selig and co-workers, and a full report of the chemical and analytical aspects of their intriguing graphite intercalation compounds is given in the comprehensive review by Selig and Ebert (*5*).

It is perhaps relevant that in both the noble gas fluorides and the halogen fluorides, semiionic bonding (*108*) occurs. Thus, KrF_2 can be represented as a resonance hybrid of the canonical forms $(FKr)^+F^-$ and $F^-(KrF)^+$. Similarly, BrF_3 can be represented as the resonance hybrid of the canonical forms $[FBr(F)]^+F^-$ and $F^-[Br(F)F]^+$. These semi-ionic bonds bestow large multipole moments on these molecules. It is likely that they produce appreciable π-cloud perturbation in the carbon layers, which results in substantial multipole–multipole attraction between guest and host. Further insight must clearly await more detailed structural studies than have appeared so far.

REFERENCES

A. "Proceedings of the Franco American Conference on Intercalation Compounds of Graphite" (F. L. Vogel and A. Hérold, eds.). Elsevier Sequoia, Lausanne, 1977; reprinted from *Mater. Sci. Eng.* **31,** 1–352 (1977).

B. "Physics and Chemistry of Materials with Layered Structures, Vol. 6, Intercalated Layered Materials" (F. A. Levy, ed.). Reidel Publ., Dordrecht, Netherlands, 1979.

C. *Proc. Int. Conf. Layered Mater. Intercalates, Nijmegen, Neth., 1979; Physica B (Amsterdam)* **99B** (1980).

1. P. Schauffaütl, *J. Prakt. Chem.* **21,** 155 (1841).

2. M. Armaud and P. Touzain, ref. A, pp. 319–329.

3. A. Hérold, ref. A, pp. 1–16; ref. B, pp. 323–421.

4. L. B. Ebert, *Annu. Rev. Mater. Sci.* **6,** 182 (1976).

5. H. Selig and L. B. Ebert, *Adv. Inorg. Chem. Radio Chem.* **23,** 281 (1980).

6. J. E. Fischer, ref. B, pp. 481–532.

7. Ref. A.

8a. *Synth. Met.* **2** (1980).

8b. *Synth. Met.* **3** (1981).

9a. J. G. Hooley, ref. A, pp. 17–24.

9b. J. G. Hooley, *in* "Preparation and Crystal Growth of Materials with Layered Structures" (R. M. A. Lieth, ed.), pp. 1–33. Reidel Publ., Dordrecht, Netherlands, 1977.

10a. A. R. Ubbelohde, *Synth. Met.* **1,** 13 (1979).

10b. A. R. Ubbelohde, ref. A, pp. 341–350.

11. F. Jeitner, E. Nedopil, and O. Vohler, *Ber. Dtsch. Keram. Ges.* **41,** 135 (1964).

12. A. R. G. Brown, W. Watt, R. W. Powell, and R. R. Tye, "Industrial Carbon and Graphite." Soc. Chem. Ind., London, 1958.
13. P. R. Wallace, *Phys. Rev.* **71,** 622 (1947). A recent elaboration is given by A. Charlier, F. DuJardin, M. F. Charlier, and J. P. Decruppe, *Bull. Soc. Chim. Fr.* No. 3/4, p. 1 (1979).
14a. J. C. Slonczewski and P. R. Weiss, *Phys. Rev.* **109,** 272 (1958).
14b. J. W. McClure, *Phys. Rev.* **108,** 612 (1957).
14c. G. S. Painter and D. E. Ellis, *Phys. Rev. B* **1,** 4747 (1970).
14d. R. R. Haering, *Can. J. Phys.* **36,** 352 (1958).
15. J. W. McClure, *IBM J.* July, p. 255 (1964); M. S. Dresselhaus and J. G. Mauroides, *IBM J.* July, p. 262 (1964).
16. A. R. Ubbelohde, *Carbon* **14,** 1 (1976).
17. S. A. Solin, ref. C, pp. 443–452.
18. L. B. Ebert and H. Selig, ref. A, pp. 177–181.
19. N. Bartlett, R. N. Biagioni, B. W. McQuillan, A. S. Robertson, and A. C. Thompson, *J.C.S. Chem. Commun.* p. 200 (1978).
20. N. Bartlett, B. McQuillan, and A. S. Robertson, *Mater. Res. Bull.* **13,** 1259 (1978).
21. F. J. Salzano and S. Aronson, *J. Chem. Phys.* **45,** 2221 (1966).
22. F. R. M. McDonnell, R. C. Pink, and A. R. Ubbelohde, *J. Chem. Soc.* p. 191 (1951).
23. J. J. Lander and J. Morrison, *Surf. Sci.* **6,** 1 (1967).
24. S. M. Heald and E. A. Stern, *Phys. Rev. B* **17,** 4069 (1978).
25. S. M. Heald and E. A. Stern, *Synth. Met.* **1,** 249 (1979–1980).
26. D. A. Young, *Carbon* **15,** 373 (1977); A. S. Bender and D. A. Young, *J. Solid State Phys.* **5,** 2163 (1972); J. D. Hibbs and D. A. Young, *Chem. Phys. Lett.* **53,** 361 (1978).
27. N. Daumas and A. Hérold, *C. R. Hebd. Seances Acad. Sci., Ser. C* **268,** 373 (1969).
28. J. M. Thomas, G. R. Millward, R. F. Schlogl, and H. P. Boehm, *Mater. Res. Bull.* **15,** 671 (1980).
29. S. A. Safran and D. R. Hamann, *Phys. Rev. Lett.* **42,** 1410 (1979); and *Physica B* (*Amsterdam*) **99B,** 469 (1980); S. A. Safran, *Phys. Rev. Lett.* **44,** 937 (1980); and ref. *8a*, p. 1.
30. A. R. Ubbelohde, *Proc. R. Soc., London, Ser. A* **327,** 289 (1972).
31. C. Zeller, A. Denestein, and G. M. I. Foley, *Rev. Sci. Instrum.* **50,** 602 (1979).
32. C. Zeller, G. M. T. Foley, E. R. Falardeau, and F. L. Vogel, ref. A, p. 255.
33. Y. N. Novikov and M. E. Volpin, *Russ. Chem. Rev.* **40,** 733 (1971).
34. Refs. A, B, and 4.
35. S. C. Jain and K. S. Krishnam, *Proc. R. Soc., London, Ser. A* **213,** 143 (1952).
36. D. Guérard, M. Chaabouni, P. LaGrange, M. El Makrini, and A. Hérold, *Carbon* **18,** 257 (1980).
37a. M. El Makrini, D. Guérard, P. LaGrange, and A. Hérold, *Physica B* (*Amsterdam*) **99B,** 481 (1980).
37b. R. D. Shannon, *Acta Crystallogr., Sect. A* **32,** 751 (1976).
38. G. S. Parry and D. E. Nixon, *Nature* (*London*) **216,** 909 (1967); D. E. Nixon and G. S. Parry, *J. Phys. Chem.* **2,** 1732 (1969).
39. A. Hérold, ref. B, pp. 361–369 contains an excellent summary and literature references.
40. D. Billaud, A. Hérold, and F. L. Vogel, *Mater. Sci. Eng.* **45,** 55 (1980).
41. G. Furdin, D. Billaud, A. Hérold, C. Zeller, and A. Marchard, *J. Solid State Chem.* **34,** 361 (1980).
42. D. Billaud and A. Hérold, *C. R. Hebd. Seances Acad. Sci., Ser. C* **281,** 305 (1975).
43. P. LaGrange, M. El Makrini, D. Guérard, and A. Hérold, *C. R. Hebd. Seances Acad. Sci., Ser. C* **289,** 288 (1979).
44. D. Guérard, P. LaGrange, and A. Hérold, ref. A, p. 29; A. Hérold and P. LaGrange, ref. A, p. 33.

45. M. Nomine and L. Bonnetain, *J. Chem. Phys.* **50,** 1731 (1969); C. Minh-Duc, C. Mai, R. Riviére, and J. Gole, *J. Chim. Phys.* **6,** 986, 991 (1972).

46a. W. Rudorff and E. Z. Schulze, *Z. Anorg. Allgem. Chem.* **277,** 156 (1954).

46b. P. Delhaes, J. C. Rouillon, J. P. Manceau, D. Guérard, and A. Hérold, *J. Phys., Lett. (Orsay, Fr.)* **37,** L127 (1976).

47. A. R. Ubbelohde, *Nature (London)* **232,** 43 (1971).

48. E. McRae, D. Billaud, J. F. Mareche, and A. Hérold, *Physica B (Amsterdam)* **99B,** 489 (1980).

49. J. J. Murray and A. R. Ubbelohde, *Proc. R. Soc., London, Ser. A* **312,** 361 (1969).

50. D. Guérard, G. M. T. Foley, M. Zanini, and J. E. Fischer, *Nuovo Cimento Soc. Ital. Fis. B* **35,** 410 (1977).

51. F. J. DiSalvo, S. A. Safran, R. C. Haddon, and J. Fischer, *Phys. Rev. B* **20,** 4883 (1979); S. A. Safran and F. J. DiSalvo, *Phys. Rev. B* **20,** 4889 (1979).

52. C. Ayache, E. Bonjour, R. Lagnier, and J. E. Fischer, *Physica B (Amsterdam)* **99,** 547 (1980); M. G. Alexander, D. P. Goshorn, D. Guérard, P. LaGrange, M. El Makrini, and D. G. Orr, ref. *8a*, p. 203.

53. See J. E. Fischer, ref. B, pp. 505–507, for a summary of this work.

54. T. Ohno, K. Nakao, and H. Kamimura, *J. Phys. Soc. Jpn.* **47,** 1125 (1979); H. Kamimura, K. Nakao, T. Ohno, and T. Inoshita, *Physica B (Amsterdam)* **99B,** 201 (1980).

55. W. Rudorff and U. Hofmann, *Z. Anorg. Chem.* **238,** 1 (1938).

56. W. Rudorff, *Z. Phys. Chem., Abt. B* **45,** 42 (1940).

57. A. R. Ubbelohde, *Proc. R. Soc. London, Ser. A* **309,** 297 (1969).

58. A. R. Ubbelohde, *Proc. R. Soc. London, Ser. A* **321,** 445 (1971).

59. P. Touzain, *Synth. Met.* **1,** 3 (1979).

60. H. Fuzellier, J. Melin, and A. Hérold, ref. A, pp. 91–94.

61. N. Bartlett, E. M. McCarron, B. W. McQuillan, and T. E. Thompson, *Synth. Met.* **1,** 221 (1980).

62. N. Bartlett and R. N. Biagioni, to be published; R. N. Biagioni, Ph.D. Thesis, University of California, October 1981, LBL Reprint 11760.

63. D. E. Nixon, G. S. Parry, and A. R. Ubbelohde, *Proc. R. Soc. London, Ser. A* **291,** 324 (1966).

64. For references on CrO_3 intercalation, see ref. 5; see also W. Metz and H. Meyer-Spasche, *Synth. Met.* **1,** 53 (1979–1980).

65. N. Bartlett, *Angew. Chem., Int. Ed. Engl.* **7,** 433 (1968).

66. J. L. Beauchamp, *J. Chem. Phys.* **64,** 929 (1976).

67. R. N. Compton, *J. Chem. Phys.* **66,** 4478 (1977).

68. P. Plurien, J. Chatelet, M. Luce, and P. Rigny, *Int. Symp. Fluorine Chem., 7th, Santa Cruz, Calif.* Abstr. P-18 (1973).

69. P. Barberi and N. Bartlett, to be published.

70. P. M. George and J. L. Beauchamp, *Chem. Phys.* **36,** 345 (1979).

71. B. K. Annis and S. Datz, *J. Chem. Phys.* **66,** 4468 (1977).

72. A. A. Opalovskii, Z. M. Kuznetsova, Y. N. Chigagov, A. S. Nazarov, and A. A. Uminskii, *Russ. J. Inorg. Chem.* **19,** 1134 (1974).

73. J. Binenboym, H. Selig, and S. Sarig, *J. Inorg. Nucl. Chem.* **38,** 2313 (1976).

74. T. E. Thompson, E. M. McCarron, and N. Bartlett, ref. *8b*, p. 255.

75. E. M. McCarron, Y. J. Grannec, and N. Bartlett, *J.C.S. Chem. Commun.* p. 890 (1980).

76. E. M. McCarron and N. Bartlett, *J.C.S. Chem. Commun.* p. 404 (1980).

77. E. R. Falardeau, Y. M. I. Foley, C. Zeller, and F. L. Vogel, *J.C.S. Chem. Commun.* p. 389 (1977).

78. L. B. Interrante, R. S. Markiewicz, and D. W. McKee, *Synth. Met.* **1,** 287 (1980).

79. A. P. Altshuller, *J. Am. Chem. Soc.* **77,** 6187 (1955).
80. J. L. Bills and F. A. Cotton, *J. Phys. Chem.* **64,** 1477 (1960).
81. A. D. Cohen, Fr. Patent 2,291,151 (1975).
82. D. D. Gibler, Ph.D. Thesis, Princeton University 1973.
83. Values from A. D. F. Toy, *Comp. Inorg. Chem.* **2,** 420, 428 (1973).
84. E. Rudzitis, E. H. Van Devemter, and W. N. Hubbard, *J. Chem. Thermodyn.* **2,** 221 (1970).
85. A. A. Woolf, *J. Fluorine Chem.* **15,** 533 (1980).
86. P. A. G. O'Hare and W. N. Hubbard, *J. Phys. Chem.* **70,** 3353 (1966).
87. V. Munch and H. Selig, *Synth. Met.* **1,** 407 (1979/80).
88. N. Bartlett, E. M. McCarron, T. Mallouk, and F. Okino, to be published.
89. L. C. Hsu, H. Selig, M. Rabinovitz, I. Agranat, and S. Sarig, *J. Inorg. Nucl. Chem. Lett.* **11,** 601 (1975).
90. E. R. Falardeau, L. R. Hanlon, and T. E. Thompson, *Inorg. Chem.* **17,** 301 (1978).
91. H. Selig, M. J. Vasile, F. A. Stevie, and W. H. Sunder, *J. Fluorine Chem.* **10,** 299 (1977).
92. J. E. Fischer, *J.C.S. Chem. Commun.* p. 544 (1978).
93. M. J. Moran, J. E. Fischer, and W. R. Salanck, *J. Chem. Phys.* **73,** 629 (1980).
94. *NBS Tech. Note* (*U.S.*) No. 270-3 (1969).
95. G. Furdin, M. LeLaurain, E. McRae, J. F. Mareche, and A. Hérold, *Carbon* **17,** 329 (1979).
96. A. I. Popov, *in* "Halogen Chemistry" (V. Gutmann, ed.), Vol. 1, pp. 225–260. Academic Press, New York, 1967.
97. R. Vangelisti and A. Hérold, *C. R. Hebd. Seances Acad. Sci., Ser. C* **276,** 1109 (1973).
98. H. D. B. Jenkins, R. R. Pratt, G. Mairesse, P. Barbier, and J. P. Wignacourt, *Inorg. Chem.* **19,** 2796 (1979). The gas phase chloride ion affinity should be 357 kJ not 257 kJ. See also H. D. B. Jenkins, *Inorg. Chem.* **15,** 241 (1976); R. C. Gearhart, J. D. Beck, and R. H. Wood, *Inorg. Chem.* **14,** 2513 (1975).
99. B. Bach and A. R. Ubbelohde, *Proc. R. Soc. London, Ser. A* **325,** 437 (1971).
100. B. Bach and A. R. Ubbelohde, *J. Chem. Soc. A* p. 3669 (1971).
101. A. J. Freeman, *J.C.S. Chem. Commun.* p. 193 (1968); J. G. Hooley, M. W. Bartlett, B. V. Liengone, and J. R. Sams, *Carbon* **6,** 681 (1968).
102. J. R. Ritsko and E. J. Mele, ref. C, pp. 425–429.
103. C. M. Cook and W. E. Dunn, *J. Phys. Chem.* **65,** 1505 (1961).
104. D. Cubicciotti, *J. Chem. Phys.* **34,** 2189 (1961).
105. H. D. B. Jenkins and R. P. Thakur, *J. Chem. Educ.* **56,** 576 (1979).
106. E. Stumpp, ref. A, pp. 53–59.
107. H. Schäfer, *Angew. Chem., Int. Ed. Engl.* **15,** 713 (1976).
108. C. A. Coulson, *J. Chem. Soc.* p. 1442 (1964).

3

Sheet Silicate Intercalates: New Agents for Unusual Chemical Conversions*

J. M. THOMAS

I. INTRODUCTION

In the context of intercalation the most attractive features possessed by certain kinds of naturally occurring sheet silicates are the following. (a) They freely incorporate interlamellar water to a greater or lesser degree depending on both the chemical driving force (for example, P_{H_2O} if uptake proceeds from the vapor) and the nature of the interlamellar cation. (b) They may also equally readily take up (instead of or in addition to the water) interlamellar organic species of many different kinds, sometimes in the form of essentially neutral molecules, sometimes as ionized entities (e.g., amino acids), and

* This chapter is dedicated to the memory of Dr. S. H. Graham.

INTERCALATION CHEMISTRY

ISBN 0-12-747380-7

sometimes in both states. (c) They may be converted from the hydrophilic to the organophilic state, for example, simply by exposing the parent water-rich intercalate (when the interlamellar inorganic cation is hydrated) to an appropriate organic reagent such as a primary amine. (d) They exhibit strong acidity. More often than not, Brønsted acidity predominates either because of the influence of the strong internal electric fields exerted on the interlamellar water molecules (which leads to detachment of protons by dissociation) or because of the additional influence of certain hydrated interlamellar cations, notably Al^{3+} (for example, $M^{n+}(H_2O)_m \rightarrow [M(H_2O)_{m-1}OH]^{(n-1)+} + H^+$). The sheet silicate may also function as a Lewis acid in that there are identifiable sites (for example, "structure" Fe^{3+} ions) which function as electron-accepting centers. (e) They may be prepared in a variety of cation-exchanged forms, in which the original exchangeable cation (generally Na^+ in the naturally occurring clay minerals) is replaced, as desired, by one (or possibly more) of a whole range of inorganic or organic cations. When these ions are bulky and multiply charged, they serve as rather rigid, cross-linking rods or bridges. (f) They can be converted [see property (e)] to even more rigid, but open structures, less prone to intumescence upon uptake of organic or inorganic guests and also capable of exhibiting molecular sieve and related properties.

There are several distinct kinds of sheet silicates that freely form intercalates, well-known examples being montmorillonite, hectorite, vermiculite, and kaolinite, and in the context of solid-state and synthetic chemistry, and also topochemistry, the supreme advantage offered by these sheet silicates is that organic intercalates may be formed which give rise to several novel types of chemical reactions, the exploitation of which for synthetic purposes has only just begun. We may legitimately talk of "clay catalysts" in which special advantage is taken of the microenvironment of the intercalate intermediates. Abundant scope for tailoring these intercalates exists. Moreover, now that there are numerous powerful analytical and structural probes for exploring the identity of parent and intermediate sheet silicate compounds, we are aware of many more catalytic possibilities than even the substantial number that have already come to light.

In this chapter particular attention is paid to the reactions that have been explored by my colleagues and I, initially (since 1972) at the University College of Wales, Aberystwyth, and later (since 1978) at the University of Cambridge. Since 1976 we have also collaborated with Professor Purnell and Dr. Ballantine's group at the University College of Swansea. No attempt has been made to be comprehensive in coverage of all relevant background literature, since numerous previous reviews have served that purpose. Instead, we highlight some of the rather remarkable organic conversions that have been discovered using essentially smectite (i.e., montmorillonite- and hectorite-based) intercalates.

We begin by recalling some of the essential structural details, as well as some bold and rather incomplete statements of fact as to which sheet silicates do and which do not freely form intercalates. A summary of some of the structural techniques that have helped to elucidate the nature and properties of relevant organic intercalates is also given. But the main body of the chapter is devoted to a classification and description of the substantial

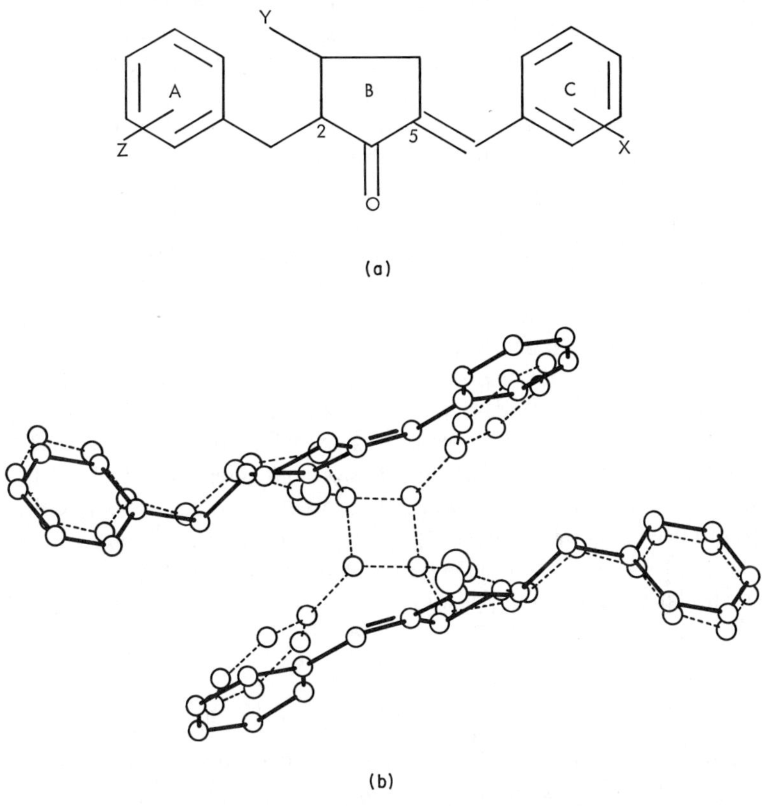

Fig. 1. 2-Benzyl-5-benzylidenecyclopentanone (BBCP) crystallizes in the space group *Pbca*. The monomer (a) packs in the crystalline state in such a way that neighboring molecules form incipient pairs (solid lines in b) so close together that, when the parent crystal is irradiated with UV light, dimer molecules are produced [by formation of cyclobutane rings (dashed lines in b)] in such a way that the single-crystal monomer is converted to the single-crystal dimer without phase separation and with smooth but minute changes in unit-cell parameters and unit-cell volume. In BBCP, X = Y = Z = H. [From Jones *et al.* (*4*); Nakanishi *et al.* (*5*).]

number of novel organic reactions that my colleagues and I have discovered during the course of our studies.

In some respects sheet silicate intercalate chemistry may be regarded as another manifestation of what is now designated (*1–5*) "crystal engineering." In organic solid-state chemistry, this involves selecting molecules according to their size, shape, or stereochemistry and then "persuading" them to crystallize into certain desired crystallographic space groups. Within those crystals, by appropriate stimulation, say with UV light, diffusionless intermolecular reactions are allowed to proceed. The secret is to select the parent molecules and crystal structure in such a way as to lead to one (topochemically dictated) product. The number of possible reaction coordinates "open" to the reactant species, which themselves are "fixed" in one (or perhaps two) specific conformational states, is strictly limited so that the solid-state reaction that ensues is clean and well-defined and generally yields crystalline products in very high yield. An example of such a tailored reaction (and product) is shown in Fig. 1, which refers to the cyclopentanone derivative BBCP (2-benzyl-5-benzylidenecyclopentanone).

In a few of the reactions of organic sheet silicates that we have investigated, reaction does indeed proceed in the solid state of the parent material (e.g., conversion of the 4,4′-diamino-*trans*-stilbene intercalate of montmorillonite to gaseous aniline). In the overwhelming majority of cases that we have studied to date, however, the reactions are effected in contact with solvent from which the guest molecules are initially introduced into the host silicate to form the intermediate intercalate. In some cases, reactions turn out to be "single pass," in the sense that once the crucial intercalate intermediate is formed and converted to product it is difficult to sustain the reaction by forming more of the intermediate. In many cases, however, the parent sheet silicate host may be continuously reused, and intercalates can form and react in specific fashion for long periods.

II. STRUCTURAL FEATURES

Of the various sheet silicates that are well known to form intercalates with guest organic molecules, there are two main structural types that concern us here: two-layered sheets, often termed kandites, and the three-layered sheets that embrace the smectites and vermiculites.

Kaolinite and Dickite. Regarded as polytypic variants of kaolinite itself, these are the best examples of the kandites. They have idealized formulas $Al_2Si_2O_5(OH)_4$ and may be conveniently regarded as the condensation–polymerization product of $Al(OH)_6$ "octahedral" (O) sheets with "tetrahedral" (T) sheets of $Si_2O_3(OH)_2$ (see Figs. 2 and 3). The exterior, basal faces of crystallites of kaolinite and dickite are therefore of two distinct kinds. One face exposes oxide ions (attached to Si^{4+} ions), and the other hydroxyls, which are attached to the Al^{3+} sublattice. There appears to be little (although by no means negligible) likelihood that a fraction of the cations (Al^{3+} and Si^{4+}) in the kandites would be replaced by ions of lower valency, so that these sheet silicates tend to be electrically neutral. (That substitution can occur; e.g., iron may replace aluminum in some kaolinites, as demonstrated by combined Mössbauer spectroscopy and electron microscopy.)

Serpentines. These sheet silicates have ideal composition $Mg_3Si_2O_5(OH)_4$ and are structurally similar to kaolinites. They merely have Mg^{2+} ions entirely replacing the Al^{3+} ions (i.e., $3Mg^{2+}$ for each pair of Al^{3+} ions) in the octahedral sheets.

Smectites, Vermiculites, and Micas. Electrically neutral structures, these silicates consist of three-layered sheets (symbolized in Fig. 2b) and have idealized formula $Al_2(Si_4O_{10})(OH)_2$ (otherwise known as *pyrophillite*) or

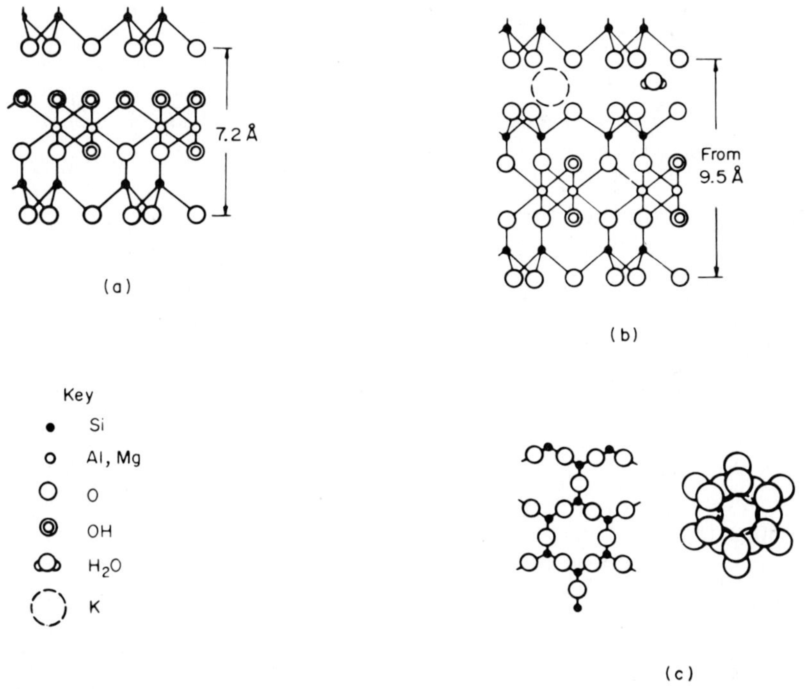

Fig. 2. Two-layered (a) and three-layered (b) variants of sheet silicates. In part (a) one of the exposed faces consists of oxide ions which are part of hexagonal sheets of linked SiO_4 tetrahedra. In part (b) both exposed faces consist of such SiO_4 tetrahedra (see text). Part (c) shows the hexagonal (6-ring) depressions viewed from above the "SiO_4" layer.

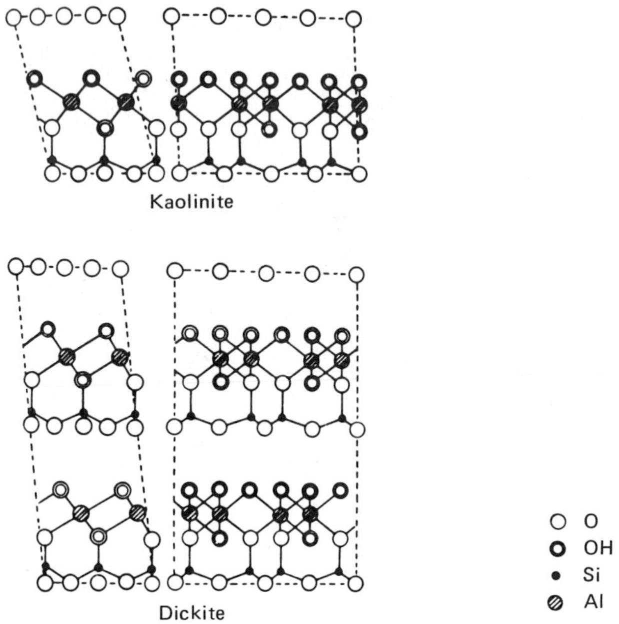

Fig. 3. The crystal structure of kaolinite and dickite viewed along [010]. Contrast this structure with that of vermiculite (see text) in Fig. 4.

$Mg_3(Si_4O_{10})(OH)_2$ (otherwise known as *talc*) depending on whether the central cationic sites in the octahedral network are occupied by Al^{3+} or Mg^{2+}. The former (pyrophillite) has a so-called dioctahedral structure (i.e., only two-thirds of all possible octahedral sites are occupied), whereas the latter (talc) has a trioctahedral structure since the Mg^{2+} ions occupy all three such sites in the unit cell. Whereas the kandite structures can be symbolized by TO, the micas are examples of TOT structures.

If substitution of Si^{4+} by Al^{3+} occurs in the tetrahedral networks, the three-layered sheets take on a surplus negative charge. When one-quarter of the Si^{4+} ions in pyrophillite are replaced by Al^{3+} and the compensating cations that must be incorporated interlamellarly are K^+, the result is *muscovite mica*, idealized formula $KAl_2(AlSi_3O_{10})(OH)_2$. Likewise, when one-quarter of the Si^{4+} ions in talc are replaced by Al^{3+} and the interlamellar compensating cations are again K^+, the compound is *phlogopite mica*, idealized formula $KMg_3(AlSi_3O_{10})(OH)_2$. Neither of these micas shows much tendency to take up guest organic species, presumably because of the strong electrostatic forces that have to be overcome to separate the three-layered sheets.

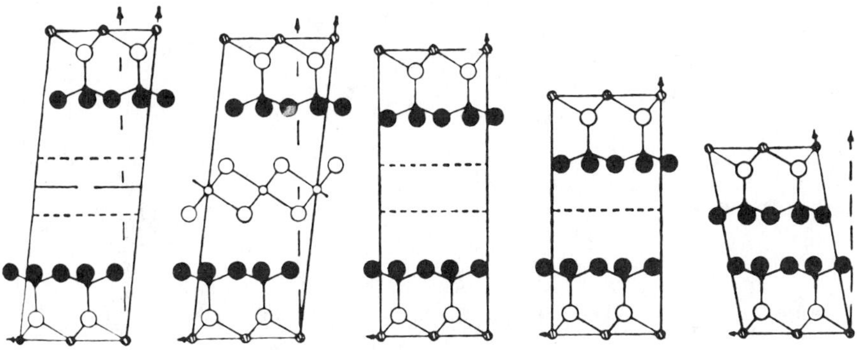

Fig. 4. Cross section of Mg-vermiculite. The sequence shows the silicate layer at various stages of dehydration: (a) the 14.81 Å complex (with water), (b) 14.36 Å complex, (c) 13.82 Å complex, (d) 11.59 Å complex, and (e) 9.02 Å phase. The dashed lines in parts a, c, and d represent the water layers. [After Barrer (*12*).]

Apart from the substitution of Si by Al in the tetrahedral network (which introduces one negative charge on the sheet for each Al), several other kinds of ions may replace Al^{3+} ions in the octahedral network, especially when those substituting ions have a radius comparable to that of Al^{3+}. Thus, Li^+, Mg^{2+}, and Li^+, Fe^{2+} may replace Al^{3+} in the octahedral layer with no alteration in the surplus negative charge on the sheet. Hence, the dioctahedral mica (e.g., muscovite) is converted to a trioctahedral mica (phlogopite). In *lepidolite mica*, in addition to the replacement of Al^{3+} by Li^+ in the octahedral sublattice, some hydroxyls are also replaced by fluoride ions. In this mica, ideal formula $K(Li_2,Al)(Si_4O_{10})(F,OH)_2$, there is isomorphous substitution in the octahedral cation sublattice as well as in the associated anions. In principle, it is generally possible to have substitution not only in these two kinds of site but also, as we have already seen, in the tetrahedral sublattice. In the *vermiculites*, for example, Mg^{2+} (Fig. 4), and to a somewhat lesser degree Fe^{2+}, can replace Al^{3+} in octahedral sites and Al^{3+} replaces Si^{4+} in the tetrahedral sites. In its most general form we write the idealized formula for vermiculite as $(Ca,Mg)_{x/2}(Mg,Fe,Al)_3[(Al,Si)_4O_{10}](OH)_2$. However, a simpler and reasonably accurate description for most vermiculites (the precise compositions of which vary according to their geological provenance) is $M_x(Mg)_3[(Al_xSi_{4-x})O_{10}](OH)_2$, where M is a monovalent, exchangeable cation, and x, as before, is the number of equivalents of exchangeable cations present.

In the *smectites* (Table I), with which we are principally concerned in this chapter, Li^+, Mg^{2+}, and Fe^{2+} (or Fe^{3+}) can replace Al^{3+} in octahedral

TABLE I

Idealized Formulas of Some Important Smectite (TOT) Sheet Silicates[a]

Mineral	Idealized formula	Cation-exchange capacity (mEq/100 gm)
Montmorillonite	$Na_x(Al_{2-x}Mg_x)(Si_4O_{10})(OH)_2 \cdot mH_2O$ (Mg^{2+} in octahedral layer can be replaced by Fe^{2+})	All in range 80–150
Saponite	$Ca_{x/2}Mg_3(Al_xSi_{4-x}O_{10})(OH)_2 \cdot mH_2O$	
Beidellite	$Na_xAl_2(Al_xSi_{4-x}O_{10})(OH)_2 \cdot mH_2O$ (tetrahedral replacement of Si^{4+} by Al^{3+})	
Hectorite	$Na_x(Mg_{3-x}Li_x)(Si_4O_{10})(OH)_2 \cdot mH_2O$	
Fluorohectorite	$(Na,Ca)_{x/2}(Mg_{3-x}Li_x)(Si_4O_{10})(OH,F) \cdot mH_2O$	
Vermiculite	$(Na,Ca)_{x/2}(Mg_3)(Al_xSi_{4-x})O_{10}(OH)_2 \cdot mH_2O$	
Pyrophyllite	$Al_2(Si_4O_{10})(OH)_2$	
Talc	$Mg_3(Si_4O_{10})(OH)_2$	
Muscovite	$KAl_2(AlSi_3O_{10})(OH)_2$	252
Phlogopite	$KMg_3(AlSi_3O_{10})(OH)_2$	240

[a] For comparison, some micas, pyrophyllite, and talc are included.

layers and Al^{3+} can replace Si^{4+} in tetrahedral layers. In synthetic smectites it is relatively easy to arrange for the OH^- ions to be replaced by F^-, and this is one of the principal features of fluorohectorite.

It is important to note that the area of a single 6-ring (Fig. 2c) in the outer part (T layer) of the micas, vermiculites, and smectites is close to 24 Å^2. The interlamellar cations naturally present in these sheet silicates can therefore be partly or totally keyed into these hexagonal (6-ring) depressions in the anhydrous, outgassed solids. It is significant that, in muscovite, phlogopite, lepidolite, and other micas that show essentially no tendency to incorporate other interlamellar entities (for example, biotite, margarite, and zinnwaldite), the area available to the interlamellar cation is ca 24 Å^2. In other words, there are no vacant 6-ring depressions. On the other hand, with the vermiculites and the smectites, because the charge density of the TOT layers is substantially smaller than that of the micas, much larger areas (for example, 35–120 Å^2) are available per interlamellar cation, so that, even in their anhydrous forms, these sheet silicates have only a fraction of their 6-ring depressions occupied. Two useful conclusions follow. First, we see why, on electrostatic grounds, it should be easier to separate the triple (TOT) layers in smectites and vermiculites than in micas. Second, when these swellable clays incorporate substantial quantities of water and/or organic "solvent," the solvated ions are no longer "pinned" to the depressions they would preferentially occupy in the anhydrous state, and a fair degree of mobility, in the

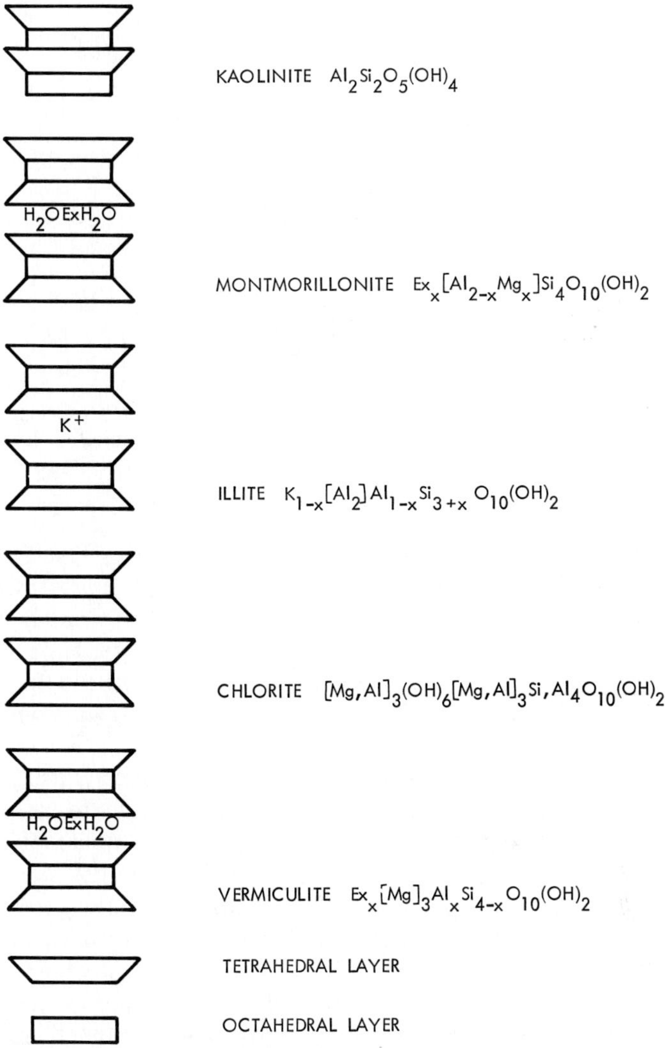

Fig. 5. A shorthand method of representing the structures of the principal members of the clay minerals. Ex, exchangeable cation.

interlamellar regions, ensues. In other words, the solvated ions will tend to "float." Likewise, bulky organic cations will not, in general, be bound to the 6-ring depressions.

A useful shorthand method of representing the structures of and interrelationships between kaolinite, montmorillonite (the most important smectite), mica, and vermiculite is shown in Fig. 5. Figure 6 shows two other

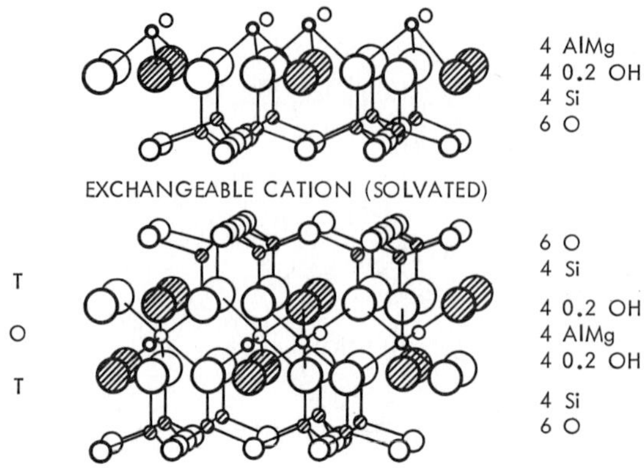

Fig. 6. The parent montmorillonite structure. The tetrahedral (T) and octahedral (O) sublattices are emphasized and the TOT notation explained. The individual layers are not regularly stacked (as in the micas). The structure is turbostratic (see text). [From Brown (*10*).]

ways in which the parent montmorillonite structure, as established by Hofmann *et al.*, Marshall, Hendricks, and MacEwan, is represented (*6–8*). There is no long-range order in the stacking of the individual layers in montmorillonite. It is a so-called turbostratic structure, in contrast to dickite, which has a two-layer repeat [compare *c*-axis repeats in kaolinite (7.37 Å) and in dickite (14.42 Å) and muscovite, which occurs with both a ca 20 Å and a ca 30 Å *c*-axis repeat (*9*, *10*)].

The cation-exchange capacity (CEC) of a sheet silicate is itself a useful guide, qualitatively and quantitatively, as far as tendency to intercalate is concerned. The unit commonly used for this purpose is milliequivalents of interlamellar cation that may be exchanged per 100 gm of silicate. For the micas (muscovite and phlogopite) the CEC is ca 250 mEq per 100 gm; it is somewhat more (ca 270) for lepidolite, rather less for vermiculites (100–170), and a good deal less (60–120) for the smectites as a family. Again, the precise values depend on the extent and nature of the isomorphous substitution, which, in turn, is governed by the manner of synthesis, whether it be in nature or in the laboratory. As a rough rule, it is useful to bear in mind that montmorillonites posses a surplus electronic charge on their TOT layers of ca 0.5–0.7 per formula unit.

Further information pertaining to structure, including tabulations of *d* spacings of various clay minerals in their "air-dried" and wet states, is given in standard references, notably the books by Grim (*11*) and Barrer (*12*). A summary of the cation-exchange capacities, expressed in the conventional units of clay mineralogy, milliequivalents per 100 gm, is given in Table II.

TABLE II

Typical Magnitudes of the Cation-Exchange Capacity of Some Sheet Silicates

Sheet silicate	Cation-exchange capacity (mEq/100 gm)
Kaolinite	3–15
Halloysite · $2H_2O$	5–10
Smectites	60–120
Vermiculites	100–140

III. TECHNIQUES FOR ELUCIDATING THE NATURE OF INTERCALATED CLAYS

In this section we briefly and selectively cite the various techniques that are used to identify and characterize intercalated clays.

Simple measurements of the amount of guest species taken up by a host clay, especially when expressed in mole percent or as a fraction of the theoretically available interlamellar space, soon affords insights as to whether genuine intercalation has taken place. Gravimetric methods, particularly thermogravimetry, in which both sorption and desorption can readily be monitored, quickly reveal whether intercalation, rather than just multilayer physical adsorption, is involved. X-ray diffraction is extremely helpful since there is usually an increase in the *c*-axis repeat distance (the interlayer spacing) when organic molecules are taken up. This simple test, although generally reliable, can be misleading: There are now many instances in which it is known on other grounds (see below) that intercalation has indeed taken place but there is very little change in *d* spacing. A carefully recorded X-ray powder diffractometer trace, in which the intensities of each peak are readily seen, does, however, offer generally reliable information as to the occurrence of intercalation.

Having established that intercalation is a reality, several methods are then, in principle, available for more precise characterization.

1. Detailed X-ray and/or neutron diffraction studies may be performed on highly oriented specimens. These yield one-dimensional Fourier maps, from which the orientation of the guest molecule in the interlamellar space may be inferred. If the intercalate is very well ordered, it may be possible to record a full three-dimensional structure of the complex. Examples are given later.

2. Infrared and other spectroscopic techniques, including X-ray-induced photoelectron spectroscopy (XPS or ESCA, electron spectroscopy for chemical analysis), X-ray photoelectron diffraction (XPD), Mössbauer spectroscopy, NMR, and ESR may also be invoked.

3. Electron microscopy may be used in certain cases. However, these techniques are used to best advantage when a stable intercalate is formed. When chemical reactions occur freely within the microenvironment of the interlamellar region, it is difficult to monitor directly the changes that accompany, or illuminate the nature of, the changes themselves. Frequently, one simply monitors the product distribution or amount of converted reactant as a function of contact time with the clay intercalate. Later sections in this chapter, particularly those describing studies of the etherification of 1-alkenes, illustrate this point. In these studies, as we shall see, mass spectrometry, gas chromatography, X-ray diffraction, infrared, and NMR were all used to piece together the nature of the chemical conversions that had taken place.

A. *d* Spacings

A selection of typical *d* spacings (in the c^* direction) for a number of cation-exchanged montmorillonites is given in Table III.

TABLE III

Typical *d* Spacings for Certain Cation-Exchanged Montmorillonites (*8*)

System[a]	Approx. formula[b]	Basal spacing (Å)
Na^+-Mont hydride	$Na_{0.5}M \cdot 3.5H_2O$	12.4
Na^+-Mont–tetrahydropyran	$Na_{0.5}M \cdot (C_5H_{10}O)_{1.3}$	14.99
Cu^{2+}-Mont hydride	$Cu_{0.25}M \cdot 3.5H_2O$	12.5
Cu^{2+}-Mont–azobenzene	$Cu_{0.25}M(C_{12}H_{10}N_2)_{0.8}$	20.5
Ni^{2+}-Mont hydride	$Ni_{0.25}M \cdot 7.0H_2O$	15.1
Ni^{2+}-Mont–tetrahydrofuran	$Ni_{0.25}M(C_4H_8O)_{2.3}$	14.50
Sr^{2+}-Mont–γ-butyrolactone	$Sr_{0.25}M \cdot (C_4H_6O_2)_{4.8}$	23.1
	$Sr_{0.25}M \cdot (C_4H_6O_2)_{3.2}$	18.3
	$Sr_{0.25}M \cdot (C_4H_6O_2)_{1.6}$	13.2
Sr^{2+}-Mont–tetrahydropyran	$Sr_{0.25}M \cdot (C_5H_{10}O)_{1.0}$	14.81
Sr^{2+}-Mont–1,4-dioxan	$Sr_{0.25}M \cdot (C_4H_8O_2)_{1.1}$	15.00

[a] Mont ≡ montmorillonite (≡ bentonite).
[b] M ≡ $(Al_{3.5}Mg_{0.5})Si_8O_{20}(OH)_4$.

B. Fourier Plots from X-Ray and Neutron Data

More revealing than the bold statement of interlayer spacing are the one-dimensional electron density maps [projected onto the z direction from the Fourier summation of the structural factors of the (00l) reflexions]. For results of the required detail, l must be as large as possible, typically greater than 10. Figure 7 shows rather well how, merely by replacing Na^+ by Sr^{2+} as an interlamellar cation, one can alter the disposition of the tetrahydropyran (THP) with respect to the silicate sheets (*1*, *8*, *13*). In both cases the THP is essentially perpendicular to the sheets, but, whereas the oxygen is nearest to the linked SiO_4 tetrahedra of the silicate sheets with Na^+ as the interlamellar cation, the molecule rotates about an axis parallel to the sheets, and by ca 60°, when Na^+ is replaced by Sr^{2+}.

It can be shown that electron density maps from X-ray intensities, and indeed the X-ray diffractograms themselves, show significant differences depending on whether an intercalated molecule such as n-hexylamine lies

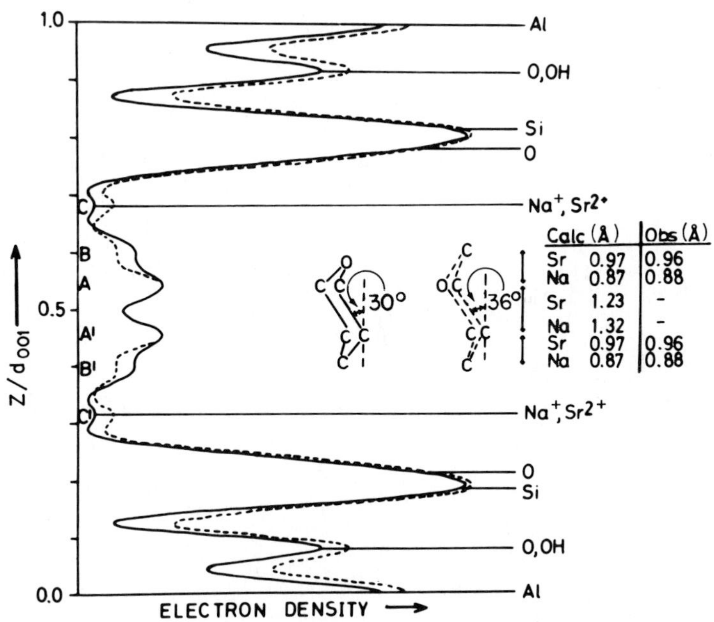

Fig. 7. Electron density distribution of Na^+-exchanged (——) and Sr^{2+}-exchanged (---) montmorillonite with intercalated tetrahydropyran. Subtle changes in the orientation of the guest molecules may result from such minor changes as the replacement of Na^+ ions by Sr^{2+} as exchangeable cations. [From Thomas (*1*); Thomas *et al.* (*8*).]

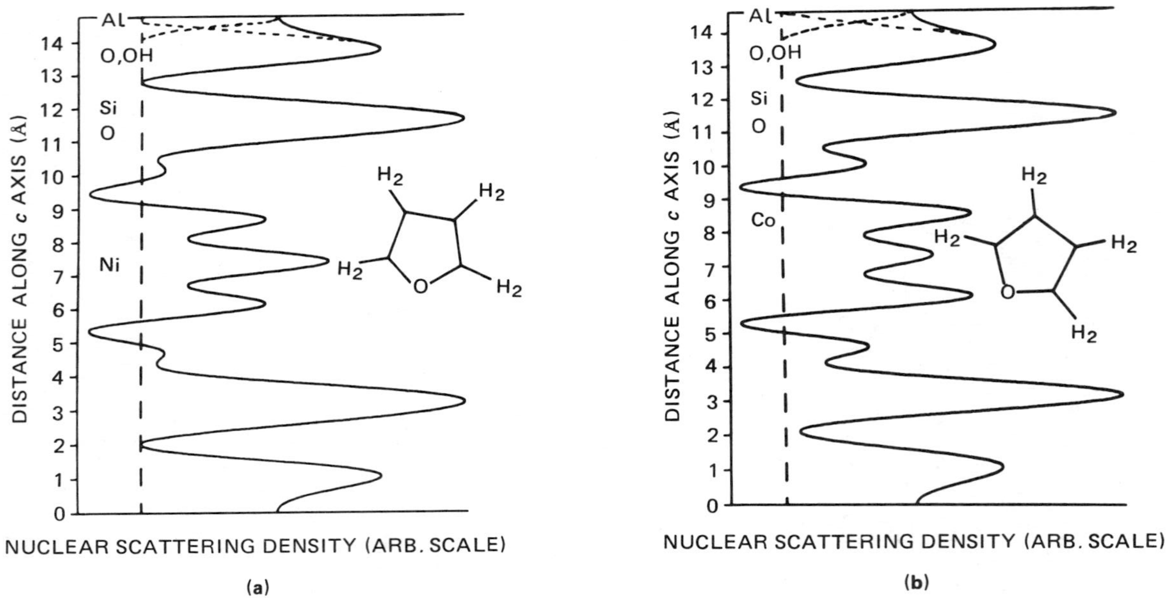

Fig. 8. Projections of nuclear scattering density for (a) Ni^{2+}-montmorillonite–tetrahydrofuran (THF) (2.3THF per formula unit) and (b) Co^{2+}-montmorillonite–tetrahydrofuran (2.2THF per formula unit). The Ni^{2+} ion is situated centrosymmetrically within the interlamellar space but the Co^{2+} ion is not. [From Thomas *et al.* (*8*).]

parallel to the sheets but with the zigzag pattern of an sp^3-bonded carbon atom perpendicular to the sheet, or whether the backbone is still parallel but with the zigzag pattern also parallel to the sheets (*14*).

When neutron diffraction is employed, the structural detail attained is greater than with X-ray diffraction since, *inter alia*, the positions of H atoms can be located more readily. This point is well illustrated (*8*) in our study of Ni^{2+}- and Co^{2+}-substituted montmorillonites into which tetrahydrofuran has been intercalated. Figure 8, which shows Fourier maps derived from neutron diffraction data, clearly indicates that, whereas the Ni^{2+} ion in the tetrahydrofuran intercalate of Ni^{2+}-exchanged montmorillonite $[(Al_{3.5}Mg_{0.5})Si_8O_{20}(OH)_4 \cdot Ni_{0.25} \cdot (C_4H_8O)_{2.3}]$ is situated centrosymmetrically in the interlamellar region, the Co^{2+} ion in a very similar intercalate $[(Al_{3.5}Mg_{0.5})Si_8O_{20}(OH)_4 \cdot Co_{0.25}(C_4H_8O)_{2.2}]$ is off-center.

C. Kinetic Data from X-Ray and Neutron Studies

Neutron diffraction studies of the stable pyridine (py) intercalate of Na^+-exchanged montmorillonite [basal spacing 14.8 Å; idealized formula $Na_{0.5}{}^+(Al_{3.5}Mg_{0.5})Si_8O_{20} \cdot (OH)_4 \cdot py \cdot 2H_2O$] showed that the guest (py) molecule is oriented perpendicular to the aluminosilicate sheets with its C—N axis inclined at 60°. From kinetic X-ray diffraction studies (*8, 15*), it proved possible to follow the dynamics of the conversion,

$$2[Na_{0.5}{}^+(Al_{3.5}Mg_{0.5})Si_8O_{20}(OH)_4] \cdot (py)_4 2H_2O \xrightarrow[+H_2O]{-py}$$

(23.3 Å *d* spacing)

$$2[Na_{0.5}{}^+(Al_{3.5}Mg_{0.5})Si_8O_{20}(OH)_4] \cdot (py)_2 \cdot 4H_2O$$

(14.8 Å *d* spacing)

at room temperature. Likewise, the intercalate first formed when γ-butyrolactone is brought into contact with Sr^{2+}-exchanged montmorillonite has a *d* spacing of 23.1 Å, corresponding to a so-called three-layered intercalate (see Fig. 9a). At room temperature this intercalate changes into a two-layered analog by liberation of γ-butyrolactone, *d* spacing of 18.3 Å. Eventually, a single-layered intercalate is formed, *d* spacing 13.2 Å. Dynamic X-ray studies yield electron density maps (Fourier plots), which indicate the ways in which the organic molecules are oriented in each of these three intercalates.

D. Three-Dimensional Crystal Structures of Intercalates

Not surprisingly, in view of the turbostratic nature of montmorillonite, it is not possible to establish by diffraction techniques a three-dimensional

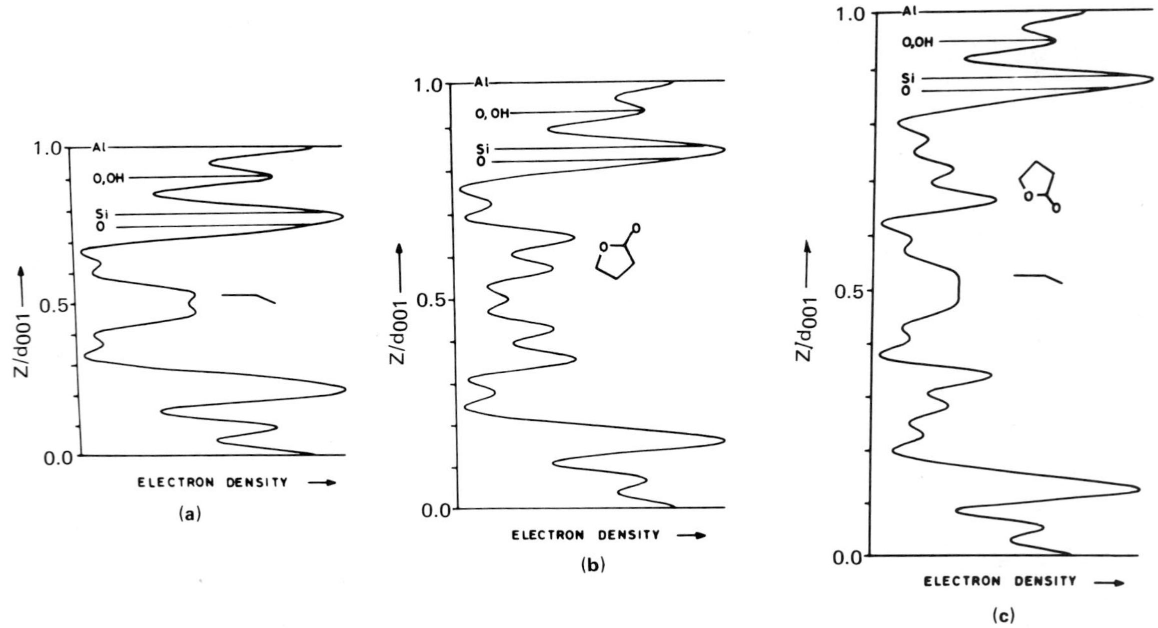

Fig. 9. Electron density projections for the three distinct "phases" of the Sr^{2+}-montmorillonite–γ-butyrolactone intercalate (1.6, 3.2, and 4.8 γ-butyrolactones per formula units in parts c, b, and a, respectively). [From Thomas *et al.* (*8*).]

structure for the intercalate; order in a direction perpendicular to the basal planes simply does not exist. For the two-layer-based (kandite) intercalates, however, the situation is rather different, since with dickite, as well as with some of its intercalates with simple organic guests, three-dimensionally ordered structures are indeed readily formed. The results obtained (*16*) with the dickite–formamide intercalate are shown in Fig. 10, from which two principal points emerge. First, rather extensive "anchoring," via five guest–host hydrogen bonds, is involved. Second, the formamide molecule is so

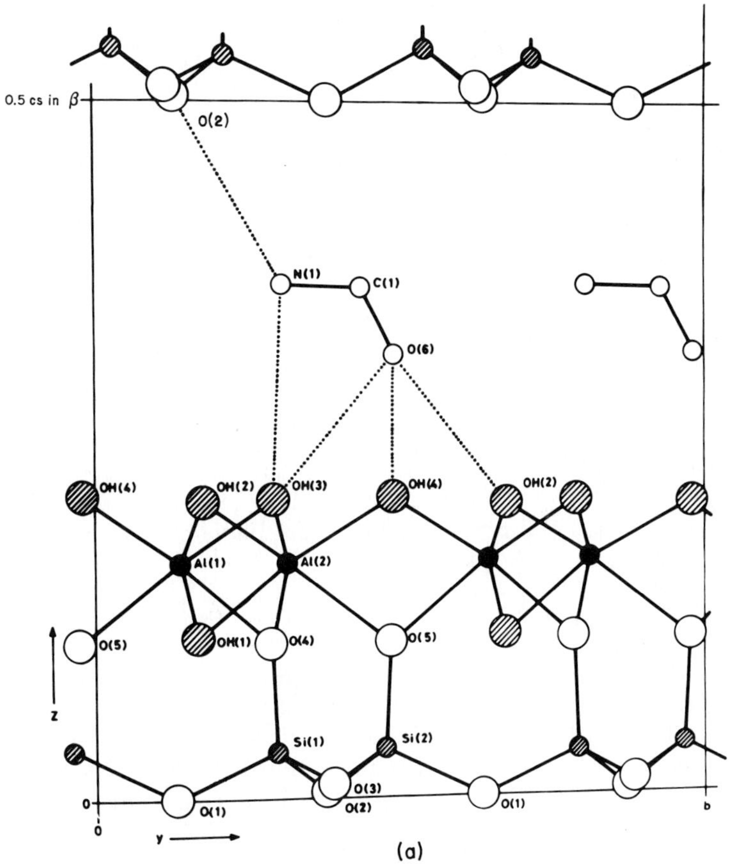

Fig. 10. (a) Projection of the structure of the dickite–formamide intercalate along the *a* axis; (b) projection of the AlO_6 octahedra of the dickite–formamide intercalate onto the *ab* plane. The hydrogen bonding of the formamide molecules to the aluminosilicate is indicated by dotted lines. [From Thomas *et al.* (*8*).] (*continued*)

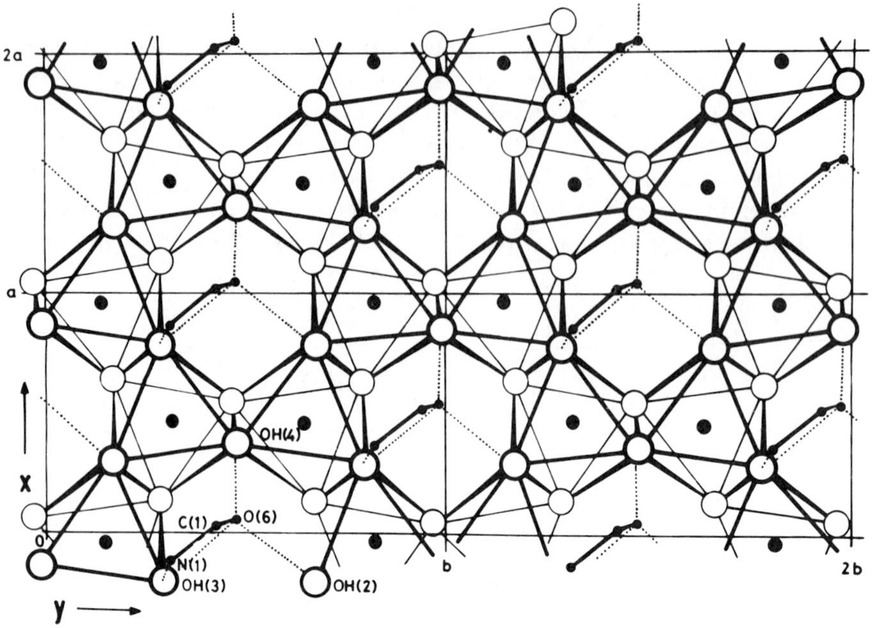

Fig. 10 (*continued*)

compact that it can nestle neatly into one of the hexagonal pockets mentioned earlier, so that, with this intercalate, no superlattice is formed and the relationship between the respective guest and host lattices is commensurate. [This is not the situation that prevails when simple guests, e.g., $FeCl_3$ or $FeCl_2$, are intercalated by graphite (*17*, *18*).]

E. Spectroscopic Studies

It is well known that infrared, visible, and UV spectroscopies yield useful information about the intercalation of organic molecules in clay minerals; see the monograph by Farmer (*19*) for more details. The lattice modes, as well as specific vibrational frequencies associated with characteristic groupings (H_2O "bending," Si—O—Si stretch, FeAlOH stretch, including distinguishing Fe^{2+}—OH—Al and Fe^{3+}—OH—Al deformations, librations of lattice hydroxyls, bending and stretching frequencies of interlamellar water or amines, protonated NH_2 group stretching frequencies, etc.), are readily detected, and their identification is usually straightforward (*20*). The presence of alkyl carbenium ions can also be detected in this way, as can "displacement" reactions which entail, for example, the elimination of most of the water in solvation shells of interlamellar cations (Ni^{2+}, Al^{3+}) when films of "wet" clay are exposed to certain amines (*21*).

Two techniques that have not yet been fully exploited for the study of clay intercalates are XPS and NMR. The technique has shown (*15*), e.g., that the pyridine attached to Na^+ ions in the interlamellar regions of the Na^+-exchanged montmorillonite intercalate with this base is retained even in vacua of $10^{-6}\ N \cdot m^{-2}$ at room temperature and that the nitrogen atoms are situated within essentially one type of environment.

By using the recently developed variation of photoelectron spectroscopy known as XPD, it is possible to differentiate among equivalent, nearly equivalent, and nonequivalent sites occupied by two (or more) elements either in a single crystal or in crystals of closely similar structure, even when the element(s) concerned comprise only a small fraction of the crystal and when the sublattice lacks both long- and short-range order [see Thomas (*22*) and Adams *et al.* (*23–25*)]. Although this technique is not well suited, as yet, to the study of intercalates in which there is considerable motion exhibited by interlamellar species, it has proved especially useful in exploring the immediate environment of ions in some layered silicates. Thus, the principal interlayer cations in the vermiculite single crystals studied by Evans, Adams, and Thomas (*25*), namely, potassium and calcium, could be entirely replaced with lead by prolonged refluxing in lead nitrate solution. The lead and calcium retain their hydration spheres, whereas potassium coordinates directly (without hydration) to the layer oxygen, as in the true micas.

We have demonstrated (*26*) that pulsed Fourier transform NMR spectroscopy can be used to assess (a) the degree of molecular freedom, (b) the composition of mixtures, and (c) the keto–enol equilibrium of certain organic species retained in the interlamellar spaces of a synthetic hectorite, idealized formula $Na_{0.66}(Li_{0.66}Mg_{5.34})Si_8O_{20}(OH)_4$. High-resolution ^1H- and ^{13}C-NMR spectra were obtained using a conventional Varian CFT-20 spectrometer at 80 and 20 MHz, respectively. Both proton and carbon spectra were obtained using a spinning sample.

Typical spectra of powdered specimens of the organic intercalates at 30°C are shown in Figs. 11 and 12. Although the proton peaks are rather broad, they are much narrower than would be obtained from solid *p*-xylene or solid γ-butyrolactone, indicating that these organic molecules have considerable freedom of motion in the interlamellar region. The two proton peaks for the xylene intercalate have relative intensities of 2:3, and the chemical shifts are very similar to the aromatic and methyl resonances in the high-resolution solution spectrum. The ^{13}C spectrum shows separate sharp absorptions for

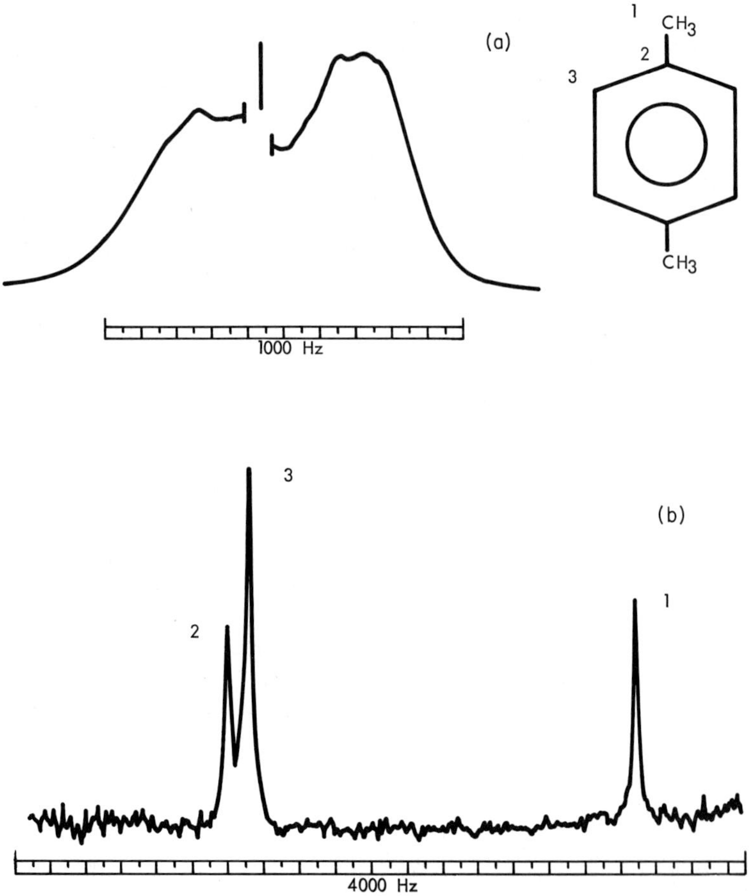

Fig. 11. (a) The ^{1}H-NMR spectrum of the *p*-xylene–hectorite intercalate at 80 MHz (50 pulses). The vertical line indicates the position of a small, sharp peak arising from HOD impurity in the D_2O lock reference, which has been deleted for clarity (see also Fig. 12a). (b) The ^{13}C-NMR spectrum of the same intercalate at 20 MHz (4000 pulses). The numbers identify the assigned carbon atoms in the aromatic molecules (inset of part a).

each of the three different types of carbon in the molecule (linewidth ca 25 Hz), and the relative chemical shifts (C-1 = 0.0 ppm, C-2 = -114.0 ppm, and C-3 = -108.5 ppm) are identical to those of the pure liquid (0.0, -114.0, and -108.5 ppm, respectively), again indicating that, motionally, the molecules are relatively free. The ^{13}C spectrum of γ-butyrolactone (Fig. 12) shows four distinct peaks, and good-quality, readily interpreted NMR spectra are obtained from hectorite intercalates of several

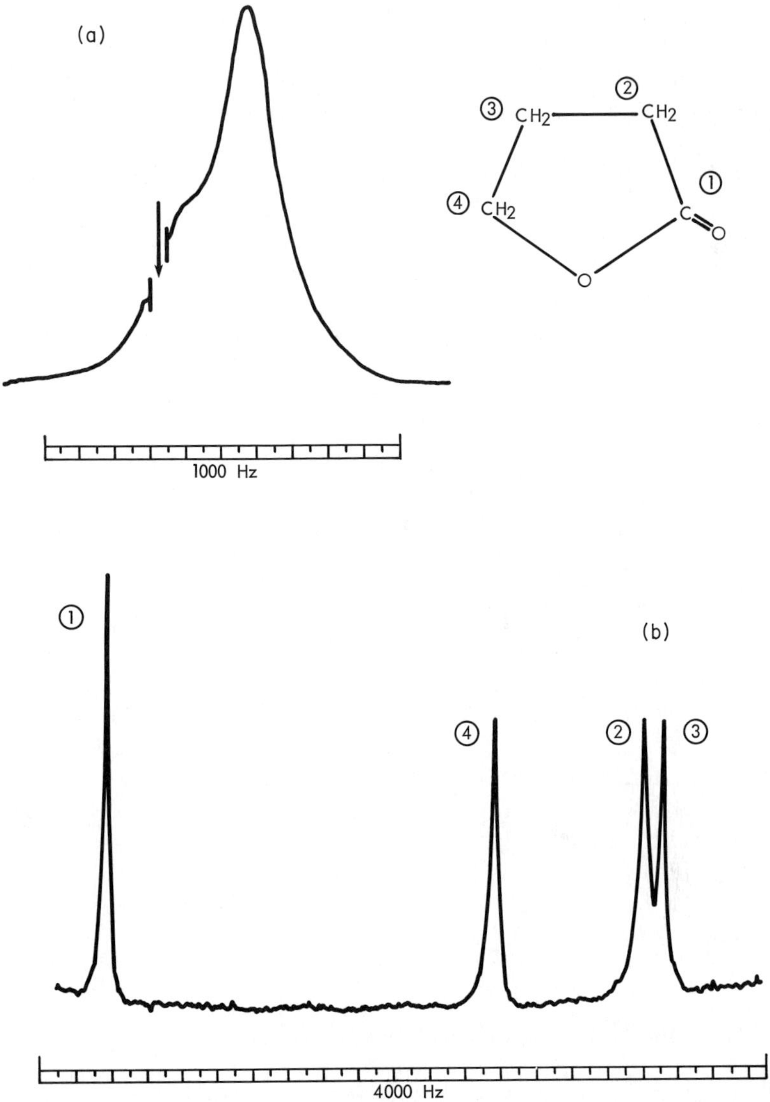

Fig. 12. (a) The ^{1}H-NMR spectrum of the γ-butyrolactone intercalate of hectorite at 80 MHz (15 pulses). (b) The ^{13}C-NMR spectrum of the same intercalate recorded at 20 MHz (16,000 pulses).

other organic molecules including tetrahydrofuran, ethylbenzene, and 1,3-pentanedione.

The composition of ethylbenzene–*p*-xylene mixed intercalates could readily be deduced from the ^{13}C-NMR spectra of the powdered solid. Moreover, the keto–enol equilibrium of 1,3-pentanedione in the interlamellar space is seen, by *in situ* NMR measurement, to be displaced in favor of the enol form.

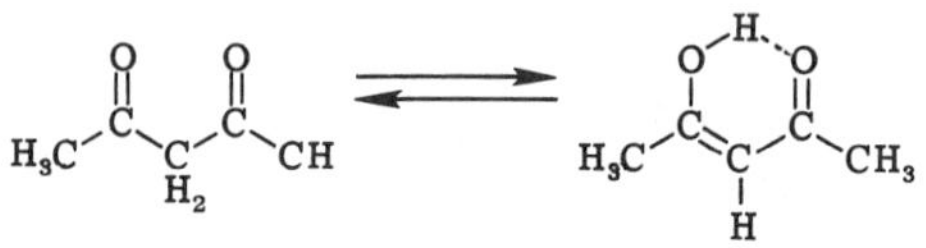

It proved readily possible with these systems to measure, by standard methods, ^{13}C spin–lattice relaxation times. For the *p*-xylene intercalate the values are C-1 = 490 msec, C-2 = 430 msec, and C-3 = 480 msec, essentially indistinguishable from one another within the accuracy of the experiment. (Compare the corresponding values of 11.8, 13.5, and 44.7 sec for the pure *p*-xylene as liquid.) These differences again point to the integration of the xylene guest into the sheet silicate host. We have also shown, and shall describe more fully elsewhere, that magic-angle spinning, as expected (*27*), enhances the resolution of both the proton and the carbon spectra of these intercalates.

Another technique that is surprisingly revealing concerning charge-transfer processes involving intercalates is Mössbauer spectroscopy (*28–30*). This approach affords direct evidence for the involvement of lattice-substituted iron (III) as an electron-accepting site within montmorillonite following intercalation of the clay by benzidine (which results in the production of benzidine radical-cations).

F. Direct Imaging by High-Resolution Microscopy

Although aluminosilicates are not, in general, as resistant to electron beam irradiation as other layered solids that readily form intercalates, it is possible to image directly the individual sheets of some clay minerals under appropriate conditions. Graphite and transition metal chalcogenides, in both their original and intercalated forms [see Thomas *et al.* (*18*) and (*31*), respectively], are rather easy to image by modern high-resolution electron microscopic techniques (*32, 33*). A high-resolution electron micrograph

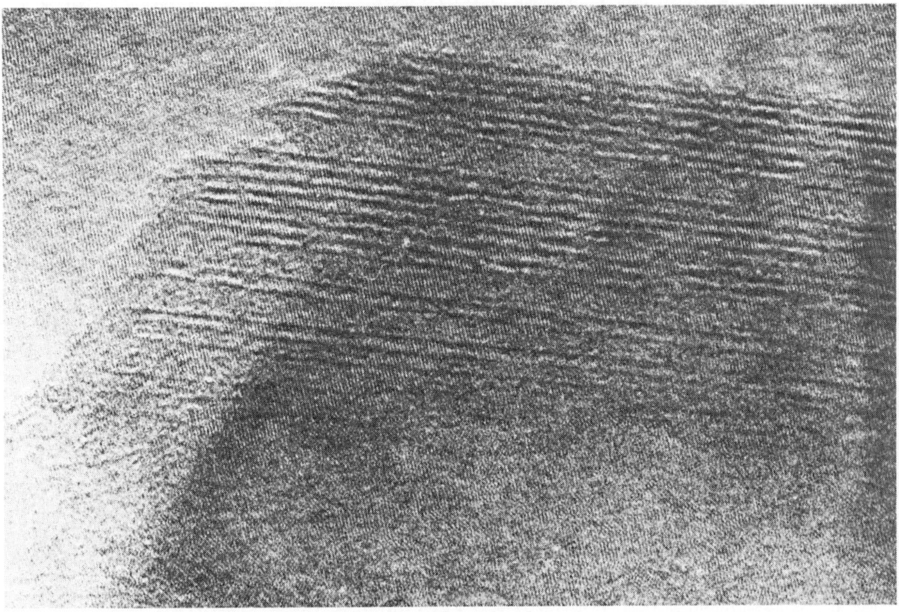

Fig. 13. High-resolution electron micrograph showing two-dimensional structural detail in a "heat-stable" Al-montmorillonite. The fringes parallel to the free surface represent the layers, spaced 14.5 Å apart. The 4.5 Å fringes also visible arise from either (110) or (020). [From Williamson (*20*).]

showing the (00*l*) spacings (14.5 Å) of a "heat-stable" aluminum-exchanged montmorillonite, as well as the 4.5 Å spacings corresponding to (110) and (020) planes, is shown (*34*) in Fig. 13. In micrographs of organomontmorillonites taken by Yoshida (*35*), regular and random interstratification is directly observed. Some of the fringes observed from these clays are seen to be deflected, a fact that is attributable to variation in the orientations of the organic molecules within individual layers, caused by inhomogeneities in the charge density in the aluminosilicate sheets. Such localized "defects" are also visible in Fig. 13; these would not have been readily manifested in X-ray diffractograms.

By way of contrast, Fig. 14 shows a typical example of a high-resolution micrograph of a randomly intercalated graphite–ferric chloride intercalate. Unlike the sheet silicates, graphite as a host structure exhibits the phenomenon of "staging," in which only some of the interlamellar spaces become occupied by guest species.

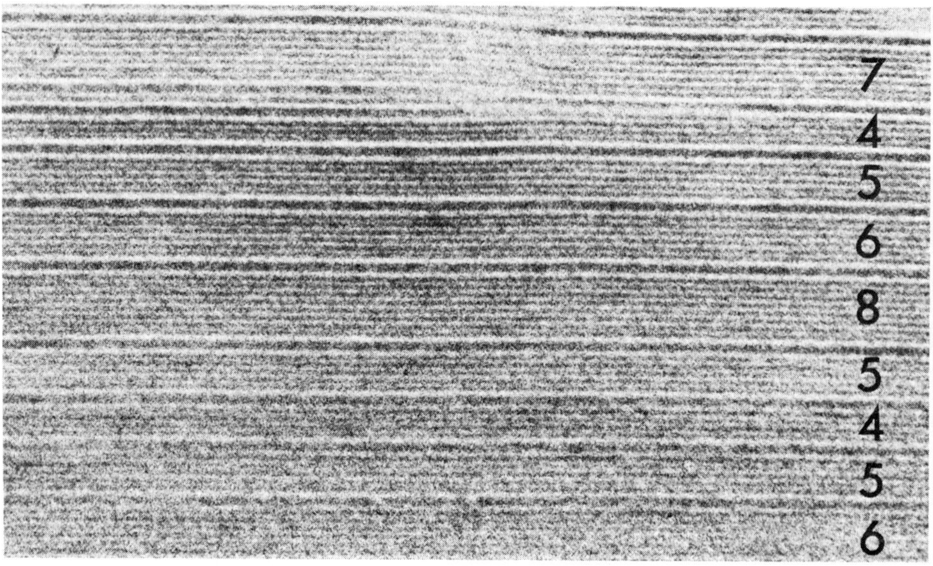

Fig. 14. High-resolution micrograph illustrating the phenomenon of staging in graphite–$FeCl_3$ intercalates. The individual $FeCl_3$ layers do not enter every available interlamellar space (which has a thickness of 3.35 Å). [From Thomas *et al.* (*18*).]

IV. UNUSUAL CHEMICAL CONVERSIONS OF ORGANIC MOLECULES USING SILICATE INTERCALATES

It is important at the outset to distinguish the many reactions that are catalyzed at clay mineral (in this context smectite or kandite) *surfaces*, on the one hand (*8*, *36–38*), from those novel organic reactions that take place *between* the individual silicate *sheets*, or at least involve the intercalation of one or more of the reactants, on the other. Whereas the former category encompasses a wide variety of examples (including the synthesis of amino acids, purines, pyrimidines, and hydrocarbons), the latter is restricted to a relatively small, but growing number of well-documented reactions. Previous work carried out by other investigators has been summarized elsewhere (*7*, *39–43*).

It is also important to recognize the *degrees of control* that are available in adjusting the silicate host before the chemical reaction. Thus, the nature of the interlamellar cation. as mentioned in Section I, may be modified by straightforward cation exchange. Furthermore, the surrounding medium in the microenvironment of the interlamellar space can be predetermined within quite broad limits. By simply allowing the cation-exchanged montmorillonite or hectorite to stand in well-defined conditions of relative humidity, a greater or lesser amount of interlamellar water may be incorporated. [This may influence both the nature and the extent of the subsequent reaction as, for example, in the etherification of alk-1-enes (see Section IV,A,6).] In addition, the water may be largely displaced by one of the major organic reactants—an amine, alcohol, or acid (see Section IV,A,7). Some organic molecules (e.g., benzene) enter the interlamellar spaces but cause no significant swelling. Moreover, it is possible to place variable quantities of both aqueous and nonaqueous interlamellar species close to the exchangeable cation.

In this section we first enumerate the reactions that my colleagues and I have discovered; the type of reaction as well as a specific example of each type is cited. In the second part of this section we consider relevant mechanistic details both with a view to rationalizing the facts so far discovered and with the intention of identifying possible future strategies.

A. Summary of the Types of Organic Reactions Favored in Sheet Silicate Intercalates

A few of the reactions described below take place when the isolated, solid intercalate is heated, and the gaseous products reveal the nature of the reation. The majority, however, involve more activated, but still quite mild conditions in which the silicate, reactant(s), and a benign solvent (for example, *n*-hexane or cyclohexane for the alkene reactants) are refluxed or heated in a bomb under self-generated pressure up to a temperature of ca 200°C for several hours. Full experimental details are given in the original references.

1. Thermally Induced Breakdown

An early example (*44*) of this kind of reaction is the conversion of diprotonated 4,4′-diamino-*trans*-stilbene–montmorillonite to aniline, the sole gaseous product, upon gentle heating:

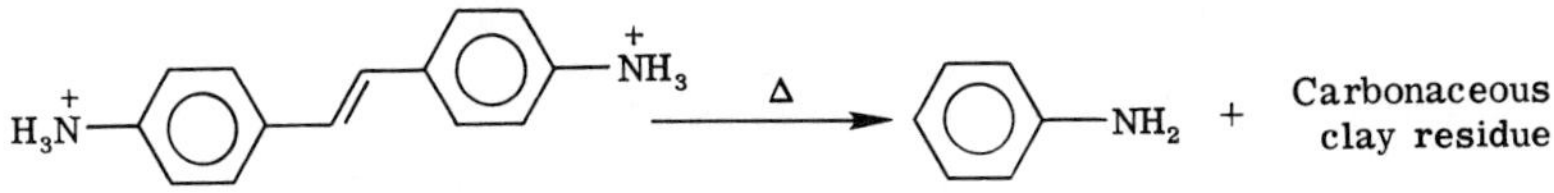

Another reaction type in this category is the conversion of esters to olefin and carboxylic acid. This process normally takes place, homogeneously, at temperatures in the vicinity of 400°C. In the presence of appropriately exchanged smectites, however, a typical ester, such as methyl cyclohexane carboxylate is rapidly and cleanly broken down to cyclohexene and acetic acid at 140°C:

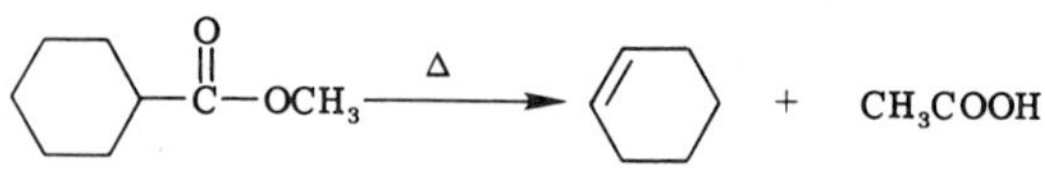

In a comparative study (*43, 45, 46*) of various aluminosilicate catalysts, it was established that, for fixed experimental conditions, the percent conversions of the methyl cyclohexane carboxylate were as follows:

Al^{3+}-Exchanged montmorillonite (14.8% interlamellar water, w/w)	98%
Ni^{2+}-Exchanged montmorillonite (14.8% interlamellar water, w/w)	78
Cu^{2+}-Exchanged montmorillonite (14.8% interlamellar water, w/w)	65
Cu^{2+}-Zeolite 13X (Faujasite-based synthetic zeolite with Si/Al ratio ca 1:4)	20
Cu^{2+}-Exchanged montmorillonite (after thermal collapse and loss of the original 7.8% water)	0
Na^{+}-Exchanged montmorillonite	0

Clearly both the nature of the cation and the water content of the silicate (as well as the nature of the framework, since the faujasite structure of Cu^{2+}-exchanged zeolite is a relatively poor catalyst) are important determinants.

2. *Thermally Induced Dimerizations*

Early in our studies of the thermal stability of dried samples of various cation-exchanged montmorillonite–hydrocarbon intercalates, we observed that appreciable quantities of dimer (and, to a lesser extent, trimer and higher oligomers) were formed by heating *in vacuo*, in the range 50°–250°C, complexes in which the hydrocarbon possessed an olefinic double bond. Thus, the Cu^{2+}-exchanged montmorillonite intercalate with *trans*-stilbene formed, on heating, small quantities of the cyclobutane (*44*) dimer as well as the

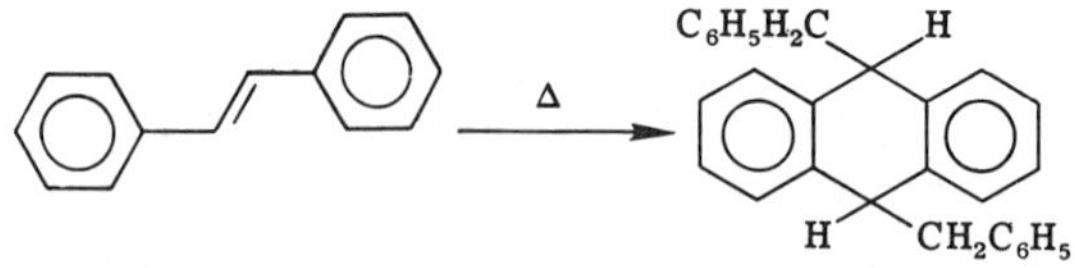

9,10-dibenzyl-9,10-dihydroanthracene (*46*). The conditions for optimal production of these dimers were not determined. With the indene intercalate

of the Cu^{2+}-exchanged clay, dimer and trimer molecules were detected mass-spectrometrically.

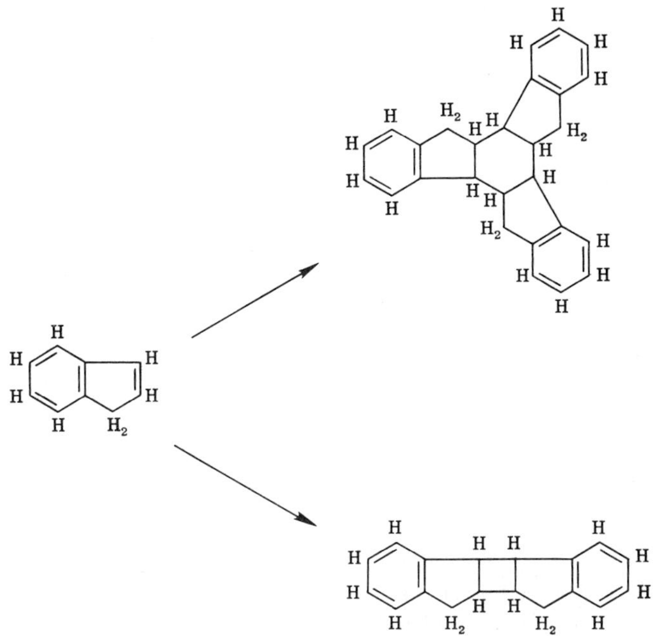

When 1,1-diphenylethylene is heated to reflux for 30 min in contact with Cu^{2+}-exchanged montmorillonite that has not been especially dried, the unusual dimer 1-methyl-1,3,3-triphenylindan is formed (*47*) in high yield:

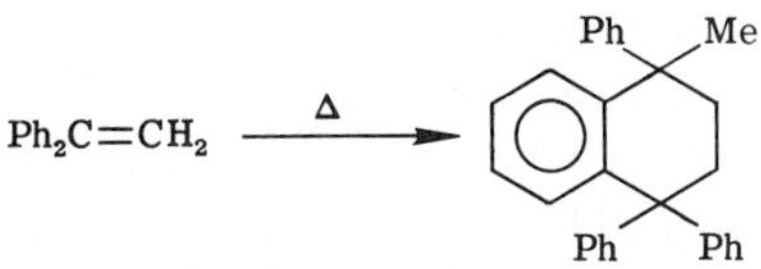

The Na^+-exchanged clay also serves as a catalyst but is less efficient than the Cu^{2+} analog.

When anethole (*43, 45*) is heated in contact with one of a variety of cation-exchanged montmorillonites, two primary dimeric products, isoanethole and the tricyclic dimer metanethole (Fig. 15), are formed. The extent of the guest monomer as well as the precise composition of the products is a function of the particular ion in the interlamellar space. A 95% conversion of the parent intercalate is achieved with both the Fe^{3+}- and the Al^{3+}-exchanged montmorillonites. In the Fe^{3+} case, for example, there is a 5:1 ratio of metanethole (50% of product) to isoanethole and ca 40% of unidentified product.

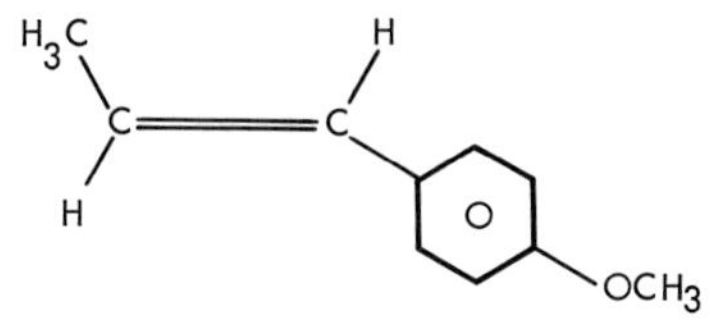

ANETHOLE

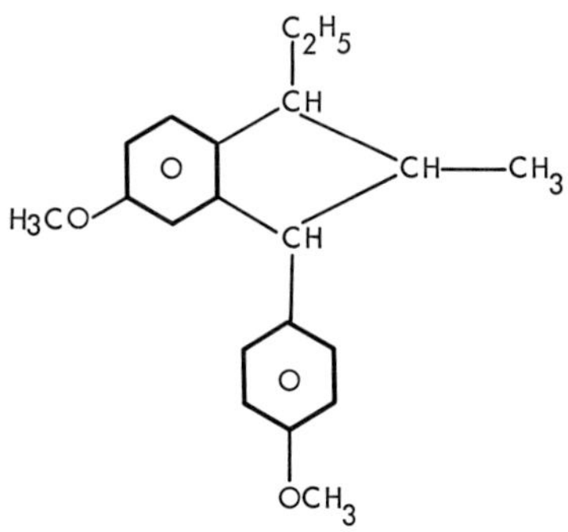

METANETHOLE

Fig. 15. The monomer anethole may be converted (see Fig. 18) to its dimer metanethole, which is tricyclic (see test). [From Davies (*43*); Davies *et al.* (*45*).]

When hectorites are used in place of montmorillonites, significant differences in behavior of both anethole and isohomogenol are noted (*48*). Again, anethole may be dimerized (to yield metanethole and isoanethole and some unidentified products), but by no means all of the interlamellar inorganic cations facilitate this reaction. Thus, Al^{3+} ions are effective, but anethole does not dimerize when Cu^{2+}-exchanged hectorite is used. Isohomogenol shows no tendency to dimerize with any cation-exchanged hectorites, in contrast to the situation with montmorillonite.

3. Reduction–Oxidation Processes

So far, we have found only one clear-cut example of this type of process, the conversion of 1,1-diphenylethylene to 1,1-diphenylethane and benzophenone using a synthetic fluorohectorite [contrast the behavior when montmorillonite is used (*48*)]. Although the mechanism of the reaction is not yet clear, we do know from X-ray studies that the unsaturated hydrocarbon enters the interlamellar spaces and also that a weight change, corresponding to a loss of water, occurs at a temperature close to that at which reaction freely ensues. The water content of the clay cannot, however, be the source of either hydrogen or oxygen incorporated into the product molecules, amounting as it does to only about 10% of the requisite quantity. With certain smectites it is conceivable that water could be abstracted from hydroxyl groups attached to octahedrally coordinated cations. This, however, is not a tenable view in the present circumstances since the hydroxyl groups are substituted solely by fluoride. The oxygen incorporated into the benzophenone is of atmospheric origin since, in an atmosphere of nitrogen and in the presence of the fluorohectorite catalyst, diphenylethylene is converted smoothly to diphenylmethane. The only source of hydrogen is the diphenylethylene itself. It could lose some hydrogen, being converted to fused-ring aromatic hydrocarbons in the process (a noncrystalline fraction of such hydrocarbon is indeed formed).

We conclude from the above that synthetic fluorohectorite clearly mediates both in a hydrogen-exchange reaction and in the atmospheric oxidation of diphenylethane and diphenylethylene.

4. Intermolecular Hydrogen Exchange

A reaction involving hydrogen exchange that is closely similar to that described in Section 3, as well as to the intermolecular reactions (*49*) that involve esterification of carboxylic acids with isomeric pentanols and hexanols, is the hydrogen exchange between two different molecules both of which are capable of being intercalated by the smectite (*50*). When 9,10-dihydroanthracene and 1,1-diphenylethylene were reacted in refluxing octane in the presence of fluorohectorite and under N_2 gas, a smooth conversion to anthracene and diphenylethane occurred (reaction was half-complete in 90 min). A slower conversion, which may readily be monitored by NMR spectroscopy, takes place in refluxing hexane.

9,10-Dihydroanthracene will not reduce *cis*- or *trans*-stilbene, diphenylacetylene, anethole, or isohomogenol under comparable conditions. But when a stronger Brønsted acid sheet silicate is used (for example, Fe^{3+}-exchanged montmorillonite), 1,2-diphenylethane is formed smoothly from *trans*-stilbene and 9,10-dihydroanthracene.

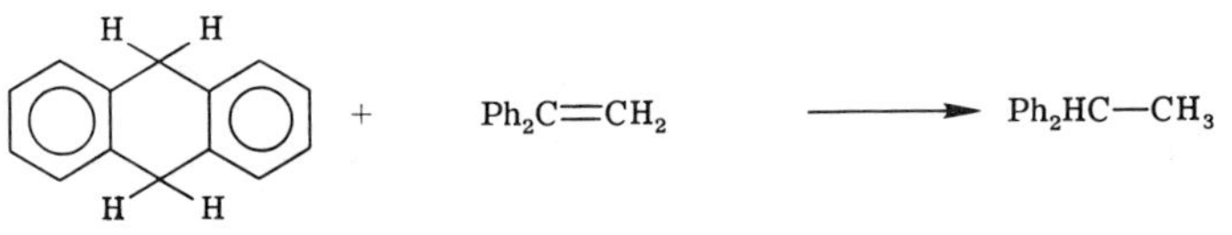

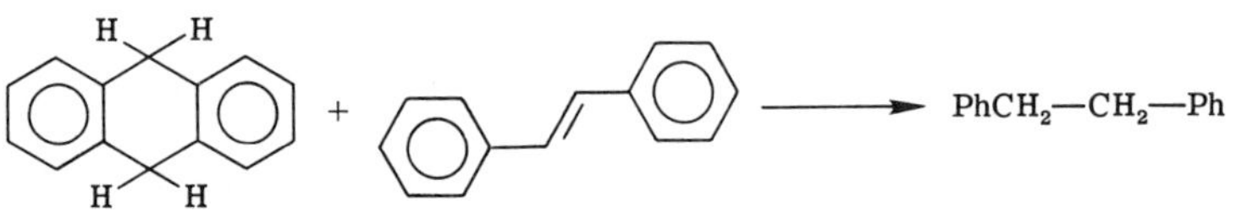

We have also established (*48*) that trimethoxystilbene takes up hydrogen even more readily than stilbene under comparable conditions. Moreover, it has been noted that, in the mixture of dimers produced when the 2-methyl-1-pentene intercalate with metal-exchanged montmorillonite is heated, some compounds with a molecular weight of 166 are generated (*48*). This indicates that some hydrogen exchange takes place after, or possibly during, the actual dimerization.

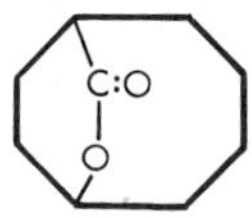

1,4–CYCLOOCTANECARBOLACTONE

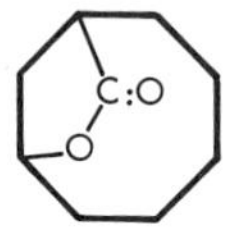

1,3–CYCLOOCTANECARBOLACTONE

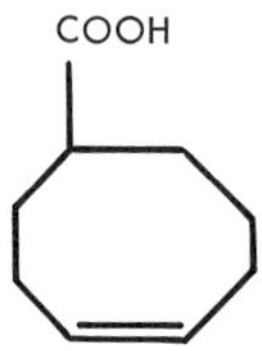

5–CARBOXYCYCLOOCTENE

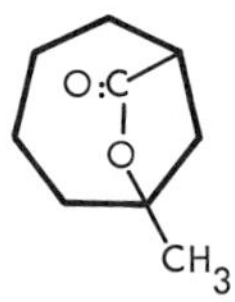

3–METHYL–3–CYCLOHEPTANECARBOLACTONE

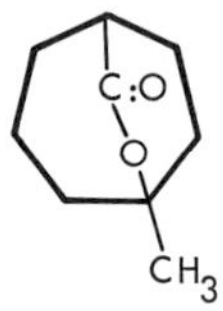

4–METHYL–4–CYCLOHEPTANECARBOLACTONE

UNKNOWN LACTONE(S)

Fig. 16. Lactones that may be produced from the parent 5-carboxycyclooctene. [From Davies (*43*); Davies *et al.* (*45*).]

Hydrogen-exchange reactions on an alumina surface, with alcohol as hydrogen donor, have been observed previously (*51*), but to our knowledge this is the first example of hydrogen exchange between hydrocarbons other than at a catalytic metal surface. (The use of alcohols as hydrogen donors with sheet silicates is precluded since, as we shall see later, they are often rapidly dehydrated or suffer other fates.)

5. *Lactonization*

Lactone formation from 5-carboxycyclooctene (Fig. 16) illustrates nicely how the product distribution may be influenced and tuned by varying the nature of the interlamellar cation (as well as the customary secondary factors, such as fractional conversion or time of reaction). In the liquid phase, at least five distinct lactones may be isolated (with sulfuric acid as the homogeneous catalyst). Although it is possible, by careful manipulation, to obtain one of these five lactones as the major product in sulfuric acid solutions, it is not possible to maximize yields of the other four. With the aid of cation-exchanged montmorillonites, on the other hand, one of these four may be obtained as the principal product. Further details are given elsewhere (*43*, *45*), but we note here that favored interlamellar cations for this purpose are Al^{3+}, Co^{2+}, and Ni^{2+}.

6. *Ether Formation from Alkenes*

Intercalation of alk-1-enes is a prerequisite for this reaction, which entails the addition of a molecule of water between two molecules of the olefin to yield the corresponding di(alk-2-yl) ether (*52*, *53*):

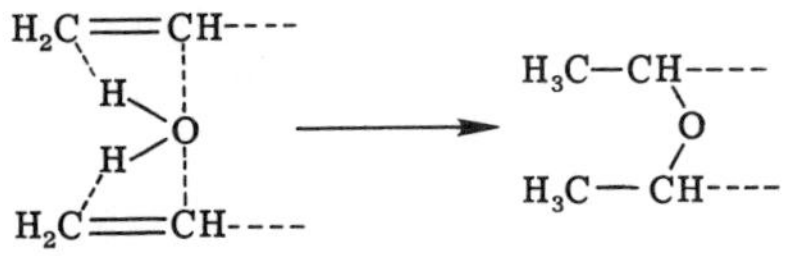

These bis-*sec*-alkyl ethers are readily formed by refluxing the appropriate alk-1-ene (for example, hex-1-ene, hept-1-ene, or oct-1-ene) in hexane solution with one of a range of cation-exchanged montmorillonites. A number of cations facilitate conversion, the most efficient being Al^{3+}, Fe^{3+}, Cr^{3+}, Cu^{2+}, and Fe^{2+}. For the Cu^{2+}-exchanged clay the optimal cation–water ratio (in the interlamellar space) is ca 1 : 12, and under these circumstances 100% conversion of usable interlamellar water to ether is achieved. [Use of other sheet structures, for example, $Cu(UO_2)_2PO_4{\cdot}6H_2O$ did not result in conversion of the alkenes to the ethers.] Essentially no ether is produced when collapsed montmorillonites (irrespective of the nature of the cation) are used in place of the deliberately expanded varieties.

The relative ease with which ethers could be prepared by incorporation of interlamellar water into two molecules of alkene prompted us to examine (*54*) the feasibility of effecting comparable ether syntheses using interlamellar methanol in place of the water. Methyl *tert*-butyl ether (MTBE), or 2-methyl-2-methoxypropene, is currently extensively used as a gasoline blending agent [see, e.g. (*55*)] and is prepared industrially using a range of rather exotic and expensive solid catalysts often under high pressure. We discovered (*54*) that Cu^{2+}-exchanged montmorillonite serves as an efficient catalyst for the reaction:

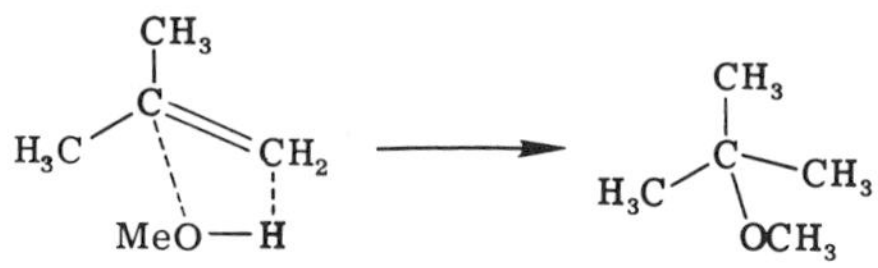

Many other alkenes form intercalates with the Cu^{2+}-exchanged montmorillonite, including 1-hexene; 1-butene; 3,3-dimethyl-1-butene; 4-methyl-1-pentene; 2-methyl-1-pentene; 2-methyl-2-pentene; 2,4,4,-trimethyl-1-pentene; and 2,4,4-trimethyl-2-pentene. However, only the last four of these add on methanol relatively easily to form methyl ethers. As noted above, interlamellar water adds freely to straight-chain alkenes to form bis-*sec*-alkyl ethers; the addition of alcohol has so far been observed only with branched alkenes to give tertiary ethers.

It seems (*56*) that aliphatic primary alcohols react preferentially, as intercalates, to give high yields of di(alk-1-yl) ethers rather than undergo competitive intramolecular dehydration to form alkenes, whereas aliphatic secondary and tertiary alcohols yield alkenes almost exclusively.

7. *Ester Formation from Alkenes*

As an extension of the principles outlined above, we have discovered (*57*) a direct route to ester formation using sheet silicate intercalates. Ethene and acetic acid react in the interlamellar region of certain cation (for example, Al^{3+})-exchanged montmorillonites to yield ethyl acetate as the sole product. Moreover, a variety of carboxylic acids readily add to C_2–C_8 alkenes at temperatures above 100°C to yield the corresponding esters in high and selective yields.

8. *Facile Production of Secondary Amines and Thioesters*

It was mentioned earlier that interlamellar water is rapidly dislodged when cation-exchanged montmorillonite is exposed to any of a large range of primary amines. We have also noted (*58a*) that ammonia is readily eliminated from molecular pairs of intercalated aliphatic primary amines, thereby generating the secondary amines in high yield:

$$2R{-}CH_2NH_2 \xrightarrow{-NH_3} R{-}CH_2{-}NH{-}CH_2{-}R$$

Likewise, H_2S is eliminated (*58b*) from thiols to form thioethers.

9. Other Reported Reactions

We have not yet reexamined the validity of claims made concerning the ease with which the following reactions allegedly occur in contact with various montmorillonites. Some of these may, for example, take place at the exterior surfaces, rather than in the interlamellar regions, of the sheet silicates.

a. Formation of nitriles, for example,

$$CH_3COOH + NH_3 \rightarrow CH_3CN$$

b. Dechlorination, for example,

$$PhCH_2Cl + Ph \cdot H \rightarrow PhCH_2Ph$$

c. Polymerizations of terpenes, for example,

Pinene → sesquiterpenes and polyterpenes

d. Alkylation of aromatic compounds, for example,

$$CH_2{=}CHR + PhH \rightarrow R{-}CH(CH_3){-}Ph$$

10. Summary of Reaction Types

Table IV summarizes the various reactions that we have cited above and that we shall now discuss from the mechanistic viewpoint.

B. Mechanistic Implications

1. General Considerations

As mentioned in the Introduction, it is well known that sheet silicates, like zeolites, may possess Brønsted or Lewis acid sites or both. At the outset of our work it was recognized that plausible overall mechanisms could be formulated for the various novel reactions we discovered in terms of one or another

TABLE IV

Reaction Types That Occur Freely in Sheet Silicate Intercalates[a]

Reaction type	Example
Thermal decomposition	Breakdown of esters to acid and hydrocarbon
Dimerization	Formation of indan and anthracene derivatives from 1,1- and 1,2-diphenylethylene
Reduction–oxidation	Conversion of 1,1-diphenylethylene to 1,1-diphenylethane and benzophenone
Hydrogen transfer	Between 9,10-dihydroanthracene and 1,1-diphenylethylene
Lactonization	Cyclooctene carboxylic acid to 1,4- and 1,3-lactones
Ether formation	Formation of di (alk-2-yl) ether from alk-1-ene; 1-alkyl-2-alkyl ethers from alk-1-enes and alk-1-ols; and di(alk-1-yl) ethers from alkan-1-ols
Ester formation	Ethyl acetate from ethylene and acetic acid
Conversion of primary to secondary amines	Elimination of NH_3 from two molecules of $R—CH_2—NH_2$
Unconfirmed reactions	Alkylation of aromatic compounds with substituted alkenes; formation of nitriles from carboxylic acids and ammonia

[a] Reactions include only those discussed in this chapter.

of these acidic entities. In recent years we have been able, using some of the techniques mentioned in Section III, to identify some intermediate species, earlier postulated to be involved in a number of the processes we discovered. For example, infrared spectroscopy has revealed the presence of certain protonated species in the interlamellar spaces, and Mössbauer spectroscopy, the existence of electron-accepting sites. However, much remains to be learned about important mechanistic aspects. What is it that favors dehydration rather than ether formation when a given alcohol undergoes interlamellar reaction with a certain cation-exchanged sheet silicate? What are the precise origins of the Lewis and Brønsted acid sites? What is it that fundamentally determines why some organic species more rapidly displace interlamellar water than do others? What are the steady-state local concentrations of the intermediates, their precursors, and decay products? More specifically, does the interlamellar cation itself function as a crucial reaction center, or does it serve merely as a primary source of reactant (e.g., protons as a result of hydrolysis)? Many of these questions have not been fully answered. Nor do we know enough about precisely which types and subtypes of organic molecules can be intercalated by given cation-exchanged sheet silicates.

It is clear, however, that protonated intermediates [e.g., alkyl carbenium (carbonium*) or oxonium ions] often play crucial roles in the reactions under discussion. The early mechanisms formulated by Whitmore (*59*), Evans and Polanyi (*60*), and Venuto and Landis (*61*) pertaining to the behavior of carbenium ions are therefore relevant here. Moreover, the relationship between the behavior of acid zeolites is strong. It is pertinent to recall not only the mode of formation of carbenium ions at zeolitic (*62*, *63*) or silica–alumina (*64*) catalysts,

$$RCH{=}CH_2 + H^+ \rightarrow R\overset{+}{C}H{-}CH_3$$

but also the feasibility of subsequent events such as isomerization, addition to further olefins, or even β-elimination,

$$R_1^+ \rightleftarrows R_2^+$$

$$R_1^+ + \;>C{=}C<\; \rightarrow R_2^+$$

$$R_1^+ \rightarrow R_2^+ + \;>C{=}C<$$

where R_1^+, R_2^+, and $R\overset{+}{C}H{-}CH_3$ are carbenium ions. It is also worth recalling the viability of hydride shift

$$R_1{-}H + R_2^+ \rightleftarrows R_1^+ + R_2{-}H$$

and of the following addition,

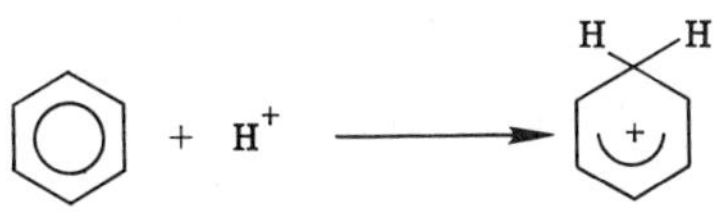

in the presence of acidic catalysts.

Not surprisingly, we recognize similar patterns of behavior between certain reactions that take place in the intracrystalline cavities of zeolites, on the one hand, and in the interlamellar regions of sheet silicates, on the other. The

* Ions formerly designated carbonium ions are now, according to the IUPAC ruling, termed carbenium ions.

flow chart below, valid for zeolites (*62*, *65*), is expected to be relevant in present contexts,

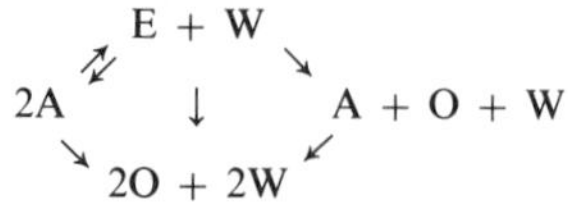

where A, E, O, and W refer, respectively, to alcohol, ether, olefin, and water.

Detailed spectroscopic, notably NMR and infrared, studies by Fripiat (*66–69*) have revealed that, whereas for the gel-type aluminosilicates the surface-protonated (organic) species are formed as a result of transfer of H^+ from a surface Brønsted site, in montmorillonite protonation results from the enhancement of the degree of dissociation of the adsorbed (or interlamellar) water due to the polarizing field of the surface (or interlamellar) cations. Fripiat *et al.* (*68*, *69*) have shown that, for methanol at acidic zeolite surfaces (where OH groups act as proton sources), $CH_3OH_2{}^+$ species possess an acidity ($pK_a \simeq -4.5$) that is almost as strong as that of the surface OH groups ($pK_a \simeq -6$). For 2-butanol on η-alumina surfaces, these workers found that the alcohol itself is the proton donor. In the clay catalysts discussed here, especially when residual water or other protonic entities are retained in the interlamellar space occupied by strongly polarizing cations, we may expect relatively high concentrations of protons.

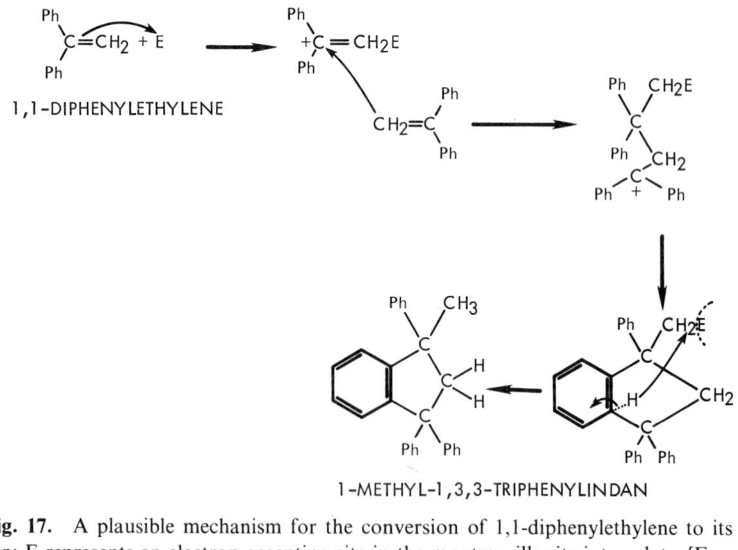

Fig. 17. A plausible mechanism for the conversion of 1,1-diphenylethylene to its indan dimer; E represents an electron-accepting site in the montmorillonite intercalate. [From Reid (*42*); Adams *et al.* (*47*).]

2. *Some Specific Examples*

a. DIMERIZATIONS. We now formulate mechanisms that account for the formation of the products described in Section IV,A,2.

Consider the dimerization of 1,1-diphenylethylene. We predicate (Fig.17) the existence of a Lewis acid site (E) on the clay (*42*).

In the dimerization of anethole (*43*, *45*) (Fig. 18) we note how the carbenium ion plays a crucial role. The initial and final steps are, respectively, the gain and loss of a proton from and to the solid catalyst (the clay).

A possible mechanism for the production of 9,10-dihydro-9,10-dibenzylanthracene from *trans*-stilbene is given in Fig. 19.

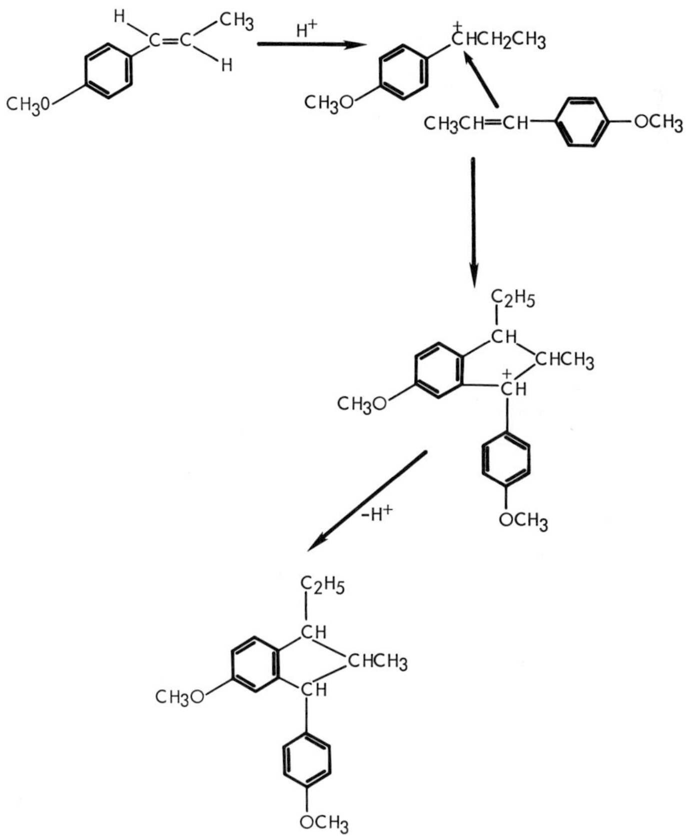

Fig. 18. A plausible mechanism for the dimerization of anethole to metanethole involving Brønsted acidity. [From Davies (*43*); Davies *et al.* (*45*).]

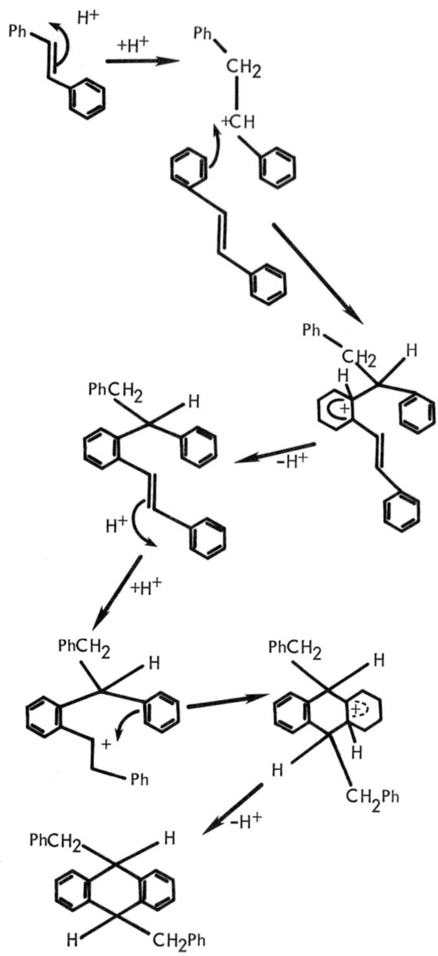

Fig. 19. A mechanism based on proton transfer for the dimerization of *trans*-stilbene to 9,10-dihydro-9,10-dibenzylanthracene. [From Davies (*43*).]

b. LACTONIZATION. Again, a plausible proton-mediated (*43*, *45*) process may be formulated (Fig. 20) for the conversion of cyclooctene-5-carboxylic acid to 4-methyl-4-cycloheptanecarbolactone.

c. ETHER FORMATION. We may account for the formation of the bis-*sec*-alkyl ethers (Section IV,A,6) from alk-1-enes first in terms of a Lewis acid intermediary, E again denoting an electron-accepting site in the montmorillonite (Fig. 21) (*42*, *53*). We may also account for the formation of the same ether in terms of proton-catalyzed processes (Fig. 22).

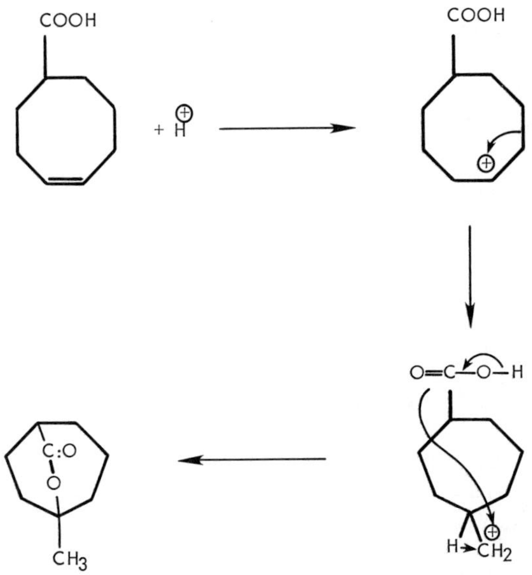

Fig. 20. A suggested mechanism involving acid catalysis for the lactonization of 5-carboxy-cyclooctene. [From Davies (*43*); Davies *et al.* (*45*).]

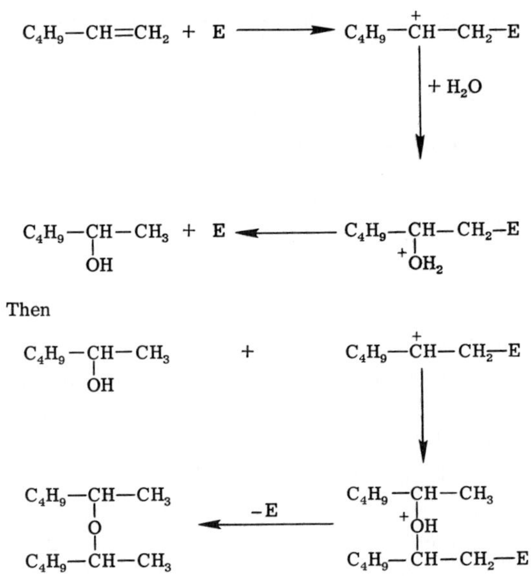

Fig. 21. A suggested mechanism involving an electron-accepting site E for the conversion of 1-hexene to bis-*sec*-hexyl ether. [From Adams *et al.* (*53*).]

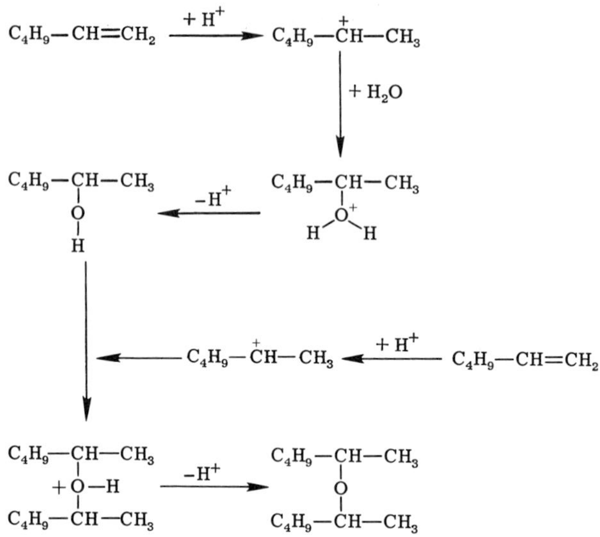

Fig. 22. Mechanism for acid-catalyzed conversion of 1-hexene to bis-*sec*-hexyl ether. [From Ballantine *et al.* (*56*).]

d. SECONDARY AMINE FORMATION. It is easy to see how protonated and unprotonated primary amines may combine to yield the secondary amine (Section IV,A,8).

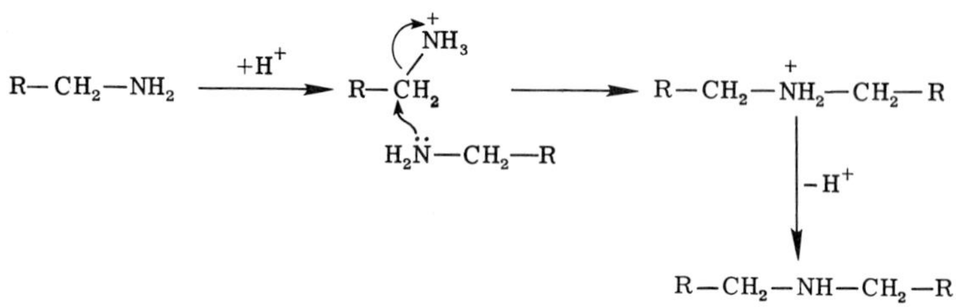

In view of what has been uncovered already in connection with the unusual organic reactions that may take place in sheet silicate intercalates, it is reasonable to speculate about the feasibility of several other novel reactions, most of which can be made to proceed at high rates and in good yield in the presence of specially tuned zeolites.

3. Possible New Reactions

a. DEHYDRATION OF ALKANOLS. In many respects, this is the least novel reaction, but mechanistically perhaps it is one of the better understood (*70*).

For example, in the formation of propene from 2-propanol, the oxonium ion is involved (*70*).

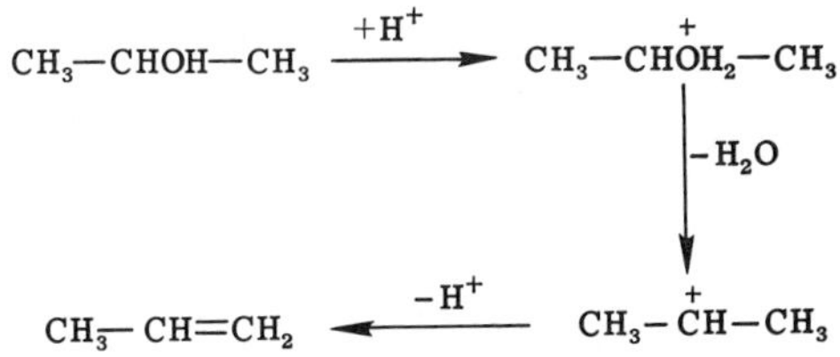

b. Conversion of Epoxides to Aldehydes or Ketones. These processes are acid-catalyzed (*60, 71*). The overall mechanism is probably as shown below, with the rearrangement possibly being a concerted process:

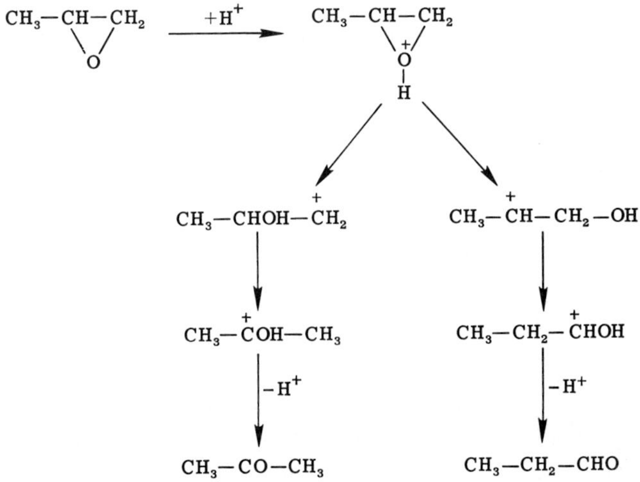

c. Conversion of Oximes to Nitriles. This reaction has been described by Landis and Venuto (*72*). It is of interest that acidic zeolite catalysts convert cyclohexanone to ϵ-caprolactam (*71*).

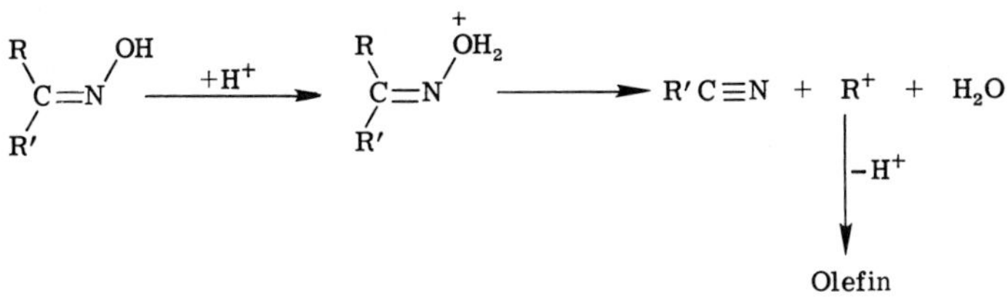

d. Ring Transformations of Heterocycles. The reactions offer scope for "tuning" of the selectivity in the sense used earlier in this chapter. Thus, depending whether or not the Brønsted acidity is pronounced, one would expect (*73*) γ-butyrolactone to yield either ω-hydroxybutyronitrile or 2-pyrrolidone:

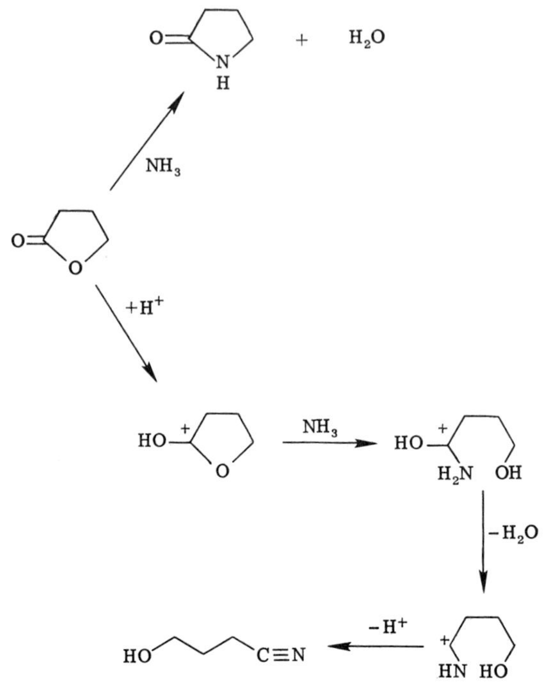

e. Carbonyl Group Reactions. Many reactions of this type have been discussed by Venuto and Landis (*74*). Thus, the condensation of formaldehyde and methyl acetate over an acidic faujasite (Y-type zeolite) at ca 350°C yielded methyl acrylate, a reaction formally akin to an aldol condensation. In the presence of an excess of aromatic compounds, bisalkylalkanes were formed as a result of alkylation. Formaldehyde and phenol, for example, give mixtures of 2,2′-, 2,4′-, and 4,4′-dihydroxydiphenylmethanes:

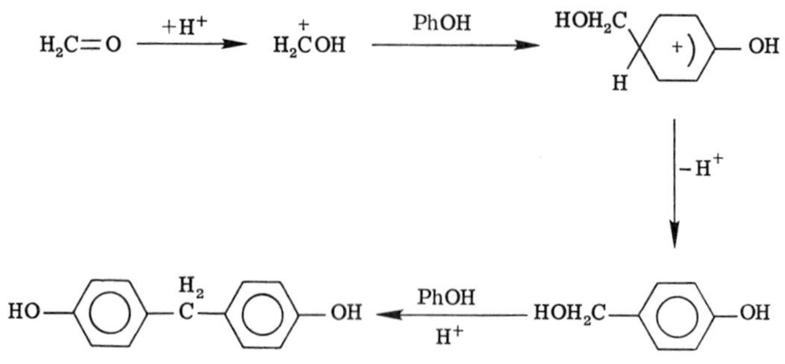

By "blocking" either the 2 or 4 position in the phenol ring with methyl groups, more selective (either exclusively para or exclusively ortho) products could no doubt be engineered.

V. FUTURE POSSIBILITIES

Very considerable scope exists for further exploitation of the structure and properties of the sheet silicates. One development initiated by Barrer (*12, 75–79*) and others is to introduce more or less permanent connecting "pillars" between the sheets. This creates a structure intermediate between the swellable clay minerals and the zeolites. Such cross-linked or pillared clays possess a so-called permanent porosity, which has the advantage of providing access to intracrystallite surfaces for reactant species up to ca 800°C (depending on the chemical stability of the pillar). Figure 13 above is an electron micrograph of one of the pillared clays prepared from the recipe given by Lussier *et al.* (*79*). Further details of these useful materials, which seem well suited for some of the bimolecular reactions described in Section IV,B, are given elsewhere (*12, 75–79*).

ACKNOWLEDGMENTS

The author acknowledges the stimulus and help provided by many present and past collaborators. Especial thanks are conveyed to Drs. D. T. B. Tennakoon, M. J. Tricker, and S. H. Graham, who were involved at the inception of this work. Later invaluable cooperation was provided by Drs. J. M. Adams, P. I. Reid, M. J. Walters, A. Bylina, and S. E. Davies. Finally, since 1976, Professor J. H. Purnell, Dr. J. A. Ballantine, and their colleagues at Swansea have played a vital part in this work. My Cambridge colleagues (Dr. W. Jones and L. J. Williamson) and I are indebted to them and to the Science Research Council and British Petroleum (since 1979) for supporting many aspects of this work.

REFERENCES

1. J. M. Thomas, *Philos. Trans. R. Soc. London* **277,** 251 (1974).
2. J. M. Thomas, S. E. Morsi, and J. P. Desvergne, *Adv. Phys. Org. Chem.* **15,** 63 (1977).
3. J. M. Thomas, *Pure Appl. Chem.* **51,** 1065 (1979).
4. W. Jones, H. Nakanishi, C. R. Theocharis, and J. M. Thomas, *J.C.S. Chem. Commun.* p. 610 (1980).
5. H. Nakanishi, W. Jones, J. M. Thomas, M. B. Hursthouse, and M. Motevalli, *J.C.S. Chem. Commun.* p. 611 (1980).
6. For comprehensive references to the work of earlier investigators, see R. M. Barrer, *in* "Non-Stoichiometric Compounds" (L. Mandelkern, ed.), p. 309. Academic Press, New York, 1964; see also refs. 7 and 8.

7. A. Weiss, *in* "Organic Geochemistry" (G. Eglinton and M. T. J. Murphy, eds.), p. 737. Springer-Verlag, Berlin and New York, 1969.
8. J. M. Thomas, J. M. Adams, S. H. Graham, and D. T. B. Tennakoon, *Adv. Chem. Ser.* No. 163, p. 298 (1977).
9. W. L. Bragg and G. F. Claringbull, "The Crystal Structure of Minerals." Bell & Sons, London, 1965.
10. G. Brown, "X-Ray Identification and Crystal Structures of Clay Minerals." Mineral. Soc., London, 1961.
11. R. E. Grim, "Clay Mineralogy," 2nd ed. McGraw-Hill, New York, 1968.
12. R. M. Barrer, "Zeolites and Clay Minerals as Sorbents and Molecular Sieves." Academic Press, New York, 1978.
13. J. M. Adams, *J.C.S. Dalton*, p. 2286 (1974).
14. L. J. Williamson and T. Rayment, personal communication.
15. J. M. Adams, J. M. Thomas, and M. J. Walters, *J.C.S. Dalton*, p. 1459 (1975).
16. J. M. Adams and D. A. Jefferson, *Acta Crystallogr., Sect. B* **32,** 1180 (1976).
17. E. L. Evans and J. M. Thomas, *J. Solid State Chem.* **14,** 99 (1975).
18. J. M. Thomas, G. R. Millward, R. Schlögl, and H. P. Boehm, *Mater. Res. Bull.* **15,** 671 (1980).
19. V. C. Farmer, "Infra Red Spectra of Minerals." Mineral. Soc., London, 1974.
20. L. J. Williamson, CPGS Thesis, Univ. of Cambridge, 1980.
21. L. J. Williamson, personal communication (1980).
22. J. M. Thomas, *Nature (London)* **279,** 755 (1979).
23. J. M. Adams, S. Evans, and J. M. Thomas, *J. Am. Chem. Soc.* **100,** 3260 (1978).
24. J. M. Adams, S. Evans, and J. M. Thomas, *J.C.S. Chem. Commun.* p. 210 (1978).
25. S. Evans, J. M. Adams, and J. M. Thomas, *Philos. Trans. R. Soc. London, Ser. A* **292,** 563 (1979).
26. C. A. Fyfe, J. M. Thomas, and J. R. Lyerla, *Angew Chem. Int. Ed.* **20,** 96 (1981).
27. J. R. Lyerla, C. A. Fyfe, and C. S. Yannoni, *J. Am. Chem. Soc.* **101,** 1351 (1979).
28. D. T. B. Tennakoon, J. M. Thomas, and M. J. Tricker, *J.C.S. Dalton*, p. 2211 (1974).
29. P. Bowen, W. Jones, J. M. Thomas, and L. J. Williamson, unpublished observations.
30. L. Rozenson and L. Heller-Kallai, *Clays Clay Miner.* **26,** 88 (1978).
31. J. M. Thomas, E. L. Evans, B. Bach, and J. LL. Jenkins, *Nature (London), Phys. Sci.* **235,** 126 (1972).
32. J. M. Thomas and D. A. Jefferson, *Endeavour* **2,** 127 (1978).
33. J. M. Thomas, *New Sci.* Aug. 21, p. 580 (1980).
34. L. A. Bursill, W. Jones, L. J. Williamson, and J. M. Thomas, unpublished observations.
35. T. Yoshida, *Clays Clay Miner.* **21,** 413 (1973).
36. D. Yoshino, R. Hayatsu, and E. Anders, *Geochim. Cosmochim. Acta* **35,** 927 (1971).
37. J. J. Fripiat and M. I. Cruz-Camplido, *Annu. Rev. Earth Planet Sci.* **2,** 239 (1974).
38. B. K. C. Theng, "The Chemistry of Clay-Organic Reactions." Hilger, London, 1974.
39. G. W. Brindley, *Reun.Hisp.-Belga Miner. Arcilla, An., Madrid, 1970,* p. 55 (1971).
40. T. J. Pinnavaia and P. K. Welty, *J. Am. Chem. Soc.* **97,** 3819 (1975).
41. D. T. B. Tennakoon, Ph.D. Thesis, Univ. of Wales, 1974.
42. P. I. Reid, Ph.D. Thesis, Univ. of Wales, 1978.
43. S. E. Davies, Ph.D. Thesis, Univ. of Wales, 1980.
44. M. J. Tricker, D. T. B. Tennakoon, J. M. Thomas, and S. H. Graham, *Nature (London)* **253,** 110 (1975).
45. S. E. Davies, J. M. Adams, S. H. Graham, and J. M. Thomas, in preparation.
46. J. A. Ballantine, J. H. Purnell, P. I. Reid, J. M. Adams, S. H. Graham, and J. M. Thomas, unpublished observations (1977).

47. J. M. Adams, S. H. Graham, P. I. Reid, and J. M. Thomas, *J.C.S. Chem. Commun.* p. 67 (1977).

48. J. M. Thomas, S. H. Graham, S. E. Davies, A. Bylina, J. M. Adams, J. A. Ballantine, and J. H. Purnell, unpublished observations (1978).

49. J. Shabtai, N. Frydman, and R. Lazar, *Proc. Int. Congr. Catal., 6th, London, 1976* **1,** 660 (1977).

50. J. M. Adams, S. E. Davies, S. H. Graham, and J. M. Thomas, *J.C.S. Chem. Commun.* p. 527 (1979).

51. G. H. Posner, A. W. Runquist, and M. J. Chapdelaine, *J. Org. Chem.* **42,** 1202 (1977).

52. J. M. Adams, J. H. Ballantine, S. H. Graham, R. J. Laub, J. H. Purnell, P. I. Reid, W. Y. M. Shaman, and J. M. Thomas, *Angew. Chem., Int. Ed. Engl.* **90,** 280 (1978).

53. J. M. Adams, J. A. Ballantine, S. H. Graham, R. J. Laub, J. H. Purnell, P. I. Reid, W. Y. M. Shaman, and J. M. Thomas, *J. Catal.* **58,** 238 (1979).

54. A. Bylina, J. M. Adams, S. H. Graham, and J. M. Thomas, *J.C.S. Chem. Commun.* p. 1003 (1980).

55. *Chem. Eng. News*, Mar. 26, p. 9 (1979).

56. J. A. Ballantine, M. Davies, J. H. Purnell, M. Rayanakorn, J. M. Thomas, and K. J. Williams, *J.C.S. Chem. Commun.* p. 427 (1981).

57. J. A. Ballantine, M. Davies, J. H. Purnell, M. Rayanakorn, J. M. Thomas, and K. J. Williams, *J.C.S. Chem. Commun.* p. 8 (1981).

58a. J. A. Ballantine, J. H. Purnell, M. Rayanakorn, J. M. Thomas, and K. J. Williams, *J.C.S. Chem. Commun.* p. 9 (1981).

58b. J. A. Ballantine, R. P. Gavin, R. M. O'Neil, J. H. Purnell, M. Rayanakorn, and J. M. Thomas, *J.C.S. Chem. Commun.* p. 695 (1981).

59. F. C. Whitmore, *Ind. Eng. Chem.* **26,** 94 (1934); *Chem. Eng. News* **26,** 668 (1947).

60. A. G. Evans and M. Polanyi, *J. Chem. Soc.* p. 252 (1947).

61. P. S. Venuto and P. S. Landis, *Adv. Catal.* **18,** 259 (1969).

62. P. A. Jacobs, "Carboniogenic Activity of Zeolites." Elsevier, Amsterdam, 1977.

63. R. Prins and G. C. A. Schuit, eds., *Chem. Chem. Eng. Catal. Processes, NATO Adv. Study Inst. Ser.* No. 39.

64. J. H. C. van Hooff, ref. 63, p. 161.

65. H. Knözinger and R. Köhne, *J. Catal.* **5,** 264 (1966).

66. J. J. Fripiat, *Ind. Chem. Belg.* **38,** 404 (1973).

67. P. Salvador and J. J. Fripiat, *J. Phys. Chem.* **79,** 1842 (1975).

68. J. J. Fripiat, *in* "Catalysis by Zeolites" (B. Imelik, ed.), p. 161. Elsevier, Amsterdam, 1980.

69. B. Stübner, H. Knözinger, J. Conrad, and J. J. Fripiat, *J. Phys. Chem.* **82,** 1311 (1978).

70. S. J. Gentry and R. Rudham, *J.C.S. Faraday I* **70,** 1685 (1974).

71. M. L. Poutsma, *in* "Zeolite Chemistry and Catalysis" (J. A. Rabo, ed.), ACS Monograph, No. 171, p. 529. Am. Chem. Soc., Washington, D.C., 1976.

72. P. S. Landis and P. B. Venuto, *J. Catal.* **6,** 245 (1966).

73. M. Hatada, M. Shimada, Y. Ono, and T. Keii, *J. Catal.* **37,** 166 (1975).

74. P. B. Venuto and P. S. Landis, *J. Catal.* **6,** 237 (1966).

75. R. M. Barrer and R. J. B. Craven, in preparation; cited in ref. 12.

76. R. J. B. Craven, Ph.D. Thesis, Univ. of London, 1976.

77. G. W. Brindley and R. E. Sempels, *Clay Miner.* **12,** 229 (1977).

78. N. Lahav, U. Shani, and J. Shabtai, *Clays Clay Miner.* **26,** 107, 116 (1978).

79. D. E. W. Vaughan, R. J. Lussier, and J. S. Magee, Jr., *Ger. Patent* 2,825,769 (1979).

4

Diffusion and Shape-Selective Catalysis in Zeolites

E. G. DEROUANE

I. INTRODUCTION

Cronstedt (*1*) coined the term "zeolite" from the Greek *zeo*, "to boil," and *litos*, "stone," to denote minerals that expel water when heated (*2*). A zeolite is a crystalline aluminosilicate whose structure contains channels and/or cages filled with exchangeable cations and water and which is of the general formula

$$M_xD_y(Al_{x+2y}Si_{n-(x+2y)}O_{2n})\cdot mH_2O$$

where M stands for a monovalent cation, H^+, Na^+, K^+, etc., and D represents a divalent cation, Ba^{2+}, Ca^{2+}, Mg^{2+}, etc. Zeolites occur naturally but can also be synthesized. Syntheses are generally performed hydrothermally.

INTERCALATION CHEMISTRY

ISBN 0-12-747380-7

When such materials are dehydrated (which makes them seem to boil), their channels and/or cages, which determine their *intracrystalline* volume, are made accessible to those molecules with dimensions and shape that permit their passage through the host structure. This unique physical aspect of zeolite structures is the reason for their designation as "molecular sieves" and accounts for their well-known use as selective adsorbents [see, e.g. (*3*)]. Molecules traveling in the intracrystalline volume of the host zeolite crystals can meet therein other molecules and/or catalytically active sites, which leads to their "activation" and further reaction or transformation. Building-in catalytic sites within such structures is the basis for molecular shape-selective catalysts. Shape-selective catalysis also includes, of course, considerations about the diffusivities of molecules through spaces of nearly molecular dimensions.

The first report on shape-selective catalysis, by Weisz and Frilette (*4*), appeared in 1960. Since then more than 130 related articles, technical papers, and patents have been published in the literature and reviewed up to 1976 in an excellent chapter by Csicsery (*5*). Many applications are found in the petroleum and chemical industries for catalytic cracking and hydrocracking [selectoforming (*6*)], olefin and paraffin isomerization, and aromatic alkylation.

Describing the molecular shape-selective physical and catalytic properties of zeolites requires (a) the classification and description of the various zeolite frameworks and pore systems, (b) a delineation of the factors that govern intracrystalline diffusion, and (c) a presentation of some relevant and striking examples of what has been named, with good reason, "molecular engineering" (*7*). Changes in pore size, nature of the cations, ratio of silicium to aluminum, and residual water content make possible, in principle, the systematic design of specific catalysts.

This chapter presents an overview of shape-selective catalysis using the three points listed above as a general outline. Recent developments, namely, those involving the use of tubular-pore zeolites including ZSM-5 (*8*) and its homologs, are emphasized. Only zeolites that are of industrial importance, such as types A, X, and Y, erionite, mordenite, faujasite, and ZSM-5, are considered in detail.

This chapter is not intended to provide an extensive review of this field but to stress some of the major points and facts that have been ascertained.

II. STRUCTURAL ASPECTS OF ZEOLITES

The structure of zeolites is interesting in many respects. Zeolites can be considered crystalline aluminosilicate polymers that are obtained by the building of a three dimensional network of AlO_4 and SiO_4 tetrahedra linked by the sharing of oxygen atoms. The geometric arrangement is also such that it excludes Al—O—Al bonds (*9*) as much as possible. The framework thus obtained contains pores, channels, and cages, or interconnected voids. As trivalent aluminum ions replace, to a given amount (depending on the nature of the zeolite), tetravalent silicium ions at lattice positions, the network bears a net negative charge, which must be compensated for by counterions (cations). The latter are mobile and may occupy various, exchange sites depending on their radius, charge, or degree of hydration, for example. They can also be replaced, to various degrees, by exchange with other cations. Because of the need to maintain electrical neutrality, there is, of course, a direct 1:1 relationship between the aluminum content of the framework and the number of positive charges provided by the exchange cations.

Considering again the general formula of zeolites, one can then view their chemical structure as shown in Table I.

A. Building and Classification of Zeolite Frameworks

Thirty-four natural zeolites are known at present, and nearly 100 synthetic molecular sieves have now been reported (*10, 11*). Their frameworks may be divided into groups, each group having a common structural subunit, i.e., a specific array of $(Al,Si)O_4$ tetrahedra. The more recent classification (*12–14*) is based on secondary building units (*9, 11*), topological relationships, and the concept of coordination sequences (*14*).

The building of a zeolite framework can be viewed in various ways. A pictorial approach is to discuss the linking of building blocks as it can be done to relate the structures of sodalite, faujasite, and A, X, and Y zeolites or to describe the structure of the synthetic ZSM-5 zeolite (*8*). A more complex and rigorous description of zeolite framework is obtained by linking finite units, which may contain up to 16 tetrahedral units (T atoms). The finite

TABLE I

Chemical Structure of Zeolites

$$\text{General formula: } \underbrace{M_xD_y}_{A}\underbrace{(Al_{x+2y}Si_{n-(x+2y)}O_{2n})}_{B}\cdot\underbrace{mH_2O}_{C}$$

A. Exchangeable cations; moving species; located at so-called exchange sites
B. Aluminosilicate three-dimensional framework, which includes well-defined pores, channels, and cages
C. Intracrystalline water, which can be removed, sometimes reversibly

units (secondary building units, SBU) are chosen such that the framework may be built from one type of unit only (*13*, *14*).

1. Sodalite, Faujasite, and A, X, and Y Zeolites

In these zeolites, the silica and alumina tetrahedra are linked together to form a cubooctahedron, as shown in Fig. 1. Two representations are given, one (Fig. 1a) in which the oxygen atoms are explicitly represented, and another (Fig. 1b) which is more formal and in which the Al or Si atoms are located at the intersection of lines that represent oxygen bridges. This cubooctahedron (octahedron truncated perpendicularly to all C_4-axes) is also called the *sodalite unit* since it is the basis for the structure of sodalite ($Na_6Al_6Si_6O_{24}{\cdot}2NaCl$). It can be seen from Fig. 1b that the sodalite unit is built from six- and four-membered rings. Sodalite is obtained by directly connecting such cubooctahedra through sharing of the four-membered rings (Fig. 1c).

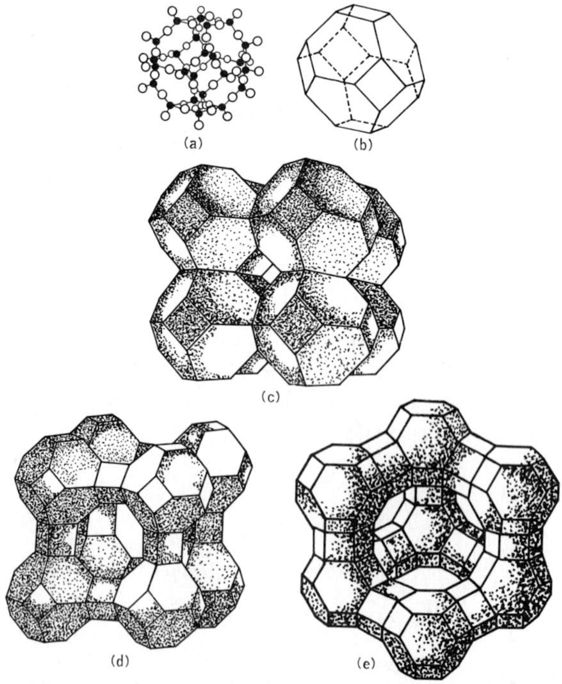

Fig. 1. Topological relationships between sodalite, faujasite, and A, X, and Y zeolites. (a) Sodalite cage structure showing individual atoms (○, oxygen; ●, aluminum or silicon). (b) Formal structure: Intersections represent Al or Si ions; lines correspond to oxygen anion bridges. (c) Stacking of cubooctahedra in sodalite. (d) Arrangement of sodalite units in type A zeolite. (e) Arrangement of sodalite units in faujasite. [Adapted from Gates *et al.* (*92*)].

If, however, the sodalite units are linked together by oxygen bridges between the four-membered rings, one obtains the framework shown in Fig. 1d, which corresponds to type A zeolite $(Na_{12}Al_{12}Si_{12}O_{48}\cdot 27H_2O)_8$. A third possibility consists of connecting the sodalite units by oxygen bridges between the six-membered rings. This leads to the structure of faujasite $[(Na_2,Ca,Mg)_{29}Al_{58}Si_{134}O_{384}\cdot 240H_2O]$, as shown in Fig. 1e. The X and Y zeolites are topologically similar to faujasite. These three materials differ chemically by their Si/Al ratios, which are 1–1.5, 2.3, and 1.5–3.0 for X zeolite, faujasite, and Y zeolite, respectively.

It can be seen from Fig. 1c to 1e that the stacking of the sodalite cages leads to larger cavities (supercages), which are connected by apertures (pores) of variable sizes (four-, eight-, or twelve-membered rings). These cavities and pores constitute the internal surface (intracrystalline volume) of these zeolites. Further attention is paid to these in Section II,C.

In the former case, frameworks are obtained by a three-dimensional stacking of closed building blocks. At variance is the case of ZSM-5, which may be considered the ideal example of a "tubular" zeolite. In contrast to faujasite and X, Y, and A zeolites, the porosity of which is essentially due to the presence of "cages," ZSM-5 (as mordenite, erionite, ferrierite, etc.) presents linear channels of nearly molecular size.

2. *ZSM-5 and Mordenite Structures*

The framework of ZSM-5 involves a new type of building block (Fig. 2a). These units can be linked together, leading to chains (Fig. 2b). The latter in turn can be combined to form sheets (Fig. 2c), which can generate the ZSM-5 framework when they are connected, leading to the formation of four- and six-membered rings. ZSM-5 then shows a unique pore structure that consists of two intersecting channel systems: one straight and the other sinusoidal and perpendicular to the former (Fig. 2c) (*8*). Both channel systems have ten-membered-ring elliptical openings (ca 5.5 Å in diameter).

The mordenite framework, as determined by Meier (*15*), is built from chains of tetrahedra cross-linked by oxygen bridges. Each Al or Si tetrahedron is, in addition, part of a five-membered ring. The chains are then interconnected to obtain the mordenite structure. The latter is schematized in Fig. 3. The mordenite pore structure consists of elliptical and noninterconnected channels parallel to the *c* axis of the orthorhombic structure. Their openings are limited by twelve-membered rings (6.7–7 Å). In the hydrogen form (Na^+ exchanged with H^+), mordenite also offers another channel system, perpendicular to the *c* axis, limited by an eight-membered ring and with opening of 2.9 × 5.7 Å.

Both ZSM-5 (typically $Na_nAl_nSi_{96-n}O_{192}\cdot 16H_2O$, with *n* smaller than 6) and mordenite $(Na_8Al_8Si_{40}O_{96}\cdot 24H_2O)$ show a very high thermal and acid

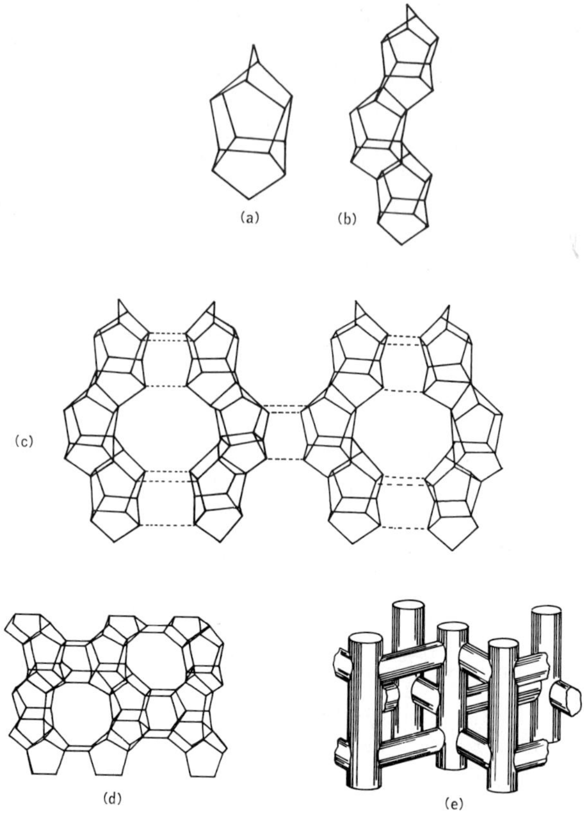

Fig. 2. Structure of ZSM-5. (a) Building block. (b) Linear chain obtained by the linkage of three building blocks. (c) Linear-chain combination. (d) Skeletal diagram of the (010) face of the ZSM-5 unit cell. The x axis is horizontal and the y axis is vertical. (e) Channel network. [Adapted from Kokotailo *et al.* (*8*).]

stability, which is believed to be due to their high Si/Al ratio and their high content of five-membered rings.

3. Classification of Zeolite Framework Types Using Coordination Sequences (14)

Assuming that the entire framework of a given zeolite can be obtained from *one* building unit only, one can derive secondary building units (*9*, *11*) (the primary building units are the Al and Si tetrahedra). These may contain up to 16 T atoms, and their three-dimensional arrangement leads ultimately to the different zeolite structures, enabling one to classify them on purely

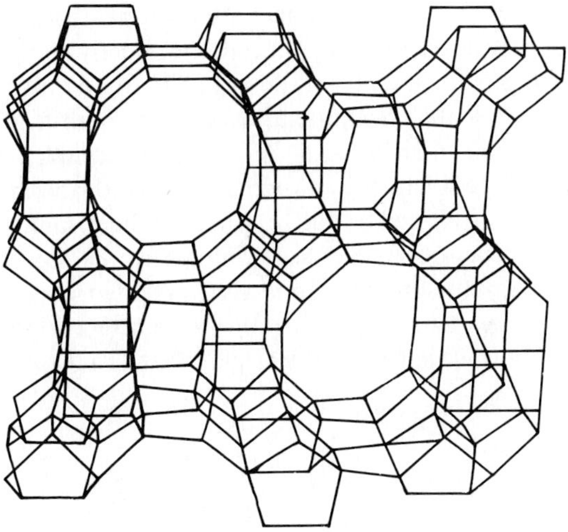

Fig. 3. Perspective view of the mordenite structure. [Adapted with permission from Meier and Olson (*13*).]

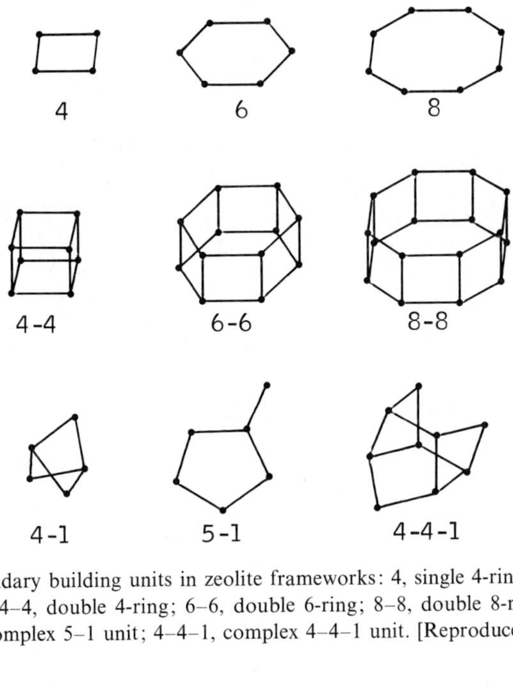

Fig. 4. Secondary building units in zeolite frameworks: 4, single 4-ring; 6, single 6-ring; 8, single 8-ring; 4–4, double 4-ring; 6–6, double 6-ring; 8–8, double 8-ring; 4–1, complex 4–1 unit; 5–1, complex 5–1 unit; 4–4–1, complex 4–4–1 unit. [Reproduced from Meier and Moeck (*14*).]

topological grounds (*13*, *14*). Secondary building units are shown in Fig. 4. These SBU's can be used to describe and construct the frameworks of the zeolites discussed previously. This is illustrated in Fig. 5 for sodalite, faujasite, ZSM-5, and type-A zeolite. Sodalite appears to be built from single 6-rings, faujasite from double 6-rings, and ZSM-5 from complex 5–1 units.

Implementing these ideas, it is possible (*13*, *14*) to obtain a classification of the zeolite structures based on the topology of their frameworks, as shown in Table II. The structure type codes conform to the recommendations of IUPAC. The maximum space group is also indicated, as are the number of topologically inequivalent T atoms (Q), the nature of the SBU, and the framework density (number of T atoms per 1000 Å^3). Framework densities

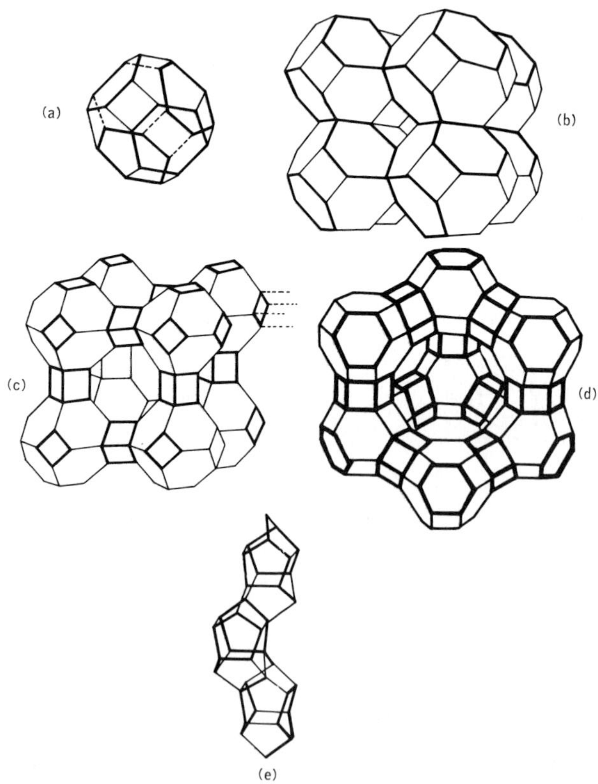

Fig. 5. Structures of (a) the sodalite cage, (b) sodalite, (c) type A zeolite, (d) faujasite, and (e) ZSM-5 showing the connections between the secondary building units. [Reproduced from Gates *et al.* (*92*) and from Kokotailo *et al.* (*8*).]

TABLE II

Topological Description of Zeolite Structures (*13*, *14*)

Type code	Name	Maximum space group	Q^a	Number of T atoms in unit cell	SBU[b]	Si/Al ratio	Framework density[c]
ABW	Li-A(BW)	$Imam$	1	8	4	1	19
AFG	Afghanite	$P6_3/mmc$	3	48	6	1	15.9
ANA	Analcime	$Ia3d$	1	48	4	2	18.6
BIK	Bikitaite	$Cmcm$	2	6	5–1	2	20.2
BRE	Brewsterite	$P2_1/m$	4	16	4	3	17.5
CAN	Cancrinite	$P6_3/mmc$	1	12	6	1	16.7
CHA	Chabazite	$R\bar{3}m$	1	36	6	2	14.6
DAC	Dachiardite	$C2/m$	4	24	5–1	3.8	17.3
EDI	Edingtonite	$P\bar{4}2_1m$	2	10	4–1	1.5	16.6
EPI	Epistilbite	$C2/m$	3	24	5–1	2	18.0
ERI	Erionite	$P6_3/mmc$	2	36	6	3	15.6
FAU	Faujasite	$Fd3m$	1	192	6–6	2.3	12.7
FER	Ferrierite	$Immm$	4	36	5–1	5	17.7
GIS	Gismondine	$I4_1/amd$	1′	16	4	1	15.4
GME	Gmelinite	$P6_3/mmc$	1	24	6	2	14.6
HEU	Heulandite	$C2/m$	5	36	4–4–1	3.5	17.0
KFI	ZK-5	$Im3m$	1	96	6–6	2.2	14.7
LAU	Laumontite	$A2/m$	3	24	6	2	17.7
LEV	Levyne	$R\bar{3}m$	2	54	6	2	15.2
LIO	Liottite	$P\bar{6}m2$	4	36	6	1	15.7
LOS	Losod	$P6_3/mmc$	2	24	6	1	15.8
LTA	Linde type A	$Pm3m$	1	24	4–4	1	12.9
LTL	Linde type L	$P6/mmm$	2	36	6	3	16.4
MAZ	Mazzite	$P6_3/mmc$	2	36	5–1	2.6	16.1
MER	Merlinoite	$I4/mmm$	1	32	8–8	2.55	16.0
MEL	ZSM-11	$I\bar{4}m2$	7	96	5–1	~30	17.7
MFI	ZSM-5	$Pnma$	12	96	5–1	~30	17.9
MOR	Mordenite	$Cmcm$	4	48	5–1	5	17.2
NAT	Natrolite	$I4/amd$	2	40	4–1	1.5	17.8
OFF	Offretite	$P\bar{6}m2$	2	18	6	3.5	15.5
PAU	Paulingite	$Im3m$	8	672	4	3.4	15.5
PHI	Phillipsite	$Cmcm$	2	16	4	3–5	15.8
RHO	Rho	$Im3m$	1	48	8–8	3	14.3
SOD	Sodalite	$I\bar{4}3m$	1	12	6	1	17.2
STI	Stilbite	$Fmmm$	4	72	4–4–1	2.6	16.9
THO	Thomsonite	$Pmma$	3	40	4–1	1	17.7
YUG	Yugawaralite	$C2/m$	2	16	4	3	18.3

[a] Number of topologically inequivalent atoms (per unit cell).
[b] Secondary building units; see Fig. 4.
[c] Framework density, number of T atoms per 1000 Å^3.

vary from 12.7 to 20.2, indicating wide variations in pore structures and intracrystalline volumes.

B. Cation Positions

A zeolite framework bears a net negative charge per unit cell equal to the number of Al tetrahedra it contains. In addition, the small Si^{4+} and Al^{3+} ions are most effectively shielded from interacting with species present in the intracrystalline volume of the zeolite because of the oxygen anions that tetrahedrally surround them. Finally, it is logical to suppose that this negative charge is partially delocalized, although this cannot be quantified, and therefore the extent of delocalization is not known.

Although retaining a certain degree of mobility, given cations locate themselves preferentially in the zeolite framework according to their degree of hydration and their preferred coordination with oxygen (*16*). Each (exchange) cation in the zeolite is virtually isolated from its analogs and, being essentially unshielded, can act with its own chemistry. The importance of exchangeable cations has been ascertained with respect to the adsorptive properties of the zeolites (*17*, *18*) and their role as adsorption centers for hydrocarbons (*19*).

Figure 6 shows the various exchangeable-cation positions that may be found in faujasite and its topological homologs (X and Y zeolites). The S_I site (which will offer an octahedral coordination) is located at the center of the hexagonal prisms connecting the sodalite cages. The S_I' site is located inside

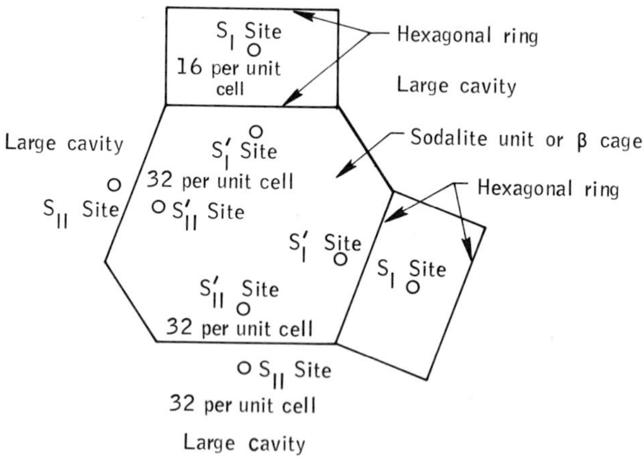

Fig. 6. Positions of exchange sites in faujasite and X and Y zeolites. [Reproduced from Gates *et al.* (*92*).]

the sodalite cage, just on the other side of the hexagonal face from the S_I site. The S_{II} site is located in the supercage, slightly beyond the plane of the unshared hexagonal face. The S_{II}' site is its counterpart, located in the sodalite cage. A site designated U corresponds to the center of the sodalite cavity.

Exchangeable-cation positions have not been determined to the same extent for the other zeolites, apart from mordenite, for which some partial data are also available (*15*).

C. Pore Structure

Pore structure (dimensions and network) varies greatly among zeolites. The lowest pore size is about 2.6 Å, and the highest is 7.4 Å. Pores may lead to linear, parallel, or interconnected channels or may give access to larger intracrystalline cavities. For all zeolites, the pore opening is determined by the free aperture of the oxygen ring that limits the pore. Maximum values for the four-, six-, eight-, ten-, and twelve-membered rings have been calculated to be 2.6, 3.6, 4.2, 6.3, and 7.4 Å, respectively (*11*).

As mentioned earlier, shape-selective reactions occur when the zeolite framework and pore structure allow only molecules of a given size and shape to reach active sites located in the intracrystalline free space. It is therefore important to characterize accurately the pore structure that is encountered in the various zeolite frameworks.

Table III describes the channel system in the zeolite frameworks. The shorthand notation is that introduced by Meier and Olson (*13*), and it includes information on the number of oxygen atoms that constitute the smallest ring determining the pore size (underlined), the free diameter of the pore (calculated from atomic coordinates for the hydrated material and assuming an oxygen radius of 1.35 Å), and the number of dimensions (number of asterisks) in which the channel runs. Channels may also be separated (|), interconnected (↔), or perpendicular (⊥). Openings smaller than six-membered rings are not considered since they give rise to "windows" more than channels because of their very small size. Openings may sometimes be elliptical, depending on the degree of tension and the flatness of the oxygen ring.

On the basis of pore size, zeolites can be classified into two groups: the large-pore zeolites, which show twelve-membered rings [such as X and Y zeolites (faujasite) and mordenite] and the shape-selective zeolites with eight- or ten-membered rings (such as ferrierite and erionite). Large-pore zeolites can accommodate aliphatic and aromatic molecules in their channels, whereas only (normal) paraffins can enter shape-selective structures. ZSM-5 and ZSM-11 zeolites are extraordinary because they admit both aliphatic and light aromatic molecules.

TABLE III

Pore Structure of Zeolites (*13*)

Type code	Name	Channel system (A)[a]
ABW	Li-A(BW)	8 3.6 × 4.0*
BIK	Bikitaite	8 3.2 × 4.9*
BRE	Brewsterite	8 2.3 × 5.0* ↔ 8 2.7 × 4.1*
CAN	Cancrinite	12 6.2*
CHA	Chabazite	8 3.6 × 3.7***
DAC	Dachiardite	10 3.7 × 6.7* ↔ 8 3.6 × 4.8*
EAB	TMA-E(AB)	8 3.7 × 4.8**
EDI	Edingtonite	8 3.5 × 3.9** ↔ 8 . . .*
EPI	Epistilbite	10 3.2 × 5.3* ↔ 8 3.7 × 4.4*
ERI	Erionite	8 3.6 × 5.2***
FAU	Faujasite (X, Y)	12 7.4***
FER	Ferrierite	10 4.3 × 5.5* ↔ 8 3.4 × 4.8*
GIS	Gismondine	(8 3.1 × 4.4 ↔ 8 2.8 × 4.9)***
GME	Gmelinite	12 7.0* ↔ 8 3.6 × 3.9**
HEU	Heulandite	8 4.0 × 5.5* ↔ (10 4.4 × 7.2* + 8 4.1 × 4.7*)
KFI	ZK-5	8 3.9*** | 8 3.9***
LAU	Laumontite	10 4.0 × 5.6*
LEV	Levyne	8 3.3 × 5.3*
LTA	Linde type A	8 4.1***
LTL	Linde type L	12 7.1*
MAZ	Mazzite	12 7.4*
MEL	ZSM-11	10 5.1 × 5.5***
MFI	ZSM-5	(10 5.4 × 5.6 ↔ 10 5.1 × 5.5)***
MOR	Mordenite	12 6.7 × 7.0* ↔ 8 2.9 × 5.7*
NAT	Natrolite	8 2.6 × 3.9** ↔ 8 . . .*
OFF	Offretite	12 6.4* ↔ 8 3.6 × 5.2**
PAU	Paulingite	8 3.9*** | 8 3.9***
RHO	Rho	8 3.9 × 5.1*** | 8 3.9 × 5.1***
STI	Stilbite	10 4.1 × 6.2* ↔ 8 2.7 × 5.7*
YUG	Yugawaralite	8 3.1 × 3.5* ↔ 8 3.2 × 3.3*

[a] For key to symbols, see text.

As mentioned earlier, zeolites can also contain cavities or cages (see, for example, the faujasite and type A zeolite structures in Fig. 2). These cavities may also accommodate molecules and, in principle, play a role in shape-selective reactions. Table IV summarizes some data for structural cavities in various zeolites (*20*). Cages smaller than 5 Å have not been considered. Here again, a wide range of cage sizes is observed. One should also be aware that channel intersections (when they occur, see Table III) offer more available space.

TABLE IV

Cavities in Zeolites

Type code	Name	Type of cavity	Free dimension (Å)
SOD	Sodalite	Sodalite cage	6.6
FAJ	Faujasite (X, Y)	Sodalite cage	6.6
		Supercage (26-hedron)	11.8
LTA	Wide type A	Sodalite cage	6.6
		Supercage (26-hedron)	11.4
GME	Gmelinite	14-Hedron	6.0 × 7.4
OFF	Offretite	14-Hedron	6.0 × 7.4
PAU	Paulingite	18-Hedron	6.6 × 10.8
		26-Hedron	11.4
KFI	ZK-5	18-Hedron	6.6 × 10.8
		26-Hedron	11.4
CHA	Chabazite	20-Hedron	6.5 × 11
ERI	Erionite	23-Hedron	6.3 × 13

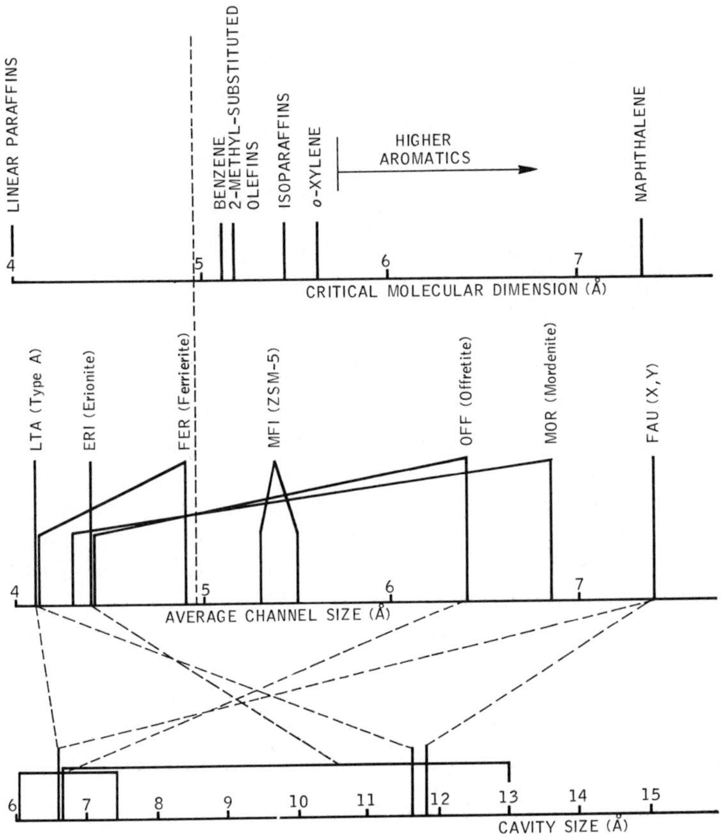

Fig. 7. Comparison of the pore structure of some major zeolites.

The zeolites that are of major industrial importance are the A, X, and Y zeolites, ZSM-5, erionite, offretite, and mordenite. Since they are the main structures that are dealt with in the following sections, it was considered appropriate to summarize their pore structure, as in Fig. 7. There, average channel sizes, cavity dimensions, and critical molecular sizes (using covalent radii for the atoms; hence, different from the values listed in Table VI) are compared.

III. INTRACRYSTALLINE DIFFUSION IN ZEOLITES

In classical diffusion theory, two regimes are defined. One is the ordinary diffusion regime, for which the molecular mean free path (of the gaseous molecules) is smaller than the pore size of the solid porous medium. When the pore size becomes smaller than the molecular mean free path, one is dealing with the Knudsen diffusion regime and the diffusivity is then inversely proportional to the pore dimension. As pointed out by Weisz (*21*), zeolites introduce a third regime, that of *configurational diffusion.* Here, diffusion migration is affected by the matching of the size, shape, and configuration of the diffusing species to the corresponding parameters for the host zeolite framework. If the molecule is large enough, its shape and configuration can change in time, eventually under the influence of the surroundings, and the diffusivity of such molecules in the configurational regime will then also be influenced by their molecular dynamics.

This type of intracrystalline diffusion plays a major role in shape-selective reactions. Indeed, the reactant molecules must reach the catalytic sites, and the product molecules must leave these sites and the intracrystalline space to be desorbed.

Because of the drastic diffusion constraints imposed in the configurational regime, one could also expect a distinction to exist between "deep" active sites and their homologs closer to the external surface of the crystallites. Some of the questions that should be raised in this respect are the following: (a) Is there an optimum size for the crystallites of a given zeolite catalyst used for a given reaction? (b) Does crystallite shape (or size) affect the diffusion behavior of molecules in zeolites? (c) Does crystallite size (or shape) affect the product distribution for a given reaction?

Crystal shape may be expected to have an effect in structures presenting diffusion anisotropy. Exchange ions may also influence the diffusion characteristics of the zeolite framework because of their size (steric hindrance), charge (modification of the local electric field), and degree of hydration (steric hindrance, location in the structure, and competitive adsorption).

To summarize, the following factors essentially govern diffusion in zeolites: (a) the geometry and dimension(s) of the channel network; (b) the size, charge, position, and concentration of the cations; (c) the size, shape, polarity, and flexibility of the penetrant molecules; and (d) the presence of impurity molecules in the channels.

A. Theory and Data Analysis

A comprehensive review by Barrer (*22*) summarizes most of the fundamental observations and their interpretations related to intracrystalline diffusion in zeolites. Evidence is presented for the effect and importance of some of the parameters that were just mentioned. Most of the data analyses are performed in a classical manner.

Statistical–thermodynamic methods have been proposed to analyze the single-component diffusion and adsorption equilibria, essentially by Ruthven (*23, 24*), Brauer *et al.* (*25*), and Fiedler and Gelbin (*26*). In the recent model of Fiedler and Gelbin molecules are either localized or adsorbed at minimum free-energy sites inside the cavities or mobile at sites near the windows that connect cavities (diffusion theoretical models have been proposed mainly for X, Y and A zeolites, to which this would apply). Two steps are necessary for a molecule to jump from cavity to cavity, i.e., a jump from the "localized" state to the "mobile" state and a jump through the window in the neighboring cavity. This model allows the fraction of "mobile" molecules to vary with the cavity occupation level. Although intrinsic diffusivity decreases at elevated temperatures if the occupation level is high because of blockage by previously adsorbed molecules, the effective overall diffusion coefficient goes through a minimum with increasing concentration to rise again at higher loadings. This behavior is governed by the probability that a cavity will have a neighbor occupied by one less molecule than itself, multiplied by the rate of change of chemical potential with respect to zeolite loading (*26*). This has been observed experimentally for various hydrocarbons in zeolite 13X (*27*). Exchangeable cations may affect this behavior (*28*).

However, experimental data are more readily analyzed by solution of the Fickian diffusion equation, which, for an ensemble of uniform spherical particles subjected to a step change in surface concentration at time zero (constant pressure), is (*29*)

$$\frac{m_t}{m_\infty} = 1 - \frac{6}{\pi^2} \sum_{n=1}^{m=\infty} \frac{1}{n^2} \exp(-n^2\pi^2 tD/r^2) \tag{1}$$

where D is the diffusion coefficient, t the time, and r the radius of the spherical particles, m_t and m_∞ being the amounts adsorbed at time t and at equilibrium, respectively.

At large times, Eq. (1) may be simplified to (*22*)

$$\ln \frac{m_\infty - m_t}{m_\infty} = \ln \frac{6}{\pi^2} - \frac{\pi^2 t D}{r^2} \tag{2}$$

whereas at small t values (initial sorption rates) it reduces to (*22*)

$$\frac{m_t}{m_\infty} = \frac{2A}{v} \sqrt{\frac{Dt}{\pi}} \tag{3}$$

Here A is the external surface area and v the volume of the adsorbent ($2A/v = 6/r$ for a sphere). In the latter case, knowledge of the shape and size of the zeolite crystals is not required since the solutions of diffusion equations for particles of different geometries all reduce to Eq. (3) at time zero.

Corrected diffusivities D_0 may be estimated from Darken's equation by assuming that the driving force for diffusion is not the concentration gradient but the difference in chemical potentials (*30*),

$$D = D_0 \left(\frac{d \ln p}{d \ln c} \right) = D_0 \gamma \tag{4}$$

where γ is evaluated from the empirical adsorption isotherms, p is pressure, and c is concentration.

Although the corrected diffusivity should be independent of sorbate concentration within the Henry's law (low-concentration) region, this is not always the case (*27, 31*), and the derivation of the diffusion activation energy E requires the definition of a parameter D_0' such that

$$D_0 = D_0'/c \tag{5}$$

and

$$D_0' = D_*' \exp(-E/RT) \tag{6}$$

Chromatographic methods have recently been proposed as an alternative to the gravimetric or volumetric methods (*32, 33*). Chromatographic methods, although attractive because of their simplicity, are bound by lower and upper limits for the diffusion coefficients. If intracrystalline diffusion is too small, the peak will not equilibrate during its passage in the column whereas, if it is too fast, the dispersion of the chromatographic peak will reflect axial dispersion and other resistances to mass transfer. Chromatographic methods also fail when the corrected diffusivities show a concentration dependence.

The NMR pulsed field gradient techniques can also be used to determine single-component self-diffusion coefficients (*34–38*). Here, the average displacement of a molecule is measured during a given time period and under equilibrium conditions; the diffusivity is obtained from Einstein's law. Large discrepancies are observed, however, between "uptake" and NMR diffusion coefficients, and this has been interpreted as evidence for a barrier resistance to adsorption (uptake) at the outer crystal surface, whereas NMR measures essentially the true mobility within the body of the adsorbent (*36*). A recent discussion of both methods (*39*), however, seems to indicate that the barrier resistance theory does not satisfactorily explain all the observed phenomena. *Intra*crystalline adsorption and self-diffusion data show considerable differences, whereas *inter*crystalline self-diffusion and uptake coefficients are in good agreement (*35*). The observed discrepancies may originate from the techniques themselves (uptake measurements are performed under non-equilibrium conditions, whereas self-diffusion coefficients are obtained at equilibrium) or from finite adsorption heat dissipation limitations.

It is shown in the following sections that diffusion coefficients and rates may depend on molecular size or type, direction of diffusion, crystallite size or shape, and the nature and distribution of the cations. This is especially important with respect to the discussion of shape-selective processes in zeolites since the interaction between diffusive flow and reactivity in porous catalysts is well established (*40–42*). It can indeed be shown that the intrinsic reaction rate in a porous medium is moderated by an effectiveness factor η which depends on a dimensionless parameter Φ (*41*, *42*),

$$\Phi = \frac{dn}{dt}\frac{1}{c}\frac{R^2}{D} \tag{7}$$

as shown in Fig. 8. dn/dt being the actual observed reaction rate, c the concentration of the applied reactant, and D its diffusivity coefficient in the particle of equivalent radius R.

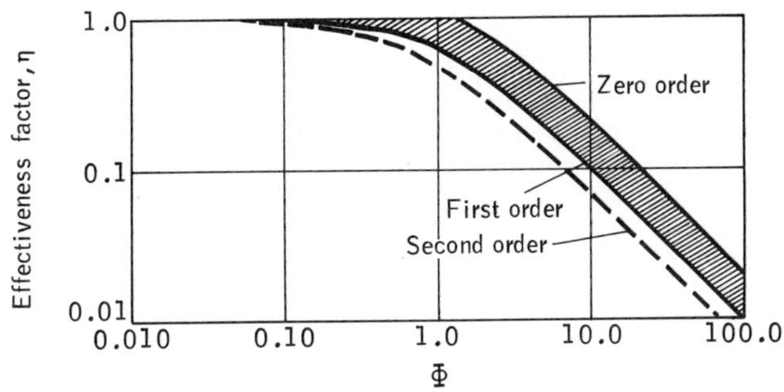

Fig. 8. Effectiveness factor (η) as a function of

$$\Phi = \frac{dn}{dt}\frac{1}{c}\frac{R^2}{D}$$

[Reproduced from Weisz, *Science*, **179**, 433–440 (1973). Copyright 1973 by the American Association for the Advancement of Science.]

The lower the diffusion coefficient, the higher will be Φ and the lower will be the value of the effectiveness factor. Diffusion coefficients may vary easily in zeolites over three to four orders of magnitude, which may result in differences in effectiveness factor of nearly two orders of magnitude.

One must also distinguish between diffusion (at nonequilibrium) and self-diffusion (at equilibrium) experiments, as briefly discussed previously. Self-diffusion may be considered a limiting case of counterdiffusion in which the two types of molecules (reactant and product) are identical.

B. Single-Component Diffusion

Most of the diffusion data that are available concern X, Y, and A zeolites (see, e.g. *27*, *43–45*). Some other information is available for erionite, gmelinite, chabbazite, and levynite (*46–48*), offretite (*49*), and mordenite (*50*, *51*, *52*).

Table V contains some typical data for the diffusion of cyclic molecules in Y (Na and H forms) zeolites. They illustrate the roles played by the critical molecular diameter, the nature of the cation, and the zeolite–adsorbate interaction.

The *critical diameter* certainly plays a major role, as seen by comparing the data for trimethylbenzene and triisopropylbenzene. As shown, the diffusion rate decreases markedly with increasing size. A similar effect was reported for the diffusion of C_1–C_4 hydrocarbons in Na-mordenite (*51*, *52*) and that of various hydrocarbons in zeolite 13X (*27*). It was also shown that propane diffused about four times faster in CaA (the Ca^{2+} form of Linde type A zeolite; pore size, 5 Å) than in NaA (pore size, 4.1 Å) (*53*).

TABLE V

Single-Component Diffusion in Y Zeolites (*44*)

Molecule	Critical diameter (Å)	Diffusion coefficient at 30°C ($D_{eff} \times 10^{17}$ m²/sec)[a] NaY	HY[b]
1,3,5-Trimethylbenzene	8.4	72	>500
2,4,6-Trimethylaniline	8.4	7.1	4.1
1,3,5-Triisopropylbenzene	9.4	0.047	12.4
1,3,5-Triisopropylcyclohexane	9.8	4.9	8.7

[a] Values at $m_t/m_\infty \rightarrow 0$.

[b] 97.7% ion exchanged to hydrogen.

Both *cis*- and *trans*-butene can diffuse in CaA, although their diffusion coefficients vary by more than two orders of magnitude because of their slightly different configuration (*54*). As previously mentioned, this is characteristic of this new configurational diffusion regime in which the diffusing species and the matrix must match each other.

Table VI lists some critical molecular sizes, which indicate the relative effect of the steric hindrance factor on molecular diffusion in zeolites.

Ion exchange also has an important effect on sorption and diffusion rates. This can occur in two ways. First, it is clear that changing the nature of the exchange cation may alter the interaction between the adsorbed molecules and the zeolite lattice. This point is examined later. Second, pore size and diffusion pathways inside the zeolite structure may depend on the nature of the cation. Indeed, channels may be more or less obstructed by cations

TABLE VI

Critical Molecular Sizes (*44*, *49*)

Molecule	Critical molecular diameter (Å)	Chain length or thickness (Å)
Methane	4.00	—
Ethane	4.6–4.9	5.26
n-Propane	4.6–4.9	6.52
n-Butane	4.6–4.9	7.78
Isobutane	5.6	—
n-Pentane	4.6–4.9	9.04
n-Hexane	4.6–4.9	10.30
n-Heptane	4.6–4.9	11.56
n-Octane	4.6–4.9	12.82
n-Nonane	4.6–4.9	14.08
n-Decane	4.6–4.9	15.34
Benzene	6.8	3.4
Toluene	6.8	3.4
p-Xylene	6.8	3.4–4.0
o-Xylene	7.4	3.4–4.0
m-Xylene	7.4	3.4–4.0
1,3,5-Trimethylbenzene	8.4	3.4–4.0
1,3,5-Triethylbenzene	9.2	3.4–4.6
1,3,5-Triisopropylbenzene	9.4	—
1,2,4,5-Tetramethylbenzene	7.4	3.4–4.0
1,2,3,5-Tetramethylbenzene	8.6	3.4–4,0
Naphthalene	7.4	3.4
Cyclohexane	6.9	4.0

depending on their size and concentration. This has been studied mainly for type A zeolite (*43*, *55*).

Diffusion of *n*-alkanes in type A zeolite occurs through the eight-membered-ring windows that connect the supercages (also referred to as α cages). The free distances between diametrically opposed oxygen ions are listed in Table VII [adapted from Quig and Rees (*43*)]. It can be clearly seen that they may vary by as much as 10% with cation and degree of hydration. For NaA, sodium ions are located in the octagonal windows, which reduces the available space. Calcium ions, however, preferably locate themselves in the six-membered windows that connect the α and β cages—hence the increase in the free pore aperture and in the diffusion coefficient for the diffusion of alkanes observed, for example, in the case of *n*-hexane (*43*).

Specific or nonspecific interactions between the adsorbate, or diffusing molecules, and the zeolite are also seen to be important. It is indeed shown in Table VI that 1,3,5-trimethylbenzene (mesitylene) and 2,4,6-trimethylaniline (mesidine), both having the same critical diameter, show very different diffusion behaviors. In NaY, mesitylene diffuses ten times faster than mesidine, the difference being more than two orders of magnitude in H-Y zeolite.

Strong electric fields exist in the pores of zeolites because of the occurrence of positive (exchange cations) and negative (framework aluminum atoms) charges. These fields affect the repartition of electrons in the adsorbed molecules (*56*, *57*) as has been proved by various techniques including the measurement of NMR chemical shifts (*57*), and result eventually in rather strong sorbent–sorbate interactions (*48*, *58*). Molecules with higher polarity are then expected to have smaller diffusion coefficients, as just shown. This

TABLE VII

Effect of Cation Nature and Degree of Hydration on Pore Openings of Type A Zeolite (*43*)

Zeolite	Pore openings (Å)[a]	
NaA, dehydrated	4.04	4.61
NaA, hydrated	4.12	4.47
66% Ca/NaA, dehydrated	4.42	5.01
66% Ca/NaA, hydrated	4.00	4.69
TlA, dehydrated	4.50	3.98
TlA, hydrated	4.19	4.51

[a] The slightly elliptical pores (windows) are characterized by two dimensions.

also explains the different mobilities in NaY of triisopropylbenzene and triisopropylcyclohexane (see Table VI).

Molecules may be classified into various groups depending on common function and/or structural features leading to a given type of adsorption behavior in zeolites (*59*). Type A molecules include the permanent gases and the alkanes. Type B molecules are compounds with π-electron structures such as the alkenes, aromatic compounds, and naphthalenes. Type D molecules possess permanent dipole groups; they may be alcohols, amines, etc. When they are of nearly similar size and structure, A molecules diffuse faster than B molecules, and the latter more rapidly than D molecules.

Pure dispersion forces may also increase the sorbate–sorbent interaction, as shown by Barrer and Clarke (*60*). Diffusion activation energies E, extrapolated at zero coverage, for n-butane, n-hexane, and n-nonane (in 32.5% Ca^{2+}, NaA) were found to be 11.3, 13.9, and 16.1 kcal/mole, respectively. The rather constant increment in E per additional methylene group, i.e., about 1 kcal/mole, confirms the idea that dispersion forces are important here; the increase in activation energy probably corresponds to the stepwise increase in the interaction that occurs with the chain in its equilibrium state, the activated state being almost unaffected by chain length.

Other parameters that may influence sorption rates and diffusion coefficient are the presence of *impurities* (*61*) and the relative effect of *crystallite shape and size distribution* (*22, 62, 63*).

C. Counterdiffusion of Liquid Hydrocarbons in Zeolites

In many applications of zeolites, molecules must diffuse in and out of the intracrystalline pore structure, and, in their catalytic uses, at least two components are present together, i.e., the reactants diffusing inward and the products diffusing outward. The counterdiffusion of the latter species may seriously affect the apparent activity and selectivity for a given reaction since counterdiffusion experiments show clearly that the diffusion flux in one direction is strongly affected by the diffusion flux in the opposite direction because of intermolecular interactions between counterdiffusing molecules.

As in single-component diffusion, the size of the diffusing molecule(s) and that of the channels or pores, the nature of the cations, the polarity of the (counter) diffusing molecule(s), and impurities are important (*64*). When molecules of nearly the same type are counterdiffused toward a common component under identical conditions, one may expect to note a neat effect of molecular size. This has indeed been reported for the counterdiffusion of various aromatic compounds with respect to cyclohexane into NaY zeolite

(*65*), as shown in Fig. 9. The effective diffusion coefficient D_{eff} varies exponentially over nearly five orders of magnitude with the critical molecular diameter [changing from 6.8 Å (toluene) to 9.2 Å (1,3,5-triethylbenzene)].

It is very interesting that molecules with a critical molecular diameter as large as 9.2 Å can still freely counterdiffuse, although the pore opening is only 7.4 Å for the Y zeolite. This reflects the three-dimensional nature of the pore system in Y zeolite and implies that the molecules may be deformed to a certain extent. Note that in a one-dimensional structure such as mordenite, molecules with dimensions close to that of the pores cannot freely counterdiffuse because they cannot pass the other molecules in the pores (*60*, *66*).

The observed variation of the diffusion rate with molecular size is similar to that observed in single-component diffusion (*44*). It can be proved that an exponential variation may be expected (*60*, *65*) when the critical molecular

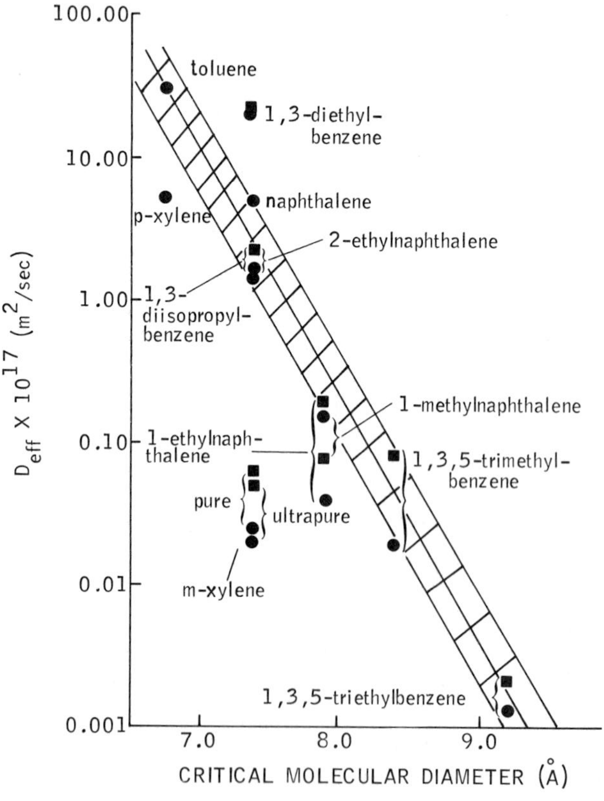

Fig. 9. Dependence of the effective diffusion coefficient on critical molecular diameter for the adsorptive counterdiffusion into cyclohexane-saturated NaY at 298 K; D_{eff} at M_t/M_∞ : ■, → 0; ●, 0.60. [Reproduced from Moore and Katzer (*65*).]

diameter is close to the pore aperture, considering the diffusion to be occurring across periodic high-energy barriers (because of electron cloud repulsion or molecular deformation) located at the pore apertures.

It must be noted finally that adsorptive and desorptive counterdiffusion rates differ, adsorption being generally faster than desorption. This could be due to the presence of an added resistance in the desorptive counterdiffusion process because of the presence of strongly adsorbed species (*60*).

D. Shape Selectivity in Adsorption and Its Implications in Shape-Selective Catalysis

A striking illustration of the close relationship that may exist between intracrystalline diffusion and shape-selective catalysis is shown in Fig. 10. Diffusion coefficients for C_3–C_{14} *n*-paraffins measured at 340°C in potassium T zeolite (*50*) are compared to the cracking product distribution from

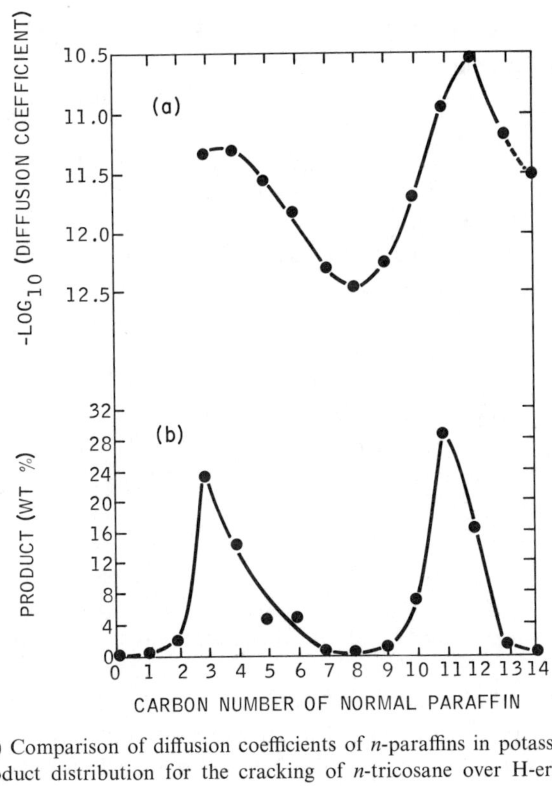

Fig. 10. (a) Comparison of diffusion coefficients of *n*-paraffins in potassium T zeolite at 340°C; (b) product distribution for the cracking of *n*-tricosane over H-erionite at 340°C. [Adapted from Gorring (*49*).]

tricosane (n-C_{23}) over H-erionite at the same temperature (*67*). The synthetic T zeolite (*68*) is primarily offretite with a minor intergrowth of erionite. However, the erionite structure is less open than that of offretite [see Table III and discussion in Gorring (*50*)], which means that erionite units in the T zeolite act as bottleneck centers and control the diffusion behavior of this material. High diffusion coefficient values in the n-C_{11}–n-C_{12} range, as compared to n-C_8 and n-C_{14}, indicate that this zeolite presents a "window of high transmittance" to molecules of a certain critical length. This has been called the "window effect."

An attractive proposal has been made to explain this unusual observation: The lengths of n-heptane (11.56 Å) and n-octane (12.82 Å) compare very well to the free length of the erionite cage (13 Å; see Table IV). Therefore, such molecules that fit very closely in the erionite hexagonally shaped cage, have low mobility (because of steric hindrance), and sit in a low-energy trap. This results in a very low diffusivity. By contrast, smaller molecules are able to move more freely and orient themselves properly with respect to the eight-membered-ring window. Larger molecules (C_9–C_{13}, with a preference for C_{12}) have to extend, partially at least, outside of one of the eight-membered rings that limit the cage and therefore also retain a high mobility because of a lower energy barrier and an imposed molecular orientation in the required direction for motion (entropy effect).

The analogy between the diffusivity data and the cracking product distribution, as compared in Fig. 10, is striking. Normally, when large-pore zeolites or aluminosilica are used as catalysts, a continuous product distribution is obtained which shows a maximum in the range C_6–C_9 (*67*). It is clear, therefore, that catalytic product distributions can be strongly altered by the superposition of a shape-selective pattern of diffusivities onto an intrinsic reaction pattern.

Related observations may be reported for ZSM-5 zeolite. This material presents the channel network shown in Fig. 2e. The channel diameter is 5.3–5.5 Å; perpendicular channels intersect (*13*), offering a maximum available space of about 7.8 Å. It was shown (*69*) that two C_3–C_5 n-alkane molecules could be absorbed per intersection as compared to only one i-C_4, i-C_5, or n-C_6 molecule (i = isoparaffin; n = normal paraffin). Further on, more molecules can be adsorbed in the channels in an "end-to-end" configuration.

It has been proposed (*70–72*) that this could explain the aromatic product cutoff at C_{10} as observed in the conversion of various hydrocarbons on ZSM-5 catalysts. Indeed, evidence was presented for a mechanism involving the "concerted" cycloaddition of a carbenium ion and of an olefin in the formation of aromatic compounds. This, of course, can occur only at the channel intersections, where at maximum, a C_{10} (two times C_5) hydrocarbon can be formed because of steric limitations. The aromatic molecules thus formed are then able to counterdiffuse (*72*) outside the zeolite because of their smaller size (as compared to an olefin dimer) and the availability of a variety of diffusion pathways (ZSM-5 does not present a unidimensional major channel network as does mordenite, for example).

In view of these observations, it is clear that sorption and diffusion play a major role in shape-selective reactions in zeolites, as discussed in the following section.

IV. SHAPE-SELECTIVE CATALYSIS IN ZEOLITES

Provided that the reactants can enter the pores (channels, cages) of a zeolite, the locus of its catalytic activity is essentially located in its intracrystalline space (*73*). Reactants have to diffuse inward to the active sites, products counterdiffuse outward to the external surface, and the environment of the active site(s) may be able to orient the reaction by its steric requirements and the presence of a high local electric field, which may affect the electronic state of the reactants. Therefore, shape-selective catalysis may be identified with molecular shape-discriminating catalytic reactions. Shape selectivity may be achieved by virtue of geometric factors, coulombic field interaction, and/or difference in diffusion rate (*7*).

The effect of geometric factors is determining whenever it is possible to divide the reactants and products into two classes of molecules, one of which cannot diffuse into the channels because of size and shape. If this restriction applies to the reactants, one refers to *charge selectivity*. Selectoforming, cracking, and hydrocracking are among the reactions that may take advantage of this factor. If, however, the geometric restriction applies to the product molecules, *product selectivity* occurs. Product selectivity has been shown to be of utmost importance in a variety of reactions conducted on ZSM-5 zeolite catalysts: The only products that appear are those that are able to pass through the channels. It also affects the deactivation behavior of the catalysts by coking: Large hydrogen-deficient organic residues (for example, polyaromatic molecules) that cannot be removed may act as "coke" precursors.

Figure 11 illustrates the role of the geometric factor. Examples of geometric factor effects are presented in the following subsections.

The magnitude of electrostatic effects also depends on the nature of both the zeolite and the reactants. Zeolites with a lower Si/Al ratio are expected to provide stronger electrostatic fields and a better discrimination between polar molecules (*7*); molecules with a high net dipole moment are more greatly affected (*74*).

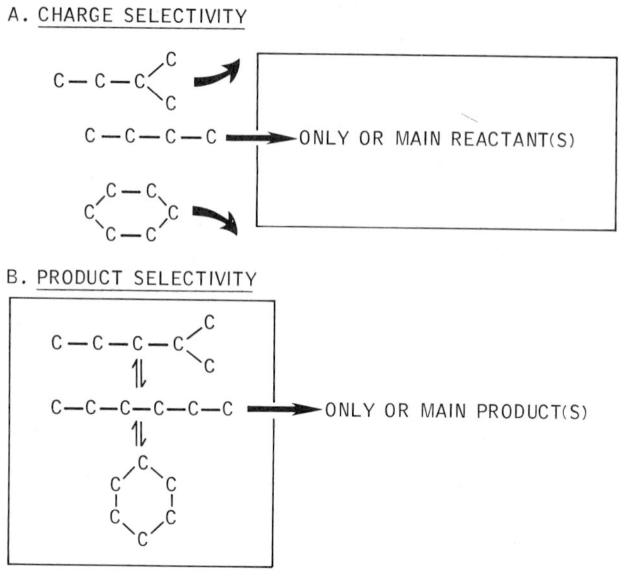

Fig. 11. Geometric effects in shape-selective catalysis. (A) Illustration of a charge selectivity effect; (b) illustration of a product selectivity effect.

The coulombic field effect arises from the electrostatic interactions between the electric field at the pore mouth(s) and the dipolar moment of the reactant molecules. Such interactions provide an energy barrier for the diffusion and may in addition orient the molecule in a favorable or unfavorable configuration with respect to the pore mouth. This is illustrated in Fig. 12 and by data for the isomerization of but-1-ene to *cis*- and *trans*-but-2-enes by type A molecular sieves (*74*) with nonobstructed pores.

The electric field $\vec{E}$ at the eight-membered-ring windows connecting the supercages (diameter ca 4.2 Å) is perpendicular to the plane of the window. Because of the electric field–dipolar moment interaction, both but-1-ene and *trans*-but-2-ene pass rather easily through the octagonal window since they are properly oriented, whereas *cis*-but-2-ene cannot. This is reflected in the product distribution, which shows trans/cis ratios of 7:14 because the trans isomer diffuses and desorbs faster than the cis isomer. Such values must be compared to the thermodynamic equilibrium value of 1.8 under the same conditions (180°C) and to experimental ratios of 0.5–1 as observed when zeolites having larger pores (X, Y, and Z) are used (*74*).

Differences in diffusion rates may also play an important role since they can be reflected in the product distribution. This is illustrated in Section III, D for the cracking of tricosane over erionite and synthetic T zeolite. In general, increasing chain length decreases the rate of diffusion (and therefore

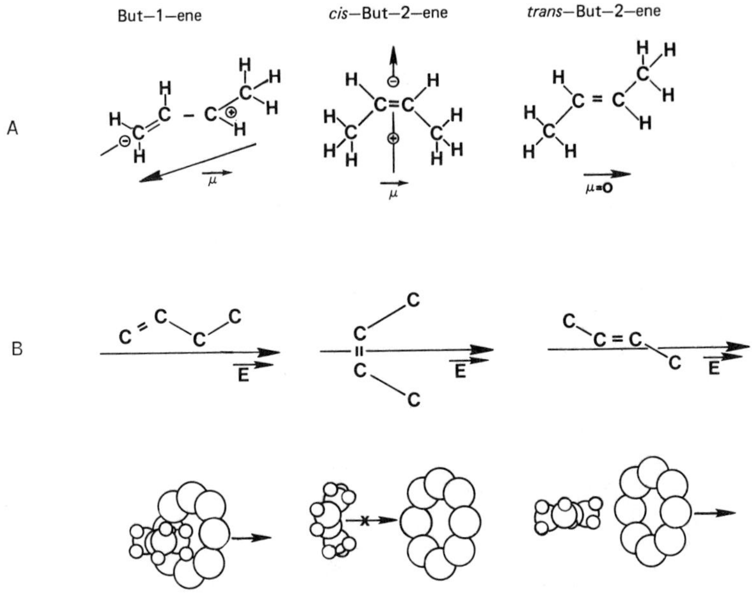

Fig. 12. Coulombic field effect in shape-selective catalysis. The isomerization of but-1-ene to *cis*- and *trans*-but-2-enes on type A zeolite. (A) Orientation of dipolar moment in molecules; (B) orientation of molecules with respect to pore mouths. [Adapted from Tempere and Imelik (*74*).]

the rate of reaction) of *n*-paraffins, as does an increase in the length and number of side chains in alkylaromatic compounds.

The combination of the former factors leads to striking differences between shape-selective and non-shape-selective catalysts. As shown in Table VIII, linear paraffins but not isoparaffins are cracked by type A zeolites, in contrast to the behavior of a silica–alumina catalyst and thermal cracking (*73*), and ethylene can be selectively removed from propylene feeds using a selective (Pt, Na)-mordenite hydrogenation catalyst (*7*) transforming ethylene into ethane.

However, one must also consider the possible role of the zeolite external surface, which normally represents about 0.5% of the total specific surface area (for particles of about 1 μm). The contribution of the external surface-active sites is difficult to predict since these may have different properties than the "internal" sites, being located in a different environment. In addition, the concentrations of the reactants (absolute and partial) may be completely different at both sites. It certainly is expected, however, that inactivating the external surface of a zeolite catalyst will increase its shape-selective character as the external surface acts as a non-shape-selective catalyst.

TABLE VIII

Demonstration of Intracrystalline Reactions in Zeolites (*7, 73*)

A. Selectivity in the cracking of C_6 paraffins
Contact time, 7 sec
Temperature, 500°C

Hydrocarbon	Conversion (%) Thermal (quartz chips)	Silica–alumina (10 wt % Al_2O_3)	Linde type 5A
n-Hexane	1.1	12.2	9.2
3-Methylpentane	<1.0	28.0	<1.0

B. Selectivity in the hydrogenation of ethylene from an ethylene–propylene feed (1:1)
Contact time, 0.18 sec
Temperature, 260°C
Pressure, 1 atm
H_2/HC, 11.5

Catalyst	Conversion (%) Ethylene	Propylene
0.2% (Pt, Na)-mordenite	28.1	<0.1
0.1% Pt on Al_2O_3	69.7	76.5

A particularly elegant demonstration of this is found in catalysts of the ZSM-5 family. ZSM-5 zeolites can be prepared with Si/Al ratios varying from 12 to nearly infinity (crystalline silica). Materials with a higher Si/Al ratio have a lower activity because of the lower number and strength of the acidic sites (*71*). It is therefore conceivable that a ZSM-5 catalyst, the crystallites of which would be terminated by a surface layer consisting of an aluminum-free isocrystalline zeolite, would show enhanced shape-selective properties. This is indeed the case, as described in the patent literature (*75*), which reports a dramatic increase (by a factor of 2) in the *p*- to *m*-xylene ratio in the products when a combined C_6–C_9 aliphatic–aromatic feed is converted over such a "coated" catalyst (316°C; H_2/hydrocarbon = 3.6; 14 atm)

Various authors have reviewed the catalytic properties of zeolites, including shape-selective catalysis (*5, 7, 21, 76–78*). We are essentially concerned in the following subsections with selected aspects of cracking and hydrocracking, alkylation of aromatic rings and coking, and general aspects of shape-selective catalysis using ZSM-5 type catalysts.

A. Cracking and Hydrocracking

An excellent and detailed review of all the aspects of catalytic cracking has been published by Venuto and Habib (*79*). Basic reaction patterns in cracking involve radical- and acid-catalyzed/carbonium-type reactions as well as hydrogen-transfer reactions (*79*, *80*). Hydrogenolysis, i.e., the scission of a C—C bond by hydrogenation, plays an important role in hydrocracking (*81*, *82*). Today, both processes take advantage of the shape-selective properties of zeolites. They are part of a series of processes aimed at producing a liquid product of high octane number, i.e., gasoline. This is schematized in Fig. 13. In the cracking operation, large petroleum molecules are converted to smaller hydrocarbons (C_2–C_{13}). Part of the cracking product to be reformed contains a large quantity of straight-chain paraffins and a relatively smaller amount of isoparaffins, olefins, naphthenes, and aromatic hydrocarbons. In the reforming step, the molecules are "reconstructed" by isomerization, dehydrogenation, dehydrocyclization, and hydrogenation reactions, without essentially changing their carbon number, leading to isoparaffins, cycloparaffins, olefins, and aromatic hydrocarbons. In the selectoforming step, C_5–C_9 *n*-paraffins (of low octane number), in combination with isoparaffins and aromatic hydrocarbons, in the reformate are selectively converted to propane using an *n*-paraffinic selective hydrocracking zeolite

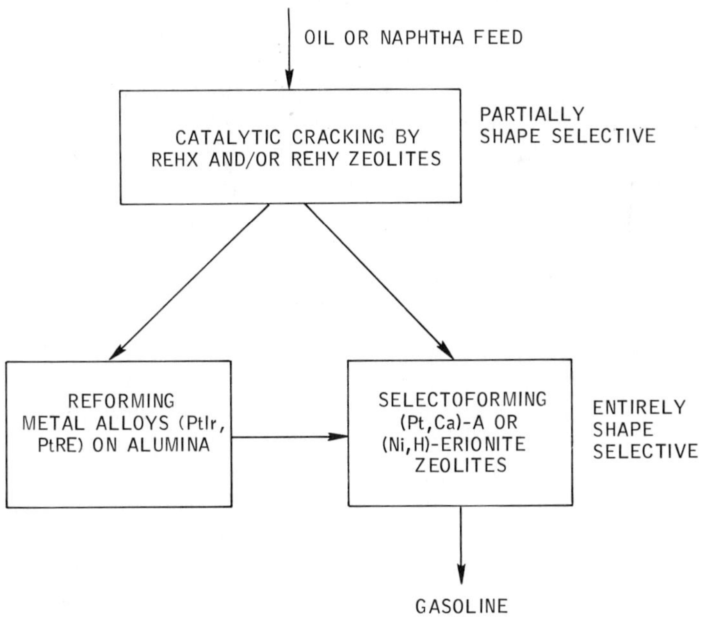

Fig. 13. Cracking and hydrocracking as part of the conversion of petroleum to gasoline.

catalyst. This yields a higher-octane gasoline-blending stock as well as more liquid petroleum gas (LPG).

1. Cracking

Molecular sieving (or shape-selective) effects in catalytic cracking by zeolites have been identified by various authors. Nace (*83*) compared the cracking activities of silica–alumina and of a REHX zeolite (rare-earth-exchanged X zeolite) for the cracking of various bulky hydrocarbons (n-$C_{16}H_{34}$, 1,3,5-triethylcyclohexane, trimethyldecalin, perhydrophenanthrene, perhydrofluorene, and perhydropyrene, listed in order of increasing size). On silica–alumina, the cracking rate increases progressively with molecular size. By contrast, the cracking rate goes through a maximum (for 1,3,5-trimethyldecalin) when an REHX catalyst is used, which can be explained only by the occurrence of diffusional limitations for the three- and four-fused-ring reactants. It is also implied that the latter may create diffusional problems for other molecules that otherwise diffuse rapidly.

Thomas and Barmby (*84*) studied the cracking of various high-boiling petroleum fractions over commercial silica–alumina, REHX, and REHY catalysts. According to them, gas-oil molecules cannot enter the zeolite pore structure and are therefore cracked on the external surface of the zeolite crystallites, as they are by silica–alumina. In the case of zeolites, however, the gasoline-range molecules thus formed may further react in the zeolite pores by secondary reactions of a hydrogen-transfer nature, explaining the higher yields in paraffins and aromatic compounds (fewer olefins). The distinction between external and internal surfaces has, however, been questioned by Eastwood *et al.* since (*85*) the higher gasoline selectivity can also be explained by a higher ratio of the hydrogen-transfer rate to that of β-scission (the latter occurs as heterolytic splitting of a C—C bond) per unit volume of zeolite.

While X and Y (faujasite-type) zeolites are the basis for commercial cracking catalysts, mordenite, erionite, and ferrierite have also attracted some attention. The occurrence of a window effect in T zeolite (which contains erionite) has been discussed previously [see Section III, D, Gorring (*50*), and Chen *et al.* (*67*)]. It is seen that the "diffusion ability pattern" superimposes itself on the primary reaction pattern, leading to a bimodal product distribution. The cracking of light n-paraffins, isoparaffins, and aromatic hydrocarbons was investigated with ferrierite (*86*). The latter shows a pore opening of 4.3×5.5 Å and is therefore expected to discriminate between the linear paraffins and the other molecules. This is indeed observed: Selectivity ratios of 20–100 for n-paraffin cracking at 400°C can be reached on deactivated catalysts, with the activation energies being 16 and 32 kcal/mole for the *n*-paraffins and isoparaffins, respectively.

This work also distinguished the relative contributions to the catalytic activity of the internal and external surfaces of the zeolite. Sorption capacity and cracking activity data are reported in Table IX for two ferrierite catalysts that differ only in their crystallite sizes. As observed by sorption of 2-methylpentane (2-MP), which cannot penetrate the structure, the external surfaces differ by a factor of 2.7, which is reflected in the 2-MP cracking rates. The internal surface, however, is not affected by crystallite size, as shown by sorption data for *n*-hexane. Therefore, the increase in the *n*-hexane cracking rate (for the smaller crystallites) must be due to a decrease in the diffusional restrictions for *n*-hexane in smaller crystallites. Hence, with this material, the selectivity toward *n*-paraffin cracking is affected by both charge selectivity and diffusion limitations.

Rollmann (*87*) did a systematic study of shape selectivity effects in various hydrocarbon reactions catalyzed by a variety of zeolites. Coking and aging behaviors were discussed at length and are dealt with in Section IV,B. An important conclusion regarding cracking is that zeolites with larger pores, such as mordenite, type L, ZSM-4 (*88*), and Y, show a preference toward the cracking of branched paraffins as expected from basic chemistry considerations (*89, 90*) if diffusional restrictions are small. The increase in cracking rate (on REHX zeolite) that has been observed for *n*-paraffins with increasing carbon number (*90*) could be due, by contrast, to higher relative contact times, because of differences in diffusion rates, or to higher concentrations of the longer-chain paraffins in the zeolite pores (*91, 92*).

TABLE IX

Effect of Crystal Size on Sorption Capacity and Cracking Activity of Ferrierite (*86*)

	Sorption capacity (μl/gm)		Cracking constant[b] (sec^{-1})		
Crystal size[a]	2-MP[d]	Hexane	2-MP[d]	Hexane	Selectivity[c]
Large	2.2	60	1.42	1.68	1.18
Small	6.0	63	3.68	5.35	1.45
Ratio	2.7	1.05	2.6	3.2	—

[a] Sizes not otherwise specified.
[b] Initial activities at 500°C.
[c] $k_{2\text{-MP}}/k_{n\text{-}C_6}$.
[d] 2-MP, 2-methylpentane.

2. *Hydrocracking*

Hydrocracking is a hydrogenolysis reaction by which a C—C bond is ruptured in the presence of hydrogen. It is particularly important for the cracking of linear paraffins, for example,

$$n\text{-}C_9H_{20} + H_2 \rightarrow C_6H_{14} + C_3H_8$$

$$n\text{-}C_6H_{14} + H_2 \rightarrow 2C_3H_8$$

since these *n*-paraffins, in equilibrium with methylparaffins, may amount to about 30% in reformate fractions. Their removal as LPG from the liquid reformate increases the octane number of the latter. Similarly, the selective hydrocracking of *n*-paraffins in jet fuel, kerosene, and heating oil improves their performance and reliability by lowering their freezing point and their viscosity.

Shape-selective catalysts that accept *n*-paraffins but not isoparaffins and other hydrocarbons must have a pore size near 5 Å (see Table VI). In addition, a hydrogenation function must be supplied by incorporating an active metal (by ion exchange and further reduction) into the zeolite structure. Eng (*93*) was the first to describe the use of 5 Å zeolites as such catalysts, Pd being the active metal. Presently, two types of systems are considered: a (Pt, Ca)-A type of zeolite and an (Ni, H)-erionite (*94*). Recently, ZSM-5-based catalysts demonstrated their superior performance (*95*), as discussed in Section IV,C.

Selectoforming, introduced by Mobil Oil Corporation, is a process by which normal paraffins in gasoline reformate fractions are hydrocracked essentially to propane (*81*, *82*). It may be considered a shape-selective hydrogenolysis. The liquid product shows an increased octane number because of the preferential conversion of its low-octane components. Because of its pore aperture (3.6 × 5.2 Å), erionite appears to be a prime candidate for hydrocracking catalysis. The conversion of C_5–C_6 naphtha is almost totally selective for *n*-paraffins, leading to a C_4-minus gas fraction with a typical composition of propane (90%), ethane (8%), and methane (1%), using a rare-earth-exchanged erionite as catalyst operating at 370°C (*96*).

Chen and Garwood (*94*, *96*, *97*) reported several excellent investigations using (Ni,H)-erionite as a hydrocracking catalyst. In a study of the hydrocracking of *n*-C_4 to *n*-C_{16} hydrocarbons, they observed a relative cracking activity pattern showing maxima at C_6 and C_{10-11}, as shown in Fig. 14 (*96*). Such a variation is attributed to a *cage effect*, analogous in essence to the *window effect* previously discussed (*50*, *67*) (Section III,D). The erionite cage structure forces interactions between specific reactant molecules and the internal surface of the catalyst by increasing their residence time in the intracrystalline space (*96*).

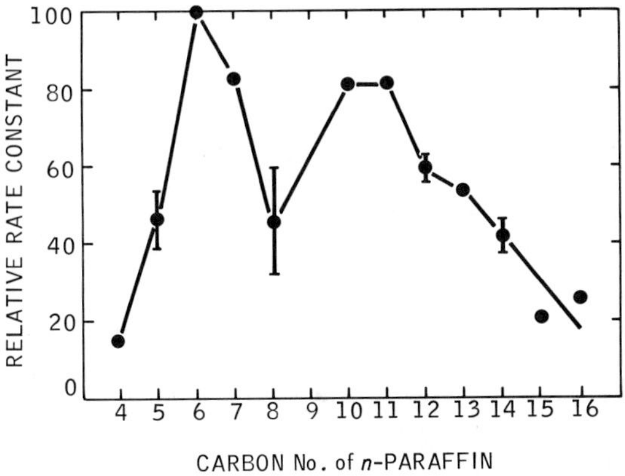

Fig. 14. Relation between the chain length of *n*-paraffins and their relative hydrocracking rate by (Ni,H)-erionite. [Reproduced from Chen and Garwood (*96*) with permission from the American Chemical Society.]

If a mixture containing C_{10}–C_{16} *n*-paraffins is used as reactant, one observes a decrease in the conversion with increasing chain for (Ni,H)-erionite, whereas the opposite behavior holds for (Ni,RE)-X zeolite. In addition, the conversion of *n*-paraffins decreases abruptly when the hydrogen pressure is lowered, whereas the conversion of isoparaffins is barely affected. Since the branched paraffins are converted on the external surface or by natural impurities and the linear paraffins *in* the zeolite, the inhibition effect observed at lower hydrogen pressure must be attributed to the formation of strongly adsorbed and bulky residual material in the pores of the catalyst (*94, 97*). These residues may eventually act as coke precursors, as discussed in Section IV,B. However, under the reaction conditions used (temp. 400°C; H_2/HC, 27:1; pressure, ca 35 atm) they are in dynamic equilibrium with the hydrogen since the inhibition effect disappears reversibly when the hydrogen pressure is increased (*93*).

B. Coke Deposition and Aging of Zeolite Catalysts

The purpose of the following discussion is to demonstrate that coking (aging, deactivation) of a zeolite is a shape-selective reaction intimately related to its pore size(s) and geometry. This requires insight into both the coking mechanism(s) and the locus of coke deposition in various zeolites, because the latter may affect the diffusion behavior of reactants and/or products (*98*).

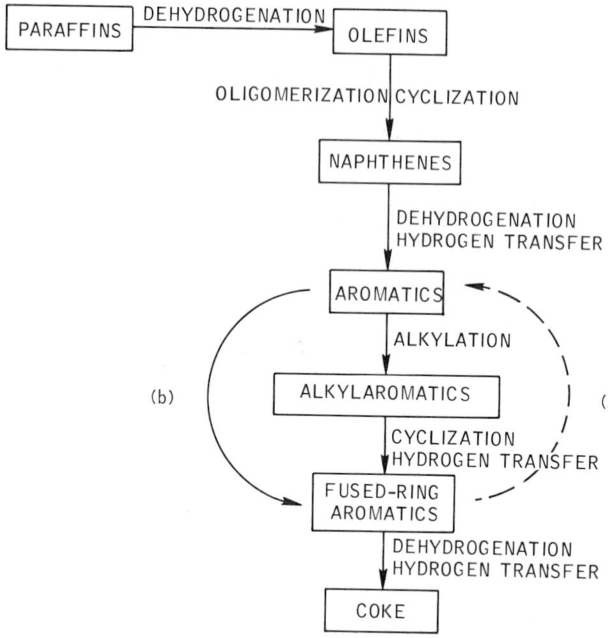

Fig. 15. Schematic representation of coke formation from aromatic compounds and olefins in zeolites. (a) Alkylation of fused-ring aromatic compounds; (b) direct condensation of aromatic rings.

Coke in zeolites, as for silicoalumina amorphous catalysts, should be considered to be a mixture of hydrogen-deficient organic residues, stemming essentially from fused-ring hydrocarbons (*99*) and characterized by a C/H ratio of 1.5–2.6 (*100*). The nature and effectiveness of various coke precursors have been discussed by Venuto (*78, 79*). Coke originates mainly from aromatic compounds (*99*) and/or olefins (*101*). Aromatic-ring alkylations and hydrogen-transfer reactions are highly important contributors to coke deposition in both cases.

A general, simplified scheme for coke deposition from aromatic compounds and/or olefins (and paraffins, if a dehydrogenating function is present) on zeolites is shown in Fig. 15. This scheme is supported by the data of Eisenbach and Gallei (*102*) (coking by hexene and hexane of Y zeolite) and Rollmann and Walsh (*103–105*) (using mixed feeds and various zeolites).

1. Shape Selectivity and Carbon Formation

Rollmann and Walsh (*87, 103–105*) have made a remarkable contribution to the understanding of coke deposition in a variety of zeolites, including the

TABLE X

Catalysts Used in Coking Rate Experiments (*104*)

Group	Si/Al[a]	Zeolite type	Hydrocarbons admitted in pores
Large-pore	2.6–54	Y L Mordenite	Aromatic compounds and paraffins
Shape-selective	3.8–110	Classical	
		Ferrierite	*n*-Paraffins only
		Erionite	
		New	
		ZSM-5 family	Paraffins and light aromatic compounds[b] only

[a] Range covered in this study. Not indicative of normal Si/Al ratio.
[b] Essentially benzene, toluene, and xylenes.

new ZSM-5 shape-selective catalyst. Shape selectivity effects were systematically investigated (*87*, *104*), as was the intrinsic mechanism, using ^{14}C-labeled compounds as radiotracers (*103*, *105*). The alkylation of aromatic compounds was found to be the initial and decisive step in coke formation over zeolites such as mordenite and type Y (*103*). The contribution of aromatic compounds to coking was also found to increase with decreasing Si/Al ratio and temperature. In a later stage, these alkylaromatic compounds lead to fused-ring products, the dehydrogenation of which results in coke formation. However, paraffins may also contribute to coke deposition (*103*) by forming precursor cycloparaffins, for example. An impressive correlation between coking activity and shape selectivity [as measured by the ratio of the cracking rates for *n*-hexane (NC6) and 3-methylpentane (3MP)] was recently proposed (*104*). Coke deposition was measured (at 427°C, 15 atm, H_2/HC, 3:1) for the cracking of a mixed feed comprising equal weights of *n*-hexane, 3-methylpentane, 2,3-dimethylbutane, benzene, and toluene. The characteristics and nature of the catalysts used are indicated in Table X.

The observed correlation is shown in Fig. 16. Care was taken to eliminate or average fluctuations or changes due to variable Si/Al ratios or crystallite sizes. The figure clearly indicates that the zeolite pore structure controls intracrystalline coking. Table XI presents other data on coke yields as compared to structural features for some selected zeolites (*87*, *105*).

The low coking activity observed for erionite and ferrierite compared to mordenite has been attributed to large restrictions on the formation of cycloparaffins, which may act as coke precursors, because of their narrow

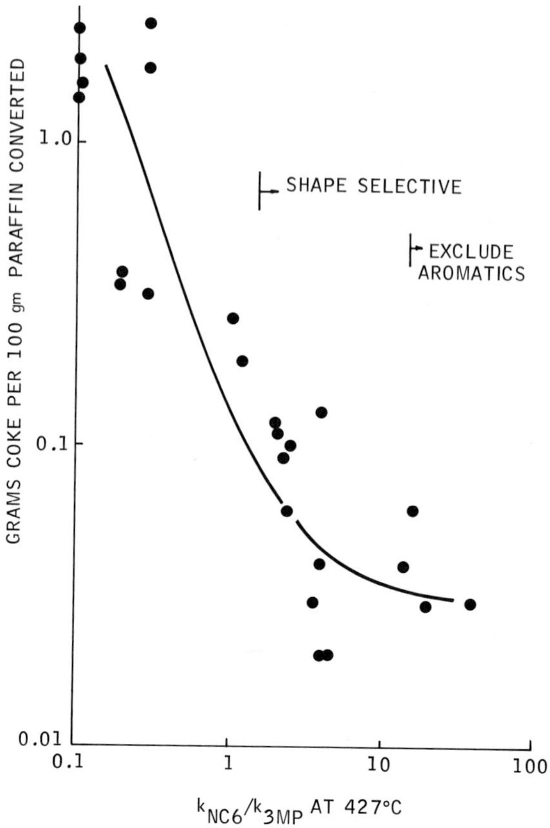

Fig. 16. Coke yield as a function of shape selectivity for the conversion of paraffins on acidic zeolite catalysts. [Reproduced from Rollmann and Walsh (*104*).]

TABLE XI

Structural Effects on the Coke Yield of Zeolites (*87, 105*)

Zeolite	Pore size (Å)[a]	Feed	Coke yield[b]
Erionite	3.6–5.2	Five-component feed[c]	0.14
Ferrierite	4.3–5.5		0.03
Mordenite	6.7–7.0	Five-component feed[c]	0.3
		Benzene + hexane[d]	~9
ZSM-5	5.1–5.6	Benzene + hexane[d]	~0.22

[a] See Table III for detailed description of pore structure.
[b] Grams per 100 gm conversion.
[c] See text for feed composition.
[d] See Walsh and Rollmann (*105*) for experimental details.

pores (*87*). As seen from Table XI, these materials have openings in the 4–5 Å range which will accept only *n*-paraffins. The higher coking activity of erionite with respect to ferrierite is probably due to the fact that ferrierite has a "pure" tubular channel network, whereas erionite also presents rather large cavities (6.3 × 13 Å), where some coking can occur. By contrast, both mordenite and ZSM-5 catalysts can accept aromatic molecules. It has been clearly demonstrated in this case that alkylaromatic molecules, when formed, cannot react further in ZSM-5 (by cyclization, dehydrogenation, further alkylation, etc.), leading to coke deposition, because of the smaller aperture of the ZSM-5 pores. This explains the low coking activity of ZSM-5 type of materials.

To summarize, our present knowledge points to the fact that the different coking activities observed for a wide range of zeolites are not due to different mechanisms and/or origins, but to structural constraints that make coking a product-selective reaction.

2. *Coking and Aging*

After coking, deactivation of the zeolite catalyst, or "aging," takes place, which may result from either of two factors (or both). First, there may be an irreversible decrease in the number of active sites that may be poisoned by organic residues. Second, coke deposition may cause *pore blockage*. In the latter case, the aging phenomenon will also depend on the structural features of the channel network(s). This is also why coking and aging must be considered distinct phenomena. The various situations that may be encountered are schematized in Fig. 17.

In mordenite, only the larger channels are accessible, and it can therefore be considered a pseudounidimensional pore network (this also holds for ferrierite). In such a system of noninterconnecting channels (I in Fig. 17), the accessibility to the active sites in a given channel is greatly decreased (Ia and Ib) or even completely blocked (Ic and Id) as soon as coke deposition occurs at some spot along the channel. Aging should therefore be rather rapid for such structures, unless coke deposition is restricted to the external surface and does not obstruct the pore mouths (*87*). Evidence for pore-mouth blocking in H-mordenite was found during the catalytic dehydration of methanol; the aging of the catalyst was observed to be independent of crystallite size.

If one deals with structures possessing interconnecting channels, such as ZSM-5 zeolites, or as a first approximation type A, X, and Y zeolites, a completely different situation is encountered. Provided that coke does *not* deposit at the channel intersections, the structure can accommodate a rather large amount of carbonaceous residue before the "circulation" of reactants and products is strongly hindered (II in Fig. 17). Type A, X, and Y zeolites are therefore able to accommodate higher coke loadings while still retaining some

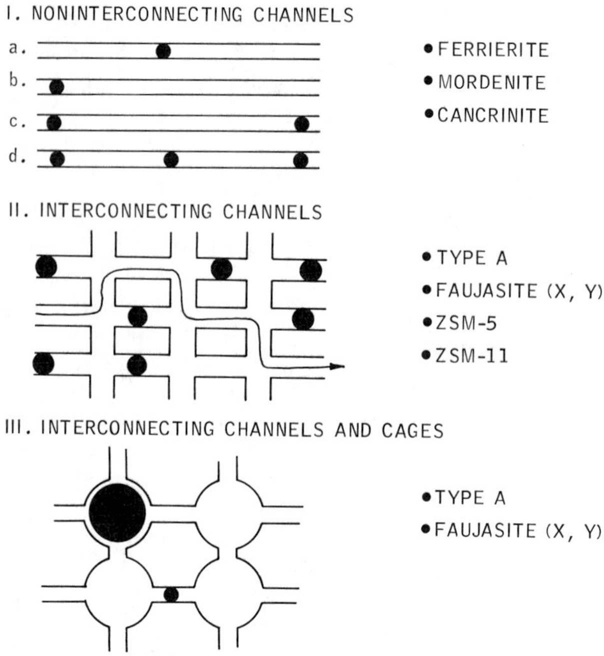

Fig. 17. Illustration of pore blockage for various structural types of channel networks.

activity. Aging should, in principle, be relatively slower than for nonintersecting channel structures.

A troublesome situation is encountered, however, in structures that are in fact interconnected cages. This is case III in Fig. 17 and is known as the "faujasite trap." Indeed, during coking, bulky deposits may develop in the larger volume offered by the cages (which may be considered channel intersections). This not only quickly deactivates the zeolite catalyst but also leads to a situation in which the catalyst cannot be regenerated under mild physical conditions (solvent extraction, desorption, etc.); C—C bonds must be broken by cracking, hydrogenation, or oxidation to allow smaller molecules to exit. This is, in fact, a reverse shape selectivity effect (*106*). It is illustrated by Fig. 18 for the isomerization–oligomerization of 1-hexene on a rare-earth-exchanged X zeolite (REX):

$$1\text{-}C_6H_{12} \xrightarrow[64°C]{REX} cis\text{- and } trans\text{-}2\text{- and } 3\ C_6H_{12} + C_{12}H_{24} + C_{18}H_{36}$$

Higher olefins ($C_{6n}H_{12n}$, n = 2–5) can be found as intracrystalline product, although they do not appear in the liquid product (*107*, *108*).

Such intracrystalline bulky material acts as a coke precursor; aging is observed because obstruction occurs in the pores and cages.

$$1\text{-}C_6H_{12} \xrightarrow[\text{1.75 hr, liquid}]{\text{REX, } 64^{\circ}\text{C}} \textit{cis-}\text{ and }\textit{trans-}\text{2- and 3-}C_6H_{12} + C_{12}H_{24} + C_{18}H_{36}$$

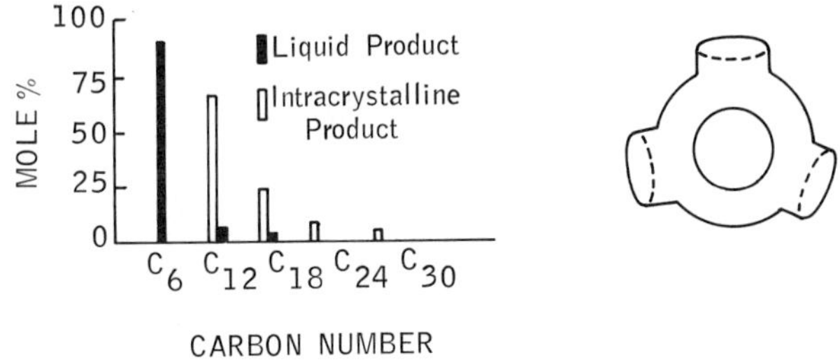

Fig. 18. Illustration of the "faujasite trap" as observed in the isomerization–oligomerization of 1-hexene. [Reproduced from Venuto (*107*) with permission from the American Chemical Society.]

C. Shape-Selective Catalysis by ZSM-5 Zeolites

Some examples of shape-selective catalytic-type reactions have already been discussed (see Sections III,D and IV,B) in which the specific catalyst is the synthetic zeolite ZSM-5. They deal with aromatization and coking.

The use of ZSM-5 zeolites as shape-selective catalysts has been described in several patents. For the present purpose, however, it is best illustrated by simple cases, such as (a) the dehydration of alcohols (*109*), (b) selectivity in the conversion of hexane isomers (*110*), (c) chain-length effects in the cracking of C_5–C_8 paraffins (*109*), (d) shape selectivity effects in aromatic alkylation (*110*), and (e) the dewaxing of high-pour gas oils (*111*). These observations are easily understood by recalling that ZSM-5, with its intermediate pore size, can accommodate at room temperature linear paraffins and isoparaffins as well as monocyclic aromatic hydrocarbons, provided that they are smaller than 1,3,5-trimethylbenzene (*110*).

Clear evidence for shape selectivity can be found by comparing the product distributions observed for the conversion of methanol, *tert*-butanol, and heptanol (*109*), as indicated in Table XII. The striking similarity between the product distributions not only suggests a common reaction pathway but is also evidence of a marked effect on shape selectivity. Six-carbon and higher aliphatic compounds as well as C_{10+} aromatic compounds are present only in small or trace amounts. Isoparaffins and isoolefins predominate. Aromatic compounds are mainly methyl-substituted. This can be rationalized by assuming that the linear C_{6+} paraffins and olefins, which diffuse more easily in such a channel network, are also more readily converted to aromatic compounds. The upper size of the aromatic compounds is in turn imposed by the ZSM-5 structure (see Section III,D).

TABLE XII

Product Distributions for the ZSM-5-Catalyzed Conversion of Various Alcohols[a]

Reactant:	Methanol	*t*-Butanol	1-Heptanol
Reaction conditions			
Temp (°C)	371	371	371
LHSV (hr^{-1})[b]	1.0	1.0	0.7
Conversion (%)	100.0	100.0	99.9
Hydrocarbon distribution (wt %)			
Methane	1.0	0.1	0.0
Ethane	0.6	0.7	0.3
Ethylene	0.5	0.5	<0.1
Propane	16.2	18.8	16.4
Propylene	1.0	1.1	0.2
i-Butane	18.7	18.4	19.3
n-Butane	5.6	8.7	11.0
Butenes	1.3	0.7	<0.1
i-Pentane	7.8	6.2	8.7
n-Pentane	1.3	1.4	1.5
Pentenes	0.5	0.2	0.1
C_{6+} aliphatics	4.3	7.6	3.0
Benzene	1.7	3.3	3.4
Toluene	10.5	11.6	14.3
Ethylbenzene	0.8	1.3	1.2
Xylenes	17.2	12.4	11.6
C_9 aromatics	7.5	6.1	5.3
C_{10} aromatics	3.3	0.4	2.9
C_{11+} aromatics	0.2	0.6	0.6

[a] From Chang and Silvestri (*109*).
[b] Liquid hourly space velocity.

Shape-selective catalysis by ZSM-5 in the hydrocracking conversion of hexane isomers or C_5–C_8 paraffins and in the alkylation of benzene and toluene have been described by Chen and Garwood (*110*). Figure 19 compares relative rate constants for the hydrocracking of C_5–C_7 paraffins (*110*). The conversion rate for *n*-hexane at 315°C and 35 atm is about twice that for 2-methylpentane and three times that for 3-methylpentane, whereas 2,3- and 2,2-dimethylbutane show very low conversions. Rates decrease for bulkier molecules, which encounter diffusion restrictions or have a lower packing density inside the zeolite because of steric factors.

Concentrating on the effect of chain length and shape (Fig. 19), one observes the following trends in the cracking rates of the paraffins: (a) $n\text{-}C_7 > n\text{-}C_6 > n\text{-}C_5$; (b) linear > 2-methyl- > 3-methyl- > dimethyl- or ethyl-. Hence, cracking activity increases with chain length as expected but

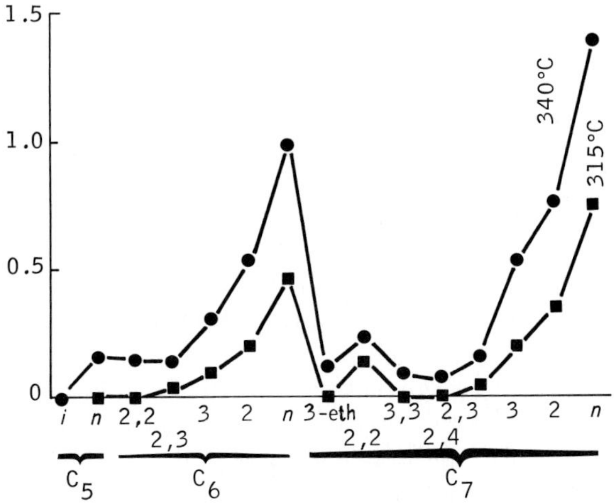

Fig. 19. Shape selectivity in the relative cracking rates of C_5–C_7 paraffins. Key: C_5—*i*: isopentane; *n*: *n*-pentane. C_6—*n*: *n*-hexane; 2: 2-methylpentane; 3: 3-methylpentane; 2,3: 2,3-dimethylbutane; 2,2: 2,2-dimethylbutane. C_7—*n*: *n*-heptane; 2: 2-methylhexane; 3: 3-methylhexane; 2,3, 2,4, 3,3, 2,2: -dimethylpentane; 3-eth: 3-ethylpentane. [Reproduced from Chen and Garwood (*110*).]

decreases for bulkier molecules, indicating that the rate is controlled by configuration effects (Section III), the pore size having essentially more importance for ZSM-5 than the channel tortuosity [in contrast to the observations for erionite, e.g., the existence of a window effect (*49*) or a cage effect (*67*)].

The intracrystalline volume of ZSM-5 also provides the necessary locus for the selective alkylation of aromatic compounds. Table XIII compares data for the alkylation of benzene and toluene (*110*). According to Stock and Brown (*112*), the ratio of the alkylation rate *k* of toluene to that of benzene

TABLE XIII

Comparative Rates of Benzene and Toluene Alkylations

	Conversion (wt %)	
	315°C	340°C
Benzene	25	41
Toluene	18	31
k (toluene)/*k* (benzene)	0.69	0.70

[a] Data from Chen and Garwood (*110*).

should be about 1.4, as confirmed experimentally for the alkylation of both these aromatic compounds by ethylene (*113*), in which case the k (toluene)/k (benzene) ratio is 1.5 at 180°C. This reflects the intrinsically higher reactivity of toluene, which should not be affected when a large-pore zeolite (REX) catalyst is used. The lower value (0.70) observed for ZSM-5 thus confirms the existence of a shape selectivity effect: The lower diffusion rate of toluene as compared to that of benzene may be responsible for the reversed order of reactivity.

The unique shape-selective properties of ZSM-5 as described earlier are the basis of the Mobil distillate dewaxing (MDDW) process (*111*). In this process, the pour point of a gas-oil distillate containing long-chain linear paraffins, isoparaffins, naphthenes, aromatic compounds, and highly branched paraffins is lowered by selectively cracking the linear paraffins and isoparaffins. These are preferentially admitted into the intracrystalline volume of the zeolite, in which the reaction essentially takes place.

Figure 20 illustrates the removal of the C_{16}–C_{28} *n*-paraffins from a midcontinent distillate (340°–390°C) by comparing typical gas chromatograms for the virgin (Fig. 20b) and processed (Fig. 20a) distillates. The corresponding pour-point lowering is from 10°C (for the virgin distillate) to −59°C (for the processed distillate). The MDDW process is of major importance because it enables one to adapt the pour or freeze point of various distillate fractions (diesel fuel, jet fuel, gasoline, heating oil) to climatic or utilization temperature requirements, thereby also optimizing the use of the hydrocarbon charge.

The former examples illustrate the specific shape-selective properties of ZSM-5 that result from its particular channel network and pore size (see Table III and Fig. 2).

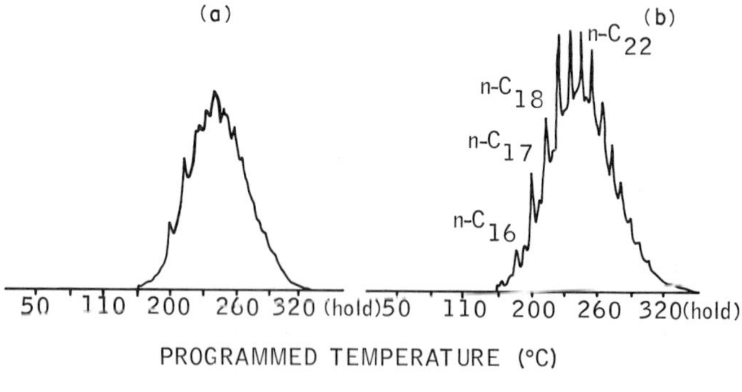

Fig. 20. Illustration of the dewaxing of a high-pour gas–oil distillate using a ZSM-5-based catalyst. (a) Processed distillate; (b) virgin distillate. [Reproduced from Chen *et al.* (*111*).]

V. CONCLUSIONS

The particular structural properties of zeolites, e.g., the presence of pores or windows, channels, and cages, with various arrangements and sizes, are intimately reflected in their catalytic and diffusional properties.

A new regime of diffusion, called "configurational diffusion," characterizes such materials. The matching of size, shape, and configuration of the diffusing species to those of the zeolite pores or channels greatly affects the diffusion behavior of several types of host molecules. Molecules that are too bulky are excluded from the intracrystalline volume, and this is the basis of the molecular sieving effect of such materials.

Shape-selective catalysis relies on the behavior of molecules through spaces of nearly molecular dimensions. Evidence for it is obtained when reaction selectivity toward either the reactants or the products is influenced by the structural properties of the zeolite catalyst. It can be due to sieving, coulombic field, or diffusional effects.

It has been found to play a major role in the conversion of hydrocarbons, which explains its great importance in the petroleum and petrochemical industry for and by which the majority of the shape-selective processes has been developed.

REFERENCES

1. A. F. Cronstedt, *K. Vetensk. Acad. Handl.* **17,** 120 (1756).
2. G. Gottardi, *in* "Natural Zeolites: Occurrence, Properties, Use" (L. B. Sand and F. A. Mumpton, eds.), p. 31. Pergamon, Oxford, 1978.
3. A. P. Bolton, *in* "Experimental Methods in Catalytic Research," Vol. II, p. 1. Academic Press, New York, 1976.
4. P. B. Weisz and V. J. Frilette, *J. Phys. Chem.* **64,** 382 (1960).
5. S. M. Csicsery, *in* "Zeolite Chemistry and Catalysis" (J. A. Rabo, ed.), ACS Monograph, No. 171, p. 680. Am. Chem. Soc., Washington, D.C., 1976.
6. N. Y. Chen, J. Maziuk, A. B. Schwartz, and P. B. Weisz, *Oil Gas J.* **18,** 154 (1968).
7. N. Y. Chen and P. B. Weisz, *Chem. Eng. Prog., Symp. Ser.* **63,** 86 (1967).
8. G. T. Kokotailo, S. L. Lawton, D. H. Olson, and W. M. Meier, *Nature* (*London*) **272,** 437 (1978).
9. W. M. Meier, *in* "Molecular Sieves," p. 10. Soc. Chem. Ind., London, 1968.
10. R. M. Barrer, *Chem. Ind.* (*London*) p. 1203 (1968).
11. D. W. Breck, "Zeolite Molecular Sieves." Wiley, New York, 1974.
12. W. M. Meier and D. H. Olson, *Adv. Chem. Ser.* No. 101, p. 155 (1971).
13. W. M. Meier and D. H. Olson, "Atlas of Zeolite Structure Types" (Special Publication by International Zeolite Association). Polycrystal Book Serv., Pittsburgh, Pennsylvania,
14. W. M. Meier and H. J. Mock, *J. Solid State Chem.* **27,** 349 (1979).
15. W. M. Meier, *Z. Kristallogr.* **115,** 439 (1961).

16. J. V. Smith, *Adv. Chem. Ser.* No. 101, p. 171 (1971).
17. F. Wolf, F. Danes, and K. Pilchowski, *Adv. Chem. Ser.* No. 102, p. 229 (1971).
18. H. W. Habgood, *Chem. Eng. Prog., Symp. Ser.* **63,** 45 (1967).
19. D. J. C. Yates, *Chem. Eng. Prog., Symp. Ser.* **63,** 56 (1967).
20. R. M. Barrer, *Adv. Chem. Ser.* No. 121, p. 1 (1973).
21. P. B. Weisz, *Chem. Technol.* **3,** 498 (1973).
22. R. M. Barrer, *Adv. Chem. Ser.* No. 102, p. 1 (1971).
23. D. M. Ruthven, *Nature (London), Phys. Sci.* **232,** 70 (1971).
24. D. M. Ruthven and R. I. Derrah, *J.C.S. Faraday I* **68,** 2332 (1972).
25. P. Brauer, A. Lopatkin, and G. Stepanez, *Adv. Chem. Ser,* No. 102, p. 97 (1971).
26. K. Fiedler and D. Gelbin, *J.C.S. Faraday I* **74,** 2423 (1978).
27. D. M. Ruthven and I. H. Doetsch, *AIChE J.* **22,** 882 (1976).
28. E. G. Derouane, J. Fraissard, J. J. Fripiat, and W. E. Stone, *Catal. Rev.* **7,** 121 (1972).
29. J. Crank, "Mathematics of Diffusion," p. 86. Oxford Univ. Press, London and New York, 1956.
30. D. M. Ruthven and K. F. Loughlin, *AIChE Symp. Ser.* **67,** 35 (1971).
31. P. E. Eberly, *Ind. Eng. Chem., Fundam.* **8,** 25 (1969).
32. P. N. Sarma and H. W. Haynes, *Adv. Chem. Ser.* No. 133, p. 205 (1974).
33. D. B. Shah and D. M. Ruthven, *AIChE J.* **23,** 804 (1977).
34. J. Kärger, M. B low, and P. Lorenz, *J. Colloid Interface Sci.* **65,** 181 (1978).
35. J. Kärger, M. B low, P. Struve, M. Kocirik, and A. Zikanova, *J.C.S. Faraday I* **74,** 1210 (1978).
36. J. Kärger, J. Caro, and M. Bülow, *Z. Chem.* **16,** 331 (1976).
37. H. Pfeifer, "NMR-Basic Principles and Progress," Vol. 7, p. 53. Springer-Verlag, Berlin and New York, 1972.
38. D. M. Ruthven and K. F. Loughlin, *Can. J. Chem. Eng.* **50,** 550 (1972).
39. D. Gelbin, *AIChE J.* **25,** 358 (1979).
40. E. W. Thiele, *Ind. Eng. Chem.* **31,** 916 (1939).
41. P. B. Weisz and C. D. Prater, *Adv. Catal. Relat. Subj.* **6,** 143 (1954).
42. P. B. Weisz, *Chem. Eng. Prog., Symp. Ser.* **55,** 29 (1959); *Science* **179,** 433 (1973).
43. A. Quig and L. V. C. Rees, *J.C.S. Faraday I* **72,** 771 (1976).
44. C. N. Satterfield and C. S. Chang, *AIChE Symp. Ser.* **67,** 43 (1971).
45. P. Salvador and R. L. Garcia Gonzales, *J. Colloid Interface Sci.* **56,** 577 (1976).
46. R. M. Barrer and I. S. Kerr, *Trans. Faraday Soc.* **55,** 1915 (1959).
47. P. E. Eberly, *Ind. Eng. Chem., Prod. Res. Dev.* **8,** 140 (1969).
48. R. M. Barrer and D. A. Ibbitson, *Trans. Faraday Soc.* **40,** 206 (1944).
49. R. L. Gorring, *J. Catal.* **31,** 13 (1973).
50. P. E. Eberly, *Ind. Eng. Chem., Prod. Res. Dev.* **10,** 433 (1971).
51. C. N. Satterfield and W. G. Margetts, *AIChE J.* **17,** 295 (1971).
52. C. N. Satterfield and A. J. Frabetti, *AIChE J.* **13,** 731 (1967).
53. E. T. Nelson and P. L. Walker, *J. Appl. Chem.* **11,** 358 (1961).
54. N. Y. Chen and P. B. Weisz, *Chem. Eng. Prog., Symp. Ser.* **63,** 86 (1967).
55. F. Wolf, F. Danes, and K. Pilchowski, *Adv. Chem. Ser.* No. 102, p. 229 (1971).
56. J. A. Rabo, V. Shoemaker, and P. E. Pickert, *Proc. Int. Congr. Catal., 3rd, Amsterdam, 1964* **2,** 1302 (1965).
57. E. G. Derouane, *in* "Reactivity of Solids" (J. Wood, O. Lindqvist, C. Helgesson, and N. G. Vannerberg, eds.), p. 559. Plenum, New York, 1977.
58. H. W. Habgood, *Chem. Eng. Prog., Symp. Ser.* **63,** 45 (1967).
59. A. V. Kiselev and A. A. Lopatkin, *in* "Molecular Sieves," p. 252. Soc. Chem. Ind., London, 1968.

60. R. M. Barrer and D. J. Clarke, *J.C.S. Faraday I* **70,** 535 (1974).
61. J. R. Katzer, Ph.D. Thesis, Mass. Inst. Technol., Cambridge, 1969.
62. D. M. Ruthven and K. F. Loughlin, *Chem. Eng. Sci.* **26,** 577 (1971).
63. D. P. Timofeev, *Adv. Chem. Ser.* No. 102, p. 247 (1971).
64. C. N. Satterfield and J. R. Katzer, *Adv. Chem. Ser.* No. 102, p. 193 (1971).
65. R. M. Moore and J. R. Katzer, *AIChE J.* **18,** 816 (1972).
66. C. N. Satterfield and W. R. Vieth, *Ind. Eng. Chem., Fundam.* **10,** 478 (1971).
67. N. Y. Chen, S. J. Lucki, and E. B. Mower, *J. Catal.* **13,** 329 (1969).
68. D. W. Breck and N. A. Acara, U.S. Patent 2,950,952 (1960).
69. J. Valyon, J. Mihalyfi, H. K. Beyer, and P. A. Jacobs, *Proc. Workshop Adsorption, Berlin, D.D.R.*, Vol. 1, 134 (1979).
70. P. Dejaifve, J. C. Vedrine, V. Bolis, J. H. C. van Hooff, and E. G. Derouane, *Prepr., Div. Pet. Chem., Am. Chem. Soc.* **24,** 286 (1979).
71. P. Dejaifve, J. C. Vedrine, V. Bolis, and E. G. Derouane, *J. Catal.* **63,** 331 (1980).
72. E. G. Derouane and J. C. Vedrine, *J. Mol. Catal.* **8,** 479 (1980).
73. P. B. Weisz, V. J. Frilette, R. W. Maatman, and E. B. Mower, *J. Catal.* **1,** 307 (1962).
74. J. F. Tempere and B. Imelik, *Bull. Soc. Chim. Fr.* p. 4227 (1970).
75. L. D. Rollmann, U.S. Patent 4,148,713 (1979).
76. J. Turkevich, *Catal. Rev.* **1,** 1 (1967).
77. P. B. Weisz, *Erdoel. Kohle, Erdgas, Petrochem.* **18,** 525 (1965).
78. P. B. Venuto, *Adv. Chem. Ser.* No. 102, p. 260 (1971).
79. P. B. Venuto and E. T. Habib, *Catal. Rev.—Sci. Eng.* **18,** 1 (1978).
80. J. A. Rabo and M. L. Poutsma, *Adv. Chem. Ser.* No. 121, p. 284 (1971).
81. N. Y. Chen, J. Mazuik, A. B. Schwartz, and P. B. Weisz, *Oil Gas J.* **66,** 154 (1968).
82. S. D. Burd, Jr. and J. Mazuik, *Hydrocarbon Process.* **51(5),** 97 (1972).
83. D. M. Nace, *Ind. Eng. Chem., Prod. Res. Dev.* **9,** 203 (1970).
84. C. L. Thomas and D. S. Barmby, *J. Catal.* **12,** 341 (1968).
85. S. C. Eastwood, C. J. Plank, and P. B. Weisz, *Proc. World Pet. Congr., 8th, Moscow* **4,** 245 (1971).
86. C. L. Kibby, A. J. Perrotta, and F. E. Massoth, *J. Catal.* **35,** 256 (1974).
87. L. D. Rollmann, *J. Catal.* **47,** 113 (1977).
88. Mobil Oil Corp., Fr. Patent 1,502,289 (1966).
89. B. S. Greensfelder, H. H. Voge, and G. M. Good, *Ind. Eng. Chem.* **38,** 1033 (1946).
90. D. M. Nace, *Ind. Eng. Chem., Prod. Res. Dev.* **8,** 31 (1969).
91. H. H. Voge, *in* "Catalysis" (P. H. Emmett, ed.), Vol. 6, p. 407. Reinhold, New York, 1958.
92. B. C. Gates, J. R. Katzer, and G. C. A. Schuit, "Chemistry of Catalytic Processes," p. 30. McGraw-Hill, New York, 1979.
93. J. Eng, U.S. Patent 3,039,953 (1962).
94. N. Y. Chen and W. E. Garwood, *Ind. Eng. Chem., Prod. Res. Dev.* **17,** 513 (1978).
95. N. Y. Chen, R. L. Gorring, H. R. Ireland, and T. R. Stein, *Oil Gas J.* **75(23),** 165 (1977).
96. N. Y. Chen and W. E. Garwood, *in* "Molecular Sieves" (W. M. Meier and J. B. Uytterhoeven, eds.), *Adv. Chem. Ser.* No. 121, p. 575, American Chemical Society, Washington, D.C., 1973.
97. N. Y. Chen and W. E. Garwood, *J. Catal.* **53,** 284 (1978).
98. P. B. Weisz, *Pure Appl. Chem.* **52,** 2091 (1980).
99. W. G. Appleby, J. W. Gibson, and G. M. Good, *Ind. Eng. Chem., Prod. Res. Dev.* **1,** 102 (1962).
100. R. C. Haldeman and M. C. Botty, *J. Phys. Chem.* **63,** 489 (1959).
101. P. B. Venuto, *in* "Catalysis in Organic Synthesis" (G. V. Smith, ed.), p. 67. Academic Press, New York, 1977.

102. D. Eisenbach and E. Gallei, *J. Catal.* **56,** 377 (1979).
103. D. E. Walsh and L. D. Rollmann, *J. Catal.* **49,** 369 (1977).
104. L. D. Rollmann and D. E. Walsh, *J. Catal.* **56,** 139 (1979).
105. D. E. Walsh and L. D. Rollmann, *J. Catal.* **56,** 195 (1979).
106. P. B. Venuto and L. A. Hamilton, *Ind. Eng. Chem., Prod. Res. Dev.* **6,** 190 (1967).
107. P. B. Venuto, *in* "Molecular Sieve Zeolites-II" (E. M. Flanigin and L. B. Sand, eds.), *Adv. Chem. Ser.* No. 102, p. 260, American Chemical Society, Washington, D.C., 1971.
108. P. B. Venuto, L. A. Hamilton, and P. S. Landis, *J. Catal.* **5,** 484 (1966).
109. C. D. Chang and A. J. Silvestri, *J. Catal.* **47,** 249 (1977).
110. N. Y. Chen and W. E. Garwood, *J. Catal.* **52,** 453 (1978).
111. N. Y. Chen, R. L. Gorring, H. R. Ireland, and T. R. Stein, *Oil Gas J.* **75(23),** 165 (1977).
112. L. M. Stock and H. C. Brown, *J. Am. Chem. Soc.* **81,** 3323 (1959).
113. P. B. Venuto, *J. Org. Chem.* **32,** 1272 (1967).

5

Intercalation Chemistry of Acid Salts of Tetravalent Metals with Layered Structure and Related Materials

G. ALBERTI

and

U. COSTANTINO

I. INTRODUCTION

Acid salts of tetravalent metals are usually obtained as amorphous precipitates, but since 1964 they have also been obtained as crystalline compounds. An extensive literature on their synthesis, structure, and properties

INTERCALATION CHEMISTRY

ISBN 0-12-747380-7

has accumulated, and this has been summarized in some recent review articles, to which the reader is referred (*1–5*). These compounds have general formula $M^{IV}(HXO_4)_2 \cdot nH_2O$ (M^{IV} = tetravalent metal; X = P, As) and can be obtained in various crystalline forms having fibrous, layered, or three-dimensional structure. The layered acid salts, in turn, may be obtained in at least two different modifications, usually known as α and γ layered forms.

The protons of the acid salts can be easily replaced by other cations; thus, acid salts are ion exchangers. Owing to their peculiar properties, such as selectivity for certain cations, high stability in strongly acidic and/or oxidizing solutions, high thermal stability, and resistance to radiation, acid salts are very attractive inorganic ion exchangers for particular uses, especially when organic resins cannot be employed because of their degradability. The first investigations on the intercalation properties of α layered acid salts were carried out by Michel and Weiss in 1965–1967 (*6, 7*). However, it must be noted that, in this period, the research efforts were essentially devoted to the synthesis and structure of new materials and to the study of their ion-exchange and ionic transport properties, so that no other papers on the intercalation of acid salts appeared until 1975. Only recently has there been increased interest in the intercalation chemistry of layered acid salts, this probably being due to a better understanding of the structure and properties of these compounds.

In this chapter, after briefly reviewing the structures of layered acid salts and the role of counterions and interlayer water, we shall examine the correlations between the structures of the host α and γ matrices and the intercalation properties.

II. ACID SALTS WITH LAYERED STRUCTURE OF THE α TYPE

In order to understand properly the intercalation chemistry of α layered acid salts, it is important to know their crystalline structures. The most extensively investigated α layered acid salt is the zirconium bis(monohydrogen orthophosphate), usually obtained as the monohydrated form, α-$Zr(HPO_4)_2 \cdot H_2O$. Its structure has been elucidated by Clearfield and collaborators (*3, 8, 9*). This compound crystallizes in the monoclinic system; and its unit-cell dimensions are a = 9.06 Å; b = 5.297 Å; c = 15.414 Å; β = 101.71°. Each layer consists of zirconium atoms lying in a plane and bridged through phosphate groups. Three oxygens of each tetrahedral phosphate are linked to three zirconium atoms so that each zirconium is octahedrally coordinated with six oxygens of six different phosphate groups.

Fig. 1. Idealized structure of the α-$[Zr_n(PO_4)_{2n}]^{2n-}$ macroanion. The fixed negative charges are balanced by cations (counterions) not shown in the figure.

The fourth oxygen of each phosphate group bears a proton. It is important to point out that the proton can be replaced with another cation without any alteration in the structure of the layer itself (*10*, *11*). Thus, each layer may be considered a planar macroanion, $[Zr_n(PO_4)_{2n}]^{2n-}$, whose negatively charged oxygens (hereafter indicated by $O^{(-)}$ or called fixed charges) are balanced by an equivalent amount of protons or other cations (mobile counterions).*

The structure of a planar macroanion is schematically shown in Fig. 1. Note that each plane of zirconium atoms is sandwiched between two sheets of phosphate groups; therefore, each macroanion possesses two planar surfaces of fixed charges. The distance between the plane passing through the baricenters of these $O^{(-)}$ and the plane of zirconium atoms is 3.15 Å. The thickness of a single macroanion is thus $2 \times 3.15 + 2r$ (where r is the radius of negatively charged oxygens). If r is taken as 1.4 Å, the thickness of the macroanion is ~9 Å.

* For this reason, hereafter the various salt forms of α-zirconium phosphate will be designated α-$[Zr(PO_4)_2]M_{2/z}$, where M is a counterion of charge z.

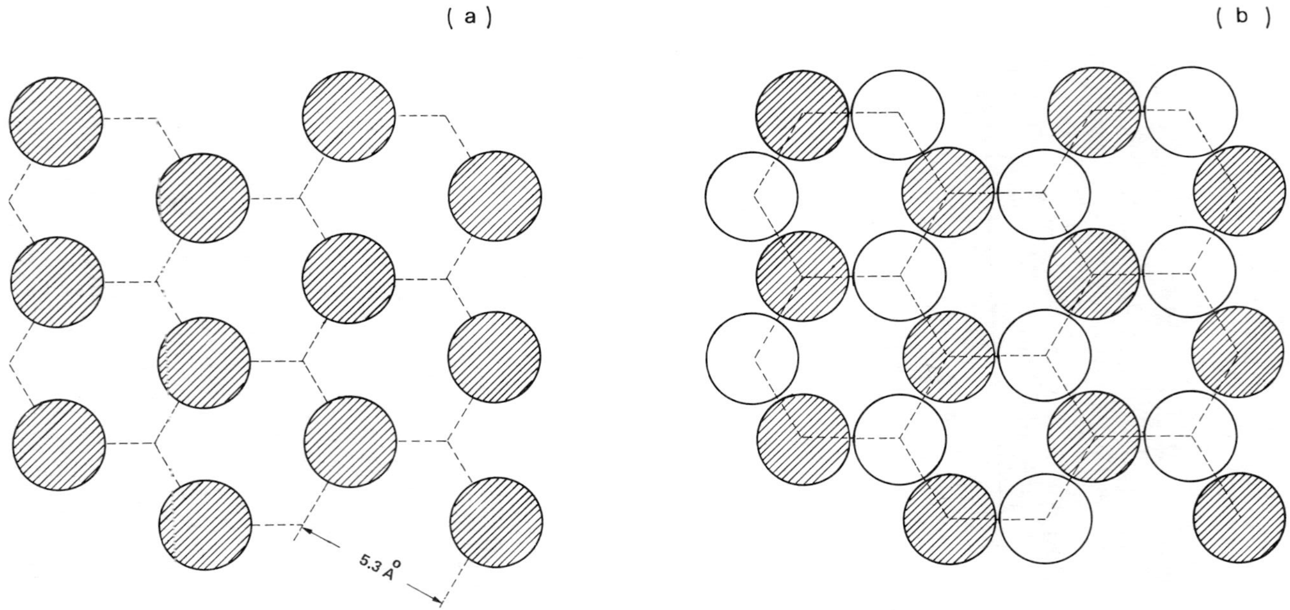

Fig. 2. (a) Arrangement of fixed charges on one side of the α-$[Zr_n(PO_4)_{2n}]^{2n-}$ macroanion. (b) Arrangement of fixed charges in the interlayer region. The figure shows their projection in a plane intermediate between and parallel to two adjacent macroanions. Striped circles refer to $O^{(-)}$ lying above and open circles refer to those lying below the intermediate plane.

The arrangement of the fixed charges in one of the two surfaces of the macroanion is schematically shown in Fig. 2a. Note that the fixed charges are arranged in a hexagonal array and that the distance between two adjacent $O^{(-)}$ is 5.3 Å. The layered crystals may be regarded as constituted by packed macroanions with interposed counterions. In the case of α-$[Zr(PO_4)_2]H_2 \cdot H_2O$, the forces holding the layers are essentially van der Waals forces (*9*, *12*). However, when the protons are replaced by other counterions, the predominant forces are probably ionic bonds of the interposed counterions with adjacent macroanions.

The interlayer distance depends on the size and hydration of the counterions. This point is examined later in more detail (Sections III,A and III,B). When the counterions are protons, the interlayer distance is 7.56 Å. The arrangement of the layers is such that each $O^{(-)}$ in the lower layer lies along a perpendicular line drawn from the zirconium atom of the upper layer (see Fig. 3). The distance between the planes passing through the

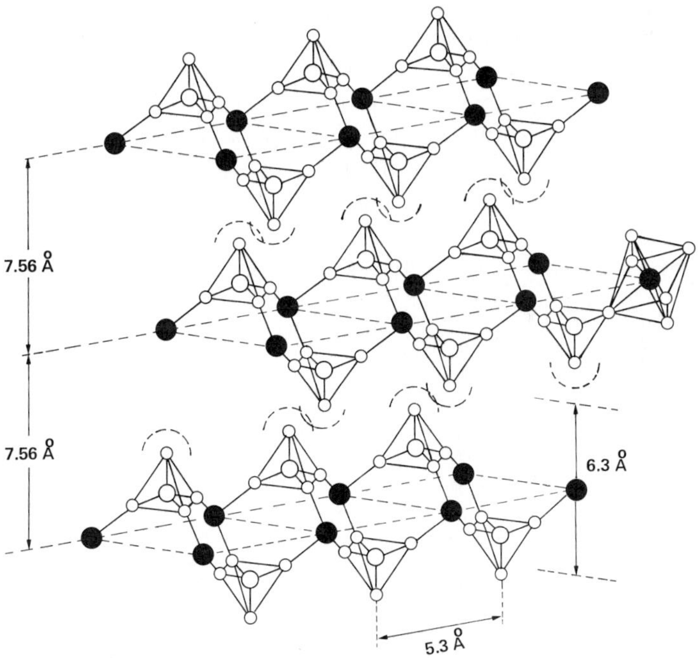

Fig. 3. Schematic drawing of the arrangement of three adjacent macroanions in α-$[Zr(PO_4)_2]H_2 \cdot H_2O$ (interlayer distance 7.56 Å). The distance between two planes passing through the baricenters of $O^{(-)}$ belonging to the two sides of the macroanion is 6.3 Å. The protons and the water are not shown. Dashed semicircles represent the approximate size of the $O^{(-)}$; ●, Zr; ○, P; o, oxygen.

baricenters of $O^{(-)}$ of two adjacent macroanions is 12.6 Å. Note that there is a certain interpenetration of $O^{(-)}$ belonging to two adjacent macroanions.

The projection of the $O^{(-)}$ in a plane intermediate between two adjacent macroanions is shown in Fig. 2b. This packing creates small zeolitic-type cavities (one for each zirconium atom), which are interconnected by windows whose maximum size is about 2.61 Å (*9*). Cations or polar molecules that are larger than the windows connecting the cavities should not diffuse between the layers. However, layered compounds are not so rigid as three-dimensional compounds; if sufficient energy is supplied, the layers may be spread apart, and large cations or large polar molecules may enter between the layers.

Several other acid salts have been obtained and studied. Some important characteristics of such compounds are listed in Table I. Note that the carboxymethanephosphonate of zirconium has an α layered structure (*13*).

In intercalation chemistry, it is very important to know the free area surrounding each active site since stoichiometric compounds can be obtained only when the cross section of the molecule to be intercalated is less than the free area of the active site. The explanation is straightforward. As the cross section of the guest molecules exceeds the free area, one molecule bonded to a site partially or completely covers adjacent active sites so that these are unavailable for bonding other molecules. The phenomenon is usually known as the "covering effect."

TABLE I

Some Characteristics of Acid Salts of Tetravalent Metals with Layered Structure of the α Type

Formula	Ion-exchange capacity (mmoles H^+/gm)	Density (gm/cm^3)	Interlayer distance (Å)	Distance between $O^{(-)}$ on the plane (Å)	Free area associated with each $O^{(-)}$ on the plane ($Å^2$)
$[Ti(PO_4)_2]H_2 \cdot H_2O$	7.76	2.61	7.56	5.0	21.6
$[Zr(PO_4)_2]H_2 \cdot H_2O$	6.64	2.72	7.55	5.3	24.0
$[Ge(PO_4)_2]H_2 \cdot H_2O$	7.08	—	7.6	—	—
$[Sn(PO_4)_2]H_2 \cdot H_2O$	6.08	3.12	7.8	—	21.4[a]
$[Pb(PO_4)_2]H_2 \cdot H_2O$	4.79	—	7.8	—	—
$[Ti(AsO_4)_2]H_2 \cdot H_2O$	5.78	3.34	7.7	—	22.2[a]
$[Zr(AsO_4)_2]H_2 \cdot H_2O$	5.14	3.39	7.71	5.4	24.7[a]
$[Sn(AsO_4)_2]H_2 \cdot H_2O$	4.80	—	7.9	—	22.6[a]
$[Zr(OOCCH_2PO_3)_2]H_2$	5.45	2.31	11.1	—	24.0[a]

[a] The value for $(a \times b)/2$ estimated by the method of Alberti *et al.* (*11*).

In α layered acid salts the free area of each active site (protons or other counterions associated with the fixed charges of the macroanions) may be easily calculated assuming that the active sites in each plane of the macroanion are arranged in the same hexagonal array as the fixed charges. In such a case, the free area is equal to $(a \times b)/2$, where a and b are the unit-cell dimensions. The value of the free area associated with each active site and the distance between two active sites in some α layered acid salts are reported in Table I.

III. INTERCALATION CHEMISTRY OF α-ZIRCONIUM PHOSPHATE

The majority of investigations on the intercalation chemistry of layered acid salts have been performed with α-zirconium phosphate. It therefore seems useful to discuss separately the results obtained with this compound. In particular, the correlations between its structure and intercalating properties are examined and discussed.

A. Interlayer Distance as a Function of the Size of Counterions

As already mentioned, the protons of the acid salts can be partially or completely replaced by other counterions. In a sense, the counterions may be regarded as charged particles intercalated between layered macroanions, and it is interesting to examine how the interlayer distance increases with an increase in size of the counterions.

Let us examine first the case of divalent counterions. As shown in Fig. 4a, the interlayer distances of the anhydrous α-$[Zr(PO_4)_2]M$ forms do not appreciably depend on the size of divalent cations and are very close to that of α-$[Zr(PO_4)_2]H_2 \cdot H_2O$. Thus, it is probable that the zeolitic-type cavities present in α-$[Zr(PO_4)_2]H_2 \cdot H_2O$ are also maintained in α-$[Zr(PO_4)_2]M$ and that the divalent cations are placed in such cavities, which have enough room to accommodate even Ba^{2+} ions (*4, 5*). As an example, the suggested arrangement of Ba^{2+} cations in α-$[Zr(PO_4)_2]Ba$ is schematically shown in Fig. 5. Also, for half-converted alkali metal ion forms, α-$[Zr(PO_4)_2]HM$, the interlayer distance is almost independent of the size of the counterions M^+ (Fig. 4a). It seems that, in this case, the large alkali metal cations are placed near the center of the cavities, whereas the small protons are very likely displaced toward the sides of the cavities (*4*). Note that this arrangement of the counterions minimizes the increase in the interlayer distance. When all the

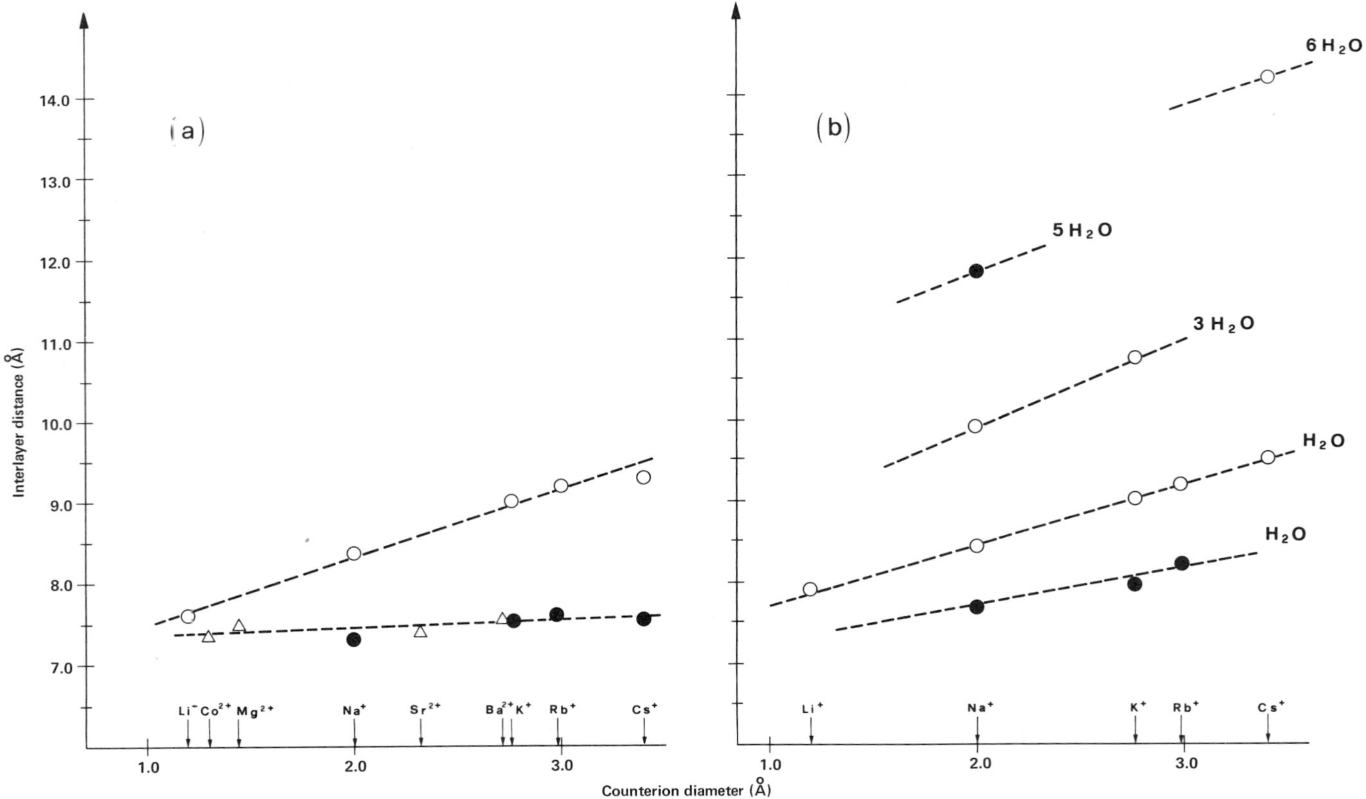

Fig. 4. Interlayer distance versus diameter of the counterions for (a) anhydrous salt forms of α-zirconium phosphate and (b) hydrated salt forms of α-zirconium phosphate: ●, α-$[Zr(PO_4)_2]HM$; ○, α-$[Zr(PO_4)_2]M_2$; △, α-$[Zr(PO_4)_2]M$.

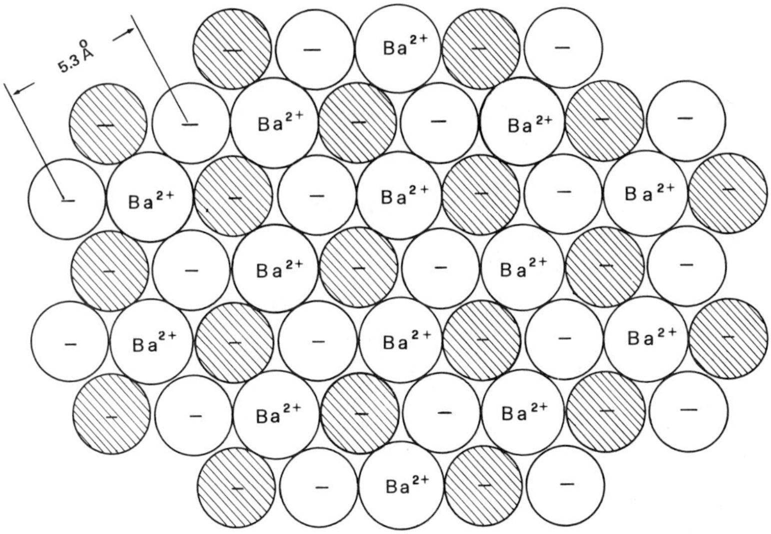

Fig. 5. Suggested arrangement of Ba^{2+} counterions in α-$[Zr(PO_4)_2]Ba$, viewed in a planar projection, as described in Fig. 2b.

protons are replaced by M^+ cations to give α-$[Zr(PO_4)_2]M_2$ forms, there is no longer sufficient room to accommodate two large cations in the same cavity. Thus, one alkali metal ion must be inserted between two fixed charges of adjacent layers, and the interlayer distance must increase regularly with increasing ionic diameter of the alkali metal ions, as experimentally found (Fig. 4a).

B. The Role of Interlayer Water

Water can be considered the most common polar molecule present in the interlayer space of acid salts (or their salt forms) because of the strong tendency of counterions to hydrate. It is therefore of the utmost importance to examine the variation of the interlayer distance as a function of the interlayer water.

Several hydrated salt forms of α-zirconium phosphate have been obtained and their interlayer distance determined (*3–5*). As an example, in Fig. 4 the interlayer distances of some anhydrous and hydrated alkali metal ion forms are plotted against the diameter of the counterions. It can be noted that the addition of one molecule of water increases the interlayer distance of half-exchanged forms and of the fully exchanged Li form, but it does not appreciably increase the interlayer distance of the other fully exchanged forms. Given the structure of α-zirconium phosphate, an explanation of these facts could be the following. As previously discussed, in the anhydrous half-exchanged forms each cavity is filled with one alkali metal ion. When one molecule of water for each cavity is added, the free volume in each cavity is no longer sufficient to accommodate the alkali metal ion and the water molecule at the same time, and the interlayer distance must increase in order to create additional room. If additional water is taken up, the interlayer distance must increase further.

Let us now examine the fully exchanged forms, $\alpha\text{-}[Zr(PO_4)_2]M_2$. In this case, the cavities have already been enlarged and the cations are displaced toward the sides of the cavities, their arrangement probably being very similar to that of the fixed charges (Fig. 2b). When one molecule of water is added, this is very likely inserted between the counterions, so that the interlayer distance does not vary appreciably. However, when two additional moles of water are taken up by the monohydrated form, this is no longer possible and the interlayer distance increases. Note that the addition of 2 moles of water per formula weight of $\alpha\text{-}[Zr(PO_4)_2]M_2 \cdot H_2O$ corresponds to the uptake of one molecule of water for each counterion. Furthermore, the fact that there is an experimental increment of about 2 Å for every two additional molecules of water seems to indicate that a monolayer of water is formed in the interlayer space of α-zirconium phosphate. In this connection, it is interesting to have a first estimation of the "free area" associated with each counterion in the interlayer space. The number of counterions is equal to the number of fixed charges of two adjacent surfaces of the macroanions, that is, $2 \times 4.12 \times 10^{14} = 8.24 \times 10^{14}$ counterions per square centimeter of interlayer area (*14*). The corresponding "free area" associated with each counterion is thus 12 Å^2. Now, the possibility of a monolayer of intercalated molecules being formed depends on the size and geometric shape of the molecules as well as their arrangement. For a fourfold planar coordination of spherical molecules, the maximum cross-sectional diameter can be estimated as $\sqrt{12} = 3.46$ Å, whereas the interstitial unavailable area is $12 - 9.4 = 2.6$ Å^2. (For sixfold planar coordination, the maximum cross-sectional diameter is slightly greater since the unavailable interstitial area is smaller.) Owing to the small size of water molecules, there is no doubt that this "free area" is sufficient to accommodate water molecules in a monolayer without covering effects, in agreement with the experimental increase in the interlayer distance. However, for large molecules such as organic molecules, this free area may be too small and, as discussed later (Section III,D), two layers of intercalated molecules (one for each adjacent macroanion) are usually present in the interlayer space of α-zirconium phosphate.

C. Uptake of Ammonia

Owing to the presence of acid ≡X—OH groups within their layers, acid salts show a high preference for ammonia. In the case of α-$[Zr(PO_4)_2]H_2 \cdot H_2O$, it has been found that one molecule of NH_3 for each ≡P—OH group is taken up, even from very dilute aqueous solutions of ammonia (*15–18*). The X-ray powder pattern of α-$[Zr(PO_4)_2]H_2 \cdot H_2O$ after the uptake of 2 moles of NH_3 is the same as that of the diammonium form, α-$[Zr(PO_4)_2](NH_4)_2 \cdot H_2O$ (interlayer distance, 9.4 Å), obtained by replacing the potassium of α-$[Zr(PO_4)_2]K_2 \cdot H_2O$ with ammonium ions (*16*). It can be concluded that the uptake of NH_3 by α-$[Zr(PO_4)_2]H_2 \cdot H_2O$ occurs via protonation of NH_3 and formation of the diammonium form, according to the process

$$\alpha\text{-}[Zr(HPO_4)_2]H_2 \cdot H_2O + 2NH_3 \rightarrow \alpha\text{-}[Zr(PO_4)_2](NH_4)_2 \cdot H_2O \tag{1}$$

It is worth noting here that the uptake of NH_3 by amorphous zirconium phosphate was successfully employed recently in kidney machines for the removal of NH_3 and NH_4^+ from human blood after previous catalytic decomposition of the urea to NH_3 and CO_2 (*19*). The uptake of NH_3 is reversible in that by heating the diammonium form at 300°C, NH_3 is completely released and the original dihydrogen form is again obtained (*20*).

D. Intercalation of Amines

Amines, being strong Brönsted bases, are expected to interact strongly with ≡X—OH groups; therefore, the intercalation of amines in layered acid salts, and in particular in α-$[Zr(PO_4)_2]H_2 \cdot H_2O$, has been extensively investigated in recent years (*6*, *21–25*). For the sake of simplicity, let us examine first the uptake of *n*-alkylmonoamines by α-$[Zr(PO_4)_2]H_2 \cdot H_2O$. The "cross-sectional area" of the alkyl chain is surely less than the "free area" of active sites (protons) on the surface of the macroanions, that is, as discussed before, 24 Å^2. Thus, all the ≡P—OH groups are available to link amines, and stoichiometric compounds with composition $[Zr(PO_4)_2]H_2 \cdot A_2$ (where A = *n*-alkylmonoamine) are expected to be formed. The problem now is to establish whether one or two layers of amines are intercalated in the interlayer space of α-zirconium phosphate—in other words, whether the *n*-alkyl chains of amines linked to the two surfaces of adjacent macroanions are, or are not, interpenetrated in the interlayer space.

It has been mentioned in relation to the uptake of water that, in order for a monolayer to be formed in the interlayer space of α-zirconium phosphate, the "cross-sectional diameter" of intercalated spherical molecules must be 3.46 Å. Even lower values are expected for an arrangement less compact than the planar fourfold coordination of spherical spheres. Thus, owing to the large "cross-sectional diameter" of their alkyl chain (4.4 Å), the monoamines would be intercalated as bilayers.

Let us now examine whether the experimental facts agree with previous expectations. The *n*-alkylmonoamines are very easily taken up by α-$[Zr(PO_4)_2]H_2 \cdot H_2O$ even from dilute solutions (*24*), and stoichiometric compounds (2 moles of monoamine per formula weight of exchanger) have indeed been obtained in all the examined cases (*6*, *22*, *24*, *25*), as expected. The interlayer distances of several α-$[Zr(PO_4)_2]H_2 \cdot A_2$ compounds have been determined (*6*, *21–25*). In order to obtain information on the number of intercalated layers in the interlayer space, it is useful to plot the interlayer distance as a function of the number of carbon atoms present in the *n*-alkyl chain. Figure 6 shows that the interlayer distance of *n*-alkylmonoamine intercalates increases linearly with the number of carbon atoms n_C in the alkyl chain. The difference in the interlayer distance for each added carbon atom (2.21 Å) can be estimated from the slope of the curve $\Delta d/\Delta n_C$. Now, if a monolayer of *n*-alkylmonoamines is formed in the interlayer space, the maximum increase in the interlayer distance for each additional carbon atom of the alkyl chain should be 1.27 Å (a value usually estimated for the *n*-alkyl chain in conformation trans,trans). In the case of an arrangement of the *n*-alkylamine not perfectly perpendicular to the plane of the macroanion, lower values are expected, depending on the inclination angle. Thus, the experimental increase of 2.21 Å for each additional carbon atom indicates that *n*-alkylmonoamines are very probably intercalated as bilayers of extended molecules (*6*, *24*, *25*). Note, however, that the experimental value is less than that expected for the addition of two carbon atoms (one carbon atom for each layer). This could be due to the fact that the *n*-alkylmonoamines are not exactly perpendicular, but are inclined with respect to the plane of the macroanion of an angle $\alpha = \arcsin(2.21/2.54) \sim 60°$. The interpenetration of *n*-alkyl chains belonging to two adjacent layers of α-zirconium phosphate does not occur because the cross-sectional area of the *n*-alkyl chains is larger than the "free area" of active sites in the interlayer space.

Since alkylamines are stronger bases than ammonia, it is probable that these molecules are present as alkylammonium ions. Note that, if the orientation of the $R{-}NH_3^+$ ion with respect to the O^- of the layers is that schematically depicted in Fig. 7, the trans,trans alkyl chains would be inclined by an angle of $\sim 55°$ with respect to the sheets; this value is very close to that found experimentally. The intercalation of *n*-alkyldiamines gives stoichiometric compounds having formula α-$[Zr(PO_4)_2]H_2 \cdot A_d$ ($A_d =$

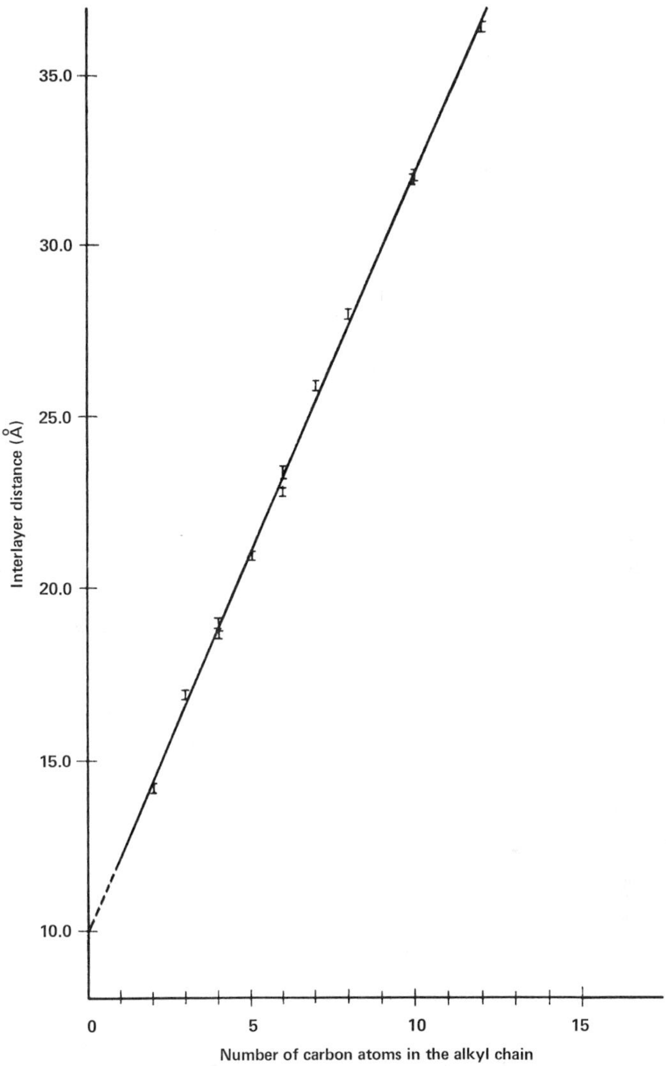

Fig. 6. Interlayer distance of alkylmonoamine intercalates of α-$[Zr(PO_4)_2]H_2$ as a function of the number of carbon atoms in the alkyl chain.

n-alkyldiamine). The interlayer distances of some of these complexes are reported in Table II. Since the guest molecule is bifunctional, it is assumed that each diamine molecule interacts with phosphate groups belonging to two adjacent layers. It is not easy to predict whether both the $—NH_2$ groups are protonated, and infrared spectroscopy would help in this respect; very likely, owing to the difference in the pK_{a_1} and pK_{a_2} values, only one of the

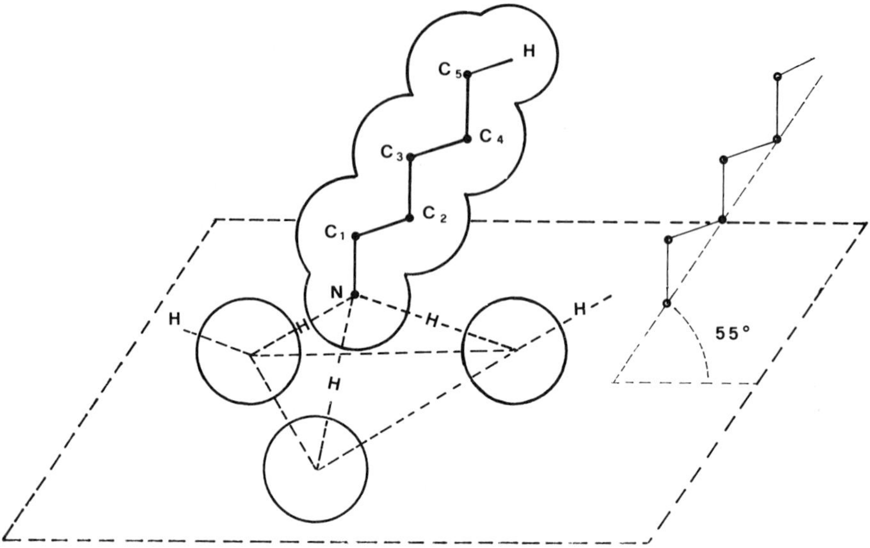

Fig. 7. Suggested arrangement for the alkylmonoamines in which the amine group interacts with three ≡P—OH groups of α-$[Zr(PO_4)_2]H_2$. Note that for this arrangement the *trans-trans* alkyl chain is inclined by an angle of ~55°.

—NH_2 groups exists as —NH_3^+ ion. It can be seen from Table II that the interlayer distances of diamine intercalates increase linearly with the increase in n_C. The increase in the basal spacing can be evaluated as 1.1 Å per each carbon atom added, as if a monolayer of alkyldiamines were present, with the trans,trans alkyl chains inclined at an angle of ~60° with respect to the plane of the layers.

Note that a monolayer could also be built with *n*-alkylmonoamines, provided that only 1 mole of amine is intercalated, giving $Zr[(PO_4)_2]H_2 \cdot A$.

TABLE II

Interlayer Distance of α-$[Zr(PO_4)_2]H_2 \cdot A_d$ and α-$[Zr(PO_4)_2]H_2 \cdot A$ Intercalates[a]

A_d	Interlayer distance (Å)	Ref.	A	Interlayer distance (Å)	Ref.
Ethylenediamine	10.7, 11.1	*24, 25*	Methylamine	9.7	*26*
Tetramethylenediamine	13.2	*24*	Ethylamine	10.6	*26, 27*
Heptamethylenediamine	15.5	*26*	Propylamine	13.0	*26*
Decamethylenediamine	19.8	*24*	Decylamine	20.5	*26*

[a] A_d, *n*-alkyldiamine; A, *n*-alkylmonoamine.

Under these conditions, the "free area" available in the interlayer region is large enough to accommodate the *n*-alkyl chains of 1 mole of monoamine per formula weight of exchanger whether the amine interacts with ≡POH groups of the same layer or is distributed among those of two adjacent layers. The values of the interlayer distance of some of these intercalation compounds, reported in Table II, are in agreement with such expectations.

Table III gives the interlayer distance and the composition of intercalation compounds of α-$[Zr(PO_4)_2]H_2 \cdot H_2O$ with various secondary and tertiary *n*-alkylamines. Note that the stoichiometric composition $[Zr(PO_4)_2]H_2 \cdot A_2$ is not reached. Taking into account that secondary and tertiary alkylamines should occupy a space at least two or three times greater than that of the primary alkylamines, the lack of stoichiometric composition, due to a covering effect, is to be expected.

It is not easy to deduce whether these molecules form a single or a double film in the interlayer region because of the numerous arrangements the alkyl chains may assume. However, it may be observed that complexes with secondary and tertiary amines having the same alkyl radical have about the same interlayer distance. Furthermore, this value is generally 1.0–1.5 Å lower than that of complexes with the corresponding primary *n*-alkylamines. This observation seems to indicate that bilayers of substituted alkylamines are present in the interlayer region. Owing to the strong interaction of amines with ≡P—OH groups, the amine intercalate compounds are very stable to heat or to washing with other polar solvents. The guest molecules

TABLE III

Interlayer Distance and Composition of Intercalates of α-Zirconium Phosphate with Substituted *n*-Alkylamines

Amine	Interlayer distance (Å)	Composition (moles amine per mole exchanger)	Ref.
Diethylamine	12.8, 12.6	1.0	*24, 27*
Dipropylamine	15.7	1.0	*24*
Dibutylamine	17.0, 17.3	0.8	*6, 24*
Dihexylamine	22.1	0.8	*6*
Dioctylamine	26.8	0.8	*6*
Didecylamine	35.3	0.8	*6*
Triethylamine	12.8, 12.6	0.5	*24, 27*
Tripropylamine	15.7	0.5	*24*
Tributylamine	17.2, 16.7	0.3	*6, 24*
Trihexylamine	19.2	0.3	*6*
Trioctylamine	26.7	0.3	*6*
Tridecylamine	31.3	0.3	*6*

are in fact lost after drying at temperatures higher than 200°C (*25*) or by treating the intercalates with strong acid solutions (*26*). On the other hand, it has been found (*28*) that the butylamine intercalate exchanges the guest molecules with Ni^{2+} or with large cations such as $Cu(NH_3)_4{}^{2+}$.

E. Mechanism of Amine Intercalation

It has been shown that the intercalation of amines in α-zirconium phosphate also occurs from dilute solutions and that the amines are protonated by the hydrogen of the phosphate groups. Thus, the intercalation process may be followed by titrating the ≡P—OH groups of the exchanger with dilute aqueous solution of amines (*22, 25, 26*). The potentiometric titration curves obtained are very useful for deducing the saturation capacity of the exchanger toward various amines at different pH values and studying the intercalation mechanism, especially if the X-ray powder patterns of samples having different loading in amines are also taken.

In Fig. 8 the potentiometric titration curves of α-$[Zr(PO_4)_2]H_2 \cdot H_2O$

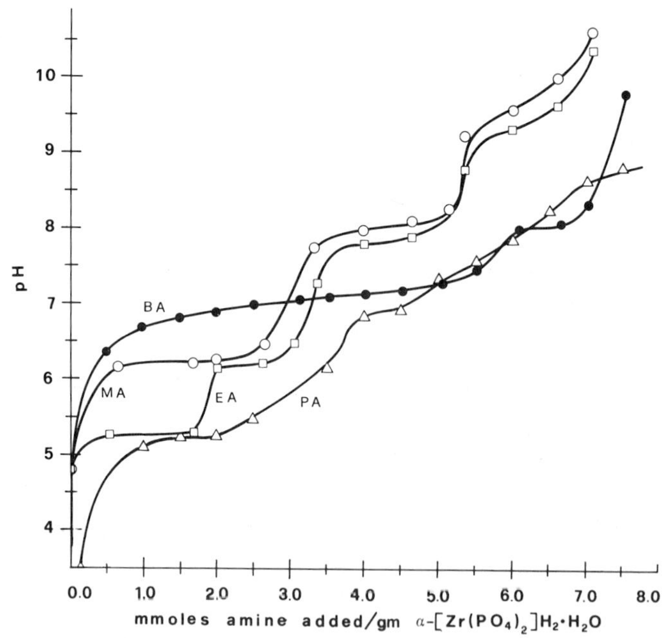

Fig. 8. Titration curves of α-$[Zr(PO_4)_2]H_2 \cdot H_2O$ with 0.1 *M* aqueous solution of the following alkylamines: MA, methylamine (*26*); EA, ethylamine (*26*); PA, *n*-propylamine (*25*); BA, *n*-butylamine (*22*).

with 0.1 *M* aqueous solution of methylamine, ethylamine, propylamine, and *n*-butylamine are shown. It can be seen that the intercalation processes occur stepwise; furthermore, the various titration curves exhibit a different number of steps, as if the exchanger displayed different ranges of acidity for each amine. These steps are not due, however, to the presence of ≡P—OH groups having different acidities but are probably related to the different energy required for the various amines to fill the interlayer region. The steps of the titration curves indicate that, before the final composition α-$[Zr(PO_4)_2]H_2 \cdot 2A$ is attained, several stable phases are formed; the molar ratio guest molecule/formula weight of host lattice may be 0.5, 1.0, 1.33, or 1.5 depending on the nature of the alkylamine (pK_a value, length of the alkyl chain) and the eventual hydration water.

Additional study is necessary to establish whether the inclusion process occurs simultaneously in all the layers, as in vermiculite (*29*), or in alternate layers, as in graphite (*30*). Recent experiments performed by the radiotracer technique seem to indicate that the Na^+/Cs^+ ion exchange in α-$[Zr(PO_4)_2]$-$HNa \cdot 5H_2O$ takes place simultaneously within each layer (*31*). Therefore, it is probable that the intercalation process occurs in a similar way. The model of coexisting phases in a single crystallite of layered acid salts (*5*) may be useful for understanding the mechanism of the intercalation. According to this model, the new intercalated phase should be initially formed in the external part of each crystal. As the intercalation proceeds, the phase boundary so formed moves toward the center of the crystal. Thus, until the original phase is present, two solid phases coexist in the same crystallite; the degree of freedom of the system (at constant temperature and pressure) is zero, and each intercalation step must occur at constant composition of the solution. The plateaus in the titration curves indeed correspond to phase transitions of intercalation steps, as confirmed by X-ray diffraction patterns (*22*, *25*, *26*). Knowledge of the compositions and of the interlayer distances of the various phases encountered during the titrations would allow us to fully describe the intercalation processes. At present, the composition is not well known, but, considering the shape of the titration curves reported in Fig. 8, the intercalation processes of methylamine (MA), ethylamine (EA), propylamine (PA), and butylamine (BA) may be schematically described by the following reactions, where the numbers in parentheses represent the interlayer distances in angstroms.

$$\underset{(7.6)}{[Zr(PO_4)_2]H_2} \xrightarrow{+MA} \underset{(9.5)}{[Zr(PO_4)_2]H_2 \cdot (MA)} \xrightarrow{+0.5MA} \underset{(12.1)}{[Zr(PO_4)_2]H_2 \cdot (MA)_{1.5}} \xrightarrow{+0.5MA} \underset{(13.3)}{[Zr(PO_4)_2]H_2 \cdot (MA)_2} \qquad (2)$$

$$
\underset{(7.6)}{[Zr(PO_4)_2]H_2} \xrightarrow{+0.5EA} \underset{(10.1)}{[Zr(PO_4)_2]H_2 \cdot (EA)_{0.5}} \xrightarrow{+0.5EA} \underset{(10.8)}{[Zr(PO_4)_2]H_2 \cdot (EA)} \xrightarrow{+0.5EA}
$$

$$
\underset{(14.7)}{[Zr(PO_4)_2]H_2 \cdot (EA)_{1.5}} \xrightarrow{+0.5EA} \underset{(16.8)}{[Zr(PO_4)_2]H_2 \cdot (EA)_2} \qquad (3)
$$

$$
\underset{(7.6)}{[Zr(PO_4)_2]H_2} \xrightarrow{+PA} [Zr(PO_4)_2]H_2 \cdot (PA) \xrightarrow{+PA} \underset{(17.3)}{[Zr(PO_4)_2]H_2 \cdot (PA)_2} \qquad (4)
$$

$$
\underset{(7.6)}{[Zr(PO_4)_2]H_2} \xrightarrow{+1.33BA} \underset{(18.2)}{[Zr(PO_4)_2]H_2 \cdot (BA)_{1.33}} \xrightarrow{0.67BA} \underset{(18.8)}{[Zr(PO_4)_2]H_2 \cdot (BA)_2} \qquad (5)
$$

Some of the processes examined deserve further consideration. In the first stages of the intercalation of butylamine or propylamine, Clearfield and Twinda (*25*) noted the presence of a phase with an interlayer spacing of 1.05 nm. This phase could arise from the presence of amines having their alkyl chains oriented parallel to the layers. Thus, according to the above-mentioned authors, in the first stages of the intercalation process the amines are accommodated in such a position because steric hindrance is absent. Then, as the loading increases, the amines are forced into a more upright position until the final arrangement of a close double layer of alkylamines is reached (*25*).

An interesting phenomenon has been observed during the titration of α-$[Zr(PO_4)_2]H_2 \cdot H_2O$ with *n*-propylamine. When the amine added is about equal to the amount necessary to saturate half of the $\equiv$P—OH groups of the exchanger, the material swells dramatically (*25*, *26*) and behaves as an amorphous solid to X-ray diffraction. However, an X-ray diffraction pattern with well-evident peaks is again obtained on the addition of further propylamine. Furthermore, when the material is treated with 0.1 *M* HCl solution, the original α-zirconium phosphate is again obtained (*26*). Since crystals of α-$[Zr(PO_4)]_2H_2 \cdot H_2O$ as large as ~ 10 mm^2 are now available (*32*), the phenomenon was followed under an optical microscope. It was found that the addition of *n*-propylamine first causes a macroscopic exfoliation of the crystals, which then assume the aspect of a gel. A similar phenomenon was also observed with methylamine, but it does not occur when the exchanger is suspended in CH_3OH and titrated with 0.1 *M* solution of the amine in methanol (*26*). Very probably, it is due to a cooperative effect of the hydrophilic character of $\equiv$P—O$^-$ and —NH_3^+ groups and the hydrophobic character of the alkyl chain. This effect seems to reach a maximum with *n*-propylamine. We point out that the marked swelling may give rise to a colloidal solution containing dispersed zirconium phosphate layers whose $\equiv$P—OH groups are half-saturated by propylamine. This system may be similar to that formed by lyotropic liquid crystals of lamellar type found in surfactant systems and may be of interest in the field of homogeneous catalysis.

F. Exchange of Ammonium and Alkylammonium Cations

The exchange of several substituted alkylammonium cations on amorphous zirconium phosphate was investigated by Amphlett and McDonald, who observed that the saturation capacity of the exchanger decreased with increasing ionic size of the cations (*33*). This effect was ascribed to the presence of cavities of different size in the amorphous material.

More recently, the exchange of some alkylammonium ions in crystalline α-zirconium phosphate was investigated by Gupta and Nowell (*27*). These authors reported the titration curves of $\alpha\text{-}[Zr(PO_4)_2]H_2 \cdot H_2O$ with ethylamine, diethylamine, and triethylamine in the presence of the respective alkylammonium chlorides. The shape of the titration curve with ethylamine is very similar to that reported in Fig. 8, but the values of the equilibrium pH are lower. In particular, the pH of the first plateau is ~3. The lowering of pH indicates that $Et{-}NH_3^+/H^+$ exchange takes place (*27*). Since the full loading of the exchanger has been achieved by adding free ethylamine, intercalation may also take place simultaneously with ion exchange. In every case, the compounds obtained by titrating the exchanger with amines in the presence of the respective alkylammonium chlorides have about the same interlayer distances as those obtained using the intercalation processes. On the other hand, it is known that $\alpha\text{-}[Zr(PO_4)_2]H_2 \cdot H_2O$ cannot exchange NH_4^+ ions, and the lack of this exchange has been related to steric hindrance (*16*). The fact that $\alpha\text{-}[Zr(PO_4)_2]H_2 \cdot H_2O$ excludes NH_4^+ ion whereas it is able to exchange, to some extent, mono-, di-, and trialkylammonium ions deserves additional investigation. Moreover, it is expected that salt forms of α-zirconium phosphate with large interlayer distances are able to exchange the metal counterions with substituted alkylammonium. Tetraalkylammonium ions are also not exchanged in $\alpha\text{-}[Zr(PO_4)_2]H_2 \cdot H_2O$, probably because of their size. However, it is possible to load the exchanger with these large ions at high pH values using the respective hydroxides. Some exchange also takes place if zirconium phosphate having a large interlayer distance, such as the polyhydrated $\alpha\text{-}[Zr(PO_4)^2]H_2 \cdot 5H_2O$ (d = 1.4 Å) or zirconium phosphate–ethanol complex (see Section III,G), is employed. In these materials the activation energy for the exchange is lowered, and the exchanger may be loaded with tetramethylammonium and tetraethylammonium up to ~25% of its ion-exchange capacity (*26*).

G. Intercalation of Alkanols and Glycols

Alkanols are much weaker proton acceptors than are amines, and the strength of their interactions with the acid ≡P—OH groups of α-[Zr(PO$_4$)$_2$]-H$_2$ · H$_2$O is also expected to be weaker. Owing to this weak interaction, alkanol intercalates are not obtained even when α-zirconium phosphate is contacted with the pure liquid. It is as if energy associated with the bond formation between the phosphate groups and the alcoholic functional group were too weak to compensate for the energy required to spread the layers apart and to overcome the activation energy of the intercalation process. Intercalation of alkanols has been achieved, however, by using salt forms of α-zirconium phosphate with large interlayer distances, e.g., α-[Zr(PO$_4$)$_2$]-HNa · 5H$_2$O (interlayer distance 1.18 Å), in order to overcome the steric hindrance to the diffusion of the alkanols in the interlayer region (*34*). The salt form is contacted with pure alkanols previously acidified with a strong mineral acid, so that the metal ions are exchanged and the overall reaction of regeneration leads to the dihydrogen forms with intercalated alkanols, e.g.,

$$\alpha\text{-}[Zr(PO_4)_2]HNa \cdot 5H_2O + x\text{ROH} + H^+ \rightarrow \alpha\text{-}[Zr(PO_4)_2]H_2 \cdot x\text{ROH} + Na^+ \quad (6)$$

The intercalation of glycols may also be easily achieved by such a procedure. The interlayer distances of the compounds obtained are reported in Fig. 9 as a function of the number of carbon atoms of the *n*-alkyl chain. It can be seen that the interlayer distance increases linearly with the increasing dimensions of the alkanol and glycol molecules, the slopes being 2.07 and 1.1 Å per carbon atom, respectively (*34*). Considerations similar to those reported for the *n*-alkylmonoamines and *n*-alkyldiamines lead us to the conclusion that alkanols are intercalated as bilayers of extended molecules inclined at an angle of ~55° with respect to the sheets, whereas glycols form a monomolecular layer. Thus, their behavior resembles that of amines, although the interactions of the respective functional groups with the ≡P—OH groups are different.

Probably the —OH groups of the *n*-alkanols are hydrogen-bonded to the ≡P—OH groups and accommodated in the half-cavities present on the layers, as the water molecule in α-[Zr(PO$_4$)$_2$]H$_2$ · H$_2$O (*9*, *12*). The weak interactions of these guest molecules with the host matrix are also revealed by the low stability of the intercalation compounds obtained. Alkanol intercalates lose the guest molecules on drying in air or on washing with polar solvents (*21*, *34*). Thus, it is not easy to determine their composition, but in analogy with the intercalates of *n*-alkylmonoamines it is likely that they contain 2 moles of guest molecule per formula weight of exchanger. Note that the methanol complex has the interlayer distance (9.3 Å) expected for a complex containing only 1 mole of guest molecule. The

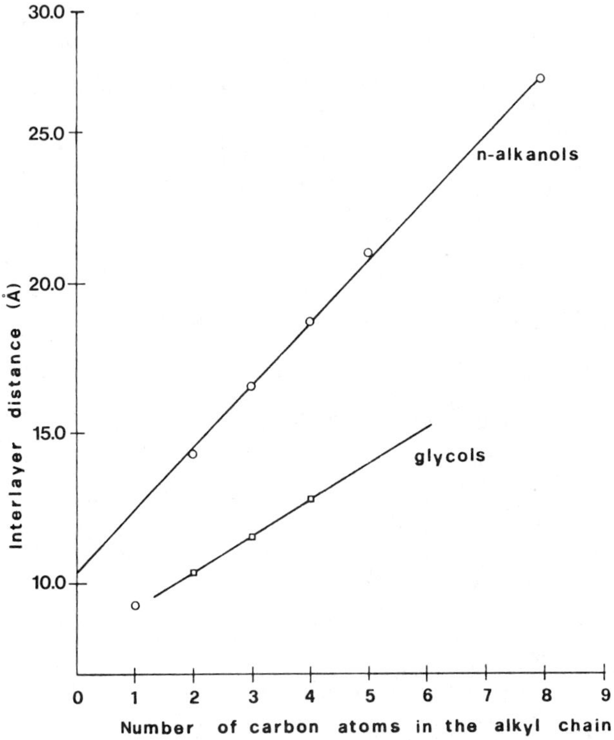

Fig. 9. Interlayer distances of *n*-alkanol and glycol intercalate compounds of α-$[Zr(PO_4)_2]H_2$ as a function of the number of carbon atoms in the alkyl chain. [Reproduced with permission from Costantino (*34*).]

stability of glycol intercalation compounds is higher than that of alkanol ones, and the composition is close to 1 mole of guest per mole of exchanger. Some long-chain or aromatic alkanols, as well as some isoalkanols, do not react as in reaction (*6*), the hydrogen form of α-zirconium phosphate being obtained. However, if an excess of these alcohols is allowed to react with the ethanol intercalate, the ethanol molecules are displaced from the interlayer spaces and the new intercalation compound is formed. By means of this procedure, other intercalation compounds have been prepared, and their interlayer distances are listed in Table IV. The values of the interlayer distance suggest that isoalkanols and benzyl alcohol also form a double layer between the sheets of the exchanger.

Regarding the intercalation mechanism, the coexistence of two intercalated phases of different composition has been detected, by X-ray diffraction, during the desorption of ethanol and ethylene glycol. It is likely

TABLE IV

Interlayer Distance of Intercalation Compounds of α-Zirconium Phosphate with Various Polar Molecules

Intercalated molecule	Interlayer distance (Å)	Ref.
Acetone	9.9	*34*
Acetylacetone	13.5	*34*
Acetonitrile	11.3	*34*
N-Methylformamide	10.5	*23*
N,*N*-Dimethylformamide	11.2, 10.7	*21, 23, 34*
Urea	9.4	*23, 34*
N,*N*-Dimethylurea	10.6	*23*
N,*N*′-Dimethylurea	9.4	*23*
Hydrazine hydrate	9.4	*23*
Piperidine	13.4	*23*
Dimethyl sulfoxide	10.8, 11.3	*21, 23*
Diethylene glycol	10.5	*21*
Octanol	26.7	*34*
Isopropanol	15.6	*34*
2-Methyl-1-propanol	17.5	*34*
3-Methyl-1-butanol	19.2	*34*
Benzyl alcohol	21.0	*34*

that the mechanism is similar to that suggested for the intercalation of amines (*34*).

H. Intercalation of Other Polar Molecules

A number of intercalation complexes of α-$[Zr(PO_4)_2]H_2 \cdot H_2O$ with various polar molecules has been prepared, and their interlayer distances are listed in Table IV. Two different methods have been used. In the first, such complexes are obtained by displacing the ethanol molecules from α-$[Zr(PO_4)_2]H_2 \cdot (C_2H_5—OH)_2$ with the pure solvent to be intercalated (*34*). In the second, α-$[Zr(PO_4)_2]H_2 \cdot H_2O$ is allowed to react for several days at high temperature with the pure liquids or with a concentrated solution of the molecules to be intercalated (*23*). The available data for each homologous series of molecules are not sufficient to permit a reliable discussion on the arrangement of these molecules in the interlayer regions. However, Behrendt *et al.* (*23*) noted that spherical molecules give intercalation compounds in which the increase in the interlayer distance is roughly equal to the diameter of the intercalated molecules minus the diameter of a water molecule. This suggests that the water molecule is displaced and a monolayer of guest molecules is formed.

Much more experimental work is needed to clarify the structural arrangement of these molecules between the layers of the exchanger. At present, we note that these materials may have interesting applications because of their large interlayer distance, the lability of the bonds between guest molecules and host matrix, and the ion-exchange and catalytic properties of zirconium phosphate, which are "tuned" by the presence of the guest molecules.

IV. INTERCALATION BEHAVIOR OF OTHER LAYERED ACID SALTS OF THE α TYPE

Apart from α-$[Zr(PO_4)_2]H_2 \cdot H_2O$, several other acid salts of tetravalent metals possessing an α layered structure have been synthesized, and some of their characteristics are reported in Table I. Since the replacement of a tetravalent atom by a different tetravalent atom does not appreciably modify the strength of the anionic field, the α layered phosphates should exhibit intercalation properties very similar to those of α-$[Zr(PO_4)_2]H_2 \cdot H_2O$. The properties of α layered arsenates should also be similar, but the weaker acidic character of the arsenate groups would lead to differences in the interaction strength between active sites and intercalated molecules. More significant differences in the intercalation behavior compared to that in α-$[Zr(PO_4)_2]H_2 \cdot H_2O$ could arise from the differences in the unit-cell dimensions. Those materials having a larger interlayer distance should possess lower steric hindrance to the diffusion of the guest molecules within the interlayer region. Furthermore, the smaller the product $a \times b$ (and hence the lower the free area associated with each phosphate and arsenate group in the plane of the sheet), the higher must be the covering effect of intercalated molecules. In this respect, α-titanium phosphate should have a higher steric hindrance and a higher covering effect than α-zirconium arsenate. The other α layered materials having unit-cell dimensions intermediate between those of the above-mentioned compounds would behave accordingly.

Very little experimental work has been performed on the intercalation chemistry of these materials. Michel and Weiss (7) reported the complexes between *n*-octylamine, dioctylamine, and trioctylamine and the phosphates and arsenates of Zr, Ti(IV), and Sn(IV). The interlayer distances of the intercalates formed by the same guest molecule and the different α layered acid salts are very similar, suggesting that the guest molecules arrange themselves in a similar way. Preliminary results (*26*) on the intercalation properties of α-$[Ti(PO_4)_2]H_2 \cdot H_2O$ toward some *n*-alkylmonoamines showed that the rates of the intercalation process are lower than those in α-$[Zr(PO_4)_2]H_2 \cdot H_2O$ and that stoichiometric compounds are obtained with greater difficulty. On the other hand, the increase in the interlayer distance of the intercalates with the increase in n_C is similar to that found in α-zirconium phosphate ($\sim$2.2 Å per carbon atom).

Additional experimental work is necessary to confirm fully the expectations based on the crystalline structure of layered acid salts.

V. ACID SALTS WITH LAYERED STRUCTURE OF THE γ TYPE

To date, only two acid salts, $[Zr(PO_4)_2]H_2 \cdot 2H_2O$ (*35*) and $[Ti(PO_4)_2]H_2 \cdot 2H_2O$ are known to have a layered structure of the γ type (*11, 35–37*). However, much experimental evidence indicates that $[Ce(AsO_4)_2]H_2 \cdot 2H_2O$ also has a γ layered structure (*38*), and it is likely that other γ layered acid salts will be prepared in the near future. The interlayer distances of some anhydrous and hydrated cationic forms of γ-zirconium phosphate and γ-titanium phosphate are listed in Table V.

A. Structure of γ Layers

Single crystals large enough for X-ray structural determination have not yet been obtained. However, considerable evidence has been collected which shows that the densities of fixed charges of γ layered acid salts are considerably higher than those of the corresponding α layered acid salts. In particular, it has been possible to calculate that the number of fixed charges per square centimeter in each of the opposite sides of the planar macroanions of γ-zirconium and titanium phosphates are 5.64×10^{14} and 6.03×10^{14}, respectively (*11*). Although the arrangement of the fixed charges is unknown, it can be shown that the covering effect in a γ compound is at least 1.3 times that of the corresponding α compound. Thus, the utilization of all the active sites is more difficult than in α layered acid salts, and the occurrence of less stoichiometric intercalated compounds is accordingly expected. On the other hand, as shown in Table V, the interlayer distances of γ layered acid salts are considerably larger than those of the α salts. The steric hindrance to the diffusion of polar molecules within the γ layered structure is accordingly smaller, and the intercalation processes are usually faster, especially in the case of molecules having a large cross-sectional area.

TABLE V

Interlayer Distance of Some Anhydrous and Hydrated Salt Forms of γ-Titanium and γ-Zirconium Phosphates

Compound	Interlayer distance (Å)	Ref.	Compound	Interlayer distance (Å)	Ref.
$[Ti(PO_4)_2]H_2$	9.2	*11, 37*	$[Zr(PO_4)_2]H_2$	9.4, 9.55	*35, 41*
$[Ti(PO_4)_2]H_2 \cdot H_2O$	11.4	*39*	$[Zr(PO_4)_2]H_2 \cdot H_2O$	11.3	*39*
$[Ti(PO_4)_2]H_2 \cdot 2H_2O$	11.6	*36, 11*	$[Zr(PO_4)_2]H_2 \cdot 2H_2O$	11.2	*35*
$[Ti(PO_4)_2]H_{1.5}Na_{0.5} \cdot H_2O$	11.0	*36*	$[Zr(PO_4)_2]HLi$	9.8	*42*
$[Ti(PO_4)_2]HNa \cdot 3H_2O$	13.2	*36*	$[Zr(PO_4)_2]HNa \cdot 2.5H_2O$	11.6	*39, 43*
$[Ti(PO_4)_2]HNa \cdot 2H_2O$	12.6	*39*	$[Zr(PO_4)_2]HNa$	10.3	*39, 42*
$[Ti(PO_4)_2HNa \cdot H_2O$	10.2	*39*	$[Zr(PO_4)_2]Na_2 \cdot 3H_2O$	12.6	*39, 43*
$[Ti(PO_4)_2]HNa$	10.1	*39*	$[Zr(PO_4)_2]Na_2 \cdot H_2O$	11.6	*39*
$[Ti(PO_4)_2]Na_2 \cdot 2H_2O$	12.8	*36*	$[Zr(PO_4)_2]Na_2$	11.7	*39*
$[Ti(PO_4)_2]Na_2 \cdot H_2O$	11.5	*39*	$[Zr(PO_4)_2]HCs$	11.9	*43*
$[Ti(PO_4)_2]Na_2$	10.5	*39*	$[Zr(PO_4)_2]Zn \cdot 4H_2O$	11.8	*44*
$[Ti(PO_4)_2](NH_4)_2 \cdot H_2O$	13.4	*40*	$[Zr(PO_4)_2]Cu \cdot 4H_2O$	12.4	*44*
$[Ti(PO_4)_2]HSr_{0.5} \cdot 3H_2O$	13.4	*36*	$[Zr(PO_4)_2]Ni \cdot 4H_2O$	12.6	*44*
$[Ti(PO_4)_2]HSr_{0.5} \cdot H_2O$	11.0	*39*	$[Zr(PO_4)_2]Co \cdot 4H_2O$	12.0	*44*
$[Ti(PO_4)_2]HSr_{0.5}$	10.7	*39*	$[Zr(PO_4)_2]Mn \cdot 4H_2O$	12.0	*44*

B. Intercalation of Polar Molecules

Few data on the intercalation behavior of γ layered acid salts are available. Behrendt *et al.* (*23*) and Yamanaka and Koizumi (*21*) reported the interlayer distance of some intercalation compounds of γ-$[Zr(PO_4)_2]H_2 \cdot 2H_2O$ with different organic molecules (Table VI). According to Yamanaka, γ-zirconium phosphate can form three types of complexes. In complexes of type I, the interaction between guest molecules (alkanols, glycols, ketones, and so on) and phosphate groups is attributed to the formation of hydrogen bonds. In type II complexes, the guest molecules (amines and other basic molecules) are protonated and thus held by ionic interactions. The complexes of type III are not true intercalation compounds since the organic molecules react with the phosphate groups (*21*) (see Section VI).

The uptake of ammonia and ammonium ions on γ-$[Ti(PO_4)_2]H_2 \cdot 2H_2O$ has been investigated (*40*). In contrast to α-$[Zr(PO_4)_2]H_2 \cdot H_2O$, which does not exchange its protons with the ammonium ions, γ-$[Ti(PO_4)_2]H_2 \cdot 2H_2O$ is a good exchanger for this cation owing to its greater interlayer distance. The NH_3 is also taken up at a high rate from aqueous solutions. The uptake of 1 mole of NH_3 per mole of exchanger does not cause an appreciable increase in the interlayer distance. Thus, it seems that ammonia replaces the water in the interlayer region.

TABLE VI

Interlayer Distance of Intercalation Compounds of γ-Zirconium Phosphate with Different Organic Molecules

Intercalated molecule	Interlayer distance (Å)	Ref.
Acetone	14.4	*21*
N-Methylformamide	15.9	*23*
N,*N*-Dimethylformamide	16.5, 15.8	*21*, *23*
Urea	13.6	*23*
N,*N*-Dimethylurea	14.1	*23*
N,*N*′-Dimethylurea	15.9	*23*
Ethyleneglycol	13.4	*21*
2-Propanol	19.4	*21*

Systematic investigations on the intercalation behavior of γ-zirconium and γ-titanium phosphates are being undertaken in our laboratory (*45*, *46*). Preliminary results indicate that, in contrast to α-$[Zr(PO_4)_2]H_2 \cdot H_2O$, the alkanol complexes are obtained by simply contacting the γ layered acid salts with the pure liquids. The different behavior of α and γ layered acid salts is probably due to the larger interlayer distances of γ materials, so that the activation energy for the diffusion of alkanols in the interlayer space is considerably lower. Amine complexes have been obtained by contacting γ layered acid salts with dilute methanolic solutions of the amine. The compositions of some γ-zirconium phosphate and γ-titanium phosphate intercalates are reported in Tables VII and VIII, respectively. The *n*-alkanols and *n*-alkylamines arrange themselves in bilayers, whereas the diamines form a monolayer of intercalated molecules, as found for α-zirconium phosphate. In addition, the inclination of the guest molecules with respect to the sheets is very similar to that found in α-$[Zr(PO_4)_2]H_2 \cdot H_2O$ since a similar increment for each additional carbon atom in the alkyl chain may be calculated from data reported in Table VII. However the interlayer distances of intercalation compounds of α- and γ-zirconium phosphate with corresponding molecules are different, the basal spacings of γ complexes always being larger than those of the α complexes. Since the thicknesses of α and γ sheets are probably almost identical and the arrangement of the guest molecules is very similar, the differences in the interlayer spacings may be attributed to a lower penetration of the terminal polar groups of the guest molecules into the surface of the γ layers compared with that in the α layers. This argument provides further evidence of the closeness of the fixed charges in γ compounds.

TABLE VII

Interlayer Distance and Composition of Intercalates of γ-Zirconium Phosphate with Alkanols and Amines

Intercalate molecule	Interlayer distance (Å)	Composition[a]
Methanol	12.7	—
Ethanol	17.2, 16.6	—
Propanol	20.2	—
Butanol	22.6	—
Pentanol	23.8	—
Hexanol[b]	24.9	—
Ethylamine	14.9	1.0
Propylamine	17.6	0.9
Butylamine	19.5	1.0
Decylamine	34	1.0
Ethylenediamine	12.4	0.53
Tetramethylenediamine	14.7	0.51
Heptamethylenediamine	18	0.55
Decamethylenediamine	22	0.5

[a] Moles intercalate per formula weight of exchanger.
[b] From Behrendt *et al.* (*23*). Other data are from Costantino (*45*, *46*).

TABLE VIII

Interlayer Distance and Composition of Intercalates of γ-Titanium Phosphate with Alkanols and Amines (*45*, *46*)

Intercalated molecule	Interlayer distance (Å)	Composition[a]
Methanol	12.7	—
Ethanol	15.7	—
Propanol	19.8	—
Butanol	21.3	—
Pentanol	23	—
Ethylamine	16.3	1.0
Propylamine	18.4	1.0
Butylamine	20.5	1.0
Decylamine	34	0.93
Tetramethylenediamine	15.0	0.53
Heptamethylenediamine	17.4	0.51
Decamethylenediamine	21.5	0.48

[a] Moles intercalate per formula weight of exchanger.

Inspection of Tables VII and VIII shows that alkylmonoamines and alkyldiamines saturate only half of the ≡P—OH groups of the γ compounds; thus, the mono- and bilayers are formed with an amount of guest molecules half that necessary for α-zirconium phosphate. This fact may also be attributed to the closeness of the ≡P—OH groups, the covering effect also being evident for the *n*-alkyl chain. In other words, the mean distance of two adjacent ≡P—OH groups in a layer of γ-zirconium or titanium phosphate should be lower than 4.3–4.9 Å which is the cross-sectional diameter of a trans,trans alkyl chain. Yamanaka and Tanaka (*47*) reported the lattice parameters of γ-$[Zr(PO_4)_2]H_2 \cdot 2H_2O$, from which those of γ-$[Ti(PO_4)_2]H_2 \cdot 2H_2O$ were also derived (*40*). Assuming that two phosphate groups are present in the $a \times b$ area, as occurs in α-$[Zr(PO_4)_2]H_2 \cdot H_2O$, the free area associated with each phosphate group may be evaluated as 18 and 17 Å^2 in γ-$[Zr(PO_4)_2]H_2 \cdot 2H_2O$ and γ-$[Ti(PO_4)_2]H_2 \cdot 2H_2O$, respectively. These values of the free area are indeed much lower than in α-zirconium phosphate (24 Å^2) and very close to the cross section of a trans, trans alkyl chain (15–19 Å^2).

VI. ORGANIC DERIVATIVES OF α- AND γ-ZIRCONIUM PHOSPHATES AND THEIR INTERCALATION BEHAVIOR

Various compounds in which organic radicals are bonded to the layered matrix of zirconium phosphate have been obtained (*10*, *13*, *48–51*). Rather than intercalates, such compounds may be considered "organic derivatives" of α- or γ-zirconium phosphates since they have the basic α or γ layered structure but with tetrahedral ≡P—OH groups replaced by ≡P—R or ≡P—OR groups (R being an organic radical). Two general methods have been used for their synthesis. In the first, proposed by Alberti and co-workers, zirconium fluoro complexes are slowly decomposed in the presence of the appropriate phosphonic acid or phosphoric ester (*10*). The organic derivatives prepared with this procedure possess the α layered structure. Maya showed that α organic derivatives may be also be obtained by direct reaction of a zirconium salt with monoalkylphosphoric esters (*48*).

In the second method, used by Yamanaka and co-workers, γ-$[Zr(PO_4)_2]$-$H_2 \cdot 2H_2O$ is allowed to react with derivatizing agents such as ethylene oxide or propylene oxide (*49*, *50*). More recently, the same authors showed that γ organic derivatives can also be prepared by exchanging the phosphate group of γ-$[Zr(PO_4)_2]H_2$ with phosphoric ester groups (*51*, *52*). The formulas, densities, and interlayer distances of some known α and γ organic derivatives are reported in Table IX. There is good agreement between the observed

TABLE IX

Formula, Density, and Interlayer Distance of Some Organic Derivatives of Zirconium Phosphate Having α or γ Layered Structure

Formula	Density (gm/cm^3)	Interlayer distance (Å)	Ref.
α-$Zr(HOCH_2PO_3)_2 \cdot H_2O$	2.20	10.1	*10*
α-$Zr(HOCH_2PO_3)_2$	2.30	9.2	*10*
α-$Zr(HOOCCH_2PO_3)_2$	2.31	11.1	*13*
α-$Zr(H_5C_6PO_3)_2$	1.89	14.7	*10*
α-$Zr(C_2H_5OPO_3)_2$	1.92	11.7	*10*
α-$Zr(C_4H_9OPO_3)_2$	1.66	15.9	*48*
α-$Zr(C_{12}H_{25}OPO_3)_2$	1.28	32.7	*48*
α-$Zr(C_{14}H_{21}OPO_3)_2$[a]	1.34	20.7	*48*
γ-$Zr(HOCH_2CH_2OPO_3)_2 \cdot H_2O$	—	18.4	*49*
γ-$Zr[(HOCH_2CH(CH_3)OPO_3)(HPO_4)]$	—	21.4	*50*
γ-$Zr(H_5C_6OPO_3)(HPO_4) \cdot 2H_2O$	—	16.4	*51*

[a] Zirconium bis(monooctylphenyl phosphate).

values of the interlayer distances of the organic derivatives and those calculated taking into account the distance between the plane of zirconium atoms and the plane of phosphorus atoms (assumed equal to 1.6 Å as in α-$[Zr(PO_4)_2]H_2$) and the lengths of the ≡P—R or ≡P—OR groups (*10, 48*). Thus, to some extent the value of the interlayer distance gives information about how the organic molecules are arranged in the solid framework. As for the intercalated alkylamine or alkanol compounds of α-$[Zr(PO_4)_2]H_2$, there is no interpenetration of organic radicals belonging to two adjacent layers.

In order to facilitate the visualization of the structure of the organic derivatives, Fig. 10 shows, as an example, the idealized structure of zirconium bis(carboxymethanephosphonate).

Apart from their fundamental aspects, the organic derivatives constitute a new class of layered materials with interesting applications. It has been found that the lauryl derivative behaves as a solid support for reverse-phase liquid chromatography (*48*), and applications in thin-layer and/or gas–solid chromatography are also expected (*10*). If an ionogenic group, such as $—SO_3H$, —COOH, or $—N(CH_3)_3Cl$, is present in the organic radical, the compound acts as an organic–inorganic ion exchanger.

Zirconium bis(carboxymethanephosphonate) is an example of such an exchanger, and its ion-exchange properties with respect to Na^+ ions have been investigated. When the protons are replaced by Na^+ ions, the interlayer distance increases from 11.1 to 14.4 Å and the phase $Zr(OOCCH_2PO_3)_2$-$Na_2 \cdot 4H_2O$ is obtained (*13*).

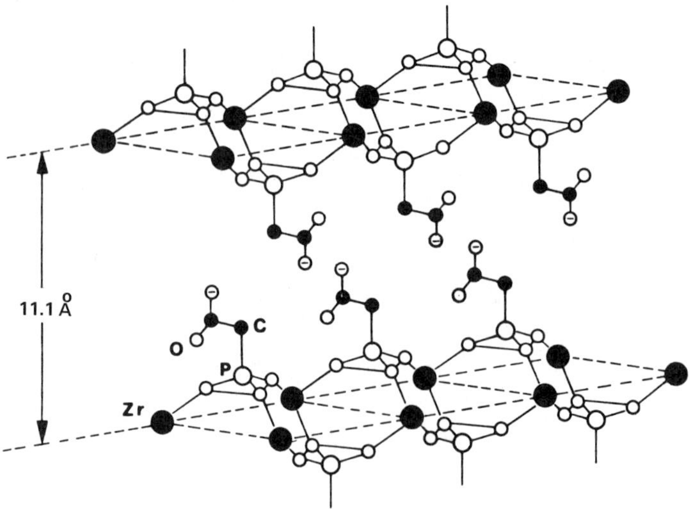

Fig. 10. Idealized crystal structure of zirconium bis(carboxymethanephosphonate). The minus sign refers to oxygens carrying the negative charges. [Reproduced with permission from Alberti *et al.* (*13*).]

Owing to their layered structure, the organic derivatives are potential intercalating agents. In this respect, the following points may be considered:

1. In many organic derivatives, the organic radicals belonging to two adjacent layers do not interpenetrate, and thus there is little steric hindrance to the diffusion of polar molecules within the interlayer region.
2. The forces holding the layers together are expected to be either weak van der Waals interactions or hydrogen bonds.
3. The free area associated with each active site in the interlayer region should be very similar to that found for α or γ layered acid salts.
4. The active sites present on the surface of the layers are the terminal groups of the organic radicals. These may vary from almost apolar methyl groups to the strongly polar —COOH groups.

The interactions between potential guest molecules and the almost apolar groups present in zirconium bis(benzenephosphonate), or in the alkyl derivatives of α-zirconium phosphate, are too weak to compensate for the energy required to spread the layers apart, and no intercalation compounds of these organic derivatives, contacted with various polar molecules, have been obtained (*26*). On the other hand, $Zr(HOCH_2PO_3)_2 \cdot H_2O$ and $Zr(HOOCCH_2PO_3)_2$ are effective intercalating agents of ammonia and amines (*26*).

TABLE X

Method of Preparation and Interlayer Distance of Some Intercalates of α-$Zr(HOCH_2PO_3)_2 \cdot H_2O$ and α-$Zr(HOOCCH_2PO_3)_2$ (*26*)

Intercalated molecule	α-$Zr(HOCH_2PO_3)_2 \cdot H_2O$ Method[a]	α-$Zr(HOCH_2PO_3)_2 \cdot H_2O$ Interlayer distance (Å)	α-$Zr(HOOCCH_2PO_3)_2$ Method[a]	α-$Zr(HOOCCH_2PO_3)_2$ Interlayer distance (Å)
Ammonia	—	—	3	13.2
Propylamine	1	18.6	2	20.5
Butylamine	1	21.5	2	22.6
Heptylamine	—	—	2	29.0
Decylamine	1	31.5	2	37.0
Benzylamine	1	20.5	—	—

[a] 1, Equilibration with pure liquid at 60°C; 2, equilibration with 0.1 *M* methanolic solution at room temperature; 3, absorption from vapor phase.

Table X reports the interlayer distances and the methods of preparation of the intercalation compounds obtained to date. It can be seen that the intercalation processes in $Zr(HOCH_2PO_3)_2 \cdot H_2O$ occur with difficulty owing to the low acidic character of its hydroxyl groups. In fact, the intercalation compounds have been obtained only after prolonged contact with the pure liquid at ~60°C (*10*, *26*). On the contrary, α-$Zr(HOOCCH_2PO_3)_2$ may be loaded with various *n*-alkylmonoamines by contacting the host compound with 0.1 *M* amine solution in methanol at room temperature. The compositions of the intercalates obtained are very close to the stoichiometric values [2 moles of alkylamines per mole of α-$Zr(HOOCCH_2PO_3)_2$]. As found for α-$[Zr(PO_4)_2]H_2$, alkylmonoamines are intercalated as a double film of extended molecules inclined at an angle of ~60° with respect to the sheets. Ammonia and alkylamines are probably protonated by the protons of the —COOH groups, and this fact accounts for the great stability of the intercalation compounds. It may be observed that, concerning their intercalation capacity, the organic derivatives may be classified as expanding or nonexpanding lamellar compounds. Their capacity for intercalating polar molecules decreases with a decrease in the polar character of their organic group, and no intercalation is obtained when this group is weakly polar or completely apolar. The method used by Yamanaka *et al.* (*49–51*) to obtain the organic derivatives seems applicable only to compounds with a γ structure, in which owing to their large interlayer distance there is an easy diffusion of the derivatizing agents or of the exchanging phosphoric esters in the interlayer region. The mechanisms of the reactions of ethylene oxide and propylene oxide with γ-[Zr(PO$_4$)$_2$]H$_2$ have been studied in detail (*49*, *50*). The ethylene oxide, taken up in the interlayer space from aqueous solution, suffers ring scission and a $\equiv$P—O—C bond is formed according to the following reaction:

$$\gamma\text{-}[Zr(PO_4)_2]H_2 + 2H_2C\overset{O}{\frown}CH_2 \leftrightharpoons \gamma\text{-}Zr(O_3P{-}O{-}CH_2{-}CH_2OH)_2 \tag{7}$$

In the case of propylene oxide only 1 mole of derivatizing agent reacts with 1 mole of γ-[Zr(PO$_4$)$_2$]H$_2$ (*50*). This effect may be attributed to the closeness of the $\equiv$P—OH groups, and it has also been observed in the exchange of the phosphate group with monophenylphosphoric ester (*51*).

The intercalation behavior of organic dèrivatives of γ-zirconium phosphate has not yet been investigated. It is known that these organic derivatives take up water molecules (*49*), but it is very probable that other polar molecules may also be intercalated.

VII. CONCLUSION

Layered acid salts of tetravalent metals represent a vast class of intercalating compounds having good chemical and thermal stability. Owing to the presence of $\equiv$X—OH acidic groups between the layers, strong interactions are especially obtained with polar molecules that are Brönsted bases. The distance among active sites of a given acid salt, [M^{IV}(XO$_4$)$_2$]H$_2$, depends on its α or γ structure and on the tetravalent metal and $\equiv$XOH tetrahedral group involved. Therefore, the intercalation properties may be studied on materials having a different density of active sites. This makes it possible to build up mono- or bimolecular films of *n*-alkyl chains having different packing densities. It is interesting that advantage could be taken of this possibility for studying the conformational transitions observed in the bilayers of phospholipids constituting biological membranes (*53*). Furthermore, owing to the structural analogies between acid salts and clays (*21*, *54*), studies on the intercalation behavior of the acid salts could provide additional insight into the factors influencing the complicated intercalation phenomena in the clay systems. Indeed, as pointed out by Behrendt *et al.* (*23*), layered acid salts "have a number of superior properties compared with other intercalating compounds." However, studies of their intercalation behavior are still in their infancy. Further investigations, especially on the structure of γ layered acid salts and on the structural disposition of the guest molecules in the interlayer region, as well as on their interaction with the active sites, are still needed to provide a deeper and more quantitative understanding of the intercalation behavior of this class of layered materials.

REFERENCES

1. C. B. Amphlett, "Inorganic Ion Exchangers." Elsevier, Amsterdam, 1964.
2. V. Veselý and V. Pekárek, *Talanta* **19,** 219 (1972).
3. A. Clearfield, G. H. Nancollas, and R. H. Blessing, *in* "Ion Exchange and Solvent Extraction" (J. A. Marinsky and Y. Marcus, eds.), Vol. V, Chap. 1. Dekker, New York, 1973.
4. G. Alberti and U. Costantino, *J. Chromatogr.* **102,** 5 (1974).
5. G. Alberti, *Acc. Chem. Res.* **11,** 163 (1978).
6. E. Michel and A. Weiss, *Z. Naturforsch., Teil B* **20,** 1307 (1965).
7. E. Michel and A. Weiss, *Z. Naturforsch., Teil B* **22,** 1100 (1967).
8. A. Clearfield and G. D. Smith, *Inorg. Chem.* **8,** 431 (1969).
9. J. M. Troup and A. Clearfield, *Inorg. Chem.* **16,** 3311 (1977).
10. G. Alberti, U. Costantino, S. Allulli, and N. Tomassini, *J. Inorg. Nucl. Chem.* **40,** 1113 (1978).
11. G. Alberti, U. Costantino, and M. L. Luciani Giovagnotti, *J. Inorg. Nucl. Chem.* **41,** 643 (1979).
12. J. Albertsson, A. Oskarsson, R. Teligren, and J. O. Thomas, *J. Phys. Chem.* **81,** 1574 (1977).
13. G. Alberti, U. Costantino, and M. L. Luciani Giovagnotti, *J. Chromatogr.* **180,** 45 (1979).
14. G. Alberti, *in* "Study week on Biological and Artificial Membranes and Desalination of Water" (R. Passino, ed.), No. 40, p. 629. Pontif. Accad. Sci. Scr. Varia, Vatican City, 1976.
15. A. Clearfield and R. A. Hunter, *J. Inorg. Nucl. Chem.* **38,** 1085 (1976).
16. G. Alberti, R. Bertrami, U. Costantino, and J. P. Gupta, *J. Inorg. Nucl. Chem.* **39,** 1057 (1977).
17. Y. Hasegawa and A. Aoki, *Bull. Chem. Soc. Jpn.* **46,** 836 (1973).
18. J. P. Gupta, N. J. Manning, and D. V. Nowell, *J. Inorg. Nucl. Chem.* **40,** 87 (1978).
19. M. Gordon, M. Popvtzer, M. Greenbaum, L. B. Marantz, M. McArthur, J. R. De Palma, and M. H. Mawell, *Dial. Renal Transplant* **5,** 86 (1968).
20. G. Alberti and P. Galli, *J. Inorg. Nucl. Chem.* **28,** 658 (1966).
21. S. Yamanaka and M. Koizumi, *Clays Clay Miner.* **23,** 477 (1975).
22. S. Yamanaka, Y. Horibe, and M. Tanaka, *J. Inorg. Nucl. Chem.* **38,** 323 (1976).
23. D. Behrendt, K. Beneke, and G. Lagaly, *Angew. Chem., Int. Ed. Engl.* **15,** 544 (1976).
24. G. Alberti, M. G. Bernasconi, M. Casciola, U. Costantino, and M. L. Luciani Giovagnotti, *Proc. Natl. Congr. Inorg. Chem., 11th, Cosenza, Italy* Pap. 1-I (1978).
25. A. Clearfield and R. M. Twinda, *J. Inorg. Nucl. Chem.* **41,** 871 (1979).
26. G. Alberti and U. Costantino, unpublished observations, 1980.
27. J. P. Gupta and D. V. Nowell, *J.C.S. Dalton,* p. 1178 (1979).
28. A. Clearfield and R. M. Twinda, *Inorg. Nucl. Chem. Lett.* **15,** 251 (1979).
29. G. F. Walker, *Nature* (*London*) **177,** 240 (1964).
30. G. H. Henning, *Prog. Inorg. Chem.* **1,** 125 (1959).
31. A. Dyer and A. M. Yusof, *J. Inorg. Nucl. Chem.* **41,** 1479 (1979).
32. G. Alberti, U. Costantino, and R. Giulietti, *J. Inorg. Nucl. Chem.* **42,** 1062 (1980).
33. C. B. Amphlett and L. A. McDonald, *Proc. Chem. Soc., London* p. 276 (1962).
34. U. Costantino, *J.C.S. Dalton* p. 402 (1979).
35. A. Clearfield, R. H. Blessing, and J. A. Stynes, *J. Inorg. Nucl. Chem.* **30,** 2249 (1968).
36. S. Allulli, C. Ferragina, A. La Ginestra, M. A. Massucci, and N. Tomassini, *J. Inorg. Nucl. Chem.* **39,** 1043 (1977).
37. E. Kobayashi, *Bull. Chem. Soc. Jpn.* **48,** 3114 (1975).
38. G. Alberti, U. Costantino, F. Di Gregorio, and E. Torracca, *J. Inorg. Nucl. Chem.* **31,** 3195 (1969).
39. A. La Ginestra and M. A. Massucci, *Thermochim. Acta* **32,** 241 (1979).

40. G. Alberti, M. G. Bernasconi, M. Casciola, and U. Costantino, *J. Inorg. Nucl. Chem.* **42,** 1637 (1980).
41. D. Dollimore, S. E. Horsley, N. J. Manning, and D. V. Nowell, *Proc. ICTA, 4th, Budapest, 1974* **1,** 647 (1975).
42. D. Dollimore, N. J. Manning, and D. V. Nowell, *Thermochim. Acta* **19,** 37 (1977).
43. A. Clearfield and J. M. Garces, *J. Inorg. Nucl. Chem.* **41,** 879 (1979).
44. A. Clearfield and J. M. Kalnins, *J. Inorg. Nucl. Chem.* **40,** 1933 (1978).
45. U. Costantino, *Proc. Natl. Congr. Inorg. Chem., 12th, Trieste* Pap. C-3 (1979).
46. U. Costantino, *J. Inorg. Nucl. Chem.* **42,** 1895 (1981).
47. S. Yamanaka and M. Tanaka, *J. Inorg. Nucl. Chem.* **41,** 45 (1979).
48. L. Maya, *Inorg. Nucl. Chem. Lett.* **15,** 207 (1979).
49. S. Yamanaka, *Inorg. Chem.* **15,** 2811 (1976).
50. S. Yamanaka, M. Tsujimoto, and M. Tanaka, *J. Inorg. Nucl. Chem.* **41,** 615 (1979.
51. S. Yamanaka and M. Hattori, *Chem. Lett.* p. 1073 (1979).
52. S. Yamanaka, H. Maeda, and M. Tanaka, *J. Inorg. Nucl. Chem.* **41,** 1187 (1979).
53. G. Lagaly, *Angew. Chem., Int. Ed. Engl.* **15,** 575 (1976).
54. A. Dyer and D. Leigh, *J. Inorg. Nucl. Chem.* **34,** 369 (1972).

6

The Intercalation Chemistry of β-Alumina

B. C. TOFIELD

I. INTRODUCTION

Since the discovery of rapid sodium ion mobility by Weber and Kummer (*1*), β-aluminas have been intensively studied. The potential use of sodium β- or β″-alumina as a solid electrolyte in the sodium–sulfur battery has given rise to a great deal of work aimed at controlling fabrication processes and sintering, optimizing the sodium ion mobility, understanding the effects of impurities on such properties, and so on. At the same time, much effort has been directed toward understanding the basic processes controlling ion mobility in β-aluminas. This has been stimulated both by the technological promise of the material and by the intrinsic complexity of the β-aluminas.

Although early work assumed a stoichiometric composition of $1Na_2O$–$11Al_2O_3$, work since 1967 has revealed that β-alumina preparations are always nonstoichiometric with excess soda, $(1 + x)Na_2O$–$11Al_2O_3$, x being

INTERCALATION CHEMISTRY

ISBN 0-12-747380-7

typically between 0.25 and 0.5. The very high sodium ion mobility (*2*) is a result of the sodium excess. To determine precisely how the nonstoichiometry is accommodated structurally is by no means straightforward, but this mechanism must be understood if the atomic details of the conduction process are to be unraveled (*3*). Several reviews of all aspects of β-alumina science and technology are available (for example *4–7*).

Alkali-ion-conducting solid electrolytes, including the β-aluminas, are not generally considered in the context of intercalation chemistry. Ion-exchange processes (*8*) do not normally permit a change in the total mobile ion composition, in contrast to the situation with electronically conducting insertion or intercalation compounds. Recent studies, however, particularly of hydrogen-containing β-aluminas, have led to a considerable increase in our understanding of many aspects of the chemistry of the conduction plane. Accurate structural work has, for example, shown the interesting behavior of both large, for example, NH_4^+, and small, for example, Li^+, ions. More profoundly, the action of hydrogen in suitable circumstances has been demonstrated to yield stoichiometric or nearly stoichiometric hydrogen β-alumina. An understanding of the chemistry of hydrogen and hydrated hydrogen β-aluminas is now being acquired and, by ion exchange, the preparation of metal β-alumina compositions, inaccessible by normal synthetic routes, has been achieved. For the first time, the properties of the stoichiometric β-aluminas are being measured and those aspects of the behavior of "normal" β-aluminas that are related directly to the presence of the nonstoichiometry are being elucidated. β-Alumina has a very rich and often subtle mirror-plane chemistry, and the advances being made in its understanding have general relevance to the study not only of alkali ion conductors but of framework ion exchangers in general.

This chapter is not intended to be exhaustive in a field that is still being defined. We do, however, attempt to review the main areas of interest in the chemistry of the β-alumina conduction plane and to outline where future work may usefully be directed. We restrict our discussion primarily to the β-alumina framework itself, where the availability of phase-pure polycrystalline material and of high-quality single crystals (*9*) has greatly facilitated both structural studies and the characterization of ion-exchange reactions. β″-Alumina (*4*), although of great interest for battery applications, has a different origin of nonstoichiometry and appears to be stabilized only by dopants such as Mg^{2+} and Li^+ in the spinel block. Also, the chemistry of β″-alumina has been less well explored than that of β-alumina; this is also true of the β-alumina structure types formed by other trivalent ions, particularly Fe^{3+} and Ga^{3+}.

In this chapter, aspects of the chemistry and structure of nonstoichiometric metal β-aluminas are discussed. The preparative routes for hydrogen-containing β-aluminas are outlined, and the relative importance of protons, water, and hydronium ions in the mirror plane is compared. Ammonium β-alumina and its thermal decomposition and the formation of hydrogen β-alumina are reviewed. Comparison is made with the preparation of hydrogen β-alumina from silver β-alumina. Finally, our state of knowledge of more stoichiometric metal β-aluminas, prepared by reverse exchange from hydrogen β-alumina, is assessed.

II. NONSTOICHIOMETRIC β-ALUMINAS

A. Cell Constants and Conductivities

The β-alumina structure is shown schematically in Fig. 1. Half a unit cell is shown. Four close-packed layers of oxygen atoms separate crystallographic

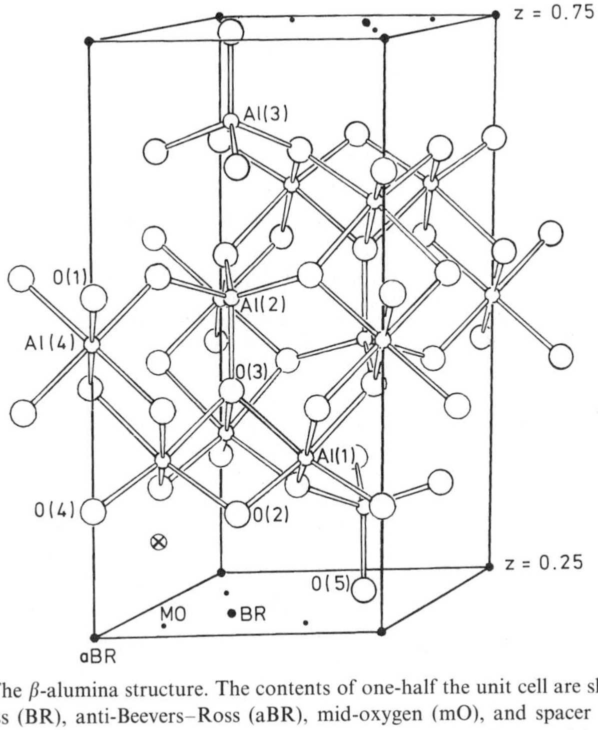

Fig. 1. The β-alumina structure. The contents of one-half the unit cell are shown, and the Beevers–Ross (BR), anti-Beevers–Ross (aBR), mid-oxygen (mO), and spacer oxygen [O(5)] positions within the mirror plane are indicated. In nonstoichiometric materials, extra cations within the conducting plane appear to be compensated by extra oxygen near mO positions. This is coordinated by aluminum atoms at interstitial sites (⊗); corresponding vacancies are observed at Al(1) positions.

mirror planes, which contain only one-quarter of the oxygen atom density. Octahedrally and tetrahedrally coordinated aluminum ions are found within the close-packed spinel-like blocks. Octahedral Al(1) and tetrahedral Al(3) ions are closest to the mirror plane. The cell is hexagonal, with the *c* axis perpendicular to the mirror planes.

The mobile ions in β-alumina are found in the conducting planes between the spinel blocks. All such ions, apart from Li^+, that have been located in crystallographic studies lie in the mirror plane itself. A neutron diffraction study of polycrystalline lithium–sodium β-alumina showed evidence for the displacement of the small Li^+ ions away from the mirror planes by about 1 Å (*10*). A single-crystal study of pure lithium β-alumina would be very useful both to confirm this result and to outline the structure of the pure material for the first time.

The linking of the spinel blocks by the spacer oxygen atoms [O(5)] provides a severe restriction on the relative displacement of the blocks parallel to the *c* axis as the mobile ion is changed. Kummer (*4*) showed, however, that there was, in general, a correlation of *c*-axis lattice constants with ion size, there being a total variation of about 2% between the values observed with the largest and smallest monovalent ions. The *a*-axis lattice constant in the basal plane, on the other hand, is almost insensitive to the nature of the mobile ion. Some *c*-axis lattice constants obtained at Harwell on Union Carbide single crystals are in good agreement with those given by Kummer (Table I). The composition of these crystals is close to $1.25Na_2O–11Al_2O_3$, very similar to

TABLE I

c-Axis Lattice Constants of β-Aluminas[a]

Ion	*c*-Axis lattice constant (Å)	
	Kummer (*4*)	This work[b]
Na^+	22.530	22.528
Ag^+	22.498	22.488
Li^+	22.570	—
K^+	22.729	22.735
NH_4^+	22.888	22.88
H^+	22.677	22.60 (from Ag^+) 22.65 (from NH_4^+)

[a] Both data sets correspond to a starting composition close to $1.25Na_2O–11Al_2O_3$.

[b] Data obtained from Union Carbide single crystals using the APEX goniometer (*66*); for compositions prepared by ion exchange, essentially complete exchange has been achieved.

that quoted by Kummer (*4*). The lattice spacing increases as the monovalent ion size increases from Na^+ to K^+ to NH_4^+, but there is a decrease from Na^+ to Ag^+. Both hydrogen β-alumina and lithium β-alumina have expanded *c*-axis lattice constants relative to sodium β-alumina.

The ionic conductivities and activation enthalpies for ion migration have been compiled by Kummer (*4*) and Kennedy (*5*). Sodium β-alumina has the highest ionic conductivity at most temperatures of interest, with a measured activation energy for conduction parallel to the basal plane of 0.13 eV (*2*) for single crystals of $1.25Na_2O–11Al_2O_3$. Some activation energies for the conductivity of several ions and room temperature conductivities are given in Table II. The trend of increasing activation energy with increasing ion size may be rationalized by assuming that the slot width of the conducting plane in β-alumina is optimal for sodium ion mobility and that the strictly limited expansion permitted as the mobile ion size is increased entails an increasing activation energy for migration.

Although this broad view is undoubtedly correct, the anomalous lattice constants of lithium and hydrogen β-aluminas, for example, warn against an oversimplified interpretation of trends such as those revealed in Tables I and II. Indeed, as discussed below, the nonstoichiometry of β-aluminas as normally prepared profoundly influences parameters such as activation energies of conductivity and lattice constants. Also, although the very low activation energy for sodium mobility results directly from the nonstoichiometry via an interstitialcy-type mechanism (*3*), within the nonstoichiometric region considerable variations in conductivity have been measured. In a comparative study (*11*) of melt-grown (at about 2000°C) Union Carbide crystals and crystals grown from Bi_2O_3 flux at a much lower temperature (about 1300°C), higher prefactors were found for the flux-grown Na^+ material and also for Ag^+, K^+, and Rb^+ crystals prepared from this by ion

TABLE II

Ionic Conductivity of β-Alumina Single Crystals

Ion	E_a (eV)	$\sigma_{25°C}$ (Ω^{-1} cm^{-1} $\times 10^{-4}$)
Na^+	0.16	140[a]
	0.13	350[b]
Ag^+	0.17	64[a]
K^+	0.29	0.65[a]
NH_4^+	0.47	0.013[c]

[a] Whittingham and Huggins (*10a*).
[b] Hooper (*2*).
[c] Hooper *et al.* (*13*).

exchange. This was in line with the greater nonstoichiometry of the flux-grown material, which contained approximately twice the excess sodium content of the melt-grown crystals. The activation energies were higher in all cases for the more nonstoichiometric crystals, however, giving lower conductivities between room temperature and several hundred degrees. Thus, the necessary introduction of increasing numbers of negative defects to compensate the sodium excess causes a decrease in the conductivity beyond a certain point.

This has been confirmed in a recent study (*12*) of $2Na_2O{-}11Al_2O_3$ prepared by the action of Na_2O_2-saturated sodium on $1.35Na_2O{-}11Al_2O_3$ at 750°C. On introducing the extra Na_2O into the conducting planes, the activation energy was raised to 0.3 eV, and the conductivity at 300°C was an order of magnitude less than was found for the less nonstoichiometric starting material.

B. Small Ions in β-Alumina

Although the *c*-axis lattice constant of hydrogen β-alumina depends on the mode of preparation (Table I), it is without doubt larger than that of nonstoichiometric sodium β-alumina. This may reflect the differing overall stoichiometry but may also be related to the absence of cations in hydrogen β-alumina in positions that normally shield trans-plane oxygen–oxygen repulsion (see below). The ionic conductivity of hydrogen β-alumina is many orders of magnitude lower than that of the alkali metal β-aluminas (*13*) because hydrogen is chemically bound to the spacer oxygen in a hydroxyl linkage (*14*, *15*). Lithium β-alumina also shows an increased lattice constant relative to sodium β-alumina ($c = 22.570$ Å, Table I); the small Li^+ ion may also be relatively inefficient at shielding the oxygen–oxygen repulsions.

Radzilowski and Kummer (*16*) refer to the extremely hygroscopic nature of fully exchanged lithium β-alumina, and recent work (*17*) reveals the care one must take when properly characterizing hygroscopic ionic conductors such as lithium β-alumina. If care is not taken to monitor, preferably by infrared spectroscopy, the presence of intercalated moisture and, if necessary, to remove it, grossly erroneous results may be obtained. Lithium ions displaced off the mirror plane can be coordinated at most by only three oxygen ions in a very asymmetric environment, and it may be presumed that this provides a favorable situation for the coordination of intercalated water molecules to establish an approximately tetrahedral environment of oxygen around the lithium ion. This would be expected to decrease the lithium ion mobility markedly, as is found (*17*, *18*).

The activation energy for Li^+ motion determined by tracer diffusion measurements [0.38 eV (*4*)] is larger than that for sodium and has been explained as arising from polarization of the small Li^+ ion off the mirror plane toward one or other of the bordering oxygen layers. Structural work (*10*) seems to confirm this displacement, as does the positive pressure dependence of Li^+ conductivity observed by Radzilowski and Kummer (*16*). No significant pressure dependence for sodium ion mobility was observed (*16*), and the conductivity of potassium β-alumina decreased with pressure, as would be expected for an oversized ion. However, conductivity studies (*18*) on fully exchanged, anhydrous lithium β-alumina indicate an activation energy of 0.29 eV and a room temperature conductivity of $3 \times 10^{-3}\ \Omega^{-1}\ cm^{-1}$. These values, comparable with those for lithium nitride (*19*), place lithium β-alumina among the best lithium ion solid electrolytes known to date.

In view of these results and in light of the possible complications arising from moisture absorption and incomplete exchange of lithium for sodium, it may not necessarily follow that the pressure dependence of conductivity reported indicates beyond doubt that Li^+ polarization is disadvantageous for fast Li^+ transport. Although, relative to sodium in β-alumina (Table II), lithium does show a decreased mobility and an increased activation energy, it may be more relevant to compare lithium β-alumina with other lithium ion conductors, especially if it is found that alkali–oxygen repulsive potentials do not scale exactly with ion size between sodium and lithium. It is by no means self-evident that the asymmetric environment of lithium in lithium β-alumina provides a higher activation energy for Li^+ motion than the four- or sixfold coordination by oxygen normally found in lithium-containing oxides. It is clear, however, that materials such as lithium β-alumina with very irregular, low lithium coordination will, in general, be metastable with respect to decomposition to compounds containing more symmetrically coordinated lithium and can thus be prepared only by ion exchange or other low-temperature routes. They will also, as discussed above, be very susceptible to moisture pickup.

C. Nonstoichiometry in β-Alumina

In an early structural determination of sodium β-alumina by Beevers and Ross, a stoichiometric composition, $NaAl_{11}O_{17}$, was established (*20*) and the sodium ion located on the Beevers–Ross sites (Fig. 1). The alternative site considered, but rejected, for the sodium occupation was the anti-Beevers–Ross position (Fig. 1). The more refined analytical and crystallographic techniques available since the discovery of fast ion conductivity in β-alumina (*1*) and the development of sodium β-alumina as the electrolyte in the sodium–sulfur battery, have clearly revealed that sodium β-alumina as normally prepared always contains excess sodium oxide. The phase diagram has been discussed by Weber [see Kummer (*4*)]. The degree of nonstoichiometry depends on the temperature and mode of preparation, but the ideal stoichiometric material is never observed. There is, indeed, a rather wide compositional range, and for commercially available, fairly pure material excess sodium oxide contents relative to the stoichiometric formula of around 25% (Union Carbide single crystals), 35% (Monofrax H polycrystalline material), and 55% (Alcoa polycrystalline β-alumina) may be observed. Although other β-aluminas, such as the potassium and rubidium compounds, can be prepared directly from the oxides, the vast majority of studies concerning the composition and nonstoichiometry of β-aluminas refer to material prepared in the sodium form, from which other compositions may be prepared by ion exchange. There is no evidence that direct preparations of other β-aluminas yield compositions outside the range observed in the sodium case.

The determination of the mechanism by which the excess sodium oxide is incorporated in the β-alumina lattice was by no means straightforwardly achieved, nor was a distinction readily made between the alternative possibilities of aluminum vacancies within the spinel blocks or oxygen interstitial atoms, presumably in the conducting plane. Thus, Whittingham and Huggins (*21*) interpreted conductivity data on silver β-alumina as indicating aluminum vacancies, and Antoine *et al.* (*22*) interpreted the EPR of manganese-doped β-alumina as supporting the presence of vacancies on Al(3) sites.

The structure of sodium β-alumina was redetermined by Peters *et al.* (*23*) using X rays and the excess sodium located within the conduction plane. The crystal studied contained 29% excess Na_2O, and only 0.75 sodium atoms per asymmetric unit were located near the Beevers–Ross positions. The extra sodium was found not on anti-Beevers–Ross sites but near mid-oxygen positions (Fig. 1). It was realized that spatial constraints precluded the occupation of more than one sodium atom per asymmetric unit in cells where the Beevers–Ross position was occupied but that the mO–mO separation was sufficient (about 2.8 Å) to allow multiple sodium occupancy. The substitution of multiply occupied cells with sodium ions on mid-oxygen positions for singly occupied cells with sodium location close to Beevers–Ross sites was thus shown to provide a mechanism for the incorporation of the excess sodium. The only other anomalous occupancy factor observed in this work was a slight underpopulation of Al(1) [11.87(7) instead of 12.0 per ideal unit cell], and this was interpreted as indicating an Al(1)-vacancy compensation mechanism.

This work was followed by an X-ray single-crystal structure refinement of silver β-alumina (*24*). Excess silver was again observed, but in this case, in contrast to the situation with the sodium compound, around 30% of the silver was located near the anti-Beevers–Ross position. It could be assumed that this resulted from a favorable covalent, almost linear O(4)—Ag—O(4) interaction. Two other silver sites were discussed, one close to the Beevers–Ross position and the other considerably displaced toward the mid-oxygen site. The distribution of silver within these two sites was difficult to establish with certainty. Roth (*24*) did not find any evidence for aluminum vacancies in the spinel block, and, although no interstitial oxygen atoms were located, bulk-density evidence was presented which slightly favored interstitial oxygen rather than aluminum-vacancy defect compensation.

The apparently correct structural mechanism by means of which excess cations are incorporated in the conduction plane was discovered by Reidinger (*25*). In a neutron diffraction study of single-crystal sodium β-alumina, difference Fourier maps perpendicular to the conducting plane showed nuclear density at an interstitial position between Al(1) and the mid-oxygen sites and with a z parameter very similar to Al(3). One such density map has so far been published (*26*). The extra scattering density represents aluminum atoms displaced from Al(1) positions (Fig. 1), and, since this initial identification of interstitial aluminum (Al_i), it has been observed in all subsequent crystallographic refinements of nonstoichiometric undoped β-aluminas. As observed by Peters *et al.* (*23*), vacancies corresponding to the displaced aluminums are found on Al(1) sites, and the concentrations of vacancies and interstitial aluminums are generally in good agreement with the excess mobile ion concentration. No single-crystal structural determinations have yet been reported on highly nonstoichiometric material with a mobile ion excess greater than 30%.

Rerefinement of the original X-ray data sets (*23*, *24*) for sodium and silver β-alumina were also in accord with the aluminum interstitial model once this defect was included (*25*). Similar behavior has since been described in potassium β-alumina (*27*), silver β-alumina (28), and ammonium β-alumina (*29*). It is assumed that the interstitial aluminums are symmetric about the mirror plane, although this cannot be confirmed from a diffraction experiment, and that they coordinate interstitial oxygen (O_i) lying within the mirror plane.

The presence of such interstitial oxygen is, however, considerably more difficult to establish than is that of the displaced aluminum. It is present at only half the interstitial aluminum concentration and, in almost all cases, lies close to a position that, in other unit cells, is occupied by metal ions. In only three cases has positive identification of the interstitial oxygen been claimed: the original neutron diffraction refinement of the sodium β-alumina (*25*), the rerefinement of the sodium β-alumina X-ray data (*23*, *25*), and a refinement of nonstoichiometric silver β-alumina (*28*). In all these refinements the maximum reduction in R factor, when the interstitial oxygen was included from a model containing only the interstitial aluminum and Al(1) vacancies, was 0.05%. Therefore, although the presence of displaced aluminum with tetrahedral coordination completed by interstitial oxygen in the mirror plane provides a convincing and plausible explanation for the mechanism of nonstoichiometry, it should be noted that definite crystallographic evidence for the presence of the interstitial oxygen is fairly marginal. In those cases where is was observed, it was necessary to constrain the occupation to that expected from the known composition.

At first sight, the deviation of the structure of nonstoichiometric β-alumina from the ideal Beevers–Ross formulation is rather minor. It is, however, a crucial part of and essential for the understanding of the intercalation chemistry of the β-aluminas. West described (*30*) the β-alumina structure from the standpoint of electrostatic bond strengths and showed that there is a net excess negative charge in the oxygen layers close to the conducting plane. Spinel block doping by lower-valent cations (for example, Mg^{2+} or Li^{+}), compensation by spinel block aluminum vacancies (which is the limiting case of lower-valent ion doping), and the interstitial defect mechanism actually observed, provide three ways of permitting the introduction of extra positive ions into the conducting plane to provide better overall electroneutrality. Nevertheless, although such considerations (*30*) are helpful for rationalizing the invariable nonstoichiometry of β-aluminas as prepared, they do not make it possible to judge why the O_i–Al_i defect is apparently always observed rather than simple aluminum-vacancy compensation. A deeper understanding of the energetics of the β-alumina structure may be provided by computer simulation methods if realistic interatomic potentials are used for all atoms. The first such calculations are now being performed and, for sodium β-alumina, confirm (*31*) the experimental observation that the interstitial O_i–Al_i defect situation with excess sodium is more stable than the ideal, stoichiometric compound. However, although one particular defect mechanism has been observed for the β-aluminas, the energy differences between the various possibilities may be fairly small, and we discuss below some other β-alumina structure types showing different structural compensation mechanisms.

The mobile ion location within the conducting plane is heavily influenced by the nonstoichiometry. As well as the excess mobile ion population itself, there are also 10% or more of cells containing interstitial oxygen–aluminum bridges. If we consider, as is reasonable, that cells containing a mobile ion on or near the Beevers–Ross position will be only singly occupied, that the alternative mobile ion location is the mid-oxygen position, and that such cells can be triply occupied by the mobile ions except where there is an interstitial oxygen defect, in which case only double occupancy is permitted, then the calculation of the maximum Beevers–Ross occupancy for a particular composition is straightforward. For 25% mobile ion excess, a maximum of 81% of cells can have Beevers–Ross occupation, and this value falls to around 62% for 50% mobile ion excess. These values imply that only 65 and 42%, respectively, of the mobile ions can be found near Beevers–Ross positions.

Table III presents fractional occupancies for several β-aluminas determined using the correct defect model. All refinements refer to compositions in the 25–30% excess range, and for sodium, potassium, and ammonium β-aluminas the similar high fractional Beevers–Ross occupation is noticeable. This suggests that for all these compounds the Beevers–Ross occupancy is maximized, implying triple occupancy ($3Na^+$ or $2Na^+ + O_i^{2-}$) for the remaining cells.

The aBR—O(4) distance, about 2.4 Å, is consistent with the sum of the ionic radii of Na^+ and O^{2-} but the aBR site is clearly less favorable for sodium occupation than the Beevers–Ross and mid-oxygen positions. The very small anti-Beevers–Ross occupancy in sodium β-alumina at room temperature given in Table III was determined by the integration of partial protections of partial Fourier syntheses (*25*), and occupancy was not found at this position in the least-squares refinement. The occupancy increases slightly with temperature as the sodium distribution becomes more smeared out, with 8% occupation being observed at 600°C (*25*). The anti-Beevers–Ross site is a bottleneck site for conductivity and long-range ion transport in the β-alumina lattice, and although the O(4)—O(4) distance is just compatible with sodium ion motion, this separation is more restricting for the larger potassium and ammonium ions, which accounts for their considerably higher activation energies of conduction (Table II). In this context, the 11% anti-Beevers–Ross occupation observed (*27*) for potassium β-alumina (Table III) is surprising; redistribution of this occupation between Beevers–Ross and mid-oxygen sites would give net contents very similar to those observed for the Na^+ and NH_4^+ materials. The anti-Beevers–Ross site is

TABLE III

Fractional Occupation of β-Alumina Mirror-Plane Sites[a]

β-Alumina	BR	BR–mO	mO	aBR	Reference
Na	0.66	—	0.30	0.04	*25*
Ag	0.25	0.44	—	0.31	*25*
Ag	0.32	0.34	—	0.34	*28*
K	0.60	—	0.29	0.11	*27*
NH_4	0.67	—	0.33	—	*29*

[a] All refinements refer to compositions in the range $(1.25–1.30)Na_2O–11Al_2O_3$. The data for Na^+, K^+, and NH_4^+ reveal maximum BR content consistent with the stoichiometry.

empty in $K_{1.62}(Al_{10.38}Mg_{0.62})O_{17}$ and $K_{1.55}(Fe_{10.61}{}^{3+}Fe_{0.31}{}^{2+}\square_{0.08})O_{17}$ as discussed below.

Thus, the crystallographic evidence collected for Na^+, K^+, and $NH_4{}^+$ β-aluminas shows that the Beevers–Ross position, or a site near this, is the favored location for the monovalent cation, implying maximum multiple occupation of mid-oxygen sites in other cells. This is clearly demonstrated in the case of ammonium β-alumina by crystallographic measurements performed after partial thermal decomposition (see below). As discussed above, the relative instability of the anti-Beevers–Ross site is explicable on steric grounds. The mid-oxygen site on the other hand is equivalent to the Beevers–Ross position from the viewpoint of nearest-neighbor oxygen atoms, but the next-nearest-neighbor Al(1) ions in the spinel block will destabilize this site with respect to the Beevers–Ross position.

The mirror-plane occupation for silver β-alumina is rather different from that for the other ions listed in Table III. The two refinements carried out since the discovery of the Al_i defect are in quite good agreement (Table III) and show that roughly one-third of the silver ions lie close to the anti-Beevers–Ross position, which is essentially empty in the other cases. The possibly favorable linear O(4)—Ag—O(4) interaction has been mentioned above. The distribution of the remaining silver ions is also rather unusual, with two sites being occupied in broadly similar concentration. One is close to the Beevers–Ross site, as is found for sodium β-alumina, and the other is near $(\frac{3}{4}, \frac{1}{4}, \frac{1}{4})$ roughly halfway between the Beevers–Ross and mid-oxygen positions. The latter site is not observed for the other ions in Table III and, as discussed below, is in fact the preferred location for silver in the β-alumina framework.

This discussion of relative site occupancies considers only the average structures determined by diffraction experiments. Although long-range order is not observed for the nonstoichiometric β-aluminas as normally prepared, considerable short-range ordering effects are expected and observed in these highly nonstoichiometric materials. Interesting work has been carried out in this area (*32, 33*), and, although short-range order models are presently only tentative, McWhan *et al.* (*32*) observed an increasing degree of frozen-in short-range order of the compensating defects as the defect concentration increased. This led to increasing short-range order among the mobile ions and hence, as discussed above, to a decreasing ionic conductivity. Both melt-grown and flux-grown crystals were studied, as were crystals doped with Co^{2+}, which, like Mg^{2+} (see below), substitutes at the Al(2) (Fig. 1) position (*34*) and decreases the interstitial defect population. A possibly more complicated set of observations for several β-aluminas containing different mobile ions and including Mg^{2+}-doped crystals has recently been given (*35*). Diffuse-scattering studies are providing a rich source of information on the short-range structural chemistry of the β-aluminas.

As already mentioned, an alternative mechanism for increasing the mobile ion content, without the introduction of extra oxygen, is doping with lower-valent ions, and Mg^{2+} has been especially investigated. For spinel block cation doping, $Al_{11-x}Mg_x$, β-alumina has been obtained (*36, 37*) for $0 \leq x \leq 0.6$ and β''-alumina for $0.64 \leq x \leq 0.74$. For a potassium β-alumina, $K_{1.62}(Al_{10.38}Mg_{0.62})O_{17}$ without oxygen defects, bond length criteria showed (*38*) Mg^{2+} to substitute preferentially at Al(2) sites in the interior of the spinel blocks. This is in accord with the behavior of β''-alumina, for which a single-crystal neutron diffraction study (*25, 39*) clearly showed dominant Mg^{2+} substitution at the Al(2) position. It is thought (*39*) that the Al^{3+} ion is too small to occupy this spinel block site without strain and that the strain can be relieved by substitution with the larger Mg^{2+} ion.

The extra alkali ions that are incorporated into the mirror plane as a result of spinel block doping and that are unaccompanied by oxygen defects might not be expected to couple very strongly to the negative defects at Al(2) in the interior of the spinel blocks, and, indeed, experiments (*40*) on Mg-doped polycrystalline sodium β-alumina showed an order of magnitude increase in the maximum room temperature conductivity at a 2 w/o MgO dopant level compared to undoped material. The sodium content corresponding to maximum conductivity agreed well with that calculated to provide compensation for all the magnesium by extra sodium, with no interstitial oxygen content being necessary. A study (*37*) of Mg-doped β-alumina crystals gave similar results. These data and those on undoped nonstoichiometric β-alumina (see above), the conductivity of which decreases with increasing nonstoichiometry (*11, 12*), clearly reveal the effect of the interstitial oxygen defects on binding and blocking the motion of sodium ions.

An alternative mechanism of increasing the mobile ion content without introducing compensating oxide ions is to dope with a higher-valent ion on the anion sublattice. Particularly, AlN–Al_2O_3 mixtures are known to yield spinel-like phases, and N^{3-} doping may be a fruitful field of study in β-aluminas. This could be achieved, for example, by incorporating AlN in the initial reaction mixture.

When the spinel block cation has variable valency, as in the ferrite analogs of β-alumina, an extra mechanism of charge compensation is possible. A recent study (*41*) of potassium β-ferrite revealed no Fe(1) defect but indicated the presence of vacancies on the Fe(2) site and the presence, according to bond length criteria, of Fe^{2+} ions in the spinel block. An overall composition of $K_{1.55}(Fe_{10.61}{}^{3+}Fe_{0.31}{}^{2+}\square_{0.08})O_{17}$ was established (*41*). Yet another defect mechanism is the location of sodium ions in the spinel blocks, which seems to occur in sodium β-gallate (*42*) and in highly sodium-rich β-alumina (*43, 44*). As expected, the sodium ions in the spinel blocks, in contrast to the ions in the conducting plane, are not readily ion-exchanged.

We have reviewed the mirror-plane structure of nonstoichiometric β-aluminas, and below we show how the compensating O_i and Al_i defects can be removed to give more stoichiometric materials. Although this is still a new and relatively poorly characterized area of β-alumina chemistry, more highly nonstoichiometric compositions are even less well studied but may also be a rich source of interesting intercalation chemistry. No crystal structure determinations have yet been reported for undoped material with greater than the 25–30% excess of mobile ions characteristic of melt-grown material, for which the Al_i—O_i—Al_i defect seems to account for all the conducting ion excess. Thus, it is not known how this situation evolves with increasing nonstoichiometry. We have shown above that the preferred Beevers–Ross position must become less occupied relative to the mid-oxygen site as nonstoichiometry increases, and this is possibly one cause of the observed change in crystal structure from β- to β″-alumina for Mg^{2+}-doped sodium β-alumina at a sodium excess greater than 0.66 (*36, 37*). However, as with nearly stoichiometric β-aluminas, possibly metastable compositions can be obtained by lower-temperature processes than by crystal growth from the melt. Nonlabile sodium in $1.82Na_2O–11Al_2O_3$ has been mentioned (*43, 44*), and two processes for reversibly incorporating extra Na_2O into sodium β-alumina of normal nonstoichiometry have been demonstrated (*12, 45*). As discussed above, the activation energy for sodium ion conductivity for a composition close to $2.0Na_2O–11Al_2O_3$, prepared from $1.25Na_2O–11Al_2O_3$ is greater (0.30 eV) than for the less nonstoichiometric precursor (0.13 eV), and the conductivity is about an order of magnitude lower, even at 300°C (*12*).

Finally, we note that all studies of ion mobility so far reported have discussed cation motion within the conduction planes. Kummer described (*4*) irradiation experiments on single crystals which showed the mobility of sodium ions in sodium β-alumina to be many orders of magnitude slower parallel to the *c*-axis than in the conduction plane. There are no studies to date that indicate the magnitude of the oxygen anion mobility in the conduction plane, although this must be involved in Na_2O incorporation (*12, 45*). Whittingham and Huggins found (*21*) no pO_2 dependence of the total conductivity of silver β-alumina up to 800°C, implying no measurable stoichiometry variation up to this temperature. Interstitial oxygen defect motion could nevertheless be a factor at some point above room temperature, especially if Al_i–Al(1) hopping were possible. Moreover, a recent Mössbauer study (*46*) of potassium β-ferrite–β-gallate solid solutions indicated oxygen uptake even at room temperature. In this case extra alkali is not involved since the extra oxygen is compensated by the oxidation of spinel block Fe(II) to Fe(III).

III. HYDROGEN-CONTAINING β-ALUMINAS

The principal mechanism by which the stoichiometry of β-aluminas may be changed involves the action of hydrogen. Considerable evidence has accumulated that, at least above 400°C, hydrogen may preferentially remove the interstitial oxygen defect as water vapor, yielding a more nearly stoichiometric material than can be obtained by normal chemical synthesis. By the use of Raman spectroscopy, a band at 250 cm^{-1} was identified (*47*) as arising from a vibrational mode of the Al(3)—O(5)—Al(3) unit. In nonstoichiometric samples a shoulder at 235 cm^{-1} was interpreted (*47*) as being an analogous vibration of the interstitial Al_i—O_i—Al_i group. This lower frequency indicates a somewhat weaker binding of the interstitial defect in the lattice compared to the regular O(5)—Al(3) structural unit, but by itself this is probably not sufficient to explain the interval of at least 250°C between the ability of hydrogen to remove interstitial oxygen and the onset of thermal decomposition of hydrogen β-alumina (*13*) with, presumably, O(5) loss. Probably, the relative lability of interstitial oxygen is associated with the ability of the interstitial aluminum atoms to return to the vacant Al(1) positions. Structural work, discussed below, on well-characterized stoichiometric or nearly stoichiometric β-aluminas shows an absence of, or certainly a significant decrease in, the interstitial aluminum population. It might follow, therefore, that the lower temperature limit of interstitial oxygen loss is set by the activation energy of motion for Al_i to Al(1), and it would be interesting to observe the effect of changing the spinel block cation (for example, Ga^{3+}, Fe^{3+}) on this decomposition temperature. By displacement into the spinel block, Al(3) atoms could also find a site with octahedral coordination, but, in contrast to the situation for Al(1) atoms, such a displaced Al(3) would face-share three O(1) atoms with a tetrahedrally coordinated Al(2) (Fig. 1), which would not provide a coulombically favorable situation.

The interstitial oxygen loss and the formation of more nearly stoichiometric β-aluminas has been reported in two ways. One involves the action of hydrogen gas on silver β-alumina (*14, 15, 48*); the other occurs via the thermal decomposition of ammonium β-alumina (*13, 49, 50*). As discussed below, the former route has successfully been used to prepare stoichiometric $DAl_{11}O_{17}$ and $AgAl_{11}O_{17}$. However, it is restricted, at least from the standpoint of conductivity and structural studies, to polycrystalline samples because of the plating-out of metallic silver that occurs during hydrogenation. The thermal decomposition of ammonium β-alumina can be studied much more conveniently using single crystals, but it appears that the formation of hydrogen β-alumina by this route is not accomplished in such a structurally well-ordered fashion as from the silver compound.

Detailed study of the structure and decomposition of single-crystal ammonium β-alumina (*13, 29, 51*), has revealed several interesting aspects of β-alumina intercalation chemistry in addition to the formation of more stoichiometric material. We discuss below the chemistry of ammonium β-alumina and outline the formation and properties of hydrogen β-aluminas. First, however, some properties of hydrogen β-aluminas prepared in sulfuric acid are discussed. During this reaction a nonstoichiometric composition seems to be maintained.

A. Action of Sulfuric Acid on Sodium β-Alumina

The formation of hydrogen-containing β-aluminas was first demonstrated by Saalfeld *et al.* (*52*), who showed that the action of hot concentrated sulfuric acid on sodium β-alumina permitted the replacement of the sodium by hydrogen-containing species.

It has become customary to describe the chemistry of the conduction planes of hydrogen-containing β-aluminas in terms of H^+, H_2O, and H_3O^+ units, and, particularly when moisture absorption has been suspected, the term hydronium (H_3O^+) β-alumina has often been used. Although infrared bands of hydrogen-containing β-aluminas prepared from sulfuric acid were assigned (*53*) to H_3O^+ vibrations or to more complex units, the presence of H_3O^+ was not confirmed by broad-line NMR (*54*). Hydrogen-containing β-alumina undergoes a reversible moisture absorption–desorption process between 200° and 300°C (*52, 55*). Below this temperature, the NMR data (*54*) did not distinguish the presence of H_3O^+ ions and separate H^+ and H_2O species; above it, isolated immobile H^+ species were strongly indicated. The latter observation agrees well with the structural data on $DAl_{11}O_{17}$ and $HAl_{11}O_{17}$ described below.

Recent infrared studies (*56*) confirm the presence of only H^+—O^{2-}(lattice) interactions in anhydrous hydrogen β-alumina, whether prepared from silver β-alumina or from ammonium β-alumina. The absorption of water from the atmosphere causes perturbation of O—H stretching vibrations as a result of weak hydrogen bonding (*56*) but the presence of relatively isolated HO^-(lattice) and H_2O units is still indicated. There thus appears to be no definite evidence for the presence of significant concentrations of H_3O^+ ions in hydrogen-containing β-aluminas, and they are almost certainly not present in the high-temperature dehydrated forms. Structural work revealing protons as well as oxygen positions is clearly urgently required for hydrated hydrogen β-aluminas. Until such studies are performed, it is proposed that H^+/H_2O is possibly a more satisfactory representation for the conduction-plane composition $H^+ + H_2O$ than is H_3O^+.

The formation, stability, and ionic conductivity of hydrogen-containing β-aluminas prepared via the sulfuric acid route have been thoroughly studied by Breiter, Farrington, and Roth (*55, 57, 58*). The acid has a fairly corrosive effect on the bulk material as well as effecting ion exchange (*58*) unless conditions are chosen with particular care (*59*); however, similar results were obtained (*55, 57, 58*) with polycrystalline and single-crystal starting materials of approximate composition $1.33Na_2O—11Al_2O_3$. Thermogravimetry of fully exchanged samples indicated an overall stoichiometry of approximately $4H_2O{\cdot}11Al_2O_3$ below about 200°C and a composition in the range $(2–2.5)H_2O{\cdot}11Al_2O_3$ above 300°C after water loss. These conclusions are somewhat different from the original results of Saalfeld *et al.* (*52*), who indicated an upper composition limit at lower temperatures of $3H_2O{\cdot}11Al_2O_3$(stoichiometric $H^+/H_2O{\cdot}Al_{11}O_{17}$), which was found to dehydrate almost completely above 300°C, yielding a composition $H_{1.0}{}^+(H_2O)_{0.1}Al_{11}O_{17}$ (that is, $1.2H_2O{\cdot}11Al_2O_3$). It is not clear whether the differences are real, reflecting perhaps different starting materials used in the two cases. The careful thermogravimetric studies carried out in the more recent work (*55, 57*) do suggest strongly, however, that in this case nonstoichiometry is maintained during the exchange.

Infrared measurements on the high-temperature ($T > 300$°C) materials would be most interesting. Although it was shown (*55*) that the polycrystalline hydrogen-exchanged material was an avid absorber of moisture at lower temperatures, which would make infrared work at room temperature difficult, work on deammoniated ammonium β-alumina single crystals (*13, 56*) shows that the use of single crystals reduces this problem very considerably. It should thus be possible to search for a band at 3548 cm^{-1}, which, in studies of hydrogen β-alumina prepared by other routes (*56*), is strongly indicated as arising from an $H—O_i$ stretching vibration. This band should be present in nonstoichiometric material with the interstitial oxygens still present, but in stoichiometric hydrogen β-alumina only a band near 3510 cm^{-1}, characteristic of the H—O(5) vibration (*56*) should be observed. These assignments will be much easier to make in high-temperature, relatively water-free material compared to the low-temperature situation for $3H_2O{\cdot}11Al_2O_3$ or $4H_2O{\cdot}11Al_2O_3$ in which these bands are strongly modified by hydrogen bonding with intercalated water molecules.

The composition $4H_2O{\cdot}11Al_2O_3$ observed (*55*) at low temperature is consistent with the replacement of every sodium ion in the starting material with a proton and a water molecule, but with the original degree of nonstoichiometry maintained. As originally pointed out (*55, 57*), the high-temperature composition indicates only partial water loss. Some possible reaction schemes, assuming non-stoichiometry is maintained, are shown in Table IV. In each case it is assumed that one proton is chemically bound to O(5), as found in anhydrous hydrogen β-alumina (see below), and that the remaining material is partitioned as water molecules and protons bonded to interstitial or other framework oxygens. Even in the low-temperature material [reaction (a)] it should be noted that, because of the presence of interstitial oxygen, the population of oxygen atoms (of water molecules) presumed to lie on or close to the Beevers–Ross site will be less than unity. Alternatively, the formulae can be recast to be consistent with the formation of stoichiometric material (*60*). The loss of only $0.5H_2O$ per formula unit around 300°C (cf Table IV) to give $3H_2O–11Al_2O_3$ would be consistent

TABLE IV

Some Possible Reactions Describing the Preparation and Thermal Decomposition of Nonstoichiometric Hydrogen-Containing β-Aluminas Prepared by the Action of Sulfuric Acid on Sodium β-Alumina

(a) $Na_{1\frac{1}{3}}Al_{11}O_{17\frac{1}{6}} \xrightarrow[300°C\ (+H_2O)]{H_2SO_4} (H^+/H_2O)_{1\frac{1}{3}}Al_{11}O_{17\frac{1}{6}} \equiv 4H_2O \cdot 11Al_2O_3$

Possible mirror-plane occupation of $4H_2O \cdot 11Al_2O_3$:

H—O(5)	1.0
H_2O	$1\frac{1}{3}$
$O_i—H_2$	$\frac{1}{6}$[a]

Maximum BR occupancy of H_2O: 0.75

(b)[b] $(H^+/H_2O)_{1\frac{1}{3}}Al_{11}O_{17\frac{1}{6}} - \frac{2}{3}H_2O \xrightarrow[T > 300°C]{Heat} H_{1\frac{1}{3}}{}^+(H_2O)_{\frac{2}{3}}Al_{11}O_{17\frac{1}{6}} \equiv 2\frac{2}{3}H_2O \cdot 11Al_2O_3$

Possible mirror-plane occupation of $2\frac{2}{3}H_2O \cdot 11Al_2O_3$:

H—O(5)	1.0
H_2O	$\frac{2}{3}$(BR)
$O_i—H_2$	$\frac{1}{6}$[a]

(c)[b] $(H^+/H_2O)_{1\frac{1}{3}}Al_{11}O_{17\frac{1}{6}} - H_2O \xrightarrow[T > 300°C]{Heat} H_{1\frac{1}{3}}{}^+(H_2O)_{\frac{1}{3}}Al_{11}O_{17\frac{1}{6}} \equiv 2H_2O \cdot 11Al_2O_3$

Possible mirror-plane occupation of $2H_2O \cdot 11Al_2O_3$:

H—O(5)	1.0
H_2O	$\frac{1}{3}$(BR)
$O_i—H_2$	$\frac{1}{6}$[a]

[a] A single O_i—H bond may be more reasonable. This would imply one-sixth of the cells containing, possibly, O(2)—H or O(4)—H bonds.

[b] The loss of between two-thirds and one H_2O seems to correspond to the results described by Roth *et al.* (*55*).

with the formation of stoichiometric $(H^+/H_2O)_{1.0}Al_{11}O_{17}$. The formula $2H_2O{\cdot}11Al_2O_3$ is equivalent to $H_{1.0}{}^+(H_2O)_{0.5}Al_{11}O_{17}$.

The thermogravimetric analysis (*55*, *57*) indicated reversible moisture absorption and desorption between room temperature and 600°C. Above this temperature, the water content diminished and an anhydrous aluminum oxide resulted by heating to 900°C. If, as seems possible, nonstoichiometry is preserved during the initial exchange, these results imply that this situation is maintained until at least 600°C, in contrast to the formation of stoichiometric hydrogen β-alumina at this temperature by the other routes. Such a difference could reflect the lack of reducing conditions, possibly necessary for the removal of interstitial oxygen as water, that pertain in the decomposition of ammonium β-alumina. There is no doubt that the details of the chemistry of this hydrogen-containing β-alumina remain poorly understood. It would be most interesting to characterize in detail on single crystals the structural evolution to 700°C of hydrogen-containing β-alumina prepared from sulfuric acid and to examine the effect of hydrogen gas on the decomposition. The decomposition of ammonium β-alumina was studied (*13*) almost entirely in hydrogen atmospheres in order to maximize the probability of interstitial oxygen loss by analogy with the results with silver β-alumina, but this has not been the case in the studies of the decomposition of material prepared by the sulfuric acid route. The effect of variations in atmosphere on the decomposition of ammonium β-alumina has not been studied in any systematic way.

In common with hydrogen β-alumina prepared by the other routes, there appears to be an irreversible thermal composition around 700°C (*55*, *57*). It has often been assumed that for hydrogen β-alumina this irreversible water loss occurs in a single step, in contrast to an irreversible partial water loss that occurs around 400°C (*37*, *54*) in hydrogen β''-Al_2O_3. However, work on the thermal decomposition of ammonium β-alumina (*13*) and also the results on hydrogen β-alumina prepared by the sulfuric acid route (*55*) reveal that this is not, in fact, the case. For hydrogen β-alumina, as well as for hydrogen β″-alumina, the loss of water is gradual, taking place in the former case between somewhere below 700°C and about 900°C.

The crystal structure of a partially reversibly dehydrated composition prepared in the earlier work (*52*) was has been reported (*61*). Oxygen was located close to the Beevers–Ross position (Fig. 1), and the composition determined from the crystal structure analysis was $1.79H_2O{\cdot}11Al_2O_3$ or $H_{1.0}{}^+(H_2O)_{0.395}Al_{11}O_{17}$, assuming a stoichiometric composition. This composition is reasonably close to that observed above 300°C in the recent work (*55*, *57*). The X-ray data were rerefined (*25*) to search specifically for interstitial aluminum and oxygen, but no evidence for them was found, which possibly confirms the more stoichiometric composition, at least in this case.

The oxygen atom of the water molecule was located exactly on the Beevers–Ross site in the later refinement, but both refinements revealed an unusual feature in that the O(5) ion was displaced off its ideal site toward the Beevers–Ross positions rather than toward the anti-Beevers–Ross positions, as is normally the case if a displacement is observed. This has the effect of lowering the O(5)—BR distance to 2.84 Å and is strongly indicative of weak hydrogen bonding between oxygen atoms on these two sites. Such an O · · · O separation is consistent with the lowering of the O(5)—H stretching frequency from 3510 cm^{-1} as observed (*56*) in anhydrous hydrogen β-alumina to around 3400 cm^{-1}, as is found after moisture pickup (*56*). Although proton positions were not determined in the structure refinement, such an interaction was indeed suggested by Kato and Saalfeld (*61*). Also, as has been pointed out (*14*), the Al(3)—O(5) distance observed (*61*) is anomalously long compared to that in most β-aluminas but is almost identical to the value observed in $DAl_{11}O_{17}$ (*14, 15*), where the presence of deuterium chemically bonded to O(5) is directly observed (see below). Similar O(5) displacements are observed in polycrystalline anhydrous $DAl_{11}O_{17}$ (*15*) and single-crystal anhydrous $HAl_{11}O_{17}$ (*51*), although in these cases, the hydrogen bond to O(2) (out of the mirror plane), which determines the O(5)—H orientation, is weaker than for the hydrated material.

The evidence, seems therefore, to support the H^+/H_2O formulation quite strongly in the case of the material studied by Kato and Saalfeld (*61*). In general, however, the chemistry and structure in the mirror plane of hydrogen-containing β-aluminas prepared by the sulfuric acid route are so far poorly characterized. Neutron diffraction studies to reveal proton positions, and detailed characterization of the thermal decomposition of well-characterized single crystals, would be particularly useful. One study (*62*) of hydrogen-containing β″-alumina prepared via the sulfuric acid route has revealed protons bonded to O(2) and O(4), although the total occupation of the mirror plane was not definitively established.

The *c*-axis lattice constants measured for hydrogen-containing β-aluminas prepared using sulfuric acid are 22.70 Å (*61*) and 22.71 Å (*55*), in very close agreement with one another and somewhat greater than observed for anhydrous hydrogen β-alumina (Table I). The larger values are probably consistent with the presence of water molecules. In contrast, a hydrogen β-alumina single crystal prepared from thermally decomposed ammonium β-alumina had $c = 22.62$ Å (*50*). In addition, this material had a short Al(3)—O(5) distance, possibly arguing against an H—O(5) interaction. However, a Beevers–Ross occupation greater than unity was observed (*50*), which probably reflected incomplete exchange and the presence of some sodium. The crystal studied may not, therefore, be characteristic of fully exchanged material. The chemistry and structure of cointercalated β-aluminas (*18*, *63*) is nevertheless proving to be an interesting area of β-alumina science (see Ref. 7).

The ionic conductivity of hydrogen-containing β-alumina single crystals prepared via the sulfuric acid route (*58*) supports the presence of nonconducting water molecules and relatively immobile protons chemically bonded to the framework or interstitial oxygen atoms. The concentration of water molecules in the conduction plane is not sufficiently high to facilitate rapid proton transfer, in contrast, possibly, to H^+/H_2O β″-alumina (*64*), and the measured conductivity therefore reflects the intrinsic proton mobility in hydrogen β-alumina, which, as would be expected, is very low (*13*). The room temperature conductivity of crystals prepared from sulfuric acid was about $10^{-11}\ \Omega^{-1}\ cm^{-1}$ (*58*). Between room temperature and 200°C an activation energy of 0.78 eV was measured (*58*) (Fig. 2). A considerable decrease in conductivity occurred as a result of water loss between 200° and 300°C, followed by a rise to 550°C with an activation energy of 1.3 eV. The conductivity at 500°C was about $10^{-6}\ \Omega^{-1}\ cm^{-1}$.

B. Ammonium β-Alumina

Tables I and II show that the *c*-axis lattice constant (22.88 Å) and the room temperature conductivity of ammonium β-alumina are, respectively, significantly higher and lower than the values for both sodium and potassium β-aluminas. Although it might be supposed that this is a straightforward reflection of the greater effective ionic radius of the ammonium ion than of K^+ and Na^+, recent work has shown (*51*) that the large *c*-axis lattice constant is a result of the nonstoichiometry rather than of the size of the $NH_4{}^+$ ion and stems from the necessity for some $NH_4{}^+$ ions to occupy unfavorable mirror-plane sites with a consequent hydrogen-bonding scheme that forces apart the spinel blocks.

Loss of the nonstoichiometric excess of NH_3 permits the optimum location of ammonium ions on Beevers–Ross sites, and the *c* axis is reduced considerably to 22.66 Å (*51*). This value is, in fact, less than that found for stoichiometric potassium β-alumina [about 22.68 Å]. Indeed, the potassium and ammonium ions are probably almost ideally sized to reside in the β-alumina lattice at the Beevers–Ross position. This favorable location, coupled with the bottleneck of these large ions at the anti-Beevers–Ross site, entails very low ionic conductivities for the stoichiometric materials. Stoichiometric potassium β-alumina is probably an ionic insulator (*65*), and the ionic conductivity of ammonium β-alumina drops by many orders of

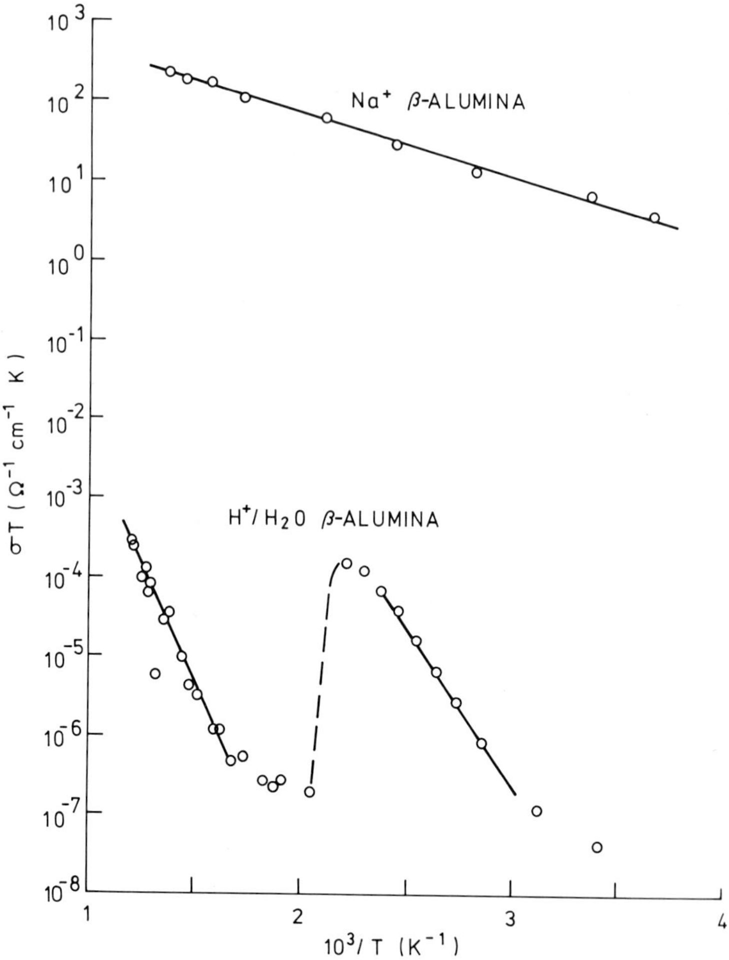

Fig. 2. Ionic conductivity of H^+/H_2O β-alumina prepared by the action of sulfuric acid on sodium β-alumina. The conductivity of the starting material is shown for comparison. [With permission from Farrington *et al.* (*58*).]

magnitude at the onset of thermal decomposition as the stoichiometric excess of NH_4^+ is removed (*13*) (Fig. 3).

An X-ray determination of the crystal structure of ammonium β-alumina was reported (*50*), and two main locations of the nitrogen atoms were found at the Beevers–Ross position and near the mid-oxygen position. Hydrogen-bonding schemes involving spinel block oxygen atoms and O(5) were proposed (*50*). The *c*-axis lattice constant was small, 22.81 Å, which

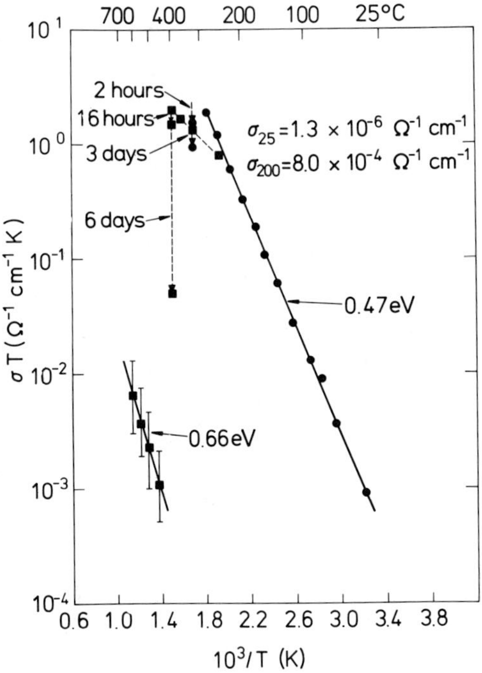

Fig. 3. Ionic conductivity of ammonium β-alumina as a function of temperature (*13*). The effect of thermal decomposition is clear. Further work has shown that the greater part of the conductivity decrease from that characteristic of nonstoichiometric ammonium β-alumina (low temperature) to that characteristic of hydrogen β-alumina (high temperature) occurs by 350°C (see text). In this later work, the conductivity near 500°C of almost fully exchanged material after complete ammonia loss is close (within an order of magnitude) to that of H^+/H_2O β-alumina in Fig. 2. Key: ●, gold electrodes; ■, platinum electrodes.

probably reflected incomplete exchange of sodium, as was indeed proposed (*50*) for the hydrogen-containing β-alumina studied in the same work (see above). In addition, a small anti-Beevers–Ross population was observed.

A detailed study of the exchange of NH_4^+ for Na^+ in molten ammonium nitrate at 210°C has since been made (*13*), and for crystals of 2–3 mm thickness in the basal plane, exposures of many weeks are required to achieve essentially complete exchange. Monitoring of the *c*-axis lattice constant across an exchanged crystal using a fine-focus high-accuracy goniometer (*66*) is an effective means of monitoring the degree of exchange. For completely exchanged Union Carbide crystals of composition $1.25[(NH_4)_2O]–11Al_2O_3$, a uniform lattice constant of 22.88 Å is observed across the crystal, but for incomplete exchange this is not the case. Some data are shown in Fig. 4. The

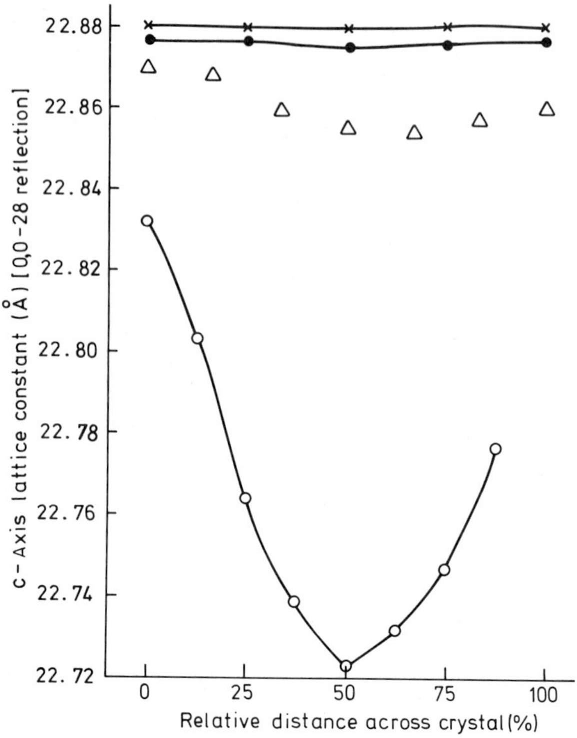

Fig. 4. *c*-Axis lattice constants for ammonium β-alumina crystals as a function of the relative distance across the crystal, showing the effect of the degree of exchange of ammonium for sodium ions (determined by chemical analysis). The data were collected using the APEX goniometer (*66*). Key: ●, 99.5% exchange, boule 1; ○, 75% exchange, boule 1; ×, 99.8% exchange boule 2; △, 88% exchange, boule 2.

degree of exchange corresponding to the curves observed was determined by chemical analysis for Na^+, Al^{3+}, and NH_4^+.

A neutron diffraction study (*29*) of a single crystal of fully exchanged ammonium β-alumina confirmed the locations of ammonium ions and the hydrogen-bonding scheme originally proposed (*50*), with the exception of anti-Beevers–Ross population, which is not observed. In the incompletely exchanged material, the observed anti-Beevers–Ross population (*50*) could be sodium ions; the increased lattice constant relative to sodium β-alumina itself would render the anti-Beevers–Ross site less unfavorable for sodium than is the case in the pure material (Table III).

The two NH_4^+ locations are shown in Fig. 5. The N(2)···O(4) hydrogen bonds are particularly short [2.664(8) Å (*29*)]. The consequences of this are that the $N(2)H_4^+$ ammonium ion is strongly hydrogen-bonded and

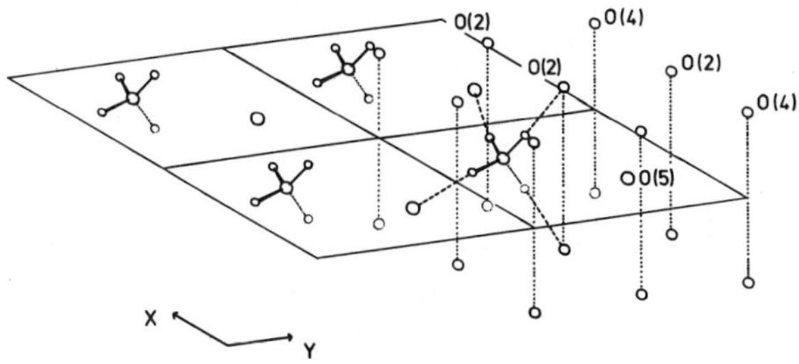

(a) N(1) Position (Beevers-Ross site)

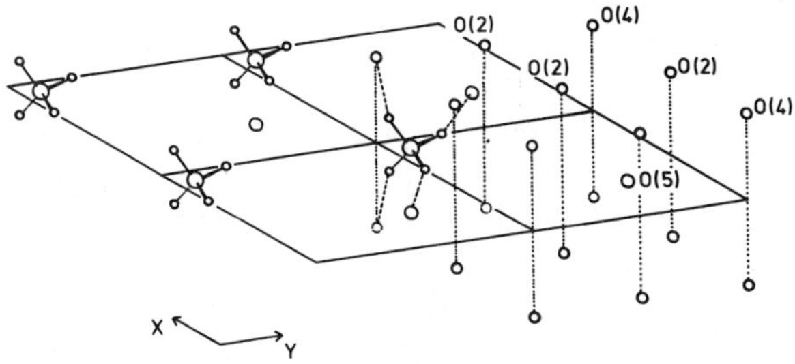

(b) N(2) Position

Fig. 5. The locations of ammonium ions in ammonium β-alumina. The $N(1)H_4^+$ ion (a) is centered at the Beevers–Ross position and is hydrogen-bonded to two O(2) ions above and below the conducting plane and to two O(5) ions in the plane. One of the rotationally equivalent orientations is shown. The $N(2)H_4^+$ ion (b) is centered between the mid-oxygen and anti-Beevers–Ross positions, allowing hydrogen bonding to two O(4) atoms above and below the plane and two O(5) ions within the plane. In contrast to the situation for $N(1)H_4^+$ ion, there is no possibility of rotational disorder. [Reproduced with permission from Newsam (*60*).]

does not take part in the conduction process, but it is unstable with respect to the $N(1)H_4^+$ ammonium ion because of the necessity of forcing apart the spinel blocks to accommodate the short hydrogen bonds.

Previous work (*49*, *50*) showed that ammonia loss occurs on heating between 300° and 600°C, with a maximum rate of weight loss near 450°C. More recent work has revealed (*13*) that this decomposition is not a simple one-stage process but that, by heating to 350°C, an intermediate phase of composition close to $(NH_4)_{1.0}H_{0.25}Al_{11}O_{17.125}$ may be obtained. The large decrease in ionic conductivity observed (*49*, *50*) on heating ammonium β-alumina in fact occurs at this stage (*13*). Loss of translucency and macroscopic fracturing of ammonium β-alumina crystals, which is observed as the initial decomposition commences between 250° and 300°C (*13*), is probably caused (*51*) by the decomposition of the unstable, but within the conduction plane immobile, $N(2)H_4^+$.

The details of the complete thermal decomposition of ammonium β-alumina are rather complex but, by annealing at 650°C or below in hydrogen gas, nearly stoichiometric hydrogen β-alumina may be obtained (*13*). A neutron diffraction study (*51*) of a single crystal obtained in this way revealed a hydrogen atom, bonded to O(5), in the conduction plane, in agreement with the results on polycrystalline $DAl_{11}O_{17}$ prepared from silver β-alumina (see below). The *c*-axis lattice constant of the crystal, 22.657(9) Å (*51*), is the same as that found for material still containing a considerable quantity of NH_4^+ (*13*, *51*) and is somewhat greater than that found for hydrogen β-alumina prepared by the silver route (Table I). The origin of this difference, which is consistently reproducible, is not yet clear, but both the structure determination (*51*) and subsequent ion-exchange experiments with alkali ions indicate that hydrogen β-alumina prepared by the thermal decomposition of ammonium β-alumina, at least in single-crystal form, is defective in some way. Possibly, complete thermal decomposition of NH_4^+ cannot be achieved without the loss of some O(5) ions also. As discussed below, this problem affects approaches to the preparation of more nearly stoichiometric β-aluminas, and electron lattice images do show some loss of structure in ammonium β-alumina crystals annealed below 650°C.

The original work (*49*, *50*) demonstrated the formation of stoichiometric hydrogen β-alumina by the removal of interstitial oxygen as water during the thermal decomposition of ammonium β-alumina, the hydrogen arising from thermal decomposition of ammonia. This is shown in Table V along with other decomposition reactions of ammonium β-alumina. By ion exchange of ammonium β-alumina equilibrated at 350°C, for example, mixed proton–alkali β-aluminas can be obtained, as is shown for silver in Table V. Such materials might be difficult to prepare by any other route. By varying the degree of decomposition, it should be possible to vary the alkali-to-hydrogen content over a wide range. In the case of reexchange with sodium, such materials might be of interest with regard to the action of water on sodium β-alumina. It is well established that water inhibits the ionic conductivity of sodium β-alumina, but the details of the precise mechanism by which this occurs are not clear (*67–69*). One possible reaction involves the formation of sodium hydroxide with incorporation of protons into the lattice:

$$y\mathrm{H_2O} + \mathrm{Na_{1+x}Al_{11}O_{17+x/2}} \rightarrow \mathrm{Na_{1+x-y}H_yAl_{11}O_{17+x/2}} + y\mathrm{NaOH}$$

This reaction is very difficult to follow directly because of the difficulties of establishing the presence of sodium hydroxide in small quantities and of achieving a homogeneous reaction with single crystals. It should be quite straightforward, however, to prepare such mixed sodium–hydrogen β-aluminas and to follow any subsequent moisture pickup by utilizing partially deammoniated β-alumina as a starting material.

TABLE V

Some Reactions of Ammonium β-Alumina and Its Decomposition Products[a]

$(NH_4)_{1+x}Al_{11}O_{17+x/2} \xrightarrow{350°C} (NH_4)_{1.0}H_xAl_{11}O_{17+x/2} + xNH_3$
$(NH_4)_{1.0}H_xAl_{11}O_{17+x/2} \xrightarrow[AgNO_3]{350°C} Ag_{1.0}H_xAl_{11}O_{17+x/2}$
$(NH_4)_{1+x}Al_{11}O_{17+x/2} \xrightarrow[T \leq 630°C,\ vac]{T \leq 650°C/H_2} HAl_{11}O_{17} + (1 + x)NH_3 + (x/2)H_2O$
$HAl_{11}O_{17} \xrightarrow[NaNO_3/450°C]{AgNO_3/350°C} MAl_{11}O_{17}$ (M = Ag, Na etc.)
$HAl_{11}O_{17} \xrightarrow[H_2]{670°C \leq T \leq 900°C} H_{1-x}Al_{11}O_{17-x/2} + (x/2)H_2O$ $(0 \leq x \leq 1)$
$(1 + x)NaAl_{11}O_{17} \xrightarrow{T \sim 1100°C} Na_{1+x}Al_{11}O_{17+x/2} + (11x/2)Al_2O_3(\alpha)$
$Al_{11}O_{16.5}(\beta) \xrightarrow{T \sim 1050°C} Al_{11}O_{16.5}(\alpha)$

[a] As discussed in the text, some of these reactions as given may be idealized.

Although the possible problem of disorder in hydrogen β-alumina prepared from ammonium β-alumina has been mentioned, the actual formation of a stoichiometric or nearly stoichiometric material is not in doubt. The infrared spectrum of ammonium-free anhydrous hydrogen β-alumina is shown in Fig. 6. As expected, the dominant feature is a sharp band near 3510 cm^{-1}, characteristic of the O(5)—H stretch. No vibrations in the 1500–1700 cm^{-1} region are observed, which is consistent with the absence of H_2O or H_3O^+ species, but an H—O(5) lattice mode is observed at 1165 cm^{-1}. In all our experiments (*13*) on hydrogen β-alumina that has been prepared from the ammonium material and that has not been heated above 650°C, a weak band near 3385 cm^{-1} is observed; the relative intensity in Fig. 6a is characteristic. This band has been assigned (*56*) to perturbed O(5)—H species; whether its presence reflects an intrinsic disorder in material prepared by this route or is simply the result of pickup of small amounts of moisture during the handling process is not yet clear.

Above 670°C in hydrogen or 650°C in vacuum (*13*), hydrogen β-alumina suffers an irreversible water loss. The *c*-axis lattice constant increases to

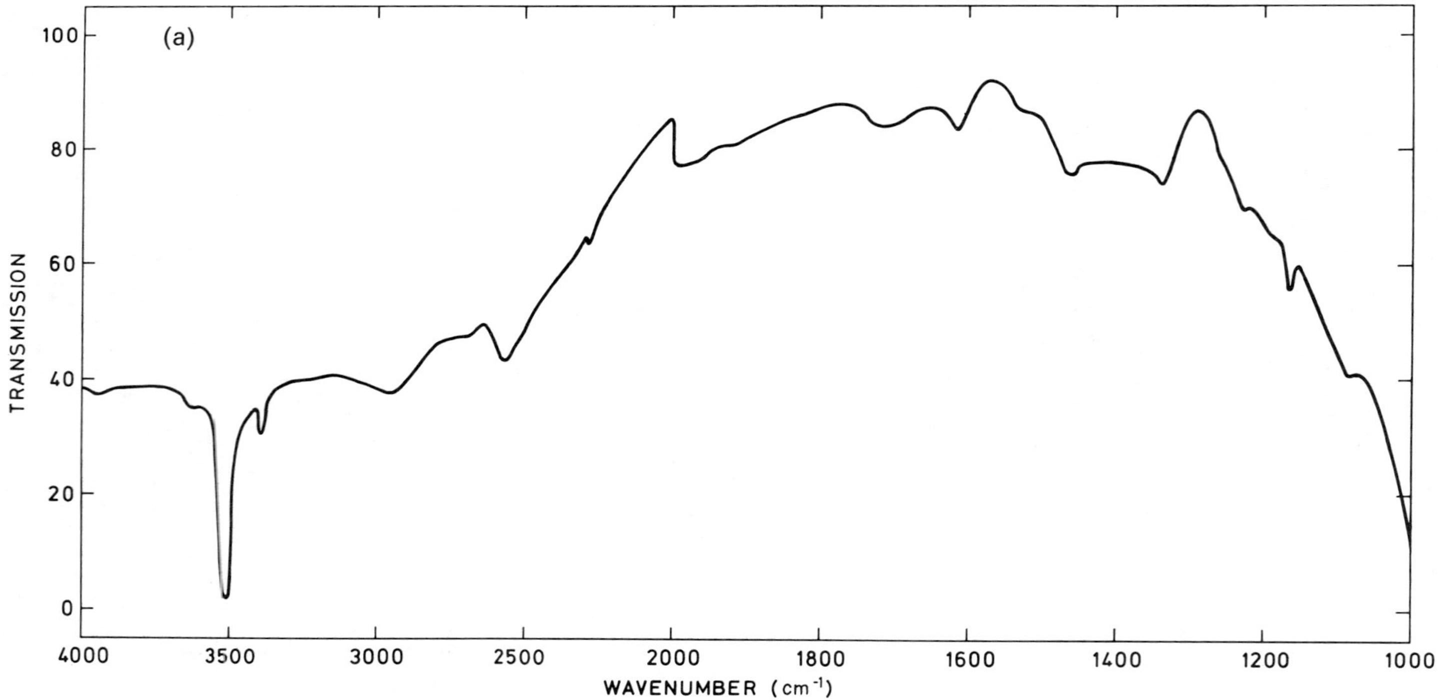

Fig. 6. Near-IR spectra of ammonium β-alumina single crystals after thermal treatments to remove all the NH_3 ($E \perp c$). (a) Spectrum of a crystal heated for 4 days in dry hydrogen at 630°C, i.e., below the irreversible decomposition temperature. (b) Spectrum of a crystal heated above this point to 800°C for 24 hr. In both cases the dominant band is that assigned to the H—O(5) stretch near 3510 cm^{-1} (*56*). Spectrum (a) is characteristic in showing a weak band near 3385 cm^{-1} arising from a perturbated H—O(5) species and related broad bands at lower frequency, possibly characteristic of H_2O. The lattice O(5)—H mode at 1165 cm^{-1} is also visible. Spectrum (b) shows no extra bands.

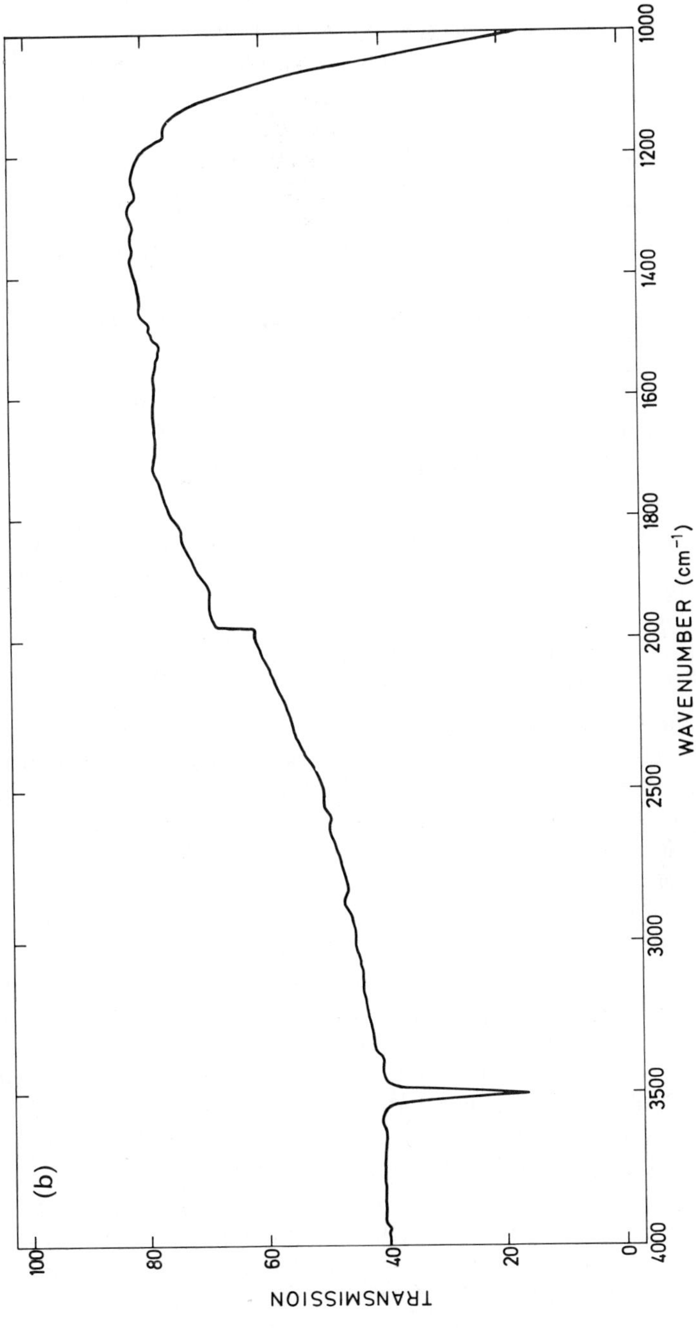
(b)
TRANSMISSION
100
80
60
40
20
0
4000
3500
3000
2500
2000
1800
1600
1400
1200
1000
WAVENUMBER (cm^{-1})

22.8–22.9 Å, and a large increase in line width indicates considerable structural disorder (*13*). In material still containing sodium, a second phase characteristic of stoichiometric sodium β-alumina (c = 22.59 Å) appears at this point. The initial decomposition step is accompanied by only partial, probably not more than 50%, water loss (Table V), and infrared spectroscopy indicates that there is still a considerable proton population for materials equilibrated at 800°C in hydrogen (Fig. 6b). The possible similarity to the decomposition observed at lower temperature in hydrogen β''-alumina has been mentioned above. The disorder in the partially decomposed β-alumina phases above the initial decomposition temperature makes structural characterization rather difficult, and none has been reported so far. However, it would be of considerable interest to determine how the β-alumina framework is modified as irreversible water loss takes place.

Water loss is essentially complete at 900°C (*13*, *55*). Structural transformation of the anhydrous metastable β-alumina, $Al_{11}O_{16.5}$, to α-Al_2O_3 (Table V) occurs at about 1050°C (*13*, *49*, *50*). For completely exchanged material, only α-Al_2O_3 lines will be observed in the diffraction pattern of a material heated to 1200°C. If residual sodium is present, β-alumina lines will also be observed. This provides a simple, albeit destructive test for the completeness of exchange in ammonium β-alumina preparations.

C. Action of Hydrogen on Silver β-Alumina

Kummer (*4*) demonstrated that hydrogen β-alumina could be prepared by the action of hydrogen on silver β-alumina at 300°C. Hydrogen incorporation into the lattice was compensated by plating-out of silver metal. Infrared spectra similar to those shown in Fig. 6 were obtained, but no details of the degree of exchange or of the structure of hydrogen β-alumina were given.

We reinvestigated this reaction at somewhat higher temperatures. With single crystals of 2–3 mm dimension in the basal plane, the rate of exchange is very slow (1–2 weeks at 550°C for complete exchange), as expected from the very low mobility of the proton bonded to O(5). Infrared data have been obtained on crystals obtained in this way, and bands at 3510 and 3548 cm^{-1} representing H—O(5) and H—O_i, respectively, have been observed (*56*). The latter disappears on complete exchange. This observation supports the other evidence which reveals that this reaction also produces a stoichiometric hydrogen β-alumina from a nonstoichiometric starting material.

With single crystals, the slowness of the exchange and the plating-out of silver within the interior of the crystal, with consequent crystal degradation, yield material unsuitable for structural study, although the c-axis lattice constant is in agreement with that determined from diffraction measurements

TABLE VI

Preparation of Hydrogen β-Alumina by the Action of Hydrogen Gas on Silver β-Alumina and Some Reactions

$$Ag_{1+x}Al_{11}O_{17+x/2} + \tfrac{1}{2}(1 + x)H_2 \xrightarrow[-550°C]{400°C} HAl_{11}O_{17} + (1 + x)Ag + (x/2)H_2O\uparrow$$

$$HAl_{11}O_{17} + (1 + x)Ag \xrightarrow[550°C]{\text{vacuum}} AgAl_{11}O_{17} + xAg + \tfrac{1}{2}H_2\uparrow$$

$$HAl_{11}O_{17} + (1 + x)Ag \xrightarrow[25°C]{HNO_3} (H + H_2O)Al_{11}O_{17}$$

$$(H + H_2O)Al_{11}O_{17} \xrightarrow[450°C]{NaNO_3} NaAl_{11}O_{17}$$

on polycrystalline material. When finely divided polycrystalline β-alumina is used, however, the reaction times are much quicker and complete exchange may be achieved in a few hours. In a closed reaction system, the evolution of water, presumably containing the original interstitial oxygen, may be directly observed as the conversion from nonstoichiometric silver β-alumina to stoichiometric hydrogen β-alumina proceeds (*14*). This reaction has been observed at 400° and 450°C. Some reactions of hydrogen β-alumina prepared from silver β-alumina are listed in Table VI.

$DAl_{11}O_{17}$ prepared in this way may be advantageously studied by neutron diffraction. Transfer to a sealed vanadium can under dry conditions avoids any problem of moisture pickup, and the exfoliated silver, which does not create an absorption problem in the case of neutron diffraction as it would with X rays, may be left on the sample. By use of the profile refinement technique (*70*), it was shown (*14*, *48*) that deuterium β-alumina prepared in this way is, within the accuracy of the structure refinement, anhydrous $DAl_{11}O_{17}$ (i.e., $D_2O{\cdot}11Al_2O_3$). The mirror-plane sites normally occupied by cations in other β-aluminas (Fig. 1) are empty, and the deuterium is bonded to O(5) in a deuteriohydroxyl linkage. This work showed that anhydrous hydrogen β-alumina is, in fact, an oxyhydroxide of aluminum with the β-alumina structure.

The deuterium location was revealed by Fourier techniques, and the scattering density in the mirror plane determined at room temperature on a second sample of $DAl_{11}O_{17}$ prepared from a starting material of stoichiometry different from that described by Tofield *et al.* (*14*, *48*) is shown in Fig. 7. This recent work (*15*) has shown that the $DAl_{11}O_{17}$ structure is

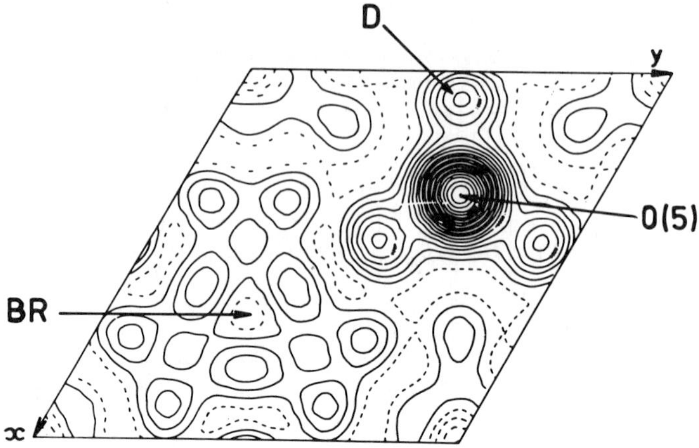

Fig. 7. Mirror-plane section of a total Fourier synthesis for $DAl_{11}O_{17}$ at room temperature. Three equivalent sites for the deuterium atom bonded to O(5) are clearly seen. The powder diffraction data were collected on the Harwell PANDA diffractometer. [Reproduced with permission from Newsam (*60*).]

maintained essentially unchanged between 4.2 K and 440°C, the maximum temperature at which it was studied.

It may be noted that the formation of stoichiometric $HAl_{11}O_{17}$ or $DAl_{11}O_{17}$ from a variety of silver β-aluminas of different stoichiometry provides rather convincing confirmation that nonstoichiometric β-alumina is indeed an intrinsic nonstoichiometric material, with interstitial oxygen stabilized by displaced Al(1) as the compensating defect, and that the nonstoichiometry does not primarily derive from the intergrowth of β and β'' structures at the unit-cell level (*71*), as has been postulated (*72*). Such chemical evidence is of a different nature from, and complementary to, the structural evidence (*25*) and is a convincing demonstration of the utility of the intercalation chemistry of β-alumina and the interest that it holds.

IV. METASTABLE β-ALUMINAS

A. More Nearly Stoichiometric Metal β-Aluminas

The fact that the interstitial oxygen of nonstoichiometric β-alumina could be removed by the action of hydrogen was discovered in independent studies of the decomposition of ammonium β-alumina (*49*, *50*) and of the structure of $DAl_{11}O_{17}$(*14*, *48*). This work and other studies of ammonium β-alumina formation and decomposition, stimulated by the work on $DAl_{11}O_{17}$ and the desire to prepare single crystals of $HAl_{11}O_{17}$, have been described above.

A path was thus opened up for the preparation of more nearly stoichiometric β-aluminas than could be prepared by direct synthesis, with the prospect of being able to measure, for the first time, the properties of intrinsic stoichiometric β-aluminas and hence to determine how the observed properties of nonstoichiometric materials were partitioned between the intrinsic β-alumina structure and the modifications due to the nonstoichiometry.

Reports on metal β-aluminas obtained by the action of molten salts on hydrogen β-aluminas prepared by both routes have been published (*65*, *73*, *74*). Although the preparation of stoichiometric or quasi-stoichiometric β-aluminas has thus been claimed (*65*, *75*), we should note that, in no case, has this been established with certainty by chemical analysis. Without doubt, metastable materials have been made which are much closer to being stoichiometric than those prepared by direct synthesis. As discussed below, however, deviations of several percent from the "ideal" stoichiometric composition may still obtain. Therefore, it is likely, but not conclusively established, that many of the properties discussed below are representative, or nearly so, of stoichiometric β-aluminas.

Several metal β-aluminas were prepared from deammoniated single crystals (*65*). For the larger ions, K^+ and Tl^+, no superlattice effects were observed (*65*), and, as discussed above, these materials are probably effectively ionic insulators, with the cation located on the Beevers–Ross site. These larger cations are appropriately sized to reside at the Beevers–Ross site with, in the stoichiometric materials, a high activation energy for ion mobility. These conclusions are strongly supported by the behavior, discussed above, of partially deammoniated ammonium β-alumina single crystals (*13*), in which the conductivity drops sharply as the ammonium content falls to the stoichiometric level and below (Fig. 3). The crystal structure of a partially deammoniated single crystal has shown (*51*) that only the Beevers–Ross sites are occupied by ammonium ions and that the $N(2)H_4^+$ sites (Fig. 5) are empty.

A more complicated situation is observed for nearly stoichiometric sodium and silver β-alumina crystals prepared via the ammonium route. A superlattice is observed in both cases with an order–disorder transition near 100 K for the sodium compound (*65*) and near 300 K for silver β-alumina. Superlattice formation indicates the displacement of metal atoms away from Beevers–Ross sites [probably accompanied by displacements of O(5) ions] resulting, essentially, from the fact that these cations are too small to reside exactly on these sites. An ordered structure indicates static cooperative displacements and probably a poor ion-conducting regime. Above the transition temperatures, instantaneous displacements still occur but ionic conduction is also likely. Results on silver β-alumina (*65*) do, indeed, show a large increase in ionic conductivity across the transition temperature.

These results imply that, at low temperatures, stoichiometric β-aluminas are no longer fast ionic conductors but that, for the smaller ions Na^+ and Ag^+, the intrinsic excitation energy to form an interstitial defect is very low so that an ionically conducting state can be reached at room temperature or below. For the latter materials, considerable differences may nevertheless be expected by comparison with the behavior of the nonstoichiometric materials, as is demonstrated below in the case of sodium β-alumina. It will be concluded that the β-alumina structure is intrinsically favorable for fast ionic conduction, at least for small cations, but that the very high conductivities, and particularly the low activation energies, observed in sodium and other β-aluminas as normally prepared, require in addition the presence of nonstoichiometry (*3*). The transition from a nonconducting to a conducting state in stoichiometric β-alumina by (at least for silver) the intrinsic excitation of a small fraction of ions to interstitial sites shows some similarity to the behavior of other intrinsic ionic conductors, particularly the fluorite compounds (*76*).

B. Silver β-Alumina

The silver compound has been the most thoroughly investigated nearly stoichiometric β-alumina. The subcell and supercell structures, determined by X-ray diffraction of single crystals, have been described (*28*). In this work, loss of the interstitial oxygen during deammoniation was deliberately not completely achieved. As discussed before, if ammonium β-alumina crystals are heated near 600°C until complete ammonia loss is achieved, the resulting hydrogen β-alumina, although qualitatively similar to material prepared by the action of hydrogen on silver β-alumina, does appear to be disordered in some way. Complete reexchange of alkali ions for protons is difficult to achieve (*75*). Whether this disorder in hydrogen β-alumina is due to some O(5) loss and/or to some defective regions appearing in the crystal, as revealed by electron diffraction lattice imaging (Fig. 8), is not clear. By incompletely decomposing the ammonium β-alumina, this problem seems to be avoided, and in the work described (*28*) no protonic species were evident in the final silver β-alumina by infrared spectroscopy.

A consequent problem, however, is that with only a 50% reduction in the interstitial aluminum content, as was the case in this work on silver β-alumina (*28*), a homogeneous, characteristic, stoichiometric material is clearly not achieved. The overall composition of the material studied, determined from the crystal structure refinement, was $0.96Ag_2O{-}11Al_2O_3$. If some residual

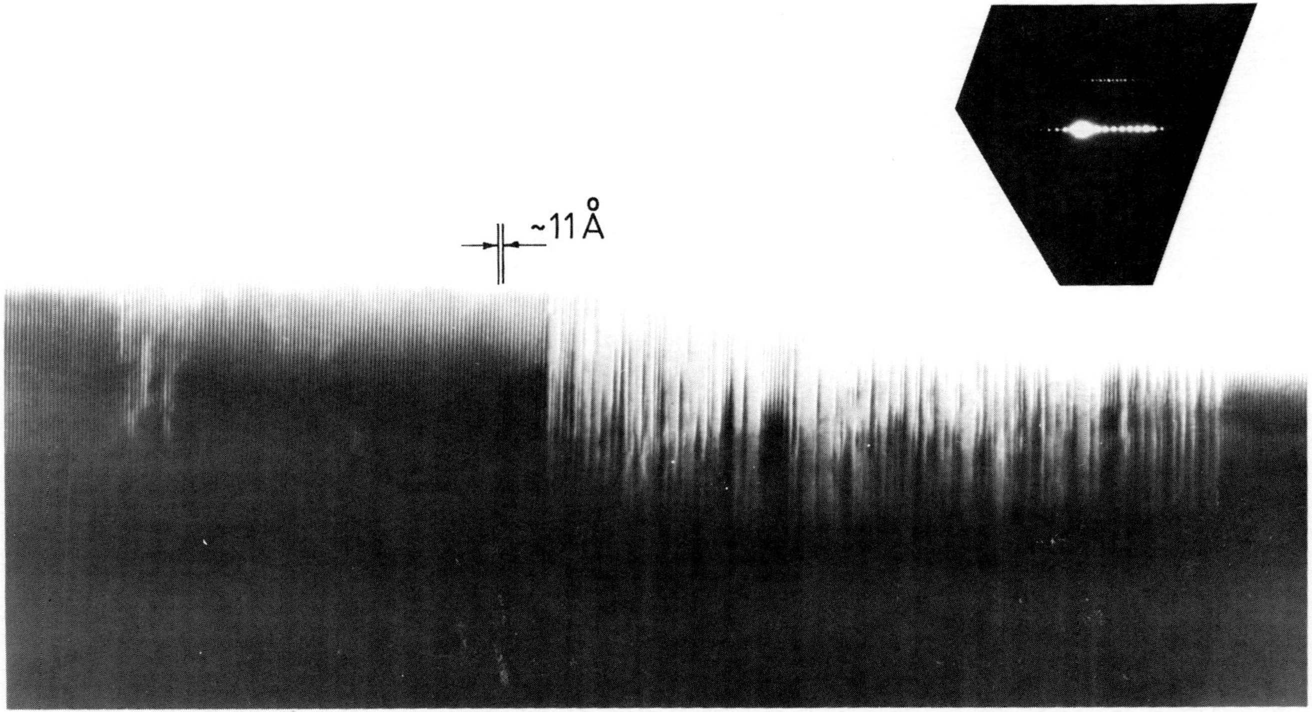

Fig. 8. Electron lattice image of a fragment of fully exchanged ammonium β-alumina single crystal after treatment at 600°C in hydrogen for 24 hr. A region showing localized onset of lattice disorder is shown, although the overall diffraction pattern indicates a nearly perfect lattice. Such disordered regions are not observed in ammonium β-alumina as prepared. (Reproduced with permission from P. T. Moseley, unpublished work.)

domains of nonstoichiometic silver β-alumina were still present, this composition would imply several percent of unit cells without any silver at all, with only about three-quarters of the cells corresponding to the ideal stoichiometric composition. The subcell structure refinement of $0.95Ag_2O$–$11Al_2O_3$ showed (*28*), roughly, a 50% decrease in the anti-Beevers–Ross population from the nonstoichiometric material, in line with the decrease in interstitial aluminum content observed, with the rest of the silver ions divided almost equally between positions close to the Beevers–Ross site and near ($\frac{3}{4}$, $\frac{1}{4}$, $\frac{1}{4}$), as found for the nonstoichiometric material.

Neutron diffraction results on apparently defect-free polycrystalline stoichiometric silver β-alumina prepared by the hydrogen gas route (*74*) indicate that the silver location in the single crystals prepared from ammonium β-alumina (*28*), may be influenced by the incomplete loss of the interstitial defects. However, although the superlattice information was obtained on similar crystals with incomplete loss of interstitial aluminum there seems no reason to suppose that this effect is not characteristic of the stoichiometric material.

The superlattice is observed (*28*) in the basal plane, with the supercell edge being $\sqrt{3}$ times the subcell lattice. In spite of the problems associated with the decomposition and reexchange of ammonium β-alumina, this route so far seems essential for obtaining the single-crystal samples necessary for effective study of the superlattice; as stated above, the hydrogen gas exchange route is suitable only for the preparation of polycrystalline samples. The superlattice is shown in detail in Fig. 9, and the cooperative displacements of silver ions close to ($\frac{3}{4}$, $\frac{1}{4}$, $\frac{1}{4}$), and of O(5) ions, which give rise to the ordering, are also shown. The order–disorder transition and the coherence length of the superlattice have been studied in detail (*78*) both for the nearly stoichiometric compound and for nonstoichiometric silver β-alumina, which shows a diffuse scattering pattern with some similarities to the superlattice but indicating only short-range order at all temperatures. The phenomenology and critical behavior of the antiferroelectric transition in silver β-alumina have also been discussed (*79*).

More recent work has shown (*73*, *74*) that, in the original study of $DAl_{11}O_{17}$ (*14*), the material studied at 550°C was not $DAl_{11}O_{17}$ but had, in fact, transformed to stoichiometric silver β-alumina with loss of hydrogen during heating under vacuum (Table VI). This is, indeed, a very straightforward preparation of stoichiometric silver β-alumina with no necessity for exposure to molten salts or moisture. The excess silver arising from the nonstoichiometric starting material is mixed with the products but causes no problem from the standpoint of structure determination by neutron diffraction and, indeed, offers (*74*) an internal standard for accurate lattice parameter determination.

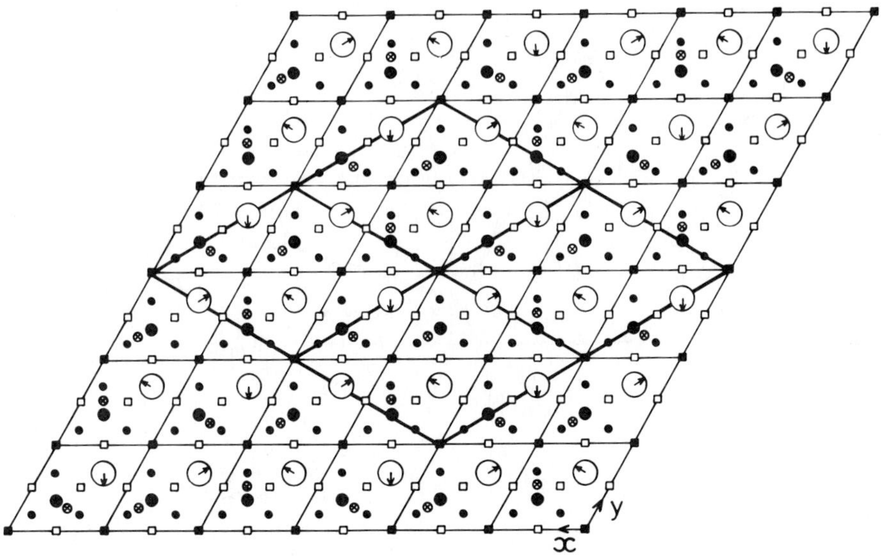

Fig. 9. Superlattice of nearly stoichiometric silver β-alumina (in heavy outline). The cooperative displacements of O(5) ions (◯) toward ordered Ag atoms (⊗) near $(\frac{3}{4}, \frac{1}{4}, \frac{1}{4})$ which lead to the superlattice are seen. Beevers–Ross (●), mid-oxygen (•), and anti-Beevers–Ross (▪) positions are indicated. The silver coordination is completed by four O(2) atoms above and below the plane; the projection on the plane of the O(2) positions is shown (□). [From Newsam and Tofield (*74*).]

The structure of this material has been determined by powder neutron diffraction at several temperatures (*74*, *80*). Of particular interest are the results at low temperature (4.2 K) and room temperature. At low temperature (*74*), within the accuracy of the crystal structure refinement carried out on the D1A diffractometer at the Institut Laue-Langevin, Grenoble (*81*), the silver was located uniquely at the site near $(\frac{3}{4}, \frac{1}{4}, \frac{1}{4})$ midway between the mid-oxygen and Beevers–Ross positions (Fig. 9). Both the Beevers–Ross and anti-Beevers–Ross sites were empty, and no interstitial aluminum was detected. Within the accuracy of the crystallographic determination, therefore, this material was ideal stoichiometric silver β-alumina. This result shows unequivocally the preferred location for silver in the β-alumina structure and indicates that, in the ordered regime, the anti-Beevers–Ross site is not occupied. At room temperature (*80*) the majority of silver was still found near the low-temperature position, but roughly 10% occupancy of the anti-Beevers–Ross site was now observed. This implies that the activation energy for intrinsic excitation to the anti-Beevers–Ross position, and presumably also for long-range silver motion, is less than 0.1 eV.

The silver coordination at 4.2 K (*74*) is shown in Fig. 10 with the anisotropic thermal ellipsoid that results from the almost ringlike coordination of

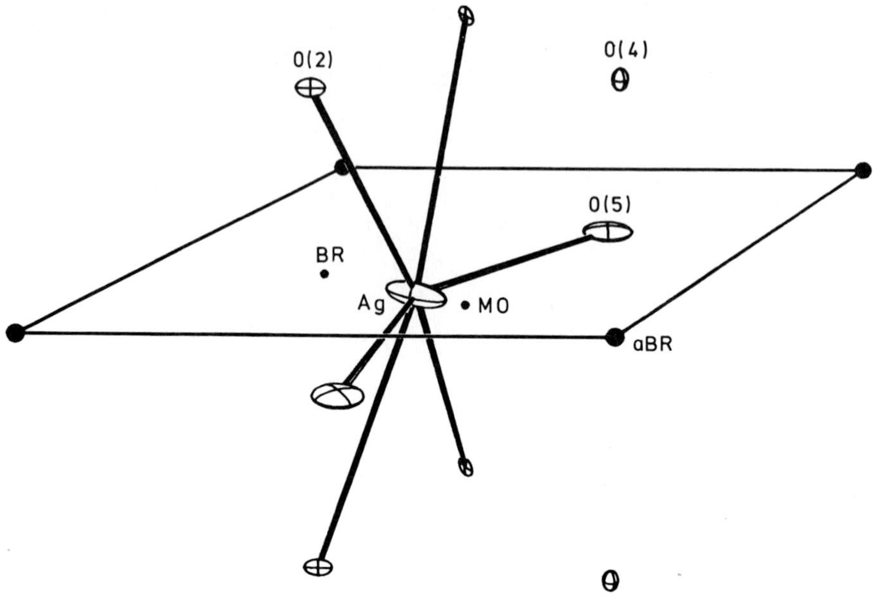

Fig. 10. Silver–oxygen coordination in stoichiometric silver β-alumina at 4.2 K. The anisotropic thermal vibration of silver reflecting the unusual coordination is clearly seen. [From Newsam and Tofield (*74*).]

oxygen around the silver. This silver location results from the desire to reduce the silver–oxygen bond distances below those obtainable for location at the Beevers–Ross site and from the ability of silver to maintain unusual coordination polyhedra. As seen from Fig. 9, silver is as close as possible to four O(2) atoms above and below the mirror plane with a reduction of about 0.1 Å below the BR—O(2) distance. Favorable displacement of two O(5) atoms within the plane, as shown in Fig. 9, gives two more, slightly shorter Ag—O distances and causes a slight shift of silver away from ($\frac{3}{4}$, $\frac{1}{4}$, $\frac{1}{4}$) toward the mid-oxygen position. The observed superlattice is, in fact, a direct consequence of this localized Ag—O interaction (*74*). For all silver ions to be favorably coordinated by two O(5) ions, as well as by O(2) ions, cooperative displacements as shown in Fig. 9 are essential. These imply the superlattice shown. As the temperature is raised and thermal excitation to the anti-Beevers–Ross position becomes possible, the driving force for such long-range ordering is decreased and the superlattice disappears, although the site near ($\frac{3}{4}$, $\frac{1}{4}$, $\frac{1}{4}$) is still, at moderate temperatures, the preferred location for silver (*80*). Superlattice reflections are only weakly visible in a powder neutron diffraction pattern, but, in the study of polycrystalline $AgAl_{11}O_{17}$, weak low-angle reflections were observed (*74*) at low temperature which

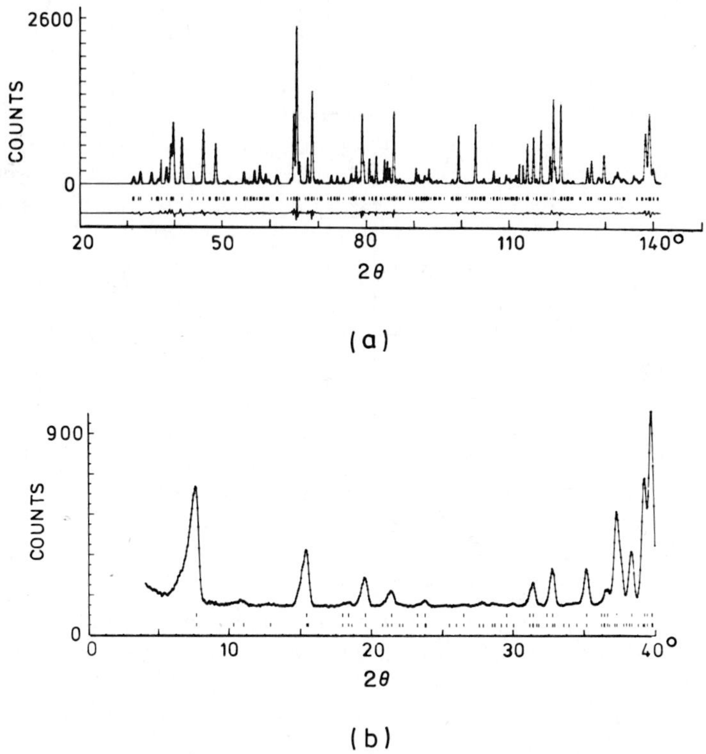

Fig. 11. Neutron diffraction pattern of $AgAl_{11}O_{17}$ at 4.2 K. (a) Observed and calculated (dots) profiles for the best refinement. The vertical bars indicate the reflection positions, and the lower trace is an ($I_{calc} - I_{obs}$) difference profile. (b) Observed low-angle pattern in more detail (only reflections above 30° 2θ were used in the profile refinement). The vertical bars indicate subcell and supercell reflection positions. Weak reflections revealing the presence of the superlattice are observable. These are not observed at room temperature and above. [From Newsam and Tofield (*74*).]

were consistent with the known superlattice (Fig. 11). These extra reflections were not observed at room temperature or above (*80*).

It is interesting that the *c*-axis lattice constant of this stoichiometric silver β-alumina is higher than that for the nonstoichiometric material (Table I), both at 4.2 K (*74*) with no anti-Beevers–Ross occupation (c = 22.513 Å) and at room temperature where there is some anti-Beevers–Ross population (c = 22.529 Å) (*80*). This *c*-axis expansion is also confirmed in the X-ray work at room temperature (c = 22.545 Å) (*28*). The room temperature values, although very similar to that for nonstoichiometric sodium β-alumina, are somewhat less than that for nearly stoichiometric sodium β-alumina at room temperature [c = 22.59 Å]. The similar differences between silver and sodium materials, in both the stoichiometric and the nonstoichiometric forms, demonstrate the effect of the silver–oxygen semicovalent interaction in pulling together the spinel blocks of the β-alumina lattice. The lower value for nonstoichiometric silver β-alumina compared to that for the stoichiometric material at room temperature indicates that population of the anti-Beevers–Ross position does not play a significant role in determining this lattice constant. On the other hand, the similar expansions for both sodium and silver β-aluminas in going from nonstoichiometric to nearly stoichiometric materials implicate the O_i—Al_i defect in reducing slightly the spinel block separation. In contrast to this expansion observed with small ions, we recall that for the larger NH_4^+ and K^+ ions, the change from nonstoichiometry to stoichiometry is manifested by a decrease in c-axis lattice constant. For these β-aluminas, this is presumably because in the non-stoichiometric materials any effect of the interstitial defect is outweighed by the lattice expansion caused by the need to locate these larger ions on unfavorable sites in the mirror plane.

C. Sodium β-Alumina

Less work has so far been reported on nearly stoichiometric β-alumina than on the silver analogue. An order–disorder transition has been observed near 100 K (*65*). The superlattice has the same dimensions as that of stoichiometric silver β-alumina, and it will be of considerable interest to determine the sodium ion location.

One study has been made (*82*) of polycrystalline nearly stoichiometric sodium β-alumina obtained by ion-exchange of hydrogen β-alumina which was prepared by the action of hydrogen gas on the nonstoichiometric silver compound. After the hydrogenation, the silver was removed by the action of cold nitric acid, and the resulting H^+/H_2O β-alumina was equilibrated in molten sodium nitrate at 450°C. A very low silver content remained after this treatment, and this was concentrated as metallic inclusions within a very few grains. The overall composition of the product, determined by chemical analysis, was $1.07Na_2O–11Al_2O_3$. The reasons for the higher stoichiometry compared to the polycrystalline silver compound discussed above are not clear but may result from incomplete decomposition during hydrogenation or, perhaps, from reintroduction of interstitial oxygen during ion exchange. As mentioned above, such a possibility has recently been demonstrated for β-ferrites (*46*). The metastability of the material is demonstrated (*82*), however, by a disproportionation occurring between 1100° and 1200°C. An exothermic transition is observed by differential thermal analysis, and samples cooled from above this temperature show lines of α-alumina in X-ray diffraction as well as β-alumina lines. A disproportionation reaction, as shown in Table V, is presumably occurring; such effects are not observed for the nonstoichiometric sodium β-alumina starting material. Similar behavior has also been reported for single-crystal sodium β-alumina prepared by the deammoniation route (*50*).

This decomposition reaction prohibits the pressing and sintering of high-density pellets to prepare conductivity specimens. However, microwave conductivity experiments performed on the powdered material above room temperature revealed ionic conductivity, but with a much higher activation energy, nearly 0.6 eV (*82*), than found for the nonstoichiometric starting material. As discussed above, these results imply that, at least above the order–disorder transition, stoichiometric sodium β-alumina will show ionic conductivity, but that this will be only moderate, at least at moderate temperatures, and that the extremely low activation energy [0.13 eV (*2*)] observed for nonstoichiometric β-alumina as normally prepared (25–30% excess Na_2O) results directly from the interstitialcy mechanism consequent upon the nonstoichiometry (*3*).

To prepare single-crystal specimens for conductivity it will be necessary to proceed via the ammonium material. To avoid problems of interpretation, as discussed above, it will be desirable to equilibrate for a necessary period below the decomposition temperature near 650°C, until the interstitial oxygen and ammonium are completely removed, and then to try to ion-exchange fully the protons for sodium ions. This will be time-consuming but certainly worthwhile.

D. Spectroscopy of Nearly Stoichiometric β-Aluminas

Although the transformation from a nonstoichiometric to a stoichiometric β-alumina entails only small changes in the overall composition, quite marked effects can be revealed by spectroscopic techniques which probe directly, on a local scale, the motions of ions in or near the conducting plane. Thus, Raman spectroscopic study (*47*) of the spinel-block vibrations of β-aluminas showed that linewidths became much sharper as the O_i–Al_i defect was removed. This was associated (*47*) with the restoration of Al_i atoms to the normal Al(1) site. Such a narrowing was not observed (*47*) as a consequence of the transitions from nonstoichiometric to stoichiometric β-gallates, suggesting (*47*) a different charge compensating mechanism and, indeed, as discussed above, nonmobile sodium may be found (*42*) within the spinel blocks of β-gallates prepared from Na_2O–Ga_2O_3 melts.

Concern about complete reexchange of hydrogen β-alumina back to metal β-aluminas has not prevented very useful information from being obtained on the low-frequency vibrational modes reflecting the attempted motions of the conducting-plane species. The far-infrared spectra (*83*) of a nonstoichiometric Union Carbide sodium β-alumina crystal and a more nearly stoichiometric sodium-containing crystal made from this via the ammonium route are compared in Fig. 12. Particularly noticeable is the dramatic increase in relative intensity and sharpness of a band at 59 cm^{-1} for the more nearly stoichiometric material, which clearly defines the attribution of this band to excitations of sodium ions at or near Beevers–Ross

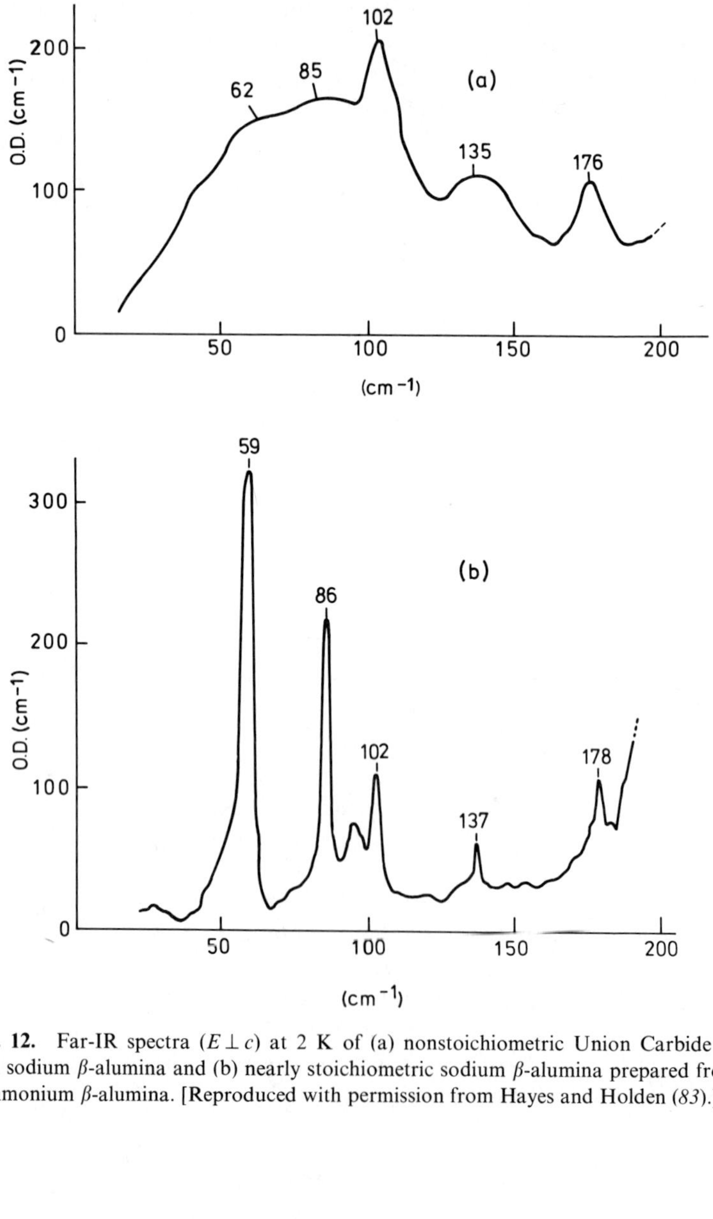

Fig. 12. Far-IR spectra ($E \perp c$) at 2 K of (a) nonstoichiometric Union Carbide single-crystal sodium β-alumina and (b) nearly stoichiometric sodium β-alumina prepared from this via ammonium β-alumina. [Reproduced with permission from Hayes and Holden (*83*).]

sites in singly occupied cells. The infrared line at 59 cm^{-1} corresponds to a band at 62 cm^{-1} observed by Raman scattering, which, in previous work on nonstoichiometric crystals only, has been correctly assigned to Beevers–Ross vibrations (*84*) and apparently incorrectly assigned to excitations of interstitial sodium pairs (*11*, *85*). The comparison (*83*) of the materials with differing stoichiometry also provides for tentative explanations of the other bands observed, and it is proposed (*83*) that the centers giving rise to the 102 cm^{-1} band make a sizable contribution to the ionic conductivity in nonstoichiometric material.

Similar conclusions are reached for potassium (*86*) and silver β-aluminas (*87*). In the former case, a band at 84 cm^{-1} is identified as being due to K^+ ions in singly occupied Beevers–Ross sites. The higher frequency than observed for sodium β-alumina indicates a stronger coupling to the lattice and agrees with the measured lower ionic conductivity (Table II). Once again, the simplification introduced by the preparation of the more nearly stoichiometric crystals has allowed a positive assignment, which was not possible in earlier work with nonstoichiometric material only (*11*, *85*), although the correct assignment was proposed in one study (*88*) of K–Sn and K–Li mixed β-aluminas. Similar conclusions concerning the assignments of stoichiometric materials have been reported (*47*).

The far-infrared spectrum of stoichiometric silver β-alumina is shown in Fig. 13. In this case, the lowest-frequency band at 21.5 cm^{-1} reflects the single-ion excitations of Ag^+ ions near $(\frac{3}{4}, \frac{1}{4}, \frac{1}{4})$. As discussed above, the best established stoichiometric or nearly stoichiometric β-alumina composition so far is silver β-alumina prepared by the thermal decomposition of silver-

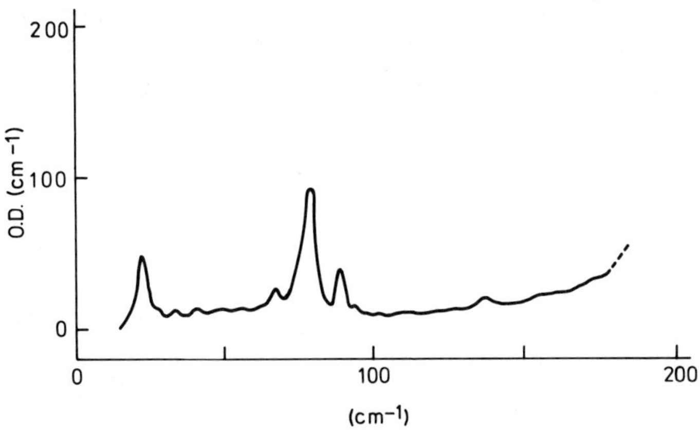

Fig. 13. Far-IR spectrum ($E \perp c$) of a stoichiometric silver β-alumina crystal prepared by in-crystal reaction (*90*) of mixed nonstoichiometric ammonium–silver β-alumina. [Reproduced with permission from Holden (*89*).]

coated deuterium β-alumina. This preparation is suitable only for polycrystalline material, but the far-infrared spectrum is in very good agreement with that obtained from nearly stoichiometric silver β-alumina crystals obtained via the ammonium route (*87*), although an out-of-plane vibration near 100 cm^{-1} gives a very strong band not seen in the single-crystal material for $E \perp c$.

The material from which the single-crystal spectrum of Fig. 13 (*89*) was taken was, however, obtained by a third route, the *in situ* decomposition of mixed nonstoichiometric ammonium–silver β-alumina. Study of this system (*90*) has revealed a feature of some mixed β-aluminas not established in previous ion-exchange studies (*8*), that of immiscibility. The study of the temperature evolution of both the chemistry and physical properties of immiscible β-aluminas coexisting in the same crystal may provide a rich new field of study in β-alumina intercalation chemistry, and an understanding of the reasons for the immiscibility will yield a deeper comprehension of the β-aluminas in general. The decomposition of the mixed, nonstoichiometric ammonium–silver crystal at 500°–600°C yields a crystal containing stoichiometric silver β-alumina in one simple stage without the need for a long and complex sequence of ion exchanges and thermal decompositions.

ACKNOWLEDGMENTS

The understanding we currently have of the fascinating details of the intercalation chemistry of β-alumina has been achieved as a result of painstaking work in several laboratories around the world. For my own understanding of the subject, I am deeply indebted to many colleagues both at Harwell and Oxford for many illuminating discussions and much valuable and expert experimentation. I acknowledge especially the contributions of J. M. Newsam, A. Hooper, C. F. Sampson, B. A. Bellamy, J. M. Butcher, L. Holden, W. Hayes, P. T. Moseley, and A. Morton.

REFERENCES

1. N. Weber and J. T. Kummer, *Proc. Annu. Power Sources Conf.* **21,** 37 (1967).
2. A. Hooper, *J. Phys. D* **10,** 1487 (1977).
3. D. Wolf, *J. Phys. Chem. Solids* **40,** 757 (1979).
4. J. T. Kummer, *Prog. Solid State Chem.* **7,** 141 (1972).
5. J. H. Kennedy, *in* "Solid Electrolytes" (S. Geller, ed.), p. 105. Springer-Verlag, Berlin and New York, 1977.
6. R. M. Dell and P. T. Moseley, *J. Power Sources,* **6,** 143 (1981).
7. P. Vashishta, J. N. Mundy, and G. K. Shenoy, eds., "Fast Ion Transport in Solids." Elsevier/North-Holland, New York, 1979.
8. Y.-F. Yao and J. T. Kummer, *J. Inorg. Nucl. Chem.* **29,** 2453 (1967).

9. L. R. Rothrock, *J. Cryst. Growth* **39,** 180 (1977).
10. B. C. Tofield and G. C. Farrington, *Nature* (*London*) **278,** 438 (1979).
10a. M. S. Whittingham and R. A. Huggins, *J. Chem. Phys.* **54,** 414 (1971).
11. S. J. Allen, Jr., A. S. Cooper, F. DeRosa, J. P. Remeika, and S. K. Ulasi, *Phys. Rev.* **17,** 4031 (1978).
12. M. A. M. Bourke, A. Hooper, P. T. Moseley, and R. G. Taylor, *Solid State Ionics* **1,** 367 (1980).
13. A. Hooper, B. C. Tofield, and C. F. Sampson, ref. 7, p. 409.
14. B. C. Tofield, A. J. Jacobson, W. A. England, P. J. Clarke, and M. W. Thomas, *J. Solid State Chem.* **30,** 1 (1979).
15. J. M. Newsam and B. C. Tofield, "Stoichiometric deuterium beta alumina: Powder neutron diffraction studies at 4.2 K, 298 K, 573 K, and 720 K," to be published.
16. R. H. Radzilowski and J. T. Kummer, *J. Electrochem. Soc.* **118,** 714 (1971).
17. R. E. Walstedt, R. S. Berg, J. P. Remeika, A. S. Cooper, B. E. Prescott, and R. Dupree, ref. 7, p. 355.
18. J. L. Briant and G. C. Farrington, *J. Electrochem. Soc.* **128,** 1830 (1981).
19. A. Hooper, T. Lapp, and S. Skaarup, *Mater. Res. Bull.* **14,** 1617 (1979).
20. C. A. Beevers and M. A. S. Ross, *Z. Kristallogr.* **97,** 59 (1937).
21. M. S. Whittingham and R. A. Huggins, *J. Electrochem. Soc.* **118,** 1 (1971).
22. J. Antoine, D. Vivien, J. Livage, J. Théry, and R. Collongues, *Mater. Res. Bull.* **10,** 865 (1975).
23. C. R. Peters, M. Bettman, J. W. Moore, and M. D. Glick, *Acta Crystallogr., Sect. B* **27,** 1826 (1971).
24. W. L. Roth, *J. Solid State Chem.* **4,** 60 (1972).
25. F. Reidinger, Ph.D. Thesis, State Univ. of New York, Albany, 1979.
26. W. L. Roth, *in* "Crystal Structure and Chemical Bonding in Inorganic Chemistry" (C. J. M. Rooymans and A. Rabenau, eds.), p. 85. North-Holland Publ., Amsterdam, 1975.
27. G. Collin, J. P. Boilot, A. Kahn, J. Théry, and R. Comès, *J. Solid State Chem.* **21,** 283 (1977).
28. J. P. Boilot, P. Colomban, G. Collin, and R. Comès, *J. Phys. Chem. Solids* **41,** 47 (1980).
29. J. M. Newsam, A. K. Cheetham, and B. C. Tofield, Ammonium Beta Alumina—the Complete Room Temperature Crystal Structure of Fully Exchanged Material, to be published.
30. A. R. West, *Mater. Res. Bull.* **14,** 441 (1979).
31. C. R. A. Catlow and J. R. Walker, *Nature* **286,** 473 (1980).
32. D. B. McWhan, P. D. Dernier, C. Vettier, A. S. Cooper, and J. P. Remeika, *Phys. Rev. B* **17,** 4043 (1978).
33. J. P. Boilot, P. Colomban, J. Théry, G. Collin, and R. Comès. *J. Phys.* (*Paris*), *Colloq. C2* **39,** 204 (1978).
34. P. D. Dernier and J. P. Remeika, *J. Solid State Chem.* **17,** 245 (1976).
35. G. Collin, P. Colomban, J. P. Boilot, and R. Comès, ref. 7, p. 309.
36. P. Colomban, Ph.D. Thesis, Univ. Pierre et Marie Curie, Paris, 1979.
37. N. Baffier, J. C. Badot, and P. Colomban, *Mater. Res. Bull.* **16,** 259 (1981).
38. G. Collin, R. Comès, J. P. Boilot, and P. Colomban, *Solid State Ionics* **1,** 59 (1980).
39. W. L. Roth, F. Reidinger, and S. J. LaPlaca, *in* "Superionic Conductors" (G. D. Mahan and W. L. Roth, eds.), p. 223. Plenum, New York, 1976.
40. J. H. Kennedy and A. F. Sammels, *J. Electrochem. Soc.* **119,** 1609 (1972).
41. J. P. Boilot, P. Colomban, G. Collin, and R. Comès, *Solid State Ionics* **1,** 69 (1980).
42. L. M. Foster, ref. 7, p. 249.
43. M. W. Breiter and G. C. Farrington, *Mater. Res. Bull.* **13,** 1213 (1978).

44. R. R. Dubin, H. S. Story, R. W. Powers, and W. C. Bailey, *Mater. Res. Bull.* **14,** 185 (1979).
45. D. Brinkmann, M. Mali, J. Roos, H. S. Story, and R. R. Dubin, ref. 7, p. 285.
46. A. T. Howe and G. J. Dudley, *J. Solid State Chem.* **30,** 157 (1979).
47. P. Colomban and G. Lucazean, *J. Chem. Phys.* **72,** 1213 (1980).
48. W. A. England, A. J. Jacobson, and B. C. Tofield, *J.C.S. Chem. Commun.* p. 895 (1976).
49. P. Colomban, J. P. Boilot, P. Chagnon, and G. Guilloteau, *Bull. Soc. Fr. Ceram.* **119,** 3 (1978).
50. P. Colomban, J. P. Boilot, A. Kahn, and G. Lucazeau, *Nouv. J. Chim.* **2,** 21 (1978).
51. J. M. Newsam, A. K. Cheetham, and B. C. Tofield, "Thermal Evolution of the Structure of Ammonium Beta Alumina," to be published.
52. H. Saalfeld, H. Matthies, and S. K. Datta, *Ber. Dtsch. Keram. Ges.* **45,** 212 (1968).
53. P. Colomban, G. Lucazeau, R. Mercier, and A. Novak, *J. Chem. Phys.* **67,** 5244 (1978).
54. G. C. Farrington, J. L. Briant, H. S. Story, and W. C. Bailey, *Electrochim. Acta* **24,** 769 (1979).
55. W. L. Roth, M. W. Breiter, and G. C. Farrington, *J. Solid State Chem.* **24,** 321 (1978).
56. W. L. Hayes, L. Holden, and B. C. Tofield, *J. Phys. C* **13,** 4217 (1980).
57. M. W. Breiter, G. C. Farrington, W. L. Roth, and J. L. Duffy, *Mater. Res. Bull.* **12,** 895 (1977).
58. G. C. Farrington, J. L. Briant, M. W. Breiter, and W. L. Roth, *J. Solid State Chem.* **24,** 311 (1978).
59. T. Nishikawa, T. Nishida, Y. Katsuyama, Y. Kawakita, and I. Uei, *Nippon Kagaku Kaishi* **6,** 1048 (1974).
60. J. M. Newsam, D.Phil. Thesis, University of Oxford (1980).
61. K. Kato and H. Saalfeld, *Acta Crystallogr., Sect. B* **33,** 1596 (1977).
62. W. L. Roth, M. Anne, and D. Tranqui, *Rev. Chim. Minér.* **17,** 379 (1980).
63. W. L. Roth and G. C. Farrington, *Electrochim. Acta* **22,** 767 (1977).
64. G. C. Farrington and J. L. Briant, ref. 7, p. 395.
65. J. F. Boilot, P. Colomban, G. Collin, and R. Comès, ref. 7, p. 243.
66. T. W. Baker, J. D. George, B. A. Bellamy, and R. Causer, *Adv. X-Ray Anal.* **11,** 359 (1968).
67. G. C. Farrington, *J. Electrochem. Soc.* **123,** 833 (1976).
68. F. G. Will, *J. Electrochem. Soc.* **123,** 834 (1976).
69. T. Kaneda, J. B. Bates, J. C. Wang, and H. Engstrom, *Mater. Res. Bull.* **14,** 1053 (1979).
70. H. M. Rietveld, *J. Appl. Crystallogr.* **2,** 65 (1969).
71. J. M. Newsam and B. C. Tofield, *Solid State Ionics* **5,** 59 (1981).
72. D. J. M. Bevan, B. Hudson, and P. T. Moseley, *Mater. Res. Bull.* **9,** 1073 (1974).
73. B. C. Tofield and J. M. Newsam, ref. 7, p. 389.
74. J. M. Newsam and B. C. Tofield, *J. Phys. C,* **14,** 1545 (1981).
75. R. Collongues, *Mater. Res. Bull.* **13,** 1281 (1978).
76. M. H. Dickens, W. Hayes, M. T. Hutchings, and C. Smith, *J. Phys. C* **12,** L97 (1979).
77. L. Holden, A. Hooper, and B. C. Tofield, unpublished observations.
78. J. P. Boilot, P. Colomban, R. Collongues, G. Collin, and R. Comès, *J. Phys. Chem. Solids* **41,** 253 (1980).
79. J. F. Gouyet, B. Sapoval, and P. Pfeuty, *J. Phys., Lett. (Orsay, Fr.)* **41,** L115 (1980).
80. J. M. Newsam, B. C. Tofield, A. K. Cheetham, and A. W. Hewat, to be published.
81. A. W. Hewat and I. Bailey, *Nucl. Instrum. Methods* **137,** 463 (1976).
82. W. Hayes, L. Holden, and B. C. Tofield, *Solid State Ionics* **1,** 373 (1980).
83. W. Hayes and L. Holden, *J. Phys. C* **13,** L321 (1980).
84. C. H. Hao, L. L. Chase, and G. D. Mahan, *Phys. Rev. B* **13,** 4306 (1976).
85. P. B. Klein, D. E. Schafer, and U. Strom, *Phys. Rev. B* **18,** 4411 (1978).
86. W. Hayes, L. Holden, and G. F. Hopper, *J. Phys. C* **13,** L317 (1980).

87. W. Hayes, L. Holden, and B. C. Tofield, *J. Phys. C* **14,** 511 (1981).
88. T. Kaneda, J. B. Bates, J. C. Wang, and H. Engstrom, ref. 7, p. 371.
89. L. Holden, D.Phil. Thesis, University of Oxford (1980).
90. B. C. Tofield, *Solid State Ionics* **5,** 295 (1981).

7

Organic and Organometallic Intercalation Compounds of the Transition Metal Dichalcogenides

ALLAN J. JACOBSON

I. INTRODUCTION

Intercalation compounds of the dichalcogenides have been extensively studied since the discovery over a decade ago that the critical temperature for the onset of superconductivity in 2H TaS_2 could be influenced and in some cases increased by the intercalation of organic molecules between the metallic dichalcogenide layers (*1*). The field has continued to grow, with added impetus provided by the discovery of charge density waves in the dichalcogenides themselves (*2*, *3*) and by the use of metal ion intercalation reactions in high-energy-density secondary batteries (*4*, *5*). The dichalcogenides of Group VI are also of considerable importance as catalysts for sulfur removal in petroleum refining (*6*), and recently there has been interest in other systems with catalytic or stoichiometric reactivity (*7*).

INTERCALATION CHEMISTRY

ISBN 0-12-747380-7

Many of the intercalation reactions of the dichalcogenides, particularly ion and solvent exchange, have parallels in other host lattices, for example, in clay minerals (*8*) and zeolites (*9*). The dichalcogenides (and graphite), however, show the widest variety of chemical reactivity in that they can both react with neutral guest species and undergo reduction to form intercalation compounds with cationic guests between negatively charged host layers. In this chapter, two classes of intercalation compounds that illustrate these two extreme types of behavior are reviewed. In the first part, the general chemistry of organic intercalation compounds is described. A comprehensive review of all organic intercalation compounds has not been attempted. Instead, emphasis has been placed on a small number of systems that have been studied in detail. Additional information can be found in a recent review article (*10*). In the second part, the recent work on organometallic systems is reviewed. The physics of intercalation compounds of the dichalcogenides is discussed in Chapter 18 and reactions with other guest species in Chapters 9 and 10.

II. STRUCTURES OF THE TRANSITION METAL DICHALCOGENIDES

The transition metal dichalcogenides of Groups IV, V, and VI have hexagonal or rhombohedral layered structures. The transition metal (M) occupies either trigonal prismatic or octahedral sites between two layers of close-packed chalcogen atoms (X). The X—M—X sandwiches are then stacked along the hexagonal *c* axis to complete the structure. Bonding within an X—M—X slab is strong, whereas bonding between slabs is weak and of the van der Waals type. Intercalation reactions occur by insertion of the guest molecule or ion into the empty layer. Different combinations of octahedral and trigonal prismatic sites and different stacking sequences of layers give rise to a number of polymorphs and polytypes. It is convenient to represent these structures by projecting the structure in the $(11\bar{2}0)$ plane and labelling the atoms $A(0, 0)$, $B(\frac{2}{3}, \frac{1}{3})$, and $C(\frac{1}{3}, \frac{2}{3})$ with respect to their translation in the xy plane from the origin at A. The notation is illustrated in Figs. 1 and 2. Uppercase letters denote chalcogen layers and lowercase letters metal layers; empty layers are indicated by /.

The most important structures that will be encountered in this chapter are shown in Fig. 2 and are labeled using the notation of Brown and Beernsten (*11*). In these projections octahedral sites can be recognized by the sequence *AbC* and trigonal prismatic by *AbA*. The simplest structure is 1T, with all octahedral metals and one X—M—X slab per unit cell. The 2Ha and 2Hc structures are the two most common polytypes of the three all-trigonal-

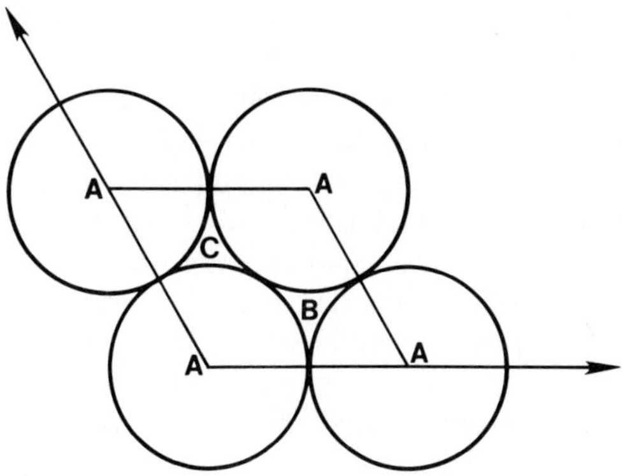

Fig. 1. Chalcogen layer illustrating the stacking sequence notation.

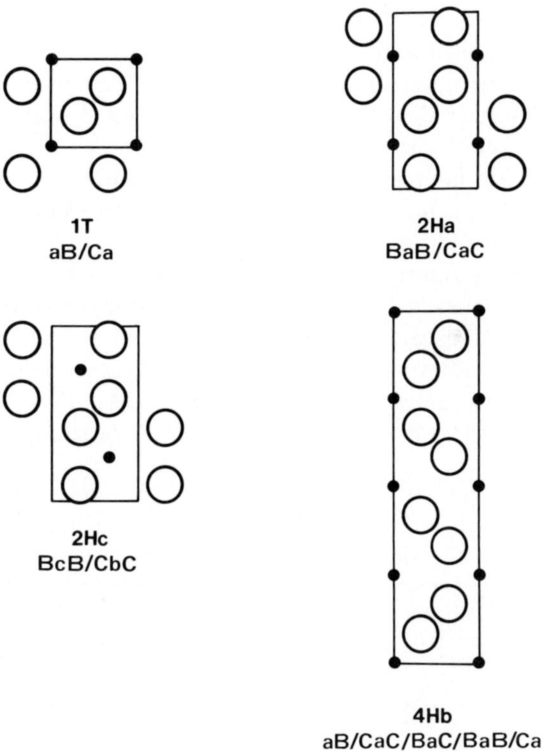

Fig. 2. The $(11\bar{2}0)$ projections of the common structures found for the transition metal dichalcogenides.

prismatic structures, having two layers per unit cell, and 4Hb is an example of a mixed octahedral/trigonal prismatic structure. The 2Ha and 2Hc structures are frequently referred to as the 2H and 2H MoS_2 structures. They differ in that in 2H the metal atoms lie directly above each other, whereas in 2H MoS_2 they are staggered. In the dichalcogenides of Group IV and vanadium the transition metals have octahedral coordination, and in those of Group VI they are in trigonal prismatic sites. Group V dichalcogenides can have either octahedral or trigonal prismatic structures. Some representative examples of the different stacking sequences are given in Table I.

In an intercalation reaction, guest species are inserted into empty sites between the layers. For small guests, such as metal cations, octahedral, tetrahedral, or trigonal prismatic sites, defined by the stacking of the chalcogen layers, are occupied. In general, the coordination of the transition metal is preserved on intercalation, but it cannot be assumed that the X—M—X stacking will remain the same. Even for small guests, e.g., Li^+, changes are observed. For example, $LiZrS_2$ has a 3R structure and not the 1T structure of the host ZrS_2 (*12*). A similar situation exists for the intercalation of large organic or organometallic molecules in which the structure is determined by guest–host bonding as well as by the steric requirements of the guest and guest–guest interactions.

In stoichiometric intercalation compounds, staging is occasionally observed, although this is much less common for the dichalcogenides than it is in graphite systems (*13*). A first-stage compound has guests between every layer, whereas in a second-stage compound every other layer is empty. Higher stages are similarly defined, although few examples of staging beyond second have been reported.

The layered dichalcogenides can all incorporate excess metal between the layers to give nonstoichiometric, metal-rich phases $M_{1+x}X_2$. The metal-rich phases are poor hosts for intercalation of other than the smallest ions because

TABLE I

Structures of the Dichalcogenides

Designation	Space group	Examples	Metal coordination
1T	$P\bar{3}m1$	Group IVB, VX_2, TaS_2, $TaSe_2$, NbS_2	Octahedral
2Ha	$P6_3/mmc$	NbX_2, TaX_2	Trigonal prismatic
2Hc	$P6_3/mmc$	Group VIB	Trigonal prismatic
4Hb	$P6_3/mmc$	TaS_2, $TaSe_2$	Trigonal prismatic and octahedral

the highly charged metal ion between the layers considerably increases the energy required to open the interlayer space for large guest molecules. In some systems the effect is subtle. For example, early failures to intercalate TiS_2 directly with pyridine or substituted pyridines have been associated with small deviations from stoichiometry. Stoichiometric TiS_2 intercalates 2,4,6-trimethylpyridine after 30*d* at 300°C, whereas no evidence for intercalation was found for $Ti_{1.001}S_2$ (*14*).

More detailed accounts of the syntheses and structures of the host dichalcogenides can be found in recent reviews (*15*).

III. ORGANIC INTERCALATION COMPOUNDS OF THE DICHALCOGENIDES

Organic intercalation compounds were first reported for amides in the semiconducting host TiS_2 (*16, 17*). Since these early reports, there has been widespread interest in both the chemistry and physics of these materials, and at this time over 240 organic intercalation compounds have been reported in the literature (*10*). A wide range of organic molecules form intercalation compounds, and the major classes are summarized in Table II. All the reactions are characterized by an expansion of the crystal lattice along the **c** direction perpendicular to the layers to an extent that may be correlated with the molecular dimensions of the guest and the stoichiometry of the final product. The host lattice may be recovered unchanged by thermal deintercalation of the organic molecule at temperatures higher than the initial reaction temperature, often by as much as 200°–300°C. Stabilities vary significantly and depend on the nature of both the guest molecule and the host lattice. The highest stabilities are observed with the host dichalcogenides 2H TaS_2, 2H NbS_2, and 1T TiS_2, and much of the literature is devoted to

TABLE II

Organic Molecules That Form Intercalation Compounds

Generic class	Examples
Amines	RNH_2, R_2NH, R_3N, $H_2N(CH_2)_nNH_2$
Phosphines	R_3P
Amides	$RCONH_2$, $CO(NH_2)_2$
Amine oxides	Pyridine *N*-oxide
Phosphine oxides	R_3PO
N-Heterocycles	Pyridine, substituted pyridines
Isocyanides	RNC

them. In contrast, 2H $NbSe_2$ does not form organic intercalation compounds, with the exception of ethylenediamine (*18*), stable at room temperature (*19*). The 2H TaS_2 structure has generally been found to react most readily and to form the most stable products and is preferred for examining new reactions.

Intercalation reactions with organic compounds are usually carried out by direct reaction of the dichalcogenide in powder form with the neat organic compound or with a benzene or toluene solution for high molecular weight systems. In some cases, reactions are facilitated by pretreatment of the dichalcogenide with ammonia or hydrazine, a procedure that may introduce other complications (see Section III,B). The progress of the reaction can often be followed qualitatively by observing the volume expansion of the solid phase. Quantitatively, reactions are studied by elemental analysis after the removal of excess unreacted guest and by the change in the X-ray powder pattern. X-ray powder patterns vary considerably in quality from system to system but in many cases are sharp and have well-defined (*hkl*) and (00*l*) reflections; the latter are usually enhanced in intensity because of preferred orientation effects. Two examples of well-ordered compounds are shown in Fig. 3A for $(\text{2-aminopyridine})_{1/2}TaS_2$ (from 2H TaS_2) (*20*) and Fig. 3B for $(\text{ethylenediamine})_{1/4}TaS_2$ (from 1T TaS_2) (*18*). The return to 1T TaS_2 on deintercalation is also shown in Fig. 3B. The increase in line width and in the

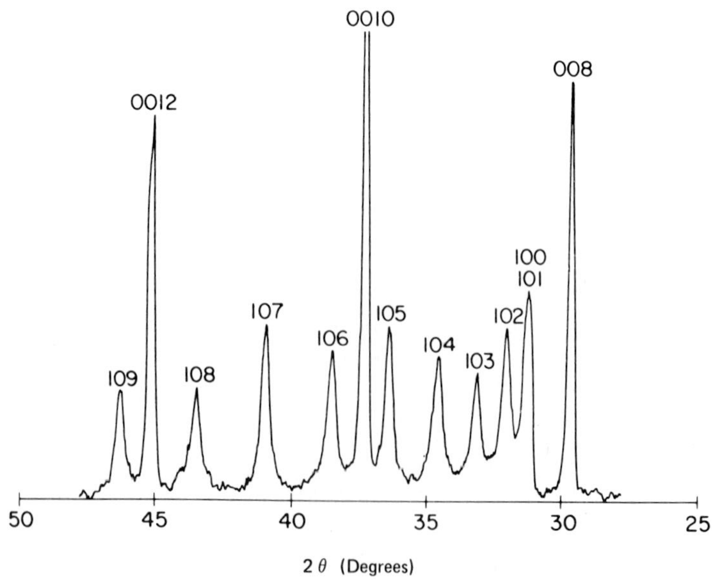

Fig. 3A. X-ray powder pattern of $(\text{2-aminopyridine})_{1/2}TaS_2$.[From Gamble *et al.* (*20*).]

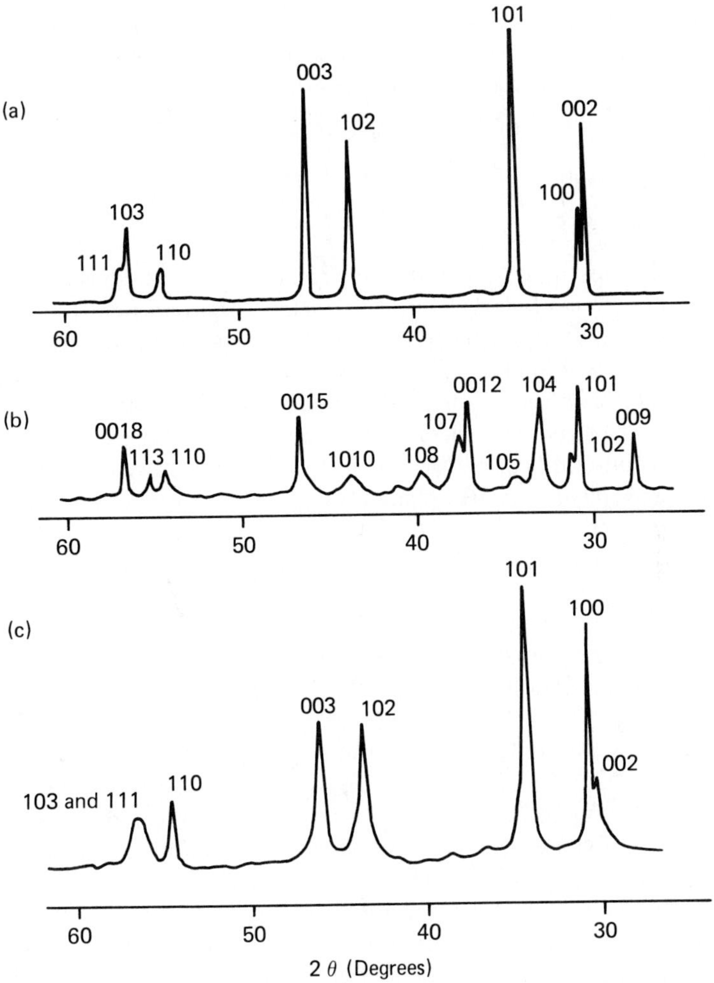

Fig. 3B. X-ray powder patterns of (a) 1T TaS_2, (b) (ethylenediamine)$_{1/4}TaS_2$, and (c) 1T TaS_2 from thermal deintercalation of (ethylenediamine)$_{1/4}TaS_2$. [From Meyer *et al.* (*18*).]

change in relative intensities of the (*hkl*) and (00*l*) reflections is indicative of stacking disorder and reduction in particle size. As the size of the guest molecule increases, disorder becomes more common; often only (*hk*0) and (00*l*) reflections are observed, and sometimes (00*l*) reflections alone are seen. In all cases, however, the most unambiguous evidence for intercalation is obtained from the X-ray data.

In the following sections two major groups of organic intercalation compounds, *n*-alkylamines and pyridines, are described in more detail. Pyridine and substituted pyridines are the most extensively studied guests and well illustrate the extent of our current understanding of structure and bonding in organic dichalcogenide systems. The *n*-alkylamines provide examples of large interlayer separations, and their chemistry parallels in some respects the well-established organic intercalation chemistry of clay minerals (*8*).

A. *n*-Alkylamines

One of the most striking intercalation reactions of the dichalcogenides is the reaction with *n*-alkylamines, $C_nH_{2n+1}NH_2$, first reported for the superconducting host 2H TaS_2 (*21*) because of the large *c*-axis expansions observed when *n* is large. A complete series of samples for $n = 1$ to 18 was prepared by direct reaction with the amine or the amine in benzene solution for $n > 12$ at 25°C for 30 days. For carbon numbers up to 4, the *c*-axis expansion remains constant, indicating that the hydrocarbon chains lie parallel to the dichalcogenide layers. As the chain length increases in this range, fewer amine groups per TaS_2 are introduced and the stoichiometry decreases. Eventually, the parallel orientation of the hydrocarbon chains gives way to a perpendicular orientation in which the stoichiometry now remains constant but the interplanar separation increases with *n*. The composition for the perpendicular orientation corresponds to $A_{2/3}TaS_2$ (A = amine), and the *c*-axis separation indicates a bilayer, as shown schematically for octadecylamine ($n = 18$) in Fig. 4. The —NH_2 groups are not shown but are thought to be adjacent to the layers and to interact with the Ta through the nitrogen lone pair. Figure 5 shows the variation in layer separation with increasing chain length. For alkylamines of intermediate chain length, $n = 5$ to 11, considerable disorder was observed and the lattice expansions could not be precisely determined. The critical temperatures for the onset of superconductivity were also determined for the same samples and are also shown in Fig. 5. A smooth decrease is observed from 4.2 to 1.8 K in the region where the stoichiometry is decreasing, but T_c is nearly constant (3 K) when the stoichiometry is constant and the interlayer separation is increasing, indicative of only weak coupling between the conduction electrons in adjacent dichalcogenide layers.

Other studies of *n*-alkylamine reactions with TaS_2 and with TiS_2 and NbS_2 have been reported (*22–24*). The direct reaction at 150°–200°C was used for TaS_2 and NbS_2, but a milder synthesis route was devised for TiS_2 and preferred for NbS_2 and TaS_2. In the indirect method, the dichalcogenide was preintercalated with ammonia or hydrazine and then reacted with amines with chain lengths up to $n = 9$ or 10. Complete reaction was observed at temperatures below 100°C in hours rather than days. Compounds of amines with chain lengths greater than 10 were prepared by displacement

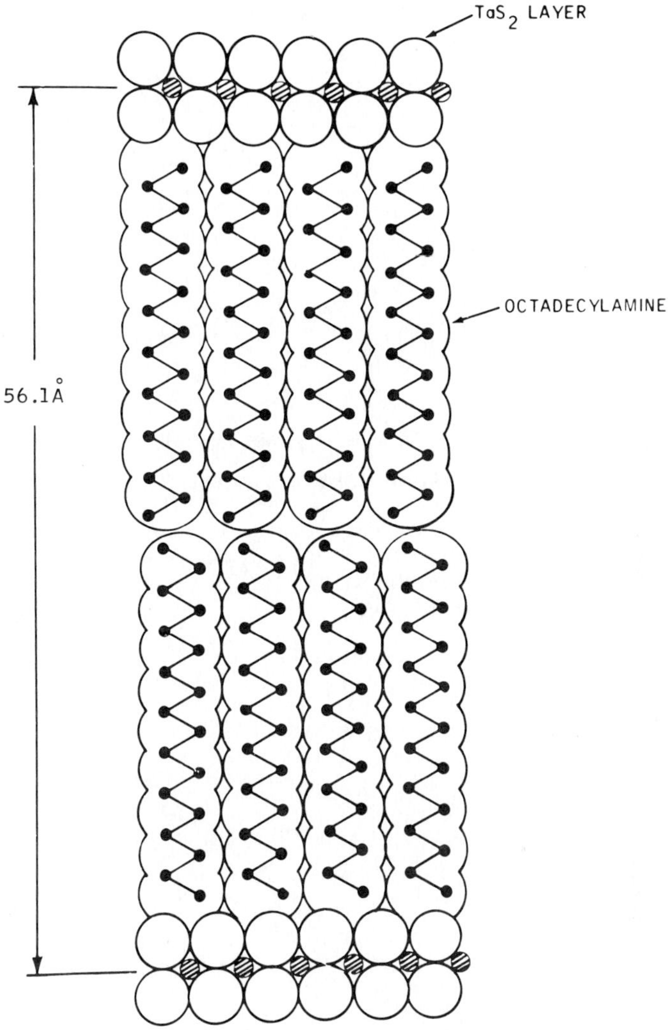

Fig. 4. Schematic representation of the structure of (octadecylamine)$_{2/3}$TaS$_2$. [After Gamble *et al. Science* **174,** 491–493 (1971); copyright 1971 by the American Association for the Advancement of Science.]

reactions of amine intercalation compounds of lower carbon numbers. The transition from a parallel orientation of amine for $n < 4$ to a perpendicular bilayer was found for all three hosts. However, on the basis of a calculation of the interlayer separation, an intermediate structure was proposed for $n = 4$ to 9 in which the amine still forms a bilayer but lies tilted at an angle of 56° to the c-axis direction. Figure 6 shows the measured c-axis spacings

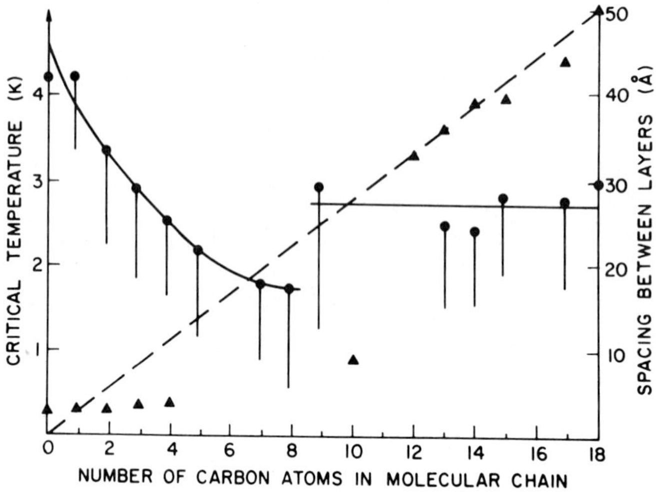

Fig. 5. Increase in the interlayer spacing (▲) and the onset temperature for superconductivity (●) as a function of n in $C_nH_{2n+1}NH_2$ for the n-alkylamine intercalation compounds of TaS_2. [From Gamble *et al. Science* **174,** 491–493 (1971); copyright 1971 by the American Association for the Advancement of Science.]

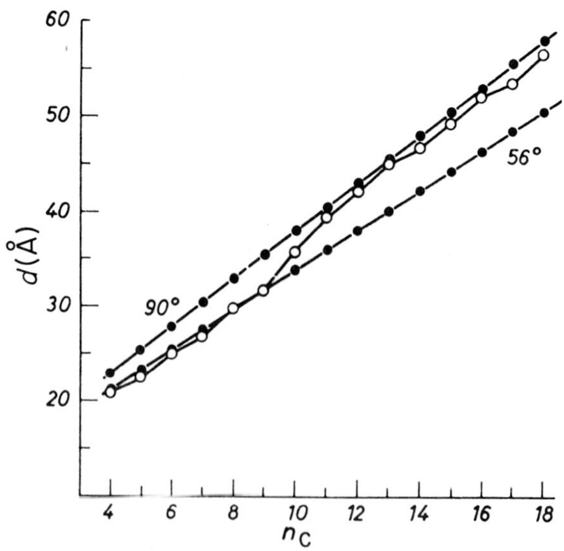

Fig. 6. Change in interlayer spacing as a function of the number of carbon atoms in the n-alkylamine intercalation compounds of TaS_2. The observed data (○) are compared with calculated results (●) for chain orientations of 90° and 56° (see text). [From Schöllhorn *et al.* (*24*).]

for TaS_2 and the calculated interlayer separations for an all-trans chain using the relation

$$d = d_{\text{host}} + 2[(n - 1)1.26 + 1.25 + 1.5 + 2.0]\ \text{Å}$$

where d and d_{host} are the c-axis spacings of the intercalation compound and the host lattice, respectively, n is the number of carbon atoms, 1.26 Å is the increase in the chain length per $—CH_2—$, 1.25 Å accounts for the C—N distance, and 1.5 and 2.0 Å are the van der Waals radii of $—NH_2$ and $—CH_3$. The data show a changeover from the 56° orientation to a 90° orientation at $n = 10$. Similar behavior was observed for TiS_2, although the changeover was found to be less sharp. Considerable disorder occurred in the region intermediate between the two orientations. In addition to structure changes with amine chain length, for a given amine both reversible phase transitions and stoichiometry changes occur on heating. For example, (octadecylamine)NbS_2 undergoes a reversible c-axis contraction from 58.6 to 46.1 Å at 52°C (*24*), and in (octadecylamine)$_{2/3}TaS_2$ a "melting" transition was observed by differential scanning calorimetry at 10°–15° above the melting point of the pure guest (*25*).

Staging of amines in the dichalcogenides does not seem to occur, although substoichiometric phases can be prepared. (Octadecylamine)$_{1/3}TaS_2$ was synthesized by direct reaction of TaS_2 with a stoichiometric amount of the amine for 4 days at 150°C (*25*). The 1:3 composition was also observed in thermogravimetric deintercalation of the 2:3 composition. However, the c-axis separation of 34 Å does not correspond to a second-stage compound, which would have a c axis of approximately 62 Å—56 Å for (octadecylamine)$_{2/3}TaS_2$ plus 6 Å for TaS_2. The result was interpreted as a structure similar to that of the fully loaded compound but collapsed by interleaving chains, which is possible because of the lower stoichiometry. Unlike octadecylamine, stearamide with the same chain length and $—CH_2NH_2$ replaced by $—CONH_2$ gives well-defined first- and second-stage compounds with $c = 57.0$ and 63.2 Å for the 2:3 and 1:3 compositions, respectively. Other than in this respect, long-chain amides (*16*, *17*, *21*, *26*) follow the same pattern as the amines: parallel orientations up to $n = 7$ and a perpendicular bilayer for $n = 18$. Reactions with amides are often slower than those with amines, and in some cases degradation of the organic compound has been observed.

The dichalcogenides of Group VI do not intercalate amines directly. However, n-alkylammonium compounds can be prepared by ion-exchange reactions of the hydrated sodium intercalation compound $Na_{0.1}(H_2O)_{0.6}MoS_2$ (*27*). The alkylammonium compounds of MoS_2 prepared in this way show lattice expansions that depend on the chain length in a very similar way to the systems described above (see Fig. 7). For n in $C_nH_{2n+1}NH_3^+$ in the

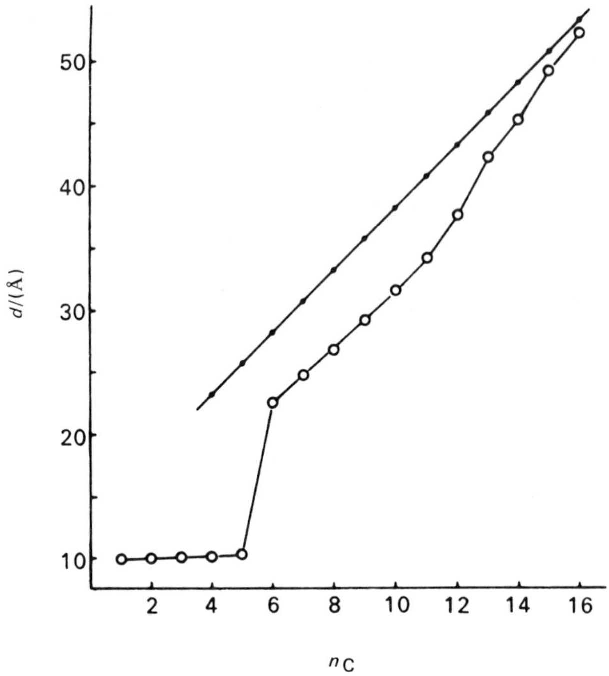

Fig. 7. Change in interlayer spacing as a function of the number of carbon atoms in the *n*-alkylammonium intercalation compounds of MoS_2. The observed data (○) are compared with calculated results (●) for a perpendicular orientation. [From Schöllhorn and Weiss (*27*).]

range 1–5 the *c*-axis spacing is constant; for $6 < n < 11$, c increases linearly with $\Delta d/n = 2.3$ Å per carbon atom, implying a bilayer tilted at 68°, and for $n > 11$, the alkylammonium cations are perpendicular to the layers. The transition from the parallel to the tilted orientation between $n = 5$ and $n = 6$ implies a layer charge of 50 Å/e, corresponding to a formulation $(RNH_3^+)_{0.15}MoS_2^{0.15-}$. Free base from the equilibrium

$$RNH_3^+ + OH^- \rightleftharpoons RNH_2 + H_2O$$

is also incorporated into the lattice for $n > 6$.

The marked similarities between this system, in which the host lattice is known to be reduced, and those prepared directly from the alkylamines and neutral hosts raises the question as to whether protonated amines are involved in both cases, particularly when the neutral host lattice has been pretreated with a reducing agent such as hydrazine. The occurrence of protonated species in the intercalation of organic bases has been considered in detail for pyridine and ammonia and is fully discussed in Chapter 10.

B. Pyridine and Substituted Pyridines

1. Synthesis

Several different synthetic techniques have been used to prepare pyridine intercalation compounds and are summarized in Table III. The most generally useful technique is the direct reaction (*21*, *28*, *29*) of dry pyridine with finely powdered (50 μm) dry host material at temperatures up to 200°C in a sealed tube. A variety of methods involving pretreatment of the host lattice with, for example, hydrazine hydrate have also been used (*28*, *29*). For single-phase stoichiometric 2H TaS_2, the direct reaction is rapid and the progress of the reaction may be followed visually. The limiting composition determined by weight uptake and elemental analysis is close to $(pyridine)_{1/2}TaS_2$. A number of preparations have given somewhat lower stoichiometries. The volume of the system decreases by 8% relative to 2H TaS_2 and liquid pyridine, only slightly greater than the decrease observed on crystallization of pyridine. The volume decrease is also observed for several other but not all substituted pyridines and is indicative of tightly packed pyridine molecules in the interlayer space. The direct reaction at 175°–200°C with dry pyridine does not give a single-phase product but a mixture of two phases with slightly different *c*-axis spacings. The two phases have *c* axes of 23.68–23.84 and 24.04–24.06 Å. One exception has been noted (*30*). A single phase with $c = 23.68$ Å was obtained, and it was suggested that the two-phase products might arise from starting TaS_2 containing stacking modifications other than 2H. However, each of the two phases can be prepared in pure form. The phase with the smaller *c* axis is obtained when the 2H TaS_2 is preintercalated with ammonia (*31*) or when small amounts of water are deliberately added to the pyridine (*32*). Electrochemical syntheses with TaS_2 as the cathode and pyridinium hydrochloride in methanol or ethanol as the electrolyte also lead to the formation of the 23.7 Å phase accompanied by hydrogen evolution (*33*). The electrochemical evidence is somewhat conflicting in that a compound with a *c* axis of 24.0 Å but a low pyridine content (0.2–0.3) has also been reported (*34*). All of these synthetic routes give more rapid reaction rates than the direct reaction under strictly anhydrous conditions, but all potentially can lead to the incorporation of protonated species along with neutral pyridine molecules (*30*, *32*, *35*).

The pure 24.04 Å phase has been obtained by addition of sulfur to the dry pyridine used in the reaction (*31*), and the results are reproducible (*36*). The 24.04 Å compound has been observed to convert to a mixture of the two phases on standing in air for several weeks (*37*). When prepared separately, the two phases show significant differences in thermal stability. The thermogravimetric analysis data are shown in Fig. 8. Loss of pyridine from

TABLE III

Synthesis of MX_2(Pyridine) Compounds

Host	Method[a]	Time (days)	Temp (°C)	a (Å)	c (Å)	Δc^{e} (Å)	Stoichiometry, x	Ref.
2H TaS_2	D1	1	200	3.326	23.70 }[b]	5.81	0.5	*1, 20, 21*
					24.04 }	5.98		
2H TaS_2	D3	1	120	3.327	23.74	5.83		*46*
2H TaS_2	I1	3	200		23.70	5.81		*31, 37*
2H TaS_2	D2	14	200		24.02	5.96		*31, 37*
2H TaS_2	D1	2	200		23.72 }[b]	5.82	0.48	*42*
					24.06 }	5.99		
2H TaS_2	D1	2	200		23.76 }[b]	5.84	0.43	*42*
					24.06 }	5.99		
2H TaS_2	D1	10 hr	100	3.32	23.68	5.80	0.49	*30*
2H TaS_2[e]	D1	1	200		23.84 }[b]	5.88		*36*
					24.06 }	5.99		
2H TaS_2[c]	D2	14	200		24.06	5.99		*36*
2H TaS_2	E		25		23.74	5.83		*33*
2H TaS_2	E		25		24.00	5.96	0.2–0.3	*34*
2H TaS_2	D1	4	175		23.68 }[b]	5.80	~0.3	*32*
					24.06 }	5.99		
2H TaS_2	I3	4	110		23.68	5.80	0.43	*32*
TaS_2[d]	D1	5	120		11.88	5.99		*29*
TaS_2[d]	I2	6	50		11.98	6.09		*29*
4Hb TaS_2	D1		150	3.365	47.384	5.917	0.5	*18*
2H TaS_2	D1	30	200	3.34	23.68	5.91	0.5	*21*
2H NbS_2	D1	3	100		? × 11.93	5.97		*29*
2H NbS_2					24.924			*42*
2H NbS_2[c]	D3		120	3.32	23.61	5.96		*40*
2H NbS_2[c]	D1	20	200	3.3	23.7	5.91		*41*
2H $NbSe_2$	E		25		24.42	5.94		*34*
$NbSe_2$	D1							*19*
2H $NbSe_2$	E		25		24.40	5.85		*34*
1T TiS_2	I1	1	90	3.426	25.13	6.02	0.5	*21*
1T TiS_2	I2	3	90		? × 11.62	5.91	0.44	*28*

[a] Synthesis method: D1, direct reaction in sealed tube; D2, direct reaction in sealed tube with pyridine saturated with sulfur; D3, refluxing pyridine; I1, direct reaction after intercalation and deintercalation of the host with ammonia; I2, reaction via the hydrazine hydrate intercalation compound; I3, pyridine and water; E, electrointercalation from pyridinium hydrochloride in methanol or ethanol.

[b] Two phases present in roughly equal amounts.

[c] Single crystals used.

[d] Polytype not given but probably 1T from c axis given as 5.89 Å.

[e] Lattice expansions calculated using the host c axis parameters 2H TaS_2(2 × 6.04 Å), 1T TaS_2 (5.85 Å), 4Hb TaS_2 (4 × 5.929 Å), 2H NbS_2 (2 × 5.945 Å), 2H $NbSe_2$ (2 × 6.272 Å), 2H $TaSe_2$ (2 × 6.346 Å), and 1T TiS_2 (5.694 Å).

[f] Detailed experimental procedures for the synthesis of 2H TaS_2 and (pyridine) TaS_2 have been given by J. F. Revelli, *Inorg. Synth.* **19,** 38 (1979).

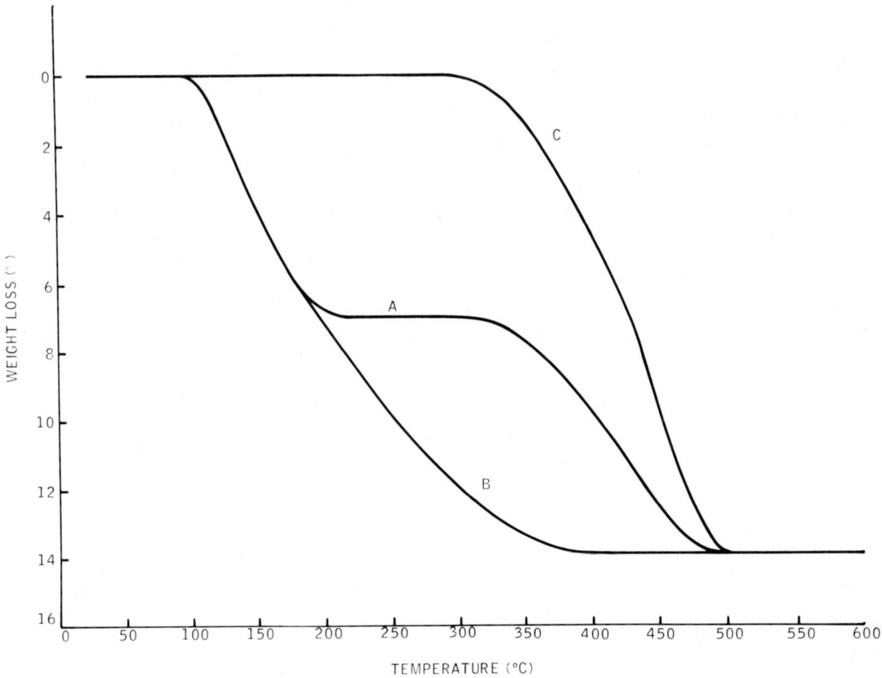

Fig. 8. Thermogravimetric analysis data for three different types of (pyridine)$_{1/2}$TaS$_2$. (A) Intercalated from pyridine directly; (B) preintercalated with ammonia and deintercalated before reaction with pyridine; (C) intercalated from pyridine saturated with sulfur. [After Thompson (*31*).]

the 23.7 Å phase begins at about 100°C, whereas no significant weight loss is observed below 300°C for the 24.04 Å compound. The thermogravimetric deintercalation of pyridine from a mixed-phase sample shows behavior characteristic of both of the two separate compounds, but the interpretation of these data is complicated by the possibility of formation of a second-stage compound in the intermediate plateau region. It has also been noted that the 24.04 Å compound, which is stable to higher temperature, typically gives a much sharper X-ray powder pattern with well-defined mixed reflections than does the 23.7 Å phase. The thermal desorption data suggest two different types of bound pyridine. Neutron diffraction studies of the exchange reaction (*38*)

$$(C_5H_5N)_{1/2}TaS_2 + 1/2C_5D_5N \rightarrow (C_5D_5N)_{1/2}TaS_2 + 1/2C_5H_5N$$

are also indicative of two different pyridine types. The reaction was followed by monitoring the relative intensities of the (002) and (004) reflections, which are sensitive to the isotope exchange. No exchange was observed with gaseous pyridine up to 100°C but was found with excess liquid C_5D_5N above 70°C, in agreement with NMR studies of pyridine mobility (*39*). Calculations of the intensities for different C_5H_5N/C_5D_5N ratios suggested that only 50% of the pyridine molecules were exchanging, indicating that half the pyridine molecules are more mobile than the other half. The neutron diffraction experiment does not distinguish between two different pyridine phases or two differently bound pyridine molecules within a single phase.

The origin of the difference in stability of the two phases is not understood at this time but may be important for an appreciation of bonding interactions in organic systems. Two-phase systems have also been reported in the reactions of 2H TaS_2 with the substituted pyridines 4-aminopyridine (23.94 and 24.80 Å) and 2-ethylpyridine (19.28 and 23.16 Å) (*20*). The syntheses of pyridine intercalation compounds of hosts other than 2H TaS_2 have been reported (Table III). Of these, the 2H NbS_2 reactions are the best documented. Powder samples (*21*, *29*) and single crystals (*39*, 40) were prepared by direct reaction and gave *c*-axis expansions in very good agreement. The *c*-axis expansion reported by Tofield and Wright (*42*) is anomalously large. Titanium disulfide was first found (*21*) not to intercalate pyridine readily by the direct reaction but only after preintercalation with ammonia. This system is very sensitive to nonstoichiometry, and highly stoichiometric material will intercalate 2,4,6-trimethylpyridine directly (*14*). Pyridine intercalation compounds of $TaSe_2$ and $NbSe_2$ have been prepared electrochemically (*32*) and also chemically for $NbSe_2$ (*19*). The compounds have lower stability than the corresponding disulfides and have been little studied. The 1T and 4Hb TaS_2 systems have also been described (*18*).

All of the pyridine intercalation compounds discussed above are first-stage, with 0.5 pyridine in every van der Waals gap. Second-stage compounds of TaS_2 may also be prepared, either by reaction of TaS_2 with a substoichiometric amount of pyridine or by reaction of 2H TaS_2 with $(\text{pyridine})_{1/2}TaS_2$ (*21*). A second-stage compound prepared by the second method at 200°C for 30 days gave a single-phase sample with a *c*-axis spacing of 18.0 Å, close to the sum of the *c*-axis separations of pure TaS_2 and first-stage $(\text{pyridine})_{1/2}TaS_2$. Ammonia also forms both first- and second-stage compounds, and it has been shown that, when the second-stage pyridine compound is contacted with ammonia at 1 atm and 23°C, ammonia is taken up into the empty van der Waals layer to give an alternately intercalated compound with the composition $(\text{pyridine})_{1/4}(NH_3)_{1/2}TaS_2$ (*43*).

The second-stage compound was observed in the initial stages of the reaction of 2H TaS_2 with C_5D_5N in a kinetic study using neutron diffraction (*38*). The reaction progress was followed by monitoring the (002) reflection of 2H TaS_2 and the (004) reflection from the first-stage compound. These two peaks coincide, but the (004) is much stronger. The data show a sigmoidal increase in first-stage intensity, indicating a nucleated process. Intermediate diffraction patterns showed, in addition, two broadened reflections at *d* spacings corresponding to the second stage. The intensity of the strongest of these two lines initially increased, reached a maximum, and then decreased as all the TaS_2 was converted to first stage. No other reflections were observed, and it was proposed that the reaction begins with second-stage formation and subsequent transformation to first stage. The positions and line widths of peaks in the intermediate diffraction patterns were analyzed in terms of a model of random stacking of blocks of first and second stage similar to treatments of the diffraction patterns of hydrated montmorillonites (*44*).

Both first- and second-stage (pyridine)TaS_2 compounds were examined by electron microscopy. Lattice images were obtained which showed clearly the repeat distances of 6, 12, and 18 Å between fringes of dark contrast corresponding to TaS_2 layers and light contrast from empty or organic layers (*45*). The electron microscope studies also revealed large numbers of lattice imperfections such as edge dislocations and stacking faults. Stacking faults were particularly prevalent in second-stage material, in which random alternations of TaS_2 and first- and second-stage regions were observed. In another electron microscope study similar results were found, although an anomalously low *c*-axis repeat of 10.1 Å was reported for first-stage (pyridine)$_{1/2}$$TaS_2$ (*46*).

2. *Structural Studies of the Pyridine Intercalation Compounds of 2H TaS_2 and 2H NbS_2*

Little is known concerning the detailed structures of the organic intercalation compounds of the dichalcogenides. Much of the reported work is concerned primarily with synthesis and X-ray lattice parameters, which, when given, are provided only as evidence that intercalation has occurred. In some cases, lattice expansions are used to deduce the orientation of the organic molecule, but such qualitative arguments are often ambiguous. Only for the cases of pyridine and ammonia have more detailed structural analyses been carried out. In this section the results for pyridine in 2H TaS_2 and 2H NbS_2 are described. In analyzing diffraction data for dichalcogenide intercalation compounds, several distinct structural aspects should be considered: the stacking sequence of the dichalcogenide layers, the orientation of the organic molecule between dichalcogenide layers, and the ordering of the organic species within and between layers. In (pyridine)$_{1/2}$$TaS_2$ the stacking sequence of the host lattice is well established. In the 2H host lattice the metal atoms lie directly above each other, and consequently the intensities of the (00*l*) reflections are weak when *l* is odd. However, on intercalation with pyridine, the intensities of the the (00*l*) lines become comparable for *l* even and *l* odd, indicating a translation of layers to a new stacking sequence (*20*). The X-ray data were more quantitatively analyzed (*47*) by calculation of (00*l*) intensities for an expanded TaS_2 lattice (ignoring the pyridine) with various layer stacking arrangements. The best agreement between the observed and calculated intensities was obtained with a translation from the *AcA*/*BcB* stacking of 2H TaS_2 to a sequence *AcA*/*AbA*. This arrangement is shown schematically in Fig. 9. The transformation brings the sulfur atoms directly above each other along the *c* axis but staggers the metal atoms.

The orientation of pyridine between the layers has been more controversial. Three distinct orientations have at various times been proposed and are shown in Fig. 10. None of the models can be distinguished from the X-ray data alone, but the model with the CN axis perpendicular to the layers (Fig. 10a) was initially suggested because this orientation apparently maximizes the interaction of the nitrogen lone pair of electrons with tantalum d orbitals. However, later work on the lattice expansions of 4-substituted pyridines (*20*) gave results incompatible with this model (Table IV). No regular *c*-axis increase was observed with ring substitution at the 4 position expected for perpendicular orientation. Consequently, the bilayer arrangement (Fig. 10c) was proposed.

The first definitive evidence as to the pyridine orientation came from a neutron diffraction study of single-crystal $(\text{pyridine})_{1/2}NbS_2$ (*39*). Samples of both pyridine and deuteropyridine intercalation compounds were pre-

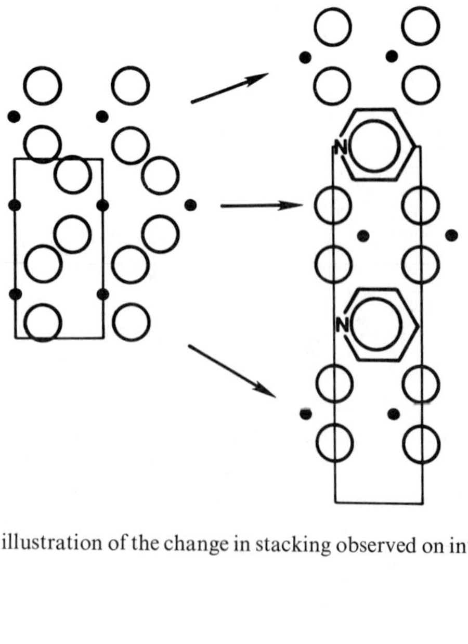

Fig. 9. Schematic illustration of the change in stacking observed on intercalation of pyridine into 2H TaS_2.

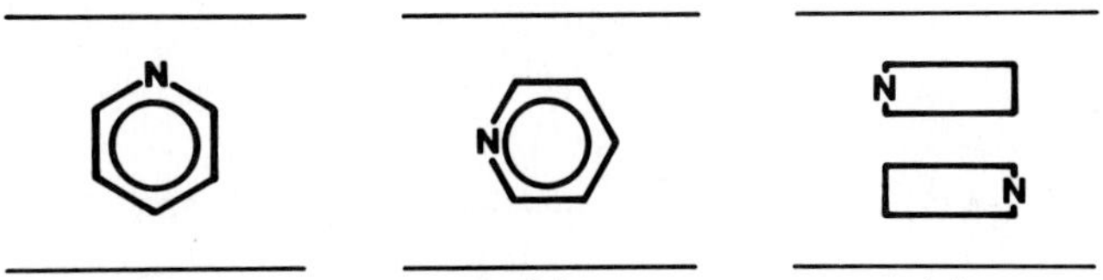

(a) (b) (c)

Fig. 10. Three models for the orientation of pyridine molecules between dichalcogenide layers.

pared from single crystals of NbS_2 by reflux with the anhydrous amines. The neutron diffraction experiment has two advantages over the previous X-ray work. The neutron intensities are more sensitive to the light atoms in the pyridine molecule, and hydrogen and deuterium have different scattering amplitudes and opposite phases for neutrons. The intensities of ten $(00l)$ reflections were measured for each crystal by film techniques, and the structure factors calculated were phased with the known NbS_2 atom parameters. One-dimensional Fourier maps were computed and are shown in Fig. 11. They reveal hydrogen atoms close to the layers identified by positive peaks for deuterium and negative peaks for hydrogen because of the opposite phases for the scattering amplitudes. Another hydrogen atom and the nitrogen were found to reside midway between the sulfur planes at $z = 1/2$. The only model that satisfactorily accounts for the data is one in which the long CN axis is parallel to the dichalcogenide layers (Fig. 10b). A similar model with respect to the nitrogen lone pair orientation had previously been proposed for $(NH_3)TaS_2$ from the NMR (*48*) and X-ray data (*49*). In a subsequent experiment (*38*) the same orientation was found for polycrystalline $(\text{pyridine})_{1/2}TaS_2$ ($c = 23.7$ Å) by a comparison of the measured in-

TABLE IV

Lattice Expansions for 4-Substituted Pyridines (*40*)

Pyridine	$c/2$ (Å)	Δc
Pyridine	11.85 }[a]	5.81
	12.02 }	5.98
4-Methylpyridine	11.76	5.72
4-Ethylpyridine	11.98	5.94
4-Isopropylpyridine	11.98	5.94
4-Propylpyridine	11.85	5.81
4-Phenylpyridine	12.31	6.27

[a] Two phases present in roughly equal amounts.

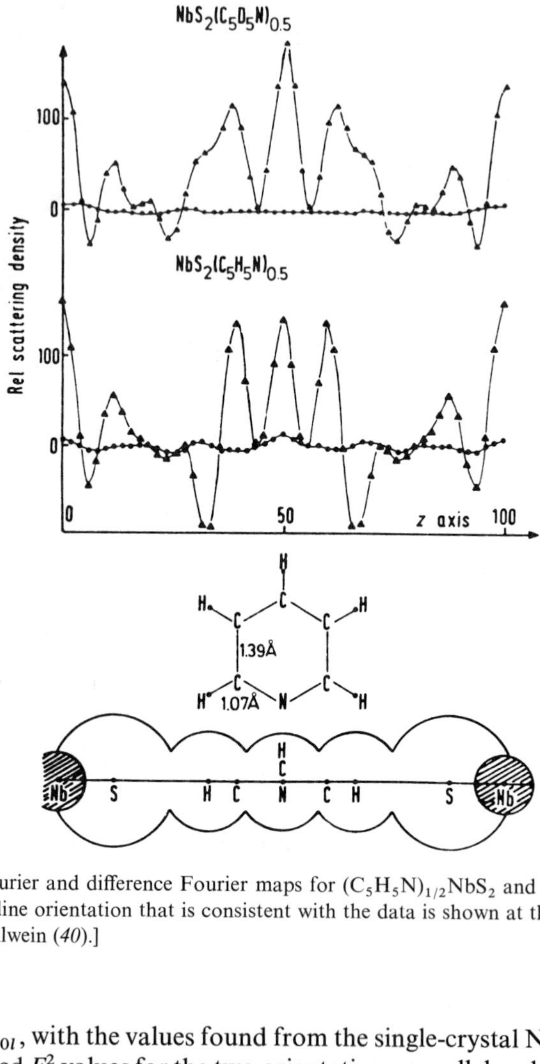

Fig. 11. Fourier and difference Fourier maps for $(C_5H_5N)_{1/2}NbS_2$ and $(C_5D_5N)_{1/2}NbS_2$ (top). The pyridine orientation that is consistent with the data is shown at the bottom. [From Riekel and Hohlwein (*40*).]

tensities, $F^2{}_{00l}$, with the values found from the single-crystal NbS_2 work and with calculated F^2 values for the two orientations, parallel and perpendicular (Fig. 12). The same orientation was proposed for the 24.06 Å phase of $(\text{pyridine})_{1/2}TaS_2$ based on results of a powder and single-crystal X-ray diffraction study (*50*). Eleven reflections (00, 14 to 00, 34) were measured using the single-crystal sample and the lower six orders from the polycrystalline phase. The structure factors were compared with calculations based on the three models, and best agreement was found for the parallel

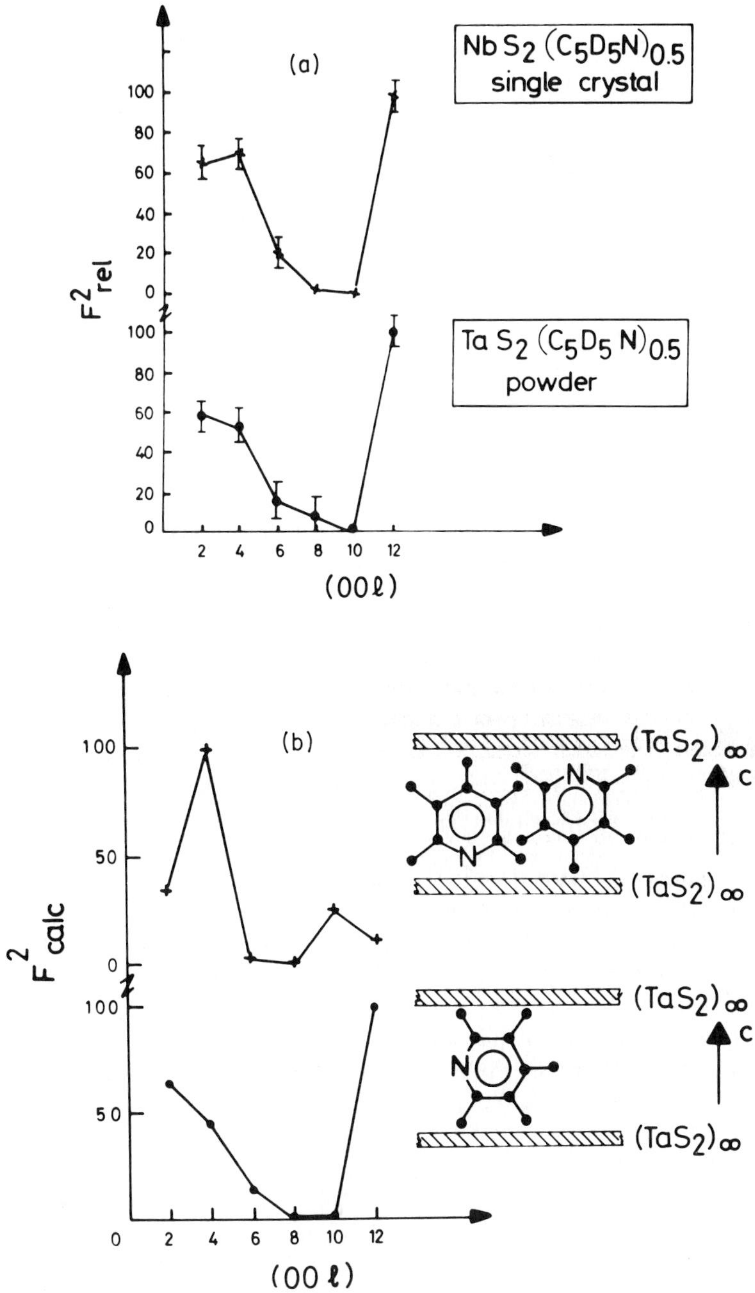

Fig. 12. (a) Comparison of the measured F^2_{00l} data for polycrystalline $(C_5D_5N)_{1/2}TaS_2$ and single-crystal $(C_5D_5N)_{1/2}NbS_2$. (b) Calculated values of F^2_{00l} for parallel and perpendicular orientations. Compare with (a). [From Riekel and Fischer (*38*).]

orientation. Because of the small X-ray scattering of pyridine relative to TaS_2, the distinction between the perpendicular and parallel orientation of the long CN axis is less well determined. The location of nitrogen in the center of the layer is surprising because, unlike the ammonia case, the distance from the center of the nitrogen to the center of a sulfur atom, 4.4 Å, is so long that it is difficult to understand the nature of the guest–host interaction. This point is further considered in the next section.

There is some evidence that the pyridine sublattice is ordered at room temperature in both $(pyridine)_{1/2}TaS_2$ and $(pyridine)_{1/2}NbS_2$. Superlattice reflections were found in both the X-ray and electron diffraction for $(pyridine)_{1/2}TaS_2$ and were indexed on a rectangular superlattice with dimensions $2a\sqrt{3} \times 13a$, where a is the TaS_2 hexagonal axis (*47*). The ideal composition requires one pyridine for every two formula units. However, the volume of the supercell could be interpreted with each pyridine occupying the area of $2\frac{1}{6}$ unit cells rather than 2, with consequently an ideal composition of $(pyridine)_{6/13}TaS_2$ or $(pyridine)_{0.46}TaS_2$. Compositions slightly less than 0.5 pyridine/TaS_2 have been noted in some preparations (*42*). Superlattice spots were also observed along a^* in single-crystal $(pyridine)_{1/2}NbS_2$ and were indexed as (101) and (201) with $a = 3.94$ Å (*38*). Some further weak reflections along a^* indicated that the true unit cell is larger. Systematic absences among the $(h0l)$ suggested that the pyridine molecules ordered in the xy plane were also registered along the c direction with every second sheet shifted by $(a/2) + (c/2)$ relative to the first sheet. A recent electron diffraction study of $(pyridine)_{1/2}NbS_2$ confirmed the enlargement of the unit cell to a rectangular lattice, $2a\sqrt{3} \times 13a$, identical to that reported for 2H TaS_2 (*41*).

A number of structural features are now well-understood for pyridine in both niobium and tantalum disulfides. The stacking sequence is known to change to *AcA*/*AbA*, and the pyridine is oriented perpendicular to the layer with the nitrogen in the center. Two- and three-dimensional ordering has been observed in some samples, but the details have not yet been worked out. The complexities of this "simple" intercalation system are illustrative of some of the difficulties involved in structural characterization of organic intercalation compounds. Further progress in this area may depend on the synthesis of more perfect, fault-free single crystals.

No detailed structural studies for substituted pyridine intercalation compounds have been made, but the lattice expansions for a large number of the known systems have been reviewed (*51*). A number of different models were proposed to account for the layer separations, but it was concluded that no one model can explain all of the data. In view of the unexpected orientation of pyridine in the 2H dichalcogenides and the marked effect of substituents both for organic and organometallic compounds (see Section IV,D), it seems unwise to offer generalizations.

C. Bonding in Organic Intercalation Compounds

The nature of the interaction between guest and host lattice in organic intercalation compounds is not completely understood. The organic intercalation compounds have been described as charge-transfer or donor–acceptor compounds in which charge is transferred from the organic molecule to the empty (Ti, Zr, Hf) or half-filled (Nb, Ta) d_{z^2} band of the dichalcogenide (*52*). The Group VI dichalcogenides with a filled low-lying d band do not directly intercalate amines. Consequently, it has been supposed that low-lying excited states in the host lattice are necessary for compound formation. Evidence from a variety of physical measurements has been cited in support of charge transfer. For example, the electric field gradient at the niobium nucleus observed in the niobium NMR of $(\text{pyridine})_{1/2}NbS_2$ was interpreted as resulting from a transfer of 0.2 electron to the conduction band (*53*). Analysis of the optical transmission spectra for pyridine compounds gave charge-transfer numbers of 0.25–0.3 electron per metal atom (*54*, *55*). Shifts in the nitrogen 1s level in the X-ray photoelectron spectra of several $NbSe_2$ and TaS_2 systems also qualitatively indicated charge transfer to the layers (*19*, 20).

For amine systems the charge transfer has been thought to arise by σ donation from the nitrogen lone pair orbital to the conduction band. Such a description requires at least some ligand basicity. An empirical correlation with basicity was found for the substituted pyridines (*20*). Molecules with pK_a values greater than 4.0 formed stable compounds with 2H TaS_2, whereas molecules with pK_a values less than 3.0 did not. In establishing correlations for low-temperature intercalation reactions, kinetics as well as thermodynamics play a role. Failure to intercalate 4-*tert*-butylpyridine after 24 days at 200°C even though its pK_a is 6.0 is probably due to steric effects on the kinetics of intercalation of this large molecule. The basicity correlation works well for amines, but a number of other systems such as amides (*16*, *17*, *21*, *26*), amine oxides (*56*), and isocyanides (*57*), which are weak bases, do form stable intercalation compounds. The isocyanides are perhaps the most extreme case in that they form very thermally stable complexes with TaS_2 but are very weak bases. The stability of these systems has been considered to arise from a combination of both σ donor and π acceptor properties of the isocyanides. The observed expansions for the 2H TaS_2 compounds with methyl, *n*-butyl, and benzyl isocyanides are consistent with an orientation in which the R—N—C axis is parallel to the layers. In this orientation π antibonding levels in the isocyanide can interact with the sulfur atoms. The argument could also be extended to phosphines, which are not as basic as amines but, like the isocyanides, have π acidity. Clearly, some σ-donor properties are required since the strong π acid CO, which is a poor σ donor, is not known to form intercalation compounds.

A difficulty with the σ lone pair donor model arises with the ammonia and pyridine intercalation compounds, which have the nitrogen lone pair midway between and parallel to the layers, precluding a direct interaction with a metal d_{z^2} orbital. For ammonia, it has been suggested that interaction with the sulfur atoms can lead to electron transfer to the condution band because of strong covalency between the transition metal and sulfur (*48*). However, it is difficult to apply this to the pyridine case because the sulfur–nitrogen distance is much greater. The vibrational spectrum measured by inelastic neutron scattering (*42*) is also not consistent with direct lone pair σ donation. To resolve this question an alternative description of the bonding in pyridine and ammonia intercalation compounds has recently been given (*30*) and is discussed together with the experimental basis in Chapter 10. In this model the bonding is described as an electrostatic interaction between negatively charged layers and cations, analogous to alkali and organometallic intercalation compounds and the bonding in clay minerals. In the case of pyridine compounds, the reaction scheme that leads to reduced TaS_2 layers and pyridinium cations is

$$2\ \text{pyridine (py)} \rightarrow \text{bipyridine (bipy)} + 2H^+ + 2e$$
$$x\text{py} + xH^+ \rightarrow x\text{pyH}^+$$
$$x\text{pyH}^+ + (0.5 - x)\text{py} + xe + TaS_2 \rightarrow (\text{pyH}^+)_x(\text{py})_{0.5-x}TaS_2$$

Some bipyridine is also found in the final intercalation compound. The product of the scheme is stabilized by ionic bonding between the guest cations and the negative layers and by hydrogen bonding between pyridinium cations and solvating pyridine. Several experimental observations can be rationalized by the ionic description: The correlation with basicity observed for the substituted pyridines is related to the stability of the protonated amine, the volume reduction on pyridine intercalation is associated with the hydrogen bonding and with the conversion of some pyridine to bipyridine, and the shifts in the nitrogen 1s level are due to protonated amines. The ionic model was also used to interpret time differential perturbed angular correlation studies of the tantalum nuclear quadrupole precession (*30*). The observed quadrupole frequency gave a charge transfer of 0.25 electron per TaS_2, and the abrupt change in the quadrupole frequency at 60 K was associated with the onset of proton exchange between neutral pyridine and pyridinium cations. Electrochemical intercalation (*30*) gave a sharp discontinuity in the voltage charge curves also at 0.25 electron per TaS_2. At this level of electron transfer each pyridinium cation has a free base partner for hydrogen bonding. It is interesting that all the experimental measurements indicate about 0.25 electron transferred to the conduction band. The difference between the ionic and covalent models rests on the presence or absence of 0.25 hydrogen ion per mole of TaS_2.

The ionic model is attractive in that it accounts for most of the experimental data and provides a single bonding description for all transition metal dichalcogenide intercalation compounds. There remain, however, a few discrepancies that must be resolved, such as the widely differing stabilities of the two (pyridine)TaS_2 phases and the observation that at 70°C only half the pyridine in (pyridine)$_{1/2}TaS_2$ can be exchanged. Nevertheless, it is clear that we are approaching an excellent understanding of one of the earliest reported organic intercalation compounds. If it is the case that pyridine, an exceptionally stable organic molecule, can be oxidatively coupled by TaS_2 to form bipyridine, then we must look for similar reactions in all of the other organic intercalation systems that have so far been studied.

IV. ORGANOMETALLIC INTERCALATION COMPOUNDS

The first organometallic intercalation compounds of the dichalcogenides were prepared by reaction of the metallocenes, cobaltocene and chromocene, with several host lattices (*58*). As in the case of the organic compounds described above, the guest organometallic compound is inserted into the interlayer space to give structures similar to that shown schematically for (cobaltocene)$_{1/4}TaS_2$ in Fig. 13. All of the organometallic compounds known to form intercalation compounds are very good reducing agents, and there is clear evidence for electron transfer from the guest molecule to the host lattice. In this respect, organometallic guests more closely resemble the alkali metals (*5*) than do organic compounds.

The presently known organometallic intercalation compounds are surveyed in Table V together with their synthesis conditions, stoichiometries, and X-ray data. It is apparent that although a large number of compounds have been prepared they represent only a very restricted group of organometallic molecules.

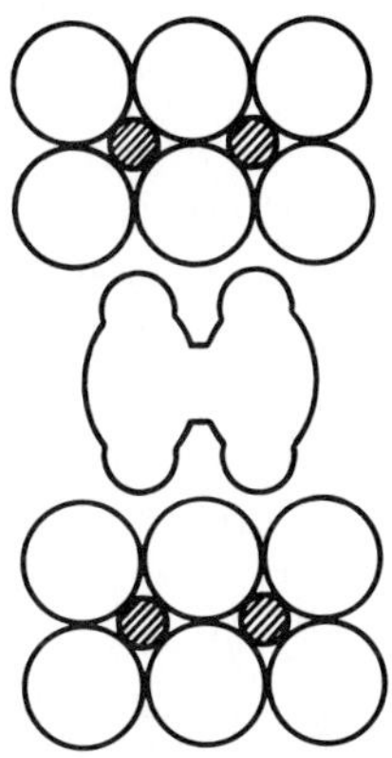

Fig. 13. Schematic illustration of the structure proposed for (cobaltocene)$_{1/4}TaS_2$.

TABLE V

Organometallic Intercalation Compounds of the Layered Dichalcogenides

Host	Organometallic guest	Reaction conditions (temp., °C; time, days)	Stoichiometry (x in G_xMX_2)	a (Å)	c (Å)	Δc (Å)	Ref.
Direct reactions							
TiS_2	Cobaltocene	23; 19	0.20			5.55	*58*
	Chromocene	100; 3	0.30			5.42	*58*
$TiSe_2$	Cobaltocene	23; 19	0.38[a]			5.52	*58*
ZrS_2	Cobaltocene	100; 4	0.27			5.35	*58*
	Chromocene	130; 3	0.25		34.31	5.61	*59*
	(η-Benzene)(η-cyclopentadienyl) chromium	130; 3	0.24		35.70	6.07	*59*
	(η-Cycloheptatrienyl)(η-cyclopentadienyl) chromium	130; 3	0.25		36.0	6.17	*59*
	(η-Cyclooctatetraene)(η-cyclopentadienyl) titanium	130; 3	0.23		36.69	6.40	*59*
	Bis(η-Benzene)chromium	130; 3	0.16		35.20	5.90	*59*
	Bis(η-Benzene)molybdenum	130; 3	0.16		34.93	5.81	*59*
	Bis(η-Toluene)molybdenum	130; 3	0.13		34.90	5.80	*59*
	Bis(η-Mesitylene)molybdenum	130; 3	0.08		34.83	5.78	*59*
	1,1′-Dimethylcobaltocene	130; 3	0.25	3.70	33.5	5.34	*7, 60*
	1,1′-Diisopropylcobaltocene	130; 3	0.15	3.65	34.7	5.74	*7, 60*
	1,1′-Di-n-butylcobaltocene	130; 3	0.13	3.67	33.5	5.34	*7, 60*

$ZrSe_2$	Cobaltocene	130; 3	0.25				*7, 60*
HfS_2	Cobaltocene	100; 5	0.38[a]			5.48	*58*
	Chromocene	200; 4	0.20			5.63	*58*
VSe_2	Cobaltocene	130; 3	0.25				*7*
$HfSe_2$	Cobaltocene	23; 19	0.31			5.53	*58*
	Chromocene	100; 7	0.20			5.56	*58*
2H TaS_2	Cobaltocene	23; 24	0.23			5.47	*58*
	Chromocene	100; 4	0.28			5.52	*58*
1T TaS_2	1,1′-Dimethylcobaltocene	130; 3	0.21	3.33	34.78	5.68	*7, 60*
	1,1′-Diisopropylcobaltocene	130; 3	0.17	3.32	36.15	6.18	*7, 60*
	1,1′-Di-*n*-butylcobaltocene	130; 3	0.13	3.37	34.31	5.53	*7, 60*
$TaSe_2$	Cobaltocene	100; 3	0.42[a]			5.45	*58*
	Chromocene	100; 4	0.25			5.57	*58*
SnS_2	Cobaltocene	100; 4	0.29			5.35	*58*
Ion-exchange reactions							
1T TaS_2	Cobaltocenium iodide		0.20	3.39	34.15	5.50	*7, 60*
	Bis(η-Benzene)chromium iodide		0.13	3.36	35.85	6.04	*7, 60*
	Bis(η-Benzene)molybdenum iodide		0.15	3.38	35.68	5.99	*5, 60*
	(η^5-Cyclopentadienyl)(η-benzene)iron hexafluorophosphate		0.16	3.38	34.90	5.72	*7, 60*
	1,1′-Dimethylcobaltocenium iodide		0.14				*7, 60*
	1,1′-Diisopropylcobaltocenium iodide		0.07				*7, 60*
	1,1′-Di-*n*-butylcobaltocenium iodide		0.05				*7, 60*
Electrointercalation							
2H TaS_2	Cobaltocenium		0.25	3.30	22.90	5.41	*61*
	Chromocenium					5.87	*61*

[a] Excess solid metallocene present.

A. Synthesis

Organometallic intercalation compounds are prepared by direct reaction. Typically, toluene solutions of the organometallic guests are heated with the solid hosts in sealed tubes at temperatures up to about 130°C. Reaction kinetics are often slow, particularly for bulky guests, and control of the dichalcogenide particle size is necessary to obtain reproducible results. Reactions at higher temperatures to increase rates are usually not feasible due to side reactions, which lead to decomposition of the organometallic compound. Of all the guests that have been investigated, cobaltocene appears to react most readily and is preferred for screening new host lattices.

Ion-exchange reactions in aqueous solution have been used successfully for TaS_2 (*60*), which is moderately stable to hydrolysis. Reactions conveniently use $Na_{1/3}(H_2O)_{2/3}TaS_2$ as a starting point and an aqueous solution of the organometallic salt and have been carried out at room temperature. The hydrated sodium intercalation compound is easily prepared by the reaction of TaS_2 with aqueous sodium dithionite (*62*). Ion-exchange reactions typically lead to lower stoichiometries and powder diffraction patterns of poorer quality than do direct reactions. This is illustrated by the data in Table V for the 1,1′-substituted cobaltocene compounds with 1T TaS_2, which were prepared by both techniques. However, in two cases, bis(η-benzene)chromium and bis(η-benzene)molybdenum with 1T TaS_2, compounds that could not be prepared by direct reaction were successfully synthesized by ion exchange. Aqueous ion exchange is clearly too restrictive a procedure for reasons of both guest and host stability, but in nonaqueous solvents reactions are limited by kinetics. For example, exchange of $[Co(\eta\text{-}C_5H_5)_2]^+I^-$ with $Na_{1/3}(H_2O)_{2/3}TaS_2$ was found to be slow in both methanol and ethanol, and no exchange was observed in tetrahydrofuran containing 18-crown-6 ether to solubilize the sodium ions (*60*).

Organometallic cations may also be intercalated from aqueous or nonaqueous solutions electrochemically (*61*). The syntheses of the chromocene and the cobaltocene intercalation compounds of 2H TaS_2 have been described (*61*), and the cobaltocene phase is identical in composition and lattice parameters to samples prepared by direct reaction. The electrointercalation technique gives additional information about intermediate compositions or nonstoichiometry. The electrochemical data for the system TaS_2-cobaltocene show that first stage $[Co(\eta\text{-}C_5H_5)_2]_{1/4}TaS_2$ is the only compound formed up to 0.25 electron per TaS_2. There is no evidence for intermediate compositions or higher stages. No evidence for a second-stage compound was found in the direct reaction of $[Co(\eta\text{-}C_5H_5)_2]_{1/4}TaS_2$ with a stoichiometric equivalent of TaS_2.

B. X-ray Diffraction

The quality of powder X-ray diffraction data for organometallic intercalation compounds varies considerably and depends on the reaction conditions and the guest species. The powder patterns show well developed ($00l$) reflections, often out to high angles, from which the interlayer separation may be determined. In many cases (see Table V) the diffraction pattern are sharp and show mixed reflections, provided that preferred orientation effects are minimized. The well-developed order is indicative of good register between adjacent layers and may be a consequence of a good fit of the organometallic sublattice to the hexagonal a axis of the host. Cobaltocene, for example, has van der Waals dimensions of 6.4–6.9 Å, which are close to $2a$ for the host disulfides. The powder diffraction pattern for a well-ordered compound, $[(\eta\text{-}C_6H_6)_2Mo]_{0.16}ZrS_2$, is shown in Fig. 14. For this compound and several others, the mixed reflections require that the c axis be tripled. The unit cell, however, is not rhombohedral since the reflections (106) and (10, 12), which do not obey rhombohedral absences, are strong. A similar effect has been observed in $LiHfS_2$ (*12*).

No detailed structures for any of the organometallic intercalation compounds have been determined.

C. Bonding

The formation of an organometallic intercalation compound involves oxidation of the organometallic guest (G) and reduction of the host lattice:

$$xG + MX_2 \rightarrow G_x{}^{+}MX_2{}^{x-}$$

Syntheses via ion exchange or electrointercalation provide indirect evidence that this bonding scheme is appropriate. Direct evidence has been obtained for cation intercalation by magnetic susceptibility measurements for $[Co(\eta\text{-}C_5H_5)_2]_{1/4}TaS_2$ and $[Cr(\eta\text{-}C_5H_4)_2]_{1/4}TaS_2$ (*63*). Both neutral cobaltocene and chromocene are paramagnetic, with magnetic moments of 1.81 and 3.15 μ_B corresponding to one and two unpaired electrons, respectively. The magnetic susceptibility χ_m of cobaltocene in TaS_2 is, however, only 8.8×10^{-5} emu/mole, somewhat less than that of 2H TaS_2 itself, and is independent of temperature down to 2 K. The loss in local moment on intercalation is consistent with complete transfer of one electron from cobaltocene to the host lattice and the formation of $[Co(\eta\text{-}C_5H_5)_2]^+$. For chromocene in TaS_2, the magnetic susceptibility obeys a Curie–Weiss law (see Fig. 15), and the Curie constant gives an effective moment of 3.76 μ_B. The observed moment is in good agreement with measurements on chro-

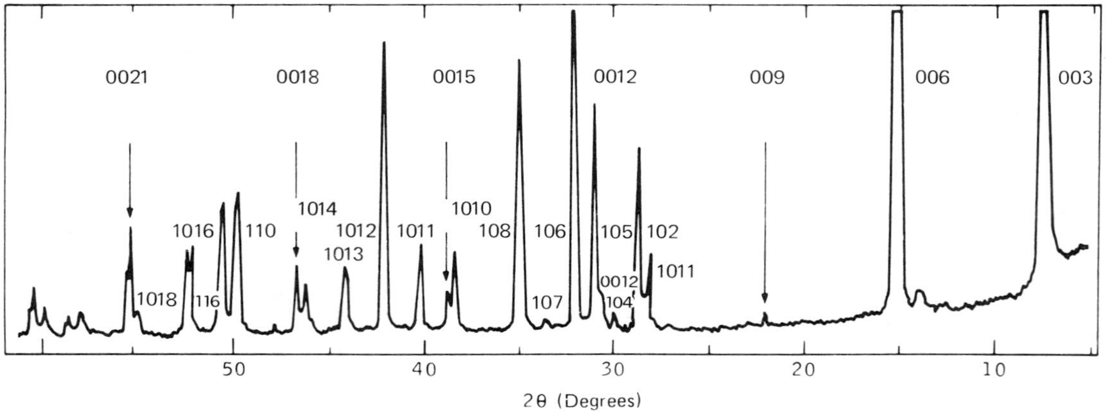

Fig. 14. Powder X-ray diffraction data for $[(\eta\text{-}C_6H_6)_2Mo]_{0.16}ZrS_2$. [From Clement *et al.* (*60*).]

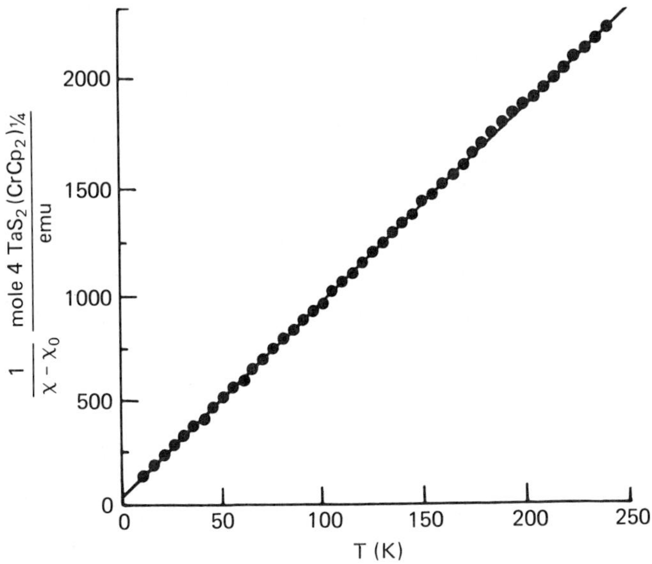

Fig. 15. Temperature dependence of the magnetic susceptibility of $[(\eta\text{-}C_5H_5)_2Cr]_{1/4}TaS_2$. [From Gamble and Thompson, *Solid State Commun.* **27**, 379 (1978); copyright 1978 by Pergamon Press, Ltd.]

mocenium salts and again indicates complete electron transfer to the host. The main motivation for the magnetic measurements was to investigate the effect of interlayer paramagnetic species on the superconducting properties of 2H TaS_2. The measured temperatures for the onset of superconductivity were 3.2 and 2.9 K for the cobaltocene and chromocene compounds, respectively. This small difference observed on insertion of a dense two-dimensional array of spins between layers suggested that the superconductivity is two-dimensional and confined to the metallic layers and that coupling between layers is not important for the stability of the superconducting state.

The ionic model for the bonding in organometallic intercalation compounds has consequences in predicting which organometallic compounds are likely to form stable intercalates. The extensive studies of the free energies of lithium intercalation reactions of the disulfides of Groups IV and V show that all of these host lattices have similar oxidizing properties. Consequently, an indication of stability can be obtained by comparing the reducing power of the guests. A qualitative scale of reducing power may be established by comparing first ionization potentials measured by photoelectron spectroscopy.

Examples of organometallic compounds that have been reacted with dichalcogenides are listed in Table VI. A survey of the known compounds shows that direct reactions occur with disulfide hosts and organometallic guests with first ionization potentials of less than 6.2 eV, provided that the

TABLE VI

First Ionization Potentials of Some Organometallic Compounds

Compound[a]	IP (eV)	Ref.
$(bnz)_2Mo$	5.52	*64*
$(tol)_2Mo$	5.32	*64*
$(mes)_2Mo$	5.13	*64*
$(bnz)_2Cr$	5.4	*65*
$(cp)_2Co$	5.56	*66*
$(cp)_2Cr$	5.71	*66*
(bnz)(cp)Cr	6.20	*64*
(cht)(cp)Cr	5.59	*64*
(cot)(cp)Ti	5.67	*64*
$(cht)_2U$	7.2	*67*
$(bnz)_2V$	5.2	*68*
$(cp)_2MoC_2H_4$	6.0	*64*
$(cp)_2MoH_2$	6.4	*64*
$(cp)_2Fe$	6.88	*69*

[a] Abbreviations: bnz, η^6-benzene; tol, η^6-toluene; mes, η^6-mesitylene; cp, η^5-cyclopentadienyl; cht, η^7-cycloheptatrienyl; cot, η^8-cyclooctatetraene.

cation formed is stable to isolation; $(\eta\text{-}C_6H_6)_2V$ has a low ionization potential (5.2 eV) but does not form a stable cation. It has also been observed that no organometallic compounds with nonparallel carbocylic rings form stable phases even when the ionization potential criterion is satisfied. Less reducing guests, for example, ferrocene (6.88 eV), can react with more oxidizing host lattices such as FeOCl. Magnetic susceptibility and Mössbauer measurements for $[(\eta\text{-}C_5H_5)_2Fe]_{0.16}FeOCl$ also show complete electron transfer from guest to host lattice (*70, 71*).

D. Orientation and Stoichiometry

The orientation of the guest molecule in the interlayer space and the stoichiometry of the intercalation compound are interrelated. No detailed crystal structures are known, and consequently orientations have been inferred by comparison of the lattice expansions along the *c* axis with estimated van der Waals dimensions for the guest molecule. Ambiguities associated with this approach have been discussed for organic compounds, and similar problems arise for organometallic molecules. Another complication is found in that the *c*-axis expansions are typically 1 Å smaller than predicted molecular dimensions, and consequently at best we can only use trends in expansions together with stoichiometries to give information about orientation.

The simple metallocenes, cobaltocene and chromocene, are nearly spherical, with estimated van der Waals dimensions of 6.4 Å parallel and 6.9 Å perpendicular to the C_5 axis. The originally proposed orientation with the C_5 axis parallel to the layers is supported by temperature-dependent proton NMR measurements (*72*). The temperature variation of the second moment (M_2) of the proton resonance is shown in Fig. 16 for $[(\eta\text{-}C_5H_5)_2Co]_{1/4}TaS_2$. On cooling from room temperature, the resonance broadens in the vicinity of 245 K and the second moment increases from 0.47 to 0.77 G^2. Further cooling produces additional broadening at 8 K (0.77 to 5 G^2). The changes in line width occur as first reorientation of the C_5 axis and subsequently C_5 spinning processes become inoperative in narrowing the resonance. A quantitative calculation of the magnitudes of the second moments for the three temperature regimes for a model with the C_5 molecular axis parallel to the layers gave very good agreement with the experimental observations.

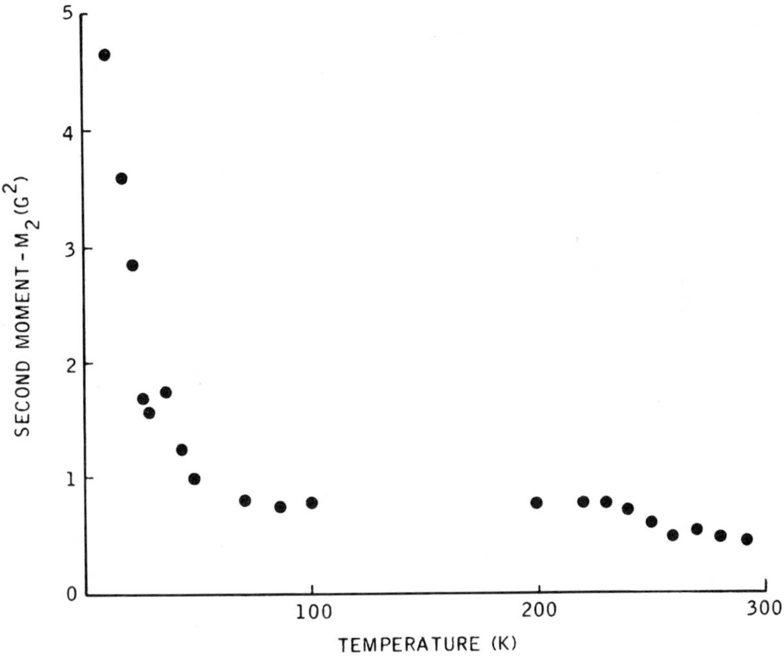

Fig. 16. Variation of the second moment of the proton NMR line width temperature for $[(\eta\text{-}C_5H_5)_2Co]_{1/4}TaS_2$. [From Silbernagel (72).]

TABLE VII

Lattice Expansions for Some Intercalation Compounds of ZrS_2

Compound	Δc (Å)
$(\eta\text{-}C_5H_5)_2Co$	5.35
$(\eta\text{-}C_5H_5)_2Cr$	5.61
$(\eta\text{-}C_6H_6)(\eta\text{-}C_5H_5)Cr$	6.07
$(\eta\text{-}C_7H_7)(\eta\text{-}C_5H_5)Cr$	6.17
$(\eta\text{-}C_8H_8)(\eta\text{-}C_5H_5)Ti$	6.40

Lattice expansions for a series of intercalation compounds with one cyclopentadienyl ring and one ring of increasing size are shown in Table VII. As the second ring size increases, the lattice expansion also increases, suggesting that for this series the principal molecular axis also lies parallel to the layers. All molecules give compounds of stoichiometry $G_{1/4}MX_2$, which is consistent with close packing of the guests onto the dichalcogenide lattice. The cross-sectional area occupied by close-packed metallocenes in the parallel orientation is about 37 $Å^2$, to be compared with the area per $4MX_2$ units of 37–40 $Å^2$. Substitution of alkyl groups in cobaltocene has been investigated for both 1T TaS_2 and ZrS_2 (Table VIII). The lattice expansion for 1,1′-diisopropylcobaltocene is significantly larger than that for the other substituents in both TaS_2 and ZrS_2. This observation suggests an orientation of the organometallic molecule with the rings parallel to the layers. The

TABLE VIII

Lattice Expansions and Stoichiometries for Substituted Metallocene and Arene Intercalation Compounds of 1T TaS_2 and ZrS_2

	1T TaS_2		ZrS_2	
	Δc (Å)	x	Δc (Å)	x
Cobaltocene	5.47	0.25	5.34	0.25
1,1′-Dimethylcobaltocene	5.68	0.21	5.34	0.25
1,1′-Di-*n*-butylcobaltocene	5.53	0.17	5.34	0.15
1,1′-Diisopropylcobaltocene	6.18	0.13	5.74	0.13
$(\eta$-Benzene$)_2$Mo			5.81	0.16
$(\eta$-Toluene$)_2$Mo			5.80	0.13
$(\eta$-Mesitylene$)_2$Mo			5.78	0.08

methyl groups point away from and the hydrogen atoms point directly toward the sulfur layers. The effective van der Waals length of the molecule is increased relative to those of the other substituted derivatives, which, if they were to adopt the same orientation, should have smaller expansions. The stoichiometries of the 1,1′-substituted cobaltocenes decrease with the size of the alkyl substituents although the 1,1′-dimethyl compounds have compositions close to 0.25. A similar decrease in stoichiometry is observed for the bisarene series (Table VIII), but here the *c*-axis expansions are closely similar. The same *c*-axis expansion for both $(benzene)_2Mo$ and $(mesitylene)_2Mo$ and the decrease in stoichiometry from 0.25 suggest that the arene series are oriented with the rings parallel to the dichalcogenide layers.

Clearly both the orientation and the stoichiometries of organometallic intercalation compounds are sensitive in a subtle way to the relative importance of guest–guest and host–guest interactions. The parallel orientation of the principal molecular axis favors π interactions with neighboring molecules as found in organic donor–acceptor complexes, and the perpendicular orientation favors interaction of the π-electron system with the sulfur layers. Failure to intercalate molecules like 1,1′-di-*tert*-butylcobaltocene or $(\eta\text{-}C_5H_5)_2MoH_2$ may reflect steric inhibition of both types of bonding interactions. However, in low-temperature intercalation reactions with large molecules, kinetic rather than thermodynamic considerations may dominate.

E. Summary

A variety of organometallic molecules with low ionization potentials react with layered transition metal dichalcogenides to give well-characterized crystalline intercalation compounds. The reactions proceed via a redox process, leading to the formation of a reduced host lattice and a guest cation. Compounds are known only for organometallic sandwich compounds with parallel rings, and stoichiometry and orientation of the guest cation are sensitive to ring substituents. These conditions for the formation of intercalation compounds have proved to be a serious restriction in the search for organometallic systems with interesting catalytic or stoichiometric reactivity. Most organometallic compounds that satisfy the apparent criteria for compound formation are inherently stable and unreactive. It remains to be seen whether the development of new synthetic procedures or the exploitation of other host lattices will lead to new compounds with which reactions of intercalated organometallic molecules can be studied. Investigations of two other host lattices, MOCl and MPS_3, have been made and are described in Chapters 8 and 12.

REFERENCES

1. F. R. Gamble, F. J. DiSalvo, R. A. Klemm, and T. H. Geballe, *Science* **168,** 568 (1970).
2. P. M. Williams, G. S. Parry, and C. B. Scruby, *Philos. Mag.* **29,** 695 (1974).
3. J. A. Wilson, F. J. DiSalvo, and S. Mahajan, *Phys. Rev. Lett.* **32,** 882 (1974).
4. M. S. Whittingham, *Science* **192,** 1126 (1976).
5. M. S. Whittingham, *Prog. Solid State Chem.* **12,** 41 (1978).
6. O. Weisser and S. Landa, "Sulfide Catalysts: Their Properties and Applications." Pergamon, Oxford, 1973.
7. R. P. Clement, W. B. Davies, K. A. Ford, M. L. H. Green, and A. J. Jacobson, *J. Mol. Catal.* **3,** 191 (1978).
8. R. G. Grimm, "Clay Mineralogy." McGraw-Hill, New York, 1968.
9. R. M. Barrer, "Zeolites and Clay Minerals." Academic Press, New York, 1978.
10. G. V. Subba Rao and M. W. Shafer, *in* "Intercalated Layered Materials" (F. Levy, ed.), p. 99. Reidel Publ., Dordrecht, Netherlands, 1979.
11. B. W. Brown and D. J. Beernsten, *Acta Crystallogr.* **18,** 31 (1965).
12. M. S. Whittingham and F. R. Gamble, *Mater. Res. Bull.* **10,** 363 (1975).
13. L. B. Ebert, *Annu. Rev. Mater. Sci.* **6,** 181 (1976).
14. A. H. Thompson, F. R. Gamble, and C. R. Symon, *Mater. Res. Bull.* **10,** 915 (1975).
15. F. Hulliger, *in* "Structural Chemistry of Layer Type Phases" (F. Levy, ed.), p. 1. Reidel Publ., Dordrecht, Netherlands, 1976.
16. A. Weiss and R. Ruthardt, *Z. Naturforsch., Teil B* **24,** 355 (1969).
17. A. Weiss and R. Ruthardt, *Z. Naturforsch., Teil B* **24,** 1066 (1969).
18. S. F. Meyer, R. E. Howard, G. R. Stewart, J. V. Acrivos, and T. H. Geballe, *J. Chem. Phys.* **62,** 11 (1975).
19. B. Bach and J. M. Thomas, *J.C.S. Chem. Commun.* p. 301 (1972).
20. F. R. Gamble, J. M. Osiecki, and F. J. DiSalvo, *J. Chem. Phys.* **55,** 3525 (1971).
21. F. R. Gamble, J. H. Osiecki, M. Cais, R. Pisharody, F. J. DiSalvo, and T. H. Geballe, *Science* **174,** 493 (1971).
22. A. Weiss and R. Ruthardt, *Z. Naturforsch., Teil B* **28,** 249 (1973).
23. A. Weiss and R. Ruthardt, *Z. Naturforsch., Teil B* **28,** 522 (1973).
24. R. Schöllhorn, E. Sick, and A. Weiss, *Z. Naturforsch.; Teil B* **28,** 168 (1973).
25. M. B. Dines, *Inorg. Chem.* **17,** 763 (1978).
26. R. Schöllhorn and A. Weiss, *Z. Naturforsch., Teil B* **26,** 172 (1973).
27. R. Schöllhorn and A. Weiss, *J. Less-Common Met.* **36,** 229 (1976).
28. R. Ruthardt, R. Schöllhorn, and A. Weiss, *Z. Naturforsch., Teil B* **27,** 1975 (1972).
29. R. Schöllhorn and A. Weiss, *Z. Naturforsch., Teil B* **27,** 1278 (1972).
30. R. Schöllhorn, H. D. Zagefka, T. Butz, and A. Lerf, *Mater. Res. Bull.* **14,** 369 (1979).
31. A. H. Thompson, *Nature (London)* **251,** 492 (1974).
32. J. W. Johnson, *Physica B + C (Amsterdam)* **99,** 141 (1980).
33. M. S. Whittingham, *J.C.S. Chem. Commun.* p. 328 (1974).
34. G. V. Subba Rao and J. C. Tsang, *Mater. Res. Bull.* **9,** 921 (1974).
35. R. Schöllhorn and H. D. Zagefka, *Angew. Chem.* **89,** 193 (1977); *Angew. Chem., Int. Ed. Engl.* **16,** 199 (1977).
36. Y. Kashihara, A. Nishida, and H. Yoshioka, *J. Phys. Soc. Jpn.* **46,** 1112 (1979).
37. A. H. Thompson, *Solid State Commun.* **13,** 1911 (1973).
38. C. Riekel and C. O. Fischer, *J. Solid State Chem.* **29,** 181 (1979); C. Riekel, *Prog. Solid State Chem.* **13,** 89 (1980).
39. S. Wada, H. Alloul, and P. Molinie, *J. Phys., Lett. (Orsay, Fr.)* **39,** L243 (1978); B. G. Silbernagel, unpublished observations (1977).

40. C. Riekel and D. Hohlwein, *J.C.S. Chem. Commun.* p. 863 (1976).
41. R. Aoki, S. Nakamura, and S. Wada, *Solid State Commun.* **34,** 71 (1980).
42. B. C. Tofield and C. J. Wright, *Solid State Commun.* **22,** 715 (1977).
43. M. B. Dines, *J.C.S. Chem. Commun.* p. 220 (1975).
44. G. W. Brindley, *in* "Crystal Structures of Clay Minerals and Their X-ray Identification" (G. W. Brindley and G. Brown, eds.), p. 125. Mineral. Soc., London, 1980.
45. H. Fernandez-Moran, M. Ohstuki, A. Hibino, and C. Hough, *Science* **1,** 4 (1971).
46. J. M. Thomas, E. L. Evans, B. Bach, and J. W. Jenkins, *Nature (London), Phys. Sci.* **235,** 126 (1972).
47. G. S. Parry, C. B. Scruby, and P. M. Williams, *Philos Mag.* **29,** 601 (1974).
48. F. R. Gamble and B. G. Silbernagel, *J. Chem. Phys.* **63,** 2544 (1975).
49. R. R. Chianelli, J. C. Scanlon, M. S. Whittingham, and F. R. Gamble, *Inorg. Chem.* **14,** 1691 (1975).
50. Y. Kashihara, A. Nishida, K. Ohshima, J. Harada, and H. Yoshioka, *J. Phys. Soc. Jpn.* **46,** 1393 (1979).
51. M. S. Whittingham, *Mater. Res. Bull.* **13,** 775 (1978).
52. J. V. Acrivos and J. R. Salem, *Philos. Mag.* **30,** 603 (1974).
53. E. Ehrenfreund, A. C. Gossard, and F. R. Gamble, *Phys. Rev. B* **5,** 1708 (1972).
54. A. R. Beal and W. Y. Liang, *Philos. Mag.* **27,** 1397 (1973).
55. A. R. Beal and W. Y. Liang, *J. Phys. C* **6,** L482 (1973).
56. R. Schöllhorn and A. Weiss, *Z. Naturforsch., Teil B* **27,** 1428 (1972).
57. M. B. Dines, *Inorg. Chem.* **17,** 762 (1978).
58. M. B. Dines, *Science* **188,** 1210 (1975).
59. W. B. Davies, M. L. H. Green, and A. J. Jacobson, *J.C.S. Chem. Commun.* p. 781 (1976).
60. R. P. Clement, W. B. Davies, K. A. Ford, M. L. H. Green, and A. J. Jacobson, *Inorg. Chem.* **17,** 2754 (1978).
61. C. Riekel, H. G. Reznik, and R. Schöllhorn, *J. Solid State Chem.* **34,** 253 (1980).
62. R. Schöllhorn, E. Sick, and A. Lerf, *Mater. Res. Bull.* **10,** 1005 (1975).
63. F. R. Gamble and A. H. Thompson, *Solid State Commun.* **27,** 379 (1978).
64. J. C. Green, B. Higginson, and S. E. Jackson, *J.C.S. Dalton,* p. 403 (1975).
65. S. Evans, J. C. Green, and S. E. Jackson, *J.C.S. Faraday II* **68,** 301 (1972).
66. S. Evans, M. L. G. Green, B. Jewitt, G. H. King, and A. F. Orchard, *J.C.S. Faraday II* **70,** 356 (1974).
67. J. P. Clark and J. C. Green, *J. Organomet. Chem.* **112,** C14 (1976)
68. S. Evans, J. C. Green, B. Higginson, and S. E. Jackson, *J.C.S. Dalton,* p. 403 (1975).
69. S. Evans, M. L. H. Green, B. Jewitt, A. F. Orchard, and C. F. Pygall, *J.C.S. Faraday II* **68,** 1847 (1972).
70. T. R. Halbert and J. C. Scanlon, *Mater. Res. Bull.* **14,** 415 (1979).
71. T. R. Halbert, D. C. Johnston, L. E. McCandlish, A. H. Thompson, J. C. Scanlon, and J. A. Dumesic, *Physica B+C (Amsterdam)* **99,** 128 (1980).
72. B. G. Silbernagel, *Chem. Phys. Lett.* **34,** 298 (1975).

8
Intercalation Chemistry of Metal Phosphorus Trichalcogenides

JACK W. JOHNSON

I. METAL PHOSPHORUS TRICHALCOGENIDES

Metal phosphorus trisulfides of the formula MPS_3 form a broad and varied class of compounds. The first metal phosphorus trisulfides (*1*, *2*) were reported before the turn of the century, but not until the systematic investigations of Klingen and Hahn (*3–7*) did the details of this interesting class of materials begin to be understood. The number of known compounds of formula MPS_3 is illustrated in Fig. 1. There are a variety of related structural types in this family. In the following sections we first discuss the structures and preparations of the host compounds and then describe the intercalation reactions they undergo and the evidence for the nature of the intercalation complexes.

A. Structure

Detailed structures from single-crystal data are known for the structural types represented by $FePS_3$ and $FePSe_3$ (*7*). $FePS_3$ has a monoclinic structure

INTERCALATION CHEMISTRY

ISBN 0-12-747380-7

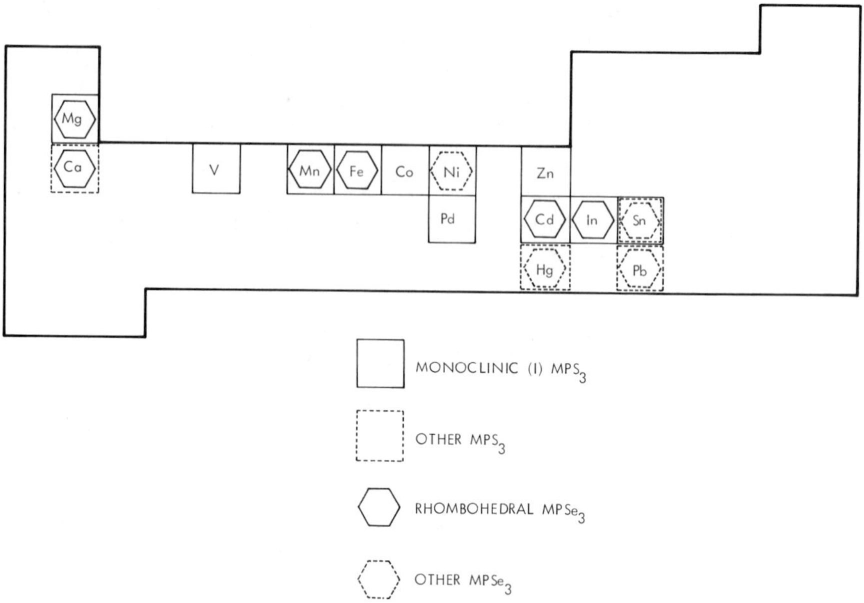

Fig. 1. Range of MPS_3 and $MPSe_3$ compounds.

that deviates only slightly from hexagonal symmetry. Two schematic representations of the structure are shown in Figs. 2 and 3. The structure can most easily be described by analogy to the layered structures of the $CdCl_2$ type. The sulfur atoms are in an essentially cubic close-packed array. Every second layer of octahedral sites between the close-packed sulfur planes is empty. The remaining layers of octahedral sites are completely filled by the transition metal ions and the phosphorus atoms as P_2 pairs. In one direction the structure is held together only by van der Waals forces, and it is in these van der Waals gaps that intercalation chemistry takes place. By rewriting the formula of the metal phosphorus trisulfides as $M_{2/3}(P_2)_{1/3}S_2$, the analogy to the transition metal disulfides is apparent. Another way of visualizing these compounds is as salts of divalent metal cations with the hexathiohypodiphosphate anion $P_2S_6^{4-}$. X-ray powder data indicate that $MgPS_3$, VPS_3, $MnPS_3$, $CoPS_3$, $NiPS_3$, $NiPSe_3$, $PdPS_3$, $ZnPS_3$, $CdPS_3$, and one phase of $SnPS_3$ all have structures very similar to that of $FePS_3$. In addition, $In_{2/3}PS_3$ has a structure like that of $FePS_3$, with a third of the metal ion sites vacant (*8*).

The structure of $FePSe_3$ is also analogous to that of the layered dichalcogenides. The individual layers are exactly analogous to those of $FePS_3$; however, in $FePSe_3$ the stacking of these layers results in hexagonal close packing of the selenium atoms and a rhombohedral lattice. This structure is

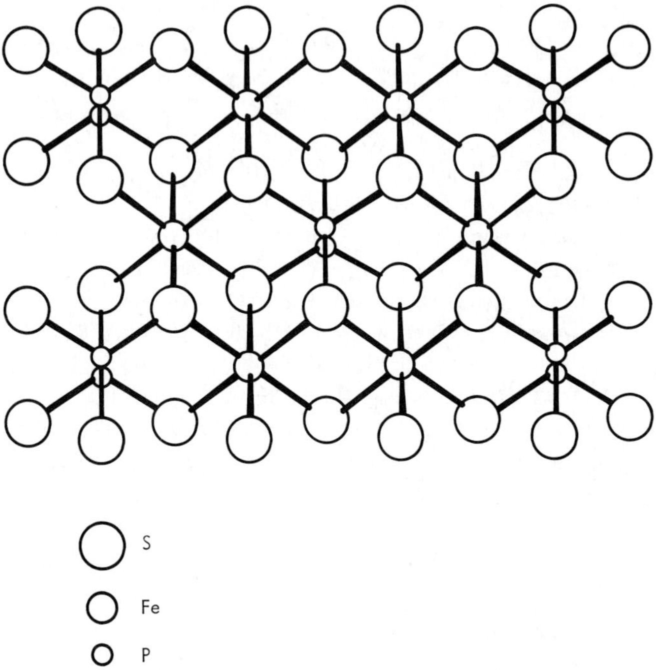

Fig. 2. Structure of $FePS_3$ viewed perpendicular to the layers.

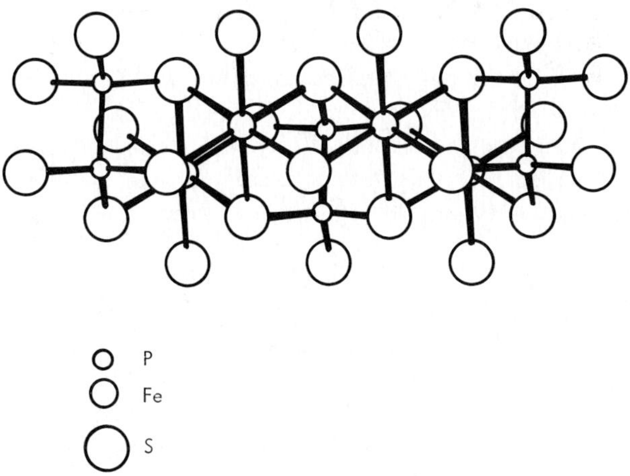

Fig. 3. Structure of $FePS_3$ viewed approximately parallel to the layers.

of the CdI_2 type, a layered structure that is exhibited by many of the transition metal dichalcogenides including TiS_2, a compound whose intercalation chemistry has been extensively studied because of its utility in secondary batteries (*9*). $MgPSe_3$, $CaMgP_2Se_6$, $MnPSe_3$, $CdPSe_3$, and one form of $SnPS_3$ all show structures isomorphous with that of $FePSe_3$ (*6*). In addition, $In_{2/3}PSe_3$ has a structure similar to that of $FePSe_3$, with one-third of the metal positions vacant (*9*).

A third structure type based on a monoclinic lattice different from that of $FePS_3$ is found in $CaPS_3$, $SnPSe_3$, $PbPS_3$, $PbPSe_3$, and a third modification of $SnPS_3$. These structures contain typical P_2S_6 or P_2Se_6 groups made up of P_2 pairs in sulfur (Se) octahedra. However, the metals are in an irregular nine-coordinate environment (*10*). Finally, both $HgPS_3$ and $HgPSe_3$ show unique lattice types in the MPS_3 family. It is likely that these other structure types are also made up of metal dications and $P_2S_6^{4-}$ anions like $FePS_3$ and $FePSe_3$, but their detailed structures are unknown. All the intercalation chemistry reported in the literature at the time of this writing has been on compounds with demonstrated layered structures of the $FePS_3$ or $FePSe_3$ type.

B. Preparation

The metal phosphorus trichalcogenides are synthesized by heating the stoichiometric quantities of the elements (or metal sulfides) in evacuated silica tubes. The optimum conditions for the reaction vary considerably among the metals. Reaction times varying from a few days to months at temperatures from 300°–800°C are required. The details of the preparation of most of the compounds are given by Klingen *et al.* (*6*). It has been pointed out (*11*, *12*) that the $FePS_3$ formed is frequently metal rich when synthesized in this fashion. Syntheses of tellurium compounds and of the phosphorus homologs have not yet been reported. The synthesis of the metal phosphorus trichalcogenides can be done more quickly by including a transporting agent such as I_2 (*13*) or Cl_2 (*14*) or by vapor sublimation in a temperature gradient. By means of these methods, large single crystals can be grown.

II. INTERCALATION CHEMISTRY OF METAL PHOSPHORUS TRICHALCOGENIDES

The intercalation chemistry of the metal phosphorus trichalcogenides can be divided conveniently into three classes according to the nature of the guest molecule. The first reported intercalation reactions were with organic amines, followed by a great deal of work on alkali metal intercalates. More recently, organometallic molecules have also been intercalated in metal phosphorus trichalcogenides. These classes are treated separately in the following sections.

A. Organoamine Intercalates

Metal phosphorus trichalcogenides react with *n*-alkylamines to form amine intercalation compounds, the reactions being similar to those of transition metal dichalcogenides (*15*). The MPS_3 hosts that have been examined are $MgPS_3$, $MnPS_3$, and $ZnPS_3$ (*16*). These three hosts react with *n*-alkylamines to give intercalation compounds that exhibit basal spacings that vary with carbon number, as demonstrated in Fig. 4. Both $MgPS_3$ and $ZnPS_3$ give products with similar basal spacings for a given amine. The increase in basal spacing in the intercalation compound that results from a lengthening of the carbon chain by one atom is 2.5 Å. This is twice the distance that an alkyl chain is lengthened by addition of a methylene group. The intercalation of $MnPS_3$ is complicated by the formation of multiple phases of different layer spacings. One of the phases formed has a layer spacing similar to that of the $MgPS_3$ and $ZnPS_3$ intercalates, but it is accompanied by other phases of larger layer spacing. Some of the guest molecules in these phases can be removed by washing the compound with acetone, resulting in a

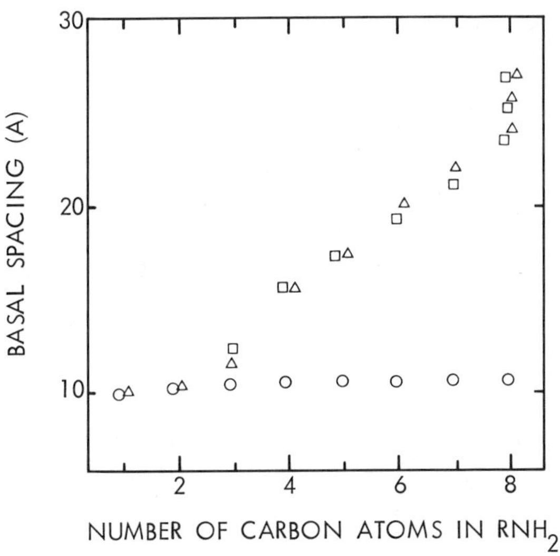

Fig. 4. *n*-Alkylamine intercalates of MPS_3 compounds. Key: □, $MgPS_3$; △, $ZnPS_3$; ○, $MnPS_3$. [From Yamanaka *et al.* (*16*).]

single phase with layer spacing of 10.5 Å for all the complexes regardless of carbon chain length. This spacing is a 4.0 Å increase over the 6.5 Å spacing of unintercalated $MnPS_3$ and corresponds to the van der Waals radius of a methyl group. From the observed lattice expansions in the $MgPS_3$ and $MnPS_3$ intercalates, the most likely configuration for the intercalated amine molecules is one in which they are in a bilayer with the bulk of the alkyl chain oriental perpendicular to the layers. In the 10.5 Å $MnPS_3$ intercalates, the alkyl chains lie parallel to the layers.

Pyridine also intercalates $MnPSe_3$, expanding the layers by 5.9 Å. When treated under similar conditions, $FePSe_3$ failed to react (*17*).

The driving force for these amine intercalation reactions is not clearly understood, nor are the details of the structural arrangement of the guest phase. Amine intercalation in the transition metal dichalcogenides has been studied extensively by a variety of workers. Schöllhorn and Zagefka (*18*) have suggested that during the amine intercalation reaction the metal disulfide lattice is reduced and the amine molecules are protonated and intercalated as ammonium cations. This model is very attractive, and its implications are still being explored in the transition metal dichalcogenides. It is likely that a similar intercalation mechanism operates in the metal phosphorus trisulfides. More detailed studies are needed in order to probe this intriguing problem in the metal phosphorus trichalcogenides.

B. Lithium Intercalates

1. Chemical Intercalation

Alkali metal cations can be intercalated into metal phosphorus trichalcogenides either by chemical or electrochemical techniques. Chemical intercalation takes place when the host material is contacted with a solution of an active organoalkali metal compound (*11*) such as *n*-butyllithium, sodium naphthalenide, or potassium naphthalenide at mild temperatures (20°–70°C). In the case of *n*-butyllithium, the reaction proceeds according to the following equation:

$$xC_4H_9Li + MPS_3 \rightarrow Li_xMPS_3 + \frac{x}{2}C_8H_{18}$$

This technique was previously used to investigate lithium intercalation in the transition metal dichalcogenides (*19*). For example, treatment of TiS_2 with excess *n*-butyllithium results in the formation of $LiTiS_2$. The metal phosphorus trichalcogenides behave in a different manner. By structural analogy to $LiTiS_2$, we would expect the product of the butyllithium–MPS_3 reaction to be $Li_{1.5}MPS_3$, with lithium cations occupying all the octahedral sites in the van der Waals gap. That which is observed varies with the nature of the metal in the host. $NiPS_3$ reacts most readily with butyllithium (*20*), giving compositions with lithium contents as high as $Li_{4.5}NiPS_3$ (*11*). Other host lattices, $FePS_3$, $FePSe_3$, and also $MnPSe_3$, react with *n*-butyllithium, whereas $CdPS_3$, $ZnPS_3$, and $MnPS_3$ do not (*20*). In contrast to the transition metal dichalcogenides, which expand upon lithium intercalation, the X-ray powder diagrams of the metal phosphorus trisulfides remain unchanged (*21*), indicating no expansion. The empty octahedral sites in the unintercalated MPS_3 compounds are the same size as the occupied Li sites in $LiTiS_2$; hence, lattice expansion is not necessary in the trisulfides. Because of this lack of evolution of structural parameters, the course of the intercalation is difficult to specify. However, at lithium contents greater than $Li_{1.2}NiPS_3$, formation of Li_2S becomes apparent in the X-ray patterns, and the lines for the $NiPS_3$ lattice broaden and diminish (*20*, *21*). Thus, the large lithium contents observed in chemical intercalation products probably result from an irreversible destruction of the host lattice to give a mixture of Li_2S and amorphous products.

The presence of lithium ions in the van der Waals gap of the host is demonstrated by X-ray examination of the products of the reaction of Li_xNiPS_3 with water (*20*). Lithium-intercalated $NiPS_3$ reacts with atmospheric moisture to yield a phase with a layer expansion of 5.8 Å. This indicates that a double layer of water molecules is solvating the intercalated lithium ions, as is observed in the hydrates of the alkali-metal-intercalated transition metal dichalcogenides (*22*). As shown in powder patterns of these hydrated samples, the lines for the unexpanded $NiPS_3$ lattice are also present, but their relative intensity decreases with increasing lithium content (see Fig. 5). This is evidence for a reaction mechanism leading to a mixture of $NiPS_3$ and $Li_{1.5}NiPS_3$ as the product of the chemical intercalation. It should be stressed that the process of chemical intercalation, carried out at room temperature, should not be expected necessarily to yield products containing the thermodynamically preferred mixture of phases.

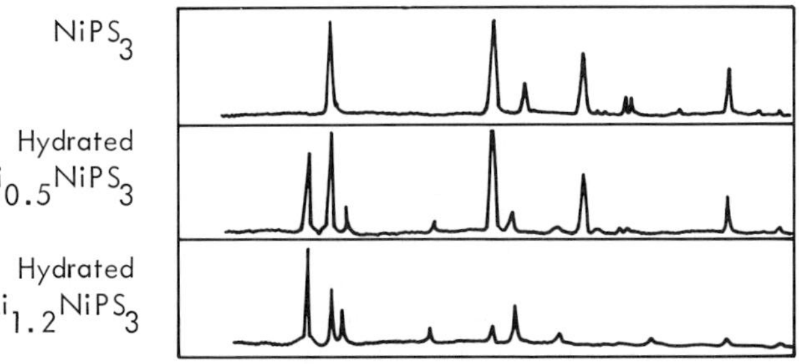

Fig. 5. Two-phase nature of Li_xNiPS_3 as demonstrated by the X-ray powder diffraction patterns of the hydrated products. [From Brec *et al.* with permission from *Inorg. Chem.* **18,** 1814 (1979); copyright (1979) American Chemical Society.]

2. *Electrochemical Intercalation*

Lithium can also intercalate into transition metal phosphorus trichalcogenides electrochemically (*11*, *21*). The intercalation reaction can be used to power a secondary (rechargeable) battery. Such a battery is schematically represented in Fig. 6. Lithium dissolves at the anode, releasing a lithium cation to the electrolyte and sending an electron through the external circuit to the $NiPS_3$ cathode. The electron enters the host matrix, and a lithium cation is intercalated from the electrolyte to balance the charge. As diagrammed, the battery produces electricity. However, if current from an external power supply drives the discharged battery backward, lithium is plated on the former anode and the Li_xNiPS_3 is deintercalated to $NiPS_3$, recharging the battery. Metal phosphorus trichalcogenides as battery cathodes are the subject of several patents (*23*).

By measurement of the open-circuit voltage of an Li–$NiPS_3$ cell as a function of discharge (Fig. 7), the mechanism of electrointercalation can be probed. For $NiPS_3$, the data have been interpreted as follows (*21*, *24*). For Li_xNiPS_3 when $0 < x < 0.5$, a single-phase region exists in which the lithium content varies continuously. For $0.5 < x < 1.5$ another such single-phase region exists. As long as x is less than 1.5, the intercalation reaction is electrochemically reversible. When the battery is discharged to x values greater than 1.5, irreversible reactions take place that result in destruction of the host lattice.

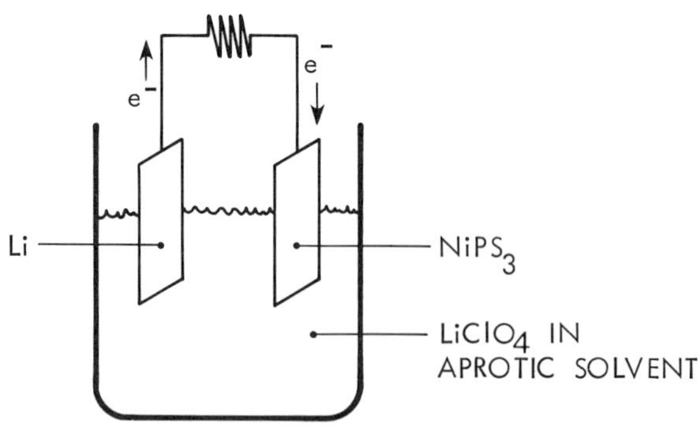

ANODE	Li	$\rightarrow Li^+ + e^-$
CATHODE	$Li^+ + NiPS_3 + e^-$	$\rightarrow Li_xNiPS_3$
CELL	$Li + NiPS_3$	$\rightarrow Li_xNiPS_3$

Fig. 6. Schematic diagram of an Li–$NiPS_3$ electrochemical cell.

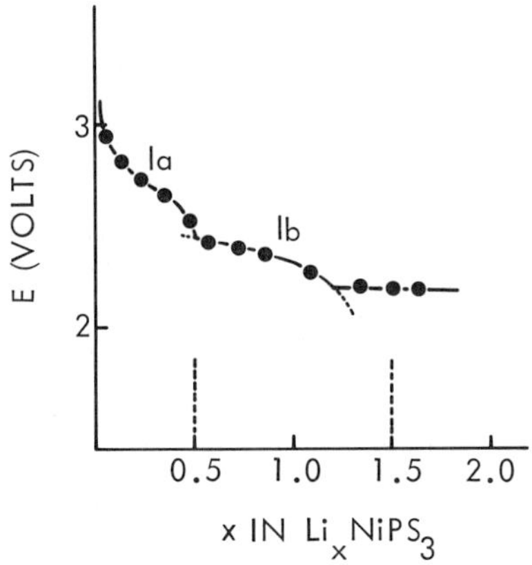

Fig. 7. Discharge curve of an Li–$NiPS_3$ electrochemical cell. [From LeMéhauté (*24*).]

The discharge current obtained depends on the electrolyte used. A differently shaped discharge curve has been observed in electrolytes incorporating dimethoxyethane in the solvent (*11*, *24*). This could be the result of intercalating solvent molecules in the coordination sphere of the lithium into the host structure (*24*).

A range of metal phosphorus trichalcogenides have been examined as cathode materials in electrointercalation cells (*21*). $NiPS_3$, $FePS_3$, and $FePSe_3$ were found to be electrochemically active, whereas $CdPS_3$, $MnPS_3$, $MnPSe_3$, $ZnPS_3$, and $In_{2/3}PS_3$ were less active or inactive. These results on lithium electrointercalation stimulated research on the physical properties of the metal phosphorus trichalcogenides and their lithium intercalation compounds in order to develop a better understanding of the intercalation process. Initial ^{7}Li NMR data (*21*) demonstrated that the lithium was present in the intercalates as Li^+ cations. Physical studies of several types have addressed the question of the fate of the associated electron; however, the answer remains unclear at this time.

3. *Physical Properties of Lithium Intercalates*

The optical properties of metal phosphorus trichalcogenides and their lithium intercalates have been studied (*20*, *25*). The host phases are broad-band semiconductors with band gaps from 1.3 to 3.5 eV, as shown in Fig. 8. It is significant that the three compounds that are most electrochemically active, $NiPS_3$, $FePS_3$, and $FePSe_3$, exhibit the lowest band gaps. In the starting materials, there is a transparent low-energy region in the optical spectrum. Upon chemical intercalation with *n*-butyllithium, the transmittance in this region decreases until the entire spectral region from 3300 to

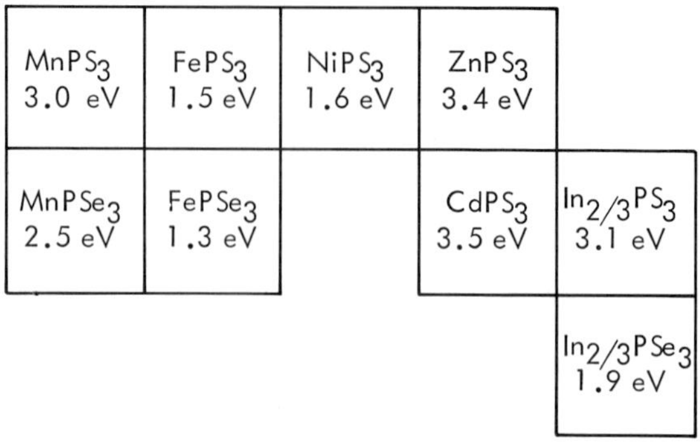

Fig. 8. Optical band gaps of MPS_3 and MPS_3 compounds. [From Brec *et al.*, with permission from *Inorg. Chem.* **18,** 1814 (1979); copyright (1979) American Chemical Society.]

500 nm is opaque. In $NiPS_3$ this process occurs very rapidly, even when dilute *n*-butyllithium solutions are used.

Measurement of the dc conductivity of intercalated crystals parallels these optical results. The resistivity of $FePS_3$ and $FePSe_3$ single crystals gradually decreases when the crystals are exposed to *n*-butyllithium. $NiPS_3$ crystals show a rapid decrease in resistivity of approximately eight orders of magnitude upon chemical intercalation. These results and the results of the optical studies suggest that the electron that accompanies the lithium cation in intercalation goes into a delocalized conduction band in these three compounds, resulting in increased electronic conductivity as well as free carrier absorption and increased reflectivity in the optical spectra. This increased conductivity allows the materials to function efficiently as cathodes in electrochemical cells. The inefficient cathode material $MnPSe_3$ showed no change in optical spectrum or dc conductivity when treated with *n*-butyllithium.

By examining the magnetic properties of the transition metal phosphorus trichalcogenides as a function of lithium intercalation, one can study the effect of the additional electrons on the electronic configuration of the transition metal ion. $FePS_3$, $FePSe_3$, $MnPSe_3$, and $NiPS_3$ are antiferromagnetic. The observed magnetic moment in the temperature regime above the Neel point indicates the presence of the M^{2+} ion in a high-spin configuration. For $FePS_3$, $FePSe_3$, and $MnPSe_3$, chemical intercalation of lithium affects the magnetic properties very little. Thus, the electronic populations of the metal 3d orbitals are not altered by the intercalation reaction. For $NiPS_3$, lithium intercalation showed a significant effect on the magnetic properties of the host. The Neel temperature remained constant, but the overall susceptibility decreased with increasing lithium content. This is the behavior one would expect if the intercalation produces a two-phase product in which one phase has magnetic properties similar to those of the unintercalated host and the other phase has a greatly reduced susceptibility.

The magnetic properties of these compounds can also be studied by NMR spectroscopy. Both ^{7}Li and ^{31}P are good NMR nuclei, and a number of metal phosphorus trichalcogenides have been studied by NMR techniques (*26, 27*). $NiPS_3$ and its lithium intercalation products have been investigated in the most detail. The chemical shift of the ^{7}Li NMR signal indicates that the lithium is intercalated as Li^+ cations throughout the concentration range $0 < x < 1.5$ in Li_xNiPS_3 (*21*). The diffusion coefficients of lithium can also be estimated. They are of the order of 10^{-14} cm^2/sec over the entire composition range. This agrees with values determined by electrochemical measurements (*24*) under equilibrium conditions. This value of the diffusion constant is four orders of magnitude lower than that of lithium in Li_xTiS_2 (*27*). However, in polarized electrochemical cells, i.e., ones with current being drawn out of them, the electrochemically estimated diffusion coefficient is of the order of 10^{-10} cm^2/sec, comparable with that of Li_xTiS_2.

The ^{31}P NMR spectra of the Li_xNiPS_3 series are very interesting (*25, 26*). Intercalation of up to 0.5 lithium per $NiPS_3$ results in no significant change in the position or intensity of the ^{31}P resonance. The compound also magnetically orders upon cooling below the Neel temperature, resulting in the disappearance of the ^{31}P NMR line. For Li contents greater than 0.5 per $NiPS_3$, a new resonance shows up in a different position, which increases in intensity with increasing lithium content at the expense of the original line. This new line does not disappear when the sample is cooled to below the Neel temperature, although the remainder of the original line still disappears. This behavior is readily explained by assuming that, in Li_xNiPS_3, when x is between 0 and 0.5 a single phase with magnetic properties the same as those of the host is present, and when x is between 0.5 and 1.5 a mixture of magnetic $Li_{0.5}NiPS_3$ and nonmagnetic $Li_{1.5}NiPS_3$ is present. Furthermore, in the domain $x < 0.5$, the lithium must be located on a specific one-third of the intralayer octahedral sites that are inversion centers for the magnetic structure.

These magnetic resonance results imply that the electron added to the host matrix in the intercalation process goes into a band that does not include either metal 3d orbitals or phosphorus orbitals. The only alternative seems to be orbitals of the sulfur atoms or possibly metal 4s orbitals. The disappearance of the magnetism in the Li_xNiPS_3 phases of $x > 0.5$ also remains unexplained.

Nuclear magnetic resonance studies on lithium-intercalated $FePS_3$ show that it behaves like $NiPS_3$, whereas the selenium compounds $NiPSe_3$,

$FePSe_3$, and $MnPSe_3$ are single phase throughout the entire lithium concentration range, without modification of the magnetic properties of the unintercalated host.

4. Remarks

It is clear from the collected work on lithium intercalation in metal phosphorus trichalcogenides that the compound that intercalates lithium most readily is $NiPS_3$, followed by $FePSe_3$ and $FePS_3$. These are the compounds with the lowest band gaps. Since intercalation results in increased conductivities, it seems that the additional electrons that accompany the lithium cations must enter a delocalized conduction band. However, the nature of this conduction band is unclear. Magnetic measurements indicate little change in the d-electron configuration of the transition metal ion. Phosphorus-31 NMR data indicate the absence of conduction electron density on the phosphorus nuclei. According to the Holmesian principle* one must conclude that the conduction band is made of sulfur wavefunctions and possibly metal 4s wavefunctions, except in the Li_xNiPS_3 for $x > 0.5$, in which the loss of paramagnetism indicates 3d participation in the conduction band.

The final chapter on lithium intercalation compounds is yet to be written. Experiments to identify clearly the fate of the electron gained by the host lattice upon intercalation are necessary. Selenium-77 NMR in Li_xNiPSe_3 has been suggested. Also, neutron diffraction measurements offer the possibility of directly locating the lithium positions in the intercalates.

Some of the results seem to be in conflict, particularly the two-phase nature of the $Li_{0.5}NiPS_3$ intercalate, as demonstrated by hydration of chemically intercalated samples, and the single-phase nature of the same composition inferred from the ^{31}P NMR measurements. The NMR estimated diffusion coefficients agree with some of the electrochemically measured ones (measured at open circuit) but are far lower than those measured in polarized electrochemical cells. Indeed, a material with a lithium diffusion coefficient as low as that estimated by the NMR measurements could not function effectively as a cathode in an electrochemical cell. One must assume that in the presence of a polarizing electric field, as in an operating electrochemical cell, an enhanced mechanism for lithium ionic diffusion must be operating when compared to diffusion in the same material at thermodynamic equilibrium. Another observation that may resolve some discrepancies is that during chemical intercalation a nonequilibrium mixture of mixed phases is often the initial product. This has been previously observed in Li_xTiS_2.

* "When you have eliminated the impossible, whatever remains, however improbable, must be the truth." S. Holmes, in A. Conan Doyle, *The Sign of the Four. VI.*

Indeed, even electrochemical intercalation in Li_xTiS_2 can lead to separate phases of differing x values under some experimental conditions (*28*).

C. Organometallic Intercalates

Another class of guest molecules that can intercalate into layered hosts consists of the metallocenes. These compounds are made of transition metal atoms sandwiched between two unsaturated carbocyclic rings. The prototypical metallocene is ferrocene, $Fe(C_5H_5)_2$, in which the iron atom is bound between two planar cyclopentadienyl rings. Variation of the transition metal and the size of the organic rings leads to a wide range of metallocene-type molecules. The first work that recognized the potential of metallocenes as guest molecules was the intercalation of transition metal disulfides by cobaltocene and chromocene, $Co(C_5H_5)_2$ and $Cr(C_5H_5)_2$ (*29*). Further studies on a range of metallocene-type guests with ZrS_2 and TaS_2 hosts established that a low ionization potential of the guest molecule was crucial for intercalation (*30*). This led to the explanation of the intercalation reaction as similar to the alkali metal intercalations of the same hosts. The organometallic molecule transfers its electron to a conduction band of the host lattice and is subsequently intercalated as a cation, causing a substantial lattice expansion. Magnetic susceptibility measurements (*31*) in cobaltocene-intercalated tantalum disulfide demonstrate the presence of the cobaltocenium cation. It has been shown that by varying the host from a transition metal disulfide to iron oxychloride, one can intercalate metallocenes with higher ionization potentials such as ferrocene, as discussed in Chapter 12.

Metal phosphorus trisulfides are also intercalated by metallocenes. Cobaltocene and dibenzenechromium intercalate into the phosphorus trisulfides of manganese, iron, nickel, zinc, and cadmium when solutions of the organometallic compound are heated at 110°–130°C with the host (*32*, *33*). The lattice expansions observed, 5.3 Å for $Co(C_5H_5)_2$ intercalates and 5.9 Å for the $Cr(C_6H_6)_2$ intercalates, agree with the expansions observed for these guests in ZrS_2 intercalates (*30*). Furthermore, the infrared spectra of the intercalation compounds show the bands expected for the metallocene cation. In addition, the absorption of the host lattices is altered. In the pure hosts, the dominant feature is a strong absorption at 560–570 cm^{-1}, which has been assigned to an asymmetric P—S stretching mode (*34*). In the cobaltocene intercalates of $MnPS_3$ (*32*), $CdPS_3$ (*33*), and $FePS_3$ (*35*) this band is split into two lines, whereas in $NiPS_3$ (*35*) it is split into three lines. This indicates that some distortion of the $P_2S_6^{4-}$ octahedron has taken place.

The location of the additional electron gained by the host lattice upon intercalation is again a perplexing problem. $CdPS_3$ is a diamagnetic, essentially transparent compound. Its cobaltocene intercalate,

$CdPS_3[Co(C_5H_5)_2]_{0.34}$, is also diamagnetic, and its optical spectrum shows weak features characteristic of the cobaltocenium cation. This means that the additional electrons gained by the host lattice upon intercalation must be localized and paired. The simplest way to explain this is to assume that the phosphorus antibonding orbitals are being occupied in selected P_2 pairs, disrupting some of the P—P bonds and reducing the phosphorus from formal valence 4 to 3. This model conflicts with that of lithium in $NiPS_3$, in which no change in the ^{31}P NMR spectrum was observed upon intercalation (*25*, *26*). However, preliminary results have shown that, when cobaltocene is intercalated in $NiPS_3$, both the position and the line width of the ^{31}P line are affected (*36*). This is consistent with the P—P bonds being the site of reduction in the metallocene intercalates.

An interesting observation has been made concerning the synthesis of the metallocene intercalates of the metal phosphorus trisulfides (*32*, *33*). Solutions of cobaltocenium cation and dibenzenechromium cation intercalate $MnPS_3$, $ZnPS_3$, and $CdPS_3$ in the absence of any additional reducing agent to give compounds that are identical to those obtained from reactions with the neutral guest molecules. Indeed, the metallocene-like cation $Fe(C_5H_5)(C_6H_6)^+$, which has no readily available neutral analog, has been intercalated by this route. In these intercalation reactions, some of the host dissolves. The additional electrons needed by the lattice to effect intercalation of the cations could be provided by oxidation of the phosphorus of the dissolving portion of the host from oxidation state 4 to 5.

D. Summary

An overview of the intercalation reactions of metal phosphorus trichalcogenides allows some general observations to be made. Neutral amines have been intercalated into the host lattices of manganese, zinc, and magnesium phosphorus trisulfides. These are materials in which the bonding is quite ionic. On the other hand, lithium intercalation proceeds most efficiently in nickel and iron phosphorus trisulfides, hosts in which there is a considerable degree of covalent nature in the bonding. Metallocenes intercalate into the whole range of hosts from ionic zinc and manganese compounds to covalent iron and nickel compounds. The details of the mechanism of intercalation and the electronic and atomic structures of the intercalation compounds probably vary from one class of guests to another. The amine intercalates have received the least attention in the literature. The driving force for the intercalation reaction and the electronic structure of the products are not known. The amine molecules could be held in the interlayer space by weak covalent forces such as hydrogen bonding to the sulfide layers or donation from lone pair orbitals to empty sulfur bands. Alternatively, the ionic model described for the amine intercalates of transition metal disulfides (*18*) could apply to the MPS_3 intercalates as well. According to this model, a fraction of the amine molecules are protonated cations and the host lattice is reduced. Hence, the intercalation reaction is a redox process as in the alkali metal intercalations. If this model is correct, it should be possible to identify chemically the oxidation product of the guest molecule either in the intercalated layer or in the reaction medium.

The lithium and the metallocene intercalation reactions of the metal phosphorus trichalcogenides are similar in that they are both clearly redox reactions. The guest molecule gives up an electron to the host and is intercalated as a cation. The fate of the electron accepted by the host is an important question, and the evidence suggests that the answer is different for lithium intercalates than it is for metallocene intercalates. In the lithium intercalates, the delocalized band structure of the covalent host accepts the electron, although the exact makeup of this accepting band is unclear. In the metallocene intercalates, of which those of the host $CdPS_3$ have been most studied, it is clear that the added electrons are in paired, localized states. One source of the differences observed in the classes of intercalation reactions might be that, in the case of both amine and metallocene intercalations, the lattice expansions are significant, resulting in well-separated layers, whereas lithium intercalation results in no lattice expansion and leaves open the possibility of interlayer electronic effects.

E. Addenda

Three papers have recently appeared by Clément and co-workers that deal with organometallic intercalates of metal phosphorus trisulfides. In the first (*37*), the infrared and Raman spectra of $MnPS_3$, $CdPS_3$, and their intercalates with $Co(C_5H_5)_2$ and $Cr(C_6H_6)_2$ are investigated in detail. The results confirm that the organometallic guest species are intercalated as cations. The P—P absorption of the host phases is identified and found to be unperturbed after intercalation, whereas the splitting of the P—S modes as previously observed (*32–35*) is confirmed. If the assignment is correct, this indicates that the additional electrons added to the host upon intercalation are not merely rupturing the P—P bond and reducing the phosphorus to the trivalent state. Investigation of the magnetic susceptibility of $MnPS_3$ and its intercalates with $Co(C_5H_5)_2$, $Cr(C_5H_5)_2$, and alkylamine are reported in a separate paper (*38*). It is stated that the antiferromagnetic interactions present in $MnPS_3$ are considerably diminished upon intercalation. This is attributed to the electrostatic effect of the positively charged interlayer species. Here and in another paper (*39*) an interesting new interpretation of the chemistry of the $MnPS_3$ intercalates is proposed. In reactions of $MnPS_3$

with aqueous solutions of organometallic salts, such as $Co(C_5H_5)_2I$, complete chemical analysis of the products shows them deficient in manganese. This leads to an explanation of the compounds as having Mn^{2+} cation vacancies within the layers, with the charge compensated by the interlayer guest cations. Thus, the intercalation compounds formed in this way have the formula $Mn_{1-x}PS_3G_{2x}(H_2O)_y$. This structure, with interlayer cation guests and intralayer cation vacancies, would involve no electron transfer. A wide range of cations, from K^+ to $(C_2H_5)_4N^+$, were shown to intercalate in this manner. The hosts $ZnPS_3$ and $CdPS_3$ reacted similarly, but the more covalent $NiPS_3$ and $FePS_3$ did not. The products resulting from direct intercalation of neutral cobaltocene into $MnPS_3$ did not show any manganese deficiency, even though its magnetic properties were similar to those of the intercalate prepared from aqueous solution (*38*). This emphasizes the previously mentioned point that intercalation into a metal phosphorus trichalcogenide can occur by different mechanisms, depending on the guest, the host, and the conditions of reaction.

REFERENCES

1. M. C. Friedel, *C. R. Hebd. Seances Acad. Sci.* **119,** 260 (1894); *Bull. Soc. Chim. Fr.* **11,** 115 (1894).
2. L. Ferrand, *Bull. Soc. Chim. Fr.* **13,** 115 (1895); *Ann. Chim. Phys.* **17,** 388 (1895); *C. R. Hebd. Seances Acad. Sci.* **122,** 621 (1896).
3. H. Hahn and W. Klingen, *Naturwissenschaften* **52,** 494 (1965).
4. W. Klingen, G. Eulenberger, and H. Hahn, *Naturwissenschaften* **55,** 229 (1968).
5. W. Klingen, G. Eulenberger, and H. Hahn, *Naturwissenschaften* **57,** 88 (1970).
6. W. Klingen, R. Ott, and H. Hahn, *Z. Anorg. Allg. Chem.* **396,** 271 (1973).
7. W. Klingen, G. Eulenberger, and H. Hahn, *Z. Anorg. Allg. Chem.* **401,** 97 (1973).
8. S. Soled and A. Wold, *Mater. Res. Bull.* **11,** 657 (1976).
9. M. S. Whittingham, *Science* **192,** 1126 (1976); M. S. Whittingham, *Prog. Solid State Chem.* **12,** 41 (1978).
10. C. D. Carpentier and R. Nitsche, *Mater. Res. Bull.* **5,** 1097 (1974).
11. A. H. Thompson and M. S. Whittingham, *Mater. Res. Bull.* **12,** 741 (1977).
12. R. Brec, D. Schleich, A. Louisy, and J. Rouxel, *Ann. Chim. (Paris)* **3,** 347 (1978).
13. R. Nitsche and P. Wild, *Mater. Res. Bull.* **5,** 419 (1970).
14. B. E. Taylor, J. Steger, and A. Wold, *J. Solid State Chem.* **7,** 461 (1973).
15. F. R. Gamble, J. H. Osiecki, M. Cais, R. Pisharody, F. J. DiSalvo, and T. H. Geballe, *Science* **174,** 493 (1971).
16. S. Yamanaka, H. Kobayashi, and M. Tanaka, *Chem. Lett.* p. 329 (1976).
17. F. Kanamaru, S. Otani, and M. Koizumi, *Int. Conf. Solid Comp. Transition Elem., 5th, Uppsala* Abstr. III-18, p. 36 (1976).
18. R. Schöllhorn and H. D. Zagefka, *Angew. Chem.* **89,** 193 (1977); *Angew. Chem., Int. Ed. Engl.* **16,** 199 (1977).
19. M. B. Dines, *Mater. Res. Bull.* **10,** 287 (1975).
20. R. Brec, D. M. Schleich, G. Ouvrard, A. Louisy, and J. Rouxel, *Inorg. Chem.* **18,** 1814 (1979).

21. A. LeMéhauté, G. Ourvard, R. Brec, and J. Rouxel, *Mater. Res. Bull.* **12,** 1191 (1977).
22. A. Lerf and R. Schöllhorn, *Inorg. Chem.* **16,** 2950 (1977).
23. A. H. Thompson and M. S. Whittingham, U.S. Patent 4,049,879 (1977); R. Brec and A. LeMéhauté, Fr. Patents 7,704,518 and 7,704,519 (1977).
24. A. LeMéhauté, *C. R. Hebd. Seances Acad. Sci., Ser. C* **287,** 309 (1978).
25. C. Berthier, Y. Chabre, and M. Minier, *Solid State Commun.* **28,** 327 (1978).
26. Y. Chabre, P. Segransan, C. Berthier, and G. Ourvard, *Fast Ion Transpt. Solids, Lake Geneva, Wis.* Abstr. PB-35 (1979).
27. M. S. Whittingham, *in* "Solid State Electrolytes" (P. Hagenmuller and W. vanGool, eds.), p. 376. Academic Press, New York, 1978.
28. A. H. Thompson, personal communication (1979).
29. M. B. Dines, *Science* **188,** 1210 (1975).
30. W. B. Davies, M. L. H. Green, and A. J. Jacobson, *J.C.S. Chem. Commun.* p. 781 (1976); R. P. Clément, W. B. Davies, K. A. Ford, M. L. H. Green, and A. J. Jacobson, *Inorg. Chem.* **17,** 2754 (1978).
31. F. R. Gamble and A. H. Thompson, *Solid State Commun.* **27,** 379 (1978).
32. R. Clément and M. L. H. Green, *J.C.S. Dalton* p. 1566 (1979).
33. J. P. Audiere, R. Clément, Y. Mathey, and C. Mazieres, *Physica B + C (Amsterdam)* **99,** 133 (1980).
34. H. Bürger and H. Falius, *Z. Anorg. Allg. Chem.* **363,** 24 (1968).
35. J. W. Johnson, unpublished observations (1978).
36. J. W. Johnson and B. G. Silbernagel, unpublished observation; (1978).
37. Y. Mathey, R. Clément, C. Sourisseau, and G. Lucazeau, *Inorg. Chem.* **19,** 2773 (1980).
38. R. Clément, J. J. Girerd, and I. Morgenstern-Badarau, *Inorg. Chem.* **19,** 2852 (1980).
39. R. Clément, *J.C.S. Chem. Commun.* p. 647 (1980).

9
Structural Aspects of Monovalent Cation Intercalates of Layered Dichalcogenides

TJIPKE HIBMA

I. INTRODUCTION

The building blocks of transition metal dichalcogenides are sandwich-like structures consisting of a layer of transition metal atoms (T) between two layers of chalcogen atoms (X). These sandwiches are only loosely bound to each other, as is demonstrated by the fact that the crystals can be easily cleaved. The weakness of the forces keeping the TX_2 units together is also the reason for the ease with which a large variety of electron donors can be introduced into the so-called van der Waals gap (Fig. 1): alkali metals, transition metals, and a broad range of inorganic and organic molecules. The electrostatic forces between the positively charged intercalate layers and

INTERCALATION CHEMISTRY

ISBN 0-12-747380-7

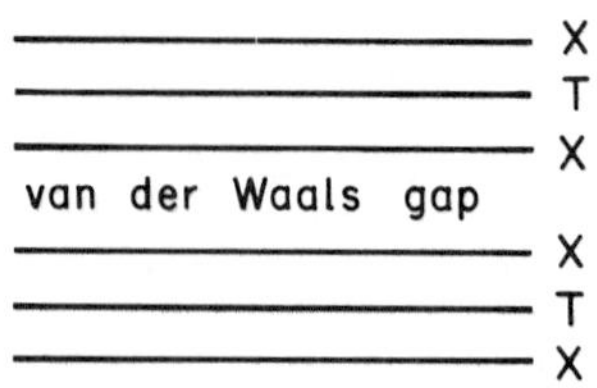

Fig. 1. Schematic picture of the host structure. Electron donors can be introduced in the so-called "van der Waals gap" between the loosely bound TX_2 sandwiches.

the negatively charged TX_2 sandwiches stabilize these layer structures. For instance, although the compound CrS_2 does not exist, the stoichiometric alkali intercalation compounds of CrS_2 have been known for a long time. In contrast to graphite, which forms intercalation compounds with electron acceptors as well as donors, intercalation compounds of acceptors are not known for the layered dichalcogenides. The reason is probably that the charge distribution over the layers would be unfavorable, because both the intercalate and the neighboring chalcogen layers would be negatively charged.

In this chapter we consider exclusively the monovalent cation intercalation compounds of the layered Group IVB, VB, and VIB transition metal dichalcogenides. These materials may be characterized as mixed conductors, i.e., good electronic as well as reasonably good ionic conductors. The electronic part of the conductivity is associated with the slabs of host material, and the ionic part with the layers of intercalate ions.

We begin the second part of the chapter with a discussion of the electronic structure of the host materials. Special attention is given to the nature of the bonds between the metal and nonmetal atoms. Two parameters are introduced to characterize the structure of these materials: (1) a geometric parameter d/a, being a measure of the deformation of the coordination polyhedra; and (2) the quantity $a/2R_{X^-}$, which is related to the ionic nature of the structure. These parameters appear to be useful in characterizing the geometric and electronic structure of the intercalation compounds as well, as shown in Sections III,B and C.

The remainder of Section III deals with the layers of intercalate ions. In many host systems, varying amounts of intercalate atoms may be introduced, and only a fraction of the available interstitial sites in the van der Waals gaps is occupied. In some instances, periodically only a fraction of all van der Waals layers is intercalated. This so-called staging phenomenon is dealt with in Section III,D. Within a single intercalate layer, the fractional occupation of equivalent lattice sites is favorable for ionic conduction. At low temperatures the ions order on the sublattice of interstitial sites, but at higher temperatures order–disorder transitions occur. The ordering of the monovalent cations and its relation to the ionic conductivity are discussed in Sections III,E and F, respectively.

II. STRUCTURE AND BONDING IN THE HOST MATERIALS

In the layered structures under consideration, the atoms of one kind are arranged on hexagonal lattice planes (Fig. 2a). In a close-packed stacking, the atoms of the next layer are shifted to either position *B* or *C*. The choice between these two possible positions has to be made for each new layer that is added to the crystal, leading to a rich variety of possible structures, differing only in the stacking sequence. The most commonly observed polytypes for layered transition metal dichalcogenides are shown in Fig. 2b. Lowercase and capital letters are used to denote the relative positions of the cations and anions, respectively. In the Introduction we characterized the host structure as a loosely bound stack of TX_2 sandwiches. Because the interactions between these TX_2 units are small, the stacking sequence of the units is not very important for most properties of these materials, and we

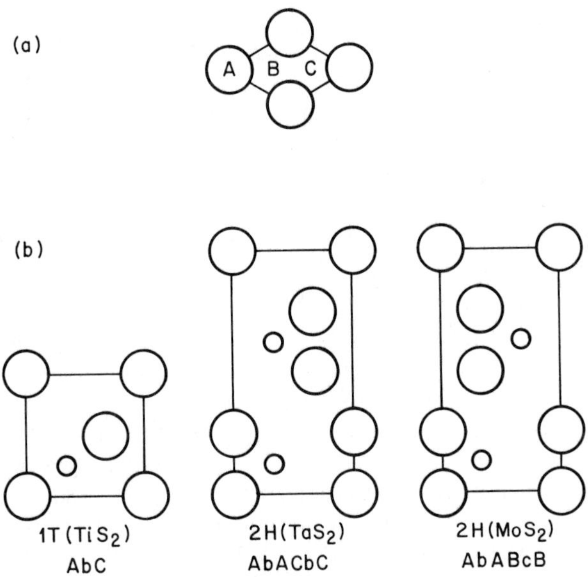

Fig. 2. (a) In a close-packed stacking, the ions can be in three different positions, indicated by *A*, *B*, and *C*. (b) Three commonly observed stacking sequences for layered transition metal dichalcogenides.

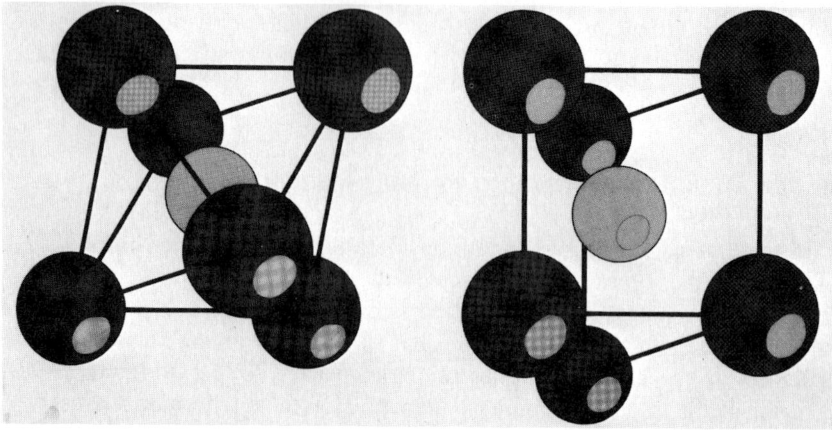

Fig. 3. The two types of coordination observed for the transition metal atoms in the layered dichalcogenides and for the alkali ions in the intercalated compounds: the trigonal prismatic and octahedral arrangements.

will ignore them. The stacking of the layers within a TX_2 slab, on the contrary, has an important influence, for instance, on the electronic band structure. There are two different possibilities, in which the transition metal atom has either a trigonal prismatic or an octahedral (trigonal antiprismatically distorted) chalcogenide coordination (Fig. 3).

The electronic structure of the layered transition metal dichalcogenides, and in particular the nature of the bond between the transition metal and chalcogen atoms, has been the subject of quite a number of papers (*1–6*). The general electron band schemes proposed by Wilson and Yoffe (*1*) and Huisman *et al.* (*2*) have been confirmed by a number of band structure calculations (*3*). The density of states as a function of energy is shown schematically in Fig. 4 for a number of representative cases. There is a wide gap between the broad bonding and antibonding s and p valence and conduction bands, which is caused by the strong covalent bonding and overlap between the metal and chalcogen s and p orbitals. In this gap the much narrower metal d bands are situated. The octahedral ligand field splits the d band into e_{2g} and t_{2g} subbands; in the trigonal prismatic case the lower t_{2g} band splits further into an upper doublet and a lower singlet, the "d_{z^2}" band. Depending on whether these bands are partly or completely filled, these materials are metals or semiconductors. Two aspects of the electronic structure of the host materials have received special attention, i.e., the ionic character of the transition metal–chalcogen bond and the factors that determine the type of coordination of the transition metals (*5, 6*). Because both aspects are of essential importance for understanding the intercalates as well, we will discuss them in somewhat more detail.

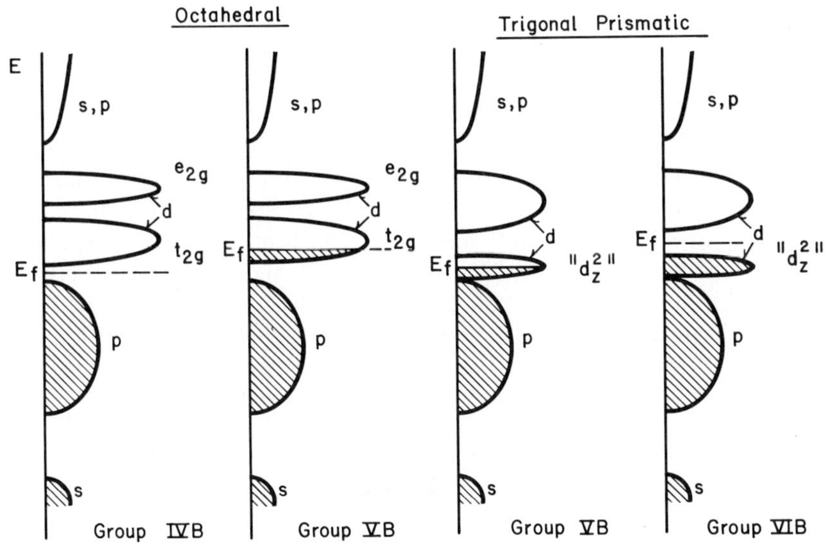

Fig. 4. Schematic representation of the band structures of the layered Group IVB, VB, and VIB transition metal dichalcogenides.

A. Ionicity of the Transition Metal–Chalcogen Bond

According to Pauling (*7*) the fractional ionic character of a single bond (or the ionicity) can be roughly estimated from the empirical expression

$$f_i = 1 - \exp[-\tfrac{1}{4}(X_A - X_B)^2] \tag{1}$$

where X_A and X_B are the electronegativities of the bonding atoms A and B. In the case of diatomic molecules the quantity f_i is directly related to the fractional charge on the atoms. As long as the number of bonds originating from an atom is equal to the covalence of that atom, expression (1) can still be used to make a rough estimate of the charge distribution in a more complicated molecule or a crystal if one adds up the contributions of the different bonds. If the number of bonds M is larger than the covalence N, the ionic character has to be calculated from (*8*)

$$1 - f_i' = (N/M)(1 - f_i) \tag{2}$$

Sometimes, f_i' is called the resonating-bond ionicity in contrast to the single-bond ionicity f_i. Equation (2) follows directly from Pauling's resonating-bond theory: The total covalence $N(1 - f_i)$ is shared by M actual bonds.

Let us apply these formulas to the TX_2 layered compounds of Groups IVB, VB, and VIB. The covalence of the chalcogen atom is 2 and its ligancy 3. If it is assumed that only four valence electrons of the transition metal participate in the bonding with the chalcogen atoms (the remaining valence electrons occupying the d-band levels), the covalence of T is 4 and its ligancy is 6. Therefore, N/M is equal to 2/3. In the most ionic situation $f_i = f_i' = 1$, and we have the charge configuration $T^{4+}(X^{2-})_2$. The most covalent case is obtained if the electronegativities of T and X are equal or if $f_i = 0$ (no charge transfer). In the latter case the ionicity of the bond differs from zero ($f_i' = 1/3$), although the T and X atoms are formally neutral. In a TX_2 compound in which effectively no charge has been transferred from the transition metal to the chalcogen atom, the bonds are nevertheless polarized in the sense that the electron pair of a bond should be partitioned between the X and T atoms in the proportion 2:1. The point to be stressed is that the ionicity of a bond should not be confused with the ionic character of the compound. It is our impression that the difference between these quantities has not been realized sufficiently in the partly controversial discussions concerning ionicity and bonding in these materials.

It has been questioned whether Eqs. (1) and (2) are very useful in a quantitative sense in more complicated systems. It was pointed out by Phillips (*9*) that a better procedure is to use two independent parameters, one for the ionic and one for the covalent character of a bond. Two such parameter schemes were successfully applied to AB compounds by Phillips (*10*) and to intermetallic compounds by Miedema *et al.* (*11*). Nevertheless, the quantity f_i of Eq. (1) may at least be considered as a reliable qualitative measure of the charge transfer. The parameter used by Miedema *et al.* (*11*) to represent the ionic character of the bond is in fact the electronegativity difference $X_A - X_B$.

The Pauling electronegativities are accurate only to within 0.1, and therefore the fractional charge-transfer parameter f_i cannot be very accurate either. Gamble (*5*) made the interesting remark that the heats of transition of the sulfides scale much more accurately with the lattice parameter a than with f_i. This suggests that the a parameter can be used to define a much more accurate scale for the fractional charge transfer. In Fig. 5 we have plotted the lattice constant a divided by $2R_{X^-}$, where R_{X^-} is the ionic radius of the chalcogenide ion (*7*), versus f_i for all the layered dichalcogenides of the Group IVB, VB, and VIB transition metals. The data used for calculating these quantities are listed in Table I. Within the accuracy of f_i, we find indeed a good linear relationship between $a/2R_{X^-}$ and f_i. The scatter in the data becomes even smaller if the electronegativities are calculated from the work function data of Miedema *et al.* (*11*). Although the bonds are predominantly covalent in all cases, the ionic radius was used to normalize the a parameter for the sulfides, selenides, and tellurides because the size of the atoms is expected to be close to the van der Waals radius in directions away from the bonds (*7*). The fact that the a parameter varies smoothly with the charge on

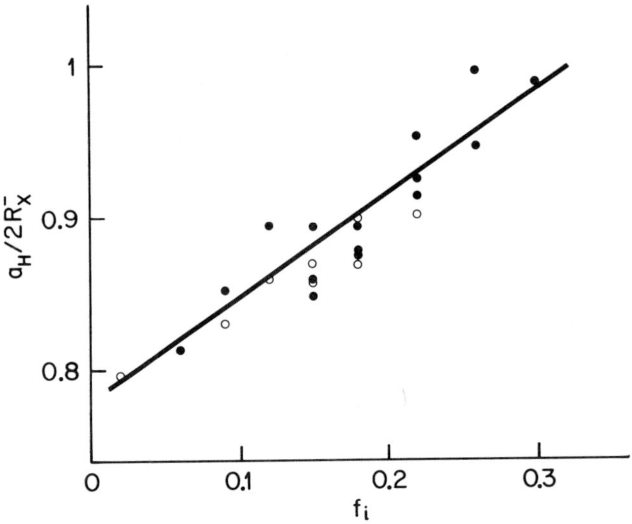

Fig. 5. Plot of the nearest-neighbor distance between the chalcogen atoms in a layer divided by twice their ionic radius versus the Pauling single-bond ionicity (or fractional charge transfer). Within the accuracy of f_i (0.1), a good linear relationship is observed.

the chalcogen atoms shows that these particles are in "contact" with each other in the hexagonal planes for all the dichalcogenides in question. In the next section we use the $a/2R_X^-$ value as a parameter representing the charge transfer.

What can be said about the actual charge on the ions? The quantity f_i is equal to about one-third for the most ionic host compounds, but, as we have argued, this value is probably not very accurate. From infrared reflectivity measurements White and Lucovsky (*12*) deduced very large effective charges for the Group IV dichalcogenides and concluded that these materials are predominantly ionic, but several authors have questioned this opinion (*5*, *6*, *13*, *14*). Band structure calculations indicate that the charge on the chalcogens never exceeds 1 (*14*), corresponding to a fractional charge transfer smaller than 0.5. The latter results are probably the most accurate.

B. Stability of the Trigonal Prismatic versus Octahedral Coordination

The stability of the trigonal prismatic coordination in molybdenite (MoS_2) was studied by Pauling (*15*) and Hultgren (*16*) as early as 1932. They pointed out that the six equivalent hybridized orbitals, which can be constructed from s, p, and d orbitals, having a trigonal prismatic configuration, have

TABLE I

Structural Data for the Host Materials

Host	Coordination[a]	f_i[b]	a_H[c] (Å)	c_H[c] (Å)	$a_H/2R_X^-$[d]	d_{TX}[e] (Å)	d_{TX}/a_H
TiS_2	O	0.22	3.407	5.695	0.925	2.426	0.712
ZrS_2	O	0.26	3.662	5.813	0.995	2.56	0.699
HfS_2	O	0.30	3.635	5.837	0.988	2.56	0.704
VS_2	O	0.18	3.221	5.755	0.875	2.35	0.730
NbS_2	T	0.18	3.31	2 × 5.945	0.899	2.47	0.746
TaS_2	T	0.22	3.315	2 × 6.05	0.901	2.48	0.748
	O	0.22	3.36	5.90	0.913	2.44	0.726
MoS_2	T	0.12	3.160	2 × 6.147	0.859	2.42	0.766
WS_2	T	0.15	3.154	2 × 6.181	0.857	2.41	0.764
$TiSe_2$	O	0.18	3.540	6.008	0.894	2.554	0.722
$ZrSe_2$	O	0.22	3.770	6.138	0.952	2.66	0.706
$HfSe_2$	O	0.26	3.748	6.159	0.946	2.66	0.710
VSe_2	O	0.15	3.359	6.108	0.848	2.495	0.743
$NbSe_2$	T	0.15	3.442	2 × 6.27	0.869	2.595	0.754
$TaSe_2$	T	0.18	3.437	2 × 6.362	0.868	2.59	0.754
	O	0.18	3.4717	6.272	0.878	2.55	0.733
$CrSe_2$	O	0.15	3.399	5.915	0.858	2.46	0.723
$MoSe_2$	T	0.09	3.288	2 × 6.460	0.830	2.49	0.772
WSe_2	T	0.09	3.286	2 × 6.488	0.830	2.51	0.764
$TiTe_2$	O	0.09	3.766	6.491	0.852	2.717	0.721
$ZrTe_2$	O	0.12	3.950	6.630	0.894	2.82	0.714
$HfTe_2$	O	0.15	3.949	6.651	0.893	2.82	0.714
VTe_2	O	0.06	3.6	6.45	0.814	2.67	0.742
$MoTe_2$	T	0.02	3.517	2 × 6.983	0.796	2.73	0.776

[a] O, octahedral; T, trigonal prismatic.

[b] Calculated from Pauling electronegativities.

[c] From Gamble (*5*), with the exception of the first-row transition metal dichalcogenides, which were taken from Wiegers (*26*).

[d] $R_{S}^- = 1.84$ Å, $R_{Se}^- = 1.98$ Å, and $R_{Te}^- = 2.21$ Å (*7*).

[e] Calculated using $z = 0.25$ for the octahedral, $z = 0.131$ for the trigonal prismatic case, if no refined values were available.

greater strength than those with an octahedral orientation. In the octahedral arrangement, on the other hand, the repulsions between the ligand atoms or ions are smaller. The octahedral coordination is therefore preferred if the bonds are predominantly ionic or if the radii of the ligand atoms are very large. Other important factors are the relative positions of the energy levels of the s, p, and d states and the number of nonbonding electrons, making some of the levels unavailable for bonding. It is difficult to quantify the above considerations. An attempt was made by Huisman *et al.*, who calculated the energy level diagram for the octahedral and trigonal prismatic TX_6 complex (*2*) in a simple Hückel-type of calculation. They found that covalent bonding stabilizes the trigonal prismatic arrangement for low d configurations (d^0, d^1, and d^2). Layered compounds of transition metal ions having a high d configuration (d^n, $n \geq 3$) indeed show exclusively the octahedral coordination, whereas the low d transition metal atoms of Groups IVB, VB, and VIB occur in both structures. Within the latter class of materials, the choice between the two possibilities is influenced by the radius ratio of the cation and anion and the fractional ionic character of the metal–chalcogen bond. This was shown by Gamble (*5*) and Madhukar (*6*), who observed that the two types of structures separate into two regions in a plot in which these two parameters are used as coordinates. Gamble defines effective radii by assuming that the size of the anion is equal to half the *a* parameter and calculating the cation radius by subtracting these effective anion radii from the experimental bond lengths. When these effective radii are used, a perfect separation is obtained for a critical value of the ratio of the cation to anion radius of 0.49, which is close to the value 0.527 for hard spheres. This result suggests that all the Group IVB, VB, and VIB layered transition metal chalcogenides would prefer a trigonal prismatic environment but that only those compounds can adopt it for which the appropriate geometric requirements are satisfied. In other words, it is mainly the hard-core repulsion that forces the chalcogenide ions to adopt the close-packed arrangement if the transition metal atoms are small.

We show in Section III,B that the intercalated materials can be analyzed along the same lines if we introduce two important modifications. First, the effective radii of Gamble are not very realistic. As was shown by Madhukar (*6*), the TX bond lengths are very close to the sum of the covalent radii, whereas, as pointed out in the last section, half the *a* parameter used by Gamble for the chalcogen radius is in fact close to the ionic radius. We propose, therefore, to use a different parameter to characterize the geometry of the polyhedra, namely, d/a, the ratio of the bond length and a. This quantity is closely related to the effective radius ratio of Gamble but has a clear geometric meaning. It measures the deformation of the trigonal prism as well as the octahedron and can be determined from experimental structural data. The second modification is the use of the quantity $a/2R_X^-$ as a second parameter, which is a measure of the charge on the chalcogen atom, as shown in the foregoing section. In Fig. 6 the d_{TX}/a versus $a/2R_X^-$ plot for the transition metal dichalcogenides is shown. The data of Table I were used. As expected, a critical value of d_{TX}/a may be defined separating the trigonal prismatic and octahedral structures. It is also possible and probably more realistic to draw a line with a considerable slope. In that case the critical

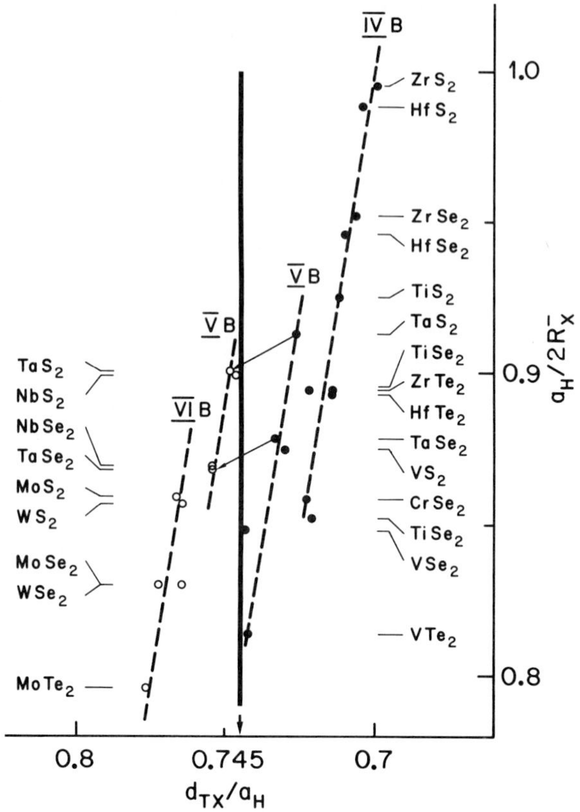

Fig. 6. Two structure parameters can be used to characterize the structures of the layered dichalcogenides: (1) d/a, the ratio of the bond length and the intralayer nearest neighbor (*nn*) distance between chalcogen atoms, which is a measure of the deformation of the TX_6 polyhedra, and (2) $a/2R_X^-$ the normalized intralayer *nn* distance between chalcogen atoms, which is taken as a measure of the fractional charge transfer (see Fig. 5). The octahedral and trigonal prismatic structures are separated on this plot by a critical d/a value [see also Gamble (*5*)]. Closed circles are octahedral, open circles are trigonal prismatic structures.

d/a ratio decreases with $a/2R_X^-$ or alternatively with the charge on the chalcogen atom. This is, in fact, very reasonable. The repulsion between the anions consists of a hard-core contribution and a more long-range coulomb contribution which obviously increases with increasing charge on the ions. A horizontal line would mean that the coulomb repulsion could be neglected, which is unrealistic. Another factor relativating the concept of a critical geometric d/a value somewhat is the observation that the transition between the different coordinations is discontinuous (or of first order) for the Nb and Ta dichalcogenides. If $a/2R_X^-$ indeed may be trusted as an accurate scale of charge transfer, it may be concluded that the ions have a slightly higher charge in the octahedral case. (Note that this could never have been deduced from a Pauling scale.) This difference in charge also causes a difference in the d/a value through a but does not account for all of the experimentally observed difference. The additional increase in d/a is due to the increased bond length in the trigonal prismatic case. Thus, the bond length seems to relax to a larger value in the trigonal prismatic case, and in the meanwhile the charge on the ions decreases somewhat, both because of the greater bond strength.

C. The van der Waals Gap

The forces keeping the TX_2 slabs together are very weak. The crystals, therefore, cleave easily, and for the same reason many of their physical properties are remarkably anisotropic. Information about the strength of these forces may be gained from infrared spectroscopy. In MoS_2 the rigid layer shear mode is observed at 32 cm^{-1} as compared to 384 cm^{-1} for the dipolar mode, showing that the interlayer forces are about 100 times weaker than the intralayer forces (*17*). Mattheiss, who explicitly considered the effect of the interactions between the TX_2 sandwiches on the electronic band structure, found that their inclusion results in a rounding off of the singularities of the band edges for the two-dimensional solutions but that the changes are not very dramatic (*14*). The name "van der Waals gap" suggests that the forces keeping the sandwiches together are van der Waals forces, but it is more likely that these forces are of a different nature. For instance, small covalent forces between the chalcogen atoms can easily be imagined through overlap of the p_z orbitals. These forces have to oppose the electrostatic repulsion between the sandwiches, which is also small.

III. THE INTERCALATED LAYERED DICHALCOGENIDES

The intercalation compounds of the layered chalcogenides have been extensively reviewed by Rouxel (*18*) and Whittingham (*19*). The interest in these materials comes from several different sides. By intercalation, the filling of the d-band levels of the host materials is changed and also the anisotropy of the electronic properties of these materials can be increased. From this point of view, intercalation may be considered a useful tool in the study of the host materials. The intercalated materials also form an extension of the class of layered materials, and some of them show very interesting electrical and magnetic properties. For instance, the transition from itinerant to localized electron behavior can be induced by a change in composition in some of the intercalation compounds of vanadium and chromium dichalcogenides (*20*). Their mixed conductor properties make these materials, in particular the lithium-intercalated substances, suitable for use as solid solution electrodes in electrochemical devices like batteries (*19*); others are superconductors (*21*).

The first intercalation compounds of the type discussed here were reported by Rudorff and Sick (*22*). From a structural standpoint, the most extensive and complete study was made by Rouxel and co-workers for the Group IVB transition metal dichalcogenides (*18*). They discovered that the lithium intercalation compound of TiS_2 is single phase over the whole composition range, whereas several different phases may be found for the other alkali metal intercalates. They also showed that the different phases occurring for the larger alkali ions may be characterized by the fractional number of intercalated van der Waals gaps (stages) and the coordination of the alkali ion, which may be octahedral or trigonal prismatic (like the coordination of the transition metal ion in the host structure). The lithium intercalates were systematically studied by Whittingham and Gamble (*23*). No fully intercalated lithium compounds were found for the molybdenum and tungsten dichalcogenides by these authors. For a number of alkali intercalates of vanadium and chromium dichalcogenides, refined structures are available from powder neutron diffraction data (*24*). Also, a number of nonstoichiometric compounds were synthesized by people from the same group (*25*, *26*). For a more complete overview of the literature on the subject, the reader is referred to the above-mentioned review articles (*18*, *19*).

A. Synthesis and Structure

The several different methods of preparation of intercalation compounds were discussed by Schöllhorn (*27*) and Rouxel (*18*). The most commonly used for the simple metal intercalation compounds are the following: (a) high-temperature synthesis from the host material and the metal or from the elements; (b) intercalation of the host material with a solution of the metal (alkali metal in liquid ammonia, butyllithium, sodium naphthalide); and (c) electrochemical intercalation. Because methods (b) and (c) are essentially room temperature methods, metastable phases may be produced and equilibration may take a very long time. Cointercalation of the solvent is also possible with these methods. For instance, cointercalated ammonia has to be removed by subsequent heat treatment (*18*). In practice, very different results are obtained. We will illustrate this for the sodium titanium sulfide system. Rouxel *et al.* (*36*) were the first to study this system. They used the liquid ammonia technique. The open-circuit voltage of an Na/Na_xTiS_2 cell

TABLE II

Comparison of the Phase Boundaries in Na_xTiS_2 Obtained by Different Authors

Ref.	Stage 3	Stage 2	Stage 1 (trigonal prismatic)	Stage 1 (trigonal antiprismatic)
18		$0.17 < x < 0.33$	$0.38 < x < 0.68$	$0.79 < x \leq 1$
29		$? < x < 0.18$	$0.35 < x < 0.58$	$0.68 < x \leq 1$
28	?	?	$0.46 < x < 0.7$	$0.80 < x \leq 1$
30	$? < x < 0.11$	$0.12 < x < 0.25$	$0.50 < x < ?$	$0.81 < x \leq 1$

as a function of composition was determined by Winn and Steele (*28*) for single crystals and by Nagelberg and Worrell (*29*) for powders. We studied the ordering of the alkali ions in this system using electrochemical intercalated TiS_2 crystals (*30*). Most of the phase boundaries given by these different authors do not agree very well, as seen from Table II. Haange *et al.* prepared two different $NaTiS_2$ phases from the elements, one of which is apparently the stable room temperature phase (*31*). This phase is very different from the phases prepared by methods (b) and (c). One of the factors that may be responsible for these differences is the nonstoichiometry of the host. A second very important factor is the slow kinetics of the reaction if method (b) or (c) is used without an appropriate heat treatment afterward. Several other examples can be given showing similar discrepancies.

B. Electronic Structure and Bonding

The more or less generally accepted picture of the alkali intercalation compounds of transition metal dichalcogenides is one in which the valence electrons of the alkali atoms have been transferred to the TX_2 sandwich, filling the lowest unoccupied d-band levels. The most important evidence for the ionic nature of the intercalation compounds is the very low activity of the alkali atoms in the intercalated dichalcogenides (or the high chemical potential of these materials with respect to the parent metals), and the very small Knight shift, indicative of a small s-electron density at the alkali nucleus (*19*). Arguments of this kind, however, should be used with great care (*32*). As discussed in more detail in the next section, the covalent bonding between the alkali and chalcogen atoms has at least to be sufficiently strong to explain the stability of the trigonal prismatic coordination of the larger alkali ions.

To our knowledge, the band structure of only one of the compounds considered here has been calculated, namely, that of $LiTiS_2$ (*33*). According to this calculation, $LiTiS_2$ should be a metal, in agreement with the experimentally observed Pauli paramagnetism. The dispersion and relative position of the d bands stay almost unchanged upon intercalation. In contrast to the d bands, the upper conduction and lower valence bands change considerably. The most dramatic change is the lowering of the conduction s band, which is caused by the mixing of the titanium and lithium s orbitals. Small changes in the valence p band are presumably due to a reduction in the overlap of sulfur p_z orbitals rather than to bonding between lithium and sulfur ions. This calculation thus supports a simple rigid band charge-transfer model for the intercalation compounds (*34*).

For the host materials, we have shown that the *a* parameter scales with the charge on the chalcogen atoms. In the intercalated materials, extra charge has a similar effect. Because the d bands, which are filled on intercalation, have only small chalcogen s and p contributions, the negative charge of the TX_2 slab is expected to be mainly on the transition metal ion, and the charge on the chalcogenide ions is not expected to increase much. In the intercalation compounds an additional increase of the *a* parameter may be caused by the attraction of the positively charged alkali ions and the negative charge on the TX_2^- units and by the intralayer repulsion of the alkali ions. The unique relation between *a* and the fractional charge transfer is thus no longer guaranteed.

In Tables III and IV most available structural data of first-stage alkali-metal-intercalated materials have been collected, and in Fig. 7 these data have been put on the same d/a versus $a/2R_X^-$ plot as the host materials. For those intercalation compounds for which the structures have not been refined, the bond lengths were taken to be the same as those of the host materials. Again a line can be drawn separating the structures with trigonal prismatically and octahedrally coordinated transition metal ions. Although no intercalation-induced change of coordination has been reported, such changes can be imagined. For instance, the trigonal prismatic Nb and Ta dichalcogenides might switch to an octahedral coordination, whereas for the Zr and Hf dichalcogenides a transition in the opposite direction seems possible. (It might well be that the coordination of Hf and Zr in $LiHfS_2$ and $LiZrS_2$ is, in fact, trigonal prismatic, because the *c* axis is tripled with respect to the host structure, showing that the stacking sequence has changed.)

More interesting, however, is the lower part of the diagram. For a hard-sphere model there also exists a lower limit for d/a in the octahedral arrangement corresponding to the situation in which the central sphere is smaller than the hole in a close-packed arrangement of the surrounding spheres. This lower limit for the d/a value is shifted by an amount $(7/12)^{1/2} - (1/2)^{1/2} \cong 0.057$ with respect to the separation between octahedral and trigonal prismatic structures. This value is even approximately correct if we allow

TABLE III

Structural Data for the Stoichiometric Intercalation Compounds MTX_2 with Octahedrally Coordinated Alkali Ions

Intercalation compound	a_I (Å)	c_I (Å)	a_I/a_H	$a_I/2R_X^{-\,a}$	$d_{TX}{}^b$ (Å)	$d_{MX}{}^b$ (Å)	$d_{TX}/a_I{}^b$	d_{MX}/a_I	Ref.
$LiTiS_2$	3.455	6.195	1.014	0.939	2.426	2.64	0.702	0.764	*19*
$LiZrS_2$	3.604	3 × 6.25	0.983	0.979	2.56	2.65	0.710	0.735	*19*
$LiHfS_2$	3.56	3 × 6.375	0.979	0.967	2.56	2.64	0.719	0.742	*19*
$LiVS_2$	3.380	6.138	1.049	0.918	2.43*; 2.35[c]	2.54*; 2.63[c]	0.719*; 0.695	0.751	*26*
$LiNbS_2$	3.342	3 × 6.421	1.001	0.908	2.47	2.55	0.739	0.763	*19*
$LiTaS_2$	3.340	2 × 6.475	1.000	0.908	2.48	2.56	0.743	0.766	*19*
$LiCrS_2$	3.464	6.037	1.065[c]	0.941	2.40*; 2.35[c]	2.62*; 2.68[c]	0.693*; 0.678	0.756	*26*
$LiTiSe_2$	3.644	6.480	1.031	0.920	2.55	2.78	0.700	0.763	*19*
$LiZrSe_2$	3.734	6.648	0.989	0.943	2.66	2.79	0.712	0.747	*19*
$LiHfSe_2$	3.715	6.642	0.993	0.938	2.66	2.77	0.716	0.746	*19*
$LiVSe_2$	3.584	6.356	1.070	0.905	2.50	2.76	0.696	0.770	*26*
$LiNbSe_2$	3.496	? × 6.772	1.013	0.883	2.60	2.68	0.742	0.767	*19*
$LiTaSe_2$	3.477	? × 6.817	1.012	0.878	2.59	2.68	0.745	0.771	*19*
$NaTiS_2$	3.47	3 × 6.87	1.018	0.942	2.43	2.88	0.699	0.830	*18*
$NaZrS_2$	3.666	3 × 6.78	1.000	0.996	2.56	2.88	0.698	0.785	*18*
$NaVS_2$	3.566	3 × 6.56	1.107	0.969	2.44*; 2.35	2.85*; 2.94	0.684*; 0.659	0.799	*26*
$NaCrS_2$	3.5544	3 × 6.497	1.094[c]	0.966	3.43*; 2.35[c]	2.82*; 2.90[c]	0.684*; 0.661	0.793	*26*
$NaTiSe_2$	3.695	3 × 7.03	1.045	0.933	2.55	3.01	0.690	0.813	*26*
$NaVSe_2$	3.735	3 × 6.88	1.115	0.943	2.59*; 2.50	2.95*; 3.04	0.693*; 0.668	0.790	*26*
$NaCrSe_2$	3.7323	3 × 6.799	1.098	0.943	2.56*; 2.46	2.95*; 3.05	0.686*; 0.658	0.790	*26*
$KZrS_2$	3.718	3 × 7.40	1.014	1.010	2.56	3.15	0.689	0.847	*18*
$KCrS_2$	3.602	3 × 7.05	1.108[c]	0.979	2.43*; 2.35[c]	3.08*; 3.16[c]	0.675*; 0.652	0.855	*26*
$KCrSe_2$	3.798	3 × 7.38	1.117	0.959	2.50	3.28	0.674	0.848	*26*

[a] $R_S^- = 1.84$ Å, $R_{Se}^- = 1.98$ Å, and $R_{Te}^- = 2.21$ Å.
[b] d_{TX} was taken from Table I. The asterisks indicate values from structure refinements.
[c] a_H was estimated from Fig. 5 as 3.25 Å.

TABLE IV

Structural Data for the Intercalation Compounds M_xTX_2 with Trigonal Prismatically Coordinated Alkali Ions

Intercalation compound	x	a_I (Å)	c_I (Å)	a_I/a_H	$a_I/2R_X^{-a}$	d_{TX}^{b} (Å)	d_{MX}^{b} (Å)	d_{TX}/a_I^{b}	d_{MX}/a_I	Ref.
Na_xTiS_2	0.55	3.433	3 × 6.98	1.008	0.933	2.426	2.88	0.707	0.841	*18*
Na_xVS_2	0.6	3.311	3 × 7.07	1.028	0.900	2.35	2.89	0.710	0.873	*26*
Na_xNbS_2	~0.6	3.363	2 × 7.26	1.016	0.913	2.47	2.86	0.734	0.850	*25*
Na_xTaS_2	~0.6	3.337	2 × 7.30	1.007	0.907	2.48	2.84	0.743	0.851	*25*
Na_xTiSe_2	0.68	3.593	3 × 7.33	1.015	0.907	2.55	3.02	0.710	0.842	*18*
	0.74	3.620	3 × 7.32	1.023	0.914	2.55	3.04	0.704	0.841	*18*
Na_xVSe_2	0.6	3.482	3 × 7.40	1.037	0.879	2.52*; 2.50	2.97*; 2.99	0.718*; 0.724	0.853	*26*
Na_xNbSe_2	~0.6	3.476	3 × 7.63	1.010	0.878	2.595	2.98	0.747	0.957	*25*
Na_xTaSe_2	~0.6	3.458	3 × 7.70	1.006	0.873	2.59	2.97	0.749	0.859	*25*
Na_xCrSe_2	~0.6	3.56	3 × 7.273	1.047	0.899	2.46	2.97	0.691	0.834	*26*
K_xTiS_2	1	3.488	3 × 7.61	1.024	0.948	2.426	3.20	0.969	0.917	*18*
K_xZrS_2	0.58	3.637	3 × 7.74	0.993	0.988	2.56	3.19	0.704	0.878	*18*
K_xNbS_2	~0.6	3.345	2 × 8.11	1.011	0.909	2.47	3.17	0.738	0.948	*25*

K_xTaS_2	~0.6	3.332	2 × 8.10	1.005	0.905	2.48	3.14	0.744	0.942	*25*
K_xMoS_2	0.4	3.204	2 × 8.29	1.014	0.871	2.42	3.18	0.755	0.993	*52*
K_xTiSe_2	0.80	3.612	3 × 8.05	1.020	0.912	2.55	3.32	0.706	0.927	*18*
	0.45	3.581	3 × 8.14	1.012	0.904	2.55	3.31	0.712	0.916	*18*
K_xZrSe_2	0.51	3.772	3 × 8.13	1.001	0.953	2.66	3.34	0.705	0.886	*18*
K_xNbSe_2	~0.6	3.480	2 × 8.52	1.011	0.879	2.595	3.30	0.746	0.948	*25*
K_xTaSe_2	~0.6	3.463	2 × 8.53	1.008	0.874	2.59	3.28	0.748	0.947	*25*
K_xCrSe_2	0.8	3.633	3 × 7.817	1.069	0.917	2.46	3.30	0.677	0.908	*26*
	0.6	3.600	3 × 7.903	1.059	0.909	2.46	3.36	0.683	0.933	*26*
Rb_xTiS_2	1	3.427	3 × 8.10	1.006	0.931	2.426	3.30	0.708	0.963	*18*
Rb_xZrS_2	0.75	3.657	3 × 8.02	0.999	0.994	2.56	3.32	0.700	0.908	*18*
Cs_xTiS_2	1	3.500	3 × 8.17	1.027	0.951	2.426	3.41	0.693	0.974	*18*
Cs_xZrS_2	0.64	3.644	3 × 8.41	0.995	0.990	2.56	3.46	0.703	0.950	*18*
K_xZrS_2[c]	0.86	3.666	9 × 7.54	0.996	0.996	2.56	3.15	0.698	0.859	*18*

[a] $R_{S^-} = 1.84$ Å, $R_{Se^-} = 1.98$ Å, and $R_{Te^-} = 2.21$ Å.
[b] d_{TX} was taken from Table I. The asterisks indicate values from structure refinements.
[c] In this compound, K has both types of coordination.

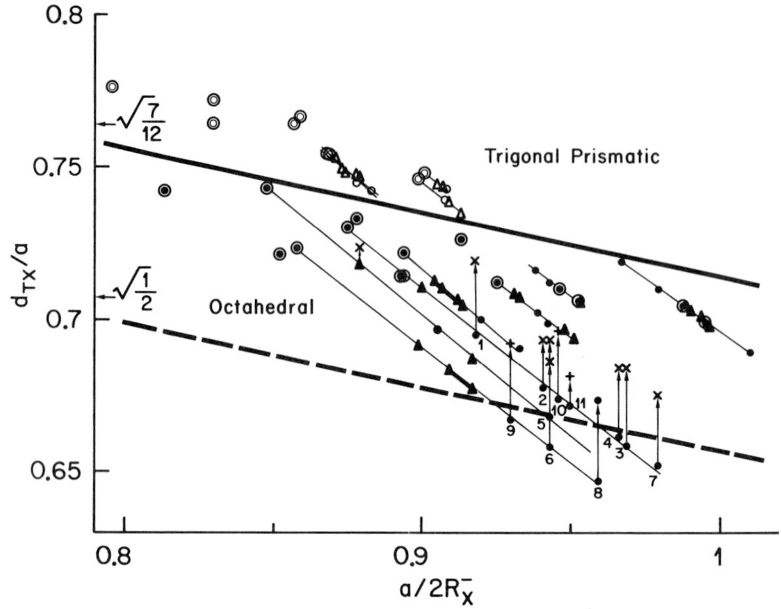

Fig. 7. Same plot as in Fig. 6 (◉, ◎ are the data from Table I), but now including the data of the intercalation compounds from Tables III (●, ○) and IV (▲, △). The closed symbols are for materials with an octahedral coordination for the transition metal atoms; the open symbols are for those with a trigonal prismatic coordination. All data were calculated using the TX bond lengths of the host (Table I). The symbols x and + indicate points calculated from the experimental bond lengths for an alkali or a silver (or copper) intercalation compound, respectively.

for an asymmetric shape of the ions and a dependence of the critical d/a value on $a/2R_X^-$. Almost all experimental points lie well above the line defined by this difference, as they should. The important point, however, is that the d/a values, calculated from the bond lengths of the host structures, of just those materials that show localized electron behavior i.e., the stoichiometric $MCrX_2$ and MVX_2 compounds, lie below the line.

This suggests that it is difficult to realize the covalent bonds in the vanadium and chromium compounds because of the relatively large size of the chalcogenide ions. For the geometrically imposed interatomic distances, which are too large, the mutual overlap between the transition metal d orbitals and the overlap between these orbitals and the chalcogen s and p orbitals is considerably reduced, causing a reduction of the bandwidths. Apparently the widths of the d bands become comparable to or smaller than the effective on-site electron–electron repulsion, resulting in the breakdown of the free electron model and the localization of the valence electrons (*20*).

Similar structural arguments may be used to explain the instability of the host materials for these elements.

C. Coordination of the Alkali Ions

In the alkali intercalation compounds the alkali ions also occur in either octahedral or trigonal prismatic coordination. As a general rule, the trigonal prismatic coordination is favored by intercalates of the large alkali ions at high alkali metal concentrations and for a large ionicity of the TX bond. Rouxel has discussed the importance of these factors (*35*) and has shown that if the ionic radius radio of the alkali and chalcogen ions is plotted against the product of the Pauling single-bond ionicities of the MX and TX bonds and the composition x, the structures showing the two different types of coordination can be separated.

That the trigonal prismatic coordination can be the more stable one for an alkali ion is not at all clear in a predominantly ionic picture, because this stability can be accounted for only by a covalent contribution to the bonding between the alkali ion and the host structure. Apparently a relatively small contribution is already sufficient. This may be plausible since the charge on the chalcogen atoms of the host materials is generally quite small and does not increase very much on intercalation, as argued in the preceding section. The electrostatic energy difference between the two coordinations, which is small at any rate, will decrease quadratically with decreasing charge on the ligand ions and will, in fact, be zero for neutral chalcogen atoms.

Obviously the bonding in the intercalation compounds is far more complicated than in the host materials. Nevertheless, we see from Fig. 8 that a d/a versus $a/2R_X^-$ plot, which was so successful in separating out the two types of coordination for the transition metal atoms in the host structure, is equally successful if applied to the sulfur coordination of the alkali ions in the intercalated materials. Bond lengths, not known from structure refinements, were calculated by assuming that the transition metal–chalcogen bonds are the same as in the host materials. These values are tabulated in Tables III and IV and are seen to be quite close to the experimental values from the alkali sulfides. Only first-stage compounds were included. It should be noted that the only intercalation compound that shows both types of coordination in the same structure, that is $K_{0.86}ZrS_2$, is very close to the line separating the two regions. The d/a value where the transition from the octahedral to the trigonal prismatic coordination occurs is much larger than the value for hard spheres and increases clearly with $a/2R_X^-$. This may be viewed as a result of the relatively small covalent contribution to the bonding between the alkali ions and the host structure.

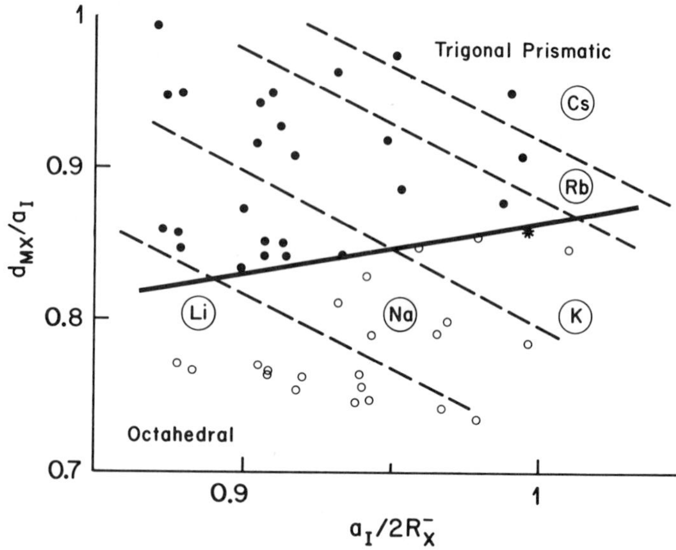

Fig. 8. Plot of d/a versus a/R_X^- for the chalcogen coordination of the alkali ions in the intercalation compounds. The data are from Tables III and IV. Closed circles and open circles indicate compounds with trigonal prismatically and octahedrally coordinated ions, respectively. The asterisk denotes the only compound showing both coordinations.

D. Stages

Rouxel and co-workers discovered that the Group IVB transition metal dichalcogenides form intercalation compounds with the larger alkali ions in which periodically only one out of two or four van der Waals layers is intercalated (*36*). A compound in which only one out of three layers is intercalated has also been reported (*30*). These so-called stages (Fig. 9) are very common for the intercalation compounds of graphite (*37*). The stability of the higher-order stages may be discussed by comparing the relevant contributions to the heat of formation of two different stages with the same overall composition. We will assume that the simple charge-transfer picture is valid, i.e., that the electronic structure of the TX_2 sandwich stays essentially unchanged on intercalation, except for the filling of the d bands. Which of the two stages will be the more stable one depends on (a) the effective interaction between the intercalate ions within the same layer and in different layers and (b) the energy required to separate the "van der Waals"-bonded layers. For stages in excess of two, the conduction electron distribution may be inhomogeneous (*38*). In that case the electrostatic contributions involving the charge on the TX_2 slabs and electronic energy differences should also be

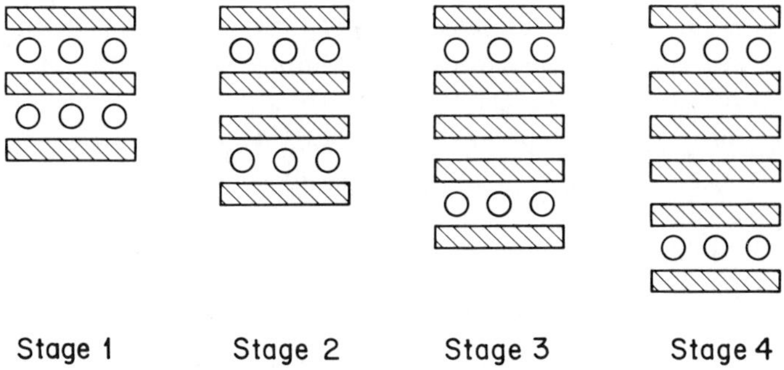

Fig. 9. For low intercalate concentrations, periodically only one out of n van der Waals gaps may be filled. Such a compound is said to be of stage n.

taken into account. Although the coordination of the alkali ions for the high-stage compounds has not been determined yet, we anticipate that it is the trigonal prismatic one on the basis of the d/a versus $a/2R_X^-$ diagram of Fig. 8. A similar deduction (with the exception of second-stage Na_xZrS_2) was made by Rouxell from his ionicity diagram (*35*). It thus appears that stages exclusively occur if the alkali ions are trigonal prismatically coordinated. The bonding between chalcogen atoms for the octahedrally and tetrahedrally coordinated cations over the van der Waals gap is not very different from the bonding in the host material, because the alkali ions are small and the chalcogen atoms still have a close-packed arrangement. Only in the case of the trigonal prismatically coordinated ions is the difference in the binding energy between the chalcogen ions in a nonintercalated and an intercalated layer apparently sufficiently large to balance the increased repulsion between the intercalate ions in the higher-stage compound. Because the forces keeping the transition metal dichalcogenide sandwiches together are presumably very small, the effective electrostatic interactions between the intercalate ions must also be small. The occurrence of stages thus shows that the screening of electrostatic interactions is very effective in these materials.

E. Ordering of the Intercalate Ions

In the preceding sections we have considered only the "average" structure of the intercalation compounds; i.e., we have tacitly assumed that the intercalate ions in the nonstoichiometric compounds are homogeneously distributed over all available equivalent sites in the van der Waals gap. This is certainly not correct. At low enough temperatures the ions will order on superlattices for certain fractional values of the composition. At finite temperatures these ordered phases will have a compositional width. The stoichiometry can be varied within certain limits by creating vacancies or adding interstitial ions. As the temperature increases, the disorder increases, and, at a critical temperature, the long-range order collapses and the system becomes disordered.

Two experimental techniques are particularly suitable for the study of the ordered structures of the intercalate ions as well as the local order in the disordered regimes, i.e., electron diffraction (*39*) and diffuse X-ray scattering (*40*). The advantage of electron diffraction is that polycrystalline materials can be used. In addition, the microcrystalline structure of the crystallites can be studied. The determination of the three-dimensional structure of a diffraction pattern and intensity measurements, on the other hand, are more easily done with X-ray techniques. If the ions are arranged in an ordered manner on a superlattice, sharp superlattice diffraction lines are observed in the diffraction patterns. Any departure from the perfect periodic order manifests as a continuous diffuse background scattering. For instance, if there are vacancies or interstitial ions in an otherwise long-range-ordered structure, a superposition of discrete superstructure reflections and a diffuse background will be observed. In the completely disordered state only a diffuse scattering pattern is present, its structure being related to the short-range order.

Depending on the coordination type of the ions, there are three possible two-dimensional sublattices for the intercalate ions in the layered dichalcogenides: a triangular lattice for the octahedrally coordinated ions, a honeycomb lattice for the trigonal prismatically coordinated ions, and a puckered honeycomb lattice for the tetrahedrally coordinated ions such as silver and monovalent copper (see Fig. 10). In the latter case the lattice also may be considered to be a superposition of two triangular lattices that are slightly shifted with respect to each other. Numerous differently ordered structures can be imagined on these sublattices (*41*). The simplest ones have an enlarged hexagonal unit cell, but structures having a lower symmetry are also possible. If in the latter case the distortions associated with this symmetry lowering are negligible, a "twinning" is expected in all equivalent directions.

Our knowledge of the ordering of the monovalent intercalate ions in transition metal dichalcogenides is very limited. Fairly extensive structural studies have been made of only two systems, that is, Cu_xNbS_2 by Boswell *et al.* (*42*), de Ridder *et al.* (*43*), and van Tendeloo *et al.* (*43*) by means of electron diffraction, and Na_xTiS_2 by our group (*30*) using X-ray techniques. An early report on Na_xTaS_2 should also be mentioned (*44*).

In Na_xTiS_2 we observed three types of superlattice patterns due to the ordering of the presumably trigonal prismatically coordinated sodium ions

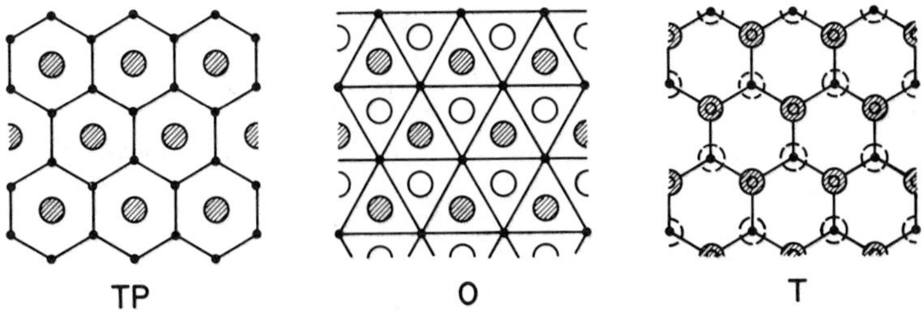

Fig. 10. Depending on the coordination of the intercalate ion, the sublattice in the van der Waals gap is (1) a honeycomb lattice (trigonal prismatic coordination, TP), (2) a triangular lattice (octahedral coordination, O), or (3) a puckered honeycomb lattice (tetrahedral coordination, T).

on a honeycomb lattice, which are denoted by A, B, and C in Fig. 11. Patterns A and B correspond to hexagonal (2×2) and $(\sqrt{3} \times \sqrt{3})$ lattices, respectively, whereas pattern C is a superposition of three rotationally equivalent rectangular $(2 \times \sqrt{3})$ lattices. The actual structures, which have also been drawn in the figure, could be determined uniquely from the unit-cell parameters, the number of ions per unit cell, and the symmetry information from systematically absent reflections. It should be noted that the schematic diffraction patterns of Fig. 11 are projections of the three-dimensional patterns and correspond, therefore, to the structure within a single layer.

The fractional compositions at which the perfectly ordered superstructures would occur have been added in Fig. 11. In reality, the phases apparently exist over a relatively wide stoichiometry range (see Table II). For compositions deviating from the stoichiometric values, the defects will cause an additional diffuse background scattering. In one case, namely, for x values slightly in excess of one-half diffuse rings were observed in addition to the superstructure pattern C and could be analyzed in terms of defects due to excess ions. From the above results it may be concluded that the sodium ions in Na_xTiS_2 are long-range-ordered at room temperature.

In Cu_xNbS_2 the copper ions are in tetrahedral positions, forming a puckered honeycomb lattice. The structure for $x = 0.5$ was studied extensively at different temperatures (*43*). Below $-30°C$ the copper ions are arranged as shown in Fig. 12. At still lower temperatures a transition to a slightly distorted arrangement occurs. The structure of Fig. 12 is closely related to the superstructure of type C in $Na_{0.5}TiS_2$. Above $-30°C$ the copper ions are disordered and the diffraction pattern is diffuse. Similar diffuse scattering patterns have been observed at other compositions (*42*) and have been interpreted in terms of a cluster model (*45*). Samples with a high copper content reveal commensurate as well as incommensurate structures, which may be described as regular arrangements of rows of copper ions interrupted by rows partially consisting of vacancies.

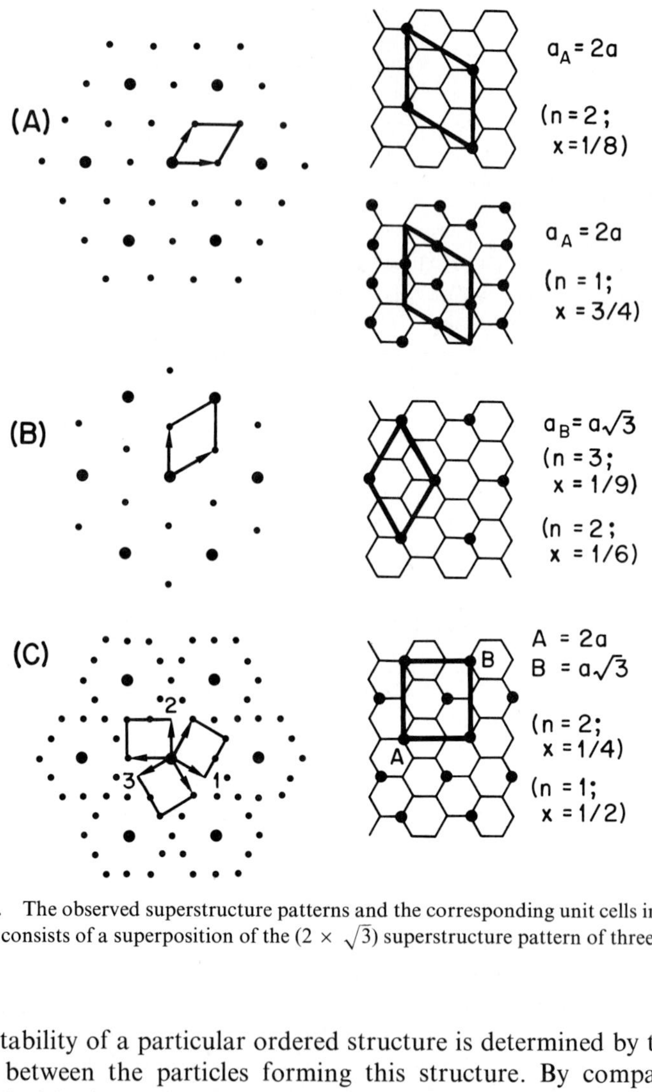

Fig. 11. The observed superstructure patterns and the corresponding unit cells in Na_xTiS_2. Pattern C consists of a superposition of the $(2 \times \sqrt{3})$ superstructure pattern of three rotational twins.

The stability of a particular ordered structure is determined by the interactions between the particles forming this structure. By comparing the observed with alternative structures (which are not observed, because they are apparently less stable), qualitative information about the interparticle interactions can be obtained. If this is done for the structures observed in Na_xTiS_2, the inequality

$$J_3 > 2J_4 > 4J_5 > 8J_6 \tag{3}$$

can be derived (*30*). Here, J_n is the effective interaction between a particle and its *n*th neighbor on a honeycomb lattice. The interaction therefore

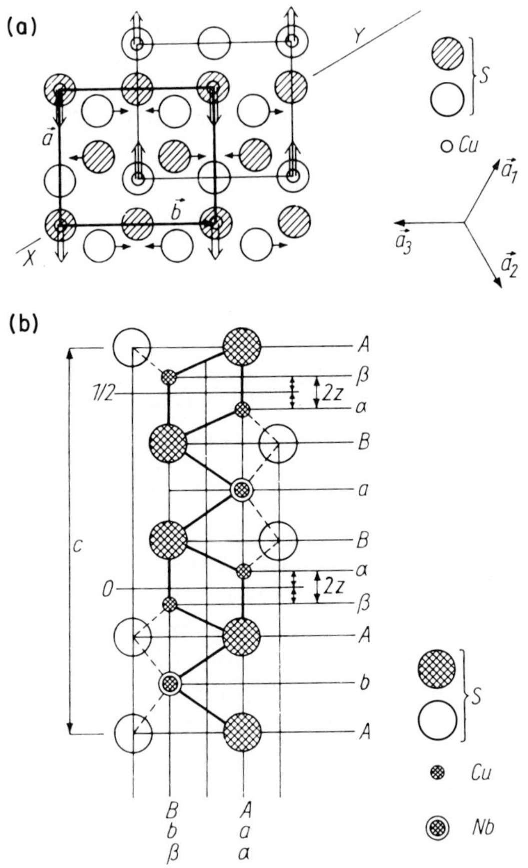

Fig. 12. Schematic representation of the structures of the β phase of Cu-intercalated NbS_2. (a) The small circles represent Cu, whereas the larger ones represent S in two levels. In this projection the Nb positions coincide partly with the Cu positions; they have not been represented in (a). One has to imagine the Nb ions in the centers of the trigonal prisms of S ions. The double arrows indicate the displacements of the Cu ions in the α phase; the single arrows indicate the simultaneous deformation of the S tetrahedra. (b) Cross section along XY of the structure according to the stacking symbol ...$[AbA\beta\alpha BaB\alpha\beta]AbA$.... The tetrahedral layers α and β are filled only for one-fourth. [After de Ridder *et al.* (*43*).]

decreases much faster with the interionic distance than according to a simple Coulomb law. This is due to a strong screening of the coulomb interaction by the valence electrons of the host. A rapidly decreasing interaction potential is expected for all the intercalation compounds, which are either metals or semiconductors with a large static dielectric constant. For semiconductors such a fast decrease with distance is expected only for a limited number of neighbors. Beyond a critical screening distance the reduction factor is constant and equals ϵ_0, the static dielectric constant (*46*).

The intercalation compounds considered here thus offer the possibility of studying the ordering of particles on simple quasi-two-dimensional lattices for a variety of compositions. Of particular interest is the lithium intercalation compound of TiS_2, because it is a single-phase system over the whole stoichiometry range. The chemical potential varies rather smoothly with x (*47*, *48*) and may be fitted approximately with the result of a mean field calculation (*49*). Nevertheless, the system appears to be long-range-ordered. Accurate measurements of the derivative of the chemical potential with respect to the composition reveal pronounced structures at well-defined fractional compositions (*47*). In addition, we have observed very weak but sharp superstructure patterns of types A and B in chemically intercalated TiS_2 crystals, suggesting long-range order for the lithium ions (*30*).

F. Ionic Conductivity and Structure

For their use as solid solution electrodes in electrochemical devices, a relatively rapid diffusion of the intercalate ions is essential. Diffusion coefficients estimated from the dynamic behavior of polarized electrochemical cells for some of the lithium and sodium compounds lie in the range between 10^{-10} and 10^{-7} cm^2/sec (*19*) at room temperature, which is relatively low and consistent with an essentially ordered nature of the alkali sublattices.

From a structural standpoint, there are mainly two factors that determine the magnitude of the ionic conductivity: (a) the ratio of the number of ions and the number of sites playing a role in the conductivity process and (b) the nature of the bottleneck connecting these sites. The barrier for ionic motion is usually at the center of the face that is common to the coordination polyhedra of neighboring sites (*50*). In the layered dichalcogenides there are three different situations, depending on the coordination of the ions. The trigonal prismatic coordination seems to be the most favorable for ionic conduction, because the trigonal prisms share rectangular faces. A rectangular arrangement of chalcogenide ions seems to be a wider bottleneck for the alkali ions than the more common triangular one. In the close-packed anion arrangement, on the other hand, there are both octahedral and tetrahedral positions, and an ion in an octahedral site can jump only to a tetrahedral site and vice versa, suggesting a relatively high effective barrier for the motion between equivalent sites. The experimental diffusion coefficients for trigonal prismatically coordinated sodium ions indeed seem to be considerably larger than those for the octahedrally coordinated ions (*19*).

The role of both types of sites in the conduction process is also apparent from X-ray data of the only layered dichalcogenide for which the dc ionic

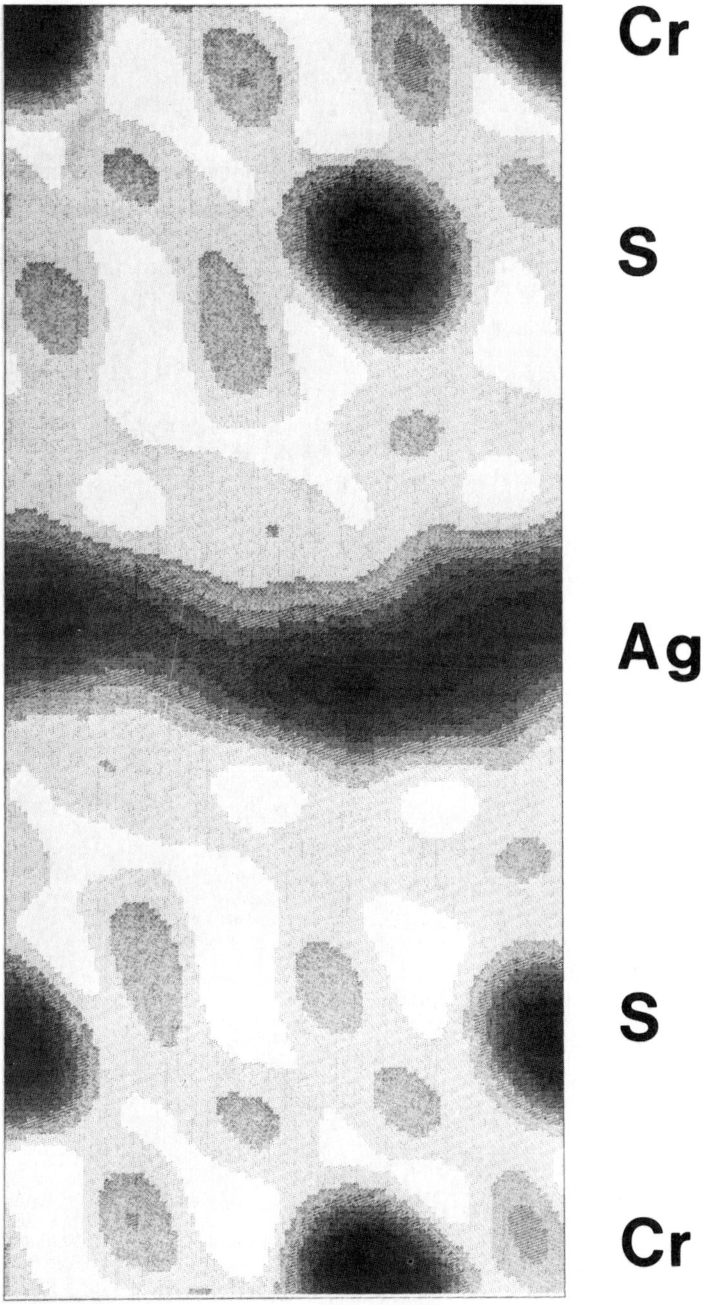

Fig. 13. Projection of the electron density of the disordered phase of $AgCrS_2$ on the (010) plane, calculated from the intensities of the ($h0l$) reflections using the signs of the $F(h0l)$ from a refinement of the projected structure. The height of the section shown is $c/3$, and the width is $a\sqrt{3}/2$. Shades increase in intensity in equal density steps; white represents small negative densities.

conductivity has been measured directly, that is, $AgCrS_2$ (*51*). At room temperature the silver ions are essentially ordered and occupy only one set of tetrahedral sites. Above the order–disorder transition temperature of 400°C, the silver ions are distributed over both sets of tetrahedral sites and the ionic conductivity is very high ($\sim 0.3\ \Omega^{-1}\ cm^{-1}$). Figure 13 shows a projection of the electron density for this material in the disordered state, which was calculated from the intensities of the ($h0l$) reflections using the signs of the scattering factors from a refinement of the projected structure. The smeared-out nature of the silver ion density in the intercalate plane is evident. In particular, the relatively large density found halfway between the tetrahedral sites means that some of the silver ions are in the octahedral holes, presumably in off-site positions. The nature of the diffuse X-ray scattering in the disordered state is also in favor of a partial off-site occupation of the octahedral sites. It consists of a number of concentric diffuse cylinders, suggesting a two-dimensional amorphous structure of the silver ions. The lack of periodic structure in the diffuse scattering pattern cannot be understood if only substitutional disorder on tetrahedral sites is considered.

ACKNOWLEDGMENTS

The author wishes to thank H. U. Beyeler, P. Brüesch, L. Pietronero, and S. Strässler for many valuable discussions.

REFERENCES

1. J. A. Wilson and A. D. Yoffe, *Adv. Phys.* **18,** 193 (1969).
2. R. Huisman, R. de Jonge, C. Haas, and F. Jellinek, *J. Solid State Chem.* **3,** 56 (1971).
3. J. L. Calais, *Adv. Phys.* **26,** 847 (1977).
4. D. W. Bullett, *J. Phys. C* **11,** 4501 (1978).
5. F. R. Gamble, *J. Solid State Chem.* **9,** 358 (1974).
6. A. Madhukar, *Solid State Commun.* **16,** 383 (1975).
7. L. Pauling, "The Nature of the Chemical Bond," 3rd ed. Cornell Univ. Press, Ithaca, New York, 1960.
8. L. Pauling, *Phys. Today* **24,** No. 2, 9 (1971).
9. J. C. Phillips, *Phys. Today* **23,** No. 2, 23 (1970).
10. J. C. Phillips, *Rev. Mod. Phys.* **42,** 317 (1970).
11. A. R. Miedema, R. Boom, and F. R. de Boer, *J. Less-Common Met.* **41,** 283 (1975).
12. R. M. White and G. Lucovsky, *Solid State Commun.* **11,** 1369 (1972).
13. A. Zunger and A. J. Freeman, *Phys. Rev. B* **16,** 906 (1977).
14. L. F. Mattheiss, *Phys. Rev. B* **8,** 3719 (1973).
15. L. Pauling, *J. Am. Chem. Soc.* **54,** 988 (1932).
16. R. Hultgren, *Phys. Rev.* **40,** 891 (1932).

17. B. L. Evans, *in* "Physics and Chemistry of Compounds with Layered Structures" (P. A. Lee, ed.), Vol. 4, p. 1. Reidel Publ., Dordrecht, Netherlands, 1976.
18. J. Rouxel, *in* "Physics and Chemistry of Compounds with Layered Structures" (F. Levy, ed.), Vol. 6, p. 201. Reidel Publ., Dordrecht, Netherlands, 1979.
19. M. S. Whittingham, *Prog. Solid State Chem.* **12,** 41 (1978).
20. C. F. van Bruggen, C. Haas, and G. A. Wiegers, *J. Solid State Chem.* **27,** 9 (1979).
21. R. F. Frindt and D. J. Huntley, *in* "Physics and Chemistry of Materials with Layered Structures" (P. A. Lee, ed.), Vol. 4, p. 403. Reidel Publ., Dordrecht, Netherlands, 1976.
22. W. Rudorff and H. H. Sick, *Angew. Chem.* **71,** 127 (1959); W. Rudorff, *Angew. Chem.* **71,** 487 (1959).
23. M. S. Whittingham and F. R. Gamble, *Mater. Res. Bull.* **10,** 363 (1975).
24. P. S. Bongert, C. F. van Bruggen, J. Koopstra, W. P. F. A. Omloo, G. A. Wiegers, and F. Jellinek, *J. Phys. Chem. Solids* **29,** 977 (1968); B. van Laar and D. J. W. Ijdo, *J. Solid State Chem.* **3,** 590 (1971); B. van Laar and F. M. R. Engelsmann, *J. Solid State Chem.* **6,** 384 (1973); F. M. R. Engelsmann, G. A. Wiegers, F. Jellinek, and B. van Laar, *J. Solid State Chem.* **6,** 574 (1973).
25. W. P. F. A. M. Omloo and F. Jellinek, *J. Less-Common Met.* **20,** 121 (1970).
26. G. A. Wiegers, *Physica* **B99,** 151 (1980).
27. R. Schöllhorn, *Physica* **B99,** 89 (1980).
28. D. A. Winn and B. C. H. Steele, *Mater. Res. Bull.* **11,** 551 (1976).
29. A. Nagelberg and V. L. Worrell, *J. Solid State Chem.* **29,** 345 (1979).
30. T. Hibma, *J. Solid State Chem.* **34,** 97 (1980); T. Hibma, *Physica* **B99,** 136 (1980).
31. R. J. Haange, A. S. A. Bos-Alberink, and C. A. Wiegers, *Ann. Chim. (Paris)* **3,** 201 (1978).
32. C. Berthier, Y. Chabre. and P. Segransan, *Physica* **B107** (1980).
33. J. V. McCanny, *Phys. Status Solidi B* **93,** K29 (1979); J. V. McCanny, *J. Phys.* **C12,** 3263 (1979).
34. J. V. Acrivos and J. R. Salem, *Philos. Mag.* **30,** 603 (1973).
35. J. Rouxel, *J. Solid State Chem.* **17,** 223 (1976).
36. J. Rouxel, M. Danot, and J. Bichon, *Bull. Soc. Chim. Fr.* p. 3930 (1971).
37. J. E. Fischer and T. E. Thompson, *Phys. Today* **31,** No. 7, 36 (1978).
38. L. Pietronero, S. Strässler, and H. R. Zeller, *Phys. Rev. Lett.* **41,** 763 (1978).
39. J. van Landuyt, *Physica* **B99,** 12 (1980).
40. A. Guinier, "X-ray Diffraction in Crystals, Imperfect Crystals and Amorphous Bodies." Freeman, San Francisco, California, 1963.
41. M. Kaburagi and J. Kanamori, *J. Phys. Soc. Jpn.* **44,** 718 (1978).
42. F. W. Boswell, A. Prodan, and J. M. Corbett, *Phys. Status Solidi A* **35,** 591 (1976).
43. R. de Ridder, G. van Tendeloo, J. van Landuyt, D. van Dyck, and S. Amelincks, *Phys. Status Solidi A* **37,** 591 (1976); G. van Tendeloo, R. de Ridder, D. van Dyck, J. van Landuyt, and S. Amelincks, *Phys. Status Solidi A* **38,** 185 (1976).
44. C. B. Carter and P. M. Williams, *Philos. Mag.* **26,** 393 (1972).
45. R. de Ridder, G. van Tendeloo, D. van Dyck, and S. Amelincks, *Acta Crystallogr. Sect. A* **32,** 216 (1976).
46. R. Resta, *Phys. Rev. B* **16,** 2717 (1977).
47. M. S. Whittingham, *J. Electrochem. Soc.* **123,** 315 (1976).
48. A. H. Thompson, *Phys. Rev. Lett.* **40,** 1511 (1978).
49. M. Armand, Ph.D. Thesis, Univ. Sci. Méd. Inst. Nat. Polytech., Grenoble, 1978.
50. S. Geller, *in* "Fast Ion Transport in Solids" (W. van Gool, ed.), p. 607. North-Holland, Amsterdam, 1973.
51. T. Hibma, *Solid State Commun.* **33,** 445 (1980).
52. R. B. Somoano, V. Hadek, and A. Rembaum, *J. Chem. Phys.* **58,** 697 (1973).

10

Solvated Intercalation Compounds of Layered Chalcogenide and Oxide Bronzes

R. SCHÖLLHORN

I. INTRODUCTION

Layered intercalation compounds have been known for a long time, and a considerable variety of host lattices that can incorporate atomic or molecular guest species have been investigated, e.g., graphite (*1*), clay

INTERCALATION CHEMISTRY

ISBN 0-12-747380-7

minerals (*2*, *3*), layered cyanides (*4*), and layered dichalcogenides (*5*). The attention paid to these compounds increased greatly approximately a decade ago, when it became established that the unusual physical properties of layered materials with electronic conductivity (anisotropic transport properties, superconductivity, cooperative lattice distortions, etc.) can be modified significantly by intercalation (*6*, *7*, *12*). This aspect resulted in a new impetus for chemical research in this area characterized by (a) an effort to understand the structure, bonding, and formation of these compounds in more detail and (b) attempts to synthesize new layered materials and intercalation compounds with interesting structures and properties. Apart from academic interest, research on these phases was stimulated by certain aspects of potential technological significance concerning the mechanical properties of these phases, their application as reversible electrodes for high-energy-density batteries (*8*), and their efficiency as homogeneous catalysts (*9*).

Considerable work has been devoted in particular to the intercalation compounds of layered transition metal dichalcogenides (*6–8*, *10–12*) for several reasons: relatively simple host structures, availability of large single crystals, favorable chemical stability, high reactivity for different types of guest species, and a wide range for modifying host lattice electronic properties, from semiconducting to metallic, depending on the choice of transition metal and chalcogen. The intercalation compounds of dichalcogenides can be divided into three categories according to the nature of the guest species: (a) compounds with metal ions or molecular ions, for example, Na^+, $[(C_5H_5)_2Co]^+$; (b) compounds with Lewis base type of molecules, for example, NH_3, pyridine; (c) compounds containing simultaneously cations and neutral polar molecules. The last group, which is of interest here, has to be described on the basis of structure and chemical reactivity in terms of a polyelectrolyte model: Mobile guest cations solvated by polar molecular species are positioned in the interlayer space between the host macroanion sheets (*7*, *11*). As a consequence of the high reactivity of these phases due to topotactic cation exchange, solvent exchange, and redox reactions by electron/ion transfer, an extraordinarily wide field is provided for the systematic synthesis of intercalation compounds. A series of structurally related layered chalcogenides has been shown to exhibit similar properties. There is also evidence now that Lewis base intercalation compounds of dichalcogenides have ionic structure and are to be considered solvated phases (*12*). Recent studies on layered oxides with electronic conductivity revealed the existence of solvated oxide bronzes with exchange and redox properties closely related to those of the layered chalcogenides.

This chapter focuses on the formation, characterization, structure, and reactivity of solvated layered chalcogenides and oxide bronzes and some related systems. The presentation of material is based essentially on the different types of host lattices and includes an introductory section giving an overview of the common characteristics of these systems.

II. GENERAL ASPECTS OF SOLVATION AND POLYELECTROLYTE PROPERTIES

Intercalation can be considered basically a reversible topotactic reaction of a solid host (Z)—which provides an interconnected system of accessible unoccupied lattice sites (□)—with ambient mobile guest species (G) that can diffuse into the solid and occupy the empty positions (*12*):

$$xG + \square_x[Z] \rightleftarrows G_x[Z] \tag{1}$$

The term "topotactic" refers to the idea that the host matrix units retain their integrity with respect to structure and composition in the course of intercalation and deintercalation. The bonding between the atoms that constitute the matrix must be strong and has to include covalent increments to ensure the stability of the structure during the solid-state reaction. Since at low temperatures the activation energy barrier for diffusion of the guest species in the solid volume inhibits intercalation, whereas at higher temperatures the matrix becomes unstable, it is apparent that intercalation reactions may proceed only at intermediate temperatures.

A considerable number of solids with layered structures satisfy the requirements discussed and undergo the type of solid-state reaction defined above. Their structure can be described in terms of stacks of layer units with partially covalent intralayer bonding which are held together in the crystal lattice essentially by weak van der Waals forces (see Section III,A). The van der Waals gap between the two-dimensional matrix units provides a system of empty lattice sites sharing polyhedral faces that can be occupied by guest species. In the course of intercalation, the interlayer spacing (distance between adjacent layers perpendicular to the layer planes) increases, the amount depending on the effective size of the guest; as a consequence of the high flexibility of layered systems, even rather large guest molecules may be accommodated by the host structure.

In the case of layered solids with electronic conductivity, it now seems to be well established for many systems that direct intercalation into the host lattice (either by chemical or by electrochemical reactions) is to be described as a reversible topotactic redox process leading to products with negatively charged host layers Z and an equivalent amount of cationic species A^+ in the interlayer space:

$$xA^+ + xe^- + Z \rightarrow A_x^+[Z]^{x-} \tag{2}$$

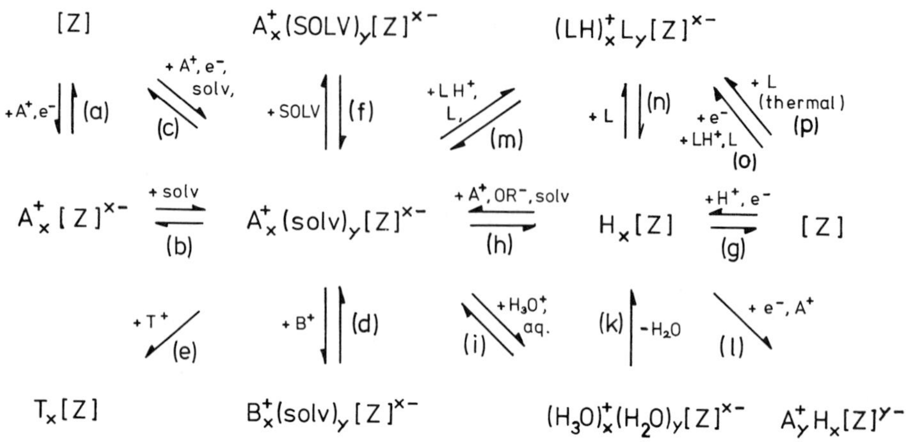

Fig. 1. Scheme of topotactic solvation, exchange, and redox reactions of layered materials with electronic conductivity. Z, host lattice, A^+, guest cation, solv, neutral polar guest molecule. See text for details.

Uncharged polar molecular species (denoted here by "solv") can be introduced by direct or indirect cointercalation or by exchange reactions:

$$xA^+ + xe^- + y(\text{solv}) + Z \rightarrow A_x^{\ +}(\text{solv})_y[Z]^{x-} \quad (3)$$

$$A_x^{\ +}[Z]^{x-} + y(\text{solv}) \rightarrow A_x^{\ +}(\text{solv})_y[Z]^{x-} \quad (4)$$

$$A_x^{\ +}(\text{solv})_y[Z]^{x-} + y(\text{SOLV}) \rightarrow A_x^{\ +}(\text{SOLV})_y[Z]^{x-} + y(\text{solv}) \quad (5)$$

A simplified scheme of reactions of layered chalcogenide and oxide systems leading to the formation of solvated phases is given in Fig. 1. We discuss below in general terms the different reaction pathways observed, which are denoted in the scheme by lowercase letters (a–p); they are all based on one of the processes described in Eqs. (2)–(5).

The empty host lattice Z may intercalate in a topotactic redox reaction (a) cations of electropositive elements (e.g., alkali and alkaline earth ions), and ternary phases $A_x^{\ +}[Z]^{x-}$ are thus obtained. These can react (b) with polar organic or inorganic molecules (such as H_2O, NH_3, and acid amides) to give solvated phases $A_x^{\ +}(\text{solv})_y[Z]^{x-}$. The same compounds may also be obtained directly from Z by chemical or electrochemical redox reaction in polar solvents (c). In addition, it is possible to intercalate via (c) solvated transition metal and molecular cations, e.g., inorganic and organic cations, radical-cations, and metal complex cations. The solvated phases exhibit polyelectrolyte character and may undergo ion-exchange reactions (d) with inorganic and organic cationic species. The extent of solvation and the solvate structure depend in the case of layered chalcogenides essentially on the interaction between solvent dipole and cation, whereas for layered oxides interactions between the solvent and the layer units may play an additional role. Ion exchange with redox-active cations or cations of low solvation energy ($T = Cu^{2+}$, Ag^+, etc.) leads to desolvated ternary phases (e). In reversion of (c), the solvated phases may be reoxidized to Z. The mobile solvate molecules can be exchanged for other polar molecular species (f); partial desolvation with the occurrence of discrete intermediate stages or complete desolvation can be performed in reversion of (b).

Cathodic reduction of Z in acid electrolytes (g) results in the formation of hydrogen compounds $H_x[Z]$, which behave as solid Brönsted acids and react with bases (h) to yield solvated phases (R = H or alkyl). Ion exchange of the latter with aqueous acids leads to hydronium compounds $(H_3O^+)_x(H_2O)_y[Z]^{x-}$, which may be converted to the hydrogen compounds by desolvation (i, k). Further cathodic reduction of $H_x[Z]$ in electrolytes (*l*) may lead either to solvated or to unsolvated phases containing H and A^+.

In homosolvated intercalation compounds $(LH^+)_xL_y[Z]^{x-}$ of Lewis bases (such as NH_3, amines, and heterocycles), cations and solvate molecules differ only by an additional proton attached to the functional group of the base. They may be obtained in several ways: ion exchange with electrolyte solutions containing protonated (LH^+) and unprotonated Lewis base (m), acid–base reaction of L with $H_x[Z]$ (n), cathodic reduction of Z in LH^+/L electrolytes (o), and direct thermal intercalation (p) of L into the host lattice Z; this process is accompanied by a molecular redox reaction (see Section III,E).

The scheme given in Fig. 1 is a simplified presentation and does not include all reactions and correlations observed so far. It is to be noted, however, that it depends on the particular type of host lattice, guest cation, and solvate species, whether all or only a part of the reactions indicated are accessible or reversible. Kinetics are usually affected by steric factors (guest dimensions), whereas the electronic factor (band structure of the host) governs the available redox range (negative layer charge density x and correlated cation density). The reactions discussed are frequently accompanied by changes in the stacking sequence of the host layers, and higher-stage compounds with a regular sequence of empty and intercalated van der Waals gaps can be observed. The occurrence of nonstoichiometry is characteristic of the class of compounds under discussion; thermodynamically, the majority of the solvated phases are to be considered metastable compounds that cannot be obtained by methods other than intercalation. In some cases, especially in the presence of protolytic molecular guest species, irreversible topochemical modification of the layer matrix by partial solvolysis may be observed on intercalation.

III. TRANSITION METAL DICHALCOGENIDES

A. Host Lattice Structures

Binary metal chalcogenides of the stoichiometry MX_2 (M = metal; X = S, Se, Te) with hexagonal layered structures are found for transition elements of Groups IVb (Ti, Zr, Hf), Vb (V, Nb, Ta), VIb (Mo, W), VIIb (Tc, Re), and VIIIb (Pt) and for the main group elements Sn and Pb. Their basic structure can be described as a stacking of two-dimensionally infinite $(MX_2)_\infty$ sheets that are held together in the crystal lattice essentially by weak van der Waals forces. The MX_2 layers consist of two anion sheets with a sheet of metal atoms in the central plane (Fig. 2); there is strong (partially covalent) bonding between the intralayer atoms M and X. A large variety of modifications are known which differ by the coordination of the metal between the anions (octahedral or trigonal prismatic, Fig. 2) and by the mode of stacking of the MX_2 layers. The electronic properties range from semiconducting to metallic behavior depending on the metal, the chalcogen, and the modification. The preparation can be achieved by thermal reaction of the elements; large single crystals are obtained by vapor-phase transport. It is essential to use experimental conditions that guarantee the formation of stoichiometric phases, since the existence of nonstoichiometric compounds $M_{1+x}S_2$ is known for many systems. In the latter case the excess metal atoms occupy positions in the van der Waals gap, which usually prevents intercalation from taking place, except in the case of small guest species, for example, Li. For details on structural aspects, the reader is referred to recent reviews (*13*).

Ternary phases A_xMX_2 (A = group Ia or IIa metal) can be prepared either by thermal methods (e.g., directly from the elements) or via intercalation. The metal atoms A are positioned in the van der Waals gap.

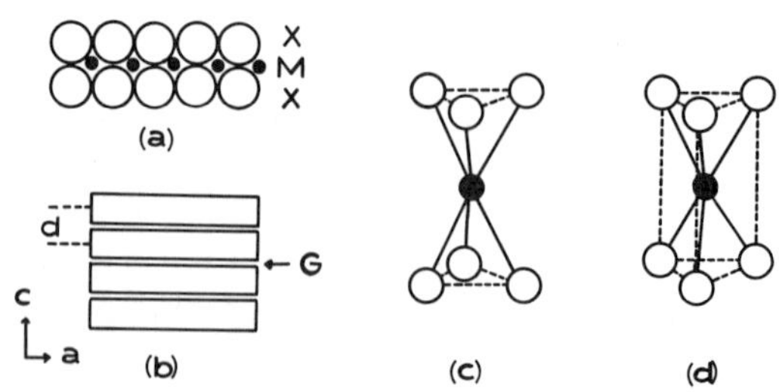

Fig. 2. Structural scheme of layered metal dichalcogenides MX_2. (a) Single layer; (b) layer stacking in crystal lattice: *d*, interlayer spacing; G, van der Waals gap; (c) octahedral coordination; (d) trigonal prismatic coordination of M.

Comprehensive reviews on the structure and properties of these phases have been published (*8*, *10*, *13*).

B. Hydrated Phases: Formation, Structure, and Reactivity

Solvated dichalcogenide compounds were first reported by Schwarz and Rüdorff (1954), who observed the uptake of water by K_xCrSe_2 (*14*), and by Rüdorff and Sick (1959), who described the simultaneous intercalation of alkali metals and ammonia into MoS_2, WS_2, and ReS_2 (*5*, *15*). Hydrates with the composition $(KOH)_{0.88}(H_2O)_{0.54}[TiS_2]$ and $K(OH)[SnS_2]$ were found by Rouxel *et al.* (1969) to form on reaction of K_xTiS_2 and $KSnS_2$ with water vapor (*16*). Gamble *et al.* (1972) obtained superconducting intercalation compounds from the reaction of aqueous alkali hydroxides with TaS_2 (*17*). Studies by Schöllhorn and Weiss (1973) on the hydration properties of ternary alkali titanium and alkali molybdenum sulfides A_xTiS_2 and A_xMoS_2 established for the reaction products the polyelectrolyte character (reversible ion and solvate exchange), the ionic structure $A_x{}^+(H_2O)_y[MX_2]^{x-}$, and the existence of two defined hydration stages dependent on cation hydration energy and water vapor pressure (*18*, *19*). Whittingham (1974) observed similar hydration behavior for intercalation compounds of dichalcogenides with aqueous NH_3 and alkali hydroxides and for hydrated phases derived from Li_xTaS_2; he proposed structural models for the position of H_2O in the interlayer space (*20*). Wiegers *et al.* (1974) reported two hydration stages of $Na_{0.6}(H_2O)_xVS_2$, for which they observed an $NaCrS_2$ type of structure (*21*). In the same year the formation of solvated phases by electrochemical and chemical reduction of dichalcogenide electrodes in aqueous electrolytes was found (*22–24*) and interpreted as reversible and topotactic redox reaction (*23*, *25*, *26*). Extensive investigations since then have established a fairly consistent picture on the properties and basic structures of the solvated layered chalcogenides.

Here, we shall first discuss hydrated phases with Group Ia and IIa cations, the behavior of which can be considered to be representative (*11*). They can be prepared directly either from the ternary phases A_xMX_2 (A = alkali or alkaline earth cation), which usually undergo spontaneous hydration (*27*),

$$A_xMX_2 + yH_2O \rightarrow A_x{}^+(H_2O)_y[MX_2]^{x-} \tag{6}$$

or by chemical and electrochemical reduction of the binary chalcogenides in the corresponding aqueous electrolytes (Fig. 3),

$$xA^+ + xe^- + yH_2O + MX_2 \rightleftarrows A_x{}^+(H_2O)_y[MX_2]^{x-} \tag{7}$$

The topotactic nature of these processes can be demonstrated by reactions of single crystals that remain single crystals after intercalation. Whereas a

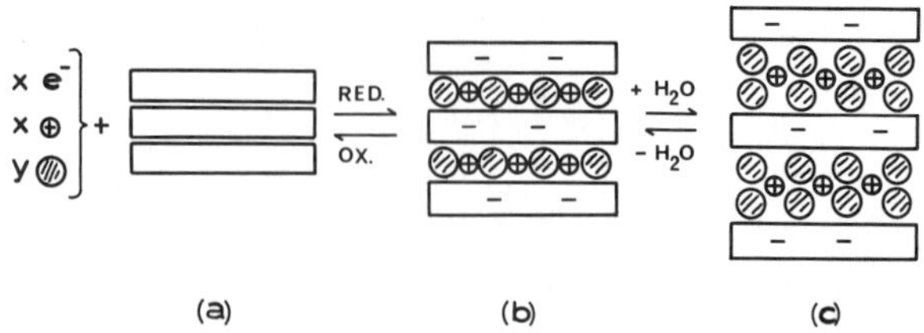

Fig. 3. Formation of hydrated phases by topotactic redox reaction in aqueous electrolytes. (a) Host lattice MX_2; (b) monolayer hydrate $A_x{}^+(H_2O)_y[MX_2]^{x-}$; (c) bilayer hydrate.

considerable change is observed for the hexagonal c-axis lattice parameter, the variation of the lattice parameters in the a/b plane is rather small and as a rule varies between 0.02 and 0.1 Å. The structure and bonding of the hydrated phases $A_x{}^+(H_2O)_y[MX_2]^{x-}$ are described in an ionic model: The $[MX_2]^{x-}$ layers represent quasi-two-dimensional macroanions with delocalized negative charges; the interlayer space is occupied by solvated cations that are highly mobile at room temperature. As a consequence, the hydrates have typical polyelectrolyte character and undergo ion- and solvent-exchange reactions. Two structurally defined hydrate stages are found in the alkali and alkaline earth series with an interlayer spacing d of ca 9 and ca 12 Å (Fig. 3). Since the thickness of one MX_2 layer equals ca 6 Å and the van der Waals diameter of H_2O amounts to ca 3 Å, these values can be interpreted reasonably in terms of water monolayer ($6 + 3 = 9$ Å) and water bilayer ($6 + 6 = 12$ Å) hydrate structures. The formation of either structure is dependent on the charge/radius ratio (e/r) of the cation, i.e., on the cation hydration energy. Figure 4 illustrates for hydrated titanium, niobium, and tantalum sulfides the dependence of the interlayer spacing values on the cation charge/radius ratio: Monovalent large cations with $e/r < 1$ exhibit monolayers of water, whereas small monovalent or bivalent cations with $e/r > 1$ have the capacity to form bilayer hydrates. The d values for Rb^+ and Cs^+ are somewhat larger than those calculated by the above considerations, since the ionic diameters are larger than the diameter of H_2O, and the interlayer spacing is thus determined by the cation size directly. In contrast to layered oxides, which show rather complex hydration behavior because of additional hydrogen bonding between solvate molecules and layers, the hydrate structures in dichalcogenides are dominated essentially by the cation/solvent dipole interaction. This is confirmed by studies on the oxidation of hydrated phases to the binary compounds in reversion of Eq. (7). After quantitative reaction, the layer charge density and the corresponding cation density are equal to zero and the layers collapse to an interlayer spacing of 6 Å; no intermediate (metastable) hydrates $MX_2 \cdot (H_2O)_y$

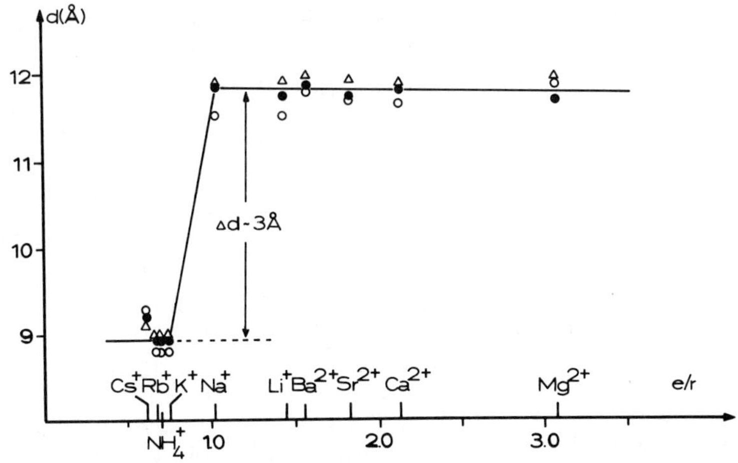

Fig. 4. Dependence of the interlayer spacing d of hydrated layered sulfides on the charge/radius ratio of the guest cation. Key: ●, $A^{n+}_{0.5/n}(H_2O)_y[TiS_2]^{0.5-}$; ○, $A^{n+}_{0.5/n}(H_2O)_y[NbS_2]^{0.5-}$; △, $A^{n+}_{0.3/n}(H_2O)_y[TaS_2]^{0.3-}$. Reproduced with permission from *Inorg. Chem.* **16,** 2950 (1977). Copyright (1977) American Chemical Society.

can be observed. This is obviously a consequence of the small energy of OH···S hydrogen bonds, i.e., a negligibly small interaction between H_2O and the chalcogenide layers.

With H_2O as the solvate, x may vary between ca. 0.1 and 0.5 depending on the host lattice (M, X). The change in interlayer spacing with changes in x, however is, very small; for cations with $e/r > 1$, even at low x values water bilayers are still stabilized by the dimensions of the cation/solvent clusters. The stoichiometry for H_2O as indicated by y depends naturally on density, charge, and size of the cations; examples of fully hydrated phases are $K_{0.33}{}^{+}(H_2O)_{0.6}[TaS_2]^{0.33-}$ and $Na_{0.3}{}^{+}(H_2O)_{1.8}[TaS_2]^{0.3-}$, which show y values that are close to those calculated from model structures. Selenides show a hydration behavior that agrees well with that of sulfides; the interlayer spacing is systematically higher, owing to the greater thickness of the layer units. The stability of the dichalcogenide layers toward solvolysis, i.e., exchange S → O, depends on the transition metal and the chalcogen but is generally rather high in neutral aqueous solutions, whereas for pH <5 and pH >8 slow hydrolysis may occur.

The change in interlayer spacing with changes in the equilibrium electrolyte concentration is rather small in the range of 10^{-5} to 4 N aqueous solutions. The stability of hydrates with H_2O bilayers, however, is dependent on the water vapor pressure: With decreasing p_{H_2O} a part of the water molecules leaves the lattice at almost constant interlayer spacing, until at a defined threshold vapor pressure a reversible bilayer hydrate → monolayer hydrate transition takes place (Fig. 5). Obviously, the cation/water clusters stabilize the structure until a critical H_2O defect concentration is reached. Complete dehydration leads to the formation of the unhydrated ternary phases A_xMX_2; for cations with higher hydration energies usually some water is retained within the lattice.

The maximum value for the negative layer charge density x of stable hydrates depends on the redox potential (electronic structure) of the host lattice, which must be higher than that of H_2/H_3O^+. If hydrates are formed from ternary alkali compounds A_xMX_2 with $x > x_{max}$(hydrate), the layers undergo partial oxidation with hydrogen formation according to

$$A_x{}^+[MX_2]^{x-} + yH_2O \rightarrow A^+_{x-n}(H_2O)_{y-n}[MX_2]^{(x-n)-} + nA^+ + nOH^- + \frac{n}{2}H_2 \quad (8)$$

For sulfides of Group IVa and Va transition metals, x_{max} varies between 0.3 and 0.5; a higher value of $x = 0.6$ was reported for the hydrated vanadium sulfides and selenides $Na_{0.6}(H_2O)_y[VX_2]$ (*21*, *28*).

In the Group VIa sulfides MoS_2 and WS_2, the d_{z^2} band of the host structure is filled up and additional electrons have to be accommodated in higher energy levels. Direct intercalation can be achieved only with strong electron

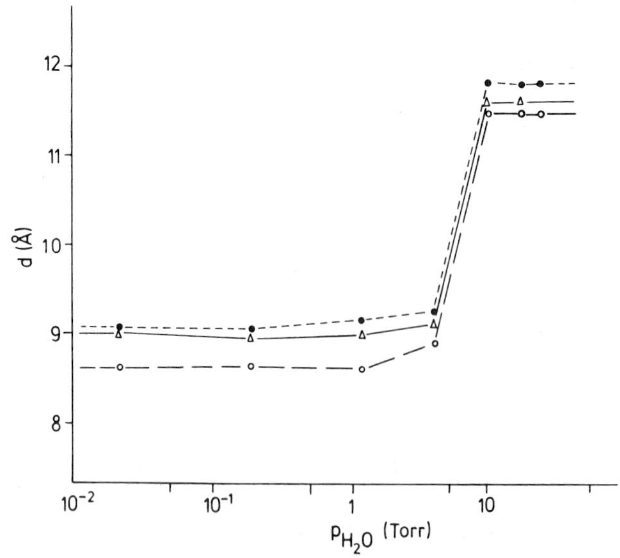

Fig. 5. Dependence of the interlayer spacing *d* of hydrated layered sulfides on ambient water vapor pressure. Key: ○, $Na^+_{0.5}(H_2O)_y[TiS_2]^{0.5-}$; ○, $Na^+_{0.5}(H_2O)_y[NbS_2]^{0.5-}$; △, $Na^+_{0.3}(H_2O)_y[TaS_2]^{0.3-}$. Reproduced with permission from *Inorg. Chem.* **16,** 2950 (1977). Copyright (1977) American Chemical Society.

donors such as alkali and alkaline earth metals (*5*). Because of the highly negative redox potential of the layers, the layer charge density after hydration is rather low, with $x \sim 0.1$, and the compounds have low stability; their solvation and ion-exchange behavior, however, is closely similar to that of Group IVa and Va chalcogenides (*19*).

For chromium only stoichiometric ternary alkali sulfides $ACrS_2$ are known (*29*), which do not hydrate spontaneously. Apparently, the redox potential of the CrS_2 layers is too high for H^+ reduction; direct intercalation is energetically unfavorable because of the high cation density, which would lead to an incomplete hydration sphere of the cations. Reaction can be achieved, however, by chemical or anodic partial oxidation, yielding the hydrates $A_{0.4}{}^+(H_2O)_y[CrS_2]^{0.4-}$; a hydrate with mixed intralayer cations $Na_{0.5}{}^+(H_2O)_y[Cr_{0.5}Ti_{0.5}S_2]^{0.5-}$ has also been reported (*25, 30, 31*).

Among the main group metals, the element tin is known to form layered chalcogenides SnX_2 with CdT_2 structure. Stoichiometric intercalation compounds $ASnS_2$ of tin disulfides with alkali metals were found to react with water to form $K(OH)SnS_2$ (*16*). In a subsequent investigation (*32*), it was shown that ionic alkali hydrates $A_x{}^+(H_2O)_y[SnS_2]^{x-}$ with $x = 0.1$–0.3 may be obtained by reduction of SnS_2 in aqueous electrolytes or by hydration of $ASnS_2$ under partial oxidation [Eq. (8)]. The metallic black compounds are unstable and decompose easily with formation of SnS. Since for tin only the oxidation states +2 and +4 are known, a mixed-valence model was proposed for the hydrated sulfides.

Besides the ternary alkali molybdenum sulfides A_xMoS_2, which are obtained by intercalation at low temperatures and have regular hexagonal lattice parameters, a second series of ternary alkali compounds A_xMoS_2 with orthorhombic lattice symmetry can be prepared at high temperatures (*33*). Their chemical reactivity and hydration behavior are closely related to those of the low-temperature phases, but the MoS_2 layers are distorted, supposedly because of the presence of Mo clusters or chains.

The close relationship between the hydrates derived from different MX_2 host structures can be demonstrated by a comparison of the Na^+ and K^+ compounds (*11*). In Table I the measured interlayer spacing values d and calculated values for the lattice expansion $\Delta d = d - d_0$ (d_0 represents the interlayer spacing of the binary unintercalated host lattice) are given for both series. It can be seen that the values for $\Delta d(Na^+)$ and $\Delta d(K^+)$ amount to ca 6 and ca 3 Å, respectively; the ratio $\Delta d(Na^+)/\Delta d(K^+)$ corresponds reasonably well to the monolayer/bilayer hydrate model.

Limited information is available on the detailed structures of the hydrates at different temperatures. For $Na_{0.5}{}^+(H_2O)_y[TiS_2]^{0.5-}$ the position of Na^+ in the midplane between the two water layers was derived from a one-dimensional Fourier projection (*18*). The structure determination of the host

TABLE I

Interlayer Spacing d and Layer Expansion Δd for Na^+ and K^+ Forms of Hydrated Layered Chalcogenides $A_x^+(H_2O)_y[MX_2]^{x-}$

M	X	$d(Na^+)$ (Å)	$\Delta d(Na^+)$ (Å)	$d(K^+)$ (Å)	$\Delta d(K^+)$ (Å)	$\frac{\Delta d(Na^+)}{\Delta d(K^+)}$
Ti	S	11.56	5.85	8.72	3.01	1.94
	Se	12.06	6.05	9.31	3.30	1.83
Nb	S	11.63	5.68	8.94	2.99	1.90
	Se	12.08	5.81	9.40	3.13	1.86
Ta	S	11.74	5.69	9.04	2.99	1.90
	Se	12.88	6.54	9.71	3.37	1.94
Cr	S	11.33	5.32	8.59	2.58	2.06
Mo	S	11.97	5.83	9.09	2.95	1.98
W	S	12.12	5.94	9.41	3.23	1.84
Re	S	12.29	6.19	9.50	3.40	1.82
Sn	S	12.64	6.74	9.29	3.39	1.99

Reproduced with permission from *Inorg. Chem.* **16**, 2950 (1977). Copyright (1977) American Chemical Society.

sublattice usually allows conclusions to be made concerning the coordination of H_2O; frequently, changes in the layer stacking relative to the binary dichalcogenide are observed. X-ray studies on the hydrates obtained from Li_xTiS_2 based on the assumption of a close packing of H_2O and S revealed trigonal prismatic coordination of H_2O for the monolayer hydrate and octahedral coordination for the bilayer phase (*20*). The monolayer hydrate of Li_xVS_2 was found to be isostructural with the titanium compound (*20*); octahedral coordination of Na^+ was observed for $Na_{0.6}(H_2O)_y[VS_2]$ with water bilayer structure (*21*). An investigation of the niobium and tantalum sulfide hydrates $K_{0.5}(H_2O)_y[NbS_2]$ and $K_{0.3}(H_2O)_y[TaS_2]$ established trigonal prismatic coordination sites for K^+ and H_2O (*34*). Recently a singel-crystal neutron diffraction study on $Na_{0.6}(D_2O)_2[VS_2]$ and $Na_{0.6}(D_2O)_2[VSe_2]$ was performed that showed the oxygen of the H_2O molecules to be in a close-packing arrangement with the chalcogen atoms; sodium was located in octahedral positions (*28, 35*). Good agreement with the data was observed for a model with one D atom of each D_2O in the plane of the oxygen atoms and the other D atom close to the sulfur layers. Information on structural aspects of the water sublattice can also be derived from ^1H NMR studies and is discussed in Section III,H.

The reaction of transition metal dichalcogenides with aqueous alkali and alkaline earth hydroxides either directly (*17, 20, 36*) or via cathodic reduction

(*24*) was reported to result in the formation of hydrated compounds $(AOH)_x(H_2O)_yMX_2$ containing alkali hydroxides intercalated along with water. Studies on stoichiometry, thermal stability (thermal gravimetric analysis, TGA), and structure revealed the dependence of the hydration stage on the intercalated cation; structural models were proposed for the position of cations and H_2O in the interlayer space (*20*). Investigations on the kinetics of the intercalation process were performed using the theory of heterogeneous reactions in solution, and second- as well as third-stage intermediates were identified (*37*).

In a study on the mechanism of this reaction for layered sulfides (*11*, *26*), it was shown that these phases are polyelectrolytes with basically similar structure and reactivity as the hydrated phases discussed above. The reduction of the dichalcogenides was explained in terms of partial hydrolysis of the chalcogenide layers by OH^- and the formation of HS^- and S^{2-} ions, respectively. The latter are able to reduce the layer lattice, being themselves oxidized to polysulfide anions and/or elemental sulfur. Synchronously, an equivalent amount of hydrated cations is taken up into the interlayer space:

$$2xOH^- + MS_2 \rightarrow MS_{2-x}O_x + xS^{2-} + xH_2O \tag{9}$$

$$xS^{2-} \rightarrow \frac{x}{8}S_8 + 2xe^- \tag{10}$$

$$xe^- + MS_{2-x}O_x \rightarrow [MS_{2-x}O_x]^{x-} \tag{11}$$

The composition of the reaction products is thus described as $A_x{}^+(H_2O)_y[MS_{2-x}O_x]^{x-}$, with negatively charged layers modified partially in composition by the replacement of sulfur with oxygen. The amount of sulfur and polysulfide $S_x{}^{2-}$ released during the reaction corresponds well with the theoretical amount calculated by the cation content of the product. Reaction with selenides proceeds in a similar way, with elemental Se_x as the reaction product. At low alkali hydroxide concentration, OH^- anions—like other anions—are not intercalated because of the unfavorable Donnan equilibrium (*26*). In concentrated hydroxide solutions, however, additional cation/anion cointercalation by alkali hydroxide can be observed (Section III,H).

Hydrated intercalation compounds $(NH_3)_x(H_2O)_yMX_2$ result from the reaction of aqueous ammonia with dichalcogenides (*20*); their formation also proceeds via the mechanism outlined in Eqs. (9)–(11), and the reaction product has to be described as $(NH_4{}^+)_x(H_2O)_y[MS_{2-x}O_x]^{x-}$. Like ammonia, a large number of other Lewis bases of suitable pK_b (aliphatic and aromatic amines, nitrogen heterocycles, etc.) can be directly intercalated at room temperature in the presence of H_2O, the products having ionic structure and sulfur-deficient layers (*26*).

C. Ion Exchange and Equilibria

The high mobility of interlayer cations and solvent molecules in hydrated dichalcogenides provides the basis for reversible ion-exchange reactions at ambient temperature,

$$xB^+ + A_x^+(H_2O)_y[MX_2]^{x-} \rightleftarrows B_x^+(H_2O)_y[MX_2]^{x-} + xA^+ \quad (12)$$

which represent a convenient and variable synthetic technique for the preparation of dichalcogenide intercalation compounds. Kinetics of the exchange reaction are dependent on the geometry of the guest cations: Under equal conditions small spherical ions undergo rather rapid exchange, whereas for molecular cations the exchange rate as a rule decreases with increasing ionic size. The ion-exchange capacity depends on the layer charge density x, the accessible range of which is governed by the electronic factor; the H_2O content is related to cation size and charge, i.e., hydration energy. Large isometric cations can frequently not be exchanged at high layer charge densities for geometric and electrostatic reasons; large, strongly anisometric cations, e.g., *n*-alkylammonium ions, may arrange with their long axis approximately perpendicular to the layer planes and may thus exchange even at high layer charge densities.

Besides the intercalation of alkali and alkaline earth ions described above, the intercalation of transition metal ions, metal complex cations with inorganic and organic ligands, organic cations, and radical-cations into dichalcogenides by ion exchange can be achieved. The interlayer spacings (ca 11.7–12.9 Å) measured for bivalent transition metal ions Mn^{2+}, Fe^{2+}, Co^{2+}, Ni^{2+}, Zn^{2+}, and $[Cr(H_2O)_5Cl]^{2+}$ with TiS_2, NbS_2, and TaS_2 as the host lattice correspond in all cases to simple bilayer hydrates with octahedral coordination of the metal cation (*38*), whereas for the trivalent ions La^{3+}, Ce^{3+}, and Cr^{3+} somewhat higher lattice spacings of 12.89, 12.91, and 11.92 Å were found (for TaS_2 as the host lattice), which may be due to the increased solvation energy of these ions and distortion of the H_2O ligand sphere (*38*, *39*). Cations of lower hydration energy, e.g., Cu^+ (Cu^{2+} is reduced to Cu^+ upon intercalation), Ag^+, and Tl^+, enter the lattice as nonhydrated ions and form ternary phases Cu_xMX_2, Ag_xMX_2, and Tl_xMX_2. Metal complex ions, e.g., $[Pt(NH_3)_4]^{2+}$, $[Fe(dipyridine)_3]^{2+}$, and $[Fe(phenanthroline)_3]^{2+}$, show interlayer spacings that depend on their molecular size; only small amounts of water are taken up into the lattice for spatial reasons and as a consequence of their low solvation energy (*38*, *39*). The extreme case is found for the intercalation of metallocenium cations by ion exchange, in which only the organometallic ions enter the lattice (*40*).

Organic amines and heterocyclic compounds can be exchanged either as protonated species or as quaternary "onium" compounds—for example,

$[(CH_3)_4N]^+$, $[(CH_3)_4P]^+$—or dye molecules, such as fuchsine and rhodamine B (*18, 19, 38, 41*). Again, the interlayer spacings are strongly dependent on the size of the molecular ions, on the hydration energy of the cations, and on potential interaction between guest cations. Conclusions on the spatial arrangement of the guest ions can be made at present only in simple cases, e.g., if heterocyclic rings are positioned in monolayers with their ring planes parallel to the MX_2 sheets. The structures observed for homosolvated phases, e.g., *n*-alkylammonium ions, are discussed in Section III,E.

Little information is available on ion-exchange reactions in nonaqueous solvents; for liquid ammonia as the solvent, the exchange $NH_4^+ \rightarrow K^+$ has been described (*42*).

Concerning ion-exchange equilibria, a study of the exchange isotherm of the hydrated system $Na^+–K^+–TaS_2$,

$$K^+_{0.3}(H_2O)_y[TaS_2]^{0.3-} + 0.3Na^+ \rightleftarrows Na^+_{0.3}(H_2O)_y[TaS_2]^{0.3-} + 0.3K^+ \tag{13}$$

has revealed a rather unusual type of selectivity for layered dichalcogenides (*43*). The pure Na^+ or K^+ phase, respectively, was equilibrated with aqueous $Na^+–K^+$ solutions. The dependence of the interlayer spacing d on the mole fraction, $\gamma_{Na} = c_{Na}/(c_{Na} + c_K)$, is given in Fig. 6a. It is seen that, for the range $0 \leq \gamma_{Na} < 0.5$, d remains almost constant and corresponds to ca 9 Å, i.e., monolayer hydrate structure. At $\gamma_{Na} \sim 0.5$, a critical threshold point is reached at which a discontinuous change appears to d = 12 Å, i.e., bilayer hydrate structure, which is found in the range $0.5 < \gamma_{Na} \leq 1$. The corresponding ion-exchange isotherm γ_{Na}(solid) versus γ_{Na}(liquid) (Fig. 6b) demonstrates again a discontinuous change: For the range $0 \leq \gamma_{Na} < 0.5$, the solid contains almost exclusively K^+, whereas for $0.5 < \gamma_{Na} \leq 1$, sodium ions are intercalated; the threshold γ value of ~0.5 correlates well

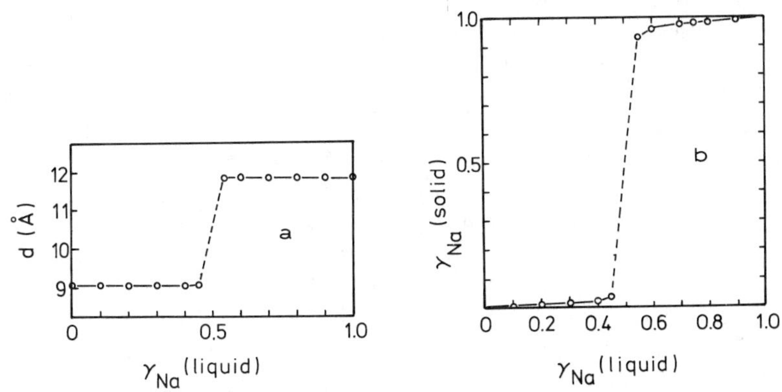

Fig. 6. Ion-exchange equilibria Na^+/K^+ for $A^+_{0.3}(H_2O)_y[TaS_2]^{0.3-}$. (a) Dependence of interlayer spacing d on the mole fraction γ_{Na}; (b) ion-exchange isotherm (298 K).

with the change in interlayer spacing shown in Fig. 6a. The phenomena encountered represent the case of a structure-induced exchange selectivity. Obviously even a partial uptake of Na^+ into the monolayer hydrate structure (or of K^+ into the bilayer structure) is energetically highly unfavorable. Since potential intermediate hydrate structures that can stabilize mixed cation forms do not exist, the structural transition must proceed in a discontinuous way. A similar behavior indicating structure-induced exchange selectivity was observed for the molybdenum sulfide system $(K/Na)_x{}^+(H_2O)_y[MoS_2]^{x-}$ (*19*).

Quantitative investigations have not been reported for other ions, but it is known qualitatively that for certain molecular ions, e.g., *n*-alkylammonium ions or pyridinium ions, in the presence of hydrated main group or transition metal ions, the equilibrium is far on the side of the organic ions, which is due to the formation of homosolvated structures (Section III,E) with particular stability.

D. Nonaqueous Solvates

The high mobility of cation and solvent molecules at room temperature in the layered polyelectrolytes also permits exchange of the solvent guest phase; a large variety of inorganic and organic polar molecules can thus be intercalated at ambient temperature:

$$A_x{}^+(\text{solv})_y[MX_2]^{x-} + y(\text{SOLV}) \rightarrow A_x{}^+(\text{SOLV})_y[MX_2]^{x-} + y(\text{solv}) \qquad (14)$$

The organic solvents are of particular interest, since they provide for systematic studies on the solvation properties a large variety of compounds with different molecular size, geometry, and functional groups of varying polarity (*11*). A convenient method is the reaction of hydrated phases with polar solvents; care has to be taken, however, to ensure quantitative reaction by repeated exchange. The preparation of nonaqueous solvate phases is also possible, e.g., by direct intercalation of solvated cations into the binary dichalcogenide via redox reactions [Fig. 1, reaction (c)] or by solvation of ternary phases A_xMX_2 [Fig. 1, reaction (b)]. The composition and structure of the compounds obtained thus depend essentially on the polarity and geometry of the solvent molecule, the hydration energy and size of the cation, and the cation density; qualitatively, the influence of changes in M and X seems to be of minor importance.

Intercalation compounds $A_x(NH_3)_yMX_2$ with ammonia as the solvent can be prepared by reaction of dichalcogenides with alkali or alkaline earth metals dissolved in liquid ammonia or by reaction of the corresponding ternary phases A_xMX_2 with gaseous or liquid ammonia (*5*, *10*, *44*).

The formation of solvated phases with organic molecules by exchange with hydrated compounds has been reported for a large series of polar organic species (alcohols; ethers; aliphatic, cyclic, and aromatic amines; phosphine oxides; sulfoxides, etc.) and for a range of different host lattices (*11*). The interlayer spacings measured for organic solvates and alkali ions in most cases considerably exceed those of the hydrated phases. Although the solvation energies of alkali and alkaline earth ions in homogeneous organic solution are notably smaller than those of the hydrates, a clear dependence of the interlayer spacing on the charge/radius ratio of the cation was established in a study on *N*-methylformamide as the solvent. For cations with $e/r < 1.33$, the interlayer spacing was 11–12 Å whereas, for $e/r > 1.33$, d values of ca 17 Å were observed. A comparison of dimethyl sulfoxide, dimethylformamide, and *N*-methylformamide solvates revealed a relationship between the dielectric constant of the pure solvent and the solvate structure (*11*).

Cointercalation of organic solvents upon reaction of dichalcogenides with lithium metal solutions in these solvents or by cathodic reduction of MX_2 electrodes in Li^+/organic electrolytes has been described for hexamethylphosphotriamide, *N*-methylformamide, propylene carbonate, dimethoxyethane, and dimethyl sulfoxide (M = Ti, Mo; X = S) (*8*, *23*, *45*, *46*). It is likely that the direct cointercalation of large organic solvent molecules is frequently prevented from taking place for kinetic reasons, which is of considerable importance for the use of dichalcogenide electrodes in Li^+/organic batteries (*8*).

The formation of colloid solutions of organic solvates $A_x{}^+(\text{solv})_y[MX_2]^{x-}$ (A = alkali cation) was observed under the influence of weak shear forces in solvents with high dielectric constant, e.g., formamide or *N*-methylformamide (*11*). The dark blue solutions obey Beer's law at low concentrations and show streaming birefringence, indicating the anisotropic character of the colloid particles. Electrophoretic investigations showed that the particles move toward the cathode, and it was concluded that the colloid is electrocratic and consists of negatively charged dichalcogenide layers surrounded by a diffuse layer of solvated alkali cations. The formation of colloid solutions was also observed by direct reaction of acid amides with TiS_2 and by electrolytic reduction of TaS_2 in H_2SO_4 and subsequent treatment of the product with surface-active agents (*47*, *48*).

The strength of the cation/solvent dipole interaction shows a wide range of variation in the dependence on cation charge and size and polarity of the solvate species. An extreme case, in which only weak physisorption occurs, is the intercalation of molecular hydrogen into K_xZrS_2, which yields $K_x{}^+(H_2)_y[ZrS_2]^{x-}$ (*49*).

E. Homosolvation and Lewis Base Intercalation

A group of solvated compounds consists of phases that are characterized by the fact that guest solvent species and guest cation are identical in composition and structure except for an additional proton or a deficient electron (e.g., radical-cation). In view of the close relationship between cation and solvent, we denote these compounds here by the term "homosolvated phases."

The intercalation compounds of *n*-alkylammonium ions $(C_nH_{2n+1}NH_3)^+$ (abbr. $R{-}NH_3{}^+$) represent a typical example of this class of compounds. The number *n* of carbon atoms in the alkyl chain may vary from 1 to 21. Their preparation can be achieved, e.g., by ion exchange of hydrated alkali phases with aqueous solutions of *n*-alkylammonium salts (*18*, *19*, *30*, *38*). At pH 5–7 these solutions contain neutral amine molecules due to the equilibrium $(R{-}NH_3)^+ + H_2O \rightleftarrows R{-}NH_2 + (H_3O)^+$. After quantitative exchange, the products have the composition $(R{-}NH_3)_x{}^+(R{-}NH_2)_y[MX_2]^{x-}$, i.e., neutral amine molecules cointercalate selectively as the solvent species. The

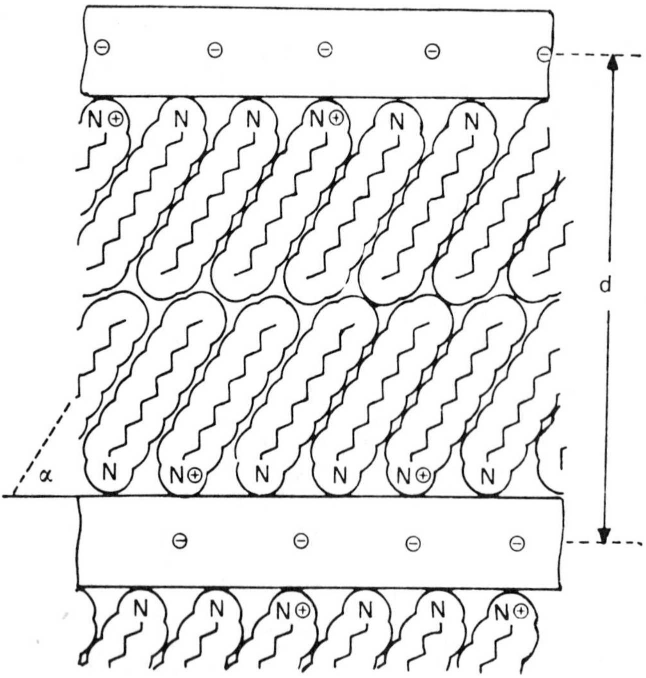

Fig. 7. Structural model of the paraffin type of arrangement of alkyl chains in homosolvated phases $(RNH_3)_x{}^+(RNH_2)_y[MX_2]^{x-}$; α, inclination angle of chain axes relative to layer planes (*19*).

exchange equilibrium A^+, $(H_2O)/RNH_3^+$, RNH_2 for $n > 6$ is far on the side of the organic intercalate due to the relatively high stability of these phases, which partially originates from entropic effects (hydrophobic bonding). For $n < 6$, the alkyl chains are arranged in bilayers parallel to the MX_2 sheet planes, whereas for $n > 6$, close-packed paraffin-type double layers are found with the chain axes perpendicular to the layer planes or inclined at an angle α (Fig. 7). The inclination angle depends on the layer negative charge density, ammonium ion concentration, pH of the solution, and temperature; upon extraction with organic solvents, the neutral amine molecules can partially or completely be removed. Alkylamine intercalation compounds may also be prepared from $R—NH_2$ and MX_2 by thermal reaction at higher temperatures (*6, 7, 50, 51*); they are likely to form ionic compounds by molecular disproportionation (see following section) via the activated hydrogen atoms at the α-carbon. Thermal phase transitions have been reported for these compounds (*51, 52*).

The interaction host lattice–guest molecule for the large class of Lewis base (L) intercalation compounds of layered dichalcogenides was interpreted originally in terms of charge transfer from the intercalated donor molecule to the acceptor host lattice (*6, 7*). Recent studies led to the hypothesis that the formation of these phases is generally correlated with molecular redox reactions and that all Lewis base intercalation compounds have ionic structure and are to be considered (homo-) solvated phases (*41, 42*). This concept has been established so far for the ammonia and pyridine intercalation compounds.

Ammonia does react at ambient temperature, e.g., with TaS_2, to give a product with the composition $TaS_2(NH_3)_{1.0}$; the lattice expansion of ca 3 Å corresponds well with the van der Waals diameter of NH_3. The initial rate of the gas–solid reaction is dominated by a nucleation period with the occurrence of higher-stage phases (*53, 54*). Deintercalation at 300 K/10^{-4} torr leads to partial removal of NH_3 with formation of $TaS_2(NH_3)_{1-x}$; a single-phase range $0 < x < 0.5$ could be established, but complete removal of NH_3 was achieved only at temperatures of >420 K (*53*). Concerning the position of NH_3 in the interlayer space, it was assumed initially that the nitrogen lone pair orbital is pointed toward the layers (Fig. 8a). Proton

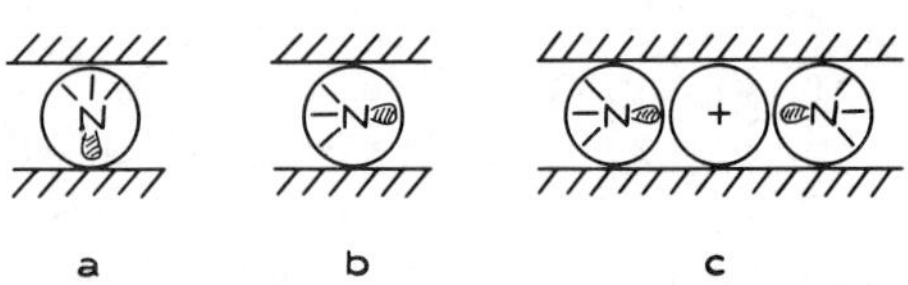

Fig. 8. Structural models proposed for the ammonia intercalation compounds MX_2-$(NH_3)_{1.0}$. Negative layer charges omitted in (c) (*42*).

NMR studies demonstrated, however, the strongly anisotropic mobility of NH_3, which was found to be oriented with the C-3 axis parallel to the layer planes, as indicated in Fig. 8b (*55*). The authors proposed an indirect donor–acceptor charge transfer $NH_3 \rightarrow TaS_2$, involving weak covalent mixing between the nitrogen lone pair orbital and electronic layer states. An investigation on the chemistry of the NH_3 intercalation revealed, however, that the reaction is associated with a redox process that implies the oxidation of a part of the guest molecules by the dichalcogenide layers with formation of dinitrogen (*42*). The reaction scheme is as follows:

$$2NH_3 \rightarrow N_2 + 6H^+ + 6e^- \tag{15}$$

$$xe^- + TaS_2 \rightarrow [TaS_2]^{x-} \tag{16}$$

$$H^+ + NH_3 \rightarrow NH_4{}^+ \tag{17}$$

The protons resulting from the oxidation step are accepted by neighboring NH_3 molecules with formation of $NH_4{}^+$, and the electrons are transferred to the TaS_2 layers. The reaction product

$$(NH_4{}^+)_x(NH_3)_{1-x}[TaS_2]^{x-}$$

with $x \sim 0.1$–0.3 has ionic structure with negative charges on the $[TaS_2]^{x-}$ macroanion layers and $NH_4{}^+$ ions solvated by neutral NH_3 molecules in the interlayer space (Fig. 8c). Exchange of $NH_4{}^+$ was demonstrated by ion exchange in liquid ammonia with K^+. Additional evidence for the ionic structure model comes from ^{181}Ta perturbed angular correlation studies that prove the presence of discrete positive charges in the interlayer space (*56*). Inelastic incoherent neutron scattering spectra of $(NH_3)_{1-x}TaS_2$ were found to exhibit bands characteristic of $NH_4{}^+$ torsional modes (*12*); the possibility of a part of the hydrogen atoms residing in intralayer positions (*57*) was dismissed by recent single-crystal neutron diffraction measurements. The unusual orientation of the NH_3 molecules in the interlayer space is thus a consequence of ion/dipole interactions (Fig. 8c). The observation of reversible partial deintercalation of NH_3 at 300 K is due to the removal and uptake of a part of the solvate molecules, while the $NH_4{}^+$ ions can leave the lattice only at elevated temperatures under decomposition into NH_3 and H_xTaS_2; the hydrogen compound is subject to further degradation at the reaction temperature.

The reaction of the nitrogen heterocycle pyridine (pyr), C_5H_5N, with, for example, TaS_2 proceeds at temperatures of ca. 400–500 K and results in the formation of $(pyr)_{0.5}TaS_2$ with an interlayer spacing of ca 12 Å. Considerations concerning interlayer spacing, guest size, and bonding originally led to two structural models with different orientation of the aromatic rings relative to the MX_2 layer planes (Fig. 9a,b); a donor–acceptor type of

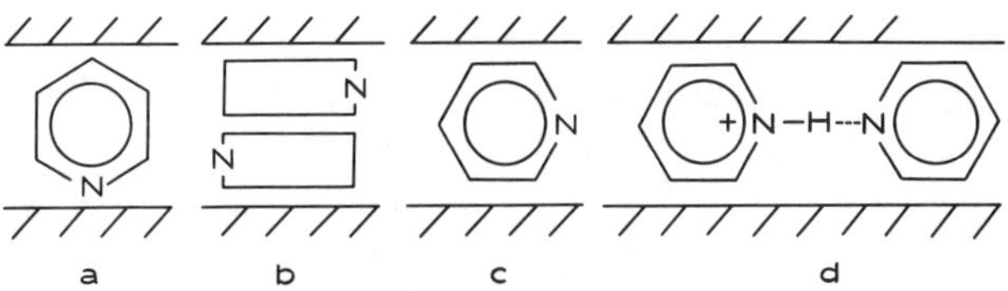

Fig. 9. Structural models proposed for the pyridine intercalation compounds MX_2-$(C_5H_5N)_{0.5}$. Negative layer charges ommitted in (d) (*41*).

charge-transfer interaction between pyridine and the host layers was proposed (*58*). A neutron diffraction study of $(pyr)_{0.5}TaS_2$ showed, however, that the C_5H_5N molecules are positioned in the interlayer space, as indicated in Fig. 9c, with the nitrogen lone pair orbitals parallel to the layer planes (*59*); this position is obviously most unfavorable for charge-transfer interaction with the TaS_2 sheets. Further investigations demonstrated that the intercalation is correlated with a redox reaction (*41*), resulting in the formation of dipyridine (pyr–pyr), NH_4C_5—C_5H_4N, protonated pyridine molecules, and negative layer charges according to

$$x(\text{pyr}) \rightarrow \frac{x}{2}(\text{pyr–pyr}) + xH^+ + xe^- \tag{18}$$

$$xH^+ + x(\text{pyr}) \rightarrow x(\text{pyrH}^+) \tag{19}$$

$$xe^- + TaS_2 \rightarrow [TaS_2]^{x-} \tag{20}$$

The product of the reaction thus has to be described as

$$(\text{pyrH}^+)_x(\text{pyr})_{0.5-2x}(\text{pyr–pyr})_{x/2}[TaS_2]^{x-}$$

with $x \sim 0.2$. The relatively high stability of the pyridine intercalation compounds must be attributed essentially to ionic bonding between the guest cations and the host macroanions. Guest/guest interactions by hydrogen bonding (Fig. 9d) and by π interaction favored by parallel stacking of the aromatic rings should also be considered. The presence of discrete ionic interlayer species is further supported by ^{181}Ta angular correlation studies; the dependence of the reactivity of substituted pyridines on the corresponding pK_b values (*58*) is a consequence of the requirement for stable protonated cation species.

Apart from the direct thermal reaction, pyridine intercalation compounds that are basically similar to the thermal product in composition and structure can be obtained in several ways (*41*): (a) ion exchange, e.g., of hydrated alkali phases with pyridinium solutions:

$$Na_x^+(H_2O)_y[MX_2]^{x-} + x(\text{pyrH}^+) + y(\text{pyr}) \rightarrow (\text{pyrH}^+)_x(\text{pyr})_y[MX_2]^{x-} + xNa^+ + yH_2O \tag{21}$$

(b) cathodic reduction of dichalcogenides in pyridinium salt electrolytes (see Section III,F); (c) acid/base reaction of pyridine with hydrogen bronzes

H_xMX_2 (see Section III,G); (d) direct reaction of aqueous pyridine solutions with dichalcogenides, which proceeds via Eqs. (9)–(11) and gives products with sulfur-deficient layers (*26*). If the solvents used in (a), (b), and (d) are not themselves strong Lewis bases that can compete with pyridine, only the latter is selectively intercalated as the guest solvent species, and homosolvated compounds are obtained. It is interesting to compare the formation after (d) with the thermal intercalation of pyridine; in both cases an external reductant is absent. The reaction with pyridine in the presence of H_2O proceeds, however, at ambient temperature, since the strong reductant HS^- is easily formed by solvolysis of the partially ionic Ta—S bonds, whereas with the exclusion of H_2O a molecular redox process [Eqs. (18)–(20)] is required with the breaking of strong covalent bonds which proceeds only at temperatures that are considerably higher.

Most direct intercalation reactions can be achieved only at relatively high temperatures (except for a few cases, e.g., NH_3) up to 500 K, and it is likely that in the case of guest molecules that are less stable than simple aromatic heterocycles or that have several reactive functional groups the reaction pathways leading to electron donation and cation formation will be far more complicated. The influence of elemental sulfur (*60*) that is present as an impurity or added to the guest phase is still unclear; small amounts of sulfur-containing side products have been detected in the system pyridine/MX_2 (*41*).

It can be concluded from the ionic structure model that the formation of Lewis base intercalation compounds should also be possible to achieve in certain cases by direct ionization of a fraction of the guest molecules to radical-cations. The requirements are a low ionization potential for L and a relatively high redox potential of the host layers. Potential guest candidates are, e.g., heterocycles or amines with extended delocalized π systems; recently, it was proposed that the thermal intercalation of *p*-phenylene-diamine into TaS_2 proceeds via direct ionization (*61*). Extremely strong electron donors, e.g., tetrakis-dimethylamino-ethylene (*62*), which behaves as a "quasi-metal-forming" stable radical-cation, should also be of interest in this respect.

The scheme of ionic homosolvated compounds may be extended formally, e.g., to metallocene and alkali metal intercalation compounds. An ionic structure has been proposed for the metallocene phases $(R_2T)_xMX_2$ (*40, 63*) (R = cyclopentadiene; T = transition metal). Although quantitative data are not available, it is likely that the ionicity depends on the ionization potential of R_2T and on the electronegativity of MX_2, and a description in static terms would thus be $(R_2T)_x{}^+(R_2T)_y[MX_2]^{x-}$. The ionicity of intercalation compounds A_xMX_2 with strongly electropositive elements depends on guest size, site symmetry, ionization potential of the guest atom, electronic structure of the host lattice, and the stoichiometric value x (*8, 10*). For low alkali content or strongly ionic bonding M—X (for example, $NaCrS_2$), the extreme case will be full ionization of A. For high x and less favorable band structure (e.g., MoS_2, WS_2), the ternary phases can be considered in a static scheme (neglecting rapid electron transfer between A^+ and A^0) to consist of cations A^+ "solvated" by neutral donor metal atoms A^0:

$$\underset{\text{Complete ionization}}{A_x^+[MX_2]^{x-}} \quad \cdots \quad \underset{\text{Partial ionization}}{A_x^+A_y^0[MX_2]^{x-}}$$

A similar scheme can be applied to graphite/alkali metal intercalation compounds which exhibit very low ionicity (*64*); heterosolvated graphite phases $A_x(\text{solv})_y[C]$ have also been reported (solv being, for example, NH_3, amines) (*5*). Recent NMR investigations present evidence for molecular redox reactions on the intercalation of graphite with strong electron acceptors (*65*), and an ionic solvate scheme with inverse polarity may describe the structure of these phases.

F. Electrochemistry and Reaction Mechanism

Initial observations on the reactivity of hydrated ternary alkali compounds established the possibility of chemical oxidation of these phases (for example, O_2 and Fe^{3+}) to the corresponding binary dichalcogenides and of chemical reduction of binary dichalcogenides in aqueous electrolytes by strong reducing agents (for example, $S_2O_4^{2-}$) to hydrated phases (*14, 15, 19, 23, 25*). The cathodic reduction of dichalcogenide electrodes in aqueous and non-aqueous electrolytes resulted in the formation of solvated and unsolvated intercalation compounds (*22–24*). In order to explain the electrochemical and chemical observations on a general basis, a concept of reversible topotactic redox reactions was put forward in which these processes are considered to be electron/ion transfer reactions leading to compounds with ionic structure in the reduced stage (*12, 23, 25, 26, 31*):

$$xe^- + xA^+ + y(\text{solv}) + MX_2 \rightarrow A_x^+(\text{solv})_y[MX_2]^{x-} \tag{22}$$

In contrast to exchange reactions, the electrochemistry of the dichalcogenides is strongly correlated with the electronic factor, i.e., the band structure of a given host lattice, which basically determines the accessible range of the layer charge density x. In protolytic electrolytes, x is, moreover, restricted by the potential of hydrogen discharge. Cointercalation of solvent molecules is again determined thermodynamically by the solvent dipole/interlayer cation interaction and the lattice energy; kinetic factors, however, frequently prevent the uptake of larger and less polar guest molecules from taking place.

The electrochemical reduction of dichalcogenides represents a general and widely variable technique for the preparation of solvated intercalation compounds; since it can be performed at ambient temperature, thermally induced side reactions are avoided. Intercalation compounds can be obtained with Group Ia and IIa cations, transition metal cations, metal complex cations, and inorganic and organic molecular cations quite similar to ion exchange.

From the appearance of reproducible potential and current steps in galvanostatic and cyclovoltammetric studies in aqueous electrolytes, the occurrence of defined reaction intermediates and of higher-stage structures

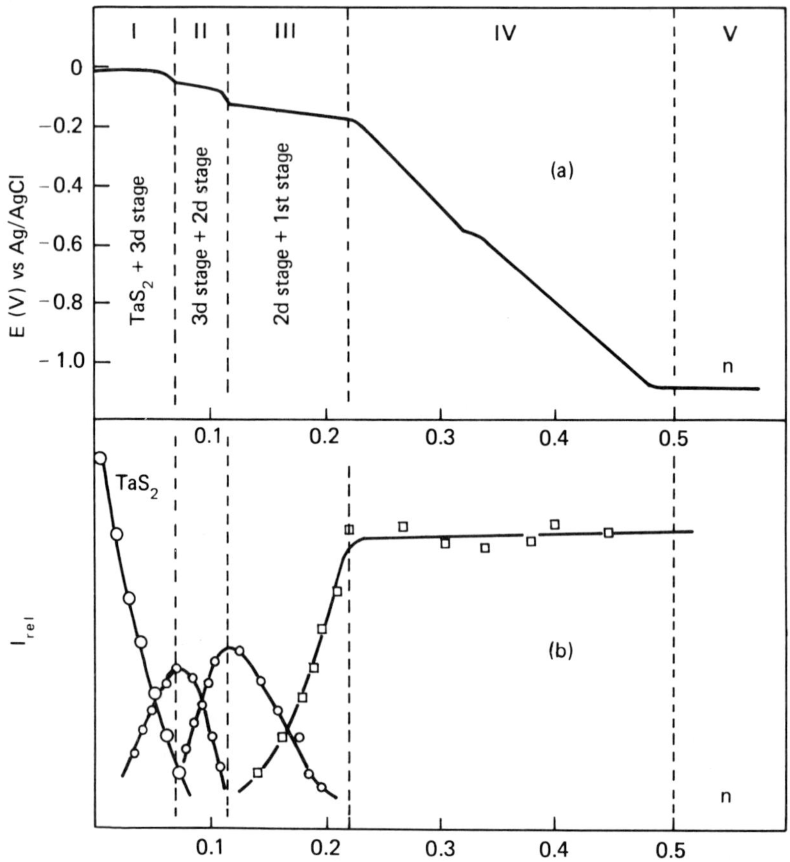

Fig. 10. Cathodic reduction of 2H TaS_2 in K^+/D_2O electrolyte. (a) Potential/charge-transfer curve (n = faraday/mole TaS_2); (b) relative intensities of characteristic diffraction lines of starting material and intermediate phases as a function of n (*39*).

was concluded (*23*). This assumption was confirmed in a recent structural investigation by neutron diffraction on the cathodic reduction of 2H TaS_2 in H_2O/K^+ electrolyte (*12*, *39*). The reaction was performed *in situ* with the dichalcogenide working electrode directly in the neutron beam during continuous galvanostatic reduction (to suppress incoherent scattering from hydrogen, D_2O was actually used as the solvent). The final reaction product was known from previous studies to have the composition $K_{0.5}{}^+(H_2O)_{0.5}[TaS_2]^{0.5-}$ with monolayers of water and an interlayer spacing of 9 Å. The intercalation enforces a change in stacking sequence; H_2O and K^+ are positioned in trigonal prismatic sites (*34*). Figure 10a shows the potential/charge-transfer curve obtained during the reduction at constant current; several characteristic potential steps appear that divide the curve into several regions (I–V). The final composition $K_{0.5}{}^+(H_2O)_{0.5}[TaS_2]^{0.5+}$ is reached after a transfer, $n = 0.5e^-/TaS_2$, at the intersection of regions IV and V, where the potential of solvent decomposition (i.e., H_2 formation) is reached (region V). Figure 10b indicates the relative intensities of characteristic (00*l*) reflections (after correction by line profile analysis) measured synchronously with the reduction by a neutron multidetector system as a function of the charge transfer n (e^-/TaS_2). It can be seen that the intensity of the TaS_2 (002) line rapidly decreases in the initial reaction period, whereas new reflections appear (and disappear) in the course of the reduction. The data analysis led to the characterization of three different two-phase regions which correspond well with the thermodynamic data of Fig. 10a:

Range I: $0 < n < 0.07$, coexistence of 2H TaS_2 plus third-stage phase ($d = 3 \times 6 + 3 = 21$ Å) with the idealized stoichiometry $K_{0.07}(H_2O)_y$-$[TaS_2]^{0.07-} \hat{=} K_{0.21}{}^+(H_2O)_{0.79}[TaS_2]_3{}^{0.21-}$, i.e., every third van der Waals gap occupied by K^+ and H_2O (Fig. 11).

Range II: $0.07 < n < 0.11$, third-stage phase plus second-stage phase ($d = 2 \times 6 + 3 = 15$ Å), stoichiometry $K_{0.11}{}^+(H_2O)_y[TaS_2]^{0.11-} \hat{=} K_{0.22}{}^+(H_2O)_{0.78}[TaS_2]_2{}^{0.22-}$ (Fig. 11).

Range III: $0.11 < n < 0.21$, second-stage phase plus first-stage phase ($d = 6 + 3 = 9$ Å), stoichiometry $K_{0.21}{}^+(H_2O)_{0.79}[TaS_2]^{0.21-}$.

In the following range IV, with $0.21 < n \leq 0.5$, the interlayer spacing remains almost constant. Because of the steep continuous increase in potential, this range is supposedly a nonstoichiometric-phase region; the reproducible small potential step close to $n = 1/3$ suggests, however, a more complicated situation with an additional intermediate stage.

The mechanism of the transition third stage → second stage is unclear as yet; third-stage phase formation has also been reported in the intercalation of dichalcogenides with alkali hydroxides and alkali metals (*10*, *37*). Uneven

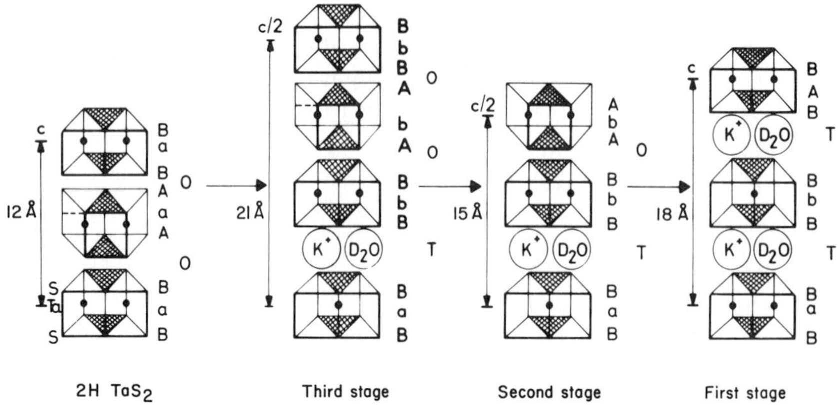

Fig. 11. Scheme of structural transformations of 2H TaS_2 on cathodic reduction in K^+/D_2O electrolyte (*39*).

stage → even stage transitions are well known from graphite chemistry and have been explained by the intergrowth of two or more stages (*66*). The interlayer cation density is apparently almost identical for the three intermediate phases in the range up to $n = 0.22$. An explanation of this could be either the assumption of a particular stability of the interlayer electrolyte phase for this composition or the possibility of electronically unstable states of first-stage phases with negative charge densities $<0.2e^-/TaS_2$.

The formation of third- and second-stage intermediates with similar phase ranges for 2H TaS_2 and 2H NbS_2 was also observed on cathodic intercalation of other hydrated mono- and divalent main group and transition metal ions (*39*). The systematic occurrence of higher-stage phases for cations of different charge and radius seems thus to favor the assumption that electronic aspects are responsible for the formation of these intermediates. Anodic oxidation was found to proceed again via first-, second-, and third-stage intermediates.

The influence of the electronic factor is also demonstrated by comparing the reaction of 1T TaS_2, which provides an octahedral environment for the intralayer cation, with 2H TaS_2, in which Ta is positioned in trigonal prismatic sites. The cathodic reduction of 1T TaS_2 in K^+/H_2O yields $K_{1/3}{}^+(H_2O)_{2/3}[TaS_2]^{1/3-}$ as the reaction product and a two-phase range, 1T TaS_2/final product, is observed for $0 < n < 1/3$. Higher-stage nuclei are present in the initial reaction period but turn over rapidly to first-stage phase; no defined intermediate phases can be isolated as compared to the 2H TaS_2 reduction (*39*).

The strong influence of kinetic factors becomes apparent on the cathodic reduction in electrolytes containing large organic cations or transition metal complex ions with large ligands and low solvation energy, for example, $(C_5H_5)_2Co^+$ or $[Fe(dipyridine)_3]^{2+}$; the potential values observed deviate strongly from equilibrium potential even at low current densities (*39*).

Cathodic reduction in aqueous electrolyte solutions containing protonated amines and heterocycles usually results in homosolvated reaction products. From aqueous pyridinium salt solutions, $(C_5H_5NH)_{0.25}{}^+(C_5H_5N)_{0.25}$-$[TaS_2]^{0.25-}$ is obtained as the final reaction product for 2H TaS_2 as the host lattice (*41, 67*). Three biphasic regions in the lower reduction range have been identified: (a) TaS_2/second-stage intermediate with the pyridine ring planes parallel to the MX_2 layers ($d = 2 \times 6 + 3 = 15$ Å), (b) second-stage phase/first-stage phase with parallel orientation of the aromatic ring planes ($d = 6 + 3 = 9$ Å), (c) first-stage phase ($d = 9$ Å)/first-stage phase ($d = 12$ Å) with the pyridine ring planes perpendicular to the layer planes. A similar reaction scheme was found for 4-methylpyridine/2H TaS_2.

As mentioned above chromium does not form stable binary chalcogenides CrX_2; the ternary stoichiometric phases $NaCrS_2$ and $KCrS_2$ were found, however, to hydrate upon partial anodic oxidation to $A_x{}^+(H_2O)_y[CrS_2]^{x-}$ (*30*). The accessible topotactic redox range in neutral electrolytes is limited in this case to $0.4 < x < 0.7$; upon oxidation beyond that range, irreversible attack of the sulfur layers is observed.

Cointercalation of solvents from organic electrolytes is discussed above (Section III,D). The cathodic reduction of 2H TaS_2 in solutions of alkali salts in liquid ammonia yielded metal/ammonia intercalates (*68*).

G. Hydrogen Bronzes

The uptake of hydrogen by layered dichalcogenides was observed originally during the cathodic reduction of TaS_2 in 0.1 *N* aqueous HCl; the hydrated product $H_xTaS_2 \cdot xH_2O$ exhibited a large interlayer spacing of ca 50 Å and was unstable (*22*). Later investigations established the existence of the following nonhydrated phases: H_xTiS_2 (*69, 70*), H_xNbS_2, H_xTaS_2 (*69, 71*), H_xNbSe_2 (*69*), H_xCrS_2 (*30*), and H_xMoS_2 (*33*), which show interlayer spacings close to those of the parent lattices. The formation of hydrogen bronzes (*12*) can be achieved either by cathodic reduction of the binary chalcogenide in strong mineral acids,

$$xH^+ + xe^- + MX_2 \rightarrow H_xMX_2 \tag{23}$$

or by reaction of hydrated phases with acids and subsequent spontaneous or enforced dehydration,

$$A_x{}^+(H_2O)_y[MX_2]^{x-} + xH_3O^+ \rightarrow (H_3O)_x{}^+(H_2O)_y[MX_2]^{x-} + xA^+ \tag{24}$$

$$(H_3O)_x{}^+(H_2O)_y[MX_2]^{x-} \rightarrow H_xMX_2 + (y + x)H_2O \tag{25}$$

Spontaneous rehydration is not observed, probably because of kinetic factors originating from essentially covalent bonding of hydrogen in the unsolvated phases. The chalcogenide bronzes show typical Brönsted acid character: upon reaction with inorganic or organic hydroxides in aqueous solutions, the corresponding hydrated phases can be obtained reversibly:

$$H_xMX_2 + xA^+ + xOH^- + yH_2O \rightleftarrows A_x{}^+(H_2O)_{y+x}[MX_2]^{x-} \tag{26}$$

In a similar way the reaction with Lewis bases leads to homosolvated ionic intercalation compounds [Eq. (27)] via partial proton transfer to the Lewis base functional group.

$$H_xMX_2 + (y + z)L \rightarrow (LH)_y{}^+L_z[H_{x-y}MX_2]^{y-} \tag{27}$$

H_xTaS_2 and H_xNbS_2 prepared by cathodic reduction of 2H TaS_2 and 2H NbS_2 are nonstoichiometric phases with a phase range of $0 < x < 0.8$. Neutron diffraction powder and single-crystal studies established a position for hydrogen in trigonal prismatic sites inside the layers at the height of the metal atoms with formation of hydrogen–metal four-center bonds (Fig. 12) (*71*). Both phases react with hydroxides and Lewis bases according to Eqs. 26 and 27. Ion exchange of the corresponding hydrated alkali phases with H_3O^+ leads to the hydronium compounds, which can be dehydrated to H_xMS_2.

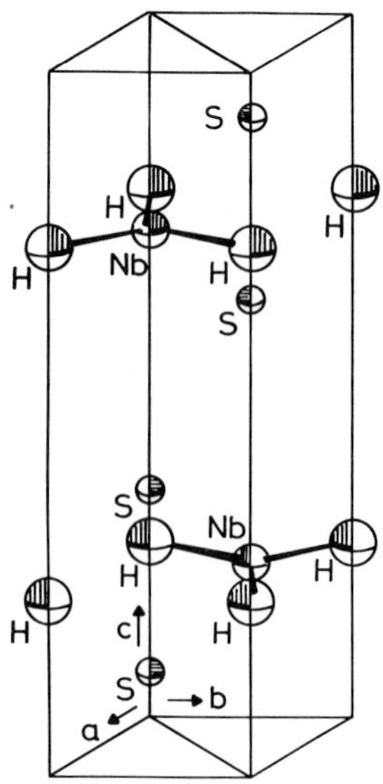

Fig. 12. Structure of $H_{0.8}NbS_2$; metal–hydrogen bonds are indicated (*71*).

Upon cathodic reduction of TiS_2 and 1T TaS_2 (both CdI_2 type of structure with octahedral positions for Ti, Ta) to the hydrated alkali phases $A_{0.3}{}^{+}(H_2O)_y[TiS_2]^{0.3-}$ and subsequent treatment with aqueous HCl/SO_2, the layers collapse instantaneously to an interlayer spacing of ca 6 Å and the hydrogen bronzes $H_{0.2}TiS_2$ and $H_{0.2}TaS_2$ are obtained, which show a chemical reactivity quite similar to that of hydrogen bronzes derived from the 2H forms described above. A neutron diffraction study on $H_{0.2}TaS_2$ showed, however, that hydrogen is located close to the van der Waals gap, approximately at the height of the sulfur layer planes between sulfur triangles (*70*).

$H_{0.3}CrS_2$ and $H_{0.5}MoS_2$ can be obtained only from the ternary alkali sulfides $ACrS_2$ and $A_{0.5}MoS_2$ prepared by thermal methods. Hydration and partial chemical or electrochemical oxidation of the latter yield the hydrated phases, which upon reaction with aqueous acids turn over directly to the hydrogen bronzes (*30*, *33*). Whereas $H_{0.3}CrS_2$ is hexagonal (a = 3.48 Å, $c/3$ = 6.01 Å), a distorted orthorhombic MoS_2 lattice was found for $H_{0.5}MoS_2$ (a = 6.47 Å, b = 11.20 Å, c = 11.98 Å). Neither compound seems to have appreciable phase ranges for hydrogen, and the corresponding binary phases are unknown. The chemical reactivity of these compounds toward hydroxides and Lewis bases is identical. however, with that of the other hydrogen bronzes described above.

H. Molecular Dynamics and Diffusion

As a consequence of the high mobility of the guest species at ambient temperatures usually encountered in solvated intercalation compounds, NMR methods are adequate for additional characterization of these phases. Nuclear magnetic resonance data may be used to obtain information on (a) the temperature-dependent dynamics of the guest species, (b) the spatial orientation of molecular guest species relative to the host lattice in the case of anisotropic mobility, and (c) chemical equilibria between the ambient guest phase and intercalated host lattice, which is of particular interest in cases in which a delicate balance exists between metastable phases that are difficult to isolate. A considerable number of appropriate atomic nuclei are available, but the investigations have concentrated so far mainly on ^1H NMR studies.

The proton exchange rate in hydrated $(NH_3)_{1/3}(H_2O)_{2/3}TaS_2$ was found to be 10^5 times faster than that in $(NH_3)_{1.0}TaS_2$ (*20*). Extensive investigations on $(NH_3)_{1.0}TaS_2$, $(NH_3)_{1.0}TiS_2$, and the ND_3 analogs between 5 and 300 K demonstrated the extreme anisotropic mobility of the guest molecules in the dichalcogenide lattice (*55*). Molecular diffusion and reorientation rates exceeded 10^5 sec^{-1} above 160 K; C-3 axis spinning continues down to at least 5 K with a rate of ca 10^8 sec^{-1}. Mechanisms of rapid diffusion of NH_3 via domain boundaries, diffusion path energy profiles, and the influence of layer charge distribution have been discussed (*8*, *72*). From the results of a study of $(NH_3)_{1.0}TaS_2$ by quasi-elastic neutron scattering, a diffusion mechanism was proposed that involves jumps of the guest species between equivalent next-nearest sites; an activation energy of 9.6 kcal/mole was measured for translational motion (*73*).

Activated ordered mobility of H_2O at 300 K was observed for monolayer hydrates $A_x^+(H_2O)_y[MS_2]^{x-}$ by ^1H NMR studies (A = alkali cation, NH_4^+; M = Ta, Nb) (*74*, *75*). The spectra obtained agree well with the calculated spectrum for an isolated two-spin system and result from the two-dimensional diffusion of water without activation of the rotation around the C-2 axis of H_2O (Fig. 13a). This anisotropic mobility is a consequence of the interaction of the negatively charged chalcogenide layers with the positively polarized hydrogen atoms of the water molecules; the negative-dipole ends of H_2O are engaged in cation/dipole interactions with neighboring cations. The transition to the rigid H_2O lattice at lower temperatures is dependent on the cation hydration energy; at higher temperatures, a single-line spectrum is found, indicating random motion.

Activation of the rotation of H_2O around the C-2 axis is observed for bilayer hydrates (*74*). The hydrated vanadium phase $Na_{0.6}(H_2O)_{3.6}VS_2$ was reported to exhibit chemical shift anisotropy due to anisotropic magnetic susceptibility; one sharp line appeared in the ^{23}Na spectrum with a chemical shift and line width identical with that of Na^+ in homogeneous aqueous solution (*35*).

The reaction of $K_{1/3}^+(H_2O)_{2/3}[TaS_2]^{1/3-}$ in aqueous electrolytes containing potassium hydroxide provides an example of an NMR study on

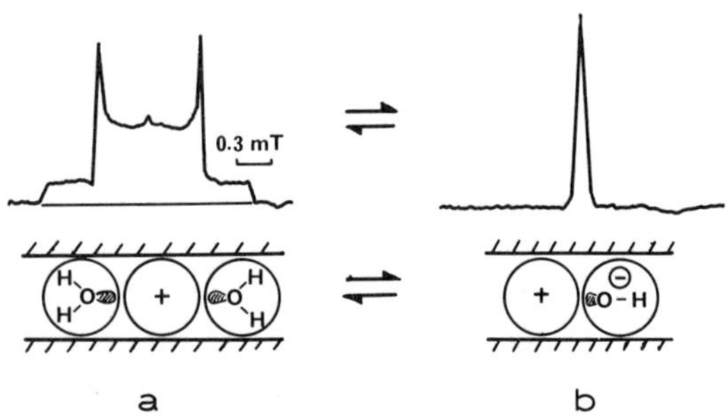

Fig. 13. Integrated ^1H NMR spectra and structural models of hydrated phases. (a) $K_{1/3}^+(H_2O)_{2/3}[TaS_2]^{1/3-}$; (b) cation/anion intercalation compound $K_{2/3}^+(OH)_{1/3}^-[TaS_2]^{1/3-}$ obtained on reversible KOH intercalation (*12*).

equilibrium states (*12*). With increasing KOH concentration (300 K) the intensity of the ^1H NMR two-spin signal of H_2O decreases reversibly at almost constant interlayer spacing, until at high concentrations only a single-line spectrum remains (Fig. 13b). Analytical data on the product obtained thus show a K/Ta ratio of ca 0.6. A single-crystal neutron diffraction study revealed that the hydrogen atoms are located in the central plane of the interlayer space. The data established that at high KOH concentration K^+ and OH^- cointercalate and a cation/anion intercalation compound $(K^+)_{2/3}(OH^-)_{1/3}[TaS_2]^{1/3-}$ is obtained. Because of the negative charge of OH^- the optimal proton position is at sites equidistant from the anion layers (Fig. 13b).

Proton NMR studies have also been applied to Lewis base intercalation compounds with heterocyclic guest species. Proton spin–lattice relaxation time studies on $TaS_2(pyr)_{0.5}$ revealed a rigid lattice for pyridine up to ca 300 K. Methyl group rotation was observed at low temperatures for 2-, 3-, and 4-picoline with an activation potential barrier of 4 kJ/mole (*76*).

It is obvious from the relatively small number of investigations performed so far that NMR studies may contribute valuable information to the characterization of solvated compounds, especially with respect to anisotropy phenomena in molecular dynamics.

I. Physical Properties

Except for the homosolvated Lewis base compounds, which are discussed in Section III,E, relatively few investigations on physical properties have been reported for the solvated phases. So far the interest has focused mainly on changes in superconducting phase transitions by incercalation.

The superconducting critical temperature T_c of 2H TaS_2 (0.7 K) was found to increase significantly to 3.3–5.3 K on intercalation with alkali hydroxides and aqueous NH_3 (*17*). This observation was confirmed by a subsequent study on TaS_2 intercalation compounds of inorganic and organic hydroxides (*36*); the highest T_c values were measured for the KOH intercalate. Similar values were obtained initially in studies on the critical temperature and the thermodynamic critical field of hydrated alkali and alkaline earth phases $A_x^+(H_2O)_y[TaS_2]^{x-}$ (*77*). In a restudy of the hydrated phases, it was shown that protolysis and solvolysis reactions may considerably affect the measurements. Investigations under improved experimental conditions demonstrated that the T_c values of the unhydrated ternary alkali phases A_xTaS_2 showed strong cation dependence, which disappeared entirely from Li to Rb after hydration; no dependence on the interlayer spacing was found (*78*). The T_c value of Cs turned out to be an exception, being considerably lower; this was explained by the size of this cation, which exceeds that of H_2O. After H_2O was replaced by the larger *N*-methylformamide molecule, the exceptional behavior of Cs disappeared. Hydration was also found to eliminate the cation dependence of T_c in the ternary alkali phases $A_{0.5}NbS_2$.

An attempt has been made to discuss the results in a rigid band approach, but it is obvious that several factors influence the superconducting properties in these phases, and additional investigations, including low-temperature structural studies, normal-state transport properties, and magnetic behavior, will be needed.

Variation of the critical temperature with the lattice expansion was reported for alkali metal/ammonia intercalation compounds $A_x(NH_3)_yTaS_2$ and for $A_x(NH_3)_yMoS_2$ (*44*, *68*).

IV. RELATED LAYERED CHALCOGENIDES

Besides the dichalcogenides with structures derived from the CdI_2 type, there is a series of ternary chalcogenides that are structurally related. These compounds contain layered matrix elements consisting of a sandwich type of array of transition metal ions and chalcogen atoms that is different in details from the CdI_2 type. The solvation, cation-exchange behavior, and redox reactions are basically similar, however, to those of the dichalcogenides. They may be divided according to composition and structure into three groups, the reactions of which are considered here. We also include a short discussion of related solvation and exchange reactions of chalcogenides with chain-type structures.

A. Tantalum Sulfide Carbide

The complex carbide Ta_2S_2C consists of van der Waals-bonded layers with a sequence ···S—Ta—C—Ta—S··· of the atoms perpendicular to the metal sulfide carbide sheets (*79*). Three hexagonal stacking modifications were reported: 1T, 2H, and random form. Intercalation reactions were first observed for Lewis base molecules (NH_3, N_2H_4, organic bases) (*80*); subsequent work described the formation of solvated phases (*81*) and of alkali metal intercalation compounds (*82*).

Chemical or electrochemical reduction of 1T Ta_2S_2C in aqueous electrolytes containing main group or transition metal ions leads to hydrated phases according to

$$xA^+ + xe^- + yH_2O \rightleftarrows A_x^+(H_2O)_y[Ta_2S_2C]^{x-} \quad (28)$$

with a layer charge density of ~ 0.2 (*81*). The hydrates obtained undergo rapid cation and solvent exchange. Figure 14 shows the interlayer spacings d measured for alkali and alkaline earth derivatives in comparison with those of the corresponding hydrates from 2H TaS_2. The average difference in d measured for the two series amounts to ca 2.5 Å, which corresponds well with the difference in thickness of the host layers: $d_0(Ta_2S_2C) - d_0(TaS_2) = 8.54 - 6.05 = 2.49$ Å. Two hydrate series are observed with lattice expansions Δd of ca 3 Å (monolayers of H_2O) and ca 6 Å (bilayers of H_2O) and a critical e/r ratio at ~ 1. Below ca 10 torr the water bilayers collapse reversibly to monolayers. Whereas hydrates of transition metal ions exhibit d values corresponding to water bilayers, the incorporation of organic ions by ion exchange or cathodic reduction leads to derivatives in which the geometry and arrangement of the cations dominate the interlayer spacing. The reduced solvated phases can be oxidized quantitatively in reversion of Eq. (*28*) to Ta_2S_2C with $d = 8.54$ Å; no metastable hydrates can be obtained for $x = 0$, which is due to the low energy of the $OH \cdots S$ hydrogen bonds. The lattice expansion values observed on replacement of H_2O by polar organic solvents are comparable to those found in the corresponding solvates of dichalcogenides.

The formation of hydrates can also be observed by the reaction of Ta_2S_2C with strongly alkaline solutions, e.g., alkali hydroxides. This process is accompanied by partial hydrolysis of the Ta_2S_2C layers under release of

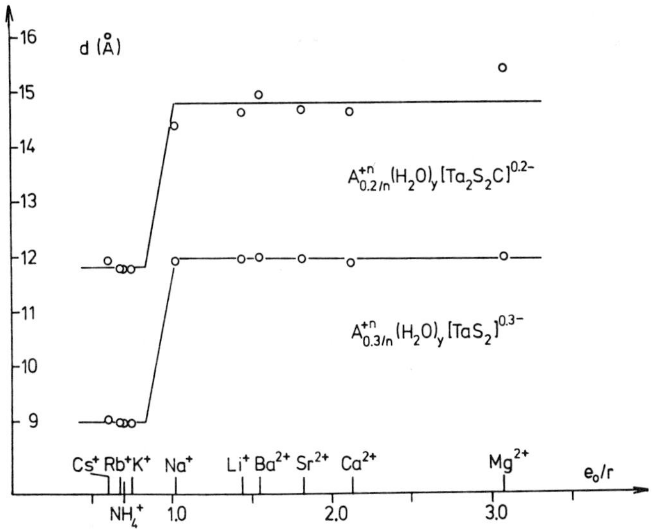

Fig. 14. Dependence of the interlayer spacing on cation charge/radius ratio for hydrated tantalum sulfide carbides and hydrated tantalum disulfides (*81*).

S^{2-}, which acts as an electron donor for the redox reaction, being itself oxidized to sulfur and polysulfides in a process similar to that described for the reaction of dichalcogenides with alkali hydroxides (Section III,B). In the reaction products $A_x^{+}(H_2O)_y[Ta_2S_{2-x}O_xC]^{x-}$, oxygen replaces the sulfur atoms lost by hydrolysis.

It is likely that the Lewis base intercalation compounds of Ta_2S_2C (e.g., with NH_3 and pyridine) are actually to be considered ionic solvated phases resulting from redox reactions similar to those established for the corresponding dichalcogenide compounds.

B. Phosphorous Sulfides

The structure of the layered ternary chalcogenides MPS_3 (M = transition metal) can be related schematically to that of $CdCl_2$: Sulfur replaces the Cl atoms, and M atoms as well as P_2 pairs occupy the octahedral metal positions: $M_{2/3}(P_2)_{1/3}S_2$ (*83*). A series of these phases has been reported to form quaternary compounds Li_xMPS_3 by intercalation of Li (*84*, *85*).

The cointercalation of solvent molecules in the course of cathodic reduction of $NiPS_3$ in Li^+/organic electrolytes has been suggested on the grounds of thermodynamic arguments (*86*); pyridine was found to intercalate into $MnPSe_3$ by thermal reaction (*87*). In a study on the solvation properties of $Na_{0.5}NiPS_3$ prepared from the reaction of $NiPS_3$ with Na in liquid ammonia, the hydrated phase $Na_{0.5}^{+}(H_2O)_y[NiPS_3]^{0.5-}$ was obtained on reaction of the Na compound with water (*67*). The hexagonal lattice parameters of the hydrate are a = 5.82 Å, $c/3$ = 12.15 Å. Rapid ion exchange was observed in aqueous electrolytes. The lattice expansions measured in the alkali/alkaline earth ion series compare well with those of the corresponding dichalcogenide monolayer and bilayer hydrates. On ion exchange with aqueous pyridinium salt solutions, the homosolvated intercalation compound $(C_5H_5NH)_x^{+}(C_5H_5N)_y[NiPS_3]^{x-}$ is formed with $x + y = 0.7$; hexagonal lattice parameters are a = 5.83 Å, $c/3$ = 12.14 Å. The lattice expansion of 5.81 Å is closely similar to that of $MnPSe_3(C_5H_5N)_{0.4}$, which amounts to 5.89 Å (*87*). $MnPSe_3(C_5H_5N)_{0.4}$ undergoes ion exchange in contact with aqueous alkali salt solutions, and hydrated phases with hexagonal lattice symmetry are obtained on quantitative exchange. The lattice expansions of the Na^+ form (a = 6.35 Å, $c/3$ = 12.63 Å) and the K^+ form (a = 6.36 Å, $c/3$ = 9.90 Å) correspond to mono-and bilayers of water (*67*).

C. Ternary Chalcogenides $A_2M_4S_6$ and $A_2M_3S_4$

The Group VIIIb transition elements Ni, Pd, and Pt form stoichiometric ternary chalcogenides $A_2Ni_3X_4$, $A_2Pd_3X_4$, $A_2Pt_3X_4$, and $A_2Pt_4X_6$ with the heavier alkali elements (A = alkali metal), which all have layer-type structures (*88*). A series of these phases has been investigated with regard to solvation, ion exchange, and redox properties (*11*, *31*, *67*).

The semiconducting potassium platinum sulfide $K_2Pt_4S_6$ (hexagonal lattice parameters a = 7.01 Å, c = 19.14 Å) has a layer structure (*88*, *89*) consisting of Pt_4S_6 sheets, which are held together by K^+ ions. The sequence of the atoms within the layers in the direction of the hexagonal c axis is ... K (S—Pt—S) K (S—Pt—S) K.... The Pt_4S_6 units contain a central layer of Pt atoms in a net of interconnected triangles; similar nets are formed above and below the Pt plane by the sulfur atoms, but one-quarter of the sulfur positions remain unoccupied. In the resulting structure, 25% of the Pt atoms are quadrivalent and have octahedral sulfur coordination, whereas the rest correspond to Pt(II) with planar coordination by S atoms. In contact with O_2-saturated water at ambient temperature, $K_2Pt_4S_6$ undergoes reversible, spontaneous hydration to $K_x^+(H_2O)_y[Pt_4S_6]^{x-}$ with the hexagonal lattice parameters a = 7.03 Å, c = 25.65 Å. The interlayer spacing changes from $d = c/3$ = 6.35 Å for $K_2Pt_4S_6$ to $d = c/3$ = 8.55 Å. In the course of hydration a part of the K^+ ions is released by oxidation; x amounts to ~1. Rapid ion exchange proceeds with main group cations, transition metal cations, and organic cations; all derivatives obtained have hexagonal structures. Alkaline earth ions, transition metal ions, and Li exhibit an interlayer spacing of d ~ 11.5 Å, whereas the other alkali ions and NH_4^+ show d values varying between 7 and 10 Å, which are strongly dependent on electrolyte concentration. This observation can be attributed to a perturbation of the cation/H_2O system in the case of ions with lower hydration energy by the large sulfur "holes" present at the surface of the Pt_4S_6 layers, which prevents a simple monolayer/bilayer hydrate correlation from occurring as observed for the dichalcogenide hydrates with main group cations.

On anodic oxidation of $K_x^+(H_2O)_y[Pt_4S_6]^{x-}$ in aqueous mineral acids Pt_4S_6 is obtained according to

$$K_x^+(H_2O)_y[Pt_4S_6]^{x-} \rightarrow Pt_4S_6 + xK^+ + yH_2O + xe^- \qquad (29)$$

The hexagonal lattice parameters of Pt_4S_6 are a = 6.98 Å, c = 13.85 Å; the interlayer spacing corresponds to $d = c/3$ = 4.62 Å. Pt_4S_6 is metastable and decomposes on heating above 200°C. Cathodic reduction of Pt_4S_6 in aqueous electrolytes results again in the formation of hydrated phases in reversion of Eq. (29).

The layered sulfides $Cs_2Pd_3S_4$, $K_2Pd_3S_4$, and $Cs_2Pt_3S_4$ are similar in structure to $K_2Pt_4S_6$ and contain large sulfur holes in the anion sheets; however, the transition metal atoms are positioned exclusively in coplanar sulfur coordination (*88*). Whereas hydration and cation exchange proceed extremely slowly with small mono- or bivalent cations, quantitative ion exchange can be achieved with *n*-alkylammonium ions at 60–80°C. $K_2Mn_3S_4$ and $Cs_2Mn_3S_4$ can exchange the alkali ions for *n*-alkylammonium ions. However, the Mn_3S_4 layers decompose rapidly by hydrolysis, which is due to the strongly ionic character of the Mn—S bonds.

Potassium nickel sulfide $K_2Ni_3S_4$ can be isolated as platelike crys tals with metallic golden luster. The structure was not reported, but coplanar coordination of nickel by the chalcogen atoms was assumed because of the weak temperature-independent paramagnetic behavior (*88*). The crystal morphology suggests a layer-type structure; this was confirmed by investigations on the solvation behavior (*31*, *67*).

$K_2Ni_3S_4$ can be indexed on an orthorhombic basis with the lattice parameters $a = 4.93$ Å, $b = 8.79$ Å, $c = 14.15$ Å, $d = c/2 = 7.07$ Å. In contact with air-saturated water, the hydrated phase $K_x^+(H_2O)_y[Ni_3S_4]^{x-}$ is formed with $x \sim 1.5$ and $y \sim 1.7$; i.e., the solvation is correlated with partial oxidation of the Ni_3S_4 layers. The lattice parameters of the hydrate are $a = 4.89$ Å, $b = 8.73$ Å, $c = 19.06$ Å, $d = c/2 = 9.53$ Å; the lattice expansion observed on hydration is approximately equivalent to a monolayer of H_2O. Water is easily replaced by polar solvent molecules, and potassium can be exchanged for mono- and divalent cations; quantitative exchange requires several days, however. All of the derivatives can be indexed by orthorhombic lattice parameters, with *a* and *b* remaining constant within ± 0.05 Å. Within the series of alkali/alkaline earth ions, two hydrate types with interlayer spacings of ca 9.5 and ca 12.5 Å are observed; the difference of ca 3 Å corresponds to one water layer. The monolayer hydrate/bilayer hydrate transition appears, however, at an e/r ratio of ca 1.7, which is considerably higher than that found for dichalcogenides; this may be related to unknown structural details of the layers.

On anodic oxidation of the hydrated K phase in K_2SO_4/H_2O, a potential step is observed after a transfer of ca $0.6e^-/Ni_3S_4$. After drying at 10^{-3} torr, the resulting product has the composition $K_{0.9}Ni_3S_4$ with the orthorhombic lattice parameters $a = 4.81$ Å, $b = 8.68$ Å, $c = 13.08$ Å, $d = c/2 = 6.5$ Å. The strong variation of *d* from $K_2Ni_3S_4$ to KNi_3S_4 suggests that sulfur holes are present in the Ni_3S_4 layer surface that can accommodate K^+ ions. KNi_3S_4 undergoes reversible solvation and ion exchange similar to $K_2Ni_3S_4$.

D. Chain-Type Host Lattices

Solvation reactions should also be expected to proceed with chalcogenides containing quasi-one-dimensional metal chalcogen units. So far two examples of reactions of ternary phases have been reported: Li_xTiS_3 and $KFeS_2$.

$KFeS_2$ forms red-violet needlelike crystals with metallic luster (*88*) that are built of infinite $[FeS_2]_\infty{}^{\infty-}$ chain anions formed by FeS_4 tetrahedra sharing opposite edges; the macroanions are held together in the lattice by potassium cations. Whereas $KFeS_2$ does not react with H_2O, suspensions of $KFeS_2$ in aqueous solutions of Mg^{2+}, Ca^{2+}, Sr^{2+}, and Ba^{2+} undergo hydration and simultaneous ion exchange (*90*):

$$KFeS_2 + 0.5A^{2+} + yH_2O \rightarrow A^{2+}_{0.5}(H_2O)_y[FeS_2]^- + K^+ \qquad (30)$$

The stability of the hydrates decreases with decreasing cation hydration energy from Ca^{2+} to Ba^{2+}. All phases are tetragonal and show a periodicity of ca. 5.6 Å in the direction of the c axis equal to that of $KFeS_2$, whereas an expansion of a is found from 7.09 Å ($KFeS_2$) to 11.3–11.5 Å upon hydration.

Transition metal trichalcogenides MX_3 are built from MX_6 trigonal prisms sharing opposite basal faces to form chains $(MX_{6/2})_\infty$. Depending on M, the structures show some layerlike characteristics as a consequence of chain/chain interaction; they are able to form metastable ternary phases Li_xMX_3 by chemical or electrochemical intercalation of Li (*8*). The uptake of organic solvent molecules, e.g., propylene carbonate, was reported for Li_xTiS_3 (*91*); the solvated phases $Li_x(solv)_y[TiS_3]$ show rather large interchain spacings with strong two-dimensional lattice disorder. On the reaction of TiS_3 with sodium in liquid NH_3 at a ratio Na/Ti = 1, expansion of the monoclinic c axis of TiS_3 for ca. 4.3 Å is measured while the chain axis b remains constant. The product composition corresponds to $Na(NH_3)_2TiS_3$. Decomposition with formation of Na_2Se is observed under identical conditions for $NbSe_3$.

V. LAYERED OXIDE BRONZES

Among the nonstoichiometric ternary oxide bronzes $A_xM_yO_z$ of transition metals with alkali metals, a large number of thermally stable phases are known to have layered structures characterized by $[M_yO_z]^{x-}$ slabs held together in the lattice by ionic bonding via the alkali cations (e.g., layered vanadates, molybdates, and cobaltates) (*92*). In addition, a few binary oxides exist with layered structures (e.g., V_2O_5, MoO_3) which are able to form nonstoichiometric alkali bronzes by topotactic redox reactions (*93–96*). The solvation, exchange, and redox properties of layered molybdate and vanadate bronzes have been studied in some detail.

Oxygen has an electronegativity that is markedly higher than that of sulfur. It is less polarizable, and metal–oxygen bonds have more ionic character than metal–sulfur bonds. Generally, the structures of layered oxides and sulfides are therefore quite different. The solvation properties of layered oxides are strongly influenced by the capacity of oxygen anions to form strong hydrogen bonds.

A. Molybdenum Oxides

The binary molybdenum(VI) oxide MoO_3 has a typical layer structure in which distorted MoO_6 octahedra share edges and vertices, respectively, to form two-dimensional MoO_3 sheets separated in the lattice by van der Waals gaps (Fig. 15). The cathodic reduction of MoO_3 in aqueous electrolytes containing main group or transition metal ions leads to the formation of metallic blue hydrated phases $A_x{}^+(H_2O)_y[MoO_3]^{x-}$ according to (*93*)

$$A_x{}^+ + yH_2O + xe^- + MoO_3 \rightleftarrows A_x{}^+(H_2O)_y[MoO_3]^{x-} \tag{31}$$

Usually several potential steps are found in the course of the reduction, indicating the presence of intermediate stages. For Na^+/H_2O, electrolyte steps occur after a transfer of $0.2e^-/MoO_3$ and $0.6e^-/MoO_3$; at $1.0e^-/MoO_3$ the potential of solvent decomposition is reached. X-ray diffraction studies showed that at the start of the reaction the interlayer spacing is already close to the value of the final reaction product. For reduction stages $>0.6e^-/MoO_3$, analytical determination of the Na^+ content indicates that protons are taken up in addition to Na^+ into the lattice, and mixed alkali/hydrogen bronzes $A_n{}^+(H_3O)_m{}^+(H_2O)_y[MoO_3]^{(n+m)-}$ are obtained (*12*). In the range of $0.5 < x \leq 1$, the redox reaction is easily reversible; quantitative reoxidation to MoO_3 is kinetically hindered and proceeds very slowly. Large

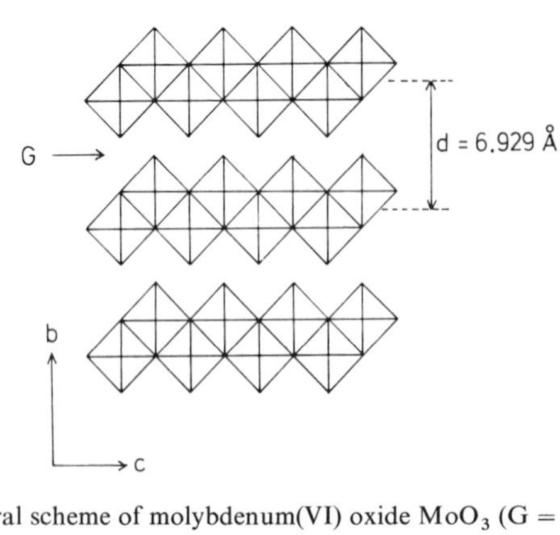

Fig. 15. Structural scheme of molybdenum(VI) oxide MoO_3 (G = van der Waals gap).

ions such as Cs^+ cannot be directly introduced into MoO_3 by cathodic reduction, but the hydrogen bronze $H_{1.0}MoO_3$ is obtained instead under these conditions. However, the Cs bronze may be prepared by cation exchange, which suggests that kinetic factors are responsible for the lack of reactivity. Similar hydrated bronzes also are formed from MoO_3 in aqueous electrolytes containing strong reducing agents, for example, $S_2O_4^{2-}$ ions, by intercalation of MoO_3 with alkali metals dissolved in liquid NH_3 and subsequent reaction with H_2O and by hydration of ternary alkali bronzes A_xMoO_3 prepared by cathodic reduction in organic electrolytes (*94*). All of the hydrated phases are subject to considerable protolysis on extended washing with H_2O. On reaction with dilute mineral acids the hydrated acid form $(H_3O)_x{}^+(H_2O)_y[MoO_3]^{x-}$ is obtained, which may again be converted to the corresponding salts by treatment with bases, e.g., alkali and alkaline earth hydroxides and quaternary ammonium and phosphonium hydroxides.

Cation exchange can be performed easily with inorganic and organic cations, and H_2O may be replaced by polar organic solvent species. All solvated MoO_3 bronzes have orthorhombic lattice symmetry like MoO_3. Changes in the c-axis parameter are systematically smaller than those for the a axis. The structural anisometry of the MoO_3 layers in the a/c plane is responsible for this phenomenon: In the direction of the c axis the MoO_6 octahedra share common edges, whereas in the direction of the a axis they share vertices only. For the hydrated alkali/alkaline earth series three groups of interlayer spacings can be distinguished: (a) large monovalent cations Rb^+, Cs^+ with d = ca 9.7 Å, (b) smaller monovalent cations Li^+, Na^+, K^+, $NH_4{}^+$ with d = ca 11.5 Å, and (c) bivalent cations with d = ca 14 Å. Although the qualitative hydration behavior is similar to that of the layered dichalcogenides, the hydrate structures are obviously more complex. The reason for this difference is the increased layer/solvent molecule interaction by O—H···O hydrogen bonds in the oxide case and the "nonplanar" surface of the MoO_3 layers, which both strongly influence the structure of the water sublattice.

At reduced water vapor pressures, reversible dehydration of the hydrated bronzes can be observed. The hydrated acid form $(H_3O^+)_{0.5}(H_2O)_y$-$[MoO_3]^{0.5-}$, which exhibits an interlayer spacing of 9.31 Å, can be dehydrated at 50°C/10^{-2} torr. X-ray powder diagrams show the product to be identical with the hydrogen bronze $H_{0.5}MoO_3$ (*97*).

The formation of hydrogen bronzes H_xMoO_3 by chemical reduction of MoO_3 in aqueous acids and the close relationship of their structure with MoO_3 were observed earlier (*97*). Cathodic reduction of MoO_3 electrodes in aqueous acids also leads to hydrogen bronzes H_xMoO_3; ^1H NMR data indicate high mobility for the H atoms in the oxide lattice (*93, 98*). X-ray and neutron diffraction studies established the structures of $H_{0.5}MoO_3$ and

$H_{0.3}MoO_3$ (*98*, *99*). The molybdenum/oxygen framework in these phases is almost unchanged as compared to MoO_3; the hydrogen atoms are located "inside" the layers at low temperatures. $H_{0.5}MoO_3$ does not hydrate spontaneously; it was found, however, that in aqueous solutions of metal hydroxides, the hydrogen bronze behaves like a Brönsted acid according to (*93*)

$$2H_{0.5}MoO_3 + A^+ + OH^- + 2nH_2O \rightarrow 2A^+_{0.5}(H_2O)_{n+0.5}[MoO_3]^{0.5-} \qquad (32)$$

Direct acid/base titrations can be performed with $H_{0.5}MoO_3$ suspensions in H_2O. The products obtained are identical in structure and chemical reactivity with corresponding hydrated phases received on cathodic reduction or ion exchange. The formation of a monoclinic sodium phase $Na_{0.94}H_{0.85}MoO_3$ was observed on the reaction of $H_{2.0}MoO_3$ with aqueous NaOH (*99*).

Hydrogen bronzes H_xMoO_3 may react directly with Lewis bases L ($L = NH_3$, hydrazine and derivatives, organic amines, heterocyclic bases, etc.) to give layered intercalation compounds $L_nH_xMoO_3$ according to (*93*, *100*)

$$nL + H_xMoO_3 \rightarrow L_nH_xMoO_3 \qquad (33)$$

The pyridine intercalation compound $(C_5H_5N)_{0.3}H_{0.5}MoO_3$ is formed on reaction of pyridine with $H_{0.5}MoO_3$ at 25°C; the orthorhombic lattice parameters are $a = 3.85$ Å, $b = 25.77$ Å, $c = 3.76$ Å, $d = b/2 = 12.88$ Å. The density of pyridine in the interlayer space calculated from the lattice expansion (5.84 Å) is close to that of liquid pyridine at room temperature. Treatment at 100°C/10^{-4} torr results in the loss of pyridine with formation of $(C_5H_5N)_{0.2}H_{0.5}MoO_3$ having a reduced interlayer spacing of $d = 11.28$ Å; reaction with excess pyridine leads back to $(C_5H_5N)_{0.3}H_{0.5}MoO_3$. To explain the host lattice/guest molecule interaction it was assumed that a partial transfer of layer protons to the lone electron pair of the nitrogen atom of pyridine takes place. The fact that for substituted pyridines the intercalation was found to be dependent on the corresponding pK_b value supports this assumption. A particularly stable situation should be expected if 50% of the pyridine rings were protonated; under these conditions, all of the heterocyclic rings would be engaged in hydrogen bonding N · · · H—N. Moreover, a position of the pyridine molecules with the ring plane perpendicular to the layer planes (compare Fig. 9d), which is compatible with the lattice expansion value of 5.84 Å, would give rise to additional π interactions between the aromatic rings, i.e., to a structure scheme that is closely related to that discussed for dichalcogenide intercalation compounds of pyridine (Section III,E). The resulting description,

$$(C_5H_5NH^+)_x(C_5H_5N)_{0.3-x}[H_{0.5-x}MoO_3]^{x-}$$

elucidates the character of a homosolvated phase for the pyridine intercalation compound.

Lewis base intercalation compounds can also be obtained by ion exchange of hydrated alkali phases with aqueous LH^+ solutions or by cathodic reduction of MoO_3 in aqueous LH^+ electrolytes. Direct thermal intercalation of MoO_3 was observed in the case of pyridine (*100*). It is likely that molecular redox reactions similar to those established for the dichalcogenides (Section III,E) are correlated with this process.

The nonstoichiometric triclinic binary oxide $Mo_{18}O_{52} \hat{=} MoO_{2.88}$ consists of MoO_3 type of layers which show regular steps originating from shear planes (*101*). $Mo_{18}O_{52}$ undergoes topotactic redox, solvation, and ion-exchange reactions quite similar to those of MoO_3 (*94*, *100*). On cathodic reduction in nonaqueous alkali electrolytes, ternary bronzes $A_xMo_{18}O_{52}$ are found which yield hydrated phases $A_x{}^+(H_2O)_y[Mo_{18}O_{52}]^{x-}$ on reaction with H_2O. Hydrated bronzes are also obtained by cathodic reduction of $Mo_{18}O_{52}$ in aqueous electrolyte solutions and by acid–base reactions of the hydrogen bronze $H_{7.7}Mo_{18}O_{52}$, which can be prepared by reduction of $Mo_{18}O_{52}$ in strong acids. Lewis base intercalation compounds can be prepared by direct reaction of L with $H_xMo_{18}O_{52}$ at 300 K or with $Mo_{18}O_{52}$ at higher temperatures.

B. Vanadium Oxides

Among the ternary vanadium oxides with electropositive metals $A_xV_2O_5$ and $A_xV_3O_8$, several phases with layered crystal structures are known (*92*). A study on exchange and solvation properties has been reported for the lithium bronze $Li_{1+x}V_3O_8$ (*103*).

$Li_{1+x}V_3O_8$ is monoclinic (lattice parameters $a = 6.68$ Å, $b = 3.60$ Å, $c = 12.03$ Å, $\beta = 107.50°$) and consists of $[V_3O_8]^{(1+x)-}$ layers held together by Li^+ ions, which are located in octahedral sites between the vanadium oxide sheets. The puckered V_3O_8 layers are formed by VO_6 octahedra and VO_5 distorted trigonal bipyramids, which share edges and corners (*102*). $Li_{1.15}V_3O_8$ reacts with aqueous acids, undergoing a change of color from grayish black to dark copper red with metallic luster (*103*). X-ray studies on single crystals show that the single-crystal nature is preserved in the reaction product. The Li^+ atoms are quantitatively exchanged for protons, and synchronously H_2O enters the interlayer space. The hydrated hydrogen bronze $H_{1.08}V_3O_8 \cdot 2.5H_2O$ is monoclinic with the lattice parameters $a = 10.27$ Å, $b = 3.61$ Å, $c = 12.42$ Å, $\beta = 112.45°$. The interlayer spacing changes during the reaction from 6.36 Å for $Li_xV_3O_8$ to 9.49 Å. Treatment at 25°C/10^{-3} torr of the hydrogen bronze leads via partial dehydration to

$H_{1.07}V_3O_8 \cdot 0.5H_2O$; the interlayer spacing changes to 6.11 Å. The hydrogen bronzes behave as solid Brönsted acids and react with aqueous bases. As a result of the high acid strength ion exchange proceeds even in neutral salt solutions with the release of protons:

$$H_x(H_2O)_y[V_3O_8] + xA^+ \rightleftarrows A_x^{\ +}(H_2O)_y[V_3O_8]^{x-} + xH^+ \qquad (34)$$

Ion exchange with inorganic and organic cations as well as solvent exchange proceed rapidly; all compounds are monoclinic. The monoclinic angle β varies depending on the cation and the solvent. For the hydrated alkali/alkaline earth ion series, three groups with different interlayer spacings can be distinguished: ca 11 Å for $e/r > 1.7$, ca 8.5 Å for $e/r < 1.1$, and an intermediate range with $d =$ ca 10 Å. The complex hydration behavior is similar to that of the hydrated MoO_3 bronzes and probably can be correlated with the V_3O_8 layer surface geometry and hydrogen bonding between the oxide layers and H_2O.

Organic Lewis bases are able to react directly with the hydrated hydrogen bronzes to intercalation compounds $L_nH_xV_3O_8$ according to

$$nL + H_x(H_2O)_y[V_3O_8] \rightarrow L_nH_x[V_3O_8] + yH_2O \qquad (35)$$

On reaction with pyridine, $(C_5H_5N)_{0.98}H_xV_3O_8$ is obtained with the monoclinic lattice parameters $a = 11.70$ Å, $b = 3.61$ Å, $c = 11.97$ Å, $\beta = 107.49°$ and an interlayer spacing of 11.16 Å. The relatively high stability observed can again be interpreted in terms of an ionic homosolvated structure due to partial transfer of lattice protons to the nitrogen lone pair of electrons at the pyridine:

$$(C_5H_5NH)_z^{\ +}(C_5H_5)_{1-z}[H_{x-z}V_3O_8]^{z-}$$

Support for this structure scheme is again provided by the observation of a strong dependence of the reactivity of substituted pyridines on the pK_b values of the guest molecules (*103*).

VI. CONCLUSIONS

The information given in the preceding pages illustrates the fascinating variety of topotactic solid-state reactions at ambient temperature exhibited by layered materials with electronic conductivity. Although there is still much work to be done on the details of reaction mechanisms and structures, the fundamental principles are now obvious. The foundation has thus been provided for a systematic approach to the tailoring of new intercalated materials with specific composition, structure, and properties by the appropriate choice of host lattice, guest species, and type of reaction. Moreover, it is apparent that a general ionic solvate bonding concept $A_x^{+}(B)_y[Z]^{x-}$ can be applied to most molecular intercalation compounds of layered chalcogenides and oxides. An ionic bonding scheme with inverse polarity is known for graphite intercalation compounds with Lewis acids. It is likely, however, that additional host lattices will be found which are able to form intercalation compounds $A_x^{-}[Z]^{x+}$ and $A_x^{-}(B)_y[Z]^{x+}$ with positively charged layer units via topotactic redox reactions. Interesting potential candidates are layered rare earth halides, e.g., PrI_2 (*13*), which exhibit metallic conductivity and are described as $[PrI_2]^{+}(e)^{-}$.

Concerning physical properties, the possibility of a systematic variation of guest cation and guest solvent species should be advantageous in identifying structure–property relationships. Solvated compounds also represent interesting model systems for the study of properties of quasi-two-dimensional electrolyte phases.

It has been established that even rather stable molecules such as pyridine may react at relatively low temperatures with layered materials via rearrangement of covalent bonds. Investigations on molecular reactions in the interlayer space, which provides a specific environment that may induce highly selective reactions, are likely to become a new area of interest with potential technological significance concerning heterogeneous catalysis and electrocatalysis.

It is apparent that the field discussed here provides attractive prospects for future research on a variety of aspects including solid-state science, molecular chemistry, interface phenomena, and electrochemistry.

REFERENCES

1. W. Rüdorff, *Adv. Inorg. Chem. Radiochem.* **1,** 223 (1959); G. R. Henning, *Prog. Inorg. Chem.* **1,** 125 (1959).
2. R. F. Grim, "Clay Mineralogy." McGraw-Hill, New York, 1968.
3. A. Weiss, *in* "Organic Geochemistry" (A. Eglinton and W. Murphy, eds.), p. 737. Springer-Verlag, Berlin and New York, 1969.
4. R. M. Barrer, *in* "Nonstoichiometric Compounds" (L. Mandelcorn, ed.), p. 268. Academic Press, New York, 1964.
5. W. Rüdorff, *Chimia* **19,** 489 (1965).
6. F. R. Gamble and T. H. Geballe, *in* "Treatise on Solid State Chemistry" (N. B. Hannay, ed.), Vol. 3, p. 84. Plenum, New York, 1976.
7. G. V. Subba Rao and M. W. Shafer, *in* "Intercalated Layered Materials" (F. Lévy, ed.), p. 99. Reidel Publ., Dordrecht, Netherlands, 1979.
8. M. S. Whittingham, *Prog. Solid State Chem.* **12,** 41 (1978).
9. L. B. Ebert, *Annu. Rev. Mater. Sci.* **6,** 181 (1976); J. M. Thomas, J. M. Adams, S. H. Graham, and D. T. B. Tennakoon, *in* "Solid State Chemistry of Energy Conversion and Storage" (M. S. Whittingham and J. B. Goodenough, eds.), Advances in Chemistry Series, No. 163, p. 298. Am. Chem. Soc., Washington, D.C., 1977.

10. J. Rouxel, *in* "Intercalated Layered Materials" (F. Lévy, ed.), p. 201. Reidel Publ., Dordrecht, Netherlands, 1979.
11. A. Lerf and R. Schöllhorn, *Inorg. Chem.* **16,** 2950 (1977).
12. R. Schöllhorn, *Angew. Chem.* **92,** 1015 (1980); R. Schöllhorn, *Angew. Chem. Int. Ed. Engl.* **19,** 983 (1980); R. Schöllhorn, *Physica* B + C **99** (*Amsterdam*), 89 (1980); R. Schöllhorn, *in* "Layered Materials and Intercalates" (C. F. van Bruggen, C. Haas, and H. W. Myron, eds.), North-Holland, Amsterdam, 1980.
13. F. Hulliger, *in* "Structural Chemistry of Layer Type Phases" (F. Lévy, ed.), Reidel Publ., Dordrecht, Netherlands, 1976; R. M. A. Lieth, ed., "Preparation and Crystal Growth of Materials with Layered Structures." Reidel Publ., Dordrecht, Netherlands, 1976.
14. H. G. Schwarz, Ph.D. Thesis, Univ. of Tübingen, 1954.
15. W. Rüdorff and H. Sick, *Angew. Chem.* **71,** 487 (1959).
16. M. Danot, A. Le Blanc, and J. Rouxel, *Bull. Soc. Chim. Fr.* p. 2670 (1969); A. Le Blanc, M. Danot, and J. Rouxel, *Bull. Soc. Chim. Fr.* p. 88 (1969); J. Rouxel, M. Danot, and J. Bichon, *Bull. Soc. Chim. Fr.* p. 3930 (1971).
17. F. R. Gamble, J. H. Osiecki, M. Cais, R. Pisharody, F. J. Di Salvo, and T. H. Geballe, *Science* **174,** 493 (1971).
18. R. Schöllhorn and A. Weiss, *Z. Naturforsch., Teil B* **28,** 711 (1973).
19. R. Schöllhorn and A. Weiss, *in* "Chemistry and Uses of Molybdenum" (P. C. H. Mitchell, ed.), p. 122. Climax Molybdenum Co., Villiers House, London, 1973; R. Schöllhorn and A. Weiss, *J. Less-Common Met.* **36,** 229 (1974).
20. M. S. Whittingham, *Mater. Res. Bull.* **9,** 1681 (1974).
21. G. A. Wiegers, R. van der Meer, H. van Heiningen, H. J. Klosterboer, and A. J. A. Alberink, *Mater. Res. Bull.* **9,** 1261 (1974).
22. M. S. Whittingham, *J. C. S. Chem. Commun.* p. 328 (1974).
23. R. Schöllhorn and H. Meyer, *Mater. Res. Bull.* **9,** 1237 (1974).
24. G. V. Subba Rao and J. C. Tsang, *Mater. Res. Bull.* **9,** 921 (1974).
25. R. Schöllhorn and A. Lerf, *Z. Naturforsch., Teil B* **29,** 804 (1974).
26. R. Schöllhorn, E. Sick, and A. Lerf, *Mater. Res. Bull.* **10,** 1005 (1975).
27. R. Schöllhorn and A. Lerf, *J. Less-Common Met.* **42,** 89 (1975).
28. A. J. A. Boos-Alberink, R. J. Haange, and G. A. Wiegers, *J. Less-Common Met.* **63,** 69 (1979).
29. W. Rüdorff and K. Stegemann, *Z. Anorg. Allg. Chem.* **251,** 376 (1943); F. Jellinek and B. van Laar, *J. Solid State Chem.* **6,** 574 (1973).
30. R. Schöllhorn, R. Arndt, and A. Kubny, *J. Solid State Chem.* **29,** 259 (1979).
31. R. Schöllhorn, *in* "Reactivity of Solids" (J. Wood, O. Lindqvist, C. Helgesson, and N. G. Vannerberg, eds.), p. 719. Plenum, New York, 1977.
32. R. Schöllhorn, W. Roer, and K. Wagner, *Monatsh. Chem.* **110,** 1147 (1979).
33. R. Schöllhorn, M. Kümpers, and D. Plorin, *J. Less-Common Met.* **58,** 55 (1978); M. Kümpers and R. Schöllhorn, *Z. Naturforsch., Teil B* **35,** 929 (1980).
34. H. A. Graf, A. Lerf, and R. Schöllhorn, *J. Less-Common Met.* **55,** 213 (1977).
35. A. J. A. Boos-Albernick, R. J. Haange, G. Stülen, and G. A. Wiegers, *Int. Conf. Solid Comp. Transition Elem., 5th, Uppsala* Abstr., p. 60 (1976).
36. G. V. Subba Rao, M. W. Shafer, and J. C. Tsang, *J. Phys. Chem.* **79,** 553 (1975).
37. G. V. Subba Rao and M. W. Shafer, *J. Phys. Chem.* **79,** 557 (1975).
38. A. Lerf, Ph.D. Thesis, Univ. of Munich, 1976.
39. C. Riekel, H. G. Reznik, and R. Schöllhorn, *J. Solid State Chem.* **34,** 253 (1980); R. Schöllhorn, *in* "Physics of Intercalation Compounds," (L. Pietronero and E. Tosatti, ed.). Springer-Verlag, New York and Berlin, 1982.

40. R. P. Clement, W. B. Davies, K. A. Ford, M. L. H. Green, and A. J. Jacobsen, *Inorg. Chem.* **17,** 2754 (1978).
41. R. Schöllhorn, H. D. Zagefka, T. Butz, and A. Lerf, *Mater. Res. Bull.* **14,** 369 (1979).
42. R. Schöllhorn and H. D. Zagefka, *Angew. Chem.* **89,** 193 (1977); *Angew. Chem., Int. Ed. Engl.* **16,** 199 (1977).
43. K. Wagner and R. Schöllhorn, *Z. Naturforsch., Teil B* (in press).
44. J. A. Woollam and R. B. Samoano, *Mater. Sci. Eng.* **31,** 289 (1977).
45. E. Bayer and W. Rüdorff, *Z. Naturforsch., Teil B* **27,** 1336 (1972).
46. J. O. Besenhard, H. Meyer, and R. Schöllhorn, *Z. Naturforsch., Teil B* **31,** 907 (1976).
47. R. Ruthardt, Ph.D. Thesis, Univ. of Munich, 1969.
48. D. W. Murphy and G. W. Hull, *J. Chem. Phys.* **62,** 973 (1975).
49. L. Christiany, H. Fuzellier, P. Lagrange, and A. Hérold, *C. R. Hebd. Seances Acad. Sci., Ser. C* **287,** 549 (1978).
50. Ar. Weiss and R. Ruthardt, *Z. Naturforsch., Teil. B* **28,** 249 (1973).
51. R. Schöllhorn and A. Weiss, *Z. Naturforsch., Teil B* **28,** 711 (1973).
52. M. B. Dines, *Inorg. Chem.* **17,** 763 (1978).
53. M. B. Dines and R. Lévy, *J. Phys. Chem.* **79,** 1979 (1975).
54. C. Riekel and R. Schöllhorn, *Mater. Res. Bull.* **11,** 369 (1976).
55. F. R. Gamble and B. G. Silbernagel, *J. Chem. Phys.* **63,** 2544 (1975); B. G. Silbernagel, M. B. Dines, F. R. Gamble, J. A. Gebhard, and M. S. Whittingham, *J. Chem. Phys.* **65,** 1906 (1976); B. G. Silbernagel and F. R. Gamble, *J. Chem. Phys.* **65,** 1914 (1976).
56. T. Butz, H. Saitovitch, A. Lerf, H. D. Zagefka, and R. Schöllhorn, *Phys. Lett. A* **67,** 74 (1978).
57. C. Riekel, *Solid State Commun.* **28,** 385 (1978).
58. F. R. Gamble, J. H. Osiecki, and F. J. Di Salvo, *J. Chem. Phys.* **55,** 3525 (1971).
59. C. Riekel, D. Hohlwein, and R. Schöllhorn, *J.C.S. Chem. Commun.* p. 863 (1976).
60. A. H. Thompson, *Nature (London)* **251,** 492 (1974).
61. A. Lerf, H. Saitovitch, A. Vasquez, and T. Butz, *Int. Conf. Solid Comp. Transition Elem., 6th*, Stuttgart, Germany, Abstr., p. 86 (1979).
62. N. Wiberg, *Angew. Chem.* **80,** 809 (1968).
63. M. B. Dines, *Science* **188,** 1210 (1975).
64. Proceedings of the Franco-American Conf. on Intercalation Compounds of Graphite, *Mater. Sci. Eng.* **31,** (1977).
65. L. B. Ebert, J. P. de Luca, A. H. Thompson, and J. S. Scanlon, *Mater. Res. Bull.* **12,** 1135 (1977).
66. A. Hérold, *C. R. Hebd. Seances Acad. Sci., Ser. C* **268,** 373 (1969).
67. H. D. Zagefka, Ph.D. Thesis, Univ. of Münster, 1980.
68. S. F. Meyer, J. V. Acrivos, and T. H. Geballe, *Proc. Natl. Acad. Sci. U.S.A.* **72,** 464 (1975).
69. D. W. Murphy, F. J. Di Salvo, G. W. Hull, J. V. Waszak, S. F. Meyer, G. R. Stewart, S. Early, J. V. Acrivos, and T. H. Geballe, *J. Chem. Phys.* **62,** 967 (1975).
70. H. G. Reznik, C. Riekel, and R. Schöllhorn, *Z. Naturforsch., Teil B* (in press).
71. C. Riekel, H. G. Reznik, R. Schöllhorn, and C. F. Wright, *J. Chem. Phys.* **70,** 5203 (1979).
72. M. S. Whittingham, *Electrochim. Acta* **20,** 575 (1975).
73. C. Riekel, A. Heidemann, B. E. F. Fender, and G. C. Stirling, *J. Chem. Phys.* **71,** 530 (1979); C. Riekel, R. Schöllhorn, I. Tomkinson, *Z. Naturforsch., Teil A* **35,** 590 (1980).
74. U. Röder, W. Müller-Warmuth, and R. Schöllhorn, *J. Chem. Phys.* **70,** 2864 (1979).
75. A. Lerf, P. Burkert, J. O. Besenhard, and R. Schöllhorn, *Z. Naturforsch., Teil B* **32,** 1033 (1977).
76. S. Wada, H, Alloul, and P. Molinié, *J. Phys., Lett. (Orsay, Fr.)* **39,** L243 (1978); M. Molitor, W. Müller-Warmuth, and R. Schöllhorn, *Ber. Bunsenges. Phys. Chem.* (in press).

77. F. Sernetz, A. Lerf, and R. Schöllhorn, *Mater. Res. Bull.* **9,** 1597 (1974); R. Schöllhorn, A. Lerf, and F. Sernetz, *Z. Naturforsch., Teil B* **29,** 810 (1974).
78. A. Lerf, F. Sernetz, W. Biberacher, and R. Schöllhorn, *Mater. Res. Bull.* **14,** 797 (1979).
79. O. Beckmann, H. Boller, and H. Nowotny, *Monatsh. Chem.* **101,** 945 (1970).
80. R. Schöllhorn and Weiss, *Z. Naturforsch., Teil B* **28,** 716 (1973).
81. R. Schöllhorn and W. Schmucker, *Z. Naturforsch., Teil B* **30,** 975 (1975).
82. R. Brec, J. Ritsma, G. Ouvrard, and J. Rouxel, *C. R. Hebd. Seances Acad. Sci., Ser. C* **281,** 531 (1975); R. Brec, J. Ritsma, G. Ouvrard, and J. Rouxel, *Inorg. Chem.* **16,** 660 (1977).
83. W. Klingen, G. Eulenberger, and H. Hahn, *Naturwissenschaften* **57,** 88 (1970).
84. A. H. Thompson and M. S. Whittingham, *Mater. Res. Bull.* **12,** 741 (1977).
85. A. Le Méhauté, G. Ouvrard, R. Brec, and J. Rouxel, *Mater. Res. Bull.* **12,** 1191 (1977).
86. A. Le Méhauté, *C. R. Hebd. Seances Acad. Sci., Ser. C* **287,** 309 (1978).
87. F. Kanamaru, S. Otani, and M. Koizumi, *Int. Conf. Solid Comp. Transition Elem., 5th, Uppsala* Abstr., p. 36 (1976).
88. W. Bronger, *in* "Crystallography and Crystal Chemistry of Materials with Layered Structures" (F. Lévy, ed.), p. 93. Reidel Publ., Dordrecht, Netherlands, 1976.
89. W. Rüdorff, A. Stössel, and V. Schmidt, *Z. Anorg. Allg. Chem.* **357,** 264 (1968).
90. H. Boller, *Monatsh. Chem.* **109,** 975 (1978).
91. D. W. Murphy and F. A. Trumbore, *J. Electrochem. Soc.* **123,** 960 (1976).
92. A. D. Wadsley, *in* "Nonstoichiometric Compounds" (L. Mandelcorn, ed.), p. 268. Academic Press, New York, 1964; P. Hagenmüller, *Comprehensive Inorganic Chemistry*, Vol. **4,** 541 (1973); C. Delmas, C. Fouassier, and P. Hagenmüller, *Mater. Sci. Eng.* **31,** 297 (1977).
93. R. Schöllhorn, R. Kuhlmann, and J. O. Besenhard, *Mater. Res. Bull.* **11,** 83 (1976).
94. J. O. Besenhard and R. Schöllhard, *J. Power Sources* **1,** 267 (1976).
95. M. S. Whittingham, *J. Electrochem. Soc.* **123,** 315 (1976).
96. D. W. Murphy, P. Y. Christian, F. J. Di Salvo, and J. N. Carides, *J. Electrochem. Soc.* **126,** 497 (1979).
97. O. Glemser and G. Lutz, *Z. Anorg. Allg. Chem.* **264,** 17 (1951); O. Glemser, G. Lutz, and G. Meyer, *Z. Anorg. Allg. Chem.* **285,** 173 (1965).
98. F. A. Schröder and H. Weitzel, *Z. Anorg. Allg. Chem.* **435,** 247 (1977).
99. P. G. Dickens, J. J. Birtill, and C. J. Wright, *J. Solid State Chem.* **28,** 185 (1979); J. J. Birtill and P. G. Dickens, *Mater. Res. Bull.* **13,** 311 (1978).
100. R. Schöllhorn, T. Schulte-Nölle, and G. Steinhoff, *J. Less-Common Met.* **71,** 71 (1980).
101. L. Kihlborg, *Ark. Kemi* **21,** 471 (1963).
102. A. D. Wadsley, *Acta Crystallogr.* **10,** 261 (1957).
103. R. Schöllhorn, F. Klein Reesink, and R. Reimold, *J.C.S. Chem. Commun.* p. 398 (1979).

11
Intercalation of Halides

JOHN D. CORBETT

The intercalation of halides represents a considerable departure from the behavior of layered disulfides of transition Groups IV and V, from which a large variety of interesting products may be formed either with molecular adducts or with cations accompanied by reduction (*1*, *2*). No examples involving normal-valent halides are known, but the layered oxyhalides best typified by FeOCl do exhibit some interesting intercalation chemistry. Reduced halides with significant metal–metal interactions (for instance, ZrCl) apparently do not form adducts with bases, and reduction accompanied by intercalation has not been reported. On the other hand layered ZrCl, ZrBr, and ThI_2 are known to exhibit significant intercalation chemistry involving reversible hydrogen uptake to form ordered, stoichiometric phases. Finally, apparently unique examples of cluster intercalation exist for hydrogen in Nb_6I_{11} and $CsNb_6I_{11}$.

I. NORMAL-VALENT HALIDES

Layered halides involving normal-valent metals are especially striking in the apparent complete absence of an intercalation chemistry. The many binary halides that occur in layered $CdCl_2$, CdI_2, $CrCl_3$, and BiI_3 structure types evidently form no such derivatives in which the layering is retained.

INTERCALATION CHEMISTRY

ISBN 0-12-747380-7

For example, $CdCl_2$ instead reacts rapidly with liquid ammonia even at $-78°C$ to form $Cd(NH_3)_6^{2+}(Cl^-)_2$. The generally lower lattice energies and higher solubilities of the halides together with the absence of any significant metal–metal interaction and band formation may all be important factors relative to the chalcogenide examples.

II. OXYHALIDES

Possibly the only examples of the intercalation of any normal-valent halide occur among the oxyhalides, particularly FeOCl. The structure of the substrate consists of distorted sheets of edge-shared octahedra of the nonmetal atoms centered about iron, the coordination polyhedra consisting of four-coordinate oxygen and two-coordinate chlorine, with the chlorine atoms defining in the outside of the layers. This material was first reported to take up pyridine (py) at 80°C to give the composition $FeOCl \cdot \frac{1}{3}py$ (*3*), the inter-sheet repeat (crystallographic *b* dimension) increasing by $\sim 70\%$ on uptake of py. The study was concerned principally with Mössbauer spectra as a function of temperature and the changes in magnetic interactions relative to that in the pure substrate, but the conductivity was found to increase by a factor of 10^7 on intercalation. Since effective coordination of py directly to the iron within the ideal FeOCl layers would appear to be sterically difficult if not impossible, the observed properties are more reasonable when considered in terms of some reduction of the layers. Intercalation has recently been accomplished with ferrocene and cobaltocene as well (*4*).

One way of accomplishing a reduction of the metal in the layered substrate would be the insertion of a countercation in the so-called "van der Waals gap" between halide layers. Although this has not been found with binary halides, it occurs for FeOCl and structurally related compounds and is also a mechanism for nonstoichiometry in the parent FeOCl. This reduction with metal was evidently first described in the patent literature for Li_xFeOCl (*5*), although the product was concluded to be inferior to Li_xTiS_2 as an anode. Armand and co-workers (*6*) described the lithiation of FeOCl as well as the isostructural VOCl and CrOCl with *n*-butyllithium in hexane (60°C, 1–8 days). For Li_xFeOCl almost no change in cell parameters (or symmetry) was observed to $x = 0.5$, and compositions to $x = 1.0$ were achieved. The X-ray powder data were of poor quality, however. On the other hand, $Li_{0.8}VOCl$ gave a sharp pattern, with a 0.25 Å decrease in the interlayer spacing after intercalation, which was attributed to the loss of a Jahn–Teller distortion of V(III) on reduction. Both this and Li_xCrOCl were decomposed by water and could not be reoxidized in $CH_3CN/LiClO_4$ after reduction electrochemically, in contrast to Li_xFeOCl. In another study

Watts and Dickens (*7*) found that solvent-free Li_xFeOCl prepared by a butyllithium route at 60°C exhibited a new unknown structure for $x > 0.3$, while at $x = 0.15$ only the *b* dimension (normal to the layers) somewhat more than doubled relative to FeOCl. However, this expansion was found to arise not from the increase in separation caused by the substantial adduct or solvent incorporation but from a supercell generated by distortion (e.g., buckling) of the layers on intercalation, the lithium cation making only a small contribution to the apparent expansion. The enthalpy of formation was also determined.

The most extensive studies of this material have been reported by Weiss and Sick (*8*) and by Meyer *et al.* (*9*). Cathodic reduction of FeOCl in DMSO/M^+ solution leads to a family of compounds $(M^+)_x(DMSO)_y[(Fe^{3+})_{1-x}(Fe^{2+})_xOCl]$, with M = Li to Cs ($x \sim 0.1$), $H^+(x \sim 0.5)$, and $NR_4{}^+(x \sim 0.05)$ in the gap between the chlorides. The amount of dimethyl sulfoxide (DMSO) taken up, presumably in solvating the intercalated cations, was inferred from the observed increase in the repeat normal to the layers. This inference is perhaps not always completely safe in view of the unit-cell expansion observed above with solvent-free Li_xFeOCl, although it is likely that polar solvents are taken up to some extent. Anodic oxidation of these intercalates was slow and incomplete.

Nonstoichiometry in FeOCl, namely, $(Fe^{2+})_{x/2}[(Fe^{3+})_{1-x}(Fe^{2+})_xOCl]$ can be induced similarly through the addition of iron during the synthesis and transport (for $0 < x < 0.2$). Reduction of stoichiometric FeOCl in aqueous systems proceeds very rapidly in many cases with concomitant hydrolysis and decomposition, but controlled topotactic reduction of single crystals takes place using iron in DMSO or aqueous NH_3, or with aqueous $FeSO_4$ or $Fe(CN)_6{}^{4-}$. The amount of solvent retained depends appreciably on the nature and concentration of the countercation, if any. Removal of solvent [H_2O, DMSO, dimethylformamide (DMF), hexamethylphosphoramide] under vacuum and its readsorption appears significant, judging from changes these processes induce in lattice dimensions. Nonstoichiometric FeOCl prepared in a dry way also takes up pyridine, DMSO, and DMF, the observed lattice repeat with pyridine being nearly that first reported (see above) for stoichiometric FeOCl. The intercalation chemistry of the metal oxyhalides is described in greater detail in Chapter 12.

III. REDUCED HALIDES WITH METAL–METAL INTERACTIONS

It seems plausible that coordination of a basic adduct to metal located in the typical layered structure with the sequence X—M—X and with weakly interacting M atoms probably must occur at every metal center or at least in a very regular way to maintain some uniformity at the metal centers and hence stability. This expectation would appear to be substantially less important for layered phases in which the metal atoms exhibit significant collective interactions and band formation, as is well known for the sulfides. This viewpoint led to the study of possible intercalates of some analogous halides, namely, ZrCl, ZrBr, and $ZrCl_2$.

Zirconium monochloride and its stacking polytype ZrBr exhibit a new type of layering: double metal layers separated by double halogen layers, as shown in Fig. 1. Strong metal–metal interactions are indicated since each

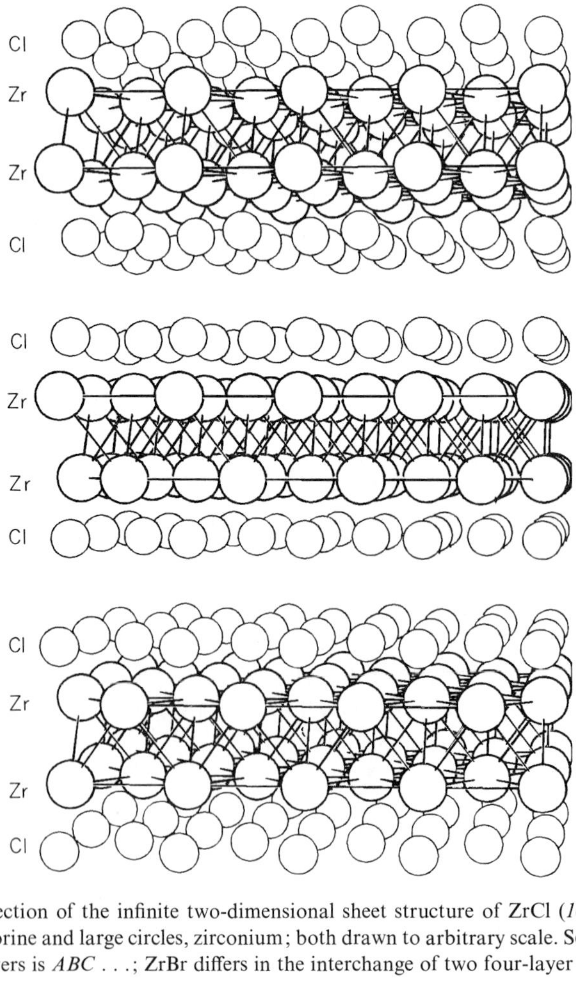

Fig. 1. Section of the infinite two-dimensional sheet structure of ZrCl (*10*). Small circles represent chlorine and large circles, zirconium; both drawn to arbitrary scale. Sequencing of the individual layers is *ABC* . . .; ZrBr differs in the interchange of two four-layer slabs.

metal in ZrCl has three other metal atoms at 3.07 Å and six at 3.43 Å, compared with twelve at an average of 3.185 Å in the metal (*10*, *11*). The phases are metallic conductors, and a strong Zr–Zr interaction is confirmed by both photoelectron spectral data and band calculations (*12*). The second type of halide substrate, $ZrCl_2$, represents a more familiar structure type: 3R MoS_2 (*13*). The material exhibits a narrow valence band in the photoelectron spectrum about 1 V below the Fermi level (E_F) and is a filled band semiconductor, as is MoS_2. However, both the ZrX and $ZrCl_2$ compounds differ in kind from the sulfides in that the metal and chlorine valence bands are separated by about 5.5 eV, indicating much less mixing of the nonmetal in the metal valence band (*14*). In any case, intercalation attempts with bases as well as $AlCl_3$ and I_2 on ZrCl and alkali metals on $ZrCl_2$ (Table I) have all been unsuccessful. Although this result may simply reflect the wrong experimental conditions, some fundamental differences from the familiar dichalcogenides are also evident. The formation of adducts with the latter may involve more or less direct coordination of a good base (or reduction) involving an empty or partially filled band on the metal, thus excluding MoS_2 with its filled band and by implication $ZrCl_2$ as well. Band calculations on ZrX indicate the presence of a substantial empty band just above E_F (*12*), but these monohalides are already relatively electron rich, perhaps making either coordination or reduction unlikely. Expectations for the intercalation of the rare earth monohalide examples of the same structure (*15*, *16*) are somewhat less clear since these compounds are formally d^2 rather than d^3. The intercalation of the layered, metallic diiodides, such as LaI_2 and CeI_2 ($MoSi_2$ type), PrI_2 and GdI_2 (various MoS_2 polytypes), and ThI_2 (4H NbS_2) (*14*), might

TABLE I

Intercalation Attempts with Molecular Adducts and Reduced Layered Halides

Substrate	Reactant	Conditions (°C; days)	Result
ZrCl, ZrBr	Pyridine (l)	115; 49	Nil
ZrBr	NH_3 (l)	23; 8	Nil
	N_2H_4 (l)	23; 4	Nil
	ϕ_3P (l)	300; 10	Nil
ZrCl	Al_2Cl_6 (g)	120; 147	Nil
	I_2 (in CCl_4)	25; 0.2	Nil
$ZrCl_2$	NH_3 (l)	−20; 0.1	
		−80; 0.7	NH_4Cl
	Na or K (NH_3)	−50; 0.7	NaCl, KCl
	Pyridine	20; 1.0	$ZrCl_3 \cdot py$

at first be thought to be more promising. However, these strong reducing agents extract protons from organic substrates readily, so that they would doubtlessly react with many of the potential intercalating species. Nearly all of these phases are known to form hydrides with elemental H_2 (*17*), and all would probably reduce protonated guest species now recognized with some layered chalcogenides.

IV. HYDROGEN IN REDUCED HALIDES

Several halides form compounds with hydrogen that qualify as intercalates within the scope of this volume, i.e., they form products that retain the basic structure of the halide substrate, with the hydrogen reversibly entering a metal–metal-bonded region more or less interstitially. Halides known to so behave are the layered ZrCl and ZrBr, ThI_2, and the novel cluster compounds Nb_6I_{11} and $CsNb_6I_{11}$. These three layered phases form new hydride phases with only narrow compositional variability at 250°–400°C, whereas Nb_6I_8 cluster units in the last two appear to take up H_2 in a one-phase reaction. Their description is subdivided according to this difference.

A. Layered Halides

The zirconium monohalides illustrated in Fig. 1 contain tighly bound double metal layers that alternate with double halogen layers. Finely powdered ZrCl and ZrBr both take up hydrogen at 100°–300°C and 50–100 torr without an induction period. Manometric studies of the adsorption–desorption process have been carried out at 250°–400°C, the temperatures necessary to generate convenient pressures, without decomposition. Typical data are shown in Fig. 2 (*18*). Both ZrCl and ZrBr form discrete phases of composition $ZrXH_{0.5}$ (lower limit) and $ZrXH_{1.0}$ (upper limit), a small nonstoichiometry meaning that a constant pressure plateau for the chloride pertains to compositions of about $ZrClH_{0.66}$ (upper limit) and $ZrClH_{0.95}$ (lower limit) at 400°C.

Although the hemihydrides $ZrXH_{0.5}$ exhibit only slight color differences from the black monohalides, the formation of the monohydrides gives a more striking change to the brass-colored ZrClH and the bronze ZrBrH. Only a slight hysteresis is observed for the adsorption and desorption cycle. An appreciable similarity between the X-ray powder patterns of the product phases and those of the parent monohalides and the ready reversibility of the hydrogenation reactions suggest that the double metal layers are retained. The hemihydride powder patterns show the greater similarities to those of the monohalides but with small splitting of many lines and the appearance of additional faint reflections. A substantially lower symmetry

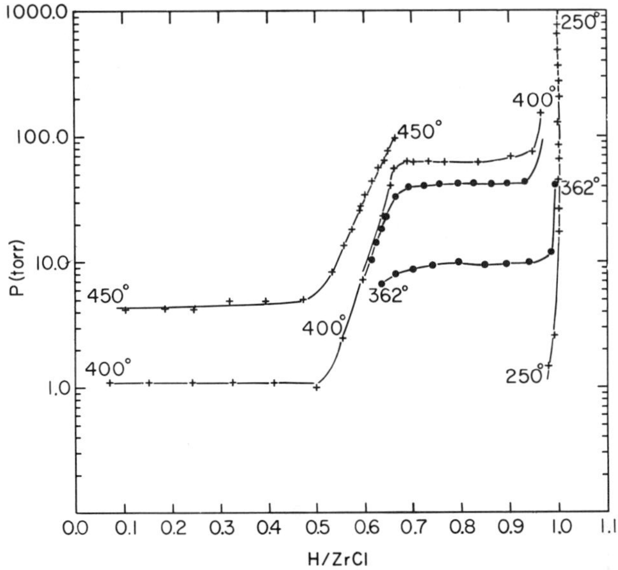

Fig. 2. Pressure–composition isotherms for ZrCl–H_2 (+, adsorption data; ●, desorption data). [Reprinted with permission from Struss and Corbett, *Inorg. Chem.* **16,** 360 (1977).]

would seem to be involved with the interstitial hydrogen ordered since the $ZrXH_{0.5}$ phases are substantially stoichiometric. The heavy atom structure of $ZrBrH_{0.5}$ has now been solved from Guinier X-ray powder data (*19*). The monoclinic cell is generated by displacement of half of a slab (one Zr and one Br layer) relative to the other half to produce zig-zag chains in which pairs of distorted metal tetrahedra share a common, long edge. The neutron diffraction pattern calculated with regular occupation of half of these tetrahedral sites is in good agreement with powder data obtained for $ZrBrD_{0.5}$ (*20*). $ZrClH_{0.5}$ has a different but presently unsolved structure.

Recently, the X-ray powder data for ZrClH and ZrBrH have been found to be consistent with a heavy-atom structure involving only one slab of four layers, which are sequenced X—Zr—Zr—X and ordered *ABAB*, in contrast to the *CBAC ABCA BCAB* sequencing for three slabs in ZrCl (Fig. 1) and a slightly different arrangement in ZrBr (*19*). This configuration results from a continuation of the intra-slab layer displacement that generates $ZrBrH_{0.5}$ from ZrBr. The overall change from ZrBr to $ZrBrH_{1.0}$ involves a cell volume increase of 2.7%, nearly all of which derives from the increased interlayer separations. The expansion presumably arises from reduced metal–metal bonding between the layers. Both the stoichiometry and the NMR properties of ZrXH are consistent with hydrogen located principally in the tetrahedral interstices.

TABLE II

Thermodynamic Properties for Reactions of Hydrogen with Some Reduced Halides, Sulfides, and Metals
(17, 18, 23)[a]

Components	$-\Delta H^0$ (kcal)	$-\Delta S^0$ (eu)
ZrX–$ZrXH_{0.5}$[b]	14.9(2)	15.4(2)
$ZrXH_{0.6}$–$ZrXH$[b]	15.5(2)	20.0(3)
Zr–ZrH (δ)	22.2	24.2
ThI_2–$ThI_2H_{0.7}$	18.4(2)	17.8(2)
$ThI_2H_{0.8}$–$ThI_2H_{1.7}$	17.9(2)	17.7(2)
Th–ThH_2	17.5	15.0
Ta_6S[c]	7.4(3)	15.8(2)
Ta (α)[c]	8.2(2)	13.7
$Nb_{21}S_8$[c]	7.1(6)	15.6(2)
Nb (α)[c]	8.8	15.5

[a] Per gram-atom of hydrogen.
[b] Average of similar values for X = Cl or Br.
[c] One-phase reaction.

The phase ThI_2 also reversibly forms two new phases, $ThI_2H_{0.7}$ and $ThI_2H_{1.77(5)}$, the color changing from gold to violet to yellow-green in the process. Only the first step appears to involve a small change in the X-ray scattering, suggesting an otherwise undefined intercalation of the single metal layers therein (*21*). Other metal-rich halides that take up hydrogen do so with apparently substantial changes in structure and so are not pertinent to the present discussion (*17*, *22*).

Table II lists enthalpy and entropy changes measured for the foregoing systems and compares these with data for hydrogen uptake in one-phase reactions with $Nb_{21}S_8$ and Ta_5S (infinite cluster chains) to 9 and 14 mole %, respectively (*22*), and in the respective metal–hydrogen systems. The magnitude of the enthalpy changes found for the halide hydrides indicates that strong interactions must be involved, comparable to those in the binary hydrides themselves and sometimes even greater (*22*). All exhibit hydrogen dissociation pressures that are higher than in the metal–hydrogen binary systems, evidently reflecting the fact that hydrogen must now compete with the nonmetal for conduction electrons and does so less effectively (*18*). Nonetheless, the four examples shown cover almost all possible ways in which alterations in ΔS^0 and ΔH^0 might achieve a higher H_2 pressure in the ternary compounds.

Manometric studies of ZrXH up to 400°C in sealed tantalum containers demonstrate that these phases are thermodynamically stable with respect to decomposition to Zr, ZrH_2, and $ZrCl_4$. However, recent studies of higher temperatures (>600°C) have shown that the compounds are only metastable at lower temperatures with respect to the decomposition

$$2ZrXH(s) \rightarrow \tfrac{1}{6}Zr_6X_{12}(s) + ZrH_2(s)$$

a reaction that in fact provides a convenient synthesis of the new cluster chloride and bromide (*24*). These cluster halides do not take up hydrogen significantly.

The parent monohalides show small, nearly temperature-independent paramagnetic susceptibilities, 1.0×10^{-4} emu/mole (corrected) at room temperature, a magnitude that is qualitatively appropriate for the Van Vleck and Pauli contributions estimated from band calculations. These values first increase slightly and then decrease by one-third on the formation of $ZrClH_{0.5}$ and ZrClH, respectively, consistent with expectations from the calculated density of states if it is assumed that the hydrogen removes the electrons from the system (*12*). As with the monohalides themselves, these hydrides show no evidence of phase transitions down to 4 K (*24*), although there is some evidence that $ZrClH_{0.5}$ may disorder at high temperature (*19*). In such a strongly reduced halide substrate as ZrX or ThI_2, there is little doubt that the electron–hydrogen interactions are, simply put, hydridic in character ($H^{\delta-}$, $0 < \delta < 1$), as in the Group IV metals themselves, with the hydrogen in effect intercalated in the electron-rich region between the double metal layers.

The hydrogen NMR properties of the four zirconium chloro- and bromohydrides and chloro- and bromodeuterides provide considerable information regarding these materials. Particularly in the chlorides the parameters characterizing the hydrogen NMR properties are determined almost exclusively by hydrogen–hydrogen interactions and by hydrogen motion. Second moments in the rigid lattice region give useful information regarding probable local structure. In view of the number of available interstitial sites, namely, 1.0 tetrahedral and 0.5 octahedral per zirconium atom, the monohydride stoichiometry suggests, and the observed second moments for ZrXH confirm, that the phase can best be understood in terms of occupancy of principally the tetrahedral sites. The relatively narrow stoichiometry range observed with the hemihydride phases suggests that an ordered arrangement must be present, and among several obvious ways of accomplishing this only some variety of parallel zig-zag chains of occupied tetrahedral positions gives respectable agreement with observed NMR properties, in agreement with the diffraction data.

The two $ZrXH_{0.5}$ phases show a two-stage line-narrowing process above about 150 K, suggesting that the higher-temperature process has the higher activation energy. Motional narrowing is complete above 350–400 K. The behavior of ZrBrH is similar to that just described, whereas ZrClH shows simple narrowing above 350 K. The observed data for the hemihydrides as a function of temperature are in excellent agreement with a model for a two-step diffusion mechanism involving protons distributed over two sublattices of tetrahedral (major occupancy) and octahedral (minor) sites (*26*, *27*).

The axially symmetric shift indicated by anisotropy of the line shape was also examined by Murphy and Gerstein (*28*, *29*), using pulse techniques to remove the homonuclear dipolar broadening and to resolve the shielding tensors. This revealed a shielding anisotropy of 102.5 ppm in ZrClH at room temperature versus the previous record of 44.8 ppm for KHF_2 (*28*).

Spin counting for all four zirconium halide hydrides gave hydrogen concentrations in excellent agreement with manometric deductions (Fig. 2). Correlation times for proton motion at room temperature were all just under 1 msec. The study further emphasizes the complex and incompletely defined hydrogen distribution in the hemihydride: Thus, $ZrClH_{0.5}$ exhibits two nonequivalent shielding tensors, whereas $ZrBrH_{0.5}$ and ZrBrH have isotropic values that are 25–60 ppm downfield from that in this chloride. The effect may reflect the different packing sequences and halide polarizations in the two hemihydrides, each metal and nonmetal in ZrBr (and $ZrBrH_{0.5}$?) having an additional second nearest neighbor across the van der Waals gap, which ZrCl (and $ZrClH_{0.5}$?) lacks. (This difference is analogous to that of the second nearest neighbors in $CdCl_2$ versus CdI_2.)

B. Hexametal Clusters

The examples in which the hydrogen atom can be put in the center of a metal octahedron in a cluster halide lack any known analogy among the chalcogenides, e.g., in Mo_6S_8. Present examples are limited to two closely related niobium iodides. [The reported small uptake of hydrogen by Mo_6Cl_{12} (*17*) evidently was caused by impurities (*24*).]

1. Nb_6I_{11}

The first example of a metal cluster compound, or any halide for that matter, containing hydrogen was $Nb_6I_{11}H$ described by Simon (*31*). The central feature of the cluster is an $Nb_6I_8^{3+}$ unit (Fig. 3) in the form of a distorted octahedron of niobium atoms exhibiting nearest-neighbor distances of 2.74–2.92 Å (versus 2.91 Å in the closely packed metal), with each face of the metal octahedron then capped by an iodine atom. [The structure has recently been better described using diffractometer data (*31*).] The iodine atoms thus ideally define a cube that is somewhat larger than the face-centered members of the metal octahedron because of metal–metal bonding, there being 19 bonding electrons delocalized within this cluster. The three

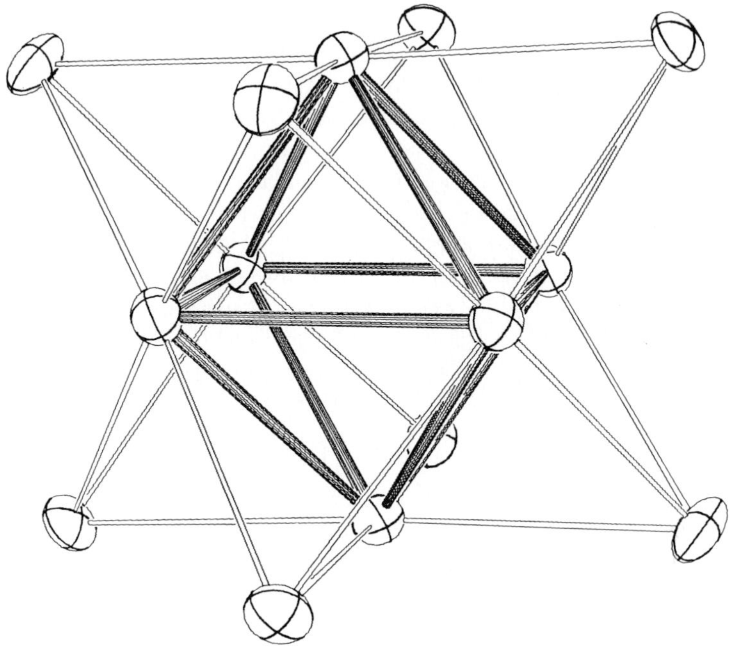

Fig. 3. Structure of the $Nb_6I_8{}^{3+}$ cluster unit in Nb_6I_{11} (*31*). The octahedron of metal atoms is shown in heavy outline.

remaining iodine atoms bridge between clusters by occuping pairs of outward-pointing positions from each apex of the metal octahedron, so that $Nb_6I_8I_{6/2}$ is a useful description.

The Nb_6I_{11} phase starts to take up hydrogen at 300°C, and a manometric study showed that this produces the monohydride composition in 10 min at 420°C and 1 atm of H_2. The reaction is one phase, so that single crystals of the parent iodide can be converted to the hydride. The hydrogen in the cluster was reported not to come off at 10^{-5} torr up to 500°C, at which point the substrate begins to decompose. [The very slow, but reversible H_2 loss at 300°C reported in another study (*17*) was most likely caused by impurities.]

The magnetic data for Nb_6I_{11} and $Nb_6I_{11}H$ are complex, but clearly the hydrogen addition has a substantial effect; in particular, the significant paramagnetism of the parent phase at low temperatures ($\sim 2\ \mu_B$) has changed to diamagnetism for the hydride (which has an even number of electrons). Both have recently been found to exhibit level-crossing transitions (*32*). Neutron diffraction data for Nb_6I_{11}, $Nb_6I_{11}H$, and $Nb_6I_{11}D$ offer good support for the assignment of hydrogen to the middle of the metal cluster (*30*). Surprisingly, the electronic contribution to the conductivity is sufficient to preclude measurements of hydrogen mobility (*31*).

The Nb_6I_{11}–$Nb_6I_{11}D$ system has been further studied by elastic as well as inelastic neutron scattering (*33*). Profile analysis of the diffraction data to $2\theta = 120°$ (with small exclusions to avoid NbO and NbD_x impurities) establishes more firmly that $Nb_6I_{11}D$ is isostructural with Nb_6I_{11}, the deuterium being well centered in the approximate octahedra. Neither the neutron or X-ray data for the deuteride are sufficiently precise to define what happens to the dimensions of the niobium cage when hydrogen is inserted, although the unit-cell volume increases by 0.2%. Inelastic neutron energy loss spectroscopy on the protide reveals an unsplit adsorption of 139 meV (1120 cm^{-1}) (FWHM = 26 meV) plus the 2ω overtone. The deuteride shows the appropriate reduction of the energy to 97 meV. The substantially harmonic adsorption of the former resolved from the known bands of an NbH_x impurity is shown in Fig. 4. With a 2.01–2.05 Å metal–hydrogen separation, the result is typical for binary hydrides of the early transition metals for hydrogen in tetrahedral sites, e.g., ZrH_2, 137; YH_2, 116; β-$NbH_{0.8}$, 120 and 168 meV; there are only a few data in the neighborhood of 55–60 meV for hydrogen with octahedral coordination in metal hydrides. However, too fine a comparison of these results with those in binary hydrides probably is not meaningful since the cluster presents a distinctly different environment for hydrogen, particularly the presence of 19 bonding electrons in the distorted octahedron versus 3.33 per octahedra in the close-packed metal assuming a uniform spatial distribution. Some of the bonding states have nodes at the body center, of course, but are still screening.

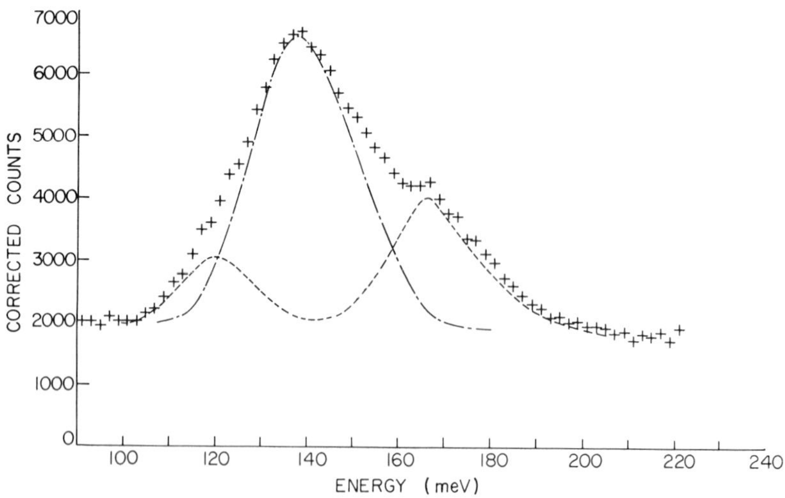

Fig. 4. Inelastic neutron energy loss spectrum of $Nb_6I_{11}H$, with the 139 meV proton feature resolved from the NbH_x impurity (*33*).

2. $CsNb_6I_{11}$

Finally, a 20-electron version of the same cluster with metal–metal distances between 2.77 and 2.94 Å is found in the solid $CsNb_6I_{11}$ (*34*). In this case three bridging iodines still link the clusters but in a more open three-dimensional trigonal array, which leaves channels for the cesium cations. Again, uptake of hydrogen to the extent of one per cluster takes place, accompanied by an increase in the unit-cell volume of 0.3%, so that the presence of an odd number of electrons (as in Nb_6I_{11}) is not critical for hydrogen uptake. Although hydrogen in these electron-rich cluster cavities formed by the more electropositive transition elements very likely acquires (overlaps with) additional delocalized electrons to produce a partial negative charge (*17*), this by no means implies that H^- is formed or that the requisite volume must therefore be available. The fundamental reasons for hydrogen uptake in these clusters but not in other octahedral metal clusters, such as Mo_6Cl_{12}, $Na_4Nb_6Cl_{18}$, Zr_6Cl_{12}, and Ta_6Cl_{15}, are not well-understood. An oxidation state below that of the binary hydrides for the respective Group III–VI metals appears to be a useful empirical requirement based on present data (*34*).

ACKNOWLEDGMENTS

The author is very appreciative of data furnished by R. Barnes, A. Simon, P. Dickens, and B. Fender in advance of publication. This chapter was written principally while the author was on leave from the Inorganic Chemistry Laboratory, Oxford University, and was supported in part by the U. S. Department of Energy, Contract W-7405-Eng-82, Office of Basic Energy Sciences.

REFERENCES

1. F. R. Gamble, *Ann. N.Y. Acad. Sci.* **313,** 86 (1978).
2. A. J. Jacobson, in this volume.
3. F. Kanamaru, M. Shimada, M. Koizumi, M. Takano, and T. Takoda, *J. Solid. State Chem.* **7,** 297 (1973).
4. T. R. Halbert and J. Scanlon, *Mater. Res. Bull.* **14,** 415 (1979).
5. M. S. Whittingham, U.S. Patent 4,049,887 (1978); *Prog. Solid State Chem.* **12,** 41 (1978).
6. M. Armand, L. Coic, P. Palvadeau, and J. Rouxel, *J. Power Sources* **3,** 137 (1978).
7. D. S. Watts and P. G. Dickens, unpublished observations (1979).
8. A. Weiss and E. Sick, *Z. Naturforsch., Teil B* **33,** 1087 (1978).
9. H. Meyer, A. Weiss, and J. O. Besenhard, *Mater. Res. Bull.* **13,** 913 (1978).
10. D. G. Adolphson and J. D. Corbett, *Inorg. Chem.* **15,** 1820 (1976).
11. R. L. Daake and J. D. Corbett, *Inorg. Chem.* **16,** 2029 (1977).
12. J. Marchiando, B. N. Harmon, and S. H. Liu, *Physica B + C (Amsterdam)* **99,** 259 (1980).

13. A. Cisar, J. D. Corbett, and R. A. Daake, *Inorg. Chem.* **18,** 836 (1979).
14. J. D. Corbett, *Adv. Chem. Ser.* No. 186, 329 (1980).
15. A. Simon, H. Mattausch, and N. Holzer, *Angew. Chem., Int. Ed. Engl.* **15,** 624 (1976).
16. K. R. Poeppelmeier and J. D. Corbett, *Inorg. Chem.* **16,** 294 (1977).
17. A. W. Struss and J. D. Corbett, *Inorg. Chem.* **17,** 965 (1978).
18. A. W. Struss and J. D. Corbett, *Inorg. Chem.* **16,** 360 (1977).
19. H. Marek, R. L. Daake, and J. D. Corbett, unpublished observations (1979, 1981).
20. R. Lapp and R. A. Jacobson, unpublished observations (1978).
21. L. J. Guggenberger and R. A. Jacobson, *Inorg. Chem.* **7,** 2257 (1968).
22. H. Imoto and J. D. Corbett, *Inorg. Chem.* **20,** 630 (1981).
23. H. F. Franzen, A. S. Khan, and D. T. Peterson, *J. Solid State Chem.* **17,** 283 (1976).
24. H. Imoto and J. D. Corbett, *Inorg. Chem.* **20,** 145 (1981).
25. F. J. DiSalvo, personal communication (1977).
26. T. Y. Hwang, D. R. Torgeson, and R. G. Barnes, *Phys. Lett. A* **66,** 137 (1979).
27. T. Y. Hwang, R. G. Barnes, R. J. Shoenberger, and D. R. Torgeson, to be published.
28. P. D. Murphy and B. C. Gerstein, *J. Magn. Reson.* **34,** 333 (1979).
29. P. D. Murphy and B. C. Gerstein, *J. Chem. Phys.* **70,** 4552 (1979).
30. A. Simon, *Z. Anorg. Allg. Chem.* **355,** 311 (1967).
31. A. Simon, personal communication (1979).
32. J. J. Finley, H. Nohl, E. E. Vogel, H. Imoto, R. E. Camley, V. Zevin, O. K. Andersen, and A. Simon, *Phys. Rev. Lett.* **46,** 1472 (1981).
33. S. A. Barrett, B. Fender, and A. Simon, unpublished observations (1978).
34. H. Imoto and J. D. Corbett, *Inorg. Chem.* **19,** 1241 (1980).

12
Intercalation Chemistry of Metal Chalcogenohalides

THOMAS R. HALBERT

I. INTRODUCTION

In a review covering the structural features of layered compounds, Hulliger (*1*) tabulated nearly 250 metal chalcogenohalides of formula MXY, the majority of which have at least one crystalline phase that can be considered layered. Representative chalcogenohalides are known for all the Group IIIA and IIIB metals, for the first row transition metals up to and including iron, for antimony and bismuth, and for nearly all the lanthanides and actinides. The metal oxychlorides form the most extensive subclass of layered

INTERCALATION CHEMISTRY

ISBN 0-12-747380-7

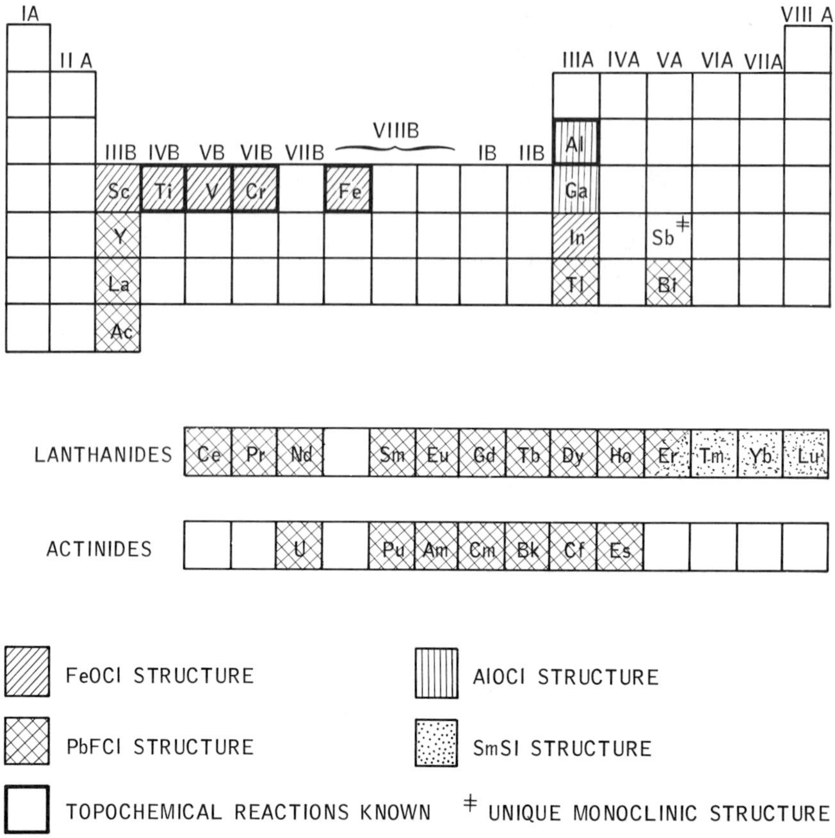

Fig. 1. Metal oxyhalides.

metal chalcogenohalides.* Known examples and their structural types are given in Fig. 1. It is apparent from this figure that, despite their wide range of existence, only a handful of the metal oxychlorides have been reported to undergo topochemical reactions. In fact, among the entire class of metal chalcogenohalides only the oxyhalides with M = Ti, V, Cr, Fe, and Al and the thio- and selenohalides with M = Al are known to be hosts for such reactions.

In the following section we begin with a brief description of the structures, physical properties, and syntheses of the layered oxyhalides, with emphasis on those that act as hosts for topochemical reactions. We then turn to the topochemical reactions themselves, which can be broken down into two

* Although there are nearly as many metal oxyfluorides, most of these compounds are not layered (*1*).

general types: intercalation reactions and topochemical substitution reactions. Known intercalation products with metal oxyhalide hosts include those with various unsolvated cationic guests (e.g., alkali metal ions, organometallic cations), with solvated cationic guests, and with neutral amine guests. Topochemical substitution products, in which the halide ion of the host is replaced by another anion, are known for amide (—NR_2) and alkoxide (—OR) anions. The structures, physical properties, and intercalation reactions of the aluminum thio- and selenohalides are less well defined and are mentioned only briefly in sections dealing with aluminum oxyhalides.

II. STRUCTURE, SYNTHESIS, AND PHYSICAL PROPERTIES OF LAYERED METAL OXYHALIDES

The layered metal oxyhalides can be conveniently grouped into four general structural types, as shown in Fig. 1 for the oxychlorides. The primary factor that controls the occurrence of a particular structure is the size of the M^{3+} cation (*1, 2*), expressed most commonly as the ratio of the M^{3+} ionic radius to the O^{2-} ionic radius ($r_{M^{3+}}/r_{O^{2-}}$). The smallest metal ions ($r_{M^{3+}}/r_{O^{2-}}$ between 0.35 and 0.44) give the AlOCl structure, in which the M^{3+} ion is four-coordinate. Metal ions of intermediate size ($r_{M^{3+}}/r_{O^{2-}}$ between 0.46 and 0.58) give the FeOCl structure, in which the M^{3+} ion is six-coordinate. Slightly larger metal ions ($r_{M^{3+}}/r_{O^{2-}}$ between 0.66 and 0.69) give the SmSI structure, in which the M^{3+} ion is seven-coordinate. Finally, the largest metal ions ($r_{M^{3+}}/r_{O^{2-}}$ between 0.69 and 0.86) give the PbFCl structure, in which the M^{3+} ion is either eight- or nine-coordinate. Although topochemical reactions are known for many of the AlOCl and FeOCl type of compounds, none have been reported for compounds in the PbFCl or SmSI structural class.

A. Metal Oxyhalides with the AlOCl Structure

There are at present four members of this group: AlOX with X = Cl, Br, or I, and GaOCl. None of these compounds has been obtained in the form of single crystals suitable for X-ray diffraction. X-ray powder studies of the microcrystalline products have, however, led to the assignment of an orthorhombic structure (*3, 4*); lattice parameters are given in Table I. In the case of GaOCl, analysis of the powder diffraction line intensities and systematic absences has produced (*4*) the detailed structural picture illustrated in Fig. 2. The space group is $Pca2_1$, with four GaOCl per unit cell. Each metal

TABLE I

Compounds with the Orthorhombic AlOCl Structure

Compound	a (Å)	b (Å)	c (Å)	Ref.
AlOCl	5.50	8.23	4.92	*3*
AlOBr	6.73	10.02	4.92	*3*
AlOI	6.86	11.03	4.92	*3*
GaOCl	5.65	8.33	5.08	*4*

ion is located in a slightly distorted tetrahedron coordinated to three O^{2-} and one Cl^-. The O^{2-} vertices of the tetrahedra are linked together in puckered layers perpendicular to the b axis; each O^{2-} is shared by three tetrahedra. The Cl^- vertices are unshared and lie on alternating sides of the Al—O net along a given row of tetrahedra. The constant c parameter in the AlOX series is evidently a result of constraints placed on the structure by strong Al—O—Al bonding in the c direction.

The first systematic investigation of synthetic routes to this class of compounds was that of Schafer and co-workers (*5, 6*), who prepared AlOCl via the sealed-tube reaction of $AlCl_3$ with a wide variety of metal oxides. The most convenient method (*3, 5*) appears to be the sealed-tube reaction of excess $AlCl_3$ with As_2O_3 at 250°C; upon completion, $AsCl_3$ and unreacted

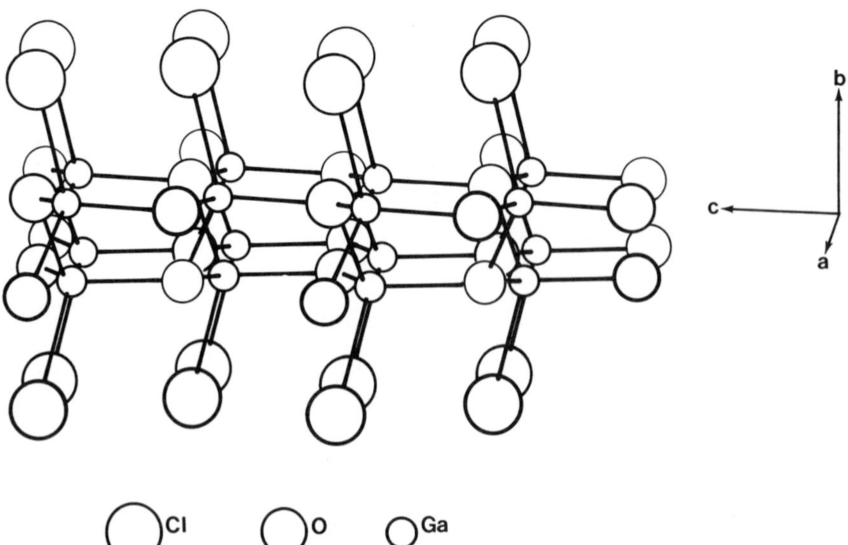

Fig. 2. Structure of GaOCl, showing part of one layer viewed 5° off a parallel to the layer.

$AlCl_3$ can be separated from the colorless, microcrystalline product by sublimation. AlOBr, AlOI, and GaOCl have been prepared by similar methods involving As_2O_3 (*3*, *7*, *8*, *9*). In each case the product is a colorless solid.

The thermal stabilities of these oxyhalides follow the order AlOCl > AlOBr ≈ GaOCl > AlOI. AlOCl and GaOCl decompose in vacuum directly to M_2O_3 and MCl_3 (at 720°C for M = Al and 410°C for M = Ga), while AlOBr and AlOI begin to decompose at 460° and 280°C, respectively, eventually forming the analogous Al_2O_3 and AlX_3 products (*3*, *7*) after passing through several intermediate steps. All three AlOX compounds react with moisture at 25°C, with rates following the order AlOI > AlOBr > AlOCl. The eventual product has been reported to be γ-AlOOH (boehmite) (*1*). GaOCl is also quite hygroscopic (*7*).

Aluminum thiohalides (AlSX, X = Cl, Br, I) have been prepared via the sealed-tube reaction of Al_2S_3 and AlX_3 (*1*, *9*, *10*). Like their AlOX counterparts, they are colorless microcrystalline powders that react readily with water at 25°C to give γ-AlOOH. Thermal decomposition in vacuum gives back Al_2S_3 and AlX_3 directly. X-ray powder diffraction patterns show only a limited number of lines, which have not been indexed. By analogy with the oxyhalides, the structure has been suggested to be orthorhombic with layers similar to those of AlOCl (*3*, *10*).

The syntheses of the corresponding selenohalides AlSeX have also been reported for X = Cl, Br, I (*11*). In these cases, however, large single crystals were obtained via vapor transport in sealed tubes containing Al_2Se_3 and excess AlX_3. X-ray precession photographs, in combination with powder diffraction data, led to the assignment of a monoclinic unit cell, with probable space group $P2_1/m$ (number 11). Unit-cell parameters are given in Table II. Although the complete structure has not been solved, the constancy of the b parameter suggests a layered structure related to that of AlOCl. The selenohalides are colorless hygroscopic solids that decompose thermally to form Al_2Se_3 and AlX_3 at 380°, 340°, and 280°C for X = Cl, Br, and I, respectively.

TABLE II

Unit-Cell Parameters of Monoclinic AlSeX Compounds (*11*)

Compound	a (Å)	b (Å)	c (Å)	β (degrees)
AlSeCl	16.33	12.87	18.17	94.0
AlSeBr	16.97	12.87	18.73	94.0
AlSeI	18.11	12.87	19.97	94.0

B. Metal Oxyhalides with the FeOCl Structure

In addition to the metal oxychlorides indicated in Fig. 1, known metal oxyhalides with the FeOCl structure include Ti, V, Cr, and In oxybromides and In oxyiodide. The thiochlorides and thiobromides of Y and all the lanthanides with $r_{M^{3+}} \leq 1.00$ Å also crystallize in this structure (*1*). Goldsztaub first determined the structure of FeOCl in 1934 (*12*). A more recent refinement (*13*) supports his early work. FeOCl crystallizes in the orthorhombic space group *Pmnm*.* The structure, illustrated in Fig. 3, is characterized by layers of distorted, edge-shared octahedra similar to those in γ-FeOOH. Each iron is coordinated to four O^{2-} and two Cl^-, with the two Cl^- in cis positions along the outside of the layers. The interlayer Cl^-–Cl^- distance of 3.680 Å is approximately twice the van der Waals radius of chlorine, consistent with the true layered nature of the compound. The lattice constants of FeOCl, along with those of other selected metal oxyhalides, are listed in Table III. The layered nature of such compounds is graphically illustrated by the lattice constants in the InOX series; whereas *a* and *c* remain nearly constant due to strong M—O—M bonding within the layers, *b* expands on going from X = Cl to Br to I. The magnitude of the expansion closely matches that expected from the increase in size of the halide ions.

Unfortunately, there seems to be no single general technique applicable to the synthesis of the oxyhalides with the FeOCl structure. Moreover, the physical properties of these compounds vary considerably. A few general observations on each individual case follow.

TiOCl. Synthetic procedures have been reviewed by Schafer *et al.* (*19*) and include the high-temperature sealed-tube reaction of $TiCl_3$ with TiO_2, Fe_2O_3, SiO_2, O_2, or H_2O, as well as the reaction of $TiCl_4$ with Ti/TiO_2 or TiO, and reduction of a $TiCl_4$/TiO_3 mixture by H_2. The best method appears to be the sealed-tube reaction of excess $TiCl_3$ with TiO_2 in a temperature gradient (650°–550°C). The product is a reflective yellow-brown crystalline solid which is stable in moist air but converts slowly to TiO_2 in boiling H_2O. In an N_2 stream, TiOCl decomposes to $TiCl_3$ and Ti_2O_3 above 700°C. Its heat of formation (ΔH_f, 298 K) is estimated to be -182 kcal/mole (*19*). TiOCl exhibits antiferromagnetic behavior; χ_m is field independent and increases with increasing temperature between 90 and 295 K, with an effective magnetic moment of 0.98 μ_B at 298 K (spin-only value for $s = 1/2$ would be 1.73 μ_B).

* *Pmnm* is a nonstandard setting of the space group *Pmmn*, number 50 (*14*). It is used here for the sake of consistency with the literature on FeOCl, which has always taken the *b* axis to be perpendicular to the layer. This convention is maintained here for all FeOCl type of compounds.

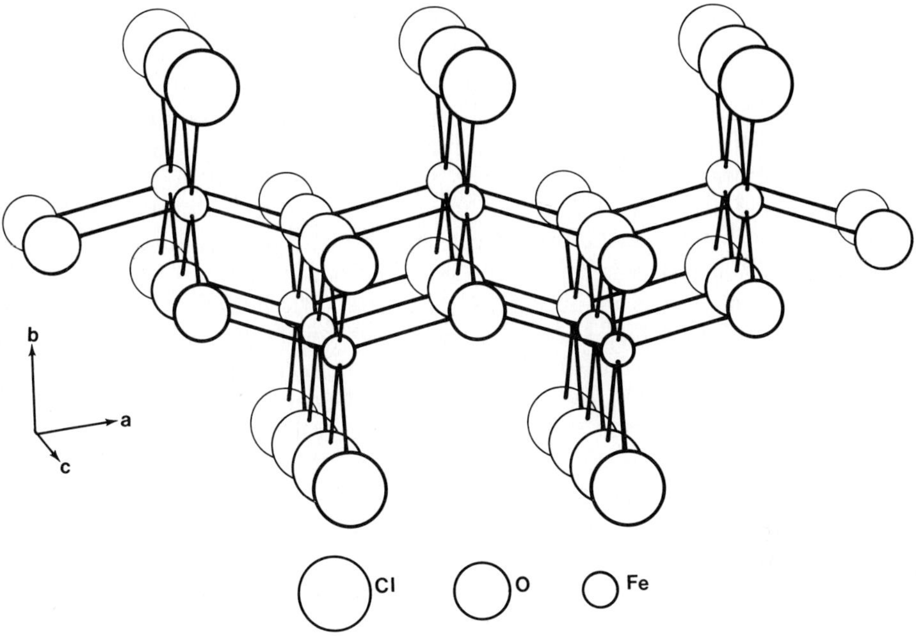

Fig. 3. Structure of FeOCl, showing part of one layer viewed 5° off *a* parallel to the layer.

TiOBr. The only reported synthesis involves the sealed-tube reaction of Ti, TiO_2, and Br_2 in a temperature gradient (650°–550°C) (*20*). The product forms reflective, red-brown flat needles. The molar susceptibility is nearly temperature independent between 90 and 295 K [$\chi_m = 450 \times 10^{-6}$, μ_B (298 K) = 1.04].

VOCl. As with TiOCl, many synthetic approaches have been reported, including disproportionation of $VOCl_2$ (*21*) and reaction of VCl_3 with Sb_2O_3

TABLE III

Unit-Cell Parameters of Selected Compounds with the FeOCl Structure

Compound	*a* (Å)	*b* (Å)	*c* (Å)	Ref.
TiOCl	3.79	8.03	3.38	*19*
VOCl	3.770(5)	7.930(6)	3.290(3)	*16, 17*
CrOCl	3.88	7.72 (7.98?)	3.20	*18*
FeOCl	3.780(5)	7.917(5)	3.302(5)	*13*
InOCl	4.065(5)	8.080(5)	3.523(5)	*15*
InOBr	4.049(5)	8.649(5)	3.611(5)	*15*
InOI	4.05(2)	9.49(2)	3.71(2)	*2*

(*22*), As_2O_3 (*16*), or V_2O_3 (*23*). Schafer, who reviewed (*23*) literature procedures from as early as 1868, favors the sealed-tube reaction of excess VCl_3 with V_2O_3 in a temperature gradient (720°–620°C). VCl_4 and $VOCl_3$ are removed from the red-brown VOCl crystals by vacuum sublimation, and VCl_3 is washed out with dimethylformamide (DMF). VOCl is stable to air and boiling water, and it decomposes at 620°C in vacuo, liberating VCl_3. Its heat of formation (ΔH_f, 298°C) is estimated to be −149 kcal/mole. VOCl is antiferromagnetic, with Neel temperature ~150 K. Single-crystal susceptibility measurements have been carried out from 4.2 to 300 K, and powder measurements up to 555 K (*16*). Above 300 K, the data fit the Curie–Weiss law. At 300 K, $\mu_B = 2.06$. The electrical resistivity of single crystals in the *a*, *c* plane is 5×10^7 Ω cm at 300 K; the optical absorption spectrum shows a weak band at 1.7 eV and a strong band beginning about 2 eV. These results suggest that VOCl is a broad-band semiconductor.

VOBr. Sealed-tube reactions of excess VBr_3 with several oxides have been carried out (*24*); As_2O_3 at 400°C gives the best results. The product is obtained as small violet crystals after washing with DMF. VOBr decomposes above 350°C in vacuum.

CrOCl. The extremely low volatility of $CrCl_3$ makes the synthesis of CrOCl relatively difficult. Schafer and Wartenpfuhl studied (*25*) the reaction of $CrCl_3$ with H_2O, Bi_2O_3, TiO_2, SiO_2, and Cr_2O_3, as well as the reaction of Cr_2O_3 with $SiCl_4$. The sealed-tube reaction of excess $CrCl_3$ with Cr_2O_3 in a temperature gradient (1040°–840°C) is recommended; however, the product must be freed of SiO_2 (from the quartz tube) by brief digestion with 50% HF. CrOCl forms as green crystals that are air and water stable. It decomposes under vacuum at 800°C to give $CrCl_3$ and Cr_2O_3. With excess $CrCl_3$, CrOCl can be vapor-transported at higher temperature; the volatile oxygen-bearing species is suggested to be CrO_2Cl_2. The heat of formation (ΔH_f, 298°C) is estimated to be −139 kcal/mole. Schafer claims that CrOCl is paramagnetic and follows the Curie–Weiss law, but the temperature range covered and actual values of χ_m are not given (*25*).

CrOBr. Danot and Rouxel were unable to prepare CrOBr directly from Cr_2O_3 and $CrBr_3$ (*26*). After examining several alternatives, they found the sealed-tube reaction of $CrBr_3$ with excess As_2O_3 at 380°C to be best. The gray violet crystals of CrOBr are purified by subliming out $AsBr_3$ and excess As_2O_3 at 330°C.

FeOCl. A review of the early literature on the synthesis and properties of FeOCl, as well as extensive original work, can be found in a series of papers by Schafer *et al.* (*27–30*). The many known synthetic procedures can be grouped into three general types: reaction of Fe_2O_3 with HCl, reaction of $FeCl_3$ with H_2O, and reaction of Fe_2O_3 with $FeCl_3$. The third reaction, first reported by Stirneman (*31*), has been commonly employed in recent work, although the temperature and ratio of reactants have been varied extensively (*13, 32–37*). A suitable procedure involves the sealed-tube reaction of a 4:3 ratio of $FeCl_3$–Fe_2O_3 at 370°C. Excess $FeCl_3$ is removed from the product by washing with dry acetone under inert atmosphere (*32*). This procedure avoids exposure of FeOCl to H_2O, which leads to hydrolysis. The extent of hydrolysis as a function of time and H_2O vapor pressure has been studied by monitoring the Cl–Fe ratio of a sample of FeOCl (*29*). In dry air, FeOCl is indefinitely stable, but, in air saturated with H_2O at room temperature, shiny dark violet FeOCl crystals are completely decomposed after 4 days. Although the product of hydrolysis appears amorphous to X rays, it has been suggested to be γ-FeOOH (lepidocrocite) (*35*). This claim seems attractive on the basis of the close structural similarity between FeOCl and γ-FeOOH.

The thermal decomposition of FeOCl in vacuum or under inert atmosphere gives Fe_2Cl_6(g) and Fe_2O_3. Schafer measured the equilibrium vapor pressure of Fe_2Cl_6(g) above FeOCl as a function of temperature; at 250°C, a vapor pressure of ~1.5 torr is produced. There have been several reports of the vapor transport of FeOCl in sealed tubes (*13, 33, 36*); the formation of $FeCl_6$(g) at elevated temperature accounts for the transport of Fe^{3+} and Cl^-. However, it seems likely that traces of H_2O are necessary to transport O^{2-} because of the low volatility of Fe_2O_3 and the absence of volatile FeO_xCl_y species.

Mössbauer spectroscopy has been applied to oriented single-crystal (*37*) and randomly ordered polycrystalline (*38–40*) samples of FeOCl. At 300 K, typical reported values of the isomer shift (with respect to iron metal) and the quadrupole splitting are 0.369 ± 0.008 mm/sec and 0.954 ± 0.010 mm/sec, respectively (*39*). These values are consistent with high-spin Fe^{3+} in an asymmetric environment. At low temperature, FeOCl orders antiferromagnetically. Ordering temperatures of 68 ± 2 K (*40*) and 92 K (*38*) have been reported; the discrepancy between these values is not understood at present. The two variable-temperature magnetic susceptibility studies available are also at variance with one another: Although the reported room temperature molar susceptibility values are reasonably close ($\chi_m = 3.86 \times 10$ and 3.26×10^{-3}), we observed (*41*) that χ_m decreases upon lowering the temperature (down to 50 K), whereas Bizette and Adam (*36*) found a steady increase in χ_m upon lowering the temperature (down to 20 K). However, this increase may be due to paramagnetic impurities. In both studies, an anomaly at ~15 K is probably due to traces of $FeCl_3$. No magnetic anomaly corresponding to the Mössbauer ordering temperature (at either 68 or 92 K) has been found. Obviously, further work is necessary to define clearly the magnetic structure of FeOCl. In particular, FeOCl samples of high purity must be prepared; the possibility of a nonstoichiometric phase $Fe_{1+x}OCl$

for $0 \leq x \leq 0.1$ (*42*) may complicate such work. Finally, the electrical resistivity of FeOCl is $\sim 10^7\ \Omega$ cm, suggesting that it may be a semiconductor (*43*).

InOX (X = Cl, Br, I). Colorless crystals of indium oxychloride and oxybromide have been prepared by the sealed-tube reaction of $In(OH)_3$ and H_2O with the appropriate indium trihalide (*15*) or by treatment of InOI with Cl_2 or Br_2 at elevated temperature (*2*). Indium oxyiodide has been prepared by passing O_2 over moist InI_3 at 285°–290°C (*2*). The resulting crystals are yellow-green, but turn red reversibly upon heating. All three oxyhalides disproportionate slowly at room temperature to In_2O_3 and InX_3; furthermore, all three are subject to slow hydrolysis. The observed order of stability is InOCl > InOBr > InOI. In contrast to transition metal oxyhalides with the FeOCl structure, indium oxyhalides are not known to undergo any topochemical reactions.

C. Metal Oxyhalides with the SmSI Structure

This structure is adopted by a relatively small number of metal chalcogenohalides (*1*): the oxychlorides of the smallest lanthanide elements (Er, Tm, Yb, Lu) and the thioiodides of some of the larger lanthanides (Ce, Pr, Nd, Sm). An individual layer consists of double sheets of M^{3+}, O^{2-} (S^{2-}), and X^-, in the order X–M–O–O–M–X. Each metal ion is found in a capped trigonal antiprism composed of a triangle of X^- on the outside of the layer and a triangle of O^{2-} in one of the O^{2-} sheets on the inside of the layer, with an additional M—O bond formed to an O^{2-} that lies above the O^{2-} triangle (in the second O^{2-} sheet) (*20*). The individual layers are joined together only by van der Waals contacts between the closely packed chloride sheets. Although one might reasonably expect to observe topochemical reactions involving this class of layered materials, there have been no reports of such reactions to date.

D. Metal Oxyhalides with the PbFCl Structure

The oxyhalides (X = Cl, Br, I) of yttrium, bismuth, and all the lanthanides crystallize in the PbFCl structure (with the exception of the oxychlorides with SmSI structure mentioned above). Other compounds that adopt this structure include many lanthanide chalcogenofluorides (chalcogen = O, S, Se, Te), several actinide oxyhalides, and bismuth oxyfluoride (*1*). In the chalcogenohalides, the PbFCl type of layers are composed of a central planar square sheet of chalcogen ions (F^- in PbFCl itself), sandwiched between sheets of halide ions. The M^{3+} ions lie on either side of the central chalcogen sheet in distorted square antiprismatic sites (coordinated to four chalcogenide ions and four halide ions). With the smaller halides

(F^-, Cl^-) an additional bond is formed from each M^{3+} to a halide in the adjacent layer. This fifth halide caps the four-halide face of the square antiprism. Thus, many of the chalcogenohalides of this class have no true van der Waals gap; e.g. in LaOCl, the *intra*layer La–Cl distance is 3.18 Å, and the *inter*layer La–Cl distance is 3.14 Å (*1*).

As with the SmSI-structure compounds, no topochemical reactions have been reported for the PbFCl-structure metal chalcogenohalides. Nevertheless, such reactions seem plausible. Even for the nine-coordinate cases, the interlayer M—X bond might be ruptured by intercalation of a Lewis base molecule that could occupy the ninth coordination site of the M^{3+} ion.

III. INTERCALATION COMPOUNDS WITH CATIONIC GUESTS

The well-developed intercalation chemistry of the transition metal dichalcogenides is dominated by cases in which the guest species is cationic and the corresponding negative charge is delocalized on the host layers. Established cases include unsolvated metal ion guests (alkali and transition metal ions), solvated metal cation guests, organic cation guests, and organometallic cation guests. For metal oxyhalide hosts, parallels are now known for each of these classes of guests, although the range of hosts is limited to the *transition* metal oxyhalides that have readily reducible M^{3+} ions. Such intercalation compounds can be prepared by either direct chemical reduction of the host or electrochemical reduction.

A. Unsolvated Metal Cations

Studies on intercalates of unsolvated metal ions in metal oxyhalides have focused almost exclusively on Li^+, the motivation being, in great part, the search for economical high-energy-density battery cathodes. Direct chemical intercalation of lithium via the *n*-butyllithium (*n*-BuLi) technique (*44*) has been reported for FeOCl (*33*, *45*, *46*), VOCl (*16*, *46*), and CrOCl (*46*), while electrochemical studies have been carried out on FeOCl (*45*, *46*), VOCl (*16*, *46*), CrOCl (*46*), and CrOBr (*46*).

The reaction of *n*-BuLi with FeOCl was first described by Whittingham in a patent on the use of metal oxyhalides as nonaqueous lithium battery cathodes (*45*). Rouxel and co-workers have investigated this reaction more extensively, using 0.1 *M* *n*-BuLi in hexane at 60°C (*33*, *46*). The lithium content of the product increases slowly during several days of exposure of FeOCl to the *n*-BuLi solution, approaching but never quite reaching one lithium per iron. Between zero and about 0.5 lithium per iron, lines in the

X-ray powder pattern of FeOCl gradually decrease in intensity without shifting position; above about 0.5, new lines assigned to "decomposition products" appear. On this basis, the existence of a poorly crystalline intercalation compound Li_xFeOCl ($0 < x \lesssim 0.5$) with lattice parameters unchanged from those of FeOCl is proposed. Possible sites for the lithium ions within the van der Waals gap are also discussed.

The poor quality of the X-ray powder spectra, the lack of any change in cell parameters upon intercalation, and the decomposition observed for $x \gtrsim 0.5$ make it impossible to determine accurately the phase boundaries for Li_xFeOCl from the above results. In fact, these results do not even conclusively demonstrate that the *n*-BuLi reaction is an *intercalation* reaction (i.e., that the basic host structure is retained). Two other lines of evidence do, however, point to this conclusion. First, the reaction of Li_xFeOCl (presumably for $x < 0.5$) with water regenerates FeOCl, although it is not known whether the oxidant is H_2O itself or O_2 (*46*). Second, a nonaqueous battery cell with FeOCl as the cathode active material and an Li metal anode can be repeatedly discharged and recharged with gradual capacity loss, suggesting that a nearly reversible intercalation reaction plays a mojor role at the cathode (*45*). Cyclic voltammetry experiments also point to nearly reversible Li^+ intercalation in FeOCl: In $LiClO_4/CH_3CN$, a well-defined reduction wave at -0.270 V versus Ag/Ag^+ is followed by a corresponding reoxidation wave at -0.080 V (*46*). However, because cyclic voltammetric experiments involve such a small amount of reduction of the host, they are of limited value in predicting whether extensive reduction (e.g., in a battery cathode or *n*-BuLi experiments) will be reversible. Furthermore, as Weiss pointed out (*47*), the observed cyclic voltammetric waves may be due to reduction and reoxidation of traces of soluble Fe^{3+} species in the electrolyte.

The reaction of *n*-BuLi with VOCl, in contrast to FeOCl, gives a crystalline phase $Li_{0.8}VOCl$. This phase exhibits relatively sharp X-ray powder diffraction lines (*16*, *46*) which differ from those of the VOCl host. The powder pattern has been indexed on the basis of an orthorhombic cell, with $a = 4.07$, $b = 7.68$, and $c = 3.31$ Å. This represents a 0.3 Å increase in the a axis and a 0.25 Å decrease in the b axis, compared to the VOCl host structure (the c axis is essentially unchanged). These changes are consistent with a flattening of the corrugated VOCl sheets; in addition, the decrease in the b axis may derive in part from the disappearance of a Jahn–Teller distortion on reduction of V^{3+} (d^2) to V^{2+} (d^3) (*16*, *46*). The intercalation of Li^+ into VOCl is not reversible: Reaction of $Li_{0.8}VOCl$ with H_2O produces LiCl and an unidentified vanadium oxide, while cyclic voltammetry in $LiClO_4/CH_3CN$ shows a reduction wave at -1.46 V versus Ag/Ag^+ reference, with no corresponding reoxidation wave (*46*).

The formation of Li_xCrOCl via the reaction of *n*-BuLi with CrOCl was reported, but neither *x* values nor X-ray powder diffraction results were given (*46*). Li_xCrOCl is said to decompose in H_2O to give LiCl and chromium oxides. Cyclic voltammetry of CrOCl shows an irreversible reduction at −1.340 V versus Ag/Ag^+ in an $LiClO_4/CH_3CN$ electrolyte. CrOBr (under similar conditions) is also irreversibly reduced, although the reduction, as might be expected, is more easily accomplished (−0.940 V versus Ag/Ag^+).

While the preceding intercalation compounds all involve Li^+ as the guest species, there is one recent report of iron cations as guest species in the van der Waals gap of nonstoichiometric iron-rich FeOCl (*42*). A phase formulated as $Fe_{x/2}^{2+}(Fe_{1-x}^{3+}Fe_x^{2+}OCl)$ for $0 < x \lesssim 0.2$ was prepared directly from $FeCl_3$, Fe_2O_3, and Fe metal at high temperature. Unit-cell parameters are identical to those of FeOCl itself. No more extensive characterization of this material has been carried out, although its reactions with various Lewis bases that solvate the interstitial iron have been investigated, as noted in the following section. In particular, it should be pointed out that no evidence for the presumed +2 oxidation state of the interlamellar iron has been presented. Mössbauer spectroscopy, magnetic susceptibility, and electrical transport studies should prove useful for gaining further understanding of the nature of such compounds.

B. Solvated Metal Cations

A variety of intercalation compounds of FeOCl have been prepared in which the guest species are alkali metal ions or iron cations solvated by water, pyridine, DMF, or dimethyl sulfoxide (DMSO) (*42, 47*). These compounds are analogous to the solvated alkali metal intercalates of transition metal dichalcogenides, which have been extensively studied (*48*).

Two general approaches have been applied to the synthesis of solvated alkali metal–FeOCl intercalates: direct chemical reduction and galvanostatic electrochemical reduction. Direct reduction of FeOCl single crystals by aqueous solutions of $M_4Fe(CN)_6$ (M = Li, Na, K, Cs) results in large expansions of the crystals perpendicular to the layers (*42*). Single crystals are necessary because of inhomogeneous reactions with powder samples and because of relatively rapid hydrolysis of microcrystalline FeOCl in H_2O. Some typical values for the increase in the *b*-axis parameter (Δb) are given in Table IV. The layer expansions in the products $M_x^+(H_2O)_y(FeOCl)^{x-}$ depend strongly on the size of the cation, the extent of the reduction ($0.05 \lesssim x \lesssim 0.28$), and the activity of H_2O in the environment (which must control the value of *y*, although no quantitative determinations of *y* have been reported).

TABLE IV

Solvated Alkali Metal Cation Intercalates of FeOCl

$M_x^+(B)_y(FeOCl)^{x-}$

M	X	B	Conditions	Layer expansion (Δb, Å)
Li	~0.14	H_2O	Under mother liquor[a]	7.76
Na	~0.14	H_2O	Under mother liquor[a]	6.25
Cs	~0.14	H_2O	Under mother liquor[a]	16.55
K	~0.14	H_2O	Under mother liquor[a]	9.61
K	~0.14	H_2O	Under distilled H_2O[a]	14.93
K	~0.14	H_2O	Under 0.1 *N* KCl[a]	9.95
K	~0.14	H_2O	Under 1 *N* KCl[a]	6.57
K	~0.14	H_2O	Under 3 *N* KCl[a]	3.56
Li	0.08–0.1	DMSO	Under DMSO[b]	9.09
Na	0.08–0.1	DMSO	Under DMSO[b]	10.61
K	0.08–0.1	DMSO	Under DMSO[b]	11.17
Rb	0.08–0.1	DMSO	Under DMSO[b]	11.61
Cs	0.08–0.1	DMSO	Under DMSO[b]	11.64

[a] Prepared by reduction with aqueous $M_4Fe(CN)_6$.
[b] Prepared by galvanostatic electrochemical reduction.

The solvating water molecules in the preceding compounds have been replaced with DMF or DMSO by soaking the solid in the organic Lewis base (*42*). However, a preferable technique that avoids possible hydrolysis of the host is galvanostatic electrochemical reduction of the FeOCl in an electrolyte composed of an alkali metal perchlorate salt dissolved in DMSO (*47*). Reduction at 0 V versus an SCE reference gives compounds of the type $M_x^+(DMSO)_y(FeOCl)^{x-}$, with $x = 0.08$ and y between 0.40 and 0.56 (y values were estimated from the volume of the van der Waals gap, assuming liquidlike density of DMSO). Layer expansions are shown in Table IV, and a typical voltage versus composition curve for $LiClO_4$ is reproduced in Fig. 4. Reductions to potentials below the first plateau (e.g., $x = 0.10$) result in smaller layer expansions, suggesting that some solvent is forced out to make room for additional cations.

In addition to alkali metal cations, solvated iron cations have been studied as guest species in FeOCl (*42*). Reduction of FeOCl with aqueous $FeSO_4$ gives a hydrated iron cation intercalate, which, like the alkali metals, exhibits a layer expansion that is sensitive to the activity of H_2O in the environment. The solvating H_2O molecules have been replaced by pyridine, DMSO, and hexamethylphosphoramide (HMPA); once again, however, detailed characterization has been hampered by the gradual hydrolysis of the host structure.

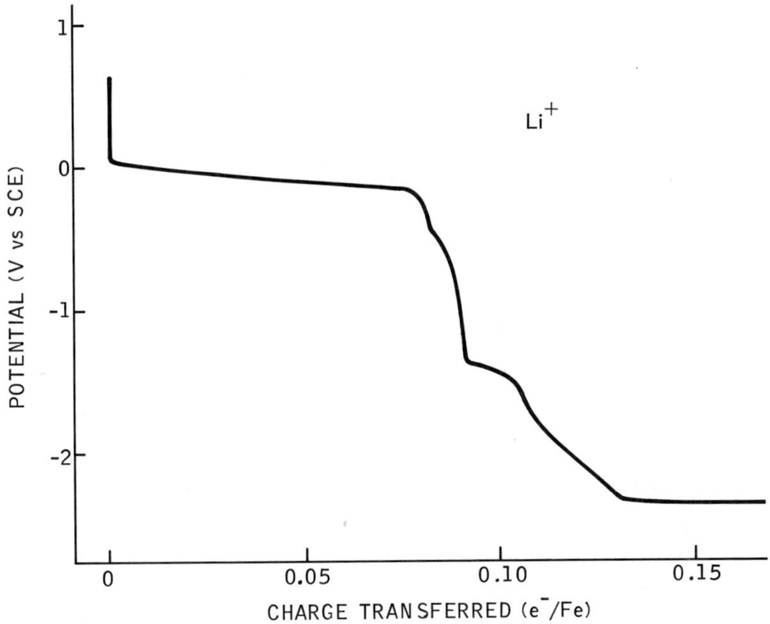

Fig. 4. Intercalation of DMSO-solvated Li^+ into an FeOCl pressed powder by galvanostatic electrochemical reduction (0.2 mA/cm^2, 0.5 *M* $LiClO_4$ in DMSO). [From Meyer *et al.* (*47*); reprinted with permission from *Mater. Res. Bull.* **13,** 1916 (1978). Copyright 1978 by Pergamon Press, Ltd.]

For the organic solvates, this problem can be circumvented by using non-stoichiometric iron-rich FeOCl prepared at high temperature (see Section III, A). Direct intercalation of pyridine, DMSO, and DMF into this material occurs readily and presumably involves solvation of the interlamellar iron ions (*42*). Finally, one other method that has been employed bears mention because of its graphic nature. If a single crystal of FeOCl under DMSO or aqueous NH_3 is touched with an iron wire, intercalation results in a swelling around the edges of the crystal, which moves concentrically toward the center of the crystal as the reaction progresses (*42*). This reaction is reminiscent of the observations of Chianelli on the *n*-BuLi intercalation of TiS_2 single crystals (*49*).

As a general rule, detailed characterizations of the structures, phase limits, and physical properties of the solvated cation intercalates have been notably lacking; for the most part, elemental analyses have not even been reported. In the cases involving H_2O solvation, this will probably remain the case, because of the metastable nature of the compounds. However, such characterization should be possible for the organic-base-solvated intercalates.

Similar solvated metal ion intercalates might also be expected with other metal oxyhalides for which reduction of the host is feasible.

C. Hydrogen Cations

The only report of the intercalation of H^+ in a metal oxyhalide involves the galvanostatic electrochemical intercalation of FeOCl in DMSO containing CF_3CO_2H (*47*). The reduction continues at a constant potential near 0.0 V versus SCE until $0.5e^-$/Fe has been passed. The resulting material exhibits a layer expansion (Δb) of 7.79 Å, suggesting that cointercalation of DMSO has occurred. There is at present no evidence bearing on the location of the H^+ ion in the structure of the product.

D. Organic Cations

The reduction of FeOCl by aqueous $FeSO_4$ in the presence of *n*-alkylammonium chlorides results in the formation of a series of intercalation compounds in which the layer expansion increases with increasing length of the alkyl chain. These compounds have been formulated as $(RNH_2)_y$-$(RNH_3{}^+)_x(FeOCl)^{x-}$. Neutral amine is present at pH $\gtrsim$ 5.8. In addition, there is at present no basis on which to rule out the cointercalation of iron cations. At low pH (<4) layer expansions vary from 3.56 Å for ethylammonium to 4.38 Å for dodecylammonium, suggesting that the alkyl chains lie flat between the layers. With *n*-hexylammonium and all larger cations up to octadecylammonium, an additional phase is observed. Above pH 4, this phase exists exclusively and exhibits a much larger layer expansion (e.g., 33.36 Å for octadecylammonium). It has been interpreted in terms of a bilayer composed of a mixture of alkylamine and alkylammonium molecules. In the bilayer, the packing density is estimated at 29 $Å^2$ per alkyl chain (*42*), close to that found in similar intercalation compounds of layered silicates (*50*).

The intercalation of tetraalkylammonium ions into FeOCl can be accomplished by galvanostatic electrochemical reduction in an $R_4N^+X^-$/DMSO electrolyte. For R = methyl and *n*-butyl, the phases $(R_4M^+)_x(DMSO)_y$-$(FeOCl)^{x-}$ have been prepared with $x \approx 0.05$. The observed layer expansions are surprisingly similar (12.45 Å for Me_4N^+ compared to 12.94 Å for Bu_4N^+).

E. Organometallic Cations

The intercalation of organometallic species into layered compounds (see Chapters 7 and 8) was first demonstrated for transition metal dichalcogenides

TABLE V

Metallocene Intercalates of Metal Oxyhalides (*32, 41*)
(Metallocene$^+$)$_x$(MOCl)$^{x-}$

Metallocene[a]	X	MOCl	b (Å)	Layer expansion (Δb, Å)
$CoCp_2$	~0.16	FeOCl	2 × 12.849	4.94
$FeCp_2$	~0.16	FeOCl	2 × 13.032	5.13
$Fe(EtMe_4Cp)_2$	~0.16	FeOCl	2 × 15.46	7.55
$CoCp_2$	~0.16	VOCl	2 × 12.77	4.86
$CoCp_2$	~0.16	TiOCl	2 × 13.16	5.12

[a] Cp = cyclopentadienyl; $EtMe_4Cp$ = ethyltetramethylcyclopentadienyl.

by Dines in 1975 (*51*). More recently, similar intercalation reactions involving bis(cyclopentadienyl) transition metal guests (metallocenes) in the transition metal oxyhalide hosts FeOCl, VOCl, and TiOCl have been reported (*32, 41*). These intercalation compounds are prepared by the direct reaction of the electron-rich metallocenes with the host compounds in inert solvents (e.g., toluene) at temperatures between 80° and 100°C.* Like the dichalcogenides, VOCl and TiOCl react only with metallocenes that have stable cations and first ionization potentials of less than 6.2 eV (e.g., cobaltocene with a first ionization potential of 5.56 eV). The more easily reduced host, FeOCl, intercalates metallocenes with higher first ionization potentials (e.g., ferrocene, 6.68 eV), illustrating that a judicious choice of layered host lattices may permit a much broader range of organometallic intercalates to be prepared.

All of the metallocene/metal oxyhalide intercalation compounds are well crystalline materials whose X-ray powder diffraction patterns can be indexed on the basis of an orthorhombic unit cell with the a and c axes nearly unchanged from those of the hosts and the b axis expanded and doubled, as indicated in Table V. Systematic absences in the powder patterns indicate that the unit cell is body-centered; the space group *Immm* is suggested by a structural model (illustrated in Fig. 5 for the cobaltocene intercalate of FeOCl) in which alternate layers of the host MOCl structure shift by one-half unit cell in the (101) direction (along the a–c diagonal). The orientation of the metallocene molecules within the van der Waals gap could not be determined directly from powder X-ray diffraction data. However, the layer

* Schafer-Stahl and co-workers (*71–75*) have now reported milder conditions for intercalation and have studied side reactions occurring at higher temperature; see Addendum.

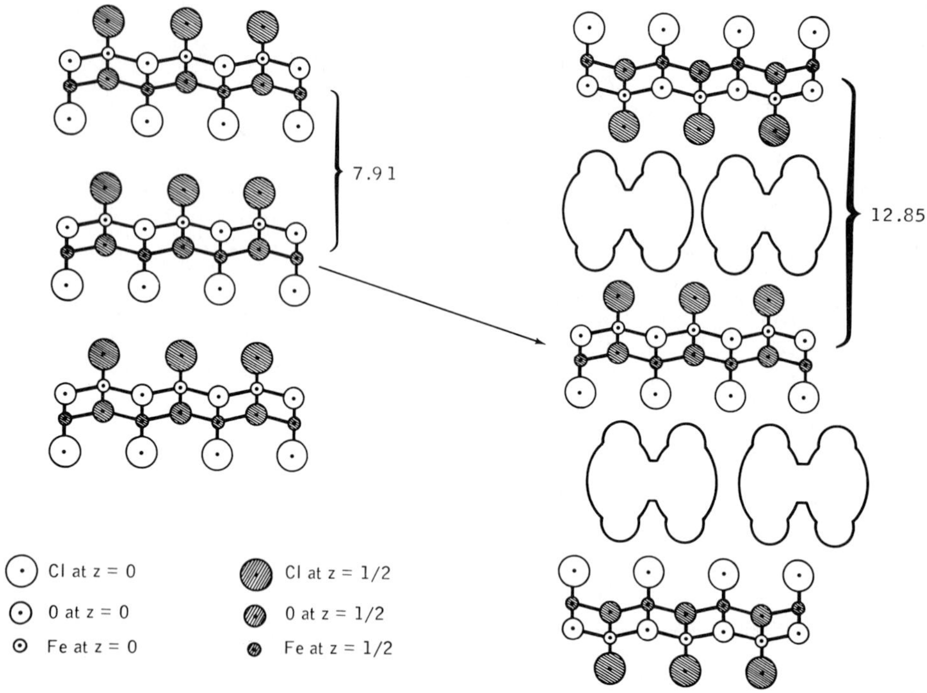

Fig. 5. Intercalation of cobaltocene into FeOCl, a schematic representation.

expansion observed for the bis(ethyltetramethylcyclopentadienyl) iron intercalate strongly suggests an orientation with the cyclopentadienyl (Cp) rings perpendicular to the layers. The stoichiometry of the metallocene intercalates (0.16 metallocene per MOCl) is near that expected for a close-packed monolayer of metallocene with each molecule occupying 41 $Å^2$.

The electronic nature of the ferrocene and cobaltocene intercalates of FeOCl has been probed by Mössbauer spectroscopy and magnetic susceptibility. The Mössbauer spectrum of the ferrocene intercalate of FeOCl at 300 K exhibits a single quadrupole-split doublet, shifted slightly from that of FeOCl itself. There is no evidence of a signal due to neutral $FeCp_2$; the expected signal for Fe^{3+} in $(FeCp_2)^+$ is masked by the doublet due to the host lattice iron. The magnetic susceptibilities of the cobaltocene and ferrocene intercalates of FeOCl have been determined between 4 and 300 K. The observed values can be interpreted on the basis of contributions from three components: a nearly temperature-independent term similar to that seen for FeOCl itself; a temperature-dependent term, obeying the Curie–Weiss law, believed to be due to localized moments generated in the host by the transfer of electrons from the metallocene; and a temperature-dependent term, obeying the Curie–Weiss law, due to the localized moment on the metallocenium cation (this term amounts to 2.4 μ_B for $FeCp_2^+$ and is zero, as expected, for diamagnetic $CoCp_2^+$) (*41*). On the basis of the above observations, an electronic picture of these intercalates emerges in which the electron-rich metallocenes are oxidized upon intercalation, and the electrons donated to the layers are essentially localized on metal sites in the host lattice.*

IV. INTERCALATION COMPOUNDS WITH "NEUTRAL" AMINE GUESTS

The reaction of a neutral amine with a metal oxyhalide typically proceeds in one of two ways. In one general case, discussed in this section, intercalation of the amine occurs with retention of the basic host structure. Known examples include the reactions of FeOCl with many *n*-alkylamines or aromatic amines (or, under certain conditions, with ammonia), the reaction of VOCl with pyridine, and the reaction of AlSeX (X = Cl, Br, I) with trimethylamine. In the other general case, discussed in the following section, topochemical substitution of the halide anions by amide anions occurs (with liberation of HX). Examples include the reactions of FeOCl with methylamine or ammonia (under certain conditions) and the reactions of most of the aluminum chalcogenohalides with ammonia or primary and secondary *n*-alkylamines.

The intercalation of FeOCl by ammonia and *n*-alkylamines was first recognized by Hagenmuller and co-workers in 1966 (*52, 53*). Table VI lists the observed compounds and lattice parameters (where reported). The reaction with NH_3 is carried out by brief immersion (~1 hr) of FeOCl in liquid NH_3; longer exposure leads to topochemical substitution (see Section V,A). The approximate doubling of the *a* and *c* lattice parameters over those of FeOCl, along with the layer expansion of 2.45 Å and the stoichiometry of $\frac{1}{2}NH_3$ per Fe, led Hagenmuller *et al.* to the conclusion that the NH_3 is coordinated to iron, occupying every second available site in a fully ordered fashion. This model suffers from the problem that the Fe^{3+} sites in the FeOCl layers are not easily accessible, and coordination of NH_3 would produce an unlikely seven-coordinate Fe^{3+}. Clarification of the bonding in this compound clearly awaits more detailed study; nevertheless, an interesting comparison can be made between $FeOCl(NH_3)_{1/2}$ and the

* More recent Mössbauer and conductivity studies, which suggest delocalization of the transferred electron density in the host lattice of FeOCl ($FeCp_2$) prepared at lower temperatures, are discussed in the Addendum.

TABLE VI

Ammonia and *n*-Alkylamine Intercalates of FeOCl[a]
(*52, 53*)

Intercalate	*a* (Å)	*b* (Å)	*c* (Å)
$FeOCl(NH_3)_{1/2}$	7.47	10.37	6.66
$FeOCl(Me_2NH)_{1/3}$	11.26	11.51	9.95
$FeOCl(Me_3N)_{1/3}$[b]			
$FeOCl(EtNH_2)_{1/3}$[b]			
$FeOCl(Et_2NH)_{1/4}$[b]			
$FeOCl(Et_3N)_{1/4}$[b]			
$FeOCl(PrNH_2)_{1/4}$[c]			

[a] Lattice parameters determined from X-ray powder diffraction patterns, indexed on the basis of orthorhombic unit cells.

[b] The spacings and approximate intensities are given in Hagenmuller *et al.* (*63*); unit-cell parameters are not given.

[c] Kikkawa *et al.* (*54*).

ammonia intercalates of transition metal dichalcogenides. In $TaS_2(NH_3)$, the observed layer expansion is 3.05 Å (*55*). The smaller expansion in the FeOCl intercalate may result from a nesting of the ammonia molecules in the corrugated FeOCl layers or from a slight flattening of the corrugation accompanying intercalation. In $TaS_2(NH_3)$, the N atom is known to be centered between the TaS_2 layers (*56*); furthermore, it has been proposed that a fraction of the ammonia (~0.1) has been converted to NH_4^+, with the remaining hydrogen bonded to the NH_4^+ and the TaS_2 layers partially reduced to balance charge (*57*). A related picture must be considered for the ammonia/FeOCl reaction. Partial ammonolysis of the FeOCl layers might give rise to a fraction of $NH_4^+Cl^-$ between the layers, with the remainder of the NH_3 hydrogen-bonded to the $NH_4^+Cl^-$, leading to the formulation of the product as $[FeOCl_{1-x}(NH_2)_x](NH_4Cl)_x(NH_3)_{0.5-2x}$. In this model, the N atom of NH_3/NH_4^+ would most likely be found near the center of the van der Waals gap, and additional NH_3 would promote further ammonolysis to $FeONH_2$, as observed (see Section V,A). Such speculation should serve primarily to promote further study.

The intercalation of *n*-alkylamines into FeOCl is much slower than the intercalation of NH_3 (*52, 54*). In liquid dimethylamine, complete reaction takes 5 days at room temperature; larger *n*-alkylamines are even slower. The products listed in Table VI are all black, crystalline solids for which stoichiometries have been obtained from elemental analyses. In general, the *n*-alkylamine intercalates of FeOCl do not readily react with additional

TABLE VII

Aromatic Amine Intercalates of Metal Oxyhalides

Compound	a (Å)	b (Å)	c (Å)	Layer expansion (Δb, Å)	Ref.
$FeOCl(pyridine)_{1/4}$		13.27		5.35	*54*
$FeOCl(pyridine)_{1/3}$	3.78(1)	13.45(2)	3.30(1)	5.53	*59*
$FeOCl(pyridine)_{1/3}$	8.12	13.60	3.28	5.68	*33*
$FeOCl(4\text{-aminopyridine})_{1/4}$		13.57		5.65	*34, 54*
$FeOCl(2,6\text{-dimethylpyridine})_{1/4}$		14.98		7.06	*34, 54*
$FeOCl(2,4,6\text{-trimethylpyridine})_{1/6}$		11.79		3.87	*34, 54*
$VOCl(pyridine)_{1/3}$	8.21	13.35	3.29	5.45	*16*

primary or secondary amine to form topochemical substitution products; the one exception is methylamine, which, surprisingly, forms such a substitution product directly, without first forming an identifiable intercalation compounds.

The intercalation of aromatic amines (e.g., pyridine) into FeOCl or VOCl has been accomplished by heating the host to temperatures between 40° and 100°C in either the liquid amine, a solution of the amine in inert solvent, or the amine vapor. Recently, such reactions have received a great deal of attention (*16*, *33*, *34*, *42*, *54*, *58–62*). The known compounds are tabulated along with lattice parameters and layer expansions (Δb) in Table VII. The stoichiometries were derived from a combination of weight gain and elemental analysis. With pyridine, the stoichiometry is variable; Koizumi reported that below 60°C, FeOCl reacts with pyridine vapor to give $FeOCl(pyridine)_{1/4}$ whereas, above 60°C, it gives $FeOCl(pyridine)_{1/3}$. The heat of reaction of pyridine with FeOCl to give $FeOCl(pyridine)_{1/3}$ is 4.09 kcal/mole (*61*). Thermogravimetric studies show that pyridine can be removed from $FeOCl(pyridine)_{1/3}$ between 150° and 250°C (*59*).

The details of structure and bonding in the aromatic amine intercalates of FeOCl and VOCl are at present a matter of some debate. On the basis of one-dimensional electron density profiles from limited X-ray powder diffraction line intensity data, Koizumi *et al.* proposed a uniform model in which the neutral amine molecules lie in the van der Waals gap with their aromatic rings perpendicular to the host layers and their nitrogen lone pairs bonding to the layers (*34*, *59*). However, as Whittingham has pointed out, this model is not at all consistent with the observed layer expansions, in which b for 2,4,6-trimethylpyridine is less than that for pyridine itself (*62*). Furthermore, whereas Koizumi *et al.* reported that the a and c unit-cell parameters of $FeOCl(pyridine)_{1/3}$ are essentially unchanged from those of FeOCl itself,

Rouxel and co-workers presented fully indexed X-ray powder diffraction data which indicated that the *a* parameter is expanded and doubled in comparison to that of FeOCl [4.06 Å × 2 for $FeOCl(pyridine)_{1/3}$ versus 3.79 Å for FeOCl]. As in the case of the dichalcogenides intercalates, any successful explanation of the existing data on metal oxyhalide intercalates will have to take into account the roles of protonated amine and water (*57*, *63*).

The physical properties of the aromatic amine intercalates of FeOCl have been studied by Koizumu and co-workers. The most striking observation is the large decrease in electrical resistivity. Although FeOCl is a poor semiconductor with a resistivity of $10^7\ \Omega\cdot$cm, single crystals of $FeOCl(pyridine)_{1/3}$ exhibit a resistivity of only 10 $\Omega\cdot$cm along the *c* axis. Resistivities of the other intercalates vary from 10^3 to $10^2\ \Omega\cdot$cm, and in all cases the thermoelectric power is negative, suggesting that the charge carriers are electrons (*34*). Mössbauer spectra of aromatic amine FeOCl intercalates have been reported in several papers, but the results are conflicting. The most recent study shows a small increase in the value of the isomer shift and a decrease in the quadrupole splitting, compared to FeOCl (*54*, *69*). Earlier reports showed essentially no differences between FeOCl and the intercalates (*58*–*60*). Low-temperature Mössbauer spectra show that antiferromagnetic ordering occurs in $FeOCl(pyridine)_{1/3}$ below about 65 K, much like FeOCl itself. Infrared spectroscopy, electron paramagnetic resonance spectroscopy, and ultraviolet (electronic) spectroscopy have also been applied (*34*, *54*, *59*), and the results have been presented in support of a model in which electron density from the nitrogen lone pair of the amine is transferred to a conduction band within the host layers of FeOCl. Although the observed conductivity changes demand explanation, the concept of removing electrons from neutral pyridine is not intuitively pleasing.

While all of the above studies deal with FeOCl or VOCl as host lattices, isolated reports of intercalation of trimethylamine in AlSeX and AlSX (X = Cl, Br, I) have appeared (*3*, *11*). With AlSeCl, which is typical, after 80 days at room temperature under 1 atm Me_3N, weight gain indicates the formation of $AlSeCl(Me_3N)$. No details of characterization have been given.

V. TOPOCHEMICAL SUBSTITUTION REACTIONS

In Section II it was pointed out that the structure of a layered metal oxyhalide generally involves a strongly interconnected central sheet of M^{3+} and O^{2-} ions with more weakly bound outer layers of halide ions. This is an ideal structure on which to observe a topochemical substitution reaction, for it should be possible to replace the halide ions by other anions having comparable steric requirements, without disrupting the basic layered structure enforced by the central metal–oxygen sheet. A number of such reactions have indeed been noted, as detailed in the following sections.

A. Reactions with NH_3 and *n*-Alkylamines

Topochemical substitution reactions involving ammonia and primary or secondary *n*-alkylamines are known for FeOCl, VOCl, and all of the reported layered aluminum chalcogenohalides. The products, listed in Table VIII, are in all cases X-ray amorphous. It is therefore not possible to prove unambiguously that the reactions proceed in a topochemical fashion. With this caveat in mind, the individual reactions will be discussed on the chemically reasonable basis that the metal–chalcogen frameworks remain intact.

Exposure of FeOCl to gaseous ammonia at 25°C results in the uptake of 2 equivalents of ammonia over a period of several months (*35*, *52*, *53*). Powder X-ray diffraction shows the presence of NH_4Cl, which can be washed out of the product with liquid NH_3 to leave a maroon, amorphous solid that analyzes as $FeO(NH_2)$. The compound is very hygroscopic and decomposes at 70°C under 1 atm NH_3 to give α-Fe_2O_3, NH_3, and iron nitride. Infrared spectral features at 3300, 3200, and 1600 cm^{-1} have been assigned to $—NH_2$-related vibrations.

The reaction of methylamine with FeOCl proceeds even more slowly than that of ammonia; after 9 months of exposure to $MeNH_2$ with repeated grinding of the solid, X-ray powder diffraction indicates that FeOCl is gone and $CH_3NH_3{}^+Cl$ has appeared. The black, hygroscopic powder is presumed to contain $FeO(HNCH_3)$.

The reaction of VOCl with NH_3 parallels that of FeOCl; 2 equivalents of NH_3 are absorbed at room temperature over a period of 1 month. Washing with liquid ammonia leaves a black amorphous product that contains no

TABLE VIII

Topochemical Substitution Products from Reactions of Ammonia and *n*-Alkylamines with Metal Chalcogenohalides

	FeOCl	VOCl	AlOX	AlSX	AlSeX
NH_3	$FeO(NH_2)$	$VO(NH_2)$	$AlO(NH_2)$	$AlS(NH_2)$	$AlSe(NH_2)$
CH_3NH_2	$FeO(HNCH_3)$		$AlO(HNCH_3)$	$AlS(HNCH_3)$	$AlSe(HNCH_3)$
$(CH_3)_2NH$			$Al_2O_2Cl[N(CH_3)_2]$	$AlS[N(CH_3)_2]$	$AlSe[N(CH_3)_2]$
$(C_2H_5)NH_2$			$Al_2O_2Cl[HN(C_2H_2)]$		
$(C_2H_5)_2NH$			$Al_2O_2Cl[N(C_2H_5)_2]$		

Cl^- and analyzes as $VO(NH_2)$ (*64*). By deuterium labeling studies (using ND_3), infrared absorptions at 3200 and 1630 cm^{-1} have been assigned to the —NH_2 group. A band at 420 cm^{-1} assigned to V—Cl stretching disappears in the product. Combined thermogravimetric and differential thermal analysis studies show that $VO(NH_2)$ decomposes at 160°C to V_2O_3 and $V(NH_2)_3$; $V(NH_2)_3$ decomposes further at 260°C to give VN and $2NH_3$. Magnetic susceptibility studies show $VO(NH_2)$ to be an antiferromagnet with T_N near room temperature and $\mu_{eff} = 2.79\ \mu_B$ at room temperature.

In comparison to the above cases, the reactions of aluminum oxyhalides with NH_3 and *n*-alkylamines are very rapid (*3, 8, 9, 65*). For example, AlOCl absorbs 2 equivalents of NH_3 at room temperature to give $AlO(NH_2)$ and NH_4Cl in less than 48 hr. The relatively high rates of these reactions have been ascribed to more rapid attack of the amine on Al^{3+}, facilitated by the open nature of the AlOX lattice; the volumes are 47.3 Å^3/FeOCl and 55.4 Å^3/AlOCl (*52*). The product of NH_3 reaction with AlOX (X = Cl, Br, I) can be freed of NH_4X by washing with liquid ammonia, leaving $AlO(NH_2)$ as a white, hygroscopic powder. This powder exhibits characteristic NH_2 infrared bands at 3150, 1605, and 1430 cm^{-1} (*66*). It decomposes in vacuum at 120°C to give Al_2O_3, AlN, and NH_3. A kinetic study of the AlOCl/NH_3 reaction has been carried out (*65*), leading to the assertion that the rate shows first-order dependence on AlOCl and overall second-order behavior. The idea of first-order dependence on AlOCl concentration is inconsistent with the fact that AlOCl is a solid with activity defined as unity.*

The reactions of aluminum oxyhalides with ammonia proceed until all the halide has been displaced. The same is true with methylamine. However, as indicated in Table VIII, dimethylamine, ethylamine, and diethylamine replace only half the halide, these stoichiometries being supported by elemental analyses and determinations of the amount of amine taken up. As with $AlO(NH_2)$, the products are white hygroscopic powders. The rates of reaction for the various oxyhalides AlOX generally follow the order X = I > Br > Cl. No reaction is observed between tertiary amines and aluminum oxyhalides.

The thio- and selenohalides of aluminum form analogous topochemical substitution products with ammonia, methylamine, and dimethylamine (see Table VIII) (*3, 9, 11*), although the materials prepared at room temperature initially contain 2 equivalents excess amine, which can be removed by heating

* If AlOCl in NH_3 is treated as a solution (i.e., NH_3 has free access to all Al sites via unhindered diffusion in the van der Waals gaps), then a kinetic expression such as the one given may be valid. This point, however, has not been demonstrated; the observed dependence of the rate of NH_3 absorption on time may, for example, arise simply from diffusion in crystallites of different size during the reaction.

under vacuum. The thio compounds are yellow hygroscopic powders; the seleno compounds are brown hygroscopic powders.

B. Reactions with Alcohols

Whereas the topochemical substitution of halide by amide or alkylamide groups is known for a large number of metal oxyhalides, topochemical substitution by alkoxide ions has been reported only in FeOCl (*34*, *67–69*). Such substitutions were first reported in 1976 by Koizumi and co-workers, who found that reaction of the aromatic amine intercalates of FeOCl with methanol at 100°C in sealed tubes produces the new layered compound $FeO(OCH_3)$, which has a powder X-ray diffraction pattern distinct from that of either FeOCl itself or the starting amine intercalates (*34*, *67*). More recently, a simpler synthesis has been reported in which FeOCl is treated directly with sodium methoxide in methanol; this method gives a brown crystalline product with the same X-ray powder diffraction pattern as the earlier material (*68*). Elemental analysis, although disturbingly low in carbon and hydrogen, shows no Cl and has been interpreted in terms of the formula $FeO(OCH_3)$. Under the same conditions, VOCl fails to react with $NaOCH_3$.

The infrared spectrum of $FeO(OCH_3)$ shows a strong band at 1050 cm^{-1} assigned to a C—O stretching vibration. The solid is reasonably stable in air but decomposes at 300°C to give γ-Fe_2O_3. The Mössbauer spectrum of $FeO(OCH_3)$ exhibits essentially the same isomer shift as FeOCl, but the quadrupole splitting is reduced substantially (from 0.92 to 0.60 mm/sec) (*67*). The powder X-ray diffraction pattern has been indexed on the basis of an orthorhombic unit cell with $a = 3.83$, $b = 9.97$, and $c = 3.99$ Å. Compared to the values for FeOCl, this represents an expansion of 2.06 Å in b and 0.69 Å in c (a is essentially unchanged). The increase in the b axis is consistent with the replacement of Cl^- by $^-OCH_3$ in the FeOCl structure. However, the increase in the c axis is not readily understood on this basis, and a more detailed structural analysis would be warranted.

In an extension of the above work, the reaction of ethylene glycol with $FeOCl(4\text{-aminopyridine})_{1/3}$ has recently been studied (*69*). Powder X-ray diffraction of the initial product, still wet with ethylene glycol, shows broad lines indicating a layer spacing of 14.5 Å. However, washing with acetone and heating at 110°C results in improved crystallinity and a reduction in layer spacing to 10.89 Å. Elcmental analysis fits well with the formula $FeO(O_2C_2H_4)_{1/2}$. An interesting question arises as to whether the two alkoxide groups of an individual glycolate dianion bind to the same Fe—O layer or to adjacent layers (thus linking the layers together). Although no direct evidence is available, the layer expansion is more consistent with the former view, in which both ends of the glycolate bind to the same layer (*69*).

The Mössbauer spectrum, infrared spectrum, and thermal behavior of $FeO(O_2C_2H_4)_{1/2}$ are very similar to those of $FeO(OCH_3)$.

To date, there have been no reports of the use of other alcohols in similar reactions, although it seems reasonable to expect this general type of reaction to be much wider in scope.

C. Reactions with Alkali Metal/Liquid Ammonia

The treatment of FeOCl with lithium dissolved in liquid ammonia results in the formation of $H_2(g)$, LiCl, and a magnetically separable powder of formula $LiFeO(NH_2)\cdot NH_3$. Ammonia can be removed from this solid by heating to 80°C under an argon stream, leaving a black ferrimagnetic solid that analyzes as $LiFeO(NH_2)$ (*70*). Whether this reaction is truly a topochemical substitution is an open question, because no structural information is available; it is included here only because of the obvious relation to the independent reactions of FeOCl with lithium or with NH_3.

VI. ADDENDUM

After the initial completion of this chapter, a number of pertinent articles appeared in the literature. Several of these articles, dealing with organometallic guests (*71–75*) and alkanol guests (*76*), will be discussed briefly here.

Schafer-Stahl and Abele have conducted an extensive investigation of the synthesis and physical properties of metallocene intercalates of FeOCl (*71–75*). The key synthetic observation is that, by use of 1,2-dimethoxyethane as the solvent, intercalation of FeOCl by metallocenes of V, Cr, Mn, Fe, Co, and Ni can be carried out at room temperature. Layer expansions of 5.18 Å are observed in the Cr, Fe, Co, and Ni cases, and stoichiometries range from FeOCl $(CrCp_2)_{0.14}$ to FeOCl $(CoCp_2)_{0.18}$ (*72*). These values are in reasonable agreement with previous observations on metallocene intercalation in FeOCl (*32, 41*). With $MnCp_2$, a complicating side reaction occurs, producing $FeCp_2$ by extraction of iron from the FeOCl and leading to exfoliation of the host crystals. With VCp_2, a side reaction also occurs, producing VCp_2Cl in solution (*73*). Intercalation of $(Me_5Cp)_2Fe$ yields an intercalate with a 7.1 Å layer expansion, in keeping with the 7.55 Å expansion previously observed for $(EtMe_4Cp)_2Fe$ (*41*) and the resulting conclusions regarding orientation of the guest molecules (see Section III,E).

The significance of the solvent is particularly clear in the reaction of $FeCp_2$ with FeOCl. Whereas in toluene temperatures of 80°C or higher are necessary to produce a reasonable rate of intercalation, in 1,2-dimethoxyethane (DME) room temperature is adequate. In fact, Schafer-Stahl observed, at >80°C in DME, some decomposition of the intercalate, resulting in partially chlorinated cyclopentadienyl rings and the appearance of Fe_2O_3 in the X-ray powder pattern (*75*). When $FeCp_2$ is intercalated from the gas phase (in an evacuated sealed tube containing no solvent), $FeCp_2{}^+FeCl_4{}^-$ is reported to be observable in the X-ray powder diffraction pattern. Above 110°C, crystalline α-Fe_2O_3 is formed (*74*, *75*).

The 4 K Mössbauer spectrum of FeOCl $(FeCp_2)_{1/6}$ prepared at room temperature shows a six-line feature attributed to antiferromagnetically ordered iron sites in the host lattice, and a single line assigned to $FeCp_2{}^+$, nicely confirming the hypothesis that $FeCp_2$ is oxidized upon intercalation (*74*). The electrical conductivity is $\sim 10^4$ greater than that of FeOCl itself and increases reversibly with increasing temperature up to 80°C. Above 80°C, a slow, irreversible decrease in conductivity begins. Thus, Schafer-Stahl suggests that the electron donated by the guest resides in the conduction band of FeOCl. This conclusion contrasts with the suggested localization of electron density in the intercalate prepared in toluene at 80°C (see Section III,E).

The first report of an *intercalation* reaction of FeOCl involving alcohols was made by Weiss and Choy (*76*). By treatment of an *n*-decylammine intercalate of FeOCl with *n*-decyl alcohol at room temperature, a crystalline solid was obtained with layer spacing of 34.6 Å. Other alkanol intercalates were prepared by exchanging the *n*-decanol in solutions of alcohols with even carbon numbers from C_2 to C_{18}. Products with stoichiometry of $FeOCl(alkanol)_{0.5}$ were obtained and characterized primarily by X-ray powder diffraction. For the long-chain alcohols, bilayers of guest species with the alkyl chains perpendicular to the layers are suggested from the layer spacings; for the shorter-chain alcohols, the results are more difficult to explain. Heating the samples to 110°C results in a reversible contraction of the interlayer space, whereas heating to above 130°C results in the expulsion of HCl and leads to the formation of topochemically substituted products believed to be $FeO(OR)_nCl_{1-n}$ with $n \sim 0.5$. In light of this work, it is unclear under what conditions the topochemical substitution reactions of alcohols discussed in Section V,B will occur and when only clean intercalation reactions will occur.

REFERENCES

1. F. Hulliger, *in* "Physics and Chemistry of Materials with Layered Structures" (F. Levy, ed.), Vol. 5, pp. 254–270. Reidel Publ., Dordrecht, Netherlands, 1976.

2. P. L. Goggin, I. J. McColm, and R. Shore, *J. Chem. Soc. A* p. 1004 (1966).

3. P. Hagenmuller, J. Rouxel, J. David, A. Colin, and B. Le Neindre, *Z. Anorg. Allg. Chem.* **323,** 1 (1963).

4. A. Hardy and A.-M. Hardy, *C. R. Hebd. Seances Acad. Sci.* **256,** 3477 (1963).
5. H. Schafer, C. Goser, and L. Bayer, *Z. Anorg. Allg. Chem.* **263,** 87 (1950).
6. H. Schafer, E. Wittig, and W. Wilborn, *Z. Anorg. Allg. Chem.* **297,** 48 (1958).
7. P. Hagenmuller and A.-M. Hardy, *C. R. Hebd. Seances Acad. Sci.* **256,** 1784 (1963).
8. P. Hagenmuller, J. Rouxel, and B. Le Neindre, *C. R. Hebd. Seances Acad. Sci.* **252,** 282 (1961).
9. P. Hagenmuller, J. Rouxel, J. David, and A. Colin, *C. R. Hebd. Seances Acad. Sci.* **253,** 667 (1961).
10. P. Hagenmuller and J. Rouxel, *C. R. Hebd. Seances Acad. Sci.* **259,** 1859 (1960).
11. P. Palvadeau, and J. Rouxel, *Bull. Soc. Chim. Fr.* **8,** 2698 (1967).
12. S. Goldsztaub, *C. R. Hebd. Seances Acad. Sci.* **198,** 667 (1934).
13. M. D. Lind, *Acta Crystallogr., Sect. B* **26,** 1058 (1970).
14. "International Tables for X-ray Crystallography," Vol. 1. Kynoch Press, Birmingham, England, 1969.
15. H. E. Forsberg, *Acta Chem. Scand.* **10,** 1284 (1956).
16. J. P. Venien, P. Palvadeau, D. Schleich, and J. Rouxel, *Mater. Res. Bull.* **14,** 891 (1979).
17. A. Haase and G. Brauer, *Acta Crystallogr.*, Sect. *B* **31,** 2521 (1975).
18. H. E. Forsberg, *Acta Chem. Scand.* **16,** 777 (1962).
19. H. Schafer, F. Wartenpfuhl, and E. Weise, *Z. Anorg. Allg. Chem.* **295,** 268 (1958).
20. H. G. von Schnering, M. Collin, and M. Hasscheider, *Z. Anorg. Allg. Chem.* **387,** 137 (1972).
21. P. Ehrlich and H.-J. Siefert, *Z. Anorg. Allg. Chem.* **301,** 282 (1959).
22. V. T. Kalinnikov, A. I. Morozov, V. G. Lebedev, O. D. Ubozhenko, and M. N. Volkov, *Russ. J. Inorg. Chem.* **17,** 353 (1972).
23. H. Schafer and F. Wartenpfuhl, *J. Less-Common Met.* **3,** 29 (1961).
24. C. Levayer and J. Rouxel, *C. R. Hebd. Seances Acad. Sci., Ser. C* **268,** 167 (1969).
25. H. Schafer and F. Wartenpfuhl, *Z. Anorg. Allg. Chem.* **308,** 282 (1961).
26. M. Danot and J. Rouxel, *C. R. Hebd. Seances Acad. Sci., Ser. C* **262,** 1879 (1966).
27. H. Schafer, *Z. Anorg. Allg. Chem.* **259,** 265 (1949).
28. H. Schafer, *Z. Anorg. Allg. Chem.* **260,** 279 (1949).
29. H. Schafer, F. Wittig, and M. Jori, *Z. Anorg. Allg. Chem.* **287,** 61 (1956).
30. H. Schafer, *Z. Anorg. Allg. Chem.* **264,** 250 (1951).
31. E. Stirneman, *Neues Jahrb. Mineral., Geol., Palaeont., Beilageband, Abt. A* **52,** 334 (1925).
32. T. R. Halbert and J. C. Scanlon, *Mater. Res. Bull.* **14,** 415 (1979).
33. P. Palvadeau, L. Coic, J. Rouxel, and J. Portier, *Mater. Res. Bull.* **13,** 221 (1978).
34. S. Kikkawa, F. Kanamaru, and M. Koizumi, *React. Solids* [*Proc. Int. Symp.*], *8th, Goeteborg, 1976* p. 731 (1977).
35. P. Hagenmuller, J. Rouxel, and J. Portier, *C. R. Hebd. Seances Acad. Sci.* **254,** 2000 (1962).
36. H. Bizette and A. Adam, *C. R. Hebd. Seances Acad. Sci., Ser. B* **275,** 911 (1972).
37. R. W. Grant, H. Wiedersich, R. M. Housley, G. P. Espinosa, and J. O. Artman, *Phys. Rev. B* **3,** 678 (1971).
38. R. W. Grant, *J. Appl. Phys.* **42,** 1619 (1971).
39. R. H. Herber and Y. Maeda, *Physica B+C (Amsterdam)* **99,** 352 (1980).
40. E. Kostiner and J. Steger, *J. Solid State Chem.* **3,** 273 (1971).
41. T. R. Halbert, D. C. Johnston, L. E. McCandlish, A. H. Thompson, J. C. Scanlon, and J. A. Dumesic, *Physica B+C (Amsterdam)* **99,** 128 (1980).
42. A. Weiss and E. Sick, *Z. Naturforsch., Teil B* **33,** 1087 (1978).
43. F. Kanamaru and M. Koizumi, *Jpn. J. Appl. Phys.* **13,** 1319 (1974).
44. M. B. Dines, *Mater. Res. Bull.* **10,** 287 (1975).
45. M. S. Whittingham, U.S. Patent 4,049,887 (1977).
46. M. Armand, L. Coic, P. Palvadeau, and J. Rouxel, *J. Power Sources* **3,** 137 (1978).

47. H. Meyer, A. Weiss, and J. O. Besenhard, *Mater. Res. Bull.* **13,** 913 (1978).
48. A. Lerf and R. Schollhorn, *Inorg. Chem.* **16,** 2950 (1977).
49. R. R. Chianelli, *J. Cryst. Growth* **34,** 239 (1976).
50. G. Lagaly, S. Fitz, and A. Weiss, *Clays Clay Miner.* **23,** 45 (1975).
51. M. B. Dines, *Science* **188,** 1210 (1975).
52. P. Hagenmuller, J. Portier, and B. Barbe, *Vietnam. Chim. Acta,* p. 59 (1966).
53. P. Hagenmuller, J. Portier, B. Barbe, and P. Bouclier, *Z. Anorg. Allg. Chem.* **355,** 209 (1967).
54. S. Kikkawa, F. Kanamaru, and M. Koizumi, *Bull. Chem. Soc. Jpn.* **52,** 963 (1979).
55. F. R. Gamble, H. H. Osiecki, M. Cais, R. Pisharody, F. J. DiSalvo, and T. Geballe, *Science* **174,** 493 (1971).
56. F. R. Gamble and B. G. Silbernagel, *J. Chem. Phys.* **63,** 2544 (1975).
57. R. Schollhorn and H. D. Zagafka, *Angew. Chem., Int. Ed. Engl.* **16,** 199 (1977).
58. F. Kanamaru, M. Shimada, M. Koizumi, M. Takano, and M. Takada, *J. Solid State Chem.* **7,** 297 (1973).
59. F. Kanamaru, S. Yamanaka, and M. Koizumi, *Chem. Lett.* p. 373 (1974).
60. F. Kanamaru and M. Koizumi, *Jpn. J. Appl. Phys.* **13,** 1319 (1974).
61. S. Yamanaka, T. Nagashima, and M. Tanaka, *Thermochim. Acta* **19,** 236 (1977).
62. M. S. Whittingham, *Mater. Res. Bull.* **13,** 775 (1978).
63. J. W. Johnson, *Physica B+C (Amsterdam)* **99,** 141 (1980).
64. N. I. Vorobev, V. V. Pechkovskii, and L. V. Kobets, *Zh. Neorg. Khim.* **19,** 3 (1974).
65. J. Portier and J. Grannec, *Bull. Soc. Chim. Fr.* p. 798 (1964).
66. A. Novak and J. Portier, *C. R. Hebd. Seances Acad. Sci.* **258,** 2582 (1964).
67. S. Kikkawa, F. Kanamaru, and M. Koizumi, *Inorg. Chem.* **15,** 2195 (1976).
68. S. Son, S. Kikkawa, F. Kanamaru, and M. Koizumi, *Inorg. Chem.* **19,** 262 (1980).
69. S. Kikkawa, F. Kanamaru, and M. Koizumi, *Inorg. Chem.* **19,** 259, (1980).
70. A. Lecerf, M. Rault, J. Portier, and G. Villers, *Bull. Soc . Chim. Fr .* p . 1208 (1965).
71. H. Stahl, *Inorg. Nucl. Chem. Lett.* **16,** 271 (1980).
72. H. Schafer-Stahl and R. Abele, *Angew. Chem., Int. Ed. Engl.* **19,** 477 (1980).
73. H. Schafer-Stahl and R. Abele, *Z. Anorg. Allg. Chem.* **465,** 147 (1980).
74. H. Schafer-Stahl, *Mater. Res. Bull.* **15,** 1091 (1980).
75. H. Schafer-Stahl and R. Abele, *Mater. Res. Bull.* **15,** 1157 (1980).
76. A. Weiss and J. H. Choy, *Z. Naturforsch., Teil B* **35,** 157 (1980).

13
Hydrogen-Containing Materials

K. H. J. BUSCHOW

and

H. H. van MAL

I. INTRODUCTION

The hydrogen atom is the smallest and simplest atom in the periodic table. It has the capacity to form compounds with a large variety of elements. Starting at the right-hand side of the periodic table, hydrogen, when combined with the halogens, forms strong acids such as HF and HI. Examples of compounds formed with elements occurring somewhat more to the left are H_2O, NH_3, CH_4, and GeH_4. Furthermore, there are a multitude of hydrides in which hydrogen is combined with metals. These comprise hydrides of

INTERCALATION CHEMISTRY

ISBN 0-12-747380-7

several d transition metals but also the hydrides of metals of Groups I, II, IIIa, IVa, and Va. Comparing the electronic properties of compounds such as HI and CsH, one may state that in the former compound the hydrogen atoms are positively charged (H^+), whereas in the latter they adopt a negative charge (about $0.7e^-$). This difference in charge state is accompanied by an even larger difference in radius of the H atoms (ions). Because of their capacity to reduce their radii to that of a proton (in extreme cases) under favorable circumstances, hydrogen atoms can possess extremely high diffusion rates in some solid materials. Even within the group of metal hydrides, hydrogen exists in a wide variety of states. It has become customary to classify the metal hydrides according to the nature of the metal–hydrogen bond (1). The elements of Group Ia and IIa form so-called ionic or saline hydrides. Examples of covalent hydrides are GeH_4, BeH_4, AsH_3, CuH, and $LiAlH_3$. Most of the hydrides are metallic, however. They comprise the hydrides of the metals of Groups IIIa, IVa, and Va, including the rare earth elements, the actinide elements, and the few representatives of the d transition metals (Pd, Ni, and Cr). The metallic hydrides also include the extremely large group of ternary hydrides formed by H_2 absorption in intermetallic compounds. Examples of such ternary hydrides are $LaNi_5H_6$, TiFeH, and $ThFe_3H_5$. Determining the state of the hydrogen in the metallic hydrides still presents some difficulties. Considerable experimental effort has been made to decide between the hydridic (or anionic) model and the protonic model (*2*). In the hydridic model the two 1s electron states of the hydrogen are situated below the Fermi energy (E_F) of the metal. In the hydride both of these levels are occupied. The extra electron is provided by the conduction band. In this model the hydrogen is present as H^-. In the protonic model E_F is lower than the two 1s electron states of the hydrogen. In this case the hydrogen donates an electron to the conduction band upon entering the lattice and is present as H^+. The difficulties mentioned stem from the fact that some of the experimental data (mostly on rare earth hydrides) can be described in terms of the hydridic model, whereas there also seems to be substantial experimental evidence in favor of the protonic model. Actually, the difference between the two models is less pronounced than described above if the hydrogen in the hydride is looked upon as a proton screened by various amounts of s and d electrons. Depending on the screening charge, a whole gamut of situations is possible in which the two situations described above take extreme positions (3).

A particularly interesting property of some of the metallic hydrides is their capacity to sorb hydrogen gas reversibly. Examples of the corresponding metals are Pd, Nb, V, and Ta. The hydrogen-to-metal atomic ratio in the binary hydrides is only moderate, however. Considerably larger quantities of H_2 gas can be absorbed by Ti, Zr, Hf, Sc, Y, the alkaline earth metals, and the rare earth metals (R). The binary hydrides of these metals are characterized, however, by a rather high stability. This, in fact, implies a very low equilibrium H_2 pressure so that, at room temperature, one can consider the H_2 absorption in these metals to be virtually irreversible. This rules out, among other things, most technological applications. More favorable properties in this respect can be expected from intermetallic compounds in which the strongly hydrogen-attracting metal is combined with one or more metals that have no or only small hydrogen-attracting power. Several investigations of these compounds have shown that such a combination can shift the equilibrium pressure to a more convenient range without greatly reducing the hydrogen-to-metal atomic ratio (*1*, *4–6*). In the context of this book, we focus our attention primarily on these ternary hydrides. The reason for this is that the reversible hydrogen absorption in these materials is accompanied by hardly any structural change. During the hydrogen absorption process the lattice expands in order to create sufficient space to accommodate the H atoms. After desorption the volume is restored to the original proportions.

Before discussing the details of hydrogen sorption, we will give a brief description in the next section of intermetallic compounds, their formation, their phase relationships, and their crystal structures. For a description of the physical properties of most of the intermetallic compounds discussed in this chapter, the reader is referred to extensive reviews on this subject (*7–9*).

II. INTERMETALLIC COMPOUNDS

Intermetallic compounds can be prepared by melting the parent metals together in an oxide crucible (for instance, Al_2O_3, ThO_2, or MgO) using either a resistance furnace or an inductance furnace. If the purity requirements are more stringent, levitation melting or arc melting can be used to avoid the possibility of contamination by reaction with the crucible material. These methods are not applicable if the vapor pressure of one or both of the parent metals is high at elevated temperatures (for instance, Mg, Zn, or Cd). In these cases the intermetallic compounds can be prepared by sealing the starting materials into tantalum or molybdenum capsules and heating in a protective atmosphere at temperatures sufficiently high for the reaction to occur. Very often annealing is required after the samples are melted. This is discussed below.

It is not always sufficiently realized that intermetallic compounds are different from ordinary solid solution alloys. In the former the atoms of the parent elements are not distributed at random over the atomic sites in the crystal lattice but are restricted to particular crystallographic sites, which are different for each kind of atom. Examples of such atomic arrangements are given in Fig. 1. Inspection of these schematic representations shows that for

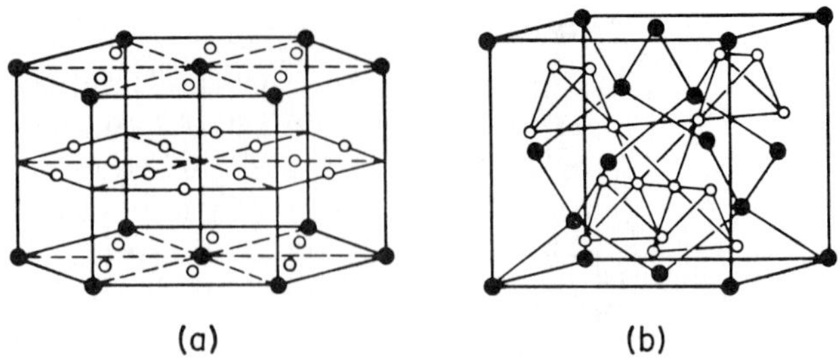

Fig. 1. Schematic representation of the crystal structures of $CaCu_5$ (a) and $MgCu_2$ (b). Only one unit cell is given for the cubic $MgCu_2$ type. The representation of the hexagonal $CaCu_5$ type contains three unit cells.

each structure type there is a fixed ratio between the number of the atoms of the two participating metals. This implies that intermetallic compounds have a well-defined stoichiometric composition.

A typical phase diagram for the combination of two metals is shown in Fig. 2. It can be seen that six intermetallic compounds can be obtained by melting Ce and Co together. The atomic proportions required correspond to the vertical lines shown in the diagram. The crystal structures of $CeCo_2$ and $CeCo_5$ are of the $MgCu_2$ and $CaCu_5$ types shown in Fig. 1. The crystal structures of $CeCo_3$, Ce_2Co_7, and Ce_2Co_{17} can be derived in a simple manner from the structure of $CeCo_5$. A brief discussion of the substitution

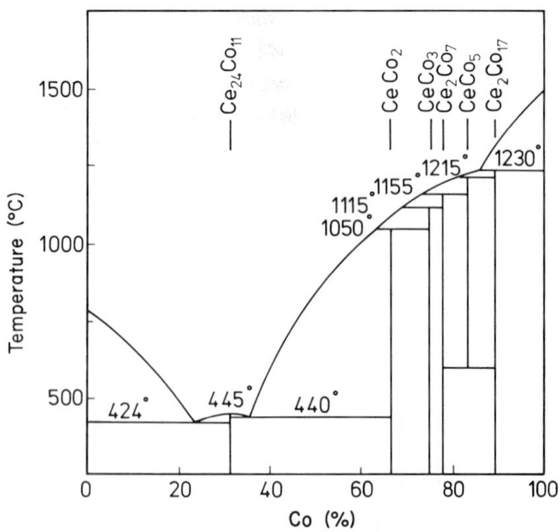

Fig. 2. Phase diagram of the binary alloy system Ce–Co.

scheme leading to these structures is given elsewhere (*7*). The crystal structure of $Ce_{24}Co_{11}$ is more complex (*10*).

It can be inferred from the phase diagram that most of the compounds have a peritectic melting point, corresponding to the horizontal lines in the phase diagram. This has serious consequences with respect to the formation of single-phase compounds. If, for instance, a melt of a composition Ce_2Co_{17} is allowed to cool, the first crystallization occurs at about 1300°C and consists of Co crystals. These Co crystals react with the melt (being richer in Ce due to the Co precipitation) at 1230°C to form the compound Ce_2Co_{17}. Practical cooling rates are too high, however, to lead to completion of the peritectic reaction, since at lower temperatures the negligible diffusion of the metal atoms inhibits further conversion. Vacuum annealing of the solidified alloy is necessary in order to obtain single-phase materials. The annealing temperature has to be sufficiently high to make diffusion of atoms possible, but it has to be lower than the corresponding peritectic melting temperature. In the case of $CeCo_5$ it has to be above 600°C since $CeCo_5$ decomposes into the two adjacent phases below this temperature (see Fig. 2). The compound $Ce_{24}Co_{11}$ melts congruently. It forms directly from the melt, and vacuum annealing is not required. Similar situations exist in other binary alloy systems. Phase diagrams corresponding to intermetallic compounds considered in this chapter can be found elsewhere (*11–14*).

The combination of two metals does not in all cases lead to the formation of intermetallic compounds. No compounds are formed, for instance, when rare earth elements are melted together with Nb, Mo, Ta, or W. Whether or not the combination of two metals leads to the formation of intermetallic compounds depends on whether the heat of formation (with respect to the pure parent metals as standard states) is negative or positive. Miedema (*15*) proposed a simple model by means of which these heats of formation can be estimated in terms of interactions between the neighboring individual atoms. Values of the formation enthalpies pertaining to intermetallic compounds of various concentrations are listed in the book "Simple Rules for Alloying" (*16*).

Although Miedema's model can give a reliable answer to the question of whether or not compound formation will occur, it does not further specify the number of compounds formed in a given binary system, nor does it give information about the composition or the crystal structure of these compounds. Predictions of the latter quantities are extremely difficult, however, since they depend on often subtle differences in free energy between the compounds possible in a small concentration range. The absolute values of the free energies (provided that these energies are negative) are not that important in this respect. It is quite possible that compounds with relatively low free energies do not exist, whereas compounds with appreciably higher values do. However, these problems fall outside the scope of this book.

III. HYDROGEN SORPTION IN INTERMETALLIC COMPOUNDS

The hydrogen sorption in metal–hydrogen systems is most conveniently studied by means of pressure–composition (P–x) isotherms. In general these isotherms look like those shown in Fig. 3 and have the following features.

The absorption of the first quantity of hydrogen gas gives rise to a relatively strong increase in the hydrogen pressure. This region refers to the original metal phase that is able to absorb a small fraction of hydrogen gas without changing its crystal structure (α phase). The next stage involves a region where the hydrogen pressure is concentration independent and where the phase is in equilibrium with the hydride phase (β phase). The pressure corresponding to the equilibrium $\alpha + H_2 \rightleftharpoons \beta$ is often referred to as the plateau pressure. A further increase in x, after the α phase has completely been converted to the β phase, is again accompanied by a strong rise in P. In some cases a second hydride (β') with a higher H content than β exists. In

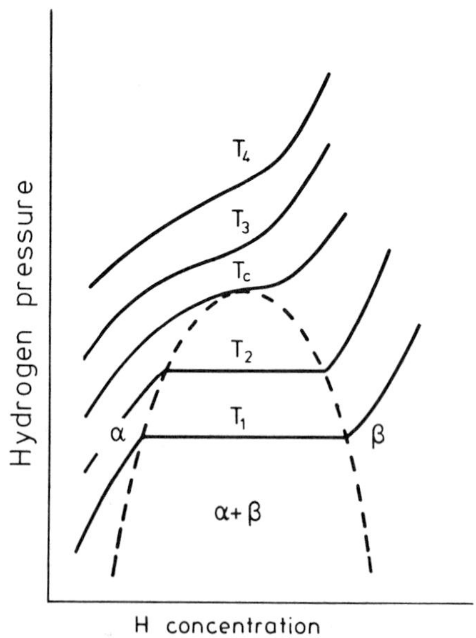

Fig. 3. Schematic representation of pressure–composition isotherms of metal–hydrogen systems. At temperatures below a critical temperature T_c, separate concentration regions exist between the α phase and the β phase. These two regions are separated by a two-phase region, indicated in the figure by a dashed line. For temperatures higher than T_c, there is no longer a flat portion in the isotherms.

these cases one observes a second "plateau" pressure due to the equilibrium $\beta + H_2 \rightleftharpoons \beta'$. These β hydrides can be looked upon as ternary intermetallic compounds in which one of the components consists of hydrogen.

The concentration region represented by the horizontal part of the isotherm can be regarded as a miscibility gap. It originates from the fact that the partial heat of solution in the intermetallic compound decreases (becomes more negative) with increasing H concentration. Alternatively, one could say that the process of solution becomes more exothermic and that it is energetically more favorable for the H atoms to be concentrated locally than to be distributed in a uniform manner throughout the metal. At higher temperature the influence of entropy counteracts this tendency to form two separate phases, so that the extent of the miscibility gap can be expected to decrease until finally the gap disappears. This is schematically represented in Fig. 3, which shows a critical pressure (P_c) occurring at a critical temperature (T_c), above which the isotherms have no horizontal portion. The absence of a flat region in the isotherm does not, therefore, necessarily mean that the sample is inhomogeneous or contains large portions of other intermetallic compound phases. In this chapter we restrict ourselves to situations in which T_c and P_c are sufficiently high for a well-defined two-phase equilibrium to exist.

Only in relatively few cases is there a spontaneous absorption of hydrogen gas by a piece of intermetallic compound when it is introduced into a hydrogen atmosphere. In most cases hydrogen absorption takes place only if the material has undergone a so-called activation treatment. The activation process can be considered as consisting of two steps (*17*). In the first step the thin surface oxide film is removed so that direct contact becomes possible between the H_2 and the metallic surface. The first step of the activation of TiFe consists of heating the alloy in H_2 to a few hundred degrees Centigrade. In many compounds based on rare earth metals, like $LaNi_5$ or the Ce–Co compounds mentioned above, such a heat treatment is not necessary. Activation can be performed by applying H_2 pressures considerably in excess of the corresponding plateau pressures. The second step of the activation consists of the conversion of the entire sample for the first time into the hydride. Microscopic examination of the hydrogenation process in TiFe has shown (*17*) that the hydride initially grows as an envelope around the partially activated particles of the compound, with the hydride phase penetrating as a series of fine platelets. The volume increase associated with the hydride phase leads to stresses exerted by the hydride envelope on the unhydrided core. This causes cracks, providing new surfaces for hydriding. In intermetallic compounds that are very brittle, the first hydrogenation already breaks the sample down into an assembly of tiny particles. For instance, in $LaNi_5$ the particle size is a broad distribution peaked at 4 μm (*18*). After 20 charging–decharging cycles the distribution became somewhat more symmetric, with the peak still at 4 μm. No further changes were observed, even after 200 cycles.

The presence of second-phase particles, dispersed in the matrix of the intermetallic compounds, can promote as well as retard activation. Second-phase particles composed of a compound richer in the hydrogen-attracting element than the main phase will in general promote activation since they can provide sites of easy hydride nucleation. They may also act as canals by means of which the hydrogen enters the main phase. Furthermore, if the presence of the second-phase particles increases the brittleness of the alloy, the major activation cracks will form more easily and activation will thus proceed much faster (*17*).

The sorption process mentioned in connection with Fig. 3 (i.e., after activation) is not strictly reversible. The occurrence of a sorption hysteresis has been observed on several occasions (*18–23*). Some of the results obtained by Kuijpers and van Mal (*20*) on the hydride $SmCo_5H_x$ are schematically represented in Fig. 4. These authors found that the sorption hysteresis (ΔP) increases with temperature but that the ratio $\Delta P/P$ remains nearly constant. They relate the hysteresis to the volume expansion and postulate it to be proportional to the hydrogen-to-metal ratio (H/M) of the corresponding hydride. The nature of the sorption hysteresis has also been studied by Lundin and Lynch (*21*). These authors argue that in the absorption isotherms the α and the β phases are compressively strained, whereas in the desorption

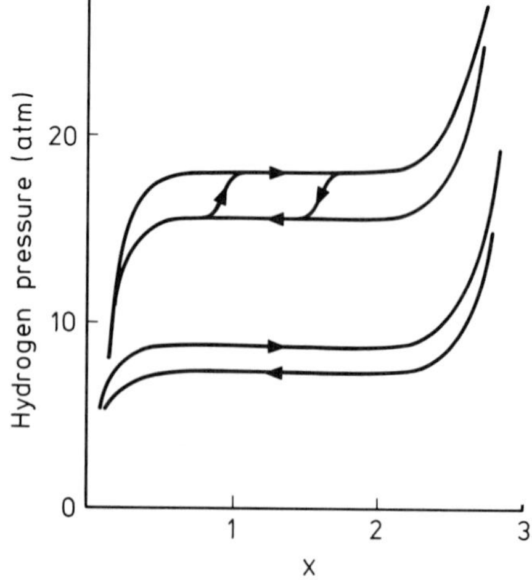

Fig. 4. Schematic representation of the absorption and desorption isotherms in $SmCo_5H_x$ at two different temperatures.

isotherms both phases are more or less strain-free. Owing to the microplastic deformation of the α and β phases in the absorption process the hysteresis is irreversible (see the minor hysteresis loops in Fig. 4). This means that the high-pressure absorption isotherm cannot be regarded as representing a metastable condition. In order to account thermodynamically for two separate paths in the absorption and desorption processes, Lundin and Lynch modified the phase rule by incorporating a strain term:

$$F = C - P + 2 + \pi(\text{strain})$$

In some respects this analysis is analogous to that used by Scholtus and Hall (*24*) to explain the hysteresis in the Pd–H system. Lundin and Lynch provide more directly a rationale pertaining to the pressure differential, in particular to that of the plateau region. According to these authors the plateau pressure increases as the size of the interstitial holes in the uncharged intermetallic compounds decreases. They note that the interstitial sites in either the α or β phases that are compressively strained during absorption are relatively smaller than the sites in either the unstrained α or β phases during desorption. The result is that the pressure observed in the absorption isotherms is higher than in the desorption isotherms. Although the explanation in terms of compressive strain seems to be generally valid, we feel that some caution is required in relating the pressure difference between the two isotherms to strain-induced changes in interstitial hole sizes, since Busch *et al.* (*25*) showed that the linear relationship between the plateau pressure of various RCo_5H_x and RNi_5H_x systems (R = rare earth metal) cannot be explained by means of elastic deformation of the interstitial volume.

It is primarily the reduced thermodynamic efficiency that makes the presence of hysteresis undesirable when it comes to technological applications. In addition, the thermodynamic efficiency can be reduced by the presence of impurities. As an example, we show in Fig. 5 the results obtained by Sandrock (*17*) on pure TiFe (solid line) and TiFe containing about 1 wt % of oxygen. It is seen that the effect of the small oxygen contamination is a substantial reduction in the absorption capacity. Very similar to the effect of impurities is the effect of a nonstoichiometric composition. A certain number of intermetallic compounds are not true "line" compounds as in Fig. 2 but exist over a small concentration range (homogeneity range). In $LaNi_5$, for instance, a small fraction of the La atoms can be replaced by a pair of Ni atoms without changing the crystal structure. The overall Ni concentration can then become as high as $LaNi_{5.5}$ without the presence of a second phase. The effect of the latter off-stoichiometric composition is to reduce the absorption capacity only slightly but to increase the plateau pressure considerably [up to 9 atm in $LaNi_{5.5}$ compared to 3 atm in stoichiometric $LaNi_5$ (*26*)].

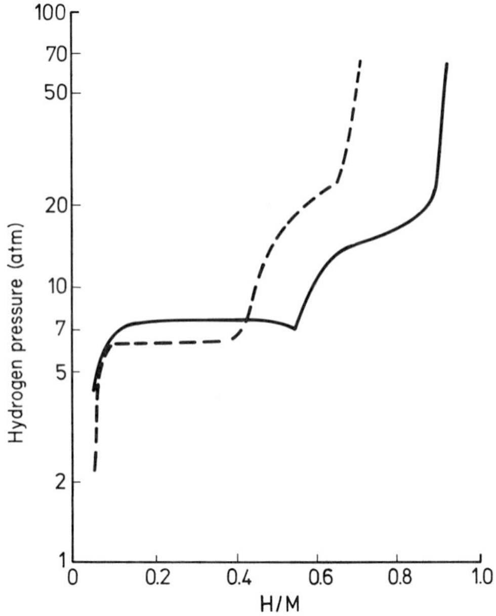

Fig. 5. Effect of an alloy oxygen content of about 1 wt % on the hydrogen desorption isotherm of TiFe at 40°C. [After Sandrock (*17*).]

It has already been mentioned that structural changes accompanying the hydrogen sorption are of minor importance. To illustrate this point we have compiled in Table I structural data obtained by means of X-ray diffraction on the various Ce–Co compounds occurring in Fig. 2 as well as on the corresponding hydrides. The hexagonal compound $CeCo_5$ undergoes a slight orthorhombic deformation upon hydrogen absorption (*27*). The relationship between the crystallographic axes in the basis plane before and after charging can be inferred from Fig. 6. This basis plane is similar to that of the $CaCu_5$ type of unit cell in Fig. 1. The compounds Ce_2Co_7 and $CeCo_3$ retain their crystal structure upon hydrogen absorption, although it can be seen from Table I that there is a considerable expansion of the unit-cell volume in the *c* direction (*23*). The X-ray diagrams of $CeCo_2$ and $Ce_{24}Co_{11}$ did not contain diffraction peaks after H_2 absorption. This indicates that long-range lattice periodicity has been lost. This atomic disordering is closely connected to the fact that the ternary hydrides are metastable under certain conditions of H_2 pressure and temperature. This point is discussed in more detail in the following section.

Because of the small atomic scattering power of the H atoms, no conclusive information on the crystallographic position of these atoms in the hydrides can be obtained by means of X-ray and neutron diffraction. For a complete

TABLE I

Composition, Equilibrium Pressure, and Structural Data of Various Compounds of Ce and 3d Metals[a]

		Lattice constants (Å)					
		Before charging		After charging			
Hydride	P_{eq} (atm)	*a*	*b*	*a*	*b*	*c*	Structure type
$CeCo_5H_{2.55}$[b]	1.55	4.922	4.019	8.726	5.105	8.146	$CaCu_5$
$Ce_2Co_7H_6$[c]	0.5	4.940	24.46	4.949	29.69		Ce_2Ni_7
$CeCo_3H_4$[c]	0.2	4.955	24.75	4.956	32.69		$PuNi_3$
$CeCo_2H_4$[c]	$<10^{-5}$	7.162					$MgCu_2$

[a] The hydride of $CeCo_5$ has a crystal structure that derives from the $CaCu_5$ type by means of a slight orthorhombic distortion (see Fig. 6) and a doubling of the *c* axis.

[b] Data of Kuijpers (*19*).

[c] Data of van Essen and Buschow (*23*).

structure determination one has therefore to rely on neutron diffraction studies made on the deuterium equivalents of the ternary hydrides. Until now only a few such studies have been undertaken. Kuijpers and Loopstra (*28*) investigated several hydrides of the RCo_5 family. Some of their results are schematically represented in Fig. 7. The relation with the original $CaCu_5$ type of structure has already been discussed in connection with Fig. 6. The crystal structure of $LaNi_5D_6$ has been studied simultaneously by several investigators (*29–31*). A schematic representation of the $LaNi_5D_6$ unit cell is shown in Fig. 8. With respect to the original $LaNi_5$ unit cell, hydriding has led to a volume increase of about 25% and the symmetry is lowered from $P6/mm$ to $P31\,m$.

Certain interesting facts have emerged from the neutron diffraction studies on these $CaCu_5$ type of compounds. If one leaves out of consideration the slight distortions that lower the symmetry, one can say that there are three octahedral interstitial positions and six tetrahedral interstitial position available per unit cell RCo_5 or RNi_5. The octahedral sites are surrounded by two R atoms and four 3d atoms and are located close to the basal plane of the $CaCu_5$ type of unit cell (i.e., at $z \approx 0$). These sites are fully occupied in $LaNi_5D_6$ and only partially occupied in the RCo_5 deuterides. The tetrahedral sites are surrounded by two R atoms and two 3d atoms and are located close to the planes at $z = 1/2$. These sites are partially occupied in the RCo_5 deuterides as well as in $LaNi_5D_6$. If both types of site were fully occupied, the resulting hydride would be of a composition $LaNi_5H_9$ or RCo_5H_9. It is interesting that a hydride approaching this composition was obtained by

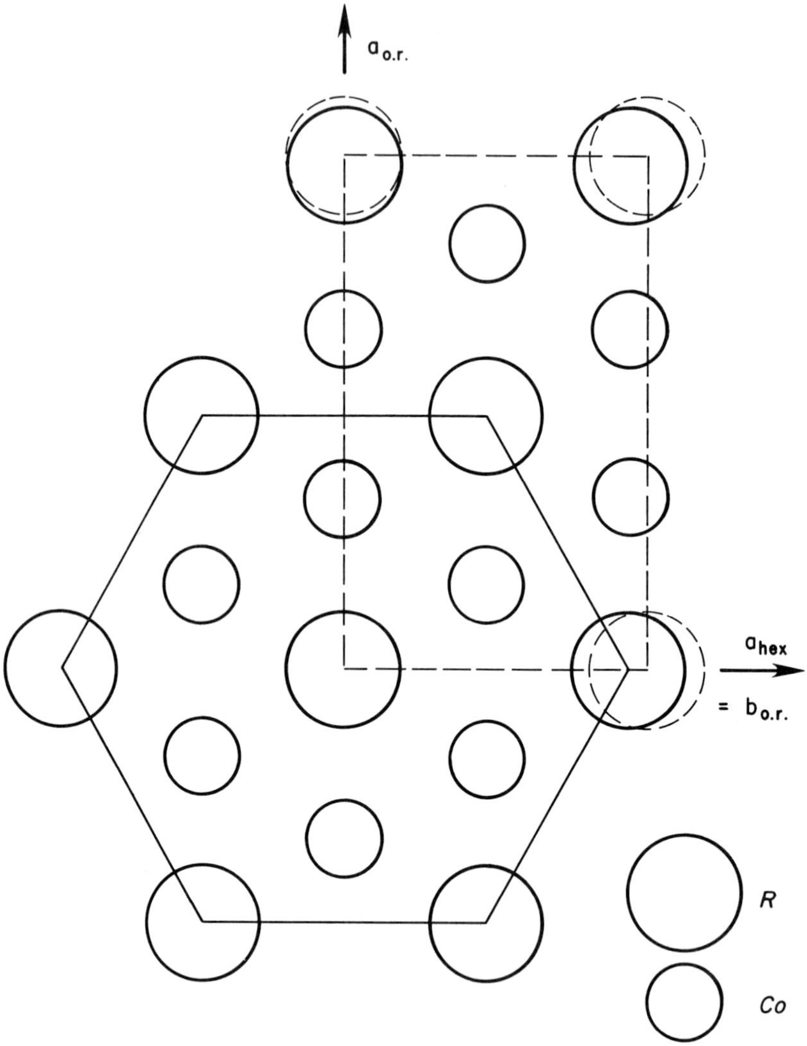

Fig. 6. Orthorhombic *a*,*b* axes in RCo_5 hydrides with respect to hexagonal axes of uncharged RCo_5. The orthorhombic *c* axis of the hydride is parallel to the hexagonal *c* axis. [After Kuijpers (*19*).]

Steward *et al.* in $LaCo_5$ at H_2 pressures higher than 1000 atm (*32*). In addition to the two interstitial sites mentioned, there are some additional tetrahedral sites in this structure type (at $z \approx 1/4$ and $z \approx 3/4$) that are surrounded exclusively by (four) 3d atoms. Neither of these sites is occupied by hydrogen. Results quite similar in this respect have also been obtained by means of

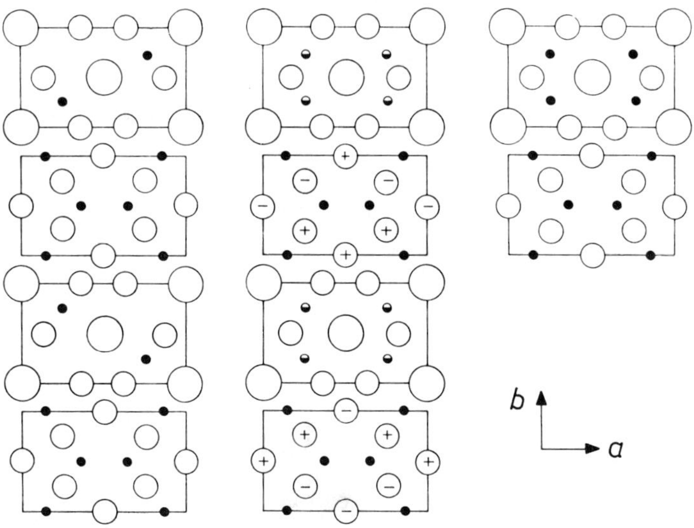

Fig. 7. Schematic representation of various sections perpendicular to the c axis in the orthorhombic crystal structures of $CeCo_5D_{2.55}$ (at $z = 0, \frac{1}{4}, \frac{1}{2}, \frac{3}{4}$), $PrCo_5D_{2.9}$ (at $z = 0, \frac{1}{4}, \frac{1}{2}, \frac{3}{4}$), and $PrCo_5D_{3.6}$ (at $z = 0, \frac{1}{2}$). The relationship between the orthorhombic structure and the original hexagonal $CaCu_5$ type of structure is shown in Fig. 6. Note that in the hydride of $CeCo_5$ the unit cell is twice as large in the c direction. [After Kuijpers (*19*).]

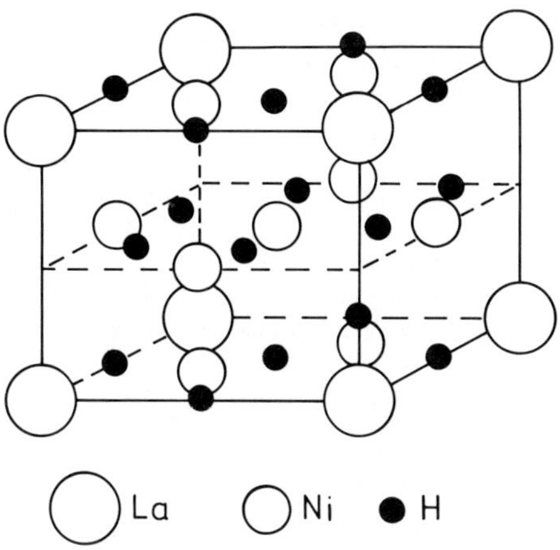

Fig. 8. Schcmatic representation of the unit cell of $LaNi_5D_6$. The hydrogen positions at $x = \frac{1}{2}$ are occupied only for 50% in a statistical manner (see text).

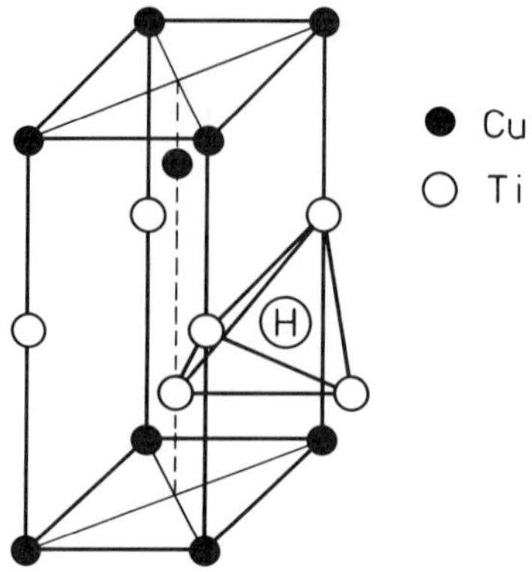

Fig. 9. Crystal structure of TiCuH. The structure of uncharged TiCu is body-centered tetragonal (B11 type) with two formula units per unit cell. [After Santoro *et al.* (*37*).]

neutron diffraction studies of the hydrides $LaNi_3D_2$ (*29*), $ZrNiD_3$ (*33*), Th_2AlH_4 (*34*), $ZrMn_2D_3$ (*35*), TiFeD (*35*, *36*), and TiCuD (*37*).

In all these cases the hydrogen or deuterium atoms occupy interstitial sites (mostly tetrahedral) that are surrounded by a large fraction of the strongly hydrogen-attracting parent metal. This is particularly pronounced in Th_2AlH_4 and TiCuH, in which the nearest neighbors of the H atoms are exclusively atoms of the strongly hydrogen-attracting component. The structure of TiCuH is shown in Fig. 9.

Not in all cases is the structural arrangement of the metal atoms roughly similar in the uncharged and charged situation. Examples are known in which the absorption of hydrogen leads to well-defined structural changes. Some results obtained on equiatomic compounds are compiled in Table II. It is seen that the structural transformations can entail a lowering of the crystal symmetry as well as an increase in symmetry.

TABLE II

Changes in Crystal Structure Observed in Some Equiatomic Intermetallic Compounds

Compound	Structure[a]	Hydride	Structure[a]	Ref.
ZrCo	CsCl (c)	$ZrCoH_{2.6}$	CrB (o.r.)	*38*, *39*
HfCo	CsCl (c)	$HfCoH_{3.2}$	CrB (o.r.)	*39*
EuPd	CrB (o.r.)	$EuPdH_3$	CsCl (c)	*40*
YbNi	FeB (o.r.)	$YbNiH_x$	CsCl (c)	*57*

[a] c, cubic; o.r., orthorhombic.

IV. THERMODYNAMIC CONSIDERATIONS

The plateau pressure associated with the equilibrium $\alpha + \frac{1}{2}xH_2 \rightleftharpoons \beta$ depends strongly on the temperature. The results obtained on $PrCo_5H_x$ shown in Fig. 10 may serve to illustrate this point. It is seen that, starting from room temperature, a rise in temperature of about 100° pushes the pressure from 0.5 atm to above 30 atm.

The plateau pressure of the hydride AB_nH_m, formed by means of the reaction

$$AB_n + \tfrac{1}{2}mH_2 \rightleftharpoons AB_nH_m \tag{1}$$

can be regarded as a measure of its stability. According to the van't Hoff equation the corresponding plateau pressure can be approximated in a limited temperature range ($T \ll T_c$) by means of the Arrhenius relation

$$\ln P_{H_2} = \Delta H/RT - \Delta S/R \tag{2}$$

where R represents the gas constant and ΔH (ΔS) are the standard enthalpy (entropy) of formation of the hydride AB_nH_m from AB_n and H_2 gas. This equation shows that $\ln P_{H_2}$ is expected to vary linearly with $1/T$. From the slope of a plot of $\ln P_{H_2}$ versus $1/T$ it is therefore possible to obtain ΔH, and

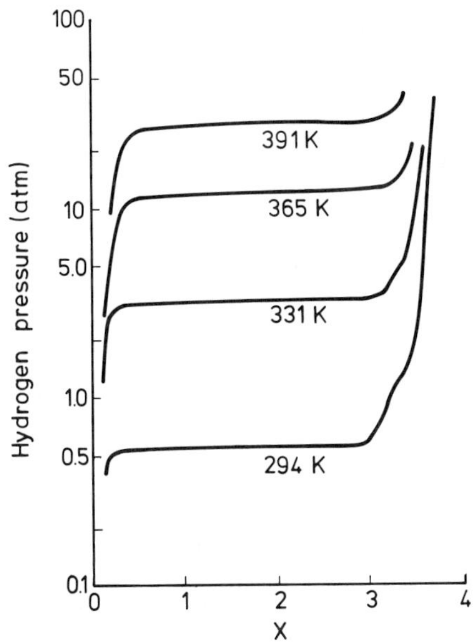

Fig. 10. Desorption isotherms of the system $PrCo_5H_x$ at various temperatures. [Data from Kuijpers (*19*).]

from the intercept with the vertical axis it is possible to derive ΔS. Figure 11 is an example of such a plot and has been made by means of the data shown in Fig. 10. The ΔH and ΔS values derived from this plot are 38.5 kJ/mole and 126.1 J/K mole H_2. It is not always completely justifiable to identify the ΔH (ΔS) value obtained in this way as the formation enthalpy (entropy) of the ternary hydride AB_nH_m of Eq. (1). The reason for this is that in most systems there is not only a substantial solubility of hydrogen in the intermetallic compound but also an appreciable deviation of the hydride phase from the stoichiometric composition. It is therefore more appropriate to identify ΔH and ΔS with the enthalpy and entropy changes associated with the reaction $\alpha + \frac{1}{2}xH_2 \rightleftharpoons \beta$ $(x < m)$.

There is yet another complication. This can be illustrated by means of Fig. 3, where one can see that, close to the critical temperature in particular, the composition of both the α phase and the β phase varies considerably with temperature. Plateau pressures measured at different temperatures do not correspond, therefore, to the same reactants and reaction products. It has been argued by Flanagan (*42*) that, in spite of this feature, conditions may be

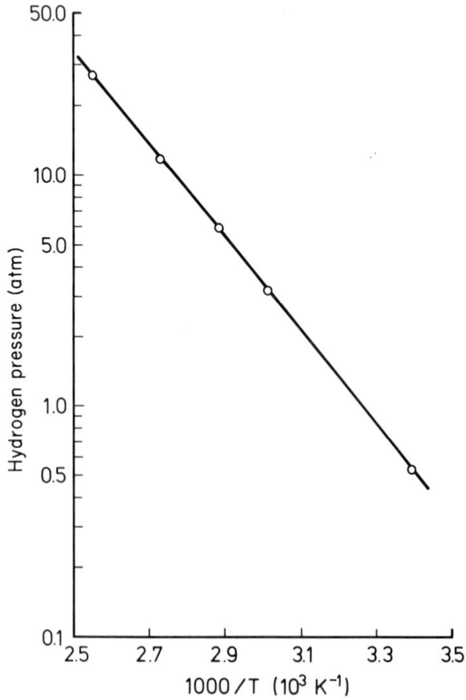

Fig. 11. Plot of ln P_{H_2} versus $1/T$ corresponding to the data shown in Fig. 10.

such that the linear relationship between ln P_{H_2} and $1/T$ can still be expected to hold. Values of ΔH and ΔS obtained from a linear ln P_{H_2} versus $1/T$ plot made on the basis of plateau pressures at temperatures where the compositional changes of the α and β phases are quite distinct have to be interpreted, therefore, with some caution. This is particularly true if the compositional dependence of the curve that defines the two-phase region in Fig. 3 is asymmetric about the critical composition (*42*).

We have listed in Table III some experimental hydrogen sorption characteristics obtained on a number of representative intermetallic compounds. Here we have restricted ourselves to the ternary hydrides in equilibrium with hydrogen gas. Details regarding the rate of hydrogen sorption are discussed in references *52–56* and will be briefly mentioned later. Inspection of the ΔH and ΔS values given in the table shows that the former values vary considerably, whereas there are only small differences in the latter. It will be clear, therefore, that the quantity ΔS is less suitable for characterizing the tendency of a given intermetallic compound to absorb hydrogen. The values of ΔS given in the table are rather close to the entropy of hydrogen gas (about 130 J/K mole H_2 at 1 atm and room temperature). The near constancy of the ΔS values can therefore be attributed to the fact that in all the hydrogenation reactions the entropy of the hydrogen as a gas is lost upon entering the metal (*6*).

Miedema (*15*) has developed a model by means of which the formation enthalpies of intermetallic compounds can be calculated (predicted) on the basis of the electronegativities of the starting materials and the corresponding electron densities at the Wigner–Seitz cell boundaries. The latter quantity can be obtained from the bulk moduli. This model has been extended to the calculation of the formation enthalpy of ternary hydrides (*6*). With respect to the uncharged compound AB_n and H_2 gas, the formation enthalpy of the hydride AB_n can be written as

$$\Delta H(AB_nH_m) \rightleftharpoons \Delta H(AH_x) + \Delta H(BH_{m-x}) - \Delta H(AB_n) \tag{3}$$

The choice of x is somewhat arbitrary. Since the concentration of the less hydrogen-attracting metal B in a given hydride is higher (n larger), the value of x will also be higher (*6*). The first term on the right-hand side of Eq. (3) represents the heat of formation of "virtual" binary hydrides of the A elements. These heats of formation can be derived from the experimental values of the actually existing binary hydrides after taking account of the deviation in composition of the virtual hydrides. The second term on the right-hand side is small. Together with the third term, it can be estimated using the model (*15, 16*) for the binary compounds and hydrides. The contribution of the third term can have small as well as large values. For a series of isostructural compounds AB_n, in which the compound A and the composition n are fixed

TABLE III

Examples of Hydrogen-Absorbing Intermetallic Compounds and Corresponding Hydrogen Sorption Parameters[a]

Compound	x	P_{H_2} (atm)	ΔH (kJ/mole H_2)	ΔS (J/K mole H_2)	Ref.
$LaNi_5$	5	2.9 (40°C)	31.0	108.9	*18*
$LaCo_5$	3.25	0.04 (20°C)	45.2	127.8	*19*
$CeCo_5$	2.55	1.55 (20°C)	40.0	136.6	*19*
$PrCo_5$	2.85	0.51 (20°C)	38.5	126.1	*19*
$SmCo_5$	2.50	3.30 (20°C)	32.7	121.5	*19*
Ce_2Ni_7	4	0.2 (50°C)	—	—	*23*
Ce_2Co_7	6	0.5 (50°C)	—	—	*23*
Gd_2Co_7	2.6	0.03 (75°C)	57.5	—	*43*
Er_2Co_7		0.44 (50°C)	39	—	*43*
Th_2Co_7	3.5	0.25 (40°C)	—	—	*50*
$CeNi_3$	3	0.09 (50°C)	—	—	*23*
$CeCo_3$	4	0.2 (50°C)	—	—	*23*
$GdCo_3$	2	0.44 (150°C)	51.1	—	*44*
$GdRu_2$	3	—	60.3	125.7	*41*
$GdRh_2$	3.2	—	49.4	134.0	*41*
$LaNi_2$	2	—	54.5	—	*45*
$ErFe_2$	2	10^{-5} (20°C)	—	—	*58*
ZrV_2	5.3	10^{-8} (20°C)	199.7	—	*48*
$ZrCr_2$	4	0.01 (20°C)	46.1	98	*48*
$ZrMn_2$	3.6	0.01 (20°C)	53.2	121.5	*48*
$TiCr_2$	3	—	23	—	*49*
$TiMn_2$	3	—	23	—	*49*
LaNi	3.84	—	104.7	—	*46*
TiCu	1	—	75	113	*47*
ZrCo	0.9	—	84.2	133.7	*38*
HfNi	3.2	0.02 (50°C)	—	—	*39*
LiPt	0.7	—	134	149	*51*
LiPd	0.8	—	75.3	97.9	*59*
Mg_2Ni	4	—	64.5	122.3	*22*
Zr_2Cu	1.2	0.2 (800°C)	120	90	*60*

[a] The compounds are arranged in order of increasing concentration of the strongly hydrogen-attracting component in AB_nH_x. The plateau pressures (first plateau) given in the third column are those closest to room temperature available in the literature.

quantities, the first term is furthermore a constant, so that the enthalpy effect for hydride formation is opposite to the enthalpy effect for compound formation. This leads to the rule of reversed stability (*6*): "The less stable the original intermetallic compound, the higher is the tendency to form a hydride, i.e., the lower is the equilibrium pressure." It can furthermore be derived by means of Eq. (3) that for fixed B and n in AB_n the equilibrium pressure is expected to increase in the series La to Lu and to be about equal for A = Ti and Hf but lower for A = Zr. For fixed A the equilibrium pressure decreases in the direction Ni to Mn, and for fixed A and B it increases with increasing *n*. There is, in general, a satisfactory agreement with experimental observations, although examples are known that are less easily interpreted in terms of Eq. (3) (*6, 61*). It will be clear that, to the extent that the H concentration varies within a series of isotypic ternary hydrides, the above rules will be obeyed less strictly. Nevertheless, they can serve as rules of thumb to facilitate the future search for new intermetallic compounds able to sorb hydrogen gas in a convenient pressure range (*62*).

If one compares the free energy of formation of a given ternary hydride (from H_2 gas and the corresponding intermetallic compound) with the free energy of formation of the binary hydrides of the constituent metals, one finds that the formation of the ternary hydride is not always the energetically most favorable reaction. Using either experimental values of ΔH and ΔS or values calculated by means of Miedema's model (*15, 16, 62*), it can be shown that under pressures that are not too high and at room temperature, most of the ternary hydrides (AB_nH_m) are metastable and decompose into the corresponding hydrides AH_2 (or AH_3) and a compound richer in B (AB_p, $p > n$). If a compound richer in B does not occur in the A–B phase diagram, pure B metal will be formed instead. That ternary compounds are nevertheless formed upon hydrogenation is due to the fact that the charging with H_2 gas is usually performed at temperatures low enough to prevent diffusion of the metal atoms. This diffusion of the metal atoms is necessary in order to achieve the separation of the phases AH_2 and AB_p associated with the decomposition of the ternary hydride. However, when the reaction between H_2 and AB_n is allowed to proceed quickly and without sufficient possibilities for transfer of the heat of hydrogenation to the surroundings in thermal contact with the sample, it can happen that the temperature of the sample becomes high enough for diffusion to take place (*63*). Very often the thermal contact of the sample with its surroundings is such that a sufficiently high temperature is reached only during a rather short period. The phase separation process will then not be able to reach completion. As a result, considerable atomic disorder will be present after charging. In general one can look upon the ternary hydride as an alloy supersaturated with respect to the phases AH_2 and AB_p. Precipitation of these phases will most likely proceed by means of a nucleation and growth mechanism and thus be strongly temperature dependent. Insufficient growth of the precipitates can lead to particles too small to be observed by X-ray diffraction. This would explain the absence of diffraction lines after charging, as reported in several places in the literature (*64, 65*). In cases in which one of the reaction products after decomposition is ferromagnetic, whereas the original ternary hydride is not, it is possible to follow the stage of the decomposition process after repeated charging and desorption cycles rather closely by means of magnetic measurements (*38*, *66–69*). In these measurements the magnetic contribution of the ferromagnetic phase will show up even though the decomposition has not yet proceeded to a stage where the precipitated ferromagnetic phase can be discerned by means of standard X-ray diffraction.

The linear relationship often observed between the logarithm of the plateau pressure and the unit-cell volume of the uncharged compound (*25*, *70*) is a great help in finding a suitable hydride for an application that requires a given plateau pressure. Results compiled by Gruen *et al.* (*70*) are shown in Fig. 12. These authors note that this volume dependence of ln *P* can

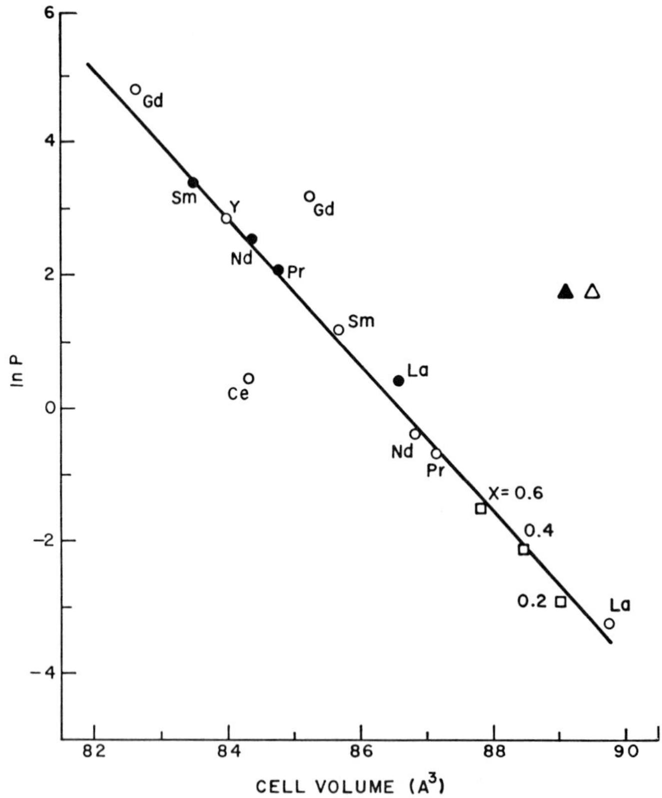

Fig. 12. Dependence of the equilibrium pressure (ln *P*) on the unit-cell volume of various $CaCu_5$ type of compounds, as found by Gruen *et al.* (*70*). Included are data on $LaNi_{4.5}Pt_{0.5}$ (solid triangle) and $LaNi_4Pd$ (open triangle). Key: ○, RCo_5; ●, RNi_5; □, $LaCo_{5-5x}Ni_{5x}$ (*50*).

be traced back to the term $\Delta H(AB_n)$ of Eq. (3). Such a volume dependence of $\Delta H(AB_n)$ is expected since this term depends on the electron density mismatch ΔN_{ws} at the Wigner–Seitz atomic cell boundaries between A and B atoms. Changes in ΔN_{ws} reflect mainly the lanthanide contraction of the rare earth component [N_{ws} values for Co and Ni are about the same (*15*)]. Gruen *et al.* (*70*) also note that the linear relationship breaks down if the non-rare earth partner elements differ too much in character. In these cases differences in N_{ws} [being proportional to $(B/V_m)^{1/2}$] between these elements may be the result of differences in the bulk modulus B rather than of differences in molar volume V_m. Nor is the mentioned relationship expected to hold in comparisons of compounds AB_n in which the B elements differ appreciably in electronegativity parameter [on Miedema's scale (*15*)]. This is most convincingly shown by results obtained on $LaNi_5$ in which Ni is partly replaced by Pt or Pd. These substitutions lead to an increase in P, even though the unit-cell volumes increase (see Fig. 12). Attempts have been made by Lundin *et al.* (*71*) to correlate the stability of the ternary hydrides (or ln P) with the tetrahedral hole sizes in the pure starting materials. It is not clear whether such a correlation has a physical basis since the differences in ln P are explained by tetrahedral hole radii differences of only several thousandths of Å, whereas it is known that hydrogen absorption gives rise to lattice expansions about two orders of magnitude larger. If the hole size in the uncharged $CaCu_5$ type of compounds RCo_5 and RNi_5 were really significant, one would expect that the correlation would also hold for compounds in which the R component has a valence different from 3. This is not the case. It is also not clear why compounds in which the R components have opposite valence deviations, as in $CeCo_5(Ce^{4+})$ and $CaNi_5(Ca^{2+})$, sometimes show deviations in the same directions from the mentioned correlation, whereas $CeCo_5$ and $ThCo_5$ (Ce and Th both tetravalent) deviate in opposite directions. The correlation is furthermore seen to break down if La in $LaNi_5$ is partly replaced by Zr. The arguments given by Lundin *et al.* (*71*) that the correlation is not expected to include $Zr_{0.2}La_{0.8}Ni_5$ because $ZrNi_5$ is not of the $CaCu_5$ type seems inappropriate since $Zr_{0.2}La_{0.8}Ni_5$ does have the $CaCu_5$ type of structure. Presumably the ln P hole size correlation is merely fortuitous and more or less reflects the volume dependence of the heat of formation $\Delta H(AB_n)$ in Eq. (3) in a way that is similar to that discussed by Gruen *et al.* (*70*). It is expected to hold for compounds in which the A components as well as the B components are similar in character, such as in RCo_5, RNi_5, and pseudo-binary compounds based on these. If more dissimilar elements are included, such as in $LaNi_4Pd$ or $LaNi_{4.5}Pt_{0.5}$, both ln P and the hole size increase with respect to $LaNi_5$, and the correlation breaks down. Further arguments against the hole size concept are the fact that the linear dependence between ln P and the hole size cannot be explained by elastic deformation of the interstitial hole size and that the volume expansion per hydrogen in a given series of compounds is not related to the associated plateau pressures (*25*).

We conclude this section by mentioning that it is highly unlikely that the tendency toward hydrogen absorption is governed by the presence of a partially empty d band in the uncharged compounds. In the past the presence of an open d band in metals such as Pd, Nd, and V has often been taken to be intimately connected with the capability of these metals to absorb hydrogen, with the hydrogen giving its electron to the d band of the metal. From the experimental standpoint, one could say that the high hydrogen absorption power of compounds such as Th_2Al (*34*), HfAl (*39*), and La_2Mg_{17} (*72*) does not point to the presence of an open d band as being the driving force behind H_2 absorption. The compound $LaNi_5$ and its hydride resemble most closely the Pd–H system in that there is a rather high d electron density of states, corresponding to an almost filled d band. The following argument shows that the presence of the partially filled 3d band in, for instance, $LaNi_5$ is not a prerequisite for the H_2 absorption. In Ni metal the number of unoccupied 3d band states is 0.6 electron per Ni. This number will be even smaller in $LaNi_5$, and we will regard it, therefore, as an upper limit. If the 3d band filling were really important, the maximum hydrogen absorption capacity would be $LaNi_5H_3$. Numerous experiments have shown that $LaNi_5$ is able to absorb more than twice this amount. Further evidence against the necessity of a d band filling upon hydrogen absorption can be derived from a study of the accompanying changes in magnetic properties. In the case of intermetallic compounds based on 3d elements, the changes in magnetic properties show that in some cases the H_2 take-up leads to a further depletion rather than a completion of the 3d band [see Buschow and de Châtel (*73*), and papers cited therein].

On the basis of enthalpy arguments one could say that the driving force behind H_2 absorption in intermetallic compounds is the presence of a metal as one of the components that is itself strongly hydrogen attracting; i.e., its binary hydride has a strongly negative enthalpy of formation. Perusal of the literature shows that, indeed, only intermetallic compounds satisfying this requirement have been reported to be hydrogen absorbers. As mentioned in Section III, it follows from neutron diffraction studies that the H atoms try to occupy positions as close to the strongly hydrogen-attracting parent metal as possible. This can be regarded as a further demonstration of the driving force being the high stability of the binary hydride. Further evidence comes from Mössbauer spectroscopy on A nuclei in A, AB_n, AH_m, and AB_nH_m showing that, from the electronic viewpoint as well, the situation in the ternary hydrides comes very close to that in the corresponding binary hydrides (*74–76*).

Before discussing some technical applications in the following section, we wish to mention that an important quantity is the rate of absorption and desorption. This rate is in general dependent on the diffusion of hydrogen in the ternary hydrides. For a thermally induced jump process, the temperature dependence of the diffusion coefficient D is given by an Arrhenius-type equation:

$$D = D_0 \exp(-E_a/kT)$$

Experimental values for the constant D_0 and the activation energy E_a can most reliably be obtained by means of nuclear magnetic resonance or by quasi-elastic neutron scattering. For more details the reader is referred to the literature (*52–56*).

V. TECHNICAL APPLICATIONS

A basic property of hydrides is the storage of hydrogen gas to a density that often exceeds the density of liquid hydrogen. The partial hydrogen densities in some hydrogen compounds are compared in Table IV. In practice, the hydrogen densities in the metal hydrides are substantially smaller since the hydrides are usually present in powder form (see Section III).

A number of applications of metal hydrides are listed below. Some of these have been described in the literature on the basis of binary hydrides. It will be

TABLE IV

Hydrogen Contents of Various Hydrogen Compounds[a]

Compound	M (kg/kmole)	ρ (kg/m^3)	N_H (10^{-28} m^{-3})	wt %	ρ_H (kg/m^3)	Ref.
H_2O	18	1000	6.7	11	111	—
Liquid CH_4	16.0	425	6.3	25	105	—
Liquid H_2	2	71	4.2	100	71	—
TiH_2	49.9	3800	9.2	4	153	*77*
ZrH_2	93.2	5610	7.3	2.1	122	*77*
YH_2	90.9	3958	5.7	2.2	95	*77*
LaH_2	140.9	5120	4.4	1.4	73	*78*
LaH_3	141.9	5350	6.5	2.1	108	*78*
$LaNi_5H_6$	438.5	6225	5.3	1.4	88	*79, 80*
$TiFeH_{1.95}$	105.7	5470	6.2	1.85	101	*81*

[a] Included are values of the molecular weight (M), the density of the hydrides (ρ), the number (N_H) of hydrogen atoms per unit volume, the weight percentage of hydrogen (wt %), and the partial hydrogen density (ρ_H).

clear that similar or even better performances can be obtained by using ternary hydrides that have the optimal hydrogen sorption characteristics required for a given application.

A. Hydrides as Moderator

The application of hydrides as a moderator for high-temperature mobile reactors is based on the large hydrogen density and high thermal stability of some particular metal–hydrogen systems (*77*, *82*).

B. Neutron Generator

In a sealed-off neutron tube, a titanium layer saturated with deuterium and tritium is used as a target. Accelerated D and T ions impinge on the target and the DT reaction yields a high flux of neutrons per second (*83*).

C. Hydrogen Getters

To bind gases liberated from the enclosed materials and from the walls of evacuated vessels after having been sealed off, getters such as Th–Ce–Al are used (*84*).

D. Hydrogen Pressure Regulator with High Absorption Rate

The hydrogen pressure inside a vessel can be varied over many orders of magnitude by varying the temperature of a layer of finely divided titanium dust deposited on the wall of a nickel cylinder (*85*).

E. Cold Accumulator

Hydrogen evaporating from a liquid hydrogen bath is absorbed by $LaNi_5$, the pressure over the liquid bath thus being kept constant. The liquid hydrogen bath can be used as a refrigerator until all the liquid is evaporated. Hydrogen can then be recondensed to make the unit ready for operation again (*18*).

F. Separation of Deuterium from Hydrogen

For metal–hydrogen absorbers in general, hydrides are found to be more stable than deuterides. For $LaNi_5$ at 40°C this is not the case (*18*). For V, Nb, and $V_{0.8}Nb_{0.2}$, the deuterides are also more stable than the hydrides. This inverse isotope effect was observed by Wiswall and Reilly (*86*), who mention a pressure ratio of 3 at 30°C. For TiNi a pressure ratio of 10 was found for the deuteride with respect to the hydride (*87*). This effect can be used to obtain an enriched gas phase of deuterium.

G. Hydrogen Storage

Hydrogen to be used as a fuel can be stored in the following ways: (a) as a compressed gas in cylinders, (b) as liquid hydrogen at about 20 K, and (c) chemically bound in hydrogen compounds or hydrides. In Table V the energy content per unit weight is given for a few hydrides, liquid hydrogen, gasoline, and compressed hydrogen. Also given are the temperatures at which the pressure will be near 1 atm.

When the use of $LaNi_5$ is considered as a source of hydrogen fuel for internal combustion engines, it is found that, for equal energy content, a $LaNi_5$ hydride reservoir will be about 34 times heavier than a full gasoline tank and will also be 8.5 times larger in volume. It is recognized, however, that the most important factors determining the future use of hydrogen as a fuel (including its transport and storage) are cost and safety.

A cost comparison of different energy carriers such as hydrogen, methane, methanol, and electricity for space heating, air conditioning, and automative applications has been made by Davitian (*88*). The comparison (made in 1974) shows that hydrogen is economically less attractive when other synthetic fuels or electricity are available in sufficient quantity.

In addition, safety aspects are rather unfavorable for liquid hydrogen. On the other hand, it appears from several studies (*22, 89*) that metal hydrides are much less hazardous than most common volatile fuels. Two points have to be kept in mind, however: (a) Some deformation of the containers might

TABLE V

Energy Content of Various Hydrogen-Containing Systems[a]

Energy carrier	Temp (°C) for 1 atm	gm H_2/kg hydride	kJ/kg hydride
$FeTiH_{1.9} \rightarrow FeTiH_{0.1}$	−10	18	2,560
$VH_{2.0} \rightarrow VH_{0.95}$	10	20	2,850
$Mg(Ni)H_{1.85} \rightarrow Mg(Ni)H_{0.1}$	300	70	10,000
$LaNi_5H_{6.0} \rightarrow LaNi_5H_{0.5}$	10	13	1,850
Liquid H_2	−250	1,000	140,000
Gasoline	—	—	48,000
High-pressure cylinder H_2	—	9	1,280

[a] Expressed as kilojoules per kilogram hydrogen.

occur, caused by expansion during the absorption process (for $LaNi_5$ 25% volume expansion), and (b) it is necessary to have a blow-off valve, since accidental heating of a container may cause the pressure to become dangerously high.

H. Hydrogen Purification

To obtain hydrogen gas of very high purity, membranes of palladium–silver alloys are commonly used. Only hydrogen gas is capable of diffusing through them. The fact that palladium–silver tubes are less susceptible to embrittlement than most metals reacting with hydrogen makes them suitable for use in membranes. These diffusion cells are rather expensive, and the process is fairly slow. More favorable in this respect is purification by means of absorption–desorption cycles based on metal hydrides operating at room temperature.

Wenzl and Klatt (*90*) have shown that, by selective absorption of hydrogen in FeTi crystals, hydrogen gas of technical purity (99.9%) can be converted to suprapure hydrogen (99.9999%). At the same time a source of pure hydrogen gas with a certain storage capacity is created. The pressure of the very pure hydrogen gas can be controlled by controlling the temperature.

It was reported by Reilly and Wiswall (*91*) that AB_5 compounds (see Table III) can still form hydrides when subjected to gas mixtures with a composition like that resulting from the steam-reforming of hydrocarbons followed by CO shifts and methanation steps. They concluded that CO_2, even when present in large concentrations, does not interfere with the H_2 take-up. Small amounts of CO, on the other hand, have an inhibiting effect. The inhibition can be diminished by raising the temperature or substituting Cu partly for Ni in $LaNi_5$. In our experiments we found that $LaNi_5$ (previously activated) exposed to CO would thereafter absorb hydrogen gas only very slowly, if at all. Saturated $LaNi_5$ hydride, exposed to CO, would subsequently desorb hydrogen gas only very slowly, even at higher temperatures. If part of the Ni in $LaNi_5$ is replaced by Cu, this inhibition effect is reduced.

I. Thermal Compression

The fast rate of absorption and desorption, the constancy of the plateau pressure (when the temperature is kept constant), the large absorption capacity, and the favorable pressure range of $LaNi_5$ hydride have been exploited to build a thermal absorption compressor for hydrogen gas (*18*, *80*). Since thermal energy is used for the work of compression, the compressor may be driven by waste heat or solar energy.

The working principle is indicated in Fig. 13. Hydrogen gas is absorbed at a low temperature level and is desorbed at a high temperature level. Despite the dead volume (at least 60% for a container with $LaNi_5$ hydride in powdered form), a high pressure ratio can still be reached because of the large absorption capacity of $LaNi_5$.

In a prototype (Fig. 14) three containers were mounted in a vacuum enclosure, each containing about 335 gm $LaNi_5$ and having its own electric heating and water cooling. This multiple system was used to obtain a nearly steady flow of 10 mg/sec of hydrogen at 45 atm and 160°C with a heat input

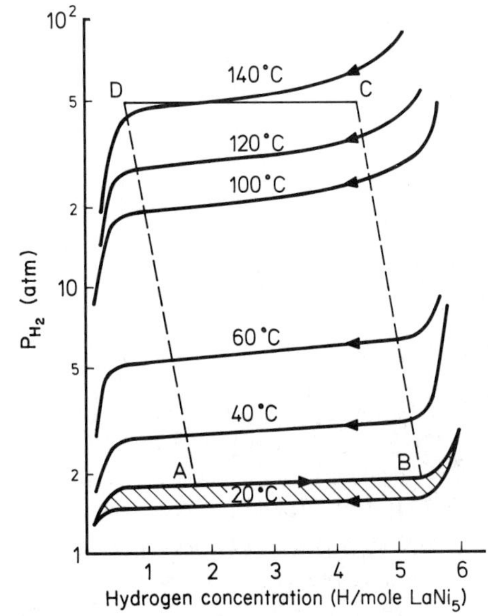

Fig. 13. Pressure–composition isotherms at various temperatures of the $LaNi_5$–hydrogen system. The hysteresis is indicated only for the absorption and desorption isotherm at 20°C. The working principle of the compressor is shown schematically. The path *AB* is the absorption process at temperature T_a and pressure P_a. The path *BC* is the heating of the $LaNi_5H_{2m}$ to temperature T_b (some desorption takes place because of the dead volume between the $LaNi_5$ particles, indicated by the slope of *BC*). The path *CD* is the desorption process at temperature T_b and pressure P_b. The path *DA* is the cooling of the $LaNi_5$ (there is some absorption of hydrogen gas from the dead volume). After the *m* moles of hydrogen gas have been driven out at T_b and P_b into a volume closed by a movable piston, a valve is closed to isolate this gas vessel. The hydrogen in this vessel is first cooled to T_a at the constant pressure P_b and is then expanded reversibly to P_a. Thereafter, the valve is reopened, and the hydrogen is again absorbed by the $LaNi_5$ (path *AB*). (See also Fig. 15.)

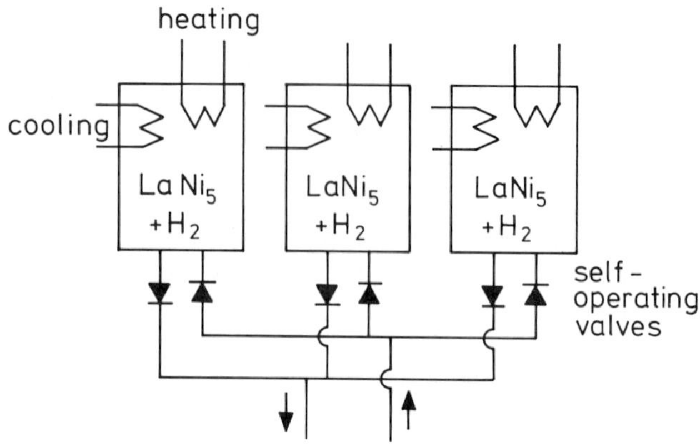

Fig. 14. Schematic diagram of an $LaNi_5$ hydride thermal absorption compressor. Hydrogen is compressed from 4 to 45 atm, with thermal energy at ~160°C and a heat sink at 15°C.

of about 1 kW. The intake pressure was about 4 atm at 15°C. The maximum attainable efficiency of the cycle is given by the relation

$$\eta = W/Q_b = [RT_a \ln(P_b/P_a)]/\Delta H_b$$

where ΔH_b is the heat of reaction for the desorption process.

To describe the cycle we follow the systematics of Fig. 13. The path *AB* represents the absorption process of *m* moles of hydrogen characterized by P_a and T_a as constants. The heat of reaction Q_a is dissipated to the cooling water. The path *BC* represents the heating of the $LaNi_5$ hydride (including its container) from T_a to T_b, whereby the pressure changes from P_a to P_b. The heat supplied to the system during this trajectory is denoted by $Q_{a,b}$. The path *CD* is the desorption process characterized by P_b and T_b as constants. The heat of reaction Q_b is supplied to the system. Finally, the path *DA* represents the cooling of the $LaNi_5$, including its container, from T_b to T_a, the cooling of the hydrogen gas from T_b to T_a, and the reversible expansion of the gas from P_b to P_a. The heat exchange with the surroundings is denoted by $Q_{b,a} + Q_{a,a}$. The various contributions are listed in Table VI, where the heat of the absorption (desorption) reaction ΔH_a (ΔH_b) refers to 1 mole of $LaNi_5$.

It is assumed that heat exchange is ideal (infinitely small ΔT's). Furthermore, the dead volume of the $LaNi_5$ containers is disregarded since the heat of desorption for the hydrogen molecules that occupy the dead volume during heating can theoretically be recovered during the cooling period.

The cycle is represented in a T–S diagram in Fig. 15. We have followed the same systematics as in Fig. 13 and Table VI. Path *AB* represents the

TABLE VI

Energy Contributions in the Thermal Compression of *m* Moles of Hydrogen Gas by 1 Mole of $LaNi_5$ and the Reversible Expansion of the Gas[a]

Trajectory	Characterization	Q_s	Q_H	W
a	P_a, T_a constant	$-m\,\Delta H_a$	0	$-mRT_a$
a,b	$P_a \rightarrow P_b$; $T_a \rightarrow T_b$	$C(LaNi_5H_{2m})(T_b - T_a)$	0	0
b	P_b, T_b constant	$m\,\Delta H_b$	0	mRT_b
b,a	P_b constant; $T_b \rightarrow T_a$	$-C(LaNi_5)(T_b - T_a)$	$-mC_p(H_2)(T_b - T_a)$	$-mR(T_b - T_a)$
a,a	$P_b \rightarrow P_a$; T_a constant	0	$mRT_a \ln(P_b/P_a)$	$mRT_a \ln(P_b/P_a)$

[a] Heat exchange between the surroundings and solid $LaNi_5$ or $LaNi_5$ hydride is indicated by Q_s, heat exchange between surroundings and hydrogen gas is represented by Q_H, and the work exerted on or done by the system is indicated by W. The heat capacity of $LaNi_5$ or $LaNi_5H_{2m}$, including the container, is indicated by C.

absorption process (the entropy of the system is reduced). Path BC represents the heating of the $LaNi_5H_{2m}$ from T_a to T_b, which is accompanied by an entropy increase,

$$dQ_{a,b}/T = C(LaNi_5H_{2m})\ln(T_b/T_a)$$

when the heat capacity C is assumed to be temperature independent. Path CD represents the desorption process giving an increase in entropy of the system, and path DA' is the cooling of the $LaNi_5$ and hydrogen gas, accompanied by a decrease in entropy,

$$dQ_{b,a}/T = [C(LaNi_5) + mC_p(H_2)]\ln(T_b/T_a)$$

During the isothermal expansion from P_b to P_a, there is an increase in entropy equal to $mR\ln(P_b/P_a)$, which is indicated in Fig. 15 as the path $A'A$.

The heat capacity of the hydride $C(LaNi_5H_{2m})$ will in general not be equal to the sum of the heat capacities of the intermetallic compound $C(LaNi_5)$ and that of the hydrogen gas $C(H_2)$. This follows from the fact that the hydrogen absorbed is expected to contribute only to the high-frequency part of the spectrum of lattice modes (*92, 93*). Therefore, only a small contribution to the heat capacity will be the result. The heat of reaction ΔH_b will therefore be different from ΔH_a:

$$m(\Delta H_b - \Delta H_a) = \int_{T_a}^{T_b} \Delta C_p\, dT$$

An irreversibility in the cycle is the hysteresis between the equilibrium pressures for absorption and desorption (see Figs. 4 and 13). The hydrogen is absorbed at a pressure $P_a + \Delta P$ at temperature T_a. It has already been

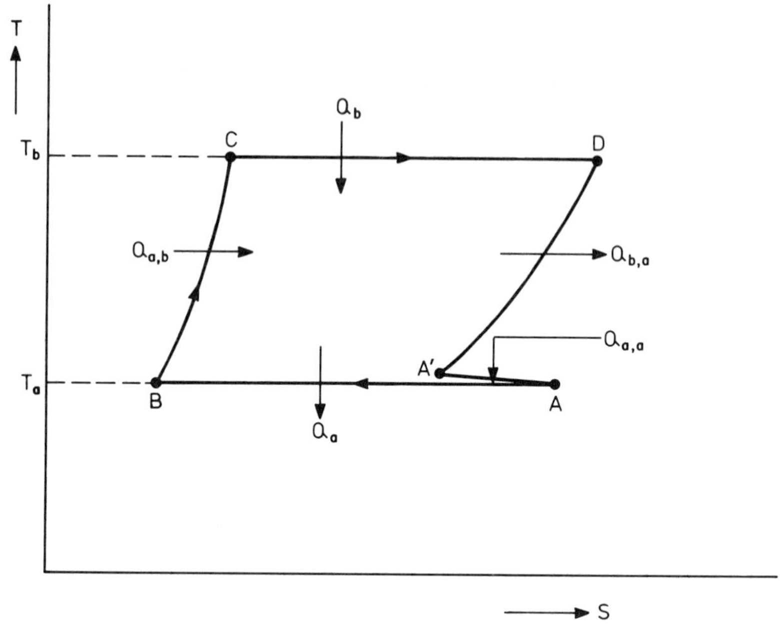

Fig. 15. A T–S diagram representing the reversible thermal compression and expansion of m moles of hydrogen gas by an $LaNi_5$ hydride absorption compressor. The systematics are the same as those in Fig. 13 with the exception of the introduction of A'. The path DA' represents the cooling of the $LaNi_5$ and the cooling of the hydrogen gas in a separate vessel at constant pressure P_b from temperature T_b to T_a. The path $A'A$ represents the reversible expansion of the hydrogen gas from P_b to P_a, during which the heat $Q_{a,a} = mRT_a \ln(P_b/P_a)$ is supplied from the surroundings to the gas.

mentioned in Section III that $\Delta P/P$ is nearly constant for different temperatures. The effect of the hysteresis is that the efficiency becomes

$$\eta_1 = [RT_a \ln(P_b/P_a) - RT_a \ln(1 + \Delta P_a/P_a)]/\Delta H_b$$

External factors, however, have a considerable influence on the efficiency. The most important is probably the ΔT effect, which is the finite temperature difference necessary for the transport of heat at a finite rate to the $LaNi_5$ powder at T_b and from the $LaNi_5$ powder at T_a.

J. Heat Pumps

The household refrigerator is a well-known example of a heat pump. Heat from the inside low-temperature level is transferred to the outside high-temperature level. The medium used is a Freon-based liquid, which flows to an evaporator inside the refrigerator. The recondensation takes place, after compression, outside the refrigerator, while condensation heat is transferred to the surroundings. The compressor is driven by electric power.

The principle of a thermomechanical engine and of a heat pump is illustrated schematically in Fig. 16a and b, respectively. Let us suppose that an amount of heat equal to Q_h is supplied to the system at a high temperature level T_h (Fig. 16a), and that an amount Q_m is removed at a medium temperature level T_m. The result is mechanical work $A = Q_h - Q_m$.

From the second law of thermodynamics, it can be shown that $Q_h/T_h = Q_m/T_m$; thus, $A = Q_h(T_h - T_m)/T_h$. Let us furthermore suppose that the mechanical work is used to drive a refrigerator. Heat Q_l at a low temperature level T_l is transferred to the temperature level T_m, where Q_m' is dissipated to the surroundings.

With $A = Q_m' - Q_l$ and $Q_m'/T_m = Q_l/T_l$, it follows that the refrigeration capacity is

$$Q_l = Q_h(1 - T_m/T_h)/(T_m/T_l - 1).$$

The heat produced at the temperature level T_m is

$$Q_m + Q_m' = Q_l(1 - T_l/T_h)/(1 - T_l/T_m)$$

Let us now look at heat pumps based on metal hydride systems. The temperature dependence of the plateau pressures of nine different hydrides is schematically represented in Fig. 17. By the use of absorption and desorption cycles of two different hydrides, the thermal energy at T_h can be used directly to pump heat from T_l to T_m. This is illustrated in Fig. 18a. Consider two metal hydrides A and B in open communication with each other. Hydride A is less stable than Hydride B. Hydride A reaches a given equilibrium pressure P_2 at a substantially lower temperature (T_m) than B (T_h). When B, saturated with hydrogen, is heated to T_h, the hydrogen pressure rises to P_2 and the hydrogen desorbs. The hydrogen is absorbed again by hydride A, which therefore increases in temperature until reaching T_m corresponding to P_2.

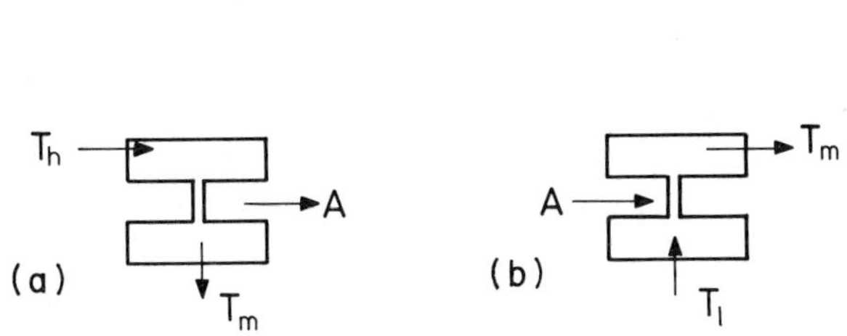

Fig. 16. (a) Thermomechanical engine. Heat Q_h is supplied to the system at temperature T_h. Heat Q_m is removed from the system at temperature T_m. Mechanical work A drives a refrigerator. (b) Heat pump. Mechanical work A drives a refrigerator. Heat Q_l is taken at low temperature level T_l and is transferred to medium temperature level T_m.

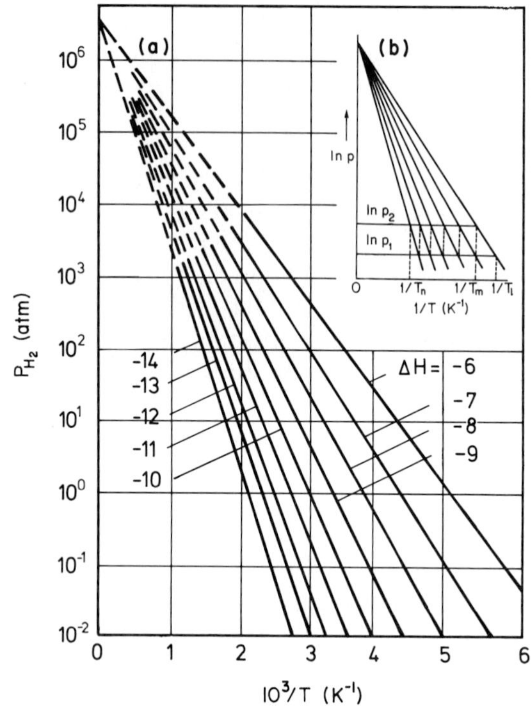

Fig. 17. Schematic representation of the dependence of the plateau pressure on the inverse absolute temperature ($\ln p = -\Delta S/R + \Delta H/RT$) of various metal–hydrogen systems. The lines converge to the point $\ln p = -\Delta S/R$, where the value $\Delta S = -30$ cal/K mole H_2 is taken for the entropy change. It must be emphasized that this picture is only schematic, since ΔS is generally not independent of temperature and differs for different metal–hydrogen systems (see Section III). The ΔH values indicated in the figure are in kilocalories per mole H_2. Inset: If a metal–hydrogen system (e.g., hydride A) is chosen to operate between the temperatures T_l and T_m, the corresponding pressures are P_1 and P_2. These pressures P_1 and P_2 consequently determine the temperatures $T_1, T_2, \ldots, T_n$.

The heat of absorption ΔH_A is dissipated at T_m. When nearly all the hydrogen from B has been desorbed, thermal contact between system B and the heat source is broken and B is brought into contact with the environment at temperature T_m. The temperature in B now drops, and hydrogen is absorbed. The hydrogen pressure decreases to P_1, and the desorption from A causes the temperature to drop to T_l. Simultaneously, A takes up heat from the source at T_l. The heat of reaction ΔH_B for the absorption by system B is dissipated to the environment at T_m. When all the hydrogen has been absorbed by B, the system B is coupled once more to the heat source and the cycle is repeated.

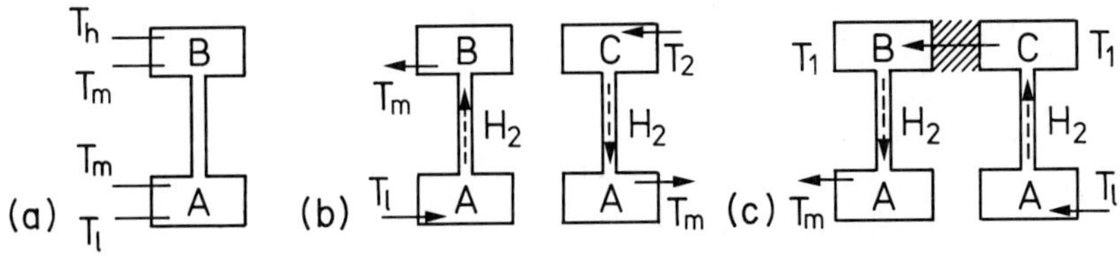

Fig. 18. (a) Single-stage heat pump (B, A) consisting of two metal–hydrogen systems A and B in communication with each other. System A operates between the temperatures T_l and T_m, with corresponding pressure P_1 and P_2. System B operates between the temperatures T_m and T_1 ($T_h = T_1$) determined by the pressures P_1 and P_2. (b) Two-stage heat pump (B, A) combined with (C, A). The hydride C operates between the temperatures T_1 and T_2 ($T_h = T_2$) determined by the pressures P_1 and P_2 (inset, Fig. 17). Hydride A of stage (C, A) is identical with hydride A of stage (B, A). Thermal energy at temperature T_2 is supplied to hydride C; the hydrogen desorbs and is absorbed by system A. The heat of absorption is used for heating purposes at the ambient temperature T_m. At the same time, system B absorbs hydrogen from another hydride A. The latter extracts heat from a low-temperature source at T_l. The heat developed by B is used for heating at the ambient temperature T_m (see text). (c) The hydrides B and C are brought into thermal contact. At temperature T_1 hydride C absorbs hydrogen and the heat of absorption is transferred to B. Hydride B desorbs at T_1, and hydride A of stage (B, A) absorbs at T_m. The heat of absorption is used for heating purposes at the ambient temperature T_m.

The heat dissipated to the environment at T_m is $\Delta H_A + \Delta H_B$. The heat input at T_h is ΔH_B. The efficiency is therefore

$$\eta = 1 + \Delta H_A/\Delta H_B$$

The principle of a two-stage heat pump is schematically indicated in Fig. 18b. The operation is as follows. Heat is supplied to C, forming part of the second stage (C, A), at T_2, and hydrogen is desorbed. The heat of reaction is equal to ΔH_C. The hydrogen is absorbed by A, and ΔH_A is dissipated at T_m. Then C is decoupled from the heat source and coupled to B, which is at a temperature T_m (Fig. 18c). The hydride is thus heated by C, which cools off. When C starts to absorb hydrogen, its heat of absorption ΔH_C passes to B. (Since B has a different heat of reaction, its total amount must be larger by a factor $\Delta H_C/\Delta H_B$ than that of hydride C.) While B and C are both at T_1, the hydride A of stage (C, A) desorbs hydrogen and extracts the heat of reaction ΔH_A from the low-temperature source at T_1. The hydride A of stage (B, A) absorbs hydrogen and dissipates the heat ΔH_A to the surroundings at T_m. Finally, B is decoupled from C and cools to T_m, starts to absorb hydrogen, and dissipates ΔH_B to the surroundings T_m. The overall efficiency is

$$\eta = 1 + \Delta H_A/\Delta H_B + \Delta H_A/\Delta H_C$$

The schematic representation (Fig. 17) of the temperature dependence of the equilibrium pressure of various hydrides makes it much easier to find the characteristic ΔH_n values of the n diffrrent metal–hydrogen systems required for an optimum multistage heat pump. Such a heat pump can in theory have the Carnot efficiency. The heat pump operates between the temperatures T_h ($T_h = T_1$ or T_2 or T_3, etc.), T_m, and T_l, where T_h is the temperature at which heat is supplied, T_m is the temperature to which heat is to be pumped (e.g., room temperature), and T_l is the temperature of the cold source from which heat is extracted. The ΔT's for heat exchange at the various temperature levels are taken to be infinitely small in our discussion.

From the inset of Fig. 17 it follows that

$$\Delta H_A = R \ln(P_2/P_1)/(1/T_m - 1/T_l)$$

$$\Delta H_B = R \ln(P_2/P_1)/(1/T_h - 1/T_m)$$

The efficiency of a single-stage heat pump is therefore

$$\eta = 1 + \Delta H_A/\Delta H_B = (1 - T_l/T_h)/(1 - T_l/T_m)$$

which is equal to η_c (efficiency of a Carnot cycle). The efficiency of a two-stage heat pump is

$$\eta = 1 + \Delta H_A/\Delta H_B + \Delta H_A/\Delta H_C = (1 - T_l/T_h)/(1 - T_l/T_m)$$

which is also equal to η_C. It is attained only if the temperatures satisfy the relation

$$T_h = T_m(T_m/T_l)$$

or in general

$$T_n = (T_{n-1})^2/T_{n-2}$$

In the foregoing discussion we have assumed ideal heat exchange and ideal metal–hydrogen systems. Regarding the latter we know that the heat of formation is in general not independent of temperature. As mentioned in Section III, there is also a hysteresis which is manifested as a pressure difference between absorption and desorption. The influence of these effects still awaits investigation.

Probably the most important factor reducing the efficiency is the temperature gradient necessary for the heat exchange with the hydride powders. Very thin powder layers with a large area for heat transfer are required. This imposes constraints on the construction of these heat pumps. In this respect the HYCSOS project of Sheft *et al.* (*94*) is worth mentioning. The unit described is a two-hydride heat pump concept of sufficient size to be able to demonstrate the feasibility of heat-transfer processes on the basis of metal hydrides.

K. Electrochemical Cells with Metal Hydrides

Earl and Dunlop (*95*) have used $LaNi_5$ hydride for the chemical storage of hydrogen in a nickel/hydrogen battery. The constant plateau pressure and the favorable pressure range make $LaNi_5$ especially suitable for this application. The reaction in the electrochemical cell is

$$NiOOH + \tfrac{1}{2}H_2 \underset{\text{charge}}{\overset{\text{discharge}}{\rightleftharpoons}} Ni(OH)_2$$

During the charging period, hydrogen pressure would normally increase from 3 to 33 atm, but because the hydrogen is stored in the $LaNi_5$ the operating pressure is only a few atmospheres. The hydrogen desorbs from the $LaNi_5$ during electrical discharge, and the nickel hydroxide is reformed.

A comparison of an NiOOH/Cd and an $NiOOH/H_2$ cell has been made by Font and Goualard (*96*). The pratical energy density of the two types of cell is given as 30 and 60 Wh/kg, respectively. For $NiOOH/LaNi_5$ an energy density of 45 Wh/kg may be expected (*97*).

The next step is to use the metal hydride itself as the negative electrode in the battery (*99–102*). The electrode can consist of sintered powder particles of, for example, titanium–nickel alloys (*100*), $LaNi_4Cu$ (*97*), or other ternary compounds based on $LaNi_5$ (*101*, *102*). A survey of the expected practical energy densities is given in Table VII.

Electrochemical power generation does not suffer from the limitations of the Carnot cycle. The conversion can be achieved with a much higher efficiency than with thermomechanical engines. Some novel devices have been proposed by Videm (*103*). One is a system for supplying a house with heat and electric power at the same time. It consists of an electrochemical cell with a metal hydride anode and an oxygen electrode. The other system is a

TABLE VII

Comparison of the Practical Energy Density in Some Electrochemical Cells, $LaNi_5$ Electrode and Gasoline

Electrochemical cell	Practical energy density (Wh/kg)
NiOOH/Cd	30
$NiOOH/H_2$	60
$NiOOH/LaNi_5H_6$	45
$LaNi_5$ electrode	400
Lead acid	22
Gasoline	16,000

hydrogen/hydride–air storage battery, which can be charged and discharged like a normal secondary cell, but which has the advantage of also being chargeable with hydrogen gas.

Characteristic features of hydrogen-absorbing alloys to be used as electrodes in aqueous solutions comprising requirements such as chemical stability and kinetics are described by van Rijswick (*104*).

L. Utility Load Leveling and Thermal Storage

Apart from the two-hydride heat pump, the most spectacular application of metal hydrides is probably that as a utility load leveling system (*22, 105*). Although the essential features do not comprise more than the large-scale storage of hydrogen (6.4 kg H_2), the project (*105*) as a whole is impressive. Hydrogen is produced electrolytically. After compression from 5 to 35 atm it is stored as iron titanium hydride. Subsequently, the hydride is decomposed to supply the hydrogen to a fuel cell, where electrical energy is again generated. The objective is to produce hydrogen during off-peak hours and to use the stored hydrogen for conversion to electricity in the peak hours.

A thermal storage system in conjunction with solar energy was suggested by Libowitz (*98*). Solar heat is used to decompose the hydride into the intermetallic compound and H_2 gas at an elevated temperature. The hydrogen is stored as a pressurized gas. The stored energy can be released by permitting the hydrogen gas to react with the previously discharged intermetallic compound.

REFERENCES

1. W. M. Mueller, J. P. Blackledge, and G. G. Libowitz, "Metal Hydrides." Academic Press, New York, 1968.

2. G. G. Libowitz, *Inorg. Chem.* **10,** 79 (1972).

3. J. Friedel, *Ber. Bunsenges. Phys. Chem.* **76,** 829 (1972).

4. H. W. Newkirk, *Lawrence Livermore Lab.* [*Rep.*] *UCRL* **UCRL-51244** (1974).

5. J. J. Reilly and R. H. Wiswall, Jr., *Brookhaven Natl. Lab.* [*Rep.*] *BNL* **BNL 50023** (1966); *Inorg. Chem.* **6,** 2220 (1967).

6. A. R. Miedema, K. H. J. Buschow, and H. H. van Mal, *Electrochem. Soc. Conf. Proc.* **77-6,** 456 (1977).

7. K. H. J. Buschow, *Rep. Prog. Phys.* **40,** 1179 (1977).

8. K. H. J. Buschow, *Rep. Prog. Phys.* **42,** 1373 (1979).

9. K. H. J. Buschow, *in* "Ferromagnetic Materials" (E. P. Wohlfarth, ed.), Vol. I, p. 298. North-Holland Publ., Amsterdam, 1979.

10. A. C. Larson and D. T. Cromer, *Acta Crystallogr.* **15,** 1224 (1962).

11. F. A. Shunk, "Constitution of Binary Alloys." McGraw-Hill, New York, 1969.

12. R. P. Elliot, "Constitution of Binary Alloys." McGraw-Hill, New York, 1965.

13. K. A. Gschneidner, "Rare Earth Alloys." Van Nostrand, Princeton, New Jersey, 1961.
14. E. M. Savitsky and V. F. Terekhova, "Metallography of Rare Earths." HAYKA Publ. Co., Moscow, 1975.
15. A. R. Miedema, *J. Less-Common Met.* **32,** 117 (1973).
16. A. R. Miedema, F. R. de Boer, and R. Boom, *in* "Crystal Structure and Chemical Bonding in Inorganic Chemistry" (C. J. M. Rooymans and A. Rabenau, eds.), p. 163. North-Holland Publ., Amsterdam, 1975.
17. G. D. Sandrock, *in* "Hydrides for Energy Storage" (A. F. Andresen and A. J. Maeland, eds.), p. 351. Pergamon, Oxford, 1978.
18. H. H. van Mal, *Philips Res. Rep., Suppl.* No. 1 (1976).
19. F. A. Kuijpers, *Philips Res. Rep., Suppl.* No. 2 (1973).
20. F. A. Kuijpers and H. H. van Mal, *J. Less-Common Met.* **23,** 395 (1971).
21. C. E. Lundin and F. E. Lynch, *in* "Hydrogen for Energy Storage" (A. F. Andresen and A. J. Maeland, eds.), p. 395. Pergamon, Oxford, 1978.
22. J. J. Reilly, *in* "Hydrides for Energy Storage" (A. F. Andresen and A. J. Maeland, eds.), p. 301. Pergamon, Oxford, 1978.
23. R. M. van Essen and K. H. J. Buschow, *J. Less-Common Met.* **70,** 189 (1980).
24. N. A. Scholtus and W. K. Hall, *J. Chem. Phys.* **39,** 868 (1963).
25. G. Busch, L. Schlapbach, and A. Seiler, *in* "Hydrides for Energy Storage" (A. F. Andresen and A. J. Maeland, eds.), p. 293. Pergamon, Oxford, 1978.
26. K. H. J. Buschow and H. H. van Mal, *J. Less-Common Met.* **29,** 203 (1972).
27. J. H. N. van Vucht, F. A. Kuijpers, and H. C. A. M. Bruning, *Philips Res. Rep.* **25,** 133 (1970).
28. F. A. Kuijpers and B. O. Loopstra, *J. Phys. Chem. Solids* **35,** 306 (1974).
29. A. F. Andresen, *in* "Hydrides for Energy Storage" (A. F. Andresen and A. J. Maeland, eds.), p. 61. Pergamon, Oxford, 1978.
30. V. V. Burnasheva, V. A. Yartis, N. V. Padeeva, S. P. Solovlev, and K. N. Somenenko, *Dokl. Akad. Nauk SSSR* **283,** 844 (1978).
31. A. Furrer, P. Fischer, A. Hälg, and L. Schlapbach, *in* "Hydrides for Energy Storage" (A. F. Andresen and A. J. Maeland, eds.), p. 73. Pergamon, Oxford, 1978.
32. S. A. Steward, J. F. Lakner, and F. Uribe, *Adv. Chem. Ser.* No. 163, p. 284 (1977).
33. S. W. Peterson, V. N. Sadana, and W. L. Korst, *J. Phys.* (*Paris*) **25,** 451 (1964).
34. J. Bergsma, J. A. Goedkoop, and J. H. N. van Vucht, *Acta Crystallogr.* **14,** 233 (1961).
35. J. J. Didisheim, K. Yvon, D. Shaltiel, and P. Fischer, *Solid State Commun.* **31,** 47 (1979).
36. P. Thompson, M. A. Pick, F. Reidinger, L. M. Corliss, J. M. Hastings, and J. J. Reilly, *J. Phys. F* **8,** L75 (1978); P. Thompson, J. J. Reilly, F. Reidinger, J. M. Hastings, and L, M. Corliss, *J. Phys. F* **9,** L61 (1979).
37. A. Santoro, A. J. Maeland, and J. J. Rush, *Acta Crystallogr.* (*B*) **34,** 3059 (1978).
38. S. J. C. Irvine and I. R. Harris, *in* "Hydrogen for Energy Storage" (A. F. Andresen and A. J. Maeland, eds.), p. 431. Pergamon, Oxford, 1978.
39. R. M. van Essen and K. H. J. Buschow, *J. Less-Common Met.* **64,** 277 (1979).
40. K. H. J. Buschow, R. L. Cohen, and K. W. West, *J. Appl. Phys.* **48,** 5289 (1977).
41. I. Jacob and D. Shaltiel, *J. Less-Common Met.* **65,** 117 (1979).
42. T. B. Flanagan, *in* "Hydrogen for Energy Storage" (A. J. Maeland and A. F. Andresen, eds.), p. 43. Pergamon, Oxford, 1978.
43. A. Goudy, W. E. Wallace, R. S. Craig, and T. Takeshita, *Adv. Chem. Ser.* No. 157, p. 212 (1978).
44. C. A. Beckman, A. Goudy, T. Takeshita, W. E. Wallace, and R. S. Craig, *Inorg. Chem.* **15,** 2184 (1976).
45. A. J. Maeland, A. F. Andresen, and K. Videm, *J. Less-Common Met.* **45,** 347 (1976).

46. G. Busch, L. Schlapbach, and T. von Waldkirch, *J. Less-Common Met.* **60,** 83 (1978).
47. A. J. Maeland, *in* "Hydrogen for Energy Storage" (A. J. Maeland and A. F. Andresen, eds.), p. 447. Pergamon, Oxford, 1978.
48. D. Shaltiel, I. Jacob, and D. Davidov, *J. Less-Common Met.* **53,** 117 (1977).
49. Y. Machida, T. Yamadaya, and M. Asanuma, *in* "Hydrogen for Energy Storage" (A. J. Maeland and A. F. Andresen, eds.), p. 329. Pergamon, Oxford, 1978.
50. K. H. J. Buschow, H. H. van Mal, and A. R. Miedema, *J. Less-Common Met.* **42,** 163 (1975).
51. B. Nacken and W. Bronger, *Z. Anorg. Allg. Chem.* **439,** 29 (1978).
52. R. C. Bowman, A. Attala, G. C. Carter, and Y. Chabre, *in* "Hydrides for Energy Storage" (A. F. Andresen and A. J. Maeland, eds.), p. 97. Pergamon, Oxford, 1978.
53. T. K. Halstead, N. A. Abood, and K. H. J. Buschow, *Solid State Commun.* **19,** 425 (1976).
54. R. G. Barnes, W. C. Harper, S. O. Nelson, D. K. Thome, and D. R. Torgeson, *J. Less-Common Met.* **49,** 483 (1976).
55. R. F. Karlicek and I. J. Lowe, *Solid State Commun.* **31,** 163 (1979).
56. E. Lebsanft, D. Richter, and J. Töpler, *J. Phys. F* **9,** 1057 (1979).
57. K. H. J. Buschow, unpublished observations (1977).
58. H. A. Kierstead, P. J. Viccaro, G. K. Shenoy, and B. D. Dunlap, *J. Less-Common Met.* **66,** 219 (1979).
59. B. Nacken and W. Bronger, *J. Less-Common Met.* **52,** 323 (1977).
60. R. Kadel and A. Weiss, *J. Less-Common Met.* **65,** 89 (1979).
61. K. H. J. Buschow and A. R. Miedema, *in* "Hydrides for Energy Storage" (A. F. Andresen and A. J. Maeland, eds.), p. 253. Pergamon, Oxford, 1978.
62. A. R. Miedema, K. H. J. Buschow, and H. H. van Mal, *J. Less-Common Met.* **49,** 463 (1976).
63. K. H. J. Buschow, *J. Less-Common Met.* **51,** 173 (1977).
64. J. Clinton, H. Bittner, and H. Oesterreicher, *J. Less-Common Met.* **41,** 187 (1975).
65. S. K. Malik and W. E. Wallace, *Solid State Commun.* **24,** 283 (1977).
66. H. C. Siegman, L. Schlapbach, and C. R. Brundle, *Phys. Rev. Lett.* **40,** 972 (1978).
67. G. Busch and L. Schlapbach, *Helv. Chim. Acta* **51,** 5 (1978).
68. L. Schlapbach, A. Seiler, and F. Stucki, *Mater. Res. Bull.* **13,** 697 (1978).
69. R. Hempelmann, D. Ohlendorf, and E. Wicke, *in* "Hydrogen for Energy Storage" (A. J. Maeland and A. F. Andresen, eds.), p. 407. Pergamon, Oxford, 1978.
70. D. M. Gruen, M. H. Mendelsohn, and I. Sheft, *Electrochem. Soc. Conf. Proc.* **77-6,** 482 (1977).
71. C. E. Lundin, F. E. Lynch, and C. B. Magee, *J. Less-Common Met.* **56,** 19 (1977).
72. S. Yajima, H. Kayano, and H. Toma, *J. Less-Common Met.* **55,** 139 (1977).
73. K. H. J. Buschow and P. F. de Châtel, *Pure Appl. Chem.* **52,** 135 (1979).
74. E. R. Bauminger, D. Davidov, I. Felner, I. Nowik, S. Ofer, and D. Shaltiel, *Physica (Amsterdam)* **86–88B,** 201 (1977).
75. F. W. Oliver, K. W. West, R. L. Cohen, and K. H. J. Buschow, *J. Phys. F* **8,** 701 (1976).
76. R. L. Cohen, K. W. West, F. W. Oliver, and K. H. J. Buschow, *Phys. Rev. B* **21,** 941 (1980).
77. J. Keinert, *Atomkernenergie* **18,** 261 (1971).
78. B. Stalinski, *Bull. Acad. Pol. Sci., Cl. 3* **3,** 613 (1935).
79. J. H. N. van Vucht, F. A. Kuijpers, and H. C. A. M. Bruning, *Philips Res. Rep.* **25,** 133 (1970).
80. H. H. van Mal, *Chem.-Ing.-Tech.* **45,** 80 (1973).
81. J. J. Reilly and R. H. Wiswall, Jr., *Inorg. Chem.* **13,** 218 (1974).
82. See Ref. 1.

83. O. Reifenschweiler, *Comm. Eur. Communities* [*Rep.*] *EUR* **EUR-4896** (1972); *Philips Res. Rep.* **16,** 401 (1961).
84. J. H. N. van Vucht, *Philips Res. Rep., Suppl.* No. 1 (1963).
85. O. Reifenschweiler, *Rev. Sci. Instrum.* **35,** 456 (1964).
86. R. H. Wiswall, Jr. and J. J. Reilly, *Inorg. Chem.* **11,** 1691 (1972).
87. H. Buchner, *in* "Hydrogen for Energy Storage" (A. F. Andresen and A. J. Maeland, eds.), p. 569. Pergamon, Oxford, 1978.
88. H. Davitian, *Proc. Intersoc. Energy Convers. Eng. Conf.* Vol. I, Am. Inst. Chem. Engrs., New York, 1976.
89. C. E. Lundin and R. W. Sullivan, *Proc. THEME Conf., Univ. Miami* (1974).
90. H. Wenzl and K. H. Klatt, "Hydrogen for Energy Storage" (A. F. Andersen and A. J. Maeland, eds.), p. 323. Pergamon, Oxford, 1978.
91. J. J. Reilly and R. H. Wiswall, Jr., *Brookhaven Natl. Lab.* [*Rep.*] *BNL* **BNL17136** (1972); E. Lebsanft, G. Arnold, K. H. Klatt, J. Töpler, and H. Wenzl, *Proc. Int. Congr. Hydrogen Met., 2nd, Paris* (1977).
92. H. E. Flotow and D. W. Osborne, *J. Chem. Phys.* **34,** 1418 (1961).
93. H. E. Flotow, R. W. Osborne, and K. Otto, *J. Chem. Phys.* **36,** 866 (1962).
94. I. Sheft, D. M. Gruen, G. J. Lemich, L. W. Carlson, A. E. Knox, J. M. Nixon, and M. H. Mendelsohn, *in* "Hydrides for Energy Storage" (A. F. Andresen and A. J. Maeland, eds.), p. 551. Pergamon, Oxford, 1978; I. Sheft, D. M. Gruen, and G. Lamich, *Argonne Natl. La.* [*Rep.*] *ANL* **ANL-79-8** (1979).
95. M. W. Earl and J. D. Dunlop, *Meet. Electrochem. Soc., Boston, Mass., 1973.*
96. S. Font and J. Goualard, *Power Sources Symp., 5th, Brighton, Eng.*, 1974.
97. P. M. Boter, *Novel Electrode Mater. Symp., Brighton, Eng., 1975.*
98. G. G. Libowitz, *Proc. Intersoc. Energy Convers. Eng. Conf.* Vol. I, p. 673, Am. Inst. Chem. Engrs., New York, 1976.
99. J. C. Barton, J. A. S. Green, and F. A. Lewis, *Nature* (*London*) **197,** 1293 (1963).
100. M. A. Gutjahr, H. Buchner, K. A. Beccu, and H. Säufferer, *Power Sources Symp., Brighton, Eng. 1972.*
101. H. Ewe, E. W. Justi, and K. Stephan, *Energy Convers.* **13,** 109 (1973).
102. A. Percheron-Guegan, L. Schlapbach, J. C. Archard, J. Sarradin, and G. Bronoël, *Bull. Soc. Chim. Fr.* (1974); Fr. Patent 7,516,160 (1975).
103. K. Videm, *in* "Hydrogen for Energy Storage" (A. F. Andresen and A. J. Maeland, eds.), p. 463. Pergamon, Oxford, 1978.
104. M. H. J. van Rijswick, *in* "Hydrogen for Energy Storage" (A. F. Andresen and A. J. Maeland, eds.), p. 261. Pergamon, Oxford, 1978.
105. J. J. Reilly, K. C. Hoffman, G. Strickland, and R. H. Wiswall, Twenty-Sixth Annual Proceedings Power Sources Conference, Atlantic City, New Jersey (1974).

14

Intercalation in Biological Systems*

W. DAVID WILSON

and

ROBERT L. JONES

* This review, which contains many of his ideas and experimental findings, is dedicated to our friend and colleague, the late Professor Edmond J. Gabbay.

INTERCALATION CHEMISTRY

ISBN 0-12-747380-7

I. EARLY INVESTIGATIONS OF THE INTERACTION OF PLANAR AROMATIC CATIONS WITH DNA

In the search for antimicrobial drugs it was found that many planar aromatic cations had significant activity. The acridine derivative quinacrine was the drug of choice against infections of the malaria parasite in World War II, and the related quinoline derivative chloroquine is still widely used as an antimalarial drug (*1*). A large number of phenanthridine derivatives such as ethidium bromide were synthesized as antitrypanosomal agents (*2*). Acridines such as proflavine were found to retard bacterial growth and induce mutations, but the results strongly depended on their structure, with both the cationic groups and the planarity and size and shape of the aromatic area being quite important for activity (*3*). The structures of these and other compounds to be discussed are shown in Fig. 1.

The growing realization that DNA was the genetic material (*4*) and the interest in DNA generated by the publication of the Watson–Crick model for the double helix (*5*) led to investigations of the interactions of several medically important planar aromatic cations with DNA [see Cavalier and Angelos (*6*)]. Dramatic changes in visible absorption and fluorescence spectra

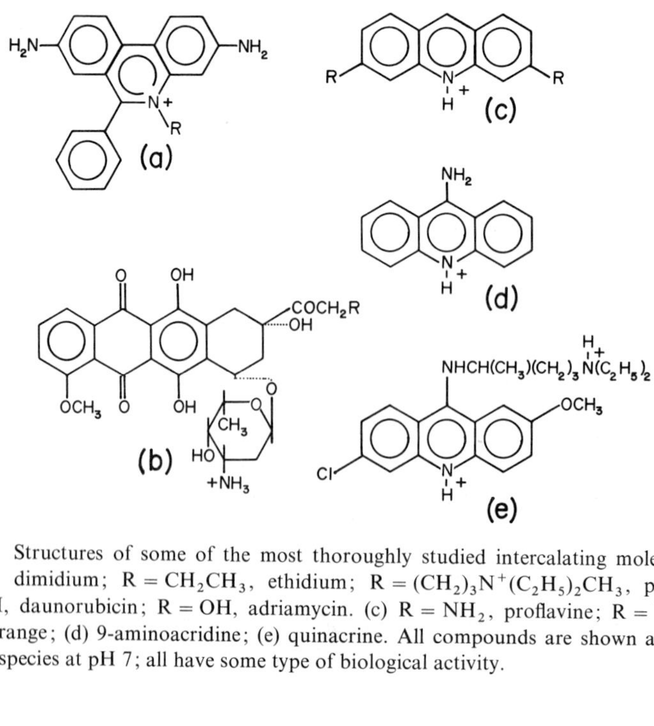

Fig. 1. Structures of some of the most thoroughly studied intercalating molecules: (a) $R = CH_3$, dimidium; $R = CH_2CH_3$, ethidium; $R = (CH_2)_3N^+(C_2H_5)_2CH_3$, propidium. (b) R = H, daunorubicin; R = OH, adriamycin. (c) $R = NH_2$, proflavine; $R = N(CH_3)_2$, acridine orange; (d) 9-aminoacridine; (e) quinacrine. All compounds are shown as the predominant species at pH 7; all have some type of biological activity.

were noted in these investigations. One of the early attempts to quantitate this type of interaction was made by Peacocke and Skerrett (*7*). They measured the amount of proflavine bound to DNA at different concentrations using spectral shifts and equilibrium dialysis. The results were plotted and analyzed according to the Scatchard equation (*8*). The plots obtained were curved, and Peacocke and Skerrett proposed that this was best explained by assuming two classes of independent, noninteracting binding sites for proflavine on DNA. The first class of sites saturated at one ligand molecule for every four to five DNA nucleotides and had a much greater binding constant than the second class of sites, which reached saturation at approximately one ligand per nucleotide. The results of a Peacocke and Skerrett spectrophotometric experiment and one of their equilibrium dialysis experiments are shown in Fig. 2. The strong binding was termed Type I binding by these authors, and they did not consider this strong interaction to cause any

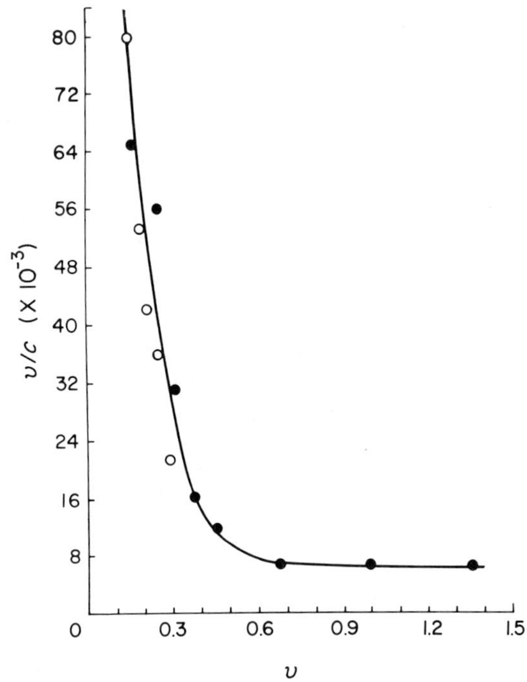

Fig. 2. A Scatchard plot using data from Peacocke and Skerrett (*7*) for the interaction of proflavine with DNA as determined spectrophotometrically (○) and by equilibrium dialysis (●); v represents moles of ligand bound per equivalent of DNA base pairs, and C represents the free-ligand concentration. The experiments were conducted at pH 6.2 and an ionic strength of 0.1.

significant structural perturbation of DNA (*7*). They thought that the ligand spectral shifts that resulted from binding could be explained by interaction with the DNA bases in the grooves of the double helix (*5*). The weaker (Type II) binding was considered to be primarily an electrostatic interaction between the polyanions of the DNA sugar–phosphate chain and a self-associated stack of the planar cations. This reasoning was based on analogy with the interaction of similar aromatic cations with polyanions (*9*). Because of the stacking of the cations in Type II binding, this interaction was found to be quite sensitive to ionic strength changes and to be unimportant at physiological conditions. This indicates that Type I binding accounts for the biological effects of proflavine and similar compounds on nucleic acids *in vivo*. Characterization of the Type I ligand–DNA complex became of utmost importance for the development of this field.

II. THE CLASSICAL INTERCALATION MODEL

In an effort to elucidate the structure of the Type I proflavine–DNA complex, Lerman (*10*) analyzed the interaction of this compound as well as acridine, acridine orange, and some simple nonplanar cations with DNA using hydrodynamic and fiber diffraction techniques. He found that the viscosity of a DNA solution was markedly increased by all the acridine derivatives, whereas simple inorganic cations and nonplanar organic cations, similar in composition to the acridines, caused slight decreases in viscosity, in agreement with previous results (*11*). The sedimentation coefficient, on the other hand, decreased with the acridines, whereas it increased with the nonplanar cations that bound to DNA. In the region of Type II binding of proflavine (*7*) the sedimentation coefficient began to increase, suggesting that Type I and Type II complexes have markedly different structures, with the Type II structure being more similar to the binding of nonplanar cations. The X-ray diffraction patterns of DNA–proflavine fibers retained the characteristic 3.4 Å meridional spots but lost the standard layer-line pattern of B-form DNA, and no new spots were found (*10*). The combined results of these experiments suggested that as a result of binding of acridine derivatives, the DNA double helix (a) became longer or some combination of longer and more stiff, (b) had a decreased mass per unit length, (c) retained its 3.4 Å spacing with the base pairs essentially perpendicular to the helix axis and (d) lost the regular repeating nature of the double helix. Within these constraints, Lerman (*12*, *13*) found that an intercalation model best described the acridine–DNA complex. Other structures, such as surface attachment of the ligand or insertion of the ligand into the double helix with displacement of a base pair, agreed with some but not all of the experimental results

(*14*). Model building studies indicated that the double helix could be extended with simple single-bond rotation to create a space for a planar aromatic molecule approximately 3.4 Å in thickness (*14*).

In a continuing investigation of this interaction Luzzati *et al.* (*15*) conducted a small-angle X-ray scattering study of the proflavine–DNA complex and found that, like pure DNA, the complex was rodlike and had a reduced mass per unit length, as expected from the intercalation model. In the next step of testing the intercalation model, Lerman (*14*) began to look at the ligand, in addition to DNA, using flow dichroism and flow-polarized fluorescence. The problem is one of fixing the orientation in solution of a plane, such as that of the acridine nucleus of quinacrine (Fig. 1), with respect to the orientation of the double helix. To do this two nonparallel and preferably nearly perpendicular vectors are needed (*16*). These can be two different transitions in absorption spectroscopy or an absorption and an emission transition in fluorescence spectroscopy. One transition or two parallel transitions can eliminate certain binding modes but can not unambiguously characterize intercalation (*14*, *16*). Since the DNA molecules are not perfectly oriented by flow, the dichroism of the complex must be compared to that of DNA alone. The dichroism of the complex was found to be slightly greater than that of DNA alone, as would be expected from its greater length, and to be the same at the DNA maximum absorption wavelength (260 nm) and at the ligand maximum, suggesting that the orientation of the acridine ring and the base pairs is the same within the complex (*14*). The flow-polarized fluorescence results also indicated that the acridine ring was oriented perpendicular to the direction of flow or to the helix axis, as predicted by the intercalation model. In a separate analysis of rates of diazotization, Lerman (*17*) found that the rates of reaction were slower when proflavine and similar ligands were bound to DNA than when they were bound to other anionic polymers. This indicated that the acridines have restricted access to the solvent when bound to DNA in a Type I complex, as predicted by the intercalation model.

These various converging lines of evidence were reviewed by Lerman (*14*), who concluded that the intercalation model was the only plausible structure for Type I binding. A unique model with the following features had evolved to explain the interaction of common acridine drugs and benzacridine carcinogens with DNA. (a) The rodlike structure, hydrogen bonding, and covalent structure of B-form DNA are retained; (b) the DNA length is increased as two adjacent base pairs are separated by 3.4 Å to create a planar space for intercalation of the acridine ring; (c) an unwinding of the double helix is required (this unwinding could not be measured and was simply a best guess based on model building; Lerman's initial estimate was 45° for unwinding), leaving the adjacent base pairs positioned more directly over each other; and (d) the acridine ring is placed in the cavity, with its long axis essentially parallel to the adjacent base pairs, in such a manner that the space created is filled as fully as possible. The size of the space created was close to that required to insert the acridine ring with its long axis oriented close to the long axes of the two adjacent base pairs. The development of this model sparked a tremendous research effort to determine how the model applied to other drugs, aromatic amino acid groups of peptides, carcinogens, and similar compounds. In this research the features of the intercalation model have undergone considerable development, but the general concepts originally postulated by Lerman have been retained.

III. DEVELOPMENT OF THE INTERCALATION MODEL

There is no attempt in this section to review the hundreds of papers that have reported on intercalating molecules since Lerman's initial work. The areas emphasized are those that resulted in new information concerning the model, but even here no exhaustive review is possible in the available space. The papers referred to are those the authors feel most clearly illustrate the general point or points under discussion.

A. The Double Helix Length Increase

Lerman (*10*, *14*) had some difficulty in the analysis of his original viscosity and sedimentation results because of the complexity of the tertiary structure of high-molecular-weight DNA in solution (*18*). Intercalation could affect both the local secondary structure of the double helix and the long-range coiling of the molecule. Cohen and Eisenberg (*19*) reasoned that this problem could be greatly simplified if shorter molecules, whose length was near or less than the persistence length of DNA in solution, were used in analyzing intercalating molecules. With these short, rodlike segments of double helical DNA, the tertiary effect due to long-range structural changes could be eliminated. In an extensive series of experimental and theoretical studies which continue to the present, Eisenberg and co-workers have characterized the hydrodynamic properties of DNA and its complexes with ligands (*20–23*). Equation (1) was derived by Cohen and Eisenberg (*19*) to relate the increase in length of the double helix on intercalation to the experimentally measured viscosity enhancement:

$$\frac{L}{L_0} = \left(\frac{[\eta]f(p)_0}{[\eta]_0 f(p)}\right)^{1/3} = 1 + \nu \qquad (1)$$

Here, L_0 is the contour length for uncomplexed DNA, L is the same quantity for the DNA–ligand complex, $[\eta]_0$ and $[\eta]$ are the intrinsic viscosities for free and complexed DNA, respectively, $f(p)_0$ and $f(p)$ are functions of the axial ratio p of free and complexed DNA, respectively, and v is the moles of bound ligand per mole of DNA base pairs. The functions $f(p)$ and $f(p)_0$ are fairly insensitive to p at high axial ratios and, because of the approximations and uncertainties in the theory and experimental methods, are frequently assumed to cancel in relating viscosity results to DNA length increases on ligand binding. Because of the experimental simplicity and accuracy of viscometric methods, they have frequently been used to analyze DNA–ligand interactions. When electronic timing is included in the experiment along with precautions to maintain temperature and eliminate dust, this method is extremely sensitive and accurate (*24*). Cohen and Eisenberg (*19*) found that the average increase in length produced by proflavine binding to DNA was approximately 80% of that predicted by Eq. (1), assuming a theoretical length increase of 3.4 Å per bound proflavine molecule. This was the first quantitative experiment to test the structural predictions of the intercalation model, and the results agreed well with predicted values.

Saucier *et al.* (*25*), also using sonicated calf thymus DNA and other similar experimental techniques, analyzed the DNA viscosity enhancement of several additional planar aromatic cations. They found that daunorubicin and 9-methoxyellipticine gave the expected increase in viscosity. Ethidium bromide, on the other hand, gave a slope of only 0.73 in a Cohen and Eisenberg type of plot, $([\eta]/[\eta]_0)^{1/3}$ versus v, compared to the theoretical slope of 1.0 (*19*). Saucier *et al.* (*25*) also found that quinacrine and proflavine gave less than the theoretical slope. This result and similar experiments with other ligands (*24–27*) indicated that the increase in DNA length produced on intercalation could vary significantly with ligand structure and was generally less than the predicted value from Eq. (1). This variation is illustrated in Fig. 3, with data from our laboratory illustrating the DNA length increase determined from viscosity measurements with sonicated DNA for binding of quinacrine, 9-aminoacridine, and ethidium bromide. The theoretical slope, assuming an ideal length increase of 3.4 Å, is illustrated in the same figure by a broken line. As with proflavine (*19*), all compounds give less than the expected length increase, with the ethidium length increase being significantly less than that for quinacrine and 9-aminoacridine. The reasons for the variation are still not completely clear (see Section V), but some modification of the classical intercalation model to include these effects was clearly necessary. Gabbay and co-workers (*27*), working with a series of methylated phenanthrolines, found that the viscosity enhancement depended on the number and position of the methyl substituents. They postulated that unsymmetric methylation of the aromatic ring gave a reduced

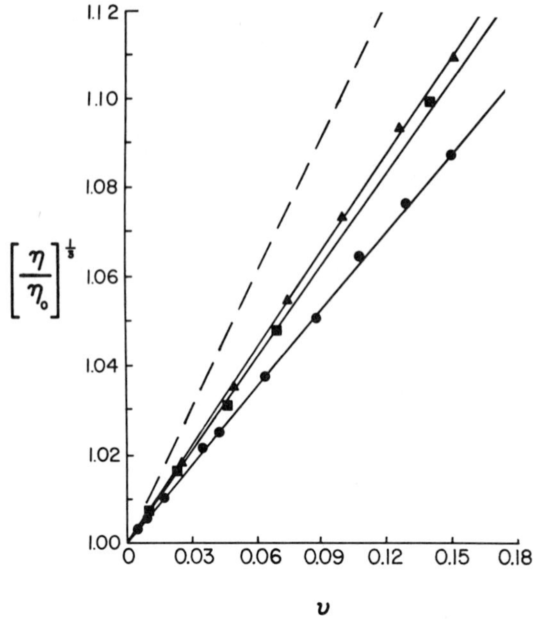

Fig. 3. Results from a viscometric titration of sonicated calf thymus DNA with quinacrine (▲), 9-aminoacridine (■), and ethidium (●), where η is the reduced specific viscosity of the DNA–ligand solution and η_0 is the reduced specific viscosity of the DNA solution before the addition of ligand. The experiments were conducted at pH 7.0 and 25°C in PIPES buffer (0.01 *M* piperazinediethanesulfonic acid, 10^{-3} *M*, EDTA, pH 7.0). The dashed line is the theoretical slope predicted by Eq. (1) for a 3.4 Å length increase in the double helix for each intercalated ligand. The reduced specific viscosity ratio η/η_0 is plotted as a function of v, the moles of ligand bound per DNA base-pair equivalent. Note that many plots in the literature are drawn as a function of the moles of ligand bound per DNA nucleotide, r, and $v = 2r$.

viscosity enhancement due to a slight bending of the double helix that occured in addition to the length increase on binding these molecules. This point is discussed in more detail in Section V.

B. Superhelical DNA: The Intercalation Unwinding Angle

As indicated in the previous section, it has generally been found that classical intercalating molecules such as proflavine give length increases in the double helix of approximately 3.4 Å. The unwinding of the double helix that results from intercalation has, however, been the subject of considerably more controversy, and as yet no standard or ideal unwinding angle is obvious. In the B-form double helix with ten base pairs per turn (360°), each base pair is rotated +36° with respect to the base pair immediately below it (*28*).

If these neighboring base pairs are separated to create a space for intercalation, this rotation is reduced and the amount of the reduction in degrees is termed the intercalation unwinding angle. Lerman revised his original postulate of 45° unwinding (a net reversed or left-handed rotation of 9°) to 36° unwinding (no net rotation) (*14*). Fuller and Waring (*29*) conducted extensive model building studies on ethidium bromide (Fig. 1) and concluded that a 12° unwinding angle was the minimum for intercalation of an aromatic ring system. In their model this unwinding angle allowed maximum separation of anionic phosphate groups and allowed the optimum interaction of the ethidium molecule with the double helix (hydrogen bonding of ethidium amino groups with the phosphate oxygens of DNA and optimum overlap of the ligand and base pair aromatic ring systems). This 12° value for ethidium became somewhat of a standard until the early 1970s, although in the same period Le Pecq and co-workers (*25*, *30*) incorrectly postulated that the double helix was actually wound by 12° when ethidium intercalates.

This controversy has now been largely eliminated by the discovery of closed circular superhelical DNA (*31*, *32*) and its use in the characterization of the intercalation reaction (*24*, *33*, *34*). As intercalating molecules are added to superhelical DNA, the double helix is unwound, and initially the natural right-handed superhelical turns are removed until the DNA has no remaining superhelical structure, at which point it is hydrodynamically equivalent to nicked circular DNA (*32*, *33*). If more of the intercalating ligand is added, the double helix continues to unwind with the formation of a left-handed superhelical structure. This is shown schematically in Fig. 4, and the predicted viscosity changes are illustrated in Fig. 5. The maximum in a viscometric titration or the minimum in a sedimentation experiment corresponds to closed circular DNA with no remaining superhelical turns.

It is obvious that, if the initial number of superhelical turns is known and if the amount of bound ligand required to remove these turns completely is determined (the amount of bound ligand required to reach the maximum in the curves in Fig. 5), the ligand unwinding angle can be calculated. Several methods have been developed for determining the number of superhelical turns in closed circular DNA (*32*, *35*, *36*), and these have been applied to circular DNA from phage, plasmids, and similar DNA samples (*32*). The other unknown, the amount of drug bound to DNA at the point where the superhelical turns are completely removed, can in principle be determined by traditional binding experiments, but this is even more difficult than usual because of the additional free-energy term involved in removal or introduction of supercoils. Vinograd and co-workers (*34*, *37*) proposed conducting several viscometric titrations at different DNA concentrations to remove the necessity of direct binding experiments and to eliminate other experimental errors (*24*). Since the DNA should contain the same number of superhelical

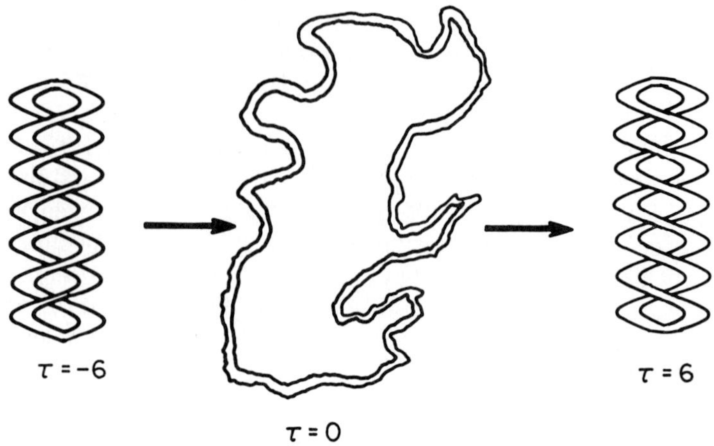

Fig. 4. Representation of the effects of the binding of an intercalating ligand to closed circular superhelical DNA. The diagram illustrates three stages of binding. On the left the DNA molecule has not bound enough ligand to remove the natural right-handed superhelical turns. In the center enough ligand has been added to remove the superhelical turns exactly (this is the maximum in a viscometric titration of superhelical DNA by an intercalating ligand). On the right more ligand has been added, and in this case unwinding of the double helix produces left-handed superhelical turns. The number of superhelical turns that the double helix makes about a superhelix axis is denoted by τ.

turns in all of the experiments, the amount of ligand bound per base pair to remove the superhelical turns completely, v', should be constant. This quantity is related to other experimentally measurable quantities through the expression

$$C_t' = v'N_t' + C' \tag{2}$$

where C_t' is the total amount of ligand and N_t' is the total nucleic acid concentration at the maximum in a viscometric titration. A plot of C_t' as a function of N_t' allows the determination of v' and C', the free ligand concentration. All primed quantities refer to those characteristic of the closed circular DNA with all superhelical turns removed, the maximum in a viscometric titration. If a standard of known unwinding angle is available, the unwinding angle of an unknown ligand can be determined using

$$\phi_k v_k' = \phi v' \tag{3}$$

where ϕ_k and ϕ are unwinding angles for a known and an unknown ligand, respectively, and v_k' refers to the known compound. Ethidium bromide has evolved as a standard for use in determinations of unwinding angles by this method. Several different types of experiments have indicated that this

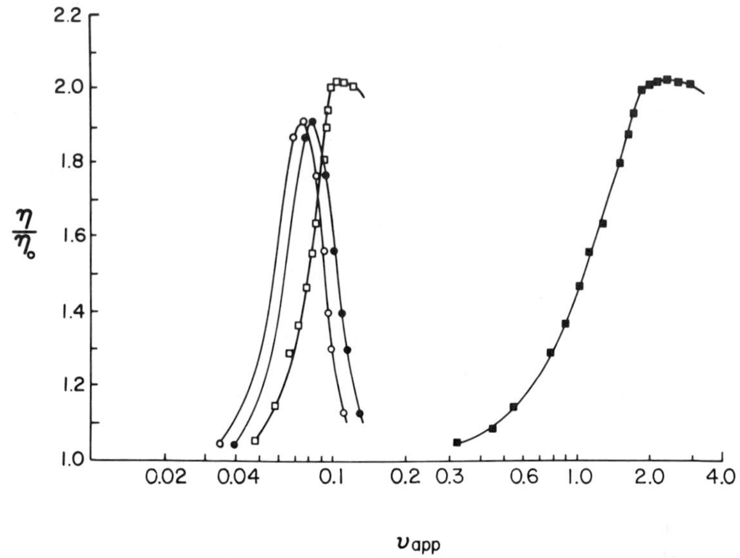

Fig. 5. A viscometric titration of closed circular superhelical Col E_1 DNA plotted as a function of ν_{app}, the total moles of ligand added per base pair equivalent of DNA (●, ethidium; ■, chloroquine; η/η_0 is defined in Fig. 3). The titrations are also shown corrected for the amount of ligand bound, ν (○, ethidium; □, chloroquine). Since ethidium and chloroquine have quite different binding constants under these experimental conditions, the corrected results are much closer than the direct experimental results. This illustrates the necessity of quantitating binding as in Eq. (2) in order to evaluate unwinding angles accurately. The experiment was conducted at 25°C in PIPES buffer (see Fig. 3) with 0.1 *M* NaCl.

molecule has an unwinding angle of 26°, in contrast to the previously assumed value of 12° (*32, 35, 36, 38*). An additional advantage of the Vinograd method is that some nicked DNA present in the sample will not change the calculated unwinding angle since both nicked and superhelical DNA have the same binding constant at the viscometric maximum at which all superhelical turns are removed. The presence of a considerable amount of nicked DNA will, however, reduce experimental accuracy since the relative change in flow times will be reduced if this species is present.

Few molecules have been quantitatively analyzed for unwinding at this time, and the influence of the various structural features that determine the unwinding angle of a ligand are still largely unknown. The 26° unwinding angle for ethidium and some related phenanthridines (*39*) is the largest unwinding angle found to date for any monomer intercalating ligand. The classical acridines originally studied by Lerman seem to give lower unwinding angles, which cluster between 17° and 20° (*24, 40*). The anthracycline drugs daunorubicin and adriamycin have been reported to have lower unwinding angles, between 10° and 12°, relative to ethidium bromide (*25, 41*). It is interesting that a large number of intercalating molecules seem to have unwinding angles near 18°. The phenanthridines, which have a very specific hydrogen-bonding interaction with DNA (*29*, *42a,b*), unwind by 8° more than this 18° value. It has also been proposed that the anthracyclines have a very specific hydrogen-bonding interaction with DNA (*43a,b*; *44*), and they unwind approximately 8° less than this common 18° value. The 26° phenanthridine value is the highest unwinding angle that has been found, and the 10° anthracycline value is the lowest for monomer intercalators.

It is tempting to speculate that the optimum unwinding of DNA is approximately 18° in the absence of other strong influences to change this interaction between the intercalated ligand and the two adjacent base pairs. This optimum could perhaps be modified by ±8° or more if the modification enhances other interactions such as specific ligand–DNA hydrogen bonding. In other words, the positive free-energy term required to move away from the optimum 18° value could be more than compensated for by other specific interactions that are optimized at some other unwinding angle. Ascertaining whether this hypothesis is true will require detailed structural and unwinding studies on intercalating molecules of varied structure. In addition, potential artifacts such as Type II binding must be carefully eliminated in all quantitative measurements of unwinding angles for these measurements to be useful in comparative analysis (*24*).

Arnott and co-workers have been quite active in developing double helical structures from fiber diffraction experiments (*45*) and in using a computerized linked-atom model building program for examining stereochemical features of nucleic acids and their complexes with ligands (*46*). In a general application of this method to the proflavine–DNA complex, they examined the stereochemical requirements for creating an intercalation space that was compatible with adjacent B-form DNA structures. With these limitations the program chose structures to optimize stacking interactions and minimize steric strain and hindrance. Because of the large number of possible variables in constructing models in this manner, some variables must be fixed at reasonable values to reduce the problem to a manageable size. With the parameters fixed by Alden and Arnott (*46*), one DNA intercalation conformation is superior to all others. In this conformation, two torsional angles were changed to trans values, and the sugar pucker was mixed at the intercalation site. An 18° unwinding angle for this general intercalation model was obtained, but in the best-fit structure this unwinding was distributed over three nucleotides rather than localized at the immediate intercalation site. Alden and Arnott point out that this long-range structural perturbation can account for neighbor exclusion binding (Section IV). A similar study has been conducted on intercalated ribonucleotide complexes (*46*). Miller carried out a similar procedure using a duplex tetranucleotide from B-form DNA and minimizing the energy of the possible intercalation conformations to obtain the optimum structure. He found two classes of acceptable intercalation sites by this procedure, and they have unwinding angles of 16° and 20°. He also found an alternating sugar pucker in the intercalation conformation. These studies reinforce the idea that there may be some optimum intercalation unwinding angle near 18° (*40*).

To assist in the visualization of this winding of native DNA, a view of two adjacent base pairs is shown in Fig. 6A, looking down at the long axis of the helix. The upper base pair is wound by 36° with respect to the lower base pair in this figure. In Fig. 6C the acridine ring system of quinacrine is shown, and in Fig. 6B the acridine ring system is inserted between the base pairs shown in Fig. 6A. As found experimentally for quinacrine (*24*, *40*) and as predicted for a general intercalation complex, the base pairs are unwound by 18° in the center diagram. The net winding of the base pairs is thus reduced from the 36° value in Fig. 6A to 18° in the complex. In this diagram the 9-amino substituent of quinacrine is inserted into the minor groove and the quinacrine acridine nitrogen is facing the major groove (*48*). Note that an excellent overlap of the base pairs and the acridine ring system (or other similar aromatic systems) is obtained in a complex with this general 18° unwinding angle. The position of the acridine chromophore of quinacrine between base pairs was selected through CPK model building experiments to minimize unfavorable steric interactions. With the complex structure shown in Fig. 6B, the chloro and methoxy substituents fit into the major groove and the 9-amino substituent is in the minor groove. All three substituents have negligible steric interference with the DNA base pairs in this conformation, which also allows excellent overlap of the acridine aromatic ring system with the adjacent DNA base pairs. Changing the angle of the acridine long axis with respect to the base pairs (which are fixed in an 18° winding angle) brings at least one of the three bulky substituents between the DNA base pairs and partially disrupts the base pair–acridine stacking interactions. The charged alkyl amino group of quinacrine has considerable freedom of motion in the minor groove in this complex as predicted by ^{13}C NMR experiments (*48*).

It should finally be emphasized that unwinding angles measured in solution with closed circular superhelical DNA are total values per bound intercalator and may be spread over several base pairs. This point is emphasized in the crystal structure of daunorubicin with a hexanucleotide DNA double helical fragment in which the base pairs adjacent the intercalated molecule have a zero unwinding angle, whereas the next base pair has a 8° unwinding angle (*44*). The total of 8° compares well with solution results that indicate an unwinding angle for daunorubicin of approximately 10° (*25*, *41*).

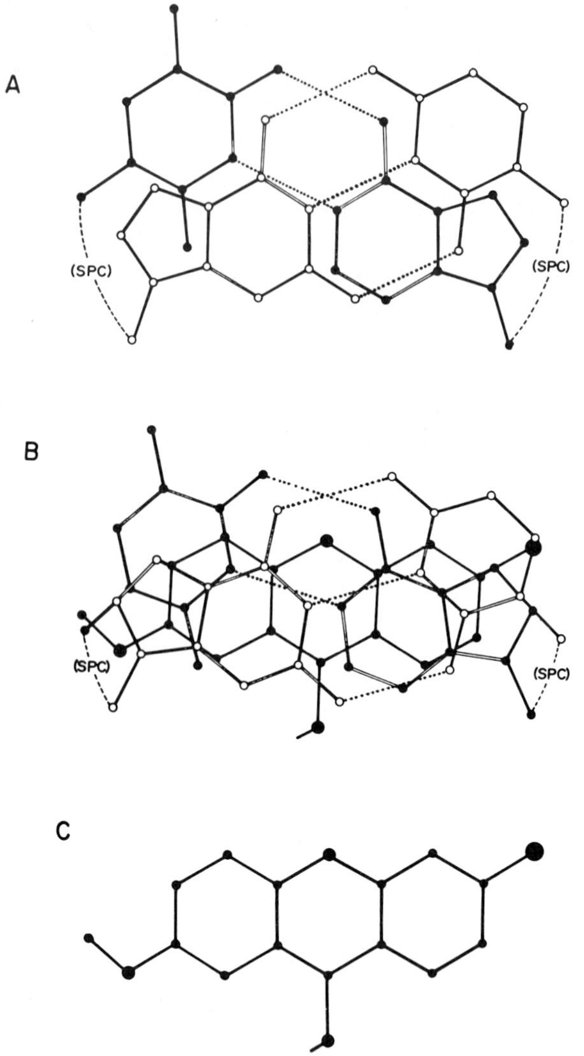

Fig. 6. (A) Two adjacent base pairs in a classical B-form Watson and Crick double helix. The top base pair (○), which is a G · C, is rotated 36° in a right-handed manner with respect to the lower base pair (●), which is an A · T. (B) The quinacrine aromatic ring system (C; see Fig. 1e) is inserted between the base pairs with an unwinding of 18°, and the quinacrine 9-amino group is oriented to place the alkylamino substituent into the minor groove. In the complex the G · C base pair has a net right-handed winding of 18° with respect to the A · T base pair. Hydrogens are not shown in the diagrams. The connecting sugar–phosphate chain (SPC) is represented by the curved dashed line.

C. Nuclear Magnetic Resonance: Complex Structures in Solution

Because of dipolar broadening, natural DNA gives very broadened spectra in both ^{1}H (*49a,b*) and ^{13}C (*48*) NMR experiments except in the denatured state. In the same manner, spectra for complexes of classical intercalating ligands with natural DNA are essentially totally broadened except in the presence of organic solvents or high salt concentrations or at high ratios of ligand to DNA, in which Type II binding is predominant (*48*, *49b*). Recent experiments with short double helical fragments of sonicated or nucleosomal DNA have produced much improved ^{1}H and ^{13}C NMR spectra (*50a–d*). A very exciting development is the preparation of homogeneous restriction fragments of DNA in sufficient quantity to perform NMR experiments (*50d*). These fragments allow high-resolution NMR studies of DNA conformation and dynamics in solution. Analysis of spin–spin and spin–lattice relaxation rates with a 12-base pair fragment, for example, indicated that single base pairs on the interior of a double helix can open for proton exchange with the solvent. At temperatures below the T_m, A·T base pairs on the interior of the helix open much faster than interior G·C base pairs (50*d*). These studies are extremely important in developing mechanisms for the opening (breathing) motions of the double helix that allow binding of an intercalating molecule.

Because of the increasing availability of Fourier transform instruments with ^{31}P capability and because of the sensitivity of ^{31}P NMR chemical shifts to phosphate bond and torsional angles (*51*), this method is beginning to be extensively applied in the study of DNA conformation in solution. Excellent ^{31}P spectra have been obtained with synthetic deoxypolynucleotides (see, for example, *52a–d*), low-molecular-weight fragments of natural DNA (see *53a–d*), and DNA complexes such as nucleosomes (*54a–d*). Jones and Wilson (*55a*) and Wilson *et al.* (*55b*) analyzed the effect of a range of intercalating compounds on the ^{31}P spectra of DNA. These compounds all caused downfield shifts and the magnitude of the shifts was most closely correlated with the unwinding angle of the intercalator. The linewidth of the DNA ^{31}P signal was also increased on binding all of the intercalators, and Wilson *et al.* (*55c*) found that these linewidth changes were quite dependent on the DNA molecular weight, temperature, and magnetic field strength. It is interesting to note that simple metal ions such as magnesium cause upfield shifts in the DNA ^{31}P signal, whereas metallointercalation reagents cause downfield shifts (*55b*). Hogan and Jardetzky (*56*) have studied the effect of ethidium on the ^{31}P NMR spectra of DNA and have obtained results which differ from those described above. A possible explanation for these differences arises from the fact that different DNA molecular weights, field strengths, and temperatures were used in the different experiments (*55c*), leading Hogan and Jardetzky (*56*) to incorrectly conclude that the DNA ^{31}P signal disappears in the presence of ethidium. As the various effects on ^{31}P chemical shifts are understood in more detail, ^{31}P NMR investigations of intercalator–DNA interactions should become even more useful, especially in the conformational analysis of DNA in aggregates such as nucleosomes. The ^{31}P technique has the advantage of being able to focus in directly on the DNA molecule in these complexes, whereas most other spectral techniques are sensitive to several molecules in the complex.

At present, far more work has been conducted on small DNA nucleotide segments, especially those that are complementary and thus have the potential to form a miniature double helix [see Krugh and Nuss (*57a*) and Patel (*57b*) for reviews of this area]. The first analysis of this type was an investigation of the stacking of ethidium with UpU and poly(U) (*58*). Most subsequent work has also utilized ethidium, along with actinomycin, and acridine derivatives with short DNA and RNA complementary segments (*57a,b*). The most fruitful experiments to date have been analyses of chemical shifts in ^{1}H NMR. These results coupled with a theoretical treatment of predicted chemical shifts based on ring current analysis strongly support the formation of intercalation complexes for both ethidium and actinomycin (*57a,b*). Using chemical shift calculations with plausible intercalated models, these experiments also allow the prediction of specific complex structures in solution.

An additional finding of interest with these short nucleotide segments is the sequence specificity exhibited by ethidium (*59a,b*), propidium iodide (*60*), and 9-aminoacridine (*61*). Ethidium shows only a slight preference for binding to G·C base pairs in native DNA (*62*), but with the short complementary nucleotides it has a strong preference for pyrimidine-(3′–5′)-purine sequences. Considering adjacent base pairs, DNA contains ten different intercalation sites (*33*), but it is impossible to analyze independent binding to each of these systematically using heterogeneous natural DNA. With synthetic dinucleotide monophosphates and longer complementary nucleotide segments, however, this type of study can be done. As mentioned above, ethidium has a marked preference for pyrimidine-(3′–5′)-purine sequences in both ribo- and deoxyribodinucleotides (for example, CpG and UpA). Extrapolating these results to heterogeneous DNA does not imply any net G·C or A·T binding specificity, since the ethidium binding preference holds for all sequences analyzed. With the potential additional influence from solvent interactions in dinucleotides, one is uncertain as to how far to extrapolate these results to natural DNA. It does seem, however, that similar results are obtained with ethidium and tetranucleotides that have been analyzed thus far. In addition, a study with six synthetic DNA samples gave results supporting the dinucleotide findings and it seems that with ethidium and actinomycin the dinucleotide binding preferences will at least be qualitatively found in natural DNA (*57a*).

The detection of sequence preferences of this type and the prediction of complex structures in solution are the strength of the NMR method. As longer complementary nucleotide segments and higher field strength instruments are used, the information obtained will be more valuable and the extrapolation to native DNA will be more straightforward, although the experimental complexity will be proportionally increased (*50d*, *61*). The development of ^{31}P NMR techniques with DNA shows promise of providing structural information even for relatively high-molecular-weight DNA (*53b*, *55c*).

D. X-ray and Model Building Studies

Lerman used fiber diffraction studies as part of his original experiments to determine the structure of the Type I binding mode (*14*). Similar studies have been conducted with other acridines (*63*), ethidium bromide (*29*), and daunorubicin (*43*). The principal result, a strong meridional reflection on the diffraction film of B-form DNA representing 3.4 Å repeat spacing in the molecule, is maintained in the diffraction patterns of DNA with intercalating ligands. All other regular features of the B-form diffraction pattern become more diffuse or disappear entirely with the complex. The diffraction patterns also vary with humidity, as for fibers of uncomplexed DNA. Bond *et al.* (*64*) obtained fiber diffraction patterns with platinum-containing intercalating ligands bound to DNA at nearly saturating levels of ligand to DNA. The electron-dense platinum provided an additional scattering center for the detection of any periodicity in the ligand distribution along the double helix. With these fibers Bond *et al.* (*64*) found the standard 3.4 Å reflection and an additional meridional line at a position corresponding to a spacing along the helix long axis of 10.2 Å. This is the expected repeat distance if one metallo-intercalation reagent is found per every other base pair. Binding studies, starting with the original work of Peacocke and Skerrett, have suggested saturation of intercalation binding sites at a level of one ligand per two base pairs. This so called "neighbor exclusion binding" was initially treated quantitatively by Crothers (*65*), and his treatment has subsequently been extended by others (see Section IV). Structural explanations for neighbor exclusion have been proposed on the basis of X-ray diffraction and model building studies discussed below.

Highly resolved structures for intercalated complexes were next obtained from single-crystal studies of dinucleotide monophosphates with intercalating ligands (*42*, *66–70*). Sobell and co-workers have been especially active in this area [see Sobell *et al.* (*71*)], frequently using iodinated derivatives of ribodinucleotide monophosphates in the crystals. The structures obtained with ethidium bromide and iodo-GpG and iodo-UpA are especially informative and have the following features. (a) The sugar puckering at the intercalation site is C3′-endo-(3′–5′)-C2′-endo; (b) changes in backbone torsional angles occur; (c) the helix axis is displaced by approximately 1.0 Å at the intercalation position; (d) a total unwinding angle of 26° (10° winding) is obtained; and (e) the base pair above the intercalated ligand is tilted approximately 8° with respect to the base pair immediately below the ligand, resulting in a bend or kink in the double helix. The 26° unwinding angle agrees with results obtained from superhelical DNA (*35*, *36*), and the mixed sugar puckering can explain neighbor exclusion binding. Crystal studies of actinomycin with deoxyguanocine (*66*) and 9-aminoacridine with 5-iodo-CpG (*67*) have provided supporting evidence for the above structural findings. Sobell and co-workers have used these results to provide a possible general model for intercalated complexes as described above and to postulate models for DNA in chromatin based on the bending obtained in the crystal structures (*71*).

Results from Sobell's laboratory with proflavine–5-iodo-GpC (*71*) and from Neidle and co-workers (*68*) with proflavine–CpG crystals and other drugs have been different from the ethidium structure described above. In the proflavine structures (*68*, *71*), sugars have C3-endo conformations, and the unwinding angle is close to zero, in contrast to the 18° unwinding angle obtained with proflavine and superhelical DNA (*33*). All of these experiments suffer, as do the small-molecule NMR experiments, from the difficulty of extrapolating results from dinucleotides to high-molecular-weight natural DNA. The results, however, have provided a wealth of ideas and models on which to develop experimental ideas as well as theoretical and structural concepts in intercalation chemistry (*68*, *71*).

X-ray studies of intercalated complexes will become even more significant as longer double helical segments are crystallized. Drew *et al.* (*72*), for example, have obtained the crystal structure of a B-form DNA dodecamer. This molecule forms an antiparallel, Watson-Crick hydrogen-bonded double helix, with close to the classical base pair winding angle and number of base pairs per turn. There is a slight propeller twisting of base pairs and some variation of the deoxyribose ring conformation. There is a slight bend (19°) in the double helix axis, which could be induced by solid state contacts. Rich and co-workers (*44*) have obtained the structure of an intercalation complex of daunorubicin with a double helical segment of d(CpGpTpApCpG). Daunorubicin molecules are intercalated between d(CpG) base pairs at the ends of the double helix. The cationic amino sugar substituent is in the minor groove and the long axis of the daunorubicin aglycone is at near right angles to the adjacent base pairs. The DNA base pairs are stacked in a relatively parallel arrangement. The net unwinding angle per intercalated daunorubicin is 8° in good agreement with the near-10° value obtained from studies with closed circular superhelical DNA (*25*, *41*). This unwinding, however, does not occur directly at the site of intercalation. The unwinding of the base pairs adjacent to the anthracycline is near 0° in the crystal, whereas the unwinding of base pairs at the site next to the intercalator is 8°. These variations, and other subtle perturbations of the double helix structure, indicate that considerable differences may be found among intercalator–DNA complexes. A common observation for both the DNA (*72*) and the daunorubicin–DNA crystal (*44*) is that the classical B-form structure, although generally correct, is an oversimplification. There is considerable variation in torsional angle, sugar conformation, base pair angle, helix axis translocation, and other similar structural features. These structural variations seem to be characteristic of DNA conformation in general. The finding that DNA can easily change to the left-handed "Z-form" double helix (*73*) is additional support for this conformational flexibility for the DNA molecule.

E. Ligand and DNA Base Pair Orientation: Dichroism Methods

In his original development of the intercalation model, Lerman used flow dichroism and flow-polarized fluorescence with heterogeneous molecular weight DNA to characterize, at least qualitatively, the orientation of the plane of an intercalated ligand (*14*). The results from small molecule X-ray and model building studies discussed above have provided renewed interest in determining the orientation of intercalated aromatic systems with respect to the base pairs of DNA in solution.

Early work on electric dichroism of DNA complexes with intercalating ligands was similar to the flow experiments which grossly analyzed the plane of the ligand with respect to the DNA base pairs. Because the DNA is incompletely oriented, even at high field, and because of the uncertainty concerning the exact orientation mechanism in electric dichroism, this technique was not widely used. Recent advances in instrumentation (adaptation of temperature-jump instruments for electric dichroism measurements) and theory (*74*) for this method as well as techniques for preparing homogeneous DNA (*74*) have increased the potential of electric dichroism for providing information about the structure of DNA and its complexes with small molecules (*74*, *75*).

One puzzling feature of electric dichroism results with DNA is that the molecule behaves as if it has a permanent dipole moment, whereas the symmetry of the double helix would seem to prevent such a dipole moment from existing along the helix axis. Crothers and co-workers [see Hogan *et al.* (*74*)] have proposed that this effect is best explained by assuming that, when the electric field is applied, an anisotropic flow of ions occurs along the helix axis. This creates an unsymmetric ion atmosphere along the helix long axis and results in orientation of the DNA molecule. They propose that, in the presence of steady-state ion flow in the electric field, the various processes that relax this ion atmosphere are not fast enough to be completely effective, and loss of symmetry of the ions surrounding the helix results. Using fairly homogeneous molecular weight DNA samples, they tested dichroism results against the predictions of the anisotropic ion flow, the permanent dipole, and the induced dipole models and found that only the flow model could explain all of the results. Previous inconsistencies in electric dichroism experiments might be due to the polydispersity of the DNA samples or the ionic strengths used. The results of experiments with several fairly homogeneous DNA samples varying in molecular weight and base pair composition suggested the somewhat surprising result that the bases in B-form DNA are inclined at an angle of approximately 73° with respect to the helix axis (*74*). It was noted that this base inclination corresponds closely to that predicted in a model for B-form DNA by Levitt (*76*). This inclination is considerably more than the 90° value of Watson and Crick's original model or Arnott and Hukin's revised calculations for B-form DNA, and its suggestion reveals how uncertain our knowledge is of the solution structure of DNA.

Crothers and co-workers applied this electric dichroism technique to intercalating drugs (*75*). They were able to determine the orientation of two nonparallel transition vectors in ethidium bromide, proflavine, and 9-aminoacridine. The results with ethidium and 9-aminoacridine are quite similar and suggest a considerable variation from perpendicularity of both the long and short axes of the ligands. With the long axis of ethidium the tilt is 23° from perpendicular, and with the short axis the twist is 10° from perpendicular. With proflavine these values are 14° and 2° for the tilt and twist, respectively. Crothers and co-workers [see Hogan *et al.* (*75*)] analyzed numerous possible experimental and theoretical artifacts as possible explanations for these somewhat unusual findings and were able to eliminate most of the possible errors based on various control experiments. Using their experimental DNA orientation from electric dichroism, they also analyzed earlier qualitative results from both flow and electric dichroism and concluded that all of these experimental findings predict a tilt and twist for intercalated ligands, which is consistent with their more quantitative results (*75*). Starting with the Levitt modified B-form DNA structure (*76*), which they prefer from previous dichroism studies (*74*), they use these tilt and twist angles to prepare a modified intercalation model that is consistent with the electric dichroism results. Their model does not involve any bending of the helix axis but does have both base pairs and the intercalated ligand oriented at an angle significantly different from 90°, in contrast to most other intercalation predictions. Their model is extremely interesting and should generate ideas for its experimental testing and refinement, but, like Levitt's model (*76*) for B-form DNA, it must be viewed as hypothetical until additional confirmatory results are available.

One of the problems of electric dichroism measurements is that the dichroism at infinite field strength must be known in order to calculate base pair and ligand orientation. Experimentally this requires extrapolation, using an appropriate theoretical equation, of $1/E$ to zero (where E is the field strength) (*74*, *77*). Crothers and co-workers extrapolated to $1/E = 0$ assuming a linear fit between dichroism and $1/E$ (*74*, *75*) for calculation of all the results discussed above. Sokerov and Weill (*77*) conducted a detailed analysis of polarized fluorescence through electric field orientation using an intercalated acridine orange-labeled DNA and disagree with the linear $1/E$ plot (*74*, *75*). They (*77*) recalculated the results of Hogan *et al.* (*74*) using a more complex extrapolation procedure and found a base tilting much less than that of Hogan *et al.* (*74*) and in good agreement with fiber diffraction studies (*45*). Additional effects of ionic strength, DNA polydispersity, and flexibility must also be considered in analyzing and comparing results from electric dichroism experiments. Sokerov and Weill (*77*) also suggested that field homogeneity may be a problem when electrodes of modified temperature-jump equipment are used in electric dichroism experiments. It seems that for electric dichroism to prove its potential as a tool in the analysis of complex formation, considerable experimental and theoretical work must still be done to resolve these points of controversy.

F. Major versus Minor Groove Binding

Although the simple classical intercalating agents such as 9-aminoacridine or proflavine do not significantly protrude into the major or minor groove in Lerman's model, other ligands can have bulky substituents which must be in one of the grooves when the planar aromatic ring system of the compound is intercalated (Fig. 1). Since many intercalating drugs possess such bulky substituents, determining which groove they are in is an important aspect of evaluating the interaction of the drug with DNA and in proposing drug modifications to enhance medicinal activity. Unfortunately, the methods for evaluating major versus minor groove binding are quite limited, and even those available have not been widely applied.

The method most frequently used is to prepare synthetic DNA or isolate a natural DNA with the major groove containing a group such as a bromine substituent or a glucose residue covalently attached to a base (*78*). This can be done by synthesizing DNA with an unusual base such as 5-bromocytidine (*79*) or by isolating a natural DNA such as that from T4 phage which contains glucose residues covalently linked to the hydroxyl group of 5-hydroxymethyl-cytosine, which replaces cytosine in T4 DNA (*78*). Results from binding studies with these modified DNA's are then compared to similar results from DNA samples without the bulky groups. If the binding is not significantly changed, it is assumed that the ligand binds in the unblocked minor groove, and, if the binding is reduced with the blocked DNA, it is assumed that the ligand binds in the major groove (*33*, *79*, *80*). Various permutations of preference for major versus minor groove are obviously possible in ligand binding and cannot be unambiguously determined by this method. For a gross determination of major versus minor groove binding this should not be a major problem.

Pigram *et al.* (*43*) conducted X-ray diffraction experiments on DNA fibers containing various amounts of the anthracycline drug daunorubicin (Fig. 1) and proposed a best-fit model of the intercalated complex which places the amino sugar substituent in the major groove. Henry (*44*) found that a modification of the sugar conformation in adriamycin from that used by Pigram *et al.* (*43*) for daunorubicin allowed a much more favorable interaction in the major groove. Patel and Canuel (*80*), however, using synthetic polydeoxynucleotides with bulky 5-iodo and 5-bromo groups, found from NMR studies that anthracyclines still bind quite well to these polymers. Since the bulky substituent in the major groove would be expected to perturb binding if the amino sugar substituent were in the major groove, they proposed reevaluation of this model. The daunorubicin-DNA crystal structure discussed earlier (Section III, D) has the amino sugar substituent in the minor groove. If this is also true for the solution complex, then all intercalators that have been analyzed in detail have their bulky substituents in the minor groove (*40*). The origin of this possible minor groove specificity is not clear at present, but it is certainly an intriguing concept for analysis of intercalator–DNA complexes. It is known, however, that naphthalene dimide intercalating molecules can place bulky cationic substituents in both the major and minor groove (*81*). These dimide molecules were designed so that if they intercalate, one bulky substituent must be moved between base pairs resulting in a final complex with substituents in opposite grooves. Viscometric analysis with linear and superhelical DNA, circular dichroism, UV-visible, and NMR spectroscopy all suggest that these molecules do bind to DNA by intercalation in the alternate groove type complex. These are, then, the first class of intercalators known to bind with bulky cationic substituents in the major groove of DNA (*81*).

Analysis of major versus minor groove binding will no doubt be quite important for the evaluation of protein–DNA interactions and for the determination of DNA structure in chromatin. Current methods for evaluating major and minor groove binding have not been rigorously tested; previously with small molecules only one intercalating class of compounds, the anthracyclines, have been proposed to lie in the major groove, and this proposal is now in some doubt (*44, 80*). Sobell *et al.* (*71*) proposed that there are mechanistically two primary classes of intercalators: those that intercalate from the minor groove and those that approach from the major groove side of the double helix. To evaluate questions of this type and other important structural aspects of DNA–ligand complexes, some precise method of evaluating major versus minor groove binding is necessary, and research in this area must have a high priority in intercalation biochemistry. The naphthalene diimides should be quite useful in future research in this area (*81*).

G. Other Methods

Additional techniques that have been applied to the analysis of DNA–ligand complexes include circular dichroism (see, for example, *82a,b*) and kinetics of binding (review in *16* and *83a*). Both of these areas have produced interesting results but both also suffer, at present, for some difficulty in data interpretation and/or experimental difficulties. The use of some temperature-jump methods in kinetic analysis (*83a,b*) has been challenged (*84*) on the basis of possible spectral changes that may occur simply because of the orientation of the DNA molecules (as in electric dichroism). This can lead to artifacts due to both orientation and randomization of DNA molecules and could affect mechanistic interpretations of ligand–DNA interactions. Stopped flow kinetic methods do not suffer from this difficulty but are generally too slow for a thorough analysis of DNA intercalator interactions. Mandal *et al.* (*85a,b*) have developed a novel flow spectrophotometric hydrogen–deuterium exchange technique for analysis of DNA dynamics. They have applied this technique to the ethidium–DNA intercalation complex to obtain both rate and equilibrium constants for the reaction (*85b*). This method allows analysis of tight binding interactions but without the limitations, discussed above, of classical T-jump and stopped flow techniques. This new method should provide considerable new information on the mechanism and dynamics of intercalator–DNA interactions. Circular dichroism is an extremely useful technique to monitor conformational changes in DNA and DNA–ligand complex but is too complex at present to interpret in terms of the exact structural changes that account for a change in the DNA circular dichroism spectrum.

Laser Raman spectroscopy also shows promise of developing into a sensitive, high-resolution method for analysis of DNA conformation and conformational changes (*86a–c*). The B to A transition (*86a,b*) and the B to Z transition (*86c*) in DNA give characteristic Raman spectral changes. Compared with CD measurements, for example, Raman spectra have more and sharper lines for high resolution structural analysis and are much less sensitive to aggregation-induced artifacts (*86a–c*). The ability to follow many different vibrational modes in a molecule means that Raman spectral analysis has the potential for a specific molecular mapping of DNA transitions. This method should provide similar specific information on the structure of some DNA complexes in future studies. The problem of ligand fluorescence presents a serious technical difficulty in the application of this method to binding analysis of fluorescent intercalating molecules.

IV. THERMODYNAMICS OF INTERCALATION

A. Nearest-Neighbor and Ligand Cooperativity

As mentioned in Section I, the early binding experiments of Peacocke and Skerrett with proflavine (*7*) (Fig. 2) gave significantly curved Scatchard plots that were resolved into a strong binding mode, Type I binding, which has been identified as intercalation, and a weaker binding interaction, Type II binding. On the basis of analogy with the work of Bradley and Wolf (*9*), Type II binding has been identified as a polyanion-induced cooperative self-association of the planar aromatic cations. Because of the multiple charge and the cooperativity of the stacking interaction involved in Type II binding, it can be eliminated by increasing concentrations of simple salts (*24*, *40*). This has facilitated the study of Type I, or intercalation, binding.

Intercalation model building experiments suggest that molecules such as proflavine can be inserted between every base pair without undue strain on the sugar–phosphate chain. Scatchard-type analyses of intercalation using Eq. (4) have consistently shown, however, that the maximum n value for intercalating molecules is 0.5 instead of 1.0:

$$v/c = K(n - v) \tag{4}$$

Here n is the maximum number of ligands bound DNA per base pair, c is the free ligand concentration, K is the observed binding constant, and v represents the moles of ligand bound per base pair under the experimental conditions. Careful Scatchard analyses of Type I binding, even for simple intercalating molecules, have also shown curvature instead of the linear behavior expected for homogeneous independent binding sites (*16*). Of the numerous possible explanations for these results (*16*), the one that seems to account best for all experimental findings has been termed "neighbor exclusion binding," as discussed in Sections II,B and III,D.

In this model each base pair in DNA is a potential binding site but, when a ligand is bound, a neighboring binding site is eliminated or in some manner becomes unable to bind a ligand. Thus, at saturation there will be only one ligand bound for every two base pairs. Several possible binding site situations are illustrated in Fig. 7. The first is the typical independent site model that frequently holds for enzyme–substrate or enzyme–inhibitor interactions and

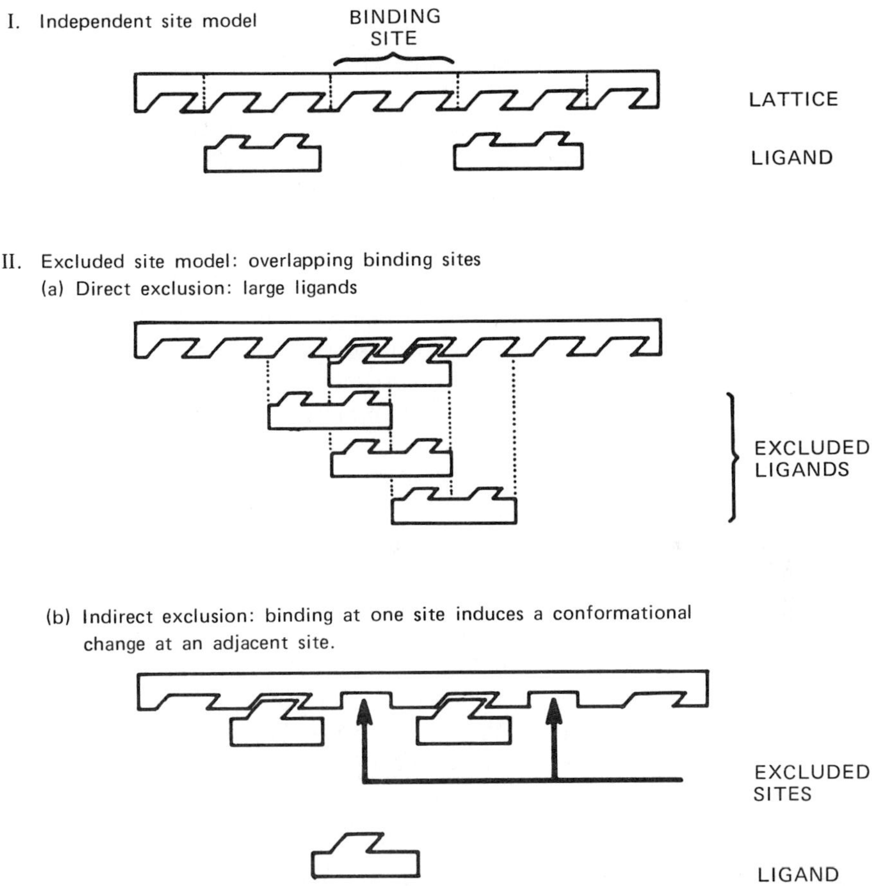

Fig. 7. Possible ligand interactions with a linear lattice such as DNA. Exclusion models involving binding site elimination through direct blockage of lattice binding sites by large ligands (IIa) and through indirect blockage caused by a conformational change induced in the lattice (double helix) by the ligand interaction (IIb) are shown. Classical independent site behavior is shown in I.

can be treated by the standard Scatchard analysis. Neighbor exclusion binding is illustrated in two possible cases: (a) The ligand physically blocks three potential binding sites, which is possible with large intercalating ligands such as actinomycin, and (b) ligand binding is excluded at a neighboring site by an induced perturbation of the DNA conformation at that site. For example, Sobell and co-workers (*71*), in their X-ray analysis of crystals of intercalating ligands with dinucleotides, found that the sugar pucker can vary symmetrically as a result of intercalation. They point out that if this is true in DNA it would automatically lead to neighbor exclusion binding. As mentioned in Section III,B, linked-atom (*46*) and potential energy calculations (*47*) designed to predict the most favorable possible intercalation conformations have also suggested that structures with alternating sugar pucker can exist for intercalation complexes.

Some early studies with synthetic bis-intercalating molecules, two intercalating aromatic rings connected by a variable-length linking chain, showed that when the chain was long enough to allow intercalation at every other base pair, both rings intercalated, in agreement with the neighbor exclusion model. However, when the chains were shorter and could allow intercalation only at adjacent base pairs, only one ring intercalated at a time (*87–89*). Lippard and co-workers (*64*), using X-ray fiber diffraction studies of a DNA fiber with an intercalated platinum metallointercalation reagent, also found that their results were best explained by a model with the platinum scattering center regularly intercalated at every other base pair, leaving neighboring sites empty.

Recent studies (see, for example, *89*; *90a,b*) with bisintercalating molecules with linking chains that are too short to allow intercalation with an empty site between intercalated ring systems have indicated that these molecules can bind with ring systems intercalated at neighboring sites in apparent violation of the neighbor exclusion principle. Bis molecules with long linking chains obey neighbor exclusion whereas compounds with very short chains give only monointercalation. Substituents on the intercalating rings also have a pronounced affect on bis- versus monointercalation (*90a*). These findings have raised questions about whether neighbor exclusion arises from a structural basis (*71*) or from a thermodynamic basis which could be overcome by bisintercalators with very high binding constants (*89*). These results do not invalidate neighbor exclusion binding by monointercalators, but they do indicate that binding of bisintercalators may be quite complex. These bis compounds may be used to exert quite different effects on DNA and to probe the energetics of neighbor exclusion.

Neighbor exclusion binding is a type of negative cooperativity, and, even if there are no direct ligand interactions or no ligand cooperativity on binding, neighbor exclusion will still give curved Scatchard plots (*65*, *91*). This type of curvature based on neighbor exclusion was first treated quantitatively by Crothers (*65*), who derived parameterized equations that could be used to calculate binding isotherms. Bauer and Vinograd (*92*) solved these equations in closed form for intercalating ligands without cooperativity and without base pair binding specificity. They found excellent fit between this equation and the binding results of ethidium bromide. They also developed equations, based on the same concept, to describe the binding of ethidium to closed circular superhelical DNA. This type of approach has been extended to include direct ligand cooperativity of homogeneous and heterogeneous binding sites (*93*; *94a,b*). McGhee and von Hipple (*95*), using a conditional probability method, derived closed-form equations to describe ligand cooperativity and site exclusion for sites of any size in the case of a ligand binding to a polymer of homogeneous lattice binding sites. The equivalent equations for these methods are interconvertible, as expected (*96*). Using the symbolism of McGhee and von Hipple, for example, noncooperative binding (no direct ligand–ligand interactions) obeys the following relationship:

$$\frac{v}{c} = K(1 - nv)\left(\frac{1 - nv}{1 - (n - 1)v}\right)^{n-1} \tag{5}$$

Here, n is the total number of binding sites eliminated when a ligand binds to the polymer, and other terms have been previously defined. When n is 2, Eq. (5) reduces to the equation of Bauer and Vinograd (*92*). With a polymer of linear binding sites such as DNA, ligand cooperative interactions can be treated in numerous ways (*16*, *95*, *96*). As a first approximation, it is useful to consider cooperative interactions only among neighboring bound ligands without intervening filled or unfilled binding sites on the linear polymer. Once again, in the symbolism of McGhee and von Hipple (*95*),

$$\frac{v}{c} = K(1 - nv)\left(\frac{(2\omega - 1)(1 - nv) + v - R}{2(\omega - 1)(1 - nv)}\right)^{n-1}\left(\frac{1 - (n + 1)v + R}{2(1 - nv)}\right)^{2} \tag{6}$$

$$R = [(1 - (n + 1)v)^2 + 4\omega v(1 - nv)]^{1/2}$$

and ω is a cooperativity parameter that is equivalent to the equilibrium constant for the reaction of transferring a bound ligand from a site with no neighboring ligands to a site with one neighboring bound ligand. When $\omega = 1$, Eq. (6) converts to Eq. (5), the noncooperative case. Positive cooperativity is defined by $\omega > 1$ and negative cooperativity by $\omega < 1$. Scatchard plots predicted from these equations are depicted in Fig. 8 for a hypothetical K value and several ω values. As can be seen, both neighbor exclusion ($\omega = 1$) and negative ligand cooperativity ($\omega < 1$) lead to curvature of Scatchard plots which could be mistakenly identified as multisite binding

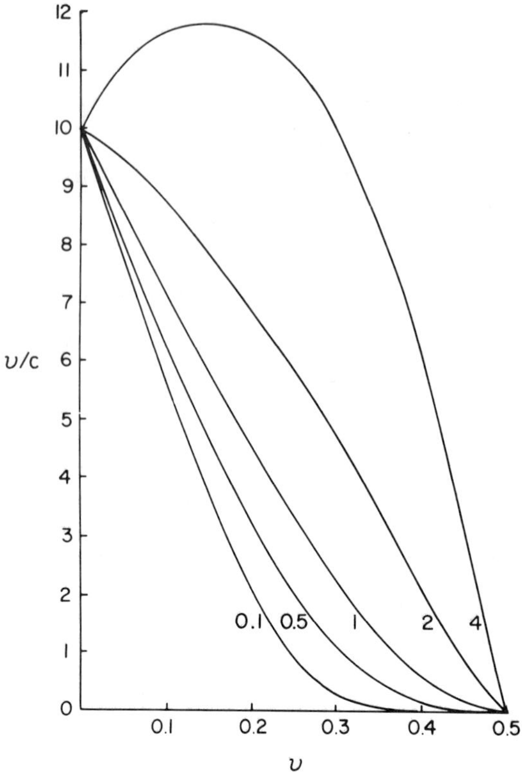

Fig. 8. Idealized Scatchard plots as predicted by Eq. (6) for different ω values. The ω value for each curve is indicated at the bottom of the figure. When $\omega = 1$, which indicates no ligand cooperativity, Eq. (6) reduces to Eq. (5). For the illustration, a K value of 10 is used for all curves. Changing K values over a fairly wide range do not dramatically affect the curve shapes.

(*95*, *96*). Positive ligand cooperativity ($\omega > 1$) in intercalation binding is also illustrated in Fig. 8.

Wilson and Lopp (*96a*) showed that negative ligand cooperativity can be important even in the interaction of simple intercalating ligands with DNA. They analyzed the interaction of the antimalarial drug quinacrine with DNA and found that for the case of nearest-neighbor cooperativity [$n = 2$ in Eq. (6)], the results required an ω value near 0.1 for fitting at all ionic strengths studied. Recent analyses of the binding of the monocationic ligands, ethidium, daunorubicin, adriamycin, and a phenazine derivative to DNA have indicated that all of these compounds have ω values near 1.0, or little if any cooperativity in their binding to DNA (*97*). A dicationic phenazine had an ω value near 0.2, indicating, as with quinacrine, pronounced negative ligand cooperativity. The negative cooperativity seems to be more a function of total ligand charge rather than compound structure with these simple intercalating molecules. This would suggest that electrostatic repulsion among bound ligands largely accounts for the negative cooperativity for these molecules.

As can be seen from Fig. 8, Scatchard plots at low ω values can give long areas that are approximately linear, and, if data in this limited ν range are analyzed by the Scatchard equation, both an incorrect K value and a grossly incorrect number of binding sites would be predicted. This may account for much of the confusion in the literature among different investigators using different number of binding sites for the same ligand or reporting a change in numbers of binding sites as conditions are changed.

The Scatchard equation, Eq. (4), has been quite successful in the treatment of protein–ligand interactions, such as enzyme–inhibitor binding, when independent binding sites are involved. With a linear duplex repeating polymer such as DNA, however, this type of model is quite artificial and, in fact, as the above discussion has shown, is incorrect. Although much work remains to be done in the development of the fine points of the neighbor exclusion model, its general validity seems to be well established by structural and thermodynamic experiments. The Scatchard equation has been applied many times to the equilibria of intercalating ligands with DNA, but it can now be seen that this is an incorrect model and can lead to significant errors in the analysis of the thermodynamics of DNA–ligand reactions. In addition, with multiple charged ligands the neighbor exclusion model with cooperativity must be used. Finally, when the ligand displays base pair specificity, which is best analyzed by the method of Müller and Crothers (*62*, *98*), this specificity must also be included in the analysis (*93*, *94*). With current experimental methods it would be difficult to obtain a unique fit to binding results for a ligand with both cooperative and specific base pair interactions. Strum and co-workers (*94a,b*) have developed a Monte Carlo method for calculation of binding isotherms for such complex situations. For complicated heterogeneous binding processes, this method offers the most straightforward approach. Strum *et al.* (*94b*) have applied this technique to the interaction of the intercalating drug tilorone to DNA and synthetic polydeoxyribonucleotides. As with other models involving heterogeneous binding, a large amount of data is required to obtain a unique fit to a particular model.

Recently Winkle and Krugh (*99*) have shown that at low saturation ratios the intercalating drugs actinomycin and daunorubicin and also some intercalating carcinogens show pronounced positive cooperativity in their interaction with DNA. The binding was analyzed using phase partition methods that allow accurate measurements at low ν values. It was only at these low ν values that the positive cooperativity was found. At higher ν values the binding curves were much like those of other intercalators in the literature (*99*). Crothers and co-workers (*100a,b*) have shown that the nonintercalating drugs distamycin and netrospin also bind cooperatively to DNA, and they suggest that the cooperativity is caused by a drug-induced allosteric transition in DNA at low v values. They also derived a quantitative method for treating this type equilibria (*100b*). Winkle and Krugh (*99*) propose that the cooperativity they see with intercalators may be due to a similar drug-induced conformational transition in DNA. The cooperativity in this case could arise due to the fact that the drug-induced conformation is not compatible with the B-form structure of DNA in solution (*100a,b*). It seems that a range of conformations are available to DNA, and shifts among these conformations can arise due to changes in solvent conditions or to complex formation with proteins or small molecules such as intercalating drugs. These cooperative binding effects are among the newest discoveries about intercalation and certainly warrant extensive study of their thermodynamics and of the conformations, which they induce in DNA.

B. Enthalpy Measurements

Because the enthalpy of ligand–DNA interactions can be determined directly by calorimetric measurements and from equilibrium measurements by van't Hoff plots, the enthalpy should be useful in characterizing intercalation reactions and in evaluating equilibrium binding models used to calculate thermodynamic results. Enthalpy determinations, however, have been rare, and the variation in reported values has been large. LePecq and Paoletti reported that the van't Hoff enthalpy of ethidium is 0 kcal below an ionic strength of 0.09, −5.2 kcal/mole between 0.09 and 0.5, and −11 kcal/mole at an ionic strength of 1 (*101*). The equilibrium data of Bresloff and Crothers (*83b*) yield a value of approximately −8.5 kcal/mole for the same reaction at an ionic strength near 1. The data of Chambron *et al.* (*102a,b*) suggest a value of approximately −5 kcal/mole for the proflavine binding reaction. Huang and Phillips (*103*) have found a ΔH° value of −5.3 kcal/mole for daunorubicin binding to DNA at an ionic strength of 0.15. McCall and Bloomfield (*104a*) found that, at an ionic strength of 0.1, ΔH° for proflavine binding to T2 bacteriophage DNA is approximately −9 kcal at both pH 5.6 and 7.6. Spectrophotometric binding experiments with the intact T2 phage yielded ΔH° values of −4.6 kcal/mole at pH 7.6 for proflavine binding to the interphage DNA. Riemer and Bloomfield (*104b*) continued the experiments with T2 DNA and investigated the effects of multications on proflavine binding at a constant ionic strength of 0.1. Their binding analysis used the neighbor exclusion model described in Section IV,A. The addition of Mg^{2+} ions, for example, resulted in a more negative ΔH° value. This change can be accounted for by the release of Mg^{2+} ions on proflavine binding as predicted by the condensation theory (Section IV,C) and by the positive enthalpy of 1–2 kcal/mole for Mg^{2+} binding. Their results generally indicate that all multiple cations have a fairly small effect on the enthalpy of proflavine binding to T2 DNA and thus cannot account for the large enthalpy difference between proflavine binding to T2 DNA versus the interphage DNA (*104b*).

Spectrophotometric binding results for dimidium bromide with DNA (*97*) are shown in Fig. 9. The solid lines of Fig. 9 are best-fit values determined using Eq. (6) with an ω value of approximately 0.8. Equilibrium constants obtained from several experiments of this type at different temperatures are graphed in a van't Hoff plot in Fig. 10. The linearity of this plot indicates that ΔH^0 is constant over this temperature range within the experimental error of the method. From the slope of the plot, ΔH^0 is approximately -6 kcal/mole. Because of the scatter of published enthalpy values, even for a single compound such as ethidium, no detailed analysis of these results is warranted at this time. It does appear that most classical intercalating molecules have ΔH° value of -7 to -8 kcal/mole for the intercalation reaction.

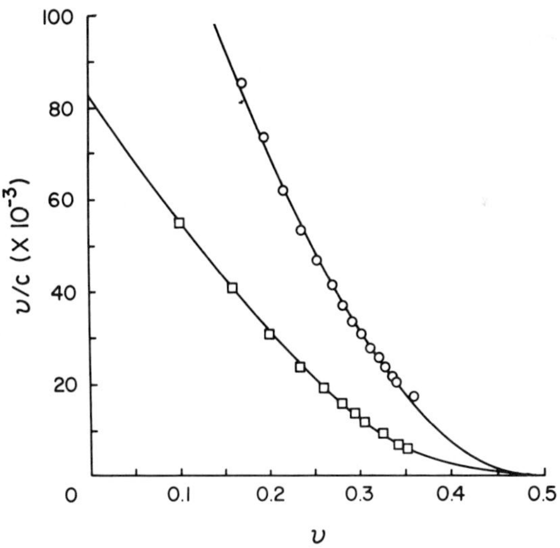

Fig. 9. Scatchard plots for the interaction of dimidium (Fig. 1) with calf thymus DNA at two different temperatures: 25°C (○) and 45°C (□). The points are experimental results, and the solid lines are plotted according to Eq. (6) with $K = 175{,}000$ at 25°C and $K = 82{,}000$ at 45°C and $\omega = 0.8$. The buffer was NaH_2PO_4, 7.5×10^{-3} *M*, EDTA, 10^{-3} *M*, NaCl, 0.1 *M* adjusted to pH 7 with NaOH. Experimental techniques have been described in more detail elsewhere [Wilson and Lopp (*96*)].

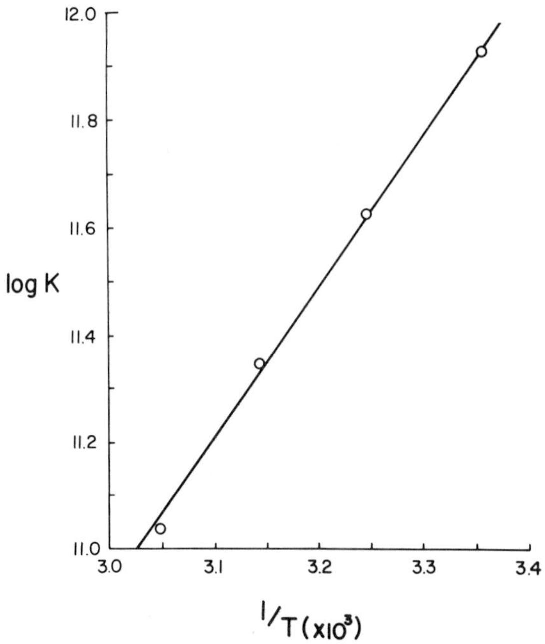

Fig. 10. A van't Hoff plot for the interaction of dimidium with calf thymus DNA. Equilibrium constants from Fig. 9 and at two other temperatures determined under the same buffer conditions are used in this plot. The linearity of the plot indicates that ΔH^0 is approximately constant over this temperature range, and the slope yields a ΔH^0 value of -5.7 kcal/mole.

Studies with a broader range of compounds under different solvent conditions are clearly necessary to test this prediction.

Extensive calorimetric measurements of intercalation enthalpies have been conducted by Crescenzi, Quadrifoglio, and their co-workers (reviewed in *105*). Since the concentrations used in these experiments are typically higher than those used in other equilibrium measurements, problems with ligand dimerization are more apparent in calorimetry. The typical problems with this type of compound, such as absorption to container surfaces, Type II binding, and corrections for the fraction of ligand bound under the conditions of the experiment, further limit the accuracy and experimental range of calorimetric measurements. Quadrifoglio *et al.* (*106*) have found with ethidium and proflavine, that ΔH^0 values are independent of the amount of ligand bound to DNA up to a v value of approximately 0.25. Above this v value, ΔH^0 apparently decreases, perhaps due to Type II binding. The ΔH^0 values are -6.2 and -6.7 kcal/mole for ethidium intercalation at ionic strengths of 0.015 and 0.1, respectively. Proflavine has ΔH^0 values of -6.7 and -7.6 kcal/mole under the same conditions (*106*). Quadrifoglio and Crescenzi (*107*) found that daunorubicin behaves in a similar manner, with a ΔH^0 value of approximately -6.5 kcal/mole at ν values less than 0.25 and steadily decreasing apparent ΔH^0 values at higher ν values. The same authors (*107*) found that the binding of actinomycin to DNA, unlike other intercalating compounds, is characterized by a positive enthalpy of binding of approximately 2 kcal/mole. They speculate that this positive ΔH^0 could be caused, at least in part, by a conformational change in the actinomycin cyclic peptide rings on binding to DNA (*107*). The enthalpy of interaction of actinomycin with deoxyguanosine 5′-monophosphate, on the contrary, is a large negative value, and this has lead Crescenzi and Quadrifoglio (*107*) to raise the question of whether actinomycin actually binds to DNA by intercalation. Other experiments with actinomycin, however, strongly support intercalation as the binding mode for this compound (*108*). Crescenzi and Quadrifoglio (*105*) also investigated the interaction of natural tetraazacyclopentazulene dyes with DNA and found that this somewhat unusual ring system results in lower negative enthalpies of binding to DNA. These compounds also have relatively constant enthalpy values below $\nu = 0.25$, but the values lie between -3 and -3.5 kcal/mole compared to ethidium, proflavine, and daunorubicin, which have calorimetric ΔH^0 values of between -6 and -7 kcal/mole. The authors speculate that this lower enthalpy could be due to a reduced stacking interaction caused by partial or nonclassical intercalation of the nonplanar compounds (*105*). Other evidence for different modes of intercalation is presented in Section V.

The limited results from enthalpy measurements up to this time indicate that most classical intercalating compounds of the type originally studied by Lerman have large negative enthalpy values of -6 to -8 kcal/mole when they bind by intercalation. Specific changes in the nonintercalating substituents on the intercalating ring such as the peptide rings of actinomycin or marked changes in the ring system such as disruption of planarity can cause significant deviations from this -6 to -8 kcal value. Calorimetric measurements using instruments with increased sensitivity and detailed van't Hoff studies using adequate binding models (Section IV,A) should remove much of the uncertainty from enthalpy measurements of intercalation and should help provide a more detailed analysis of the thermodynamics of intercalation. Because of the current uncertainty in enthalpy values, entropy values calculated using these enthalpies are especially subject to error, and we do not feel that a discussion of entropies of intercalation is warranted at the present time.

C. Ionic Effects

Since the early experiments of Peacocke and Skerrett (*7*), it has been recognized that both intercalation and outside stacking of ligands on complexation with DNA are dependent on ionic strength (*24*). Until recently there has been no generally accepted quantitative method for treating these ionic effects, which are related to other ionic-strength-dependent reactions of DNA such as thermal denaturation. The condensation theory developed by Manning (*109*) and Record and co-workers (*110*) has provided a method for the quantitative analysis and conceptual understanding of these processes. Basically DNA is treated as an infinitely long, linear molecule with an average charge density spread over the molecule (a line charge). There is a maximum value of the charge density above which any molecule will become unstable. For such a line charge there is a parameter ξ, which is determined by (*110*)

$$\xi = e^2/\epsilon kTb \tag{7}$$

where e is the protonic charge, ϵ is the bulk dielectric constant, k is Boltzmann's constant, b is the average distance between charged groups, and T is the absolute temperature.

For monovalent electrolytes the line charge will be unstable if $\xi \geq 1$ ($b \leq 7.14$ Å in DNA). B-Form DNA must then condense enough cations to lower its effective ξ value to 1 in order to be stable. The condensation is an indirect association, in contrast to a direct binding, and is essentially constant over a wide range of cation concentrations (*109–111*). Even after condensation, the DNA molecule is highly charged and can cause Debye–Hückel type screening effects for uncondensed counterions in solution. The total fraction of counterion associated per phosphate ψ is $\psi = \psi_c + \psi_s$, where ψ_c and ψ_s are the fraction counterion per phosphate condensed and screened, respectively (*112*). The ψ_s is, to a first approximation, determined by the Debye–Hückel theory, and ψ_c is directly related to ξ. For a ligand binding reaction the theory leads to Eq. (8) for the dependence of the equilibrium constant on counterion concentration:

$$\frac{\partial \log K_{\text{obs}}}{\partial \log [M^+]} = -m'\psi \tag{8}$$

Here, m' is the number of ion pairs formed in the complex and K_{obs} is the observed equilibrium constant at any salt concentration (*112*). Equation (8) assumes no conformational change in the polymer, which is not the case with intercalating ligands binding to DNA. Even if an uncharged ligand is bound to DNA by intercalation, counterions should be released just by the lowered charge density produced by lengthening the double helix (*112*). Wilson and Lopp (*96*) treat intercalation as a two-step reaction involving the intercalation conformational change [Eq. (9)] and the ligand binding step [Eq. (10)]:

$$\mathrm{D} \rightarrow \mathrm{D}^* + 2n(\psi_c - \psi_c^*)\mathrm{Na}^+ \tag{9}$$

$$\mathrm{D}^*(\mathrm{P}) + \mathrm{L} \rightarrow \mathrm{D}^*(\mathrm{PL}) + m'\psi_c^*\mathrm{Na}^+ \tag{10}$$

Here, D is native B-form DNA (B-DNA), D* is DNA in the conformation assumed in the intercalated complex (I-DNA), n is the number of base pairs per ligand involved in the intercalation conformational change [this is multiplied by 2 in Eq. (9), since there are two phosphate groups per base pair], ψ_c^* is the fraction counterion condensed per phosphate in I-DNA, D*(P) is the ligand binding site on I-DNA, D*(PL) is the intercalated complex, and m' is the number of ion pairs formed between the ligand and the DNA phosphate groups. The counting procedure in Eq. (9) is designed to determine the number of sodium ions released as a result of binding a single ligand. The counting procedure in Eq. (10) is designed to determine the number of sodium ions released as the result of direct ion pair formation. The number of phosphates involved can be different in these two equations and is different for intercalating molecules. The total reaction is conceptually viewed as two sodium ion release steps: (a) a conformation change in DNA, which increases the average b value, resulting in sodium ion release [Eq. (9)], and (b) binding of the cationic ligand to the extended DNA conformation to form m' ion pairs and release $m'\psi_c^*$ sodium ions [Eq. (10)]. The counterion dependence of the observed equilibrium constant for intercalation becomes (*96*)

$$\frac{\partial \log K_{\mathrm{obs}}}{\partial \log [M^+]} = -2n(\psi - \psi^*) - m'\psi^* \tag{11}$$

For the nearest-neighbor model (Section IV,A) n is 2, and for the intercalation geometry the average b value is 2.55 Å and ψ^* is 0.82 as compared to a ψ value of 0.88 in B-form DNA. A plot of log K_{obs} versus log $[M^+]$ with the known values for ψ and ψ^* allows the determination of the number of ion pairs m'. A plot for the dication quinacrine and a monocationic phenazine derivative is shown in Fig. 11 (*97*). As can be seen, both compounds give the expected linear plots. The number of ion pairs formed as determined from the slopes are 2.4 for quinacrine and 1.0 for the phenazine derivative, in reasonable agreement with the predicted values. Similar m' values are obtained for other mono- and dications (*97*, *112*). Errors in this method can arise from errors in K_{obs} values and in the theoretical treatment of each intercalation complex as an infinite line charge. The general applicability of

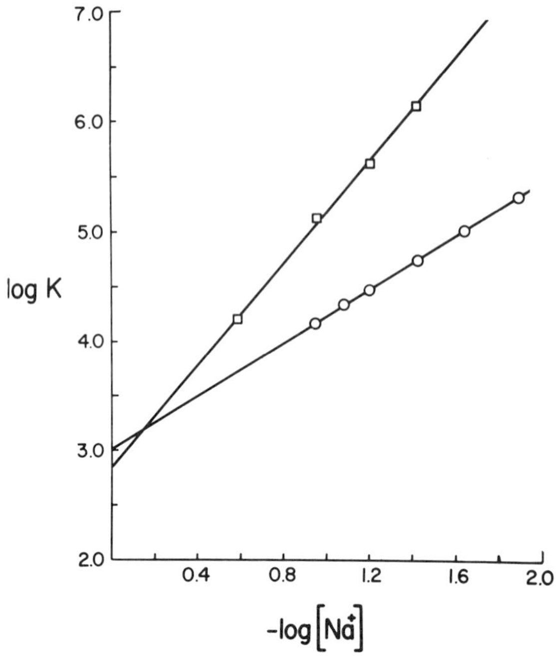

Fig. 11. Logarithms of the equilibrium constants for the interaction with calf thymus DNA of the dication quinacrine (□) and a monocation phenazine (○) versus the logarithm of the sodium ion molarity as suggested by Eq. (11). The slopes of these plots can be used to determine m', the number of ion pairs formed in the ligand–DNA complex, as discussed in the text. The excellent linearity of the plots supports the use of the ion condensation theory in the study of the interaction of intercalating ligands with DNA.

the condensation theory to ligand binding and other ionic-strength-dependent reactions of DNA seems well established (*96*, *97*, *112*).

As can be seen in Fig. 11 and as predicted by the condensation theory in Eq. (11), the dication and monocation intercalating ligand equilibrium constants are affected to different extents by salt. This has perhaps led to some confusion in the literature when relative DNA binding affinities for intercalating ligands of different charge are compared. Even less justified but perhaps more common is the comparison of binding results of ligands that have been determined at different ionic strengths. This can obviously lead to large errors, and any conclusions made from binding data at different salt concentrations will have little meaning and may be totally misleading. As a standard procedure in the analysis of ligand–DNA interactions, observed equilibrium constants should be determined at several ionic strengths and plotted as in Fig. 11. Comparison with other compounds at any ionic strength can then be made using these plots. Two precautions should be taken with this type of analysis: (a) The plot in Fig. 11 is for monovalent counterions such as Na^+ and will be different for dicationic counterions such as Mg^{2+}; and (b) the dimerization of intercalating ligands is also dependent on ionic strength, and this must be taken into consideration in quantitative work, especially at high ionic strengths (*96*). Because of possible conformational changes when other cationic ligands and macromolecules such as proteins bind to DNA, a detailed analysis of the interdependence of conformational and ionic effects on binding should be conducted.

Type II binding or outside binding through DNA-induced stacking of intercalating aromatic ring systems is known to be quite strong at low ionic strength and is essentially eliminated as ionic strength is increased (*7*, *24*). Since the stacked ligands would have a large effective cationic charge as an associated complex, this dramatic dependence on ionic strength is easily explained by Eq. (8). The associated complex would have a large m' value representative of a large number of ion pairs with DNA. At low ionic strength this would dramatically increase the free energy of the association due to the release of a large fraction of counterion. At higher ionic strengths this effect would be dramatically reduced (log K versus log $[M^+]$ would have a large negative slope), and Type II binding would be insignificant relative to intercalation, as has been observed experimentally (*24*). It should also be emphasized that much of the observed free energy of intercalation (calculated from K_{obs}) determined from experiments at low ionic strength is due to sodium ion release and not to any especially favorable interaction between the ligands and DNA. This illustrates the importance of analyzing ligands at several ionic strengths and evaluating these data by means of a plot such as that shown in Fig. 11. The dependence of intercalation on ionic strength may be the most frequently misinterpreted concept in this area of research. The application of the ion condensation theory to intercalation reactions and to ligand–DNA interactions in general should eliminate this problem.

V. NONCLASSICAL INTERCALATION

In Lerman's original model for intercalation the ligand–DNA complex was characterized by a specific unwinding angle and length increase of the double helix, as discussed in Section II. In addition it was assumed that the helix axis continued undisturbed through the intercalating molecule (e.g., the axis above and below the intercalating molecule continued just as through a base pair), and it was assumed that there was no bending of the helix as a result of intercalation (*14*). Early investigations on ligands that bind to DNA were concerned primarily with determining qualitatively whether the compound was bound by intercalation or not. As more and more intercalating compounds were discovered, especially such as quinine and chloroquine with small ring systems (*113*) and compounds such as actinomycin with unusual substituents (*108*), the more important point became an evaluation of how the various intercalated complexes were related. One of the early variations was the model of Peacocke and co-workers (*114*), which suggested that instead of the long axis of the acridine ring being parallel to the long axis of adjacent base pairs, the ligand was intercalated in a more perpendicular manner to the base pair long axis. This would allow stronger interaction of the acridine cation with the DNA phosphates and could explain the similarity of the binding of acridines to some single-stranded and double helical DNA samples (*114*).

X-ray structures with dinucleotides have cast doubt on this model, however, since in these complexes the intercalated molecule is generally more parallel to the long axis of the base pairs (see Fig. 6, for example) and close approach of the cationic groups to the phosphates seems to be of less importance than maximum overlap of the intercalated ring system with the adjacent base pairs (*68*, *71*). In addition, a *tert*-butylproflavine derivative that could not intercalate by the classical model but that could intercalate by the Peacocke model was found not to bind to DNA by intercalation (*115*).

Cohen and Eisenberg (Section III,A) devised a hydrodynamic method to evaluate quantitatively the length increase produced upon intercalation. Most compounds seem to give a smaller increase than would be expected on the basis of a 3.4 Å increase per intercalated ligand, as shown in Fig. 3. This could be interpreted as a deficiency in the theoretical or experimental methods, as a true increase in length of less than 3.4 Å per bound molecule, or as a combination of these factors. Gabbay and co-workers (*27*) synthesized a series of symmetrically and unsymmetrically methylated phenanthrolines to test these possibilities. By analyzing differentially a group of compounds of this type, errors in theoretical assumptions and experimental procedures should have a minimum effect, and molecular differences can be accentuated. They found that two pentamethyl compounds (including the quaternary *N*-methyl group) with the methyl groups spread around the ring gave the largest increase in viscosity on complex conformation of any compound in this series (*27*). The viscosity increases for these methylated compounds are considerably larger than the increase for the unsubstituted compound. The authors attributed this effect to the increased effective thickness of the ring system when extensively methylated (*27*). They found with several trimethyl derivatives, however, that the position of substitution of the methyl groups dramatically affected the viscosity increase on complex formation. A symmetrically substituted trimethyl compound gave a viscosity increase almost as large as the pentamethyl derivatives, while an asymmetrically substituted trimethyl compound gave the smallest increase of all the derivatives studied (even smaller than the unsubstituted compound). It was suggested that these results could best be explained by a nonclassical intercalation binding mechanism as follows. (a) The unsubstituted compound binds in a classical manner; (b) symmetrically substituted compounds bind in a more or less classical manner but cause larger length increases because of the greater effective thickness of their methylated intercalating ring systems; and (c) the asymmetrically substituted compounds bind by intercalation but can cause both lengthening and bending of the double helix. The net result of the latter binding mode is a smaller increase in the effective hydrodynamic helix length and viscosity than is produced in a classical binding mechanism. This type of binding model is illustrated in Fig. 12, in which the various combinations of helix length changes are illustrated.

One possibility suggested in Fig. 12 is that, if the bending of the helix is great enough, the net apparent helix length could actually drecrease as a result of nonclassical intercalation ($L_2 < L_0$). This is also supported by the work of Gabbay and co-workers with reporter molecules and peptides containing small aromatic ring systems (*116–121*). The work with reporter molecules is perhaps illustrated best by the studies of Kapicak and Gabbay (*118*), in which they synthesized several derivatives of the following structure and analyzed DNA interactions using NMR, viscometric, and thermal denaturation techniques:

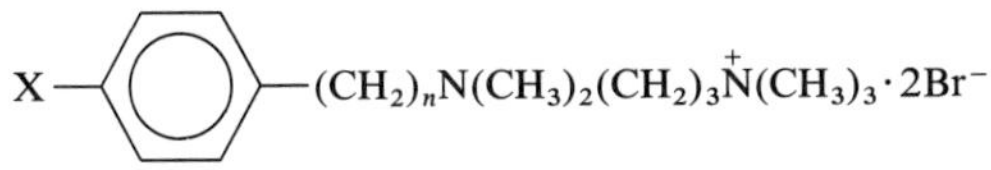

In this structure X could be H, NO_2, or CH_3, and n was varied from 1 to 4 (only the $n = 2$ derivative was made for the $X = CH_3$ derivative). These workers found that, with the DNA complex, viscosity decreased relative to uncomplexed DNA for all the compounds in which X = H. With the $X = NO_2$ compound, the complex viscosity decreased for the $n = 1$ derivative but increased for the other three nitro compounds ($n = 2, 3, 4$) and the methyl derivative. The ^1H NMR spectra of the complexes also indicated significantly more broadening for the compounds that increase in viscosity than for those that decrease in viscosity. These results were interpreted using a model such as the one in Fig. 12 with partial intercalation occurring for all the X = H and the $n = 1$, $X = NO_2$ derivatives, resulting in bending of the double helix and decreases in net length and viscosity on complex formation. With the other nitro and methyl compounds, the aromatic rings can fully inset between base pairs and more fully fill the intercalation cavity than with the X = H and $n = 1$ compounds. These compounds with $n > 1$ thus result

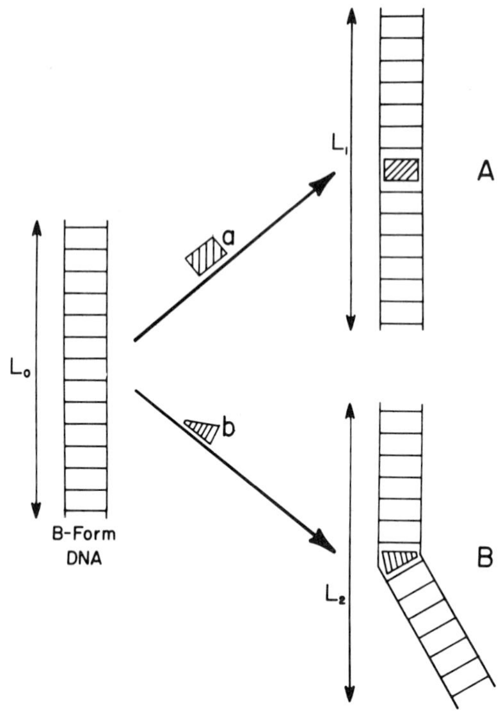

Fig. 12. Classical (A) and nonclassical (B) intercalation models for ligand–DNA interactions: L_0 is the length of a section of B-form DNA, L_1 is the length of the same section of DNA after binding an intercalating molecule (a) such as those shown in Fig. 1, and L_2 is the length of the same section of DNA after binding a nonclassical intercalating molecule (e.g., a small or nonplanar aromatic ring system or a larger aromatic ring system with an asymmetric distribution of bulky substituents). The value of L_1 is always greater than that of L_0 and L_2. However, L_2 may be greater than, equal to, or less than L_0, depending on the extent of bending or kinking caused in the double helix by interaction with the nonclassical intercalating molecule (b).

in at least slight increases in the length of the helix on complexation with DNA. Model building studies also support the conclusion of Fig. 12 (*117*).

Similar results have been obtained by Gabbay and co-workers on a broad range of peptides containing the aromatic groups of phenylalanine, tyrosine, and tryptophan (*120*, *121*), and the results have been reviewed (*117*). For example, peptides containing a lysyl residue as the N-terminal followed by an aromatic amino acid amide (for example, L-Lys-L-Phe-A, where A = amide) cause a much larger viscosity decrease than other similar peptides, such as L-Lys-L-Leu-A. These results have been interpreted in terms of partial insertion of the aromatic rings of the amino acid side chain resulting in a bending and net hydrodynamic length decrease of the double helix. A large number of longer peptides containing aromatic amino acids were also synthesized, and the results of physical studies with these compounds support a binding model in which the peptide is bound in the double helix minor groove as an extended single-stranded β chain. Whether the aromatic residue is partially inserted or does not insert depends strongly on (a) its position in the sequence and (b) its stereochemistry. As an example of (a), L-Lys-L-Phe-Gly-A binds by partial insertion, whereas L-Lys-Gly-L-Phe-A does not; and as an example of (b), L-Lys-L-Phe-A binds by partial insertion, whereas L-Lys-D-Phe-A does not (*117–121*). Because all aromatic residues in these peptides are joined to the peptide chain by a single methylene group, none are able to insert fully between DNA base pairs (*117*). This finding is similar to results with the reporter molecules of different length discussed above (*118*).

The Gabbay results have been questioned by Jacobsen and Wang (*122*), who showed that tyramine, tryptamine, and similar molecules did not unwind closed circular superhelical DNA, and by Quadrifoglio *et al.* (*123*), who found that some similar peptides gave calorimetric ΔH^0 values near zero on complexation with DNA. Nor did the latter authors (*123*) find the dramatic viscosity decrease discussed by Gabbay (*117*) and co-workers. It should be mentioned, however, that Gabbay showed that partial intercalation is quite dependent on sequence and stereochemistry of the peptide, as discussed above, and neither Jacobsen and Wang (*122*) nor Quadrifoglio *et al.* (*123*) used a series of peptides or reporter molecules which Gabbay and co-workers have shown to partially intercalate. It is quite possible that the disagreement results from an inappropriate choice of ligands for comparative study and not from any molecular difference.

The concept of Gabbay and co-workers that insertion of an aromatic group between DNA base pairs can cause bending of the double helix has received support from the X-ray analysis of intercalating drugs with dinucleotides (*71*). With ethidium bromide, for example, the phenyl and ethyl substituents are located in the minor groove and the helix is bent toward the major groove at the intercalation point (*71*). The possibility of bending the helix with larger ring systems and the steric restrictions on intercalation have been supported (*40*) using the berberinium ion (*26*) and several quinoline methanol derivatives with different numbers of bulky substituents (*124, 125*).

We have recently found that neither L-Lys-L-Phe-A nor L-Lys-D-Phe-A unwinds closed circular superhelical DNA (*126*). This result was somewhat surprising since Gabbay (*117*) has shown that these molecules have dramatically different effects on apparent helix length, although it is not clear whether partial intercalation should also significantly unwind DNA as classical intercalation does. Because of the possible implications of partial intercalation and bending of DNA for chromatic structure (*71*) and for protein–DNA recognition, this area should receive considerable future emphasis.

VI. APPLICATIONS OF INTERCALATION

A. Intercalating Drugs

Some of the early interest in intercalation arose out of the effort to understand how compounds such as proflavine, acridine orange, quinacrine, and ethidium bromide exerted medicinal and mutagenic activity (*40*). It was found that these compounds and numerous other planar aromatic cations formed a strong complex with DNA, and the complexes had the physical properties predicted by the intercalation model (Section II). Enzyme inhibition experiments both *in vitro* and *in vivo* indicated that all the compounds in this general class (Fig. 1) inhibited replication and/or transcription of DNA (*33*, *40*). A postulation for the mode of action of these compounds included formation of an *in vivo* complex between the intercalating drug and DNA followed by cell death due to the resulting metabolic disruption (*33*). There is now little question that drugs such as those listed above and the clinically important antineoplastic anthracycline compounds (*44*) include such binding as an essential part of their mode of action. Mechanistic studies on intercalating drugs attracted interest because of the importance of DNA and because its structure was known in detail for many more years than structures for other bioreceptors such as metabolic enzymes. Intercalating drugs also have a wide range of effects, including antimicrobial, antimalarial, antitrypanosomal, and antineoplastic activities, and this has increased interest in their medicinal action (*40*).

One feature that has yet to be fully explained is the specificity of the intercalating drugs. For example, how do anthracyclines distinguish antineoplastic from normal cells and, perhaps even more puzzling, how does quinacrine specifically attack the malaria parasite; ethidium, the trypanosomal parasite; anthracyclines, antineoplastic cells; and so forth? In other words, all of the compounds have been postulated to function through intercalation, but their specificity for parasitic or diseased cells varies enormously. One possible explanation for this diversity is that all of these compounds cause death if present at a sufficiently high intracellular concentration, but the difference between normal and diseased cells and the variation in the susceptibility of different diseased cells to different drugs are due to factors other than intercalation. Diseased cells might, for example, allow membrane diffusion or transport of the drug at rates that are significantly faster than those allowed by normal cells (*40*). It is interesting that resistence to drugs of this type might arise, at least in some cases, as a result of changes in cellular uptake of the drug (*127a,b*). The origin of this specificity for intercalating molecules will certainly be a very exciting research area in the future.

Other factors, including solubility, metabolic inactivation, and toxic side effects, such as the cardiotoxicity of the anthracyclines, which have nothing to do with intercalation, must also be included in any analysis of the medicinal effects of DNA binding drugs. Many of these compounds bind strongly to surfaces such as glass and, not surprisingly, interact with biopolymers, other than DNA, that can affect other cellular functions. The critical point to consider, then, is whether DNA is of primary importance in the medicinal action or whether interaction with some other bioreceptor is more important. It is, of course, not inconceivable that drugs exist which have two or more bioreceptors of approximately equal importance.

In our laboratory we are taking a differential approach to the analysis of medicinal activity of intercalating drugs in the hope that effects such as metabolic inactivation will not greatly vary through a series of drugs and that *in vivo* activities can therefore be more directly compared with *in vitro* studies on the drug–DNA complexes. One of the problems with the literature in this area is the lack of standard conditions for use in comparing the binding of drug molecules. As illustrated in Section IV,C, the binding constant of charged ligands to DNA varies considerably with ionic strength, and this must be considered when one is comparing the binding of different molecules. The dication chloroquine, for example, binds relatively strongly to DNA at very low ionic strengths but weakly at higher ionic strengths, including the physiological range (*24*, *40*). These and other results (*125*) made it unlikely that intercalation is the primary mode of action of chloroquine and other weakly binding quinolines. In fact, mefloquine, the best drug to come out of the extensive Vietnam War effort to develop new antimalarial drugs, hardly binds to DNA even at very low ionic strength (*125*). We have developed a structural model for quinoline–DNA interactions in general which indicates that, with one bulky substituent such as an aliphatic group containing a cationic nitrogen, they can bind well to DNA at least at low ionic strength. With two bulky substituents, binding is possible but somewhat reduced, while three fairly symmetrically placed bulky substituents (such as the cationic group and two trifluoromethyl groups in mefloquine) virtually eliminate binding through disruption of the stacking of the bases with the quinoline aromatic ring system (*125*).

Many of the unsolved problems involving intercalating ligands in general (Section III) will obviously also apply to intercalating drugs. Besides the areas of cell specificity and ionic effects on binding mentioned above, the following are major questions remaining to be answered in the area of DNA–intercalating drug interactions: (a) the origin of base pair binding specificity; (b) the influence of major versus minor groove binding; (c) the importance of steric and electronic substituent effects in intercalation binding; (d) how DNA conformational changes and changes in protein interactions affect medicinal activity; (e) the importance of DNA bending, kinking, etc., in drug interactions; and (f) the influence of specific interactions such as hydrogen bonds and ionic interactions in the interaction of intercalating drug molecules with DNA (*40*). Although the clinical use of intercalating drugs is frequently limited by toxicity problems, the success record of compounds of this type is undeniable, and their continued use in the near future is not in question (see (*43b*)]. Detailed research on the above questions offers the possibility of enhancing medicinal activity and decreasing toxicity if different drug molecular areas are involved in these two different biological effects. In addition, research on drug–DNA interactions has provided much of our current knowledge of DNA structure, conformational changes, and thermodynamics of ligand binding in general. This research area will no doubt continue to be one of the most active in nucleic acid chemistry.

B. Nucleic Acid Separations

With the current excitement and activity in the area of closed circular superhelical DNA and its uses in genetic engineering, methods for separating superhelical DNA from the nicked and linear molecules of the same sequence and from other contaminating molecules are quite important. CsCl density gradient techniques have been widely applied in such separations (*31–33*). Because a free-energy difference is involved in the binding of intercalating ligands to superhelical DNA relative to the nicked and linear DNA, the amount of ligand bound to the superhelical form will be different from the amount bound to the other DNA species of the same composition except at the point where the superhelical turns are exactly removed (see Fig. 4) (*92*). Since the binding of intercalating dyes such as ethidium bromide changes the density of a DNA molecule, separation of superhelical and other DNA forms can be achieved in density gradients under conditions in which the superhelical form has a different amount of dye bound than the other types of DNA (*31–33*). Ethidium is a particularly appropriate ligand for use since it does not have significant base pair binding specificity; since its fluorescence is enhanced when it binds by intercalation, it can therefore also serve as a marker for identifying the various DNA bands after centrifugation (*32*). When the DNA bands have been collected, the ligand can be removed by solvent extraction (*125*). The same technique has been used for an analytical detection of the presence of closed circular superhelical DNA and for a determination of the ratio of plasmid to chromosomal DNA from bacterial lysates (*128*). Pakroppa *et al.* (*129*) showed that quantitative separation of superhelical DNA from the nicked and linear forms can also be achieved on a hydroxyapatite column using ethidium bromide as the differentiating reagent. Hudson *et al.* (*130*) synthesized a series of phenanthridinium derivatives containing a triethylamino group attached to variable-length methylene chains in place of the quaternary ethyl group of ethidium. They found significantly improved resolution of closed circular and nicked forms from PM2 viral DNA using their derivatives instead of ethidium. An approximately linear relationship was found between the band separation of closed circular superhelical and nicked circular DNA and the mass of the dye used in the experiment. Hudson *et al.* (*130*) were able, at least qualitatively, to provide a thermodynamic explanation for this behavior and to indicate what factors should be most important in increasing the resolution in the density gradient system.

It has also been shown that closed circular and nicked DNA can be separated using agarose gel electrophoresis (*131–133*). Although this method is not generally useful for preparing superhelical DNA on a quantitative level, it is a faster method for determining analytically the purity of a superhelical DNA preparation. Dyes such as ethidium bromide can be used to vary the resolution between species in the experiment and as fluorescence stains for identifying and quantitating the DNA species present. Espejo and Lebowitz (*134*) and Delays and Jackson (*135*) have also shown that the superhelix density of a new superhelical DNA can be determined electrophoretically using an intercalating molecule of a known unwinding angle such as ethidium. These authors electrophoresed DNA samples in agarose containing different amounts of ethidium. When the superhelical species has all superhelical turns removed, it has the same electrophoretic mobility as the nicked species. At lower and higher concentrations of ethidium, it will have right- and left-handed turns, respectively, and will have a higher mobility than the nicked molecule. Since the number of superhelical turns affects the electrophoretic mobility, high-resolution electrophoresis has been used to show that natural superhelical DNA within a species consists of a population of molecules with different superhelical densities (*135*). The molecules with different superhelix density result in a series of DNA bands on electrophoresis. This has led to a technique called band counting, which allows direct determination of the unwinding angle of an intercalating molecule under a variety of conditions (*136*). Basically, a partially relaxed population of superhelical DNA molecules is produced which starts at zero superhelical turns and moves up in increments of one superhelical turn to the desired maximum number of turns that can be enzymatically adjusted

(*136*). This once again creates a series of bands on an electrophoretic gel, but now the number of superhelical turns for each band can be determined by simple counting, starting at the molecule with no superhelical turns (lowest electrophoretic mobility). The amount of intercalating molecule needed to remove the superhelical turns from most superhelical species can also be determined by monitoring its mobility, as discussed above. Since the total number of superhelical turns that were removed is known and since there are 360° per turn, the ligand unwinding angle can be calculated directly (*136*).

C. Probes of Chromatin Structure

The use of intercalating compounds as probes of chromatin structure has been reviewed by Gabbay and Wilson (*137*) and will therefore only be summarized here. The current model for chromatin structure postulates a series of DNA–protein complexes called nucleosomes which contain DNA folded in some manner (*138a,b*). These nucleosomes are then joined by sections of fairly free or uncomplexed DNA (*138a,b*). Intercalating drugs have been used to estimate what percentage of DNA in chromatin is exposed and can interact freely with the ligand (*137*). In addition, intercalating ligands can be used as probes to determine the effects of removal of specific proteins on chromatin structure and exposure and to determine the fidelity of chromatin reconstitution methods.

Binding studies to determine the number of exposed sites in chromatin versus uncomplexed DNA have generally indicated a significant reduction in binding sites in chromatin, as might be expected (*137*). The problem with these results is that they generally ignore neighbor exclusion and cooperative effects (Section IV,A) in binding, and, therefore, the quantitative assessment of binding sites and equilibrium constants must be in error. In addition, these studies show variation in work among different laboratories, which probably results from different methods of chromatin preparation, chromatin degradation or heterogeneity, and different methods of physical analysis (e.g., absorption, fluorescence, and equilibrium dialysis methods for analyzing binding) (*137*). The problem is also complicated by the results of Wu *et al.* (*139*) indicating that ethidium has very strong binding to the DNA extensively complexed with the core histones in nucleosomes. This interaction causes pronounced conformational changes in the nucleosome and should have a similar effect on fractionated chromatin (*40*, *139*).

Brodie *et al.* (*140a*) used a method that involves energy transfer between bound pairs of fluorescent ligands to measure the accessibility of chromatin DNA. The relative efficiency of energy transfer, which can be experimentally determined, between fixed amounts of ligands bound to DNA or chromatin depends on the separation distance of the energy donor and acceptor molecules along the double helix. The amount of energy transfer under these conditions will then depend on the amount of free double helix available for binding by the donor and acceptor ligands. For this method to be accurate the ligands should bind to all regions of open double helix and should not have any significant base pair binding specificity. Quinacrine is an acceptible energy donor with ethidium bromide as an acceptor since the quinacrine fluorescence emission spectrum overlaps the absorption spectrum of ethidium (*140a*). Ethidium thus quenches quinacrine fluorescence when both are bound to DNA, and the quenching efficiency depends on the distance between bound ligands or on the amount of free DNA binding sites. Quenching of the fluorescence of quinacrine bound to chromatin occurs at only about one-third of the amount of ethidium required to quench quinacrine bound to free DNA under similar conditions. Direct binding studies under similar conditions by Latt *et al.* (*140b*) also indicated that the number of binding sites on chromatin was reduced by approximately one-third compared to free DNA.

Further analysis in this area is complicated by the complexity of the binding models for purified DNA (Section III) and the additional factor of competing proteins and possible DNA conformational changes in chromatin. Future emphasis in this area should be on the application of appropriate thermodynamic models to the binding equilibria (Section IV,A) and on a more thorough investigation of ionic effects in ligand–chromatin interactions, as has been done with DNA (Section IV,C). In addition, direct binding studies using methods such as equilibrium dialysis are to be preferred in these experiments instead of indirect spectrophotometric or fluorometric analysis because of the complexity of chromatin relative to purified DNA.

D. Chromosome Banding

The intense, fairly uniform staining of chromosomes in microscopy has been known for many years and in fact accounts for the chromosome name. Caspersson and co-workers (*141*) discovered, however, that the fluorescent alkylating agent quinacrine mustard (the *N*,*N*-diethyl groups of quinacrine in Fig. 1 are replaced with *N*,*N*-2-chloroethyl groups) causes some areas of metaphase chromosomes to fluoresce brightly in ultraviolet fluorescence microscopy, whereas other areas of the chromosomes are relatively dark. It was hypothesized that the differentiation of chromosomes into fluorescent bands and dark regions was due to specific alkylation of guanine residues by the chloroethyl group of quinacrine mustard (*141*). Caspersson and co-workers (*141–143*) showed that the fluorescence microscope analysis had sufficient resolution to allow differentiation of the 23 pairs of chromosomes. Up to that time only 6 chromosome pairs could be unambiguously identified using criteria such as size, shape, and overall morphology. With the fluorescence bands, patterns specific for each chromosome pair were found which were relatively invariant among normal individuals and could therefore be used for identification. Up to 300 of these so-called Q bands can be identified by eye, and their positions soon were mapped and standardized patterns published (*144–146*).

In the course of this work, however, it was found that the guanine-specific alkylation of quinacrine mustard did not account for banding since the nonalkylating drug quinacrine also produced essentially identical fluorescence patterns (*144*). Weisblum and de Haseth (*147*) provided the next important step in the analysis of the mechanism of Q banding when they discovered that A·T-rich DNA samples enhanced the fluorescence of quinacrine, whereas DNA higher in G·C base pairs quenched quinacrine fluorescence. From these results, the proposed explanation for Q banding is that the bright bands are produced in areas that contain sequences of DNA high in A·T content, whereas the darker areas are higher in G·C base pairs. There are several problems with this explanation: (a) The heterochromatic centromeric region of chromosomes, which contains very A·T-rich satellite DNA, stains very poorly with quinacrine; (b) there has been some question as to whether the base composition of chromosomal DNA can have a sufficient variation in base content to account for Q banding; (c) there is some correlation between banding and the variation of DNA composition along the chromosomes; and (d) the proposed explanation does not include any effect for protein variation along the chromosome (*144–146*).

Comings and Drets (*148*) have analyzed the question of whether there is sufficient variation in chromosomal DNA base composition to account for banding. They found that the quinacrine fluorescence depends on the fourth power of the DNA base composition and that only approximately a 6% change in A·T content of a DNA segment could result in a 50% change in the relative fluorescence of quinacrine (*148*). The effect of chromosomal proteins was investigated quantitatively by Gottesfeld *et al.* (*149*). They isolated both euchromatin (extended) and heterochromatin (condensed) and found that the fluorescence of quinacrine bound to euchromatin was quenched much more effectively than that of quinacrine bound to heterochromatin. The DNA isolated from both types of chromatin essentially gave identically efficient fluorescence quenching and did not significantly differ in base pair composition. Chromosomal proteins alone did not significantly affect quinacrine fluorescence, but when heterochromatin was digested with pronase to remove proteins in the presence of quinacrine, the fluorescence quenching was markedly increased (*149*). The relative binding to the various chromosomal fractions was not significantly different and could not account for the differences in quenching. Circular dichroism spectra of the different fractions did indicate a possible conformational difference between euchromatin and heterochromatin (*149*). These authors (*149*) suggest, on the basis of these results, that there is a protein-induced conformational change in the DNA of heterochromatin that does not greatly change the amount of quinacrine bound to this fraction of DNA but does reduce the fluorescence quenching of bound quinacrine. In this way bands would represent regions of heterochromatin with an altered conformation of DNA and no base pair composition changes (*149*).

Disteche *et al.* (*150*) made densitometric and fluorescence intensity traces of five human chromosomes using the intercalating dyes proflavine, quinacrine, and ethidium bromide (Fig. 1). They analyzed banding in terms of variations in DNA base composition and total DNA concentration along the chromosomes. They concluded that the banding intensity obtained with compounds such as quinacrine depended primarily on the fraction of A·T base pairs along the chromosome. Secondary factors, however, such as chromatin structure, protein arrangement, and especially DNA content definitely influence the banding patterns and intensity (*150*). These results suggest that, at least, weak banding patterns should be obtainable for fluorescent intercalators having no base pair fluorescence specificity, and this hypothesis has been confirmed by Hoggard and Wilson (*151*) using a series of quinacrine derivatives that vary markedly in fluorescence sensitivity to variations in base-pair composition. Banding patterns were obtained with all of these derivatives but the brightness or resolution of the bands depended on the sensitivity of the fluorescence quantum yield of the derivative to variations in base pair composition (*151*).

In a search for other intercalating compounds that give chromosomal banding, Lin and van de Sande (*152*) found that the anthracyclines daunorubicin and adriamycin (Fig. 1) both gave banding patterns very similar to that of quinacrine. With regard to the mechanism of banding, the fluorescence quantum yields of these anthracyclines are not enhanced by A·T-rich DNA samples as is that of quinacrine. This would seem to rule out base pair composition as an important parameter in banding except for the fact that the anthracycline fluorescence was quenched less by DNA samples high in A·T than by those with a high G·C content (*153*). This leaves the possibility that anthracycline banding, now called D banding, is due to a relative difference in quenching, with areas of chromosomal DNA high in A·T giving low quenching and appearing brighter than regions of chromosomes high in G·C. As part of the same search for banding agents, van de Sande *et al.* (*154*) found that the fluorescence of the related antibiotics oligomycin, chromomycin A_3, and mithramycin is enhanced when they bind to G·C-rich regions of DNA but quenched when they bind to A·T base pairs. Since this is the opposite fluorescence behavior of that exhibited by quinacrine, these authors reasoned that, if base pair composition is largely responsible for banding, then these antibiotics should, in general, give bands that are bright where quinacrine is dark and vice versa. Indeed, such reverse banding (R banding pattern) was observed with these compounds in support of the base pair compositional contribution to banding (*154*).

As has been shown, both DNA concentration and base-pair composition explanations for banding are important. Protein composition might account for some banding, with the especially bright bands being produced in regions where DNA concentration is favorable and the A·T content is high. The reason for some A·T-rich DNA satellites not giving bright bands might then be due to interference from proteins or to sequences which cause especially pronounced quenching. Other dyes and dye mixtures, such as Giemsa, also can produce banding but usually require some pretreatment of the chromosome before strong banding patterns appear (*144–146*). The G bands produced by Giemsa occur in positions very similar to the fluorescent Q bands from quinacrine treatment but are visible without ultraviolet illumination. A typical Giemsa pattern is shown in Fig. 13. The exact mechanism for Giemsa staining is not clear at present but probably involves base pair specificity, DNA concentration, and specific protein effects (*144–146*).

Whatever the mechanism, the usefulness of the chromosomal banding technique is apparent. As mentioned above, the identification of the 23 chromosomal pairs has occurred only within the last ten years and has been due largely to the various banding techniques (*144–146*). It should be possible to develop specific dyes that are sensitive to protein content, DNA conformation, and base pair composition and that should allow detailed banding analyses of chromosome structure. Banding techniques have already greatly increased the sensitivity of chromosome analysis and now allow high-resolution detection of chromosome imperfections leading to genetic abnormalities such as malformation syndromes (*144–146*). As mentioned above, the banding patterns are generally reproducible within the members of a species, and the variation of banding among species can be a useful tool for studying the evolution of chromosomal structure. In spite of the general constancy of the banding patterns, there are, within some chromosomes, areas of considerable variation (*145*). One of the easiest to detect is the variation in size of the brightly fluorescing long arm of the male Y chromosome (*145*). This variable region can as much as triple the size of the Y chromosome. The DNA added seems to be highly repetitious and noncomplex in sequence. Individuals without the extra long arm are fertile and have male characteristics, as do those with the long fluorescence arm. This variation seems to be what may be called a normal type of polymorphism

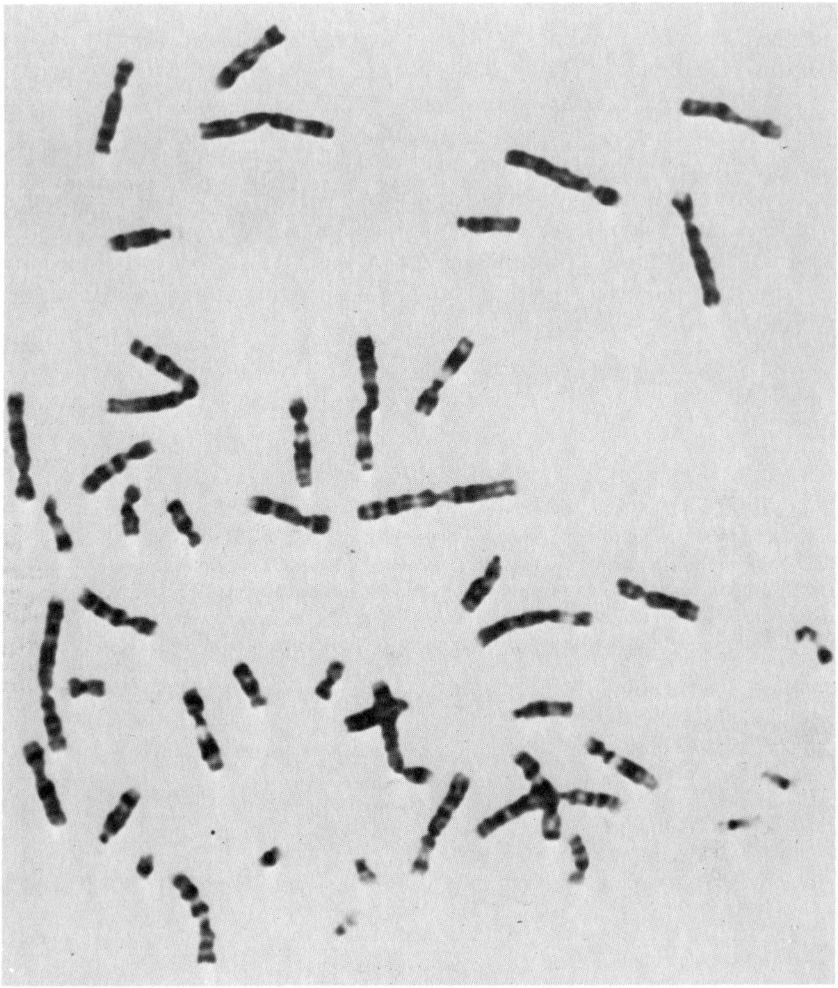

Fig. 13. A typical pattern for human chromosomes fixed with acetic acid–methanol and stained with Giemsa. The fixed chromosomes were treated with trypsin to produce banding before staining (courtesy of Ms. Margaret Hoggard).

in chromosome structure. Another variation is a chromosome area that stains homogeneously and has been discovered in a line of human neuroblastoma cells (*155*). These cells and others found in different species generally produce abnormally high levels of some type of protein, and it has been speculated that the homogeneously stained regions, which lack normal banding, are involved in the production of this protein. Rowley and co-workers have been active in applying banding techniques to analysis of chromosome abnormalities in human leukemia (reviewed in *156*). Before banding, the extent of chromosome changes involved in the transformation of a normal cell to a cancer cell was a controversial area. Using banding techniques, however, it is now apparent that specific chromosome changes are correlated with certain tumors (*156*). Chromosome translocations seem to be particularly common in the development of various human leukemias. The application of intercalating compounds, through chromosomal banding techniques, to the detection and analysis of human disease is one of the most exciting and important areas of intercalation research and should receive considerable future attention.

ACKNOWLEDGMENTS

The research reported in this review from our laboratory has been supported by The Research Corporation, the Petroleum Research Fund, administered by the American Chemical Society, and, during the writing of this review, by National Institutes of Health Grant CA 24454. We thank each of these organizations for their assistance. A continuing collaboration with Professor David Boykin has made this research on the interaction of drug molecules with DNA much more enjoyable and productive. We thank Ms. Mary Wilson and Ms. Brenda Holt for their friendly and thoughtful assistance in the preparation of the manuscript.

REFERENCES

1. E. A. Steck, "The Chemotherapy of Protozoan Diseases," pp. 23-77–23-177. U.S. Gov. Print. Off., Washington, D.C., 1972.
2. T. I. Watkins, *J. Chem. Soc.* p. 3059 (1952).
3. A. Albert, *in* "Drug Design" (E. J. Ariens, ed.), Chap. 5. Academic Press, New York, 1970.
4. O. Avery, C. MacLeod, M. McCarty, and M. McCarty, *J. Exp. Med.* **79,** 137)1944).
5. J. D. Watson and F. H. C. Crick, *Nature* (*London*) **171,** 737 (1953).
6. L. F. Cavalier and J. Angelos, *J. Am. Chem. Soc.* **72,** 4686 (1950).
7. A. R. Peacocke and J. N. H. Skerrett, *Trans. Faraday Soc.* **52,** 261 (1956).
8. G. Scatchard, *Ann. N.Y. Acad. Sci.* **51,** 660 (1949).
9. D. F. Bradley and M. K. Wolf, *Proc. Natl. Acad. Sci. U.S.A.* **45,** 944 (1959).
10. L. S. Lerman, *J. Mol. Biol.* **3,** 18 (1961).
11. L. F. Cavalieri, M. Rosoff, and B. H. Rosenberg, *J. Am. Chem. Soc.* **78,** 5239 (1956).
12. L. S. Lerman, *Proc. Natl. Cancer Conf.* **5,** 39 (1964).
13. L. S. Lerman, *Proc. Natl. Acad. Sci. U.S.A.* **49,** 94 (1963).
14. L. S. Lerman, *J. Cell. Comp. Physiol.* **64,** 1 (1964).
15. V. Luzzati, F. Mason, and L. S. Lerman, *J. Mol. Biol.* **3,** 634 (1961).
16. V. A. Bloomfield, D. M. Crothers, and I. Tinoco, Jr., "Physical Chemistry of Nucleic Acids," Chap. 7. Harper, New York, 1974.
17. L. S. Lerman, *J. Mol. Biol.* **10,** 367 (1964).

18. P. H. Lloyd, R. N. Prutton, and A. R. Peacocke, *Biochem. J.* **107,** 353 (1968).
19. G. Cohen and H. Eisenberg, *Biopolymers* **8,** 45 (1969).
20. G. Voordouw, Z. Kam, N. Borochov, and H. Eisenberg, *Biophys. Chem.* **8,** 171 (1978).
21. J. E. Godfrey and H. Eisenberg, *Biophys. Chem.* **5,** 301 (1976).
22. J. E. Godfrey, *Biophys. Chem.* **5,** 285 (1976).
23. H. Eisenberg, *in* "Basic Principles in Nucleic Acid Chemistry" (P. O. P. Ts'O, ed.), p. 173. Academic Press, New York, 1974.
24. R. L. Jones, A. C. Lanier, R. A. Keel, and W. D. Wilson, *Nucleic Acids Res.* **8,** 1613 (1980).
25. J. M. Saucier, B. Festy, and J. B. Le Pecq, *Biochimie* **53,** 973 (1971).
26. M. W. Davidson, I. Lopp, S. Alexlander, and W. D. Wilson, *Nucleic Acids Res.* **4,** 2696 (1977).
27. E. J. Gabbay, R. Scofield, and C. S. Baxter, *J. Am. Chem. Soc.* **95,** 7850 (1973).
28. S. Arnott and D. W. L. Hukins, *J. Mol. Biol.* **81,** 93 (1973).
29. W. Fuller and M. J. Waring, *Ber. Bunsenges. Phys. Chem.* **68,** 805 (1964).
30. J. Paoletti and J. B. Le Pecq, *Biochimie* **53,** 969 (1971).
31. J. Vinograd and J. Lebowitz, *J. Gen. Physiol.* **49,** 103 (1966).
32. W. R. Bauer, *Ann. Rev. Biophys. Bioeng.* **7,** 287 (1978).
33. M. J. Waring, *In* "The Molecular Basis of Antibiotic Action" (E. F. Gale, E. Cundliffe, P. E. Reynolds, M. H. Richmond, and M. J. Waring, eds.), pp. 173–277. Wiley, New York, 1972.
34. B. M. J. Révet, M. Schmir, and J. Vinograd, *Nature (London), New Biol.* **229,** 10 (1971).
35. J. C. Wang, *J. Mol. Biol.* **89,** 783 (1974).
36. D. E. Pullyblank and A. R. Morgan, *J. Mol. Biol.* **91,** 1 (1975).
37. W. Bauer and J. Vinograd, *in* "Basic Principles in Nucleic Acid Chemistry" (P. O. P. Ts'o, ed.), p. 265. Academic Press, New York, 1974.
38. W. A. Baase and W. C. Johnson, Jr., *Nucleic Acids Res.* **6,** 797 (1979).
39. L. P. G. Wakelin and M. J. Waring, *Mol. Pharmacol.* **10,** 544 (1974).
40. R. L. Jones and W. D. Wilson, *Adv. Pharmacol. Chemother.* **18,** 177–221 (1981).
41. M. J. Waring, *Prog. Mol. Subcell. Biol.* **2,** 216 (1971).
42a. C. Tsai, S. C. Jain, and H. M. Sobell, *J. Mol. Biol.* **114,** 301 (1977).
42b. S. C. Jain, C. Tsai, and H. M. Sobell, *J. Mol. Biol.* **114,** 317 (1977).
43a. W. J. Pigram, W. Fuller, and L. D. Hamilton, *Nature (London), New Biol.* **235,** 17 (1972).
43b. D. W. Henry, *in* "Cancer Chemotherapy" (A. C. Sartorelli, ed.), p. 15. Am. Chem. Soc., Washington, D.C., 1976.
44. G. J. Quigley, A. Wang, G. Ughetto, G. van der Marel, J. van Boom, and A. Rich, *Proc. Natl. Acad. Sci. U.S.A.* **77,** 6453 (1980).
45. S. Arnott, R. Chandrasekaran, and E. Selsing, *in* "Structure and Conformation of Nucleic Acids and Protein–Nucleic Acid Interactions" (M. Sudaralingam and S. T. Rao, eds.), p. 577. Univ. Park Press, Baltimore, Maryland, 1975.
46. C. J. Alden and S. Arnott, *Nucleic Acids Res.* **2,** 1701 (1975).
47. K. J. Miller, *Biopolymers* **18,** 2683 (1979).
48. M. W. Davidson, B. G. Griggs, I. G. Lopp, D. W. Boykin, and W. D. Wilson, *Biochemistry* **17,** 4220 (1978).
49a. C. C. McDonald, W. D. Phillips, and S. Penman, *Science* **144,** 1234 (1964).
49b. B. J. Blears and S. S. Danyluk, *J. Am. Chem. Soc.* **89,** 21 (1967).
50a. P. H. Bolton and T. L. James, *J. Phys. Chem.* **83,** 3359 (1979).
50b. M. E. Hogan and O. Jardetzky, *Biochemistry* **19,** 3460 (1980).
50c. R. L. Rill, P. R. Hilliard, Jr., J. T. Bailey, and G. C. Levy, *J. Am. Chem. Soc.* **102,** 418 (1980).
50d. T. A. Early, D. R. Kearns, W. Hillen, and R. D. Wells, *Biochemistry* **20,** 3756 (1981).

51. D. G. Gorenstein, *in* "Nuclear Magnetic Resonance Spectroscopy in Molecular Biology" (B. Pullman, ed.), pp. 1–15, Reidel Publ. Company, Dordrecht, Netherlands, 1978.
52a. J. Klysik, S. M. Studivant, J. E. Larson, P. A. Hart, and R. D. Wells, *Nature* (*London*) **290,** 672 (1981).
52b. R. T. Simpson, and H. Shindo, *Nucleic Acids Res.* **8,** 2093 (1980).
52c. D. J. Patel, and L. L. Canuel, *Proc. Natl. Acad. Sci. U.S.A.* **76,** 24 (1979).
52d. J. S. Cohen, J. B. Wooten, and C. L. Chatterjee, *Biochemistry* **20,** 3049 (1981).
53a. Y. H. Mariam, and W. D. Wilson, *Biochim. Biophys. Res. Commun.* **88,** 861 (1979).
53b. P. H. Bolton, and T. L. James, *J. Am. Chem. Soc.* **102,** 25 (1980).
53c. M. E. Hogan, and O. Jardetzky, *Proc. Natl. Acad. Sci. U.S.A.* **76,** 6341 (1979).
53d. H. Shindo, *Biopolymers* **19,** 509 (1980).
54a. R. I. Cotter, and D. M. J. Lilley, *FEBS Lett.* **82,** 63 (1977).
54b. N. R. Kallenbach, D. W. Appleby, and C. H. Bradley, *Nature* (*London*) **272,** 134 (1978).
54c. L. Klevan, I. M. Armitage, and D. M. Crothers, *Nucleic Acids Res.* **6,** 1607 (1979).
54d. H. Shindo, J. D. McGhee, and J. S. Cohen, *Biopolymers* **19,** 523 (1980).
55a. R. L. Jones and W. D. Wilson, *J. Am. Chem. Soc.* **102,** 7776 (1980).
55b. W. D. Wilson, R. Reddy, and L. G. Marzilli, *J. Am. Chem. Soc.*, (submitted).
55c. W. D. Wilson, R. A. Keel, and Y. H. Mariam, *J. Am. Chem. Soc.*, **103,** 6267 (1981).
56. M. Hogan and O. Jardetzky, *Biochemistry* **19,** 2079 (1980).
57a. T. R. Krugh and M. E. Nuss, *in* "Biological Applications of Magnetic Resonance" (R. G. Shulman, ed.), p. 113. Academic Press, New York, 1979.
57b. D. J. Patel, *Acc. Chem. Res.* **12,** 18 (1979).
58. G. P. Kreishman, S. I. Chan, and W. Bauer, *J. Mol. Biol.* **61,** 45 (1971).
59a. T. R. Krugh and C. G. Reinhardt, *J. Mol. Biol.* **97,** 133 (1975).
59b. C. G. Reinhardt and T. R. Krugh, *Biochemistry* **16,** 2890 (1977).
60. M. W. Davidson, B. G. Griggs, I. G. Lopp, and W. D. Wilson, *Biochim. Biophys. Acta* **479,** 378 (1977).
61. D. R. Young and N. R. Kallenbach, *Proc. Nat. Acad. Sci. U.S.A.* **77,** 6453 (1980).
62. W. Müller and D. M. Crothers, *Eur. J. Biochem.* **54,** 267 (1975).
63. D. M. Neville, Jr. and D. R. Davies, *J. Mol. Biol.* **17,** 57 (1966).
64. P. J. Bond, R. Langridge, K. W. Jennette, and S. J. Lippard, *Proc. Natl. Acad. Sci. U.S.A.* **72,** 4825 (1975).
65. D. M. Crothers, *Biopolymers* **6,** 575 (1968).
66. S. C. Jain and H. M. Sobell, *J. Mol. Biol.* **68,** 1 (1972).
67. T. D. Sakore, S. C. Jain, C. Tsai, and H. M. Sobell, *Proc. Natl. Acad. Sci. U.S.A.* **74,** 188 (1977).
68. S. Neidle, *Proc. Med. Chem.* **16,** 151 (1979).
69. A. H. J. Wang, J. Nathans, G. van der Marel, J. H. van Boom, and A. Rich, *Nature* (*London*) **276,** 471 (1978).
70. C. Tsai, S. C. Jain, and H. M. Sobell, *Proc. Natl. Acad. Sci. U.S.A.* **72,** 628 (1975).
71. H. M. Sobell, B. S. Reddy, K. K. Bhandray, S. C. Jain, T. D. Sakore, and T. P. Seshadri, *Cold Spring Harbor Symp. Quant. Biol.* **42,** 87 (1978).
72. H. R. Drew, R. M. Wing, T. Takano, C. Broka, S. Tanaka, K. Itakura, and R. E. Dickerson, *Proc. Natl. Acad. Sci. U.S.A.* **78,** 2179 (1981).
73. A. H.-J. Wang, C. J. Quigley, F. J. Kolpak, G. van der Marel, J. H. van Boom, and A. Rich, *Science* **211,** 171 (1981).
74. M. Hogan, N. Dattagupa, and D. M. Crothers, *Proc. Natl. Acad. Sci. U.S.A.* **75,** 195 (1978).
75. M. Hogan, N. Dattagupta, and D. M. Crothers, *Biochemistry* **18,** 280 (1979).
76. M. Levitt, *Proc. Natl. Acad. Sci. U.S.A.* **75,** 640 (1978).

77. S. Sokerov and G. Weill, *Biophys. Chem.* **10,** 161 (1979).
78. I. R. Lehman and E. A. Pratt, *J. Biol. Chem.* **235,** 3254 (1960).
79. R. M. Wartell, J. E. Larson, and R. D. Wells, *J. Biol. Chem.* **249,** 6719 (1974).
80. D. J. Patel and L. L. Canuel, *Eur. J. Biochem.* **90,** 247 (1978).
81. S. F. Yen, E. J. Gabbay, and W. D. Wilson, *Biochemistry* (submitted, 1981).
82a. D. G. Dalgleish, A. R. Peacocke, G. Fey, and C. Harvey, *Biopolymers* **10,** 1853 (1971).
82b. H. J. Li and D. M. Crothers, *Biopolymers* **8,** 217 (1969).
83a. D. M. Crothers, *Prog. Mol. Subcell. Biol.* **2,** 10 (1971).
83b. J. Bresloff and D. M. Crothers, *J. Mol. Biol.* **95,** 103 (1975).
84. M. Dourlent and J. F. Hogrel, *Biochemistry* **15,** 430 (1976).
85a. C. Mandal, N. R. Kallenbach, and S. W. Englander, *J. Mol. Biol.* **135,** 391 (1979).
85b. C. Mandel, S. W. Englander, and N. R. Kallenbach, *Biochemistry* **19,** 5819 (1980).
86a. S. C. Erfurth, P. J. Bond, and W. L. Peticolas, *Biopolymers* **14,** 1245 (1975).
86b. J. C. Martin and R. M. Wartell, *Biopolymers* (in press).
86c. R. M. Wartell, J. Klysek, W. Hillen, and R. D. Wells, "The Second Biomolecular Stereodynamics Meeting (Abstr.)," Albany, New York, p. 13, 1981.
87. J. B. Le Pecq, M. Lebret, J. Barbet, and B. Roques, *Proc. Natl. Acad. Sci. U.S.A.* **72,** 2915 (1975).
88. P. B. Dervan and M. M. Becker, *J. Am. Chem. Soc.* **100,** 1968 (1978).
89. B. Guagain, J. Barbet, R. Oberlin, B. P. Roques, and J. B. LePecq, *Biochemistry* **17,** 5071 (1978).
90a. L. P. G. Wakelin, M. Romanos, T. K. Chen, D. Glaubiger, E. S. Canellakis, M. J. Waring, *Biochemistry* **17,** 5057 (1978).
90b. R. G. Wright, L. P. G. Wakelin, A. Fieldes, R. M. Acheson, and M. J. Waring, *Biochemistry* **19,** 5825 (1980).
91. R. W. Armstrong, T. Kurucsev, and U. P. Strauss, *J. Am. Chem. Soc.* **92,** 3174 (1970).
92. W. Bauer and J. Vinograd, *J. Mol. Biol.* **47,** 419 (1970).
93a. A. S. Zasedatelev, G. V. Gurski, and M. V. Volkenshtein, *J. Mol. Biol.* **5,** 245 (1971).
93b. G. V. Gurski, A. S. Zasedatelev, and M. V. Volkenshtein, *J. Mol. Biol.* **6,** 479 (1972).
94a. J. Strum, *Biopolymers* **20,** 753 (1981).
94b. J. Strum, L. Schreiber, and M. Daune, *Biopolymers* **20,** 765 (1981).
95. J. D. McGhee and P. H. von Hipple, *J. Mol. Biol.* **86,** 469 (1974).
96. W. D. Wilson and I. G. Lopp, *Biopolymers* **18,** 3025 (1979).
97. W. D. Wilson, A. N. Gough, D. Gable, and R. L. Jones (manuscript in preparation).
98. W. Müller, H. Bunemann, and N. Dattagupta, *Eur. J. Biochem.* **54,** 279 (1975).
99. S. A. Winkle and T. R. Krugh, *Nucleic Acids Res.* **9,** 3175 (1981).
100a. M. Hogan, N. Duttagupta, and D. M. Crothers, *Nature (London)* **278,** 521 (1979).
100b. N. Dattagupta, M. Hogan, and D. M. Crothers, *Biochemistry* **19,** 5998 (1980).
101. J. B. LePecq and C. Paoletti, *J. Mol. Biol.* **27,** 87 (1967).
102a. J. Chambron, M. Duane, and C. Sadron, *Biochim. Biophys. Acta* **123,** 306 (1966).
102b. J. Chambron, M. Duane, and C. Sadron, *Biochim. Biophys. Acta* **123,** 319 (1966).
103. Y. M. Huang and D. R. Phillips, *Biophys. Chem.* **6,** 363 (1977).
104a. P. J. McCall and V. A. Bloomfield, *Biopolymers* **15,** 97 (1976).
104b. S. C. Riemer and V. A. Bloomfield, *Biopolymers* **18,** 1695 (1979).
105. V. Crescenzi and F. Quadrifoglio, *in* "Polyelectrolytes and Their Applications" (A. Rembaum and E. Selegny, eds.), p. 217. Reidel Publ., Company, Dordrecht, Netherlands, 1975.
106. F. Quadrifoglio, V. Crescenzi, and V. Giancotti, *Biophys. Chem.* **1,** 319 (1974).
107. F. Quadrifoglio and V. Crescenzi, *Biophys. Chem.* **2,** 64 (1974).
108. W. Müller and D. M. Crothers, *J. Mol. Biol.* **35,** 251 (1968).

109. G. S. Manning, *Q. Rev. Biophys.* **2,** 179 (1978).
110. M. T. Record, Jr., L. F. Anderson, and T. M. Lohman, *Q. Rev. Biophys.* **2,** 103 (1978).
111a. C. F. Anderson, M. T. Record, and P. A. Hart, *Biophys. Chem.* **7,** 301 (1978).
111b. M. T. Record, T. M. Lohman, and P. L. de Haseth, *J. Mol. Biol.* **107,** 145 (1976).
112. R. L. Jones and W. D. Wilson, *Biopolymers* **20,** 141 (1981).
113. F. E. Hahn, R. L. O'Brien, J. Ciak, J. L. Allison, and J. G. Olenick, *Mil. Med.* **131,** 1071 (1966).
114. A. Blake and A. R. Peacocke, *Biopolymers* **6,** 1225 (1968).
115. W. Müller, D. M. Crothers, and M. J. Waring, *Eur. J. Biochem.* **39,** 223 (1973).
116. E. J. Gabbay, B. L. Gaffney, and R. Glaser, *Ann. N.Y. Acad. Sci.* **171,** 810 (1970).
117. E. J. Gabbay, *in* "Bioorganic Chemistry" (E. E. van Tamelen, ed.), p. 33. Academic Press, New York, 1977.
118. L. Kapicak and E. Gabbay, *J. Am. Chem. Soc.* **97,** 403 (1975).
119. P. Adawadkar, W. D. Wilson, W. Brey, and E. J. Gabbay, *J. Am. Chem. Soc.* **97,** 1959 (1975).
120. E. J. Gabbay, P. Adawadkar, and W. D. Wilson, *Biochemistry* **15,** 146 (1976).
121. E. J. Gabbay, P. D. Adawadkar, L. Kapicak, S. W. Pearce, and W. D. Wilson, *Biochemistry* **15,** 152 (1976).
122. J. Jacobsen and J. C. Wang, *Biochim. Biophys. Acta* **335,** 49 (1973).
123. F. Quadrifoglio, V. Giancotti, and V. Crescengi, *FEBS Lett.* **65,** 345 (1976).
124. M. W. Davidson, B. G. Griggs, D. W. Boykin, and W. D. Wilson, *Nature* (*London*) **254,** 632 (1975).
125. M. W. Davidson, B. G. Griggs, D. W. Boykin, and W. D. Wilson, *J. Med. Chem.* **20,** 1117 (1977).
126. R. L. Jones and W. D. Wilson, unpublished observations.
127a. P. B. Macomber, R. L. O'Brien, and F. E. Hahn, *Science* **152,** 1374 (1966).
127b. C. D. Fitch, *Science* **169,** 289 (1970).
128. J. A. Meyers, D. Samchez, L. P. Elwell, and S. Falkow, *J. Bacteriol.* **127,** 1529 (1976).
129. W. Pakroppa, W. Goebel, and W. Müller, *Anal. Biochem.* **67,** 372 (1975).
130. B. Hudson, J. H. Dawson, R. Desiderio, and C. W. Mosher, *Nucleic Acids Res.* **4,** 1349 (1977).
131. F. W. Studier, *J. Mol. Biol.* **79,** 237 (1973).
132. D. A. Kaplan, R. Russo, and G. Wilcox, *Anal. Biochem.* **78,** 235 (1977).
133. W. Keller, *Proc. Natl. Acad. Sci. U.S.A.* **72,** 4876 (1975).
134. R. Espejo and Lebowitz, *Anal. Biochem.* **72,** 85 (1976).
135. R. J. Delays and D. A. Jackson, *Biochem. Biophys. Res. Commun.* **69,** 446 (1976).
136. M. Shure and J. Vinograd, *Cell* **8,** 215 (1976).
137. E. J. Gabbay and W. D. Wilson, *Methods Cell Biol.* **20,** 351 (1978).
138a. M. Zama, P. N. Bryan, R. E. Harrington, A. L. Olins, and D. E. Olins, *Cold Spring Harbor Symp. Quant. Biol.* **42,** 31 (1978).
138b. J. D. McGhee and G. Felsenfeld, *Annu. Rev. Biochem.* **49,** 1115 (1980).
139. H.-M. Wu, N. Dattagupta, M. Hogan, and D. M. Crother, *Biochemistry* **19,** 626 (1980).
140a. S. Brodie, J. Giron, and S. A. Latt, *Nature* (*London*) **253,** 470 (1975).
140b. S. A. Latt, S. Brodie, and S. H. Munroe, *Chromosoma* **49,** 17 (1974).
141. T. Caspersson, S. Farber, G. E. Foley, J. Kudynowski, E. J. Modest, E. Simonsson, V. Wagh, and L. Zech, *Exp. Cell Res.* **49,** 219 (1968).
142. T. Caspersson, L. Zech, and C. Johansson, *Exp. Cell Res.* **62,** 490 (1970).
143. T. Caspersson, L. Zech, C. Johansson, and E. J. Modest, *Chromosoma* **30,** 215 (1970).
144. D. E. Comings, *Ann. Rev. Genet.* **12,** 25 (1978).
145. H. J. Evans, *Adv. Hum. Genet.* **8,** 347 (1977).

146. S. A. Latt, *Ann. Rev. Biophys. Bioeng.* **5,** 1 (1976).
147. B. Weisblum and P. L. de Haseth, *Proc. Natl. Acad. Sci. U.S.A.* **69,** 629 (1972).
148. D. E. Comings and M. E. Drets, *Chromosoma* **56,** 199 (1976).
149. J. M. Gottesfeld, J. Bonner, G. K. Radda, and I. O. Walker, *Biochemistry* **13,** 2937 (1974).
150. C. Disteche, J. Bontemps, C. Houssier, J. Frederic, and E. Fredericq, *Exptl. Cell Res.* **125,** 251 (1980).
151. M. Haggard and W. D. Wilson (manuscript in preparation).
152. C. C. Lin and J. H. van de Sande, *Science* **190,** 61 (1975).
153. J. H. van de Sande, C. C. Lin, F. P. Johnston, and K. F. Jorgenson, *in* "Molecular Human Cytogenetics" (R. S. Sparkes, D. E. Comings, and C. F. Fox, eds.), p. 65. Academic Press, New York, 1977.
154. J. H. van de Sande, C. C. Lin, and K. F. Jorgenson, *Science* **195,** 400 (1977).
155. J. L. Biedler and B. A. Spengler, *Science* **191,** 185 (1976).
156. J. Rowley, *Annu. Rev. Genet.* **14,** 17–39 (1980).

15
Reactions in Crystallographic Shear Structures

J. S. ANDERSON

I. INTRODUCTION

Although crystallographic shear structures and intercalation compounds are usually considered to represent two very different principles of crystal chemistry, there are some close analogies and relationships between them. In both types, the underlying structural principle makes it possible to generate a very large variety of ordered structures corresponding to subtle, but tightly defined changes of composition—homologous series of intermediate compounds. The idealized structures of those compounds involve ordering at two distinct levels: Short-range order, reflecting the local composition of a crystal, which may be inhomogeneous, may be strictly preserved, whereas long-range order is not achieved. The gain in free energy of a crystal that results from perfect long-range order may be relatively small, with the result that, experimentally, crystallographic shear phases and intercalation phases may frequently be imperfectly ordered and not uniquely defined in composition. Such materials may simulate the bivariant

INTERCALATION CHEMISTRY

ISBN 0-12-747380-7

thermodynamic properties of a nonstoichiometric phase, but their crystals may be highly ordered on a short- or medium-range scale, with domains of different ordering pattern and composition coexisting. For both crystallographic shear structures and intercalation compounds, the structural principles impose a constraint on reaction processes. Changes in composition, or polymorphic transformations, leave the main fabric of the crystal unaltered while involving cooperative movements of a relatively small number of atoms on specific crystallographic sites. In those shifts of atomic positions, the principles governing short-range ordering are preserved. During the last few years, a good deal of rather detailed information has been acquired about reactions in the crystallographic shear structures at the unit-cell level, and this is surveyed in the present chapter.

II. CRYSTALLOGRAPHIC SHEAR AND THE STRUCTURE OF CRYSTALLOGRAPHIC SHEAR COMPOUNDS

A. The Crystallographic Shear Operation

The term *crystallographic shear* (CS) was originally coined to describe the relationship between the structures of transition metal oxides, particularly WO_3 and TiO_2, and the homologous series of oxygen-deficient intermediate oxides derived from them. It has proved to be a principle of wide significance whereby discrete ordered intermediate compounds can be generated in three-dimensional network structures when the stoichiometric ratio of their components is changed, without involving point defects. Structures based on crystallographic shear can be found in classes of compounds as disparate as the essentially ionic metallic oxides and the intermetallic phases (*1*).

Ionic crystals, in particular, can conveniently be regarded as being built up from coordination polyhedra (e.g., the $[MO_6]$ octahedra in TiO_2 or WO_3), centered on the cations and linked through the anions at their vertices into a three-dimensional fabric. In the ReO_3 and PdF_3 structure types (based on cubic and hexagonal packing of the anions, respectively), octahedra share all their vertices, each being common to two octahedra; in the rutile structure (which is itself related to the PdF_3 type by crystallographic shear), octahedra share a common edge, forming ribbons running parallel to the c axis. In the crystallographic shear phases derived from these types, the polyhedron linkage is locally modified and made more compact by increasing the degree of polyhedron linkage and thereby lowering the anion/cation ratio. In the derivatives of the ReO_3 type, simple vertex sharing is replaced by edge sharing; in the rutile type, edge sharing is replaced by octahedron face sharing. The coordination number of certain anions is thereby increased, but that of the cations (in most, but not all cases) remains unaltered. The resulting, locally modified structure, extending across an entire sheet of the crystal on a specific orientation, constitutes a crystallographic shear plane (CS plane); it represents a strong, completely localized fluctuation in composition, a planar singularity between two slices of virtually unmodified parent structure. The closer grouping of coordination octahedra does introduce some distortions and strain, but crystal structure determinations have shown that this attenuates fairly rapidly with distance from the CS plane. To a first approximation it may be ignored, although the strain field is important as the dominant long-range ordering force.

Some of the general characteristics of CS planes can most readily be brought out by considering crystallographic shear in the ReO_3-type structure, exemplified by the lower oxides of tungsten and molybdenum, and by the niobium oxides; the essential features are clearly displayed in a projection of the structure, as in Fig. 1. The nature of the planar boundary between two slices of unperturbed structure, which are necessarily in antiphase relation to each other, can be viewed in several ways. From one standpoint, the CS plane is equivalent to the complete elimination of a sheet of anion sites. In the parent structure, lattice planes parallel to the CS plane are of two regularly alternating kinds: planes containing both cations and anions, and planes containing anion sites only (Fig. 2). At the CS boundary, the regular ...*ABABABAB*... stacking sequence is interrupted and replaced by the sequence ...*ABAABAB*...; one anion plane is omitted. Across this interrupted stacking, the structure has been closed up by a displacement with components both normal to and parallel to the CS plane, thereby generating

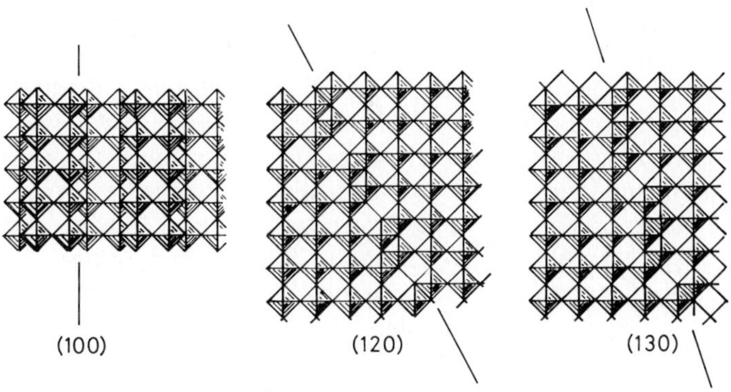

Fig. 1. Crystallographic shear in the ReO_3 structure on three orientations.

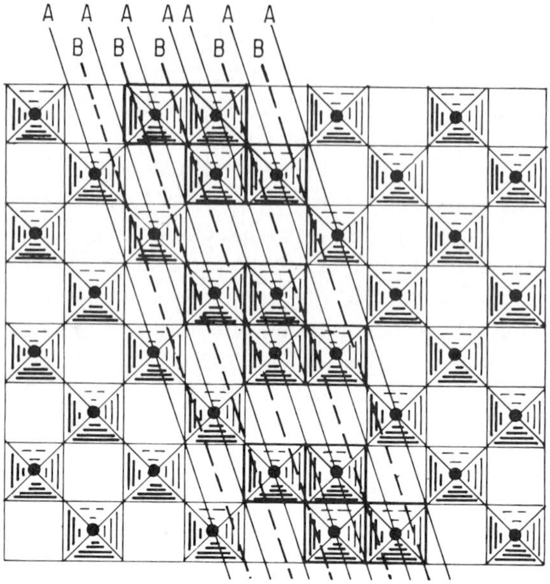

Fig. 2. Virtual elimination of a complete sheet of anion sites by crystallographic shear.

the groups of more closely linked polyhedra and producing the antiphase relation between the unmodified structure on either side. This hypothetical combination of collapse and shear can be specified by a *displacement vector* [**uvw**] operating on the *shear plane* (HKL). To a first approximation, omitting the effect of the small distortions, this displacement vector is always a perfect vector of the anion sublattice, which therefore remains invariant under the CS operation; only the cations have been displaced into a new set of equivalent sites.

In principle, this notional process of abstracting a sheet of anion sites could create CS planes on many orientations, with the same displacement vector operating on different (HKL) planes belonging to a common zone. This is indeed the case. Magneli's original work showed the existence of one homologous series of compounds M_nO_{3n-2}, typified by $W_{20}O_{58}$, with (130) CS planes involving groups of six edge-sharing octahedra, and a series M_nO_{3n-1} (for example, Mo_8O_{23}), with (120) CS planes and groups of four edge-sharing octahedra. In the polymorph R—Nb_2O_5 there are (100) CS planes, with infinite strings of edge-sharing octahedra. The displacement vector common to all of these is $\frac{1}{2}a[\bar{1}10]$; the CS planes all belong to the [001] zone, and the groups of edge-sharing octahedra seen in the projection actually form columns parallel to the zone axis.

In the derivatives of the rutile type, the columns of closely grouped (face-sharing) octahedra lying along the zone axis direction in the CS plane run obliquely through the sheared structure and are less easily visualized. The essential features are most informatively discussed from a slightly different standpoint, which is equally valid for the ReO_3 derivatives just considered. The anion sublattice of the rutile structure approximates a hexagonal close-packed arrangement, and, with only slight idealization, the (100) plane of the rutile structure could be represented by a ball model, as in Fig. 3a. There are two principal families of CS phases formed by the reduction of TiO_2 and in some double oxide systems (for example, TiO_2–Cr_2O_3) with CS planes on the $(12\bar{1})$ and the $(13\bar{2})$ orientation, respectively. Both arise from the operation of the displacement vector $[0\frac{b}{2}\frac{c}{2}]$ on planes of the $[1\bar{1}\bar{1}]$ zone. Figures 3b and c show the structure in which these oblique CS planes intersect the (100) plane. In Fig. 3d it can be seen that the same displacement vector operating on $(01\bar{1})$ produces an antiphase boundary (APB) by pure shear. If oxygen atoms were removed from the sites that are hatched in Fig. 3a, lying on a $(13\bar{2})$ anion-only lattice plane, operation of the displacement would produce the CS plane structure of Fig. 3b.

Comparison of Fig. 3b and c reveals that the stepped trace of the CS plane involves two kinds of step in the array of cations: steps at which an oxygen site has been eliminated, resulting in the rearranged, more compact linkage

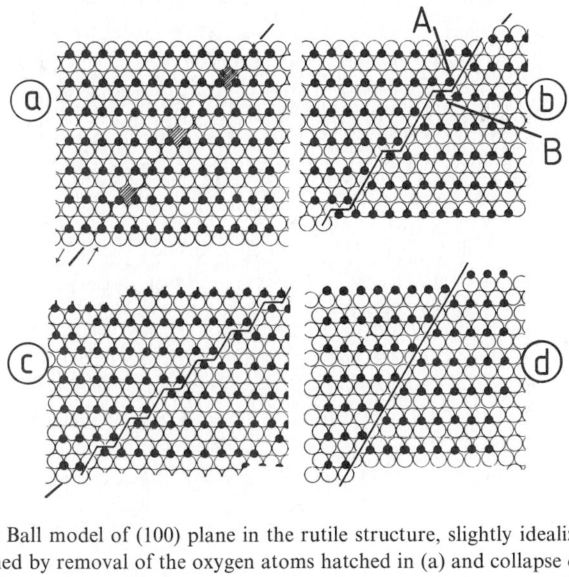

Fig. 3. (a) Ball model of (100) plane in the rutile structure, slightly idealized. (b) A $(13\bar{2})$ CS plane formed by removal of the oxygen atoms hatched in (a) and collapse of the structure. (c) A $(12\bar{1})$ CS plane involving the same displacement vector as in (b). (d) Operation of the same displacement vector on $(01\bar{1})$ produces an APB.

of coordination octahedra, and steps at which no oxygen site has been eliminated, but the cation rows have been displaced into an antiphase relation. These may be referred to as CS collapse steps and AP steps, respectively. In the $(12\bar{1})$ CS plane, each step is a collapse step; in the $(13\bar{2})$ CS plane, there is a regular 1 : 1 alternation of CS collapse steps and AP steps. Furthermore, the structural difference between these two kinds of step lies in the presence of an extra cation at each CS collapse step. In fact, either of the cations A or B (Fig. 3b) could be regarded as an interstitial atom with respect to one or other of the two slabs of unchanged rutile structure. In a sense, the CS plane could also be regarded as being formed by the ordering of interstitial atoms along an antiphase boundary into regularly spaced rods, extending along the $[1\bar{1}\bar{1}]$ zone axis direction. The $(13\bar{2})$ CS plane could be regarded as having alternate segments oriented parallel to $(01\bar{1})$ and $(12\bar{1})$, respectively. In principle, any regularly ordered sequence of these two kinds of segments would generate a CS plane, with an orientation defined by the

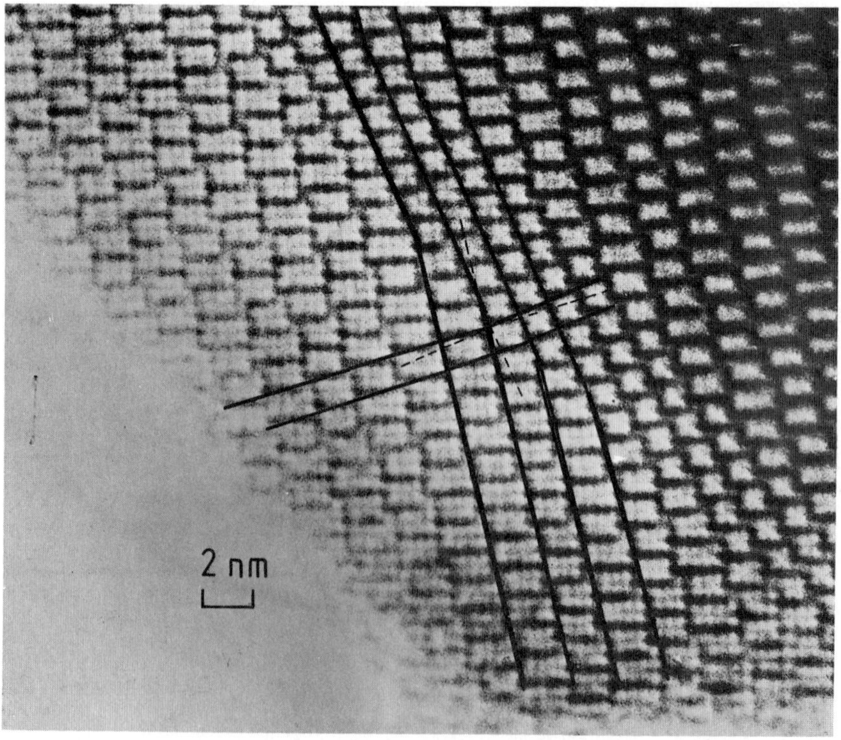

Fig. 4. Kinked CS planes in splat-cooled $TiNb_{38}O_{97}$.

indices ($p \times 01\bar{1} + q \times 12\bar{1}$) ($p,q$ integers) lying between the extremes of ($01\bar{1}$) and ($12\bar{1}$). The effect of the CS plane on the stoichiometry of the crystal (the number of anion sites eliminated) depends on the ratio of CS collapse steps to AP steps. If the two kinds of step are not regularly sequenced, the CS interface is no longer planar, but has kinks and changes of direction (Fig. 4).

B. Stoichiometric and Nonstoichiometric Crystallographic Shear Compounds

The foregoing discussion of the nature of a CS boundary can be extended to other structure types. If such planar CS boundaries recur at regular spacings, they generate a new ordered structure—a distinct chemical compound, defined by a new crystallographic unit cell, which may be regarded as a superlattice of the small subcell of the parent structure. The composition of the new unit cell will be determined by two factors: the orientation of the CS planes and the width of the slab of parent structure between successive CS planes. The latter could clearly be varied, at least hypothetically, in steps of one cation polyhedron at a time, so generating a homologous series of compounds, with a fine gradation in composition, for each possible CS plane orientation.

Since the interface between successive slabs—the structure of the CS plane—is common to all members of one homologous series, slabs representing different members of the homologous series can intergrow perfectly coherently. Disorder is possible, in which the recurrent CS planes are irregularly spaced, and the composition of a macroscopic crystal is then no longer defined. It is nonstoichiometric, but metastable with respect to a possible ordering process which would bring the CS planes into regular spacing.

C. "Swinging" CS Planes

As noted above, each orientation of CS planes should generate a family of homologous compounds. To a given chemical composition there may therefore correspond several alternative, fully ordered structures, differing in the orientation and spacing of their CS planes. Both strain energy (deformation of coordination octahedra) and cation–cation repulsion contribute to the relative stability of different structures, and, in general, ordered structures with different CS plane orientations are each stable only over certain composition ranges. Factors governing the relative stability of (120) and (130) CS compounds in the WO_{3-x} oxides have been discussed by Iguchi and Tilley (*2*). The rutile structure, with CS compounds of the general formula Me_nO_{2n-1} based on both the ($12\bar{1}$) and ($13\bar{2}$) types, is more versatile.

In the lower titanium oxides, those from Ti_4O_7 (CS plane spacing 6.2 Å) to $Ti_{10}O_{19}$ (CS plane spacing 16.3 Å) have $(12\bar{1})$ CS planes; the lower vanadium oxides V_3O_5 to V_9O_{17} also have this structure. A second family, ranging from an upper limit of $n =$ ca. 36 ($Ti_{36}O_{71}$, CS plane spacing 36.8 Å) to $Ti_{16}O_{31}$ (CS plane spacing 16.1 Å), has $(13\bar{2})$ CS planes. For reasons that will be readily apparent in the following discussion of reaction processes, changes in composition by oxidation or reduction within each of these families show strong hysteresis. The number of CS planes in a crystal must be altered, and attainment of long-range order in the spacing of CS planes is established with difficulty. Between the total compositions $TiO_{1.900}$ and $TiO_{1.937}$, however, the system is kinetically quite reversible (*3*). In that composition interval, the CS planes have to swing around, and since the cation rows in the rutile slabs in $Ti_{10}O_{19}$ and in $Ti_{16}O_{31}$ are identical in length and almost unchanged in dimensions, oxidation or reduction does not require the creation or elimination of CS planes. In principle, every possible composition between those limits could correspond to an ordered, but nonstoichiometric structure, provided that a perfectly ordered sequence of AP and CS collapse steps in the CS planes could be preserved. Whether that order is preserved is uncertain, but a continuous reorientation of CS planes is undoubtedly found in the Cr_2O_3–TiO_2 double oxides, in which there is a series of CS phases with the general formula $Cr_2Ti_{n-2}O_{2n-1}$. For low values of n (for example, $Cr_2Ti_4O_{11}$), the CS planes lie on or close to $(12\bar{1})$; for the highest values ($n = 20$–30), they are on or close to $(13\bar{2})$. Careful electron diffraction studies have shown that for intermediate Cr_2O_3/TiO_2 ratios, the CS plane orientation progressively swings around as a function of composition (*4*). Moreover, the evidence of electron microscopy is that the spacing between CS planes is extremely regular and that the CS planes are not kinked.

D. Two Sets of CS Planes: The "Block" Structures

In the foregoing discussion, the structural principle has been developed in relation to structures with a single set of CS planes. Among the niobium oxides and their derivatives is a group of compounds derived from the ReO_3 type but with two nearly orthogonal sets of CS planes with a range of possible orientations close to $(100)_R$ and $(001)_R$, respectively. As can be seen from Fig. 1, any oblique plane is necessarily stepped, with "treads" m octahedra long, parallel to $\langle 100 \rangle$, and "risers" one octahedron high between the steps. If two such sets intersect, they cut the collapsed substructure up into rectangular columns or blocks, ($m \times n$) octahedra in cross section and infinite in extension along $[010]_R$ (Fig. 5). Both the dimensions of the columns or blocks and the way in which they are linked together into a space-filling

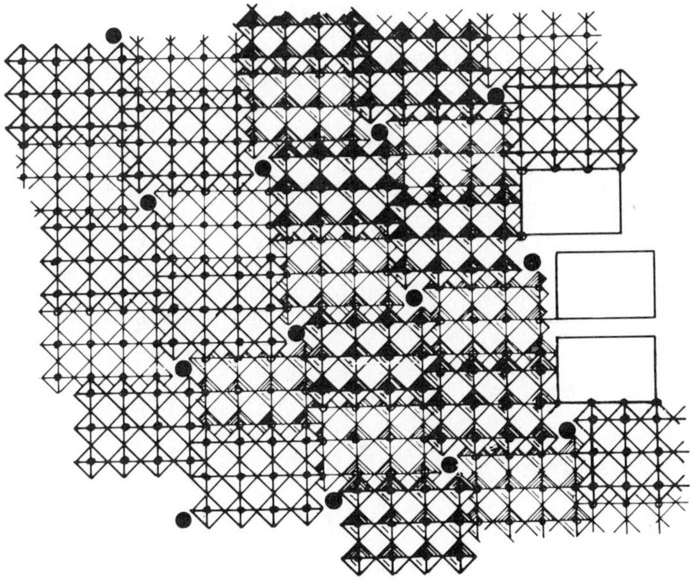

Fig. 5. The block structure principle: projection of the structure of $TiNb_{24}O_{62}$ and $Nb_{25}O_{62}$ down [010]. Block sizes and connectivity can be schematically represented by rectangular blocks drawn through the outer cation positions, as shown.

pattern are uniquely determined when the orientation of both sets of CS planes is specified. Figure 6 and Table I indicate how this hypothetical CS collapse of the ReO_3 substructure generates the structure of H—Nb_2O_5 and of other compounds, with oxygen/metal ratios ranging from $MeO_{2.333}$ ($TiNb_2O_7$) to $MeO_{2.654}$ ($W_8Nb_{18}O_{69}$) as the orientation of the CS planes changes.

As has already been seen, structures having a common CS plane orientation can intergrow coherently. In the block structures, that implies the possibility of coherence along one CS plane; the orientation of the other set of

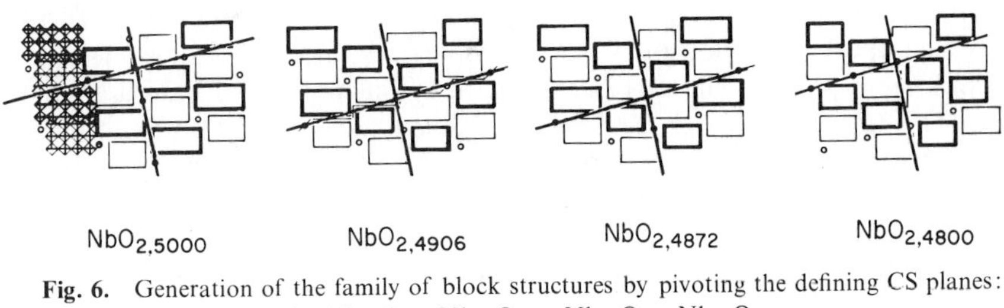

Fig. 6. Generation of the family of block structures by pivoting the defining CS planes: the regular intergrowth series $Nb_{28}O_{70}$, $Nb_{53}O_{132}$, $Nb_{39}O_{97}$, $Nb_{25}O_{62}$.

TABLE I

Generation of Block Structures from the ReO_3 Type by Double Crystallographic Shear

Compound	Block structure symbol[a]	Generating crystallographic shear operations	
		$\frac{1}{2}a[\bar{1}10]/$	$\frac{1}{2}a[011]/$
$TiNb_2O_7$	$(3 \times 3)_\infty$	(100)	$(10\bar{3})$
$Nb_{12}O_{29}$	$(4 \times 3)_\infty$	(100)	$(10\bar{4})$
$Nb_{16}O_{40}$	$(4 \times 4)_\infty$	(100)	$(10\bar{4})$
$MgNb_{14}O_{35}F_2$	$(5 \times 3)_\infty$	(100)	$(10\bar{5})$
$Nb_{22}O_{54}$	$(3 \times 3)_1 + (4 \times 3)_\infty$	(601)	$(20\bar{7})$
$Nb_{47}O_{116}$	$(3 \times 3)_1 + (4 \times 3)_\infty + (4 \times 3)_2$	(601)	$(40\overline{15})$
$Nb_{25}O_{62}$	$(4 \times 3)_2$	(601)	$(10\bar{4})$
$Nb_{53}O_{132}$	$(4 \times 3)_2 + (4 \times 3)_1 + (5 \times 3)_\infty$	(601)	$(40\overline{17})$
$Nb_{28}O_{70}$	$(4 \times 3)_1 + (5 \times 3)_\infty$	(601)	$(20\bar{9})$
$WNb_{12}O_{33}$	$(4 \times 3)_1$	(301)	$(10\bar{4})$
$W_3Nb_{14}O_{44}$	$(4 \times 4)_1$	(401)	$(10\bar{4})$
$W_5Nb_{16}O_{55}$	$(5 \times 4)_1$	(401)	$(10\bar{5})$
$W_8O_{18}O_{69}$	$(5 \times 5)_1$	(501)	$(10\bar{5})$

[a] The symbol $(m \times n)_1$ denotes a block that is related to adjacent blocks with cations at the same level across tetrahedral sites only; $(m \times n)_2$ denotes blocks paired at the same cation level via shared octahedron edges, as in Fig. 5; $(m \times n)_\infty$ denotes blocks sharing two corners in the same way, so as to form infinite ribbons.

CS planes must be altered at the junction of two structures. Compatible structures can therefore intergrow in layers, as single rows of blocks of differing dimensions and composition (so-called Wadsley defects) or in macroscopic domains. Coherently intergrown layers can recur in either an ordered or a disordered sequence. It can readily be shown that, if one CS plane orientation is held constant while the other CS plane set preserves perfect planar order as it swings around, the operation will generate a succession of ordered intergrowth phases. Such regular intergrowth compounds do, indeed, exist. If one set of CS planes is kinked, Wadsley defects are formed; if both sets of CS planes are kinked, the result must be to form a jigsaw-like intergrowth of different block sizes. Since the dimensions and the composition of the blocks are uniquely linked, a randomized arrangement of blocks indicates local fluctuations in stoichiometry, and in this way these very flexible structures accommodate nonstoichiometry. These materials have proved to be well suited for study at the atomic level by high-resolution electron microscopy, and it is through this technique that much of our understanding, both of structure and of structural transformations, has come.

III. REACTIONS IN CRYSTALLOGRAPHIC SHEAR STRUCTURES

A. One Set of CS Planes: General Considerations

1. Implications of Reaction Processes in CS Structures

Any ordered configuration of CS planes, implying slabs of parent structure of uniform width between them or columns of uniform cross section in block structures, corresponds to a definite chemical composition. Even a disordered CS or block phase can be regarded as an intergrowth of chemically defined structures on a unit-cell or microdomain scale. It follows that any reaction that changes the composition of the crystal implies an alteration in the spacing between CS planes or in their orientation; CS planes must be eliminated, or new CS planes must be introduced, or the CS planes must swing around. Furthermore, the attainment of order in disordered CS structures or transformation from one CS structure to another—processes taking place at constant composition—also requires the reorientation or the migration of CS planes.

In general terms, it can be seen that reactions in CS phases involve two kinds of processes: (a) a process that alters the chemical composition by changing either the number of CS boundaries within the crystal or their orientation; (b) a subsequent process, restoring long-range order of a new kind, appropriate for the new composition. Process (a) is inherently likely to be much faster than (b). It is highly localized, since a CS boundary is itself a localized, strong fluctuation in structure and composition. New CS planes may be introduced into, or preexisting CS planes eliminated from, a crystal at any position; changes of orientation of CS planes can involve isolated segments of CS planes. At least as a transient state, reaction necessarily produces some disorder, but the free energy of the crystal is undoubtedly minimized by complete ordering, although the decrease in enthalpy, compared with a disordered state, is probably small; the configurational entropy of an assembly of randomized CS planes is negligibly small. To transform the disordered state that is the immediate result of a reaction into an ordered state, all the CS boundaries in the crystal must be completely reshuffled.

Ordering is therefore a highly cooperative process and requires adjustments at several different levels corresponding to decreasing interaction between the structural units involved. The strongest interaction is between cations at any one step of the CS planes, occupying next-neighbor sites along the zone axis direction. In the ordered state, such an atom row must be free from jogs. If one cation jumps to a new site, it alters the potential barriers against a corresponding jump by its neighbors. The adjustment thus propagates readily along a whole rod of sites along the zone axis. This is the most facile ordering step, and much evidence from the electron microscopy of the block structures shows that ordering is usually perfect along the axis of the columns, even when long-range order is completely lacking in the arrangement of columns. Second, the CS boundaries must become planar and correctly oriented. As has been seen, this depends on establishing a regular sequence of AP and collapse steps along the boundary. To alter an existing sequence, a cooperative displacement of whole rods of sites is needed, and the rows of cations that must be shifted are separated by several interatomic spacings. In many systems that is a relatively slow process; kinked CS planes and irregular sequences of blocks may be eliminated with difficulty. Finally, the weakest interaction is between successive CS planes, which may be 20 Å or more apart (*5*). To make the width of the slab of parent structure between CS planes uniform throughout a crystal entails the bodily migration of every CS plane, possibly the cooperative movement of almost every cation in the crystal. This is therefore the most sluggish step. It is hardly to be achieved in finite annealing times if the spacing between CS planes is very long, as in the highest members of the Ti_nO_{2n-1} oxide series. As a consequence, lamellar disorder has frequently been found in CS compounds and block structures; they tend to form intergrowths with layers of several different members of a homologous series detectable by electron microscopy, although often not detectable by X-ray methods, in apparently perfect crystals. There remain inhomogeneities of composition, which could be removed only by attaining perfect long-range order.

From the foregoing discussion, three mechanistic questions emerge: (a) How are CS planes introduced and grown into the parent structure? (b) How are changes in the orientation of CS planes effected? (c) How is the migration and reshuffling of CS planes effected?

2. *The Introduction of CS Planes*

Two slightly different problems must be considered here. In the reduction of transition metal oxides, the highly localized reaction implies a net outward transport of oxygen through the crystal, or an inward transport of cations. In the formation of oxygen-deficient double oxides, oxygen atoms are added to the fabric, and there must be an inward diffusion of, e.g., trivalent cations and possibly of oxygen as well.

Considering first the reduction of TiO_2, WO_3, MoO_3, etc., some evidence comes from the introduction of individual CS planes when a specimen is strongly irradiated by the electron beam in the electron microscope. The CS planes are observed to nucleate at the edge of the crystal and advance across it. An example of such a process is shown in Fig. 7. The locus of reaction is, in effect, confined to the advancing edge of the CS plane. Since a CS plane grown in that way literally represents the elimination of a sheet of oxygen sites coextensive with its area, the boundary between growing CS plane and unreduced crystal is necessarily a dislocation with at least partial

Fig. 7. Crystallographic shear planes growing into a crystal of WO_3 under beam irradiation.

edge character; dislocation contrast has frequently been observed around the tip.

In the course of reduction carried out under less drastic conditions, it has been found that CS planes are formed at a very early stage in the reduction of the stoichiometric oxides, i.e., when the deficit of oxygen is quite small (*6*). Thus, Bursill and Hyde (*5*) found that, in a crystal with the nominal composition $TiO_{1.998}$, CS planes were already present on all orientations of the form $[13\bar{2}]$, both as isolated defects and grouped in packets. The association of CS planes in groups indicates that they exert mutual attraction, not a repulsion; within larger groups, the spacing between the interior CS planes (around 40 Å) corresponds to that of the uppermost member of the $(13\bar{2})$ family, but it is larger for the outermost CS planes of the packet. Similarly, (120) CS planes are already formed at compositions richer in oxygen than $WO_{2.99}$, in the reduction of WO_3 (*6*).

These observations are relevant both to the mechanism of formation of CS planes and to the relation between CS planes and the point defect equilibria of the oxides. It is evident that a single CS plane would effectively eliminate a very large number of point defect anion vacancies or cation interstitials, and, from the composition and the density of $(13\bar{2})$ CS planes in their material, Bursill and Hyde inferred that the residual point defect concentration in the rutile was only of the order 10^{-4}. It is conceivable, however, that the saturation concentration of point defects may be greater at high temperatures and that there is some temperature-dependent equilibrium established between point defects and CS planes. There is a considerable body of fairly concordant work on the composition/oxygen fugacity relationship and the semiconducting properties of substoichiometric TiO_{2-x} which points to titanium interstitials as the predominating defect over a wider range than indicated by Bursill and Hyde. Yagi *et al.* (*7*), from a proton channeling study, concluded that in $TiO_{1.994}$ about 90% of the stoichiometric defect must be accommodated by titanium interstitials and 10% by CS planes. As in the electron microscope work, some or all of the CS planes could have formed during the cooling from the high temperature at which the material had been prepared. Although the problem is not fully resolved, the probable situation is as represented in Fig. 8, based on the work of Baumard *et al.* (*8*). At all temperatures from 1273 to 1693 K, the conductivity versus oxygen fugacity follows a consistent pattern. At the highest oxygen pressures, with compositions close to $TiO_{2.0}$, the conductivity not only accurately conforms to the $P_{O_2}{}^{-1/5}$ relationship indicative of titanium interstitials, but is rapidly reversible, without hysteresis, when the oxygen fugacity is changed. This suggests that the equilibrium is controlled by defects with a high diffusivity, interstitial cations. At some composition

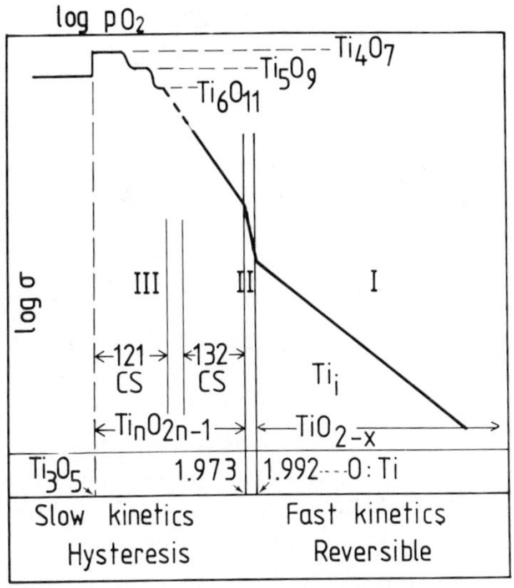

Fig. 8. Electronic conductivity, defect character, and crystal structure of TiO_{2-x} as a function of oxygen fugacity. [After Baumard (*8*).]

TiO_{2-x} (with x greater at the higher temperatures), the conductivity increases abruptly at nearly constant oxygen fugacity, corresponding to the onset of a univariant phase reaction at some equilibrium pressure. At the conclusion of this step, the composition is consistent with the conversion to CS phases of the $(13\bar{2})$ family, which is traversed and in due course converted to defined members of the $(12\bar{1})$ family, as the oxygen fugacity is lowered still further. The increasing degree of reduction is reflected in the n-type conductivity, which progressively increases with the concentration of Ti^{3+} donor centers. The significant feature is that equilibrium is attained only slowly and is subject to hysteresis in the second and third ranges of behavior; it clearly involves slower organizational processes than the simple diffusion of point defects. The model that emerges is that of point defects aggregating or ordering and being eliminated by formation of a CS plane when their concentration exceeds some limiting value. Range II, however, must imply that CS planes are present in sufficient density and grouped into domains large enough to function as a distinct "phase" in the Gibbs sense; whether isolated or possibly even finite CS planes are present in equilibrium with interstitials in range I remains unanswered.

Some evidence on this comes from very recent work by Gai and Hirsch (*9*), in which MoO_3 was studied, under equilibrium conditions of temperature and oxygen fugacity, in the high-voltage electron microscope. At rather low temperatures and for extremely small degrees of reduction, finite areas of CS plane, bounded by dislocations, were formed in the crystal. The area of CS boundary within the dislocation loop was found to diminish and finally to "dissolve" when the temperture was progressively raised. The behavior is that of a CS plane in thermodynamic equilibrium with a surrounding atmosphere of point defects (in this case oxygen vacancies), and the free energy of creation of defects proves, on analysis, to involve the interfacial energy of the CS boundary and its enclosing dislocation.

For the reaction mechanism whereby CS planes are introduced into a crystal, the basic issues are (a) the manner in which two slabs of good crystal become mutually displaced across an APB and (b) the actual locus of the reduction process. Anderson and Hyde (*10*) proposed that a CS plane is nucleated at the surface, where the reduction reaction creates a supersaturation of point defects. These locally aggregate into a vacancy disk (in WO_3 of MoO_3) or an interstitial disk, across which the structure collapses and is sheared to create the antiphase relationship. This creates a finite area with the structure of the CS interface, bounded by a dislocation loop. As reaction continues, this dislocation climbs through the crystal, extending the size of the CS interface, until it climbs out of the crystal, leaving a complete CS plane. As it climbs, it drains point defects from the surrounding structure. Furthermore, the strain associated with the dislocation raises the chemical potential of oxygen in the strained region, so that the dislocation becomes the preferred site of the reduction process. Both factors effectively confine the reaction to the periphery of the growing CS plane, and the dislocation provides a high conductivity path for the inward diffusion of oxygen vacancies.

In that mechanism, the CS plane is anchored to the point on the crystal surface at which it nucleated, and it grows only along its length. In an alternative mechanism suggested by Bursill and Hyde (*11*), the need to invoke long-range diffusion is avoided. Initiation of a CS plane is assumed to occur where a preexisting APB emerges at the free surface. Such an APB of finite area would be formed by closure of the structure around a dislocation with a Burgers' vector $\mathbf{b} = \frac{1}{2}[011]$. Removal of an oxygen atom at the resulting surface step would furnish a cation that, by executing a unit diffusion jump of $\frac{1}{2}\langle 011 \rangle$, would lodge at the adjacent step in the APB (Fig. 9), creating the configuration of a CS collapse step. A jump of one cation from this step to the next transfers this configuration from step to step along the APB. As oxygen atoms at the surface step are successively stripped from the surface, CS collapse steps are fed into the crystal and a CS plane grows along its length. The tip of this CS plane moves along the APB, which is anchored at the surface, while the point of emergence of the CS plane progressively shifts sideways as the step in the oxygen sublattice moves. If the

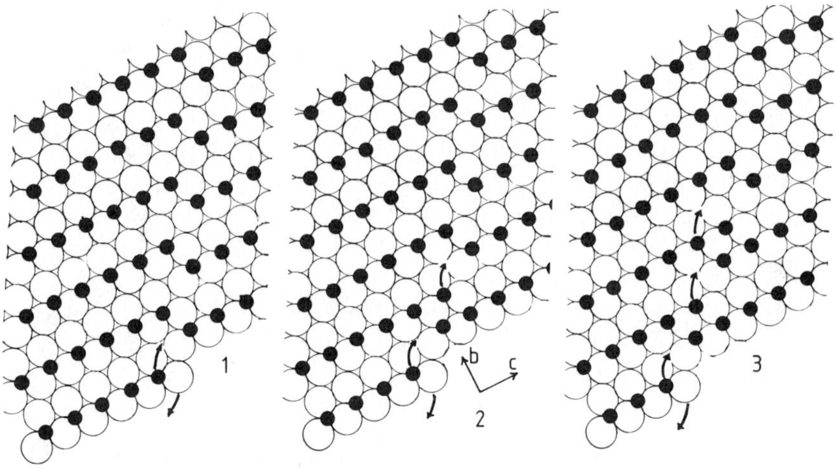

Fig. 9. The Hyde–Bursill mechanism for propagating CS planes into TiO_2 during reduction by unit hops of cations along an APB.

original dislocation climbs along its glide plane, ahead of the growing CS boundary, the process continues until the crystal is traversed. The actual reduction reaction is a surface reaction throughout, and the whole process is mediated by the unit diffusion jumps of cations at the steps of the CS plane; these involve the cooperative processes considered in the next section.

Both these mechanisms account for most of the observed facts without the assumption that large numbers of point defects are first created randomly throughout the crystal, and in some way ordered, before the CS plane is formed; either may represent the course of reaction under specific conditions. When CS planes are grown into a crystal by electron beam heating in the electron microscope, it has not been observed that they migrate laterally during their growth. Under those conditions, the Hyde–Anderson mechanism seems the more likely, but it may well be that the Hyde–Bursill mechanism operates in reactions with gaseous reductants, in which the reducing species is necessarily chemisorbed, and a gaseous product desorbed, at the surface.

The Hyde–Anderson mechanism does not seem applicable to reactions that form CS phases in double oxide systems. Our knowledge of these is largely confined to systems derived from TiO_2, which present several different kinds of CS structures of the general type Me_nO_{2n-m} (Me = Ti or some 3+ cation of suitable radius). Reference has already been made to the family $Cr_2Ti_{n-2}O_{2n-1}$, based on CS planes of the kind already considered, lying between the $(12\bar{1})$ and $(13\bar{2})$ orientations. The Fe^{3+} cation also leads to such CS planes, although no well-defined, homologous series of ordered structures is produced. Above 1450°C, however, in TiO_2 doped with Fe_2O_3, the CS planes undergo a cooperative reorientation from the $[1\bar{1}\bar{1}]$ zone to the [100] zone, with a displacement vector $\frac{1}{2}[011]$ (*12*). On the assumption that two-thirds of the octahedral sites are occupied by Fe^{3+} cations, the structure in the CS planes is effectively that of a (0001) slice of the Fe_2O_3 corundum structure. In practice, some reduction of ferric iron implies a rather larger occupancy of the octahedral sites. The somewhat smaller Ga^{3+} cation generates yet another kind of CS structure, with CS planes on (210) and the displacement vector $[\frac{1}{2}\frac{1}{4}\frac{1}{2}]$ (*13*). In this case, and for certain composition ranges, a family of fairly well ordered CS compounds of the type $Ga_4Ti_{n-4}O_{2n-2}$ ($n = 15$–23) can be formed. They can properly be included among the CS phases in that the oxygen sublattice remains invariant, whereas the composition is achieved by the typical reorganization of cation coordination polyhedra. In this case, a proportion of the Ga^{3+} cations enters tetrahedral sites, and the structure in the CS planes is effectively that of rods of the β-Ga_2O_3 structure extending along the [001] zone axis (Fig. 10). It is only for this series of double oxides that any direct evidence is available about their mechanism of formation.

In the reaction between TiO_2 and an oxide Me_2O_3, it can be assumed that the nearly close-packed oxygen sublattice of the rutile reactant is extended and that the reaction proper hinges on the diffusion of the cations. It is known that the diffusion of cations in the rutile structure is fairly rapid

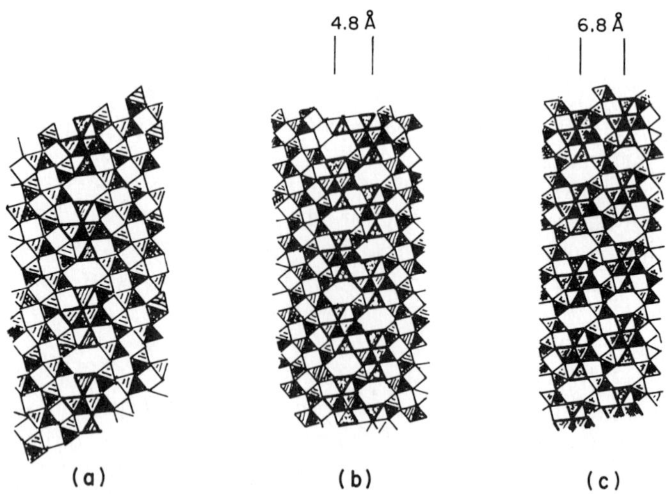

Fig. 10. (a) Structure of the CS boundary in the Ga_2O_3–TiO_2 double oxides. (b) Basic type of double-fault boundary, an intergrowth of Ga_4TiO_8. (c) Structure of the wider double-fault boundaries, elements of $Ga_4Ti_3O_{12}$, the second member of the homologous series.

and highly anisotropic at high temperatures. The nearly square channels between the ribbons of edge-sharing octahedra running down the *c*-axis provide a set of interstitial octahedral sites slightly larger than the cation octahedra; the unit diffusion jump $\frac{1}{2}[001]$ is facile. The activation energy for a jump from one channel to the next (as shown in Fig. 9 and required by the reaction mechanism considered above) is larger. In the formation of CS structures there must be a counterdiffusion process, in which three Ti^{4+} diffuse out while four Me^{3+} diffuse in.

In a detailed study of the Ga_2O_3–TiO_2 system by Stone and Bursill (*14*), using high-resolution electron microscopy at 3.2 Å resolution, the empty hexagonal tunnels seen in Fig. 10a could be unambiguously identified, and the structure assigned to the CS boundary was confirmed by a crystal structure determination of an ordered CS phase, $Ga_4Ti_{21}O_{48}$. At low Ga_2O_3 contents, these CS boundaries were replaced by isolated (210) faults with a double row of hexagonal tunnels, having the structure shown in Fig. 10b; the composition of such a defect is Ga_4TiO_8, an element of the first member ($n = 5$) of the homologous series. In addition, wider (6.8 Å) double faults are present, which have the structure shown in Fig. 10c; these are elements of the second member ($n = 7$) of the series, with the composition $Ga_4Ti_3O_{12}$. Occasional Ga_4TiO_8 defect pairs are also found between the CS planes of the ordered structures, terminating within the crystal at a dislocation. These

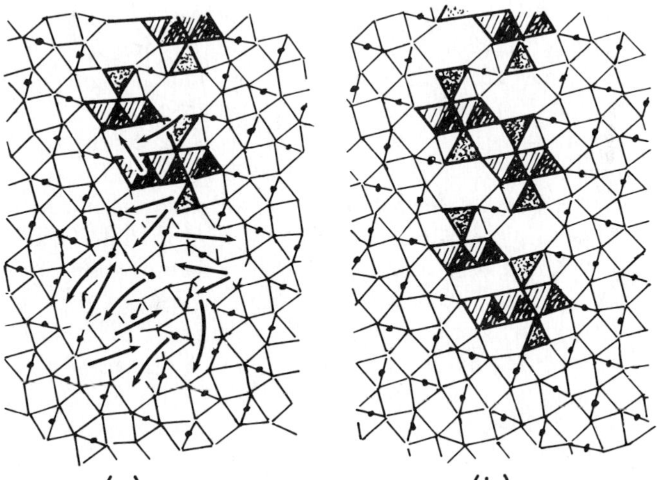

Fig. 11. Proposed mechanism for the introduction of CS planes by way of faults with the Ga_4TiO_8 structure. Arrows show the outward transport of six Ti atoms and inward transport of eight Ga atoms, to extend the CS plane by one structural unit.

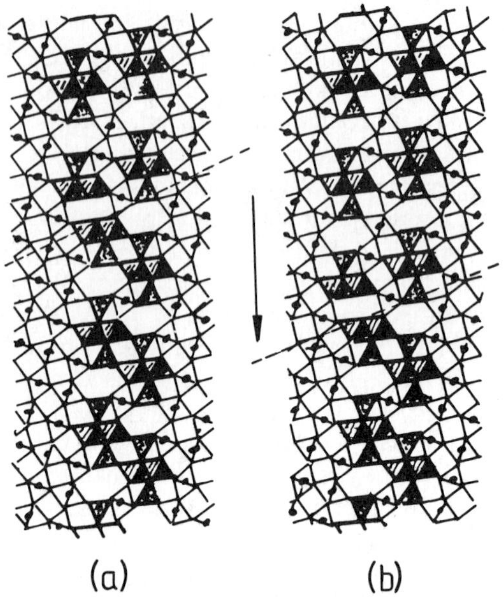

Fig. 12. Proposed mechanism for dissociating a Ga_4TiO_8 boundary (above the dashed line) by lateral migration and place exchange between Ga and Ti cations.

observations lead to a reasonable mechanism for propagating a Ga_4TiO_8 fault into the crystal. Although the structure of the dislocation core cannot be resolved in the electron microscope, Fig. 11 provides a rational model for it. If six Ti atoms leave the core and eight Ga atoms enter it by lattice diffusion, as shown, the fault is extended by one structure unit. Bursill suggests that this is the mechanism by which Ga_2O_3 enters the structure, with simultaneous shear, as the initial step in the reaction between the oxides. The wider double faults suggest the process whereby Ga_4TiO_8 faults dissociate into single CS planes, which migrate and adopt spacings appropriate to the overall composition. Figure 12 shows how such dissociation could occur by coordinated unit jumps of cations—a conservative, strain-free, homogeneous process. The unit shift of three cations, with smaller readjustments, separates the components of one unit of the Ga_4TiO_8 fault. Analogous, repeated movements can adjust the CS plane spacing, but since the CS plane is, in effect, an intergrowth of Ga_2O_3, a lateral migration of CS planes necessarily implies that there is a countercurrent exchange of places between Ti and Ga atoms.

3. The Reorientation and Reshuffling of CS Planes

Lateral migration and reorientation of CS planes have already been touched upon in the foregoing section. Both processes involve the operation of the same basic steps, so correlated as to preserve (or impose) long-range order. Those basic steps are the following:

(1) Execution of a single unit jump by one of the cations in the CS interface into an adjacent "interstitial" site. This is the primary act of any self-diffusion process. It becomes effective if the free energy is thereby lowered, so that the potential barrier becomes asymmetric, and this depends in part on the execution of similar jumps by neighboring atoms. In the case of structures derived from the ReO_3 type, the unit jump is $\frac{1}{2}\langle 110\rangle$; in rutile-based structures it is $\frac{1}{2}\langle 011\rangle$.

(2) Propagation of that unit displacement from atom to atom along a whole zone axis row of equivalent cations. As already mentioned, the strong interactions make this the most effective cooperative process, and the experimental evidence shows that good order is usually maintained along the zone axis direction during shifts of the CS planes. In ReO_3-based structures, strings of cations are linked along the [001] direction through the oxygen atoms at the vertices of their coordination octahedra, so that these oxygen atoms must also be shifted by $\frac{1}{2}\langle 110\rangle$, following the cations.

(3) As a whole zone axis row of atoms is thus displaced, further propagation of the shift from row to row along the whole CS interface.

If every cation in the CS interface executes such a unit jump, the CS interface preserves its configuration and advances by one module of spacing, e.g., by one octahedron diagonal in the ReO_3-based structures. Figure 13 illustrates this; reference to Fig. 3 will show that the same considerations apply to the advance of CS planes in the rutile structure. This is the process first formulated by Andersson and Wadsley (*15*), whereby CS planes must reshuffle and adjust their spacings to achieve long-range order.

Changing the orientation of CS planes involves the same basic steps; the cooperative processes are more complex, although, experimentally, the "swinging" of CS planes appears to be the more facile process. Change in the CS plane orientation comes from change in the relative proportions of CS collapse and AP steps in the boundary; there must be a corresponding change in the number of "pseudo-interstitial" cations, so that some inward or outward diffusion of cations along the CS plane is needed. This can be understood along the lines of the Hyde–Bursill mechanism sketched in the preceding section, and to preserve planarity there must be complete readjustment of order in the sequence of steps throughout the CS interface. As can be seen from Fig. 13b, however, the process requires that the unit shift be made by every cation in the wedge of crystal swept by the swinging CS plane. Insofar as the change in orientation arises from a reduction or other reaction that generates a net flux of cations inward from the surface, the process can be represented along the same lines as the mechanism proposed

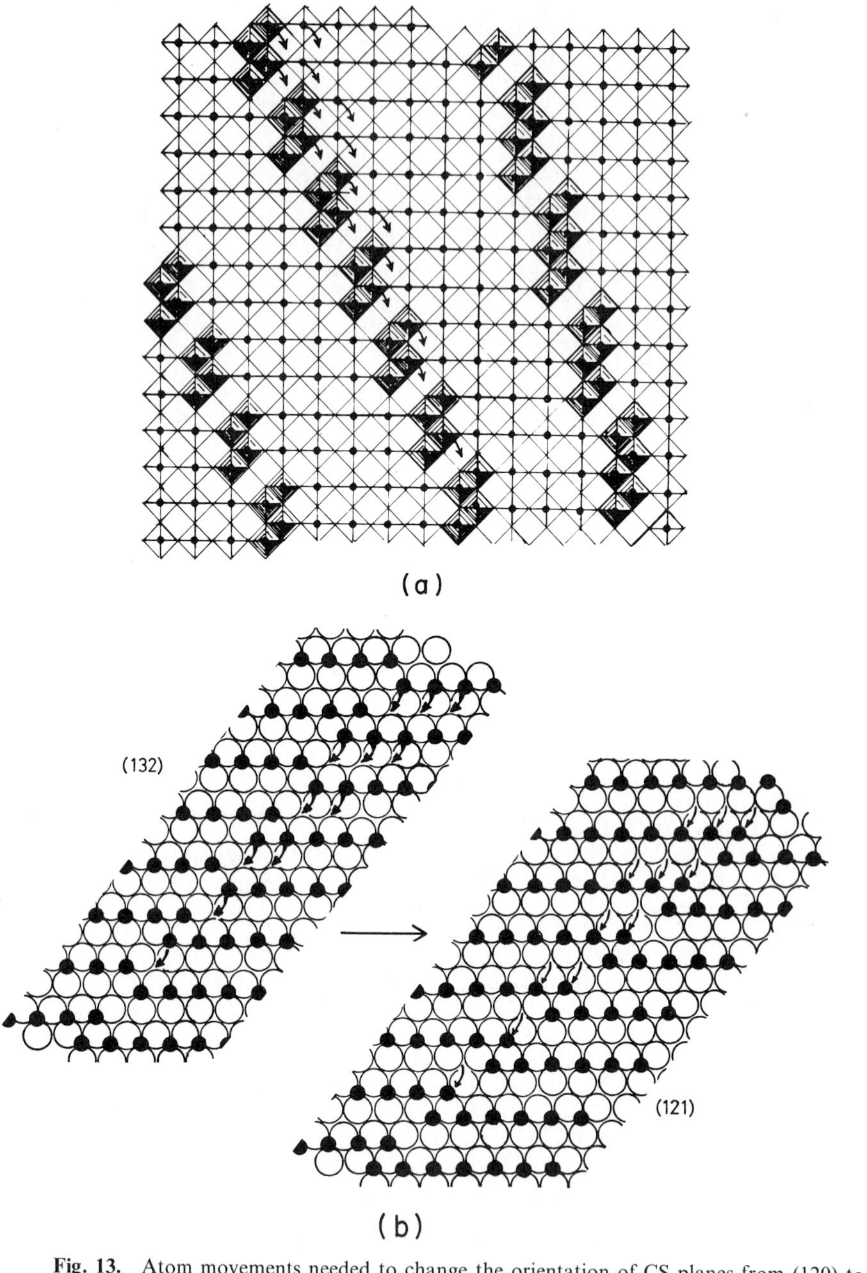

Fig. 13. Atom movements needed to change the orientation of CS planes from (120) to (130) in the ReO_3-type structure (c), and from (132) to (121) in the rutile structure (b). Arrows show the unit jumps executed by cations within a whole sector of the structure as the CS planes pivot.

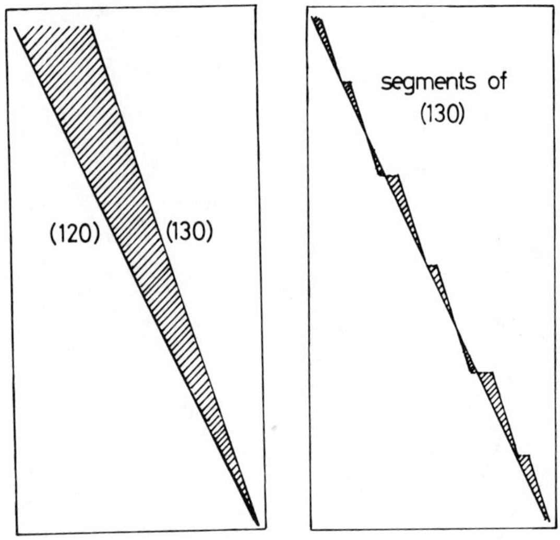

Fig. 14. Fewer strictly correlated shifts of cations are required if a CS plane pivots around in segments, creating a faulted structure. All cations in shaded area of scheme on the left must move.

by Hyde and Bursill (above). By definition, if the reaction spreads inward from the surface, the trace of the emergent CS plane must move across the surface as the CS plane pivots around; cation shifts from row to row can take place along the existing CS plane. If, however, the pivoting is initiated within the crystal, as in many reaction processes in the block structures considered below, fully cooperative shifts may be difficult to achieve, at least as a transient stage in reaction. The easier step may be to break the CS plane into segments, each of which reorients with cooperative shifts within a relatively small sector, as schematically shown in Fig. 14. Junctions between such segments are necessarily faulted; the attainment of order requires a subsequent lateral migration stage. There is evidence for such faulting and kinking of CS planes during reorientation from the electron microscope studies of Tilley (Fig. 15).

B. Reactions in the Block Structures

1. Basic Mechanisms

More detailed evidence about the processes of reaction and rearrangement has come from recent studies of the niobium oxide block structures by high-resolution lattice imaging electron microscopy. With their two sets of intersecting CS planes, these exhibit enormous flexibility and possibilities of

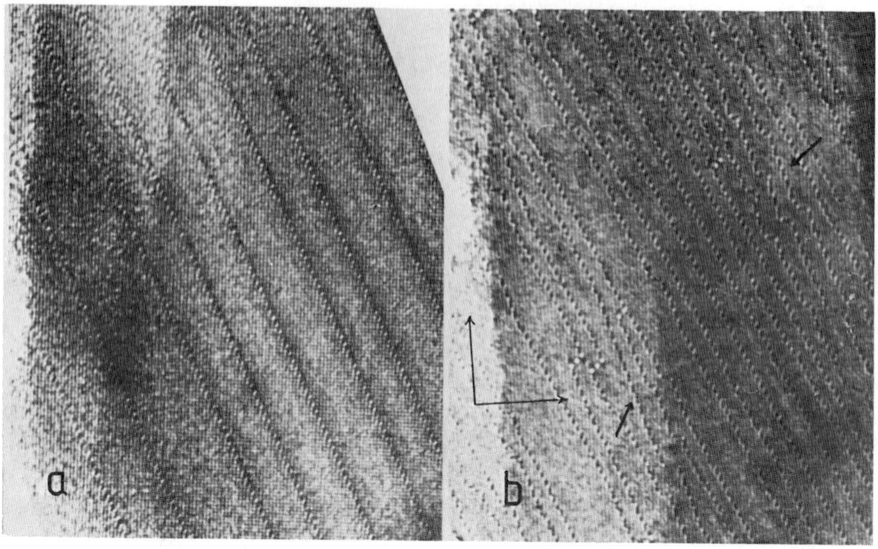

Fig. 15. (a) Almost regularly spaced, truly planar CS planes on (120) in reduced WO_3. (b) Under electron beam heating, new CS planes are introduced and changes in orientation are observable. Arrows show regions with CS planes breaking up into segments.

coherent intergrowth between columnar elements of different structures with different compositions. As noted earlier, the cross sections of these columns are determined by the orientation of the CS planes that define them, so that extended defects or jigsaw puzzle arrangements of blocks in disordered structures can be regarded as arising from the kinking of CS planes.

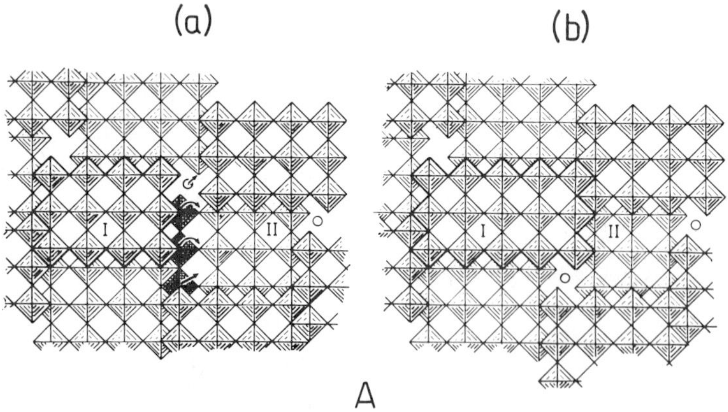

Fig. 16A. The Wadsley–Andersson mechanism for advancing CS planes. Shifts of the cations in the hatched octahedral groups in (a) change the cross sections of blocks I and II to the configuration (b).

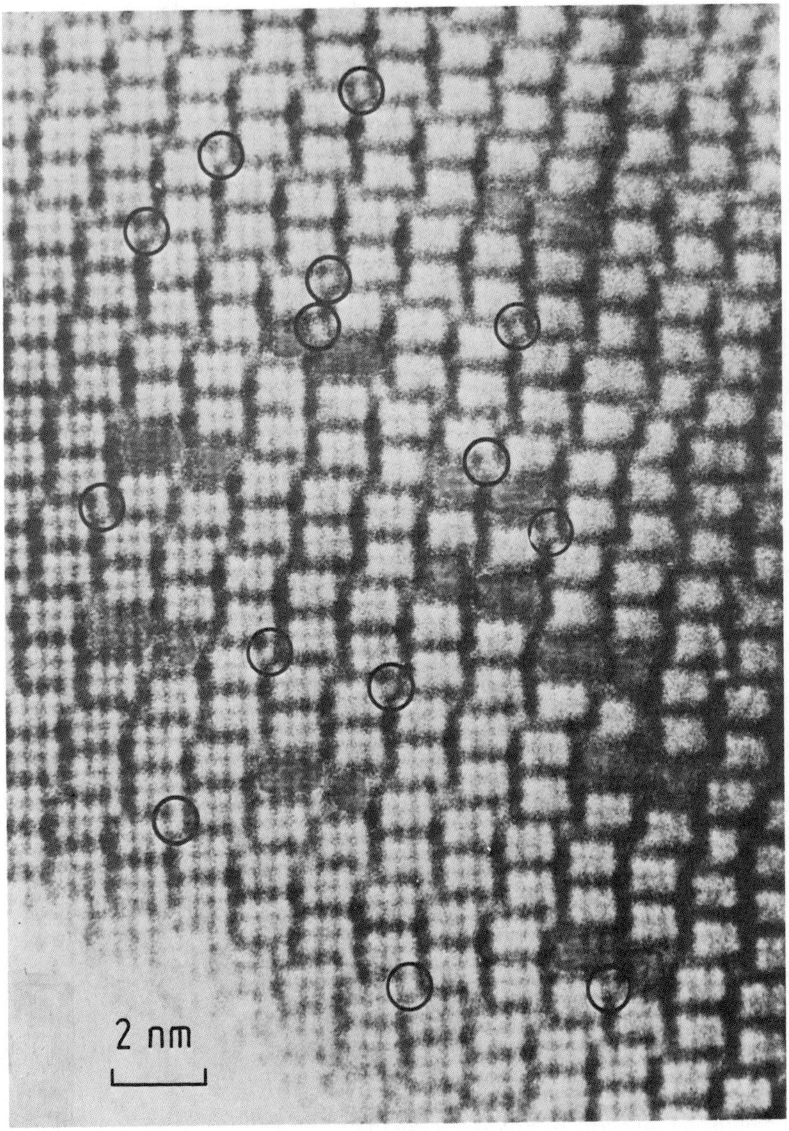

B

Fig. 16B. Disproportionation of (4 × 3) blocks in oxidized $Nb_{25}O_{62}$ exemplifies the process shown in Fig. 16A. A pair of (4 × 3) blocks is transformed into one (5 × 3) and one (3 × 3) block.

It is characteristic of reactions in these materials that the product of reactions such as the reduction of Nb_2O_5 may be formed within the parent structure in the form of a complete sheet of the new structure coherently intergrown with the parent matrix. Such a Wadsley defect is defined by a recurring double kink in one set of CS planes. In this case, the defect is a sheet of $Nb_{25}O_{62}$, the next lower oxide. As reaction proceeds, further isolated sheets or domains of $Nb_{25}O_{62}$ are formed; new CS planes must be introduced and reshuffled by processes that straighten the kinked CS planes.

In these processes, the basic step is that discussed in the previous section: the shift of atoms by a unit step of $\frac{1}{2}\langle 101 \rangle$ at the CS interface between blocks. The cross section of a pair of adjacent blocks is thereby altered (Fig. 16), without any change in composition. This transfer process, initiated at some block pair, has to be propagated (a) along a whole zone axis row of sites from level to level throughout the crystal, (b) along the whole affected edge of the blocks, (c) from block to block along a row of blocks.

In addition to that basic step, certain block structures make it possible for new anion sites to be created if stage (b) is inhibited by constraints imposed by the local structure. Figure 17a is a structure image from a crystal of $Nb_{12}O_{29+x}$ obtained at a very early stage of the homogeneous oxidation of $Nb_{12}O_{29}$ to Nb_2O_5 at a low temperature (*16*). "Ripples" in the CS planes defining the $(4 \times 3)_\infty$ block structure of $Nb_{12}O_{29}$ show where the reaction has been initiated and propagated into the crystal along a line of reaction. The structure in such a ripple is mapped schematically in Fig. 17b. It can be seen that the advance of the edge of block A has created, on one side, a channel containing cation sites of tetrahedral coordination and, on the other side, an apparent gap, 8 × 6 Å in size, which requires the assumption that the structure should be represented as in Fig. 18a or b; the coordination octahedra of the cations can be completed only if oxygen atoms are incorporated into the new anion sites thus created. This is the essence of the oxidation process, which at 150°–250°C imposes no drastic reorganization of the $Nb_{12}O_{29}$ structure.

In Fig. 17a, the resolution of structure within the blocks indicates that the stage (a) ordering has been fully achieved, at least over distances as great as the thickness of the crystal fragment imaged (of the order of 50–150 Å); if it were not, so that the interface between blocks was jogged, rows of cations would be seen as only 2 Å apart in the projection. This is less than the attainable resolution, so that the stepped interface would be imaged as an area of dark contrast 6 Å wide. There is evidence, however, from work on the transformation of unstable polymorphs of Nb_2O_5 into the stable polymorph H—Nb_2O_5 (*17*), that the propagation of the cation shifts along zone axis rows is not quasi-instantaneous, brought about by the passage of a single displacement wave, but that jogs may be frozen in as the displacement

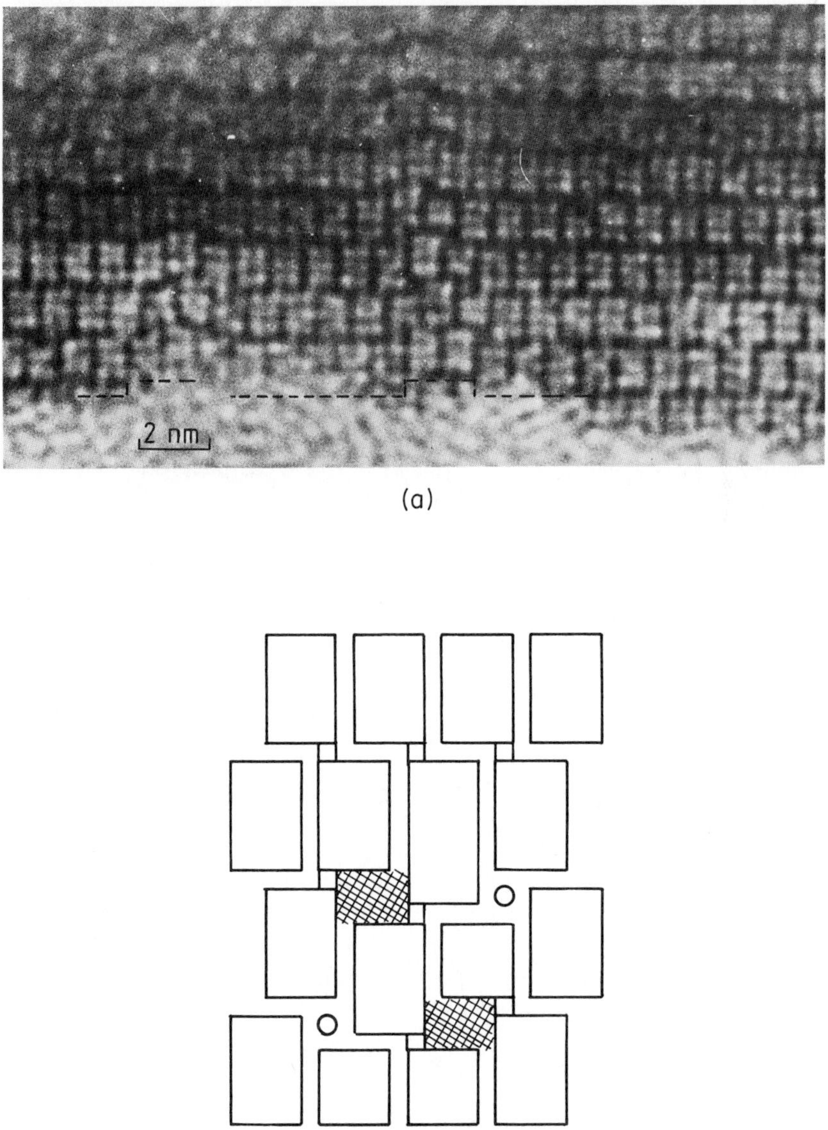

Fig. 17. (a) Rippled CS planes in $Nb_{12}O_{29+x}$ oxidized at low temperatures. (b) Schematic structure map of a ripple. The shaded area represents unresolved structure where new anion sites have been created, as in Fig. 18.

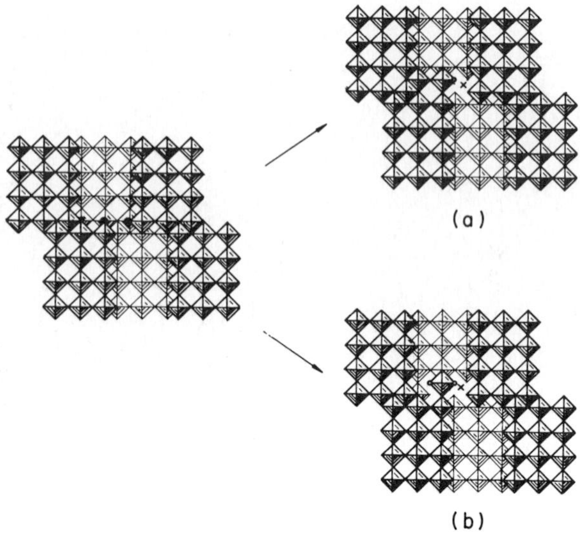

Fig. 18. Creation of new anion sites by the unit shift of cations at the edge of blocks. (a) Structure proposed by Browne (*16*); three cations shifted. (b) Structure of Iijma (*21*); two cations shifted.

propagates along a row of cation sites. In this and other polymorphic transformations between block structures, no long-range movement of atoms is required; the processes occur solely by unit shifts by cations, as discussed above. High-resolution electron micrographs from materials in the earliest stages of transformation, in which relatively few segments of CS planes have undergone a displacement, afford evidence in full harmony with the model that has been presented for the mechanism of change. If has further been found that the act of displacing the segments of CS planes appears to be initiated quite randomly at a multiplicity of sites within a crystal and that the propagation of shifts from block to block is a slower (less probable) process than propagation along a zone axis row.

2. *Formation of Wadsley Defects*

The same mechanism of reconstruction probably suffices for an interpretation of other processes, such as the formation of reaction products in the form of intergrown Wadsley defects, as can be illustrated by observations on the reduction of H—Nb_2O_5. Figure 19, a schematic structure map made from the micrograph of a slightly reduced crystal of Nb_2O_5, furnishes some suggestive evidence (*18*). The matrix structure is that of H—Nb_2O_5 in which (4×3) blocks and tetrahedral sites alternate with ribbons of (5×3) blocks,

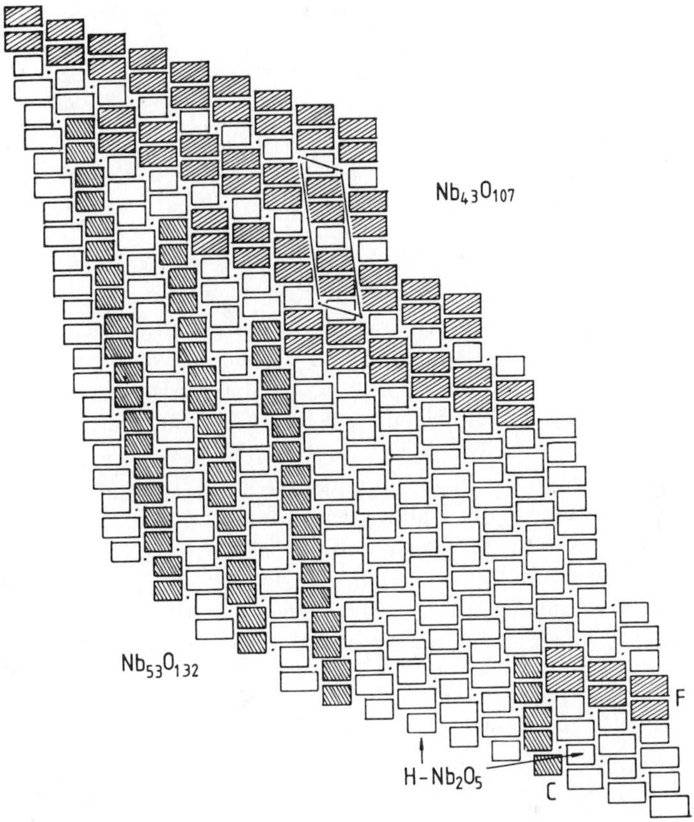

Fig. 19. Schematic structure map of a domain of new superlattice observed in the reduction of H—Nb_2O_5; C, F are Wadsley defects, terminating where they intersect and being mechanistically significant.

but in the upper part of the diagram is a new (probably metastable) superstructure with two ribbons of (5 × 3) blocks between the (4 × 3) blocks; this would have the ideal composition $Nb_{43}O_{107}$, $NbO_{2.488}$. This domain of superstructure terminates inside the crystal fragment, where it meets a domain of the 1:1 intergrowth structure $Nb_{53}O_{132}$, $NbO_{2.491}$. These domains enclose a region of H—Nb_2O_5 which is traversed, lower down, by a double fault C–F; this is the significant feature. Defect F is a short length of the double ribbon of (5 × 3) blocks that generates the $Nb_{43}O_{107}$ superstructure. It ends at a terminating row of the ultimate reduction product $Nb_{25}O_{62}$, $NbO_{2.480}$. An analysis of the atomic displacements that are needed for the interconversion of the H—Nb_2O_5 ($Nb_{28}O_{70}$), $Nb_{53}O_{132}$,

$Nb_{43}O_{107}$, and $Nb_{25}O_{62}$ structures suggests a reaction mechanism that is consistent with the foregoing discussion.

In these structures, the oxygen sublattice is almost unchanged; to convert one to another, only a few cations and fewer anions need be shifted by the $\frac{1}{2}\langle 110 \rangle$ unit step. In the reduction reaction, one oxygen atom has to be removed from the crystal, and two electrons transferred to it, to be trapped as Nb^{4+} cations, for every two formula units of $Nb_{25}O_{62}$ formed [i.e., for every two pairs of (4×3) blocks linked at the same level of the crystal: a $(4 \times 3)_2$ block pair]. As reduction becomes more extensive, new CS planes have to be created. Both of these steps are surface processes, and, in effect, the requirement is therefore to transfer niobium and oxygen outward to the crystal surface while producing the new structure within the crystal.

Figure 20 schematizes a possible role for the Wadsley defect pair C–F in this process. Starting from the perfect H—Nb_2O_5 structure, a cation shift at the edge of one $(4 \times 3)_1$ block would convert two $(4 \times 3)_1$ blocks and two $(5 \times 3)_\infty$ blocks into a $(4 \times 3)_2$ pair and a $(5 \times 3)_2$ pair, without effecting any change in composition. If, however, an itinerant electron is trapped at one of the cations in the $(4 \times 3)_2$ block pair, this becomes a true element of the $Nb_{25}O_{62}$ structure, but the complementary $(5 \times 3)_2$ block pair, which would have the composition $Nb_{31}O_{78}$, $NbO_{2.516}$, does not correspond to any viable structure, nor can such a viable structure be formed by eliminating

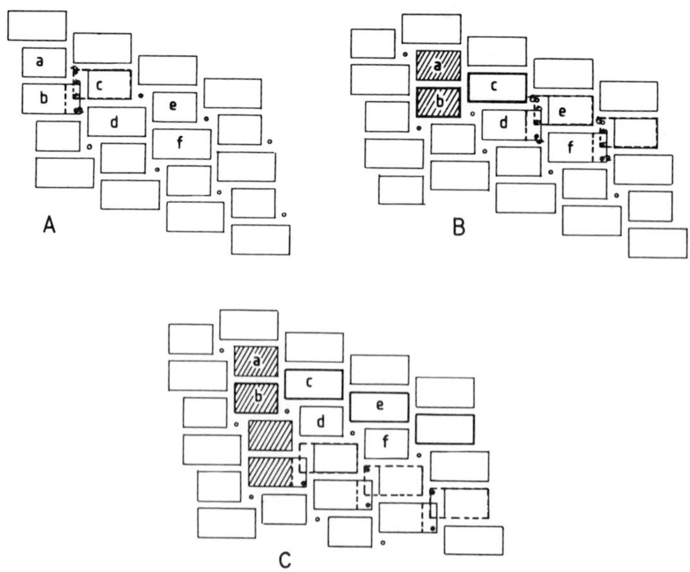

Fig. 20. Schematic mechanism for the role of the double fault C—F in the reduction reaction. A, the initiating step; B, propagation along a row of blocks; C, propagation down a file of blocks.

oxygen without also involving the adjacent structure. However, by propagating the shift of block-edge cations along the row of blocks from left to right, both oxygen and niobium can, in effect, be transferred to the outer surface, there to take part in the surface processes. As these cation shifts are repeated along the row of blocks, they automatically generate a double ribbon of (5×3) blocks, terminating at the newly formed unit of $Nb_{25}O_{62}$, as in fault F. Also produced is a stacking fault in the H—Nb_2O_5 structure below the double ribbon: a double row of $(4 \times 3)_1$ blocks. A repetition of these steps—electron trapping, coupled with the propagation of cation shifts outward—would displace the stacking fault and form another unit of $Nb_{25}O_{62}$ each time until, ultimately, an entire file of $Nb_{25}O_{62}$ structure would be formed, terminating within the crystal at the double ribbon of $(5 \times 3)_\infty$ blocks formed in the first stage. The structure seen at C–F would thus be produced. It can readily be shown that this chevron-shaped double fault can expand and climb through the crystal by the same set of operations.

On this basis, a reasonable mechanism for the formation of Wadsley defects of $Nb_{25}O_{62}$ can be framed and, *mutatis mutandis*, extended to other reactions. As in the Hyde–Bursill mechanism for introducing single CS planes, reaction is almost certainly initiated at the surface, starting with the minimum number of blocks that are needed to nucleate a chevron-shaped double fault like C–F. As reduction proceeds, the fault climbs by the cooperative processes just discussed, forming one new element of $Nb_{25}O_{62}$ structure at each step and for each electron transferred to the crystal. In general, the fault will climb right out of the crystal, thereby eliminating the double ribbon of $(5 \times 3)_\infty$ blocks and leaving a complete sheet of $Nb_{25}O_{62}$ sandwiched between two slabs of good H—Nb_2O_5 structure; the CS planes are kinked as they traverse the Wadsley defect. If the initial nucleation steps take place at random, isolated Wadsley defects or randomly stacked intergrowths will be formed. If the initiating processes effectively attack every $(4 \times 3)_1$ block in a file of H—Nb_2O_5, the result is to form an intergrown domain of $Nb_{53}O_{132}$, as in Fig. 19. The proposed mechanism does involve a very extensive set of cooperative atom movements, as indeed does any transformation in these structures that cannot be interpreted in terms of edge dislocations. The available experimental evidence indicates, however, that ordering processes, rather than the displacement of CS planes, are the sluggish steps in most reactions.

3. *Mixed Homogeneous/Heterogeneous Reactions of Block Structures*

The formation of double oxides with CS structures, discussed in Section III,A, is really a reaction of mixed type. It involves a chemical interaction between two phases at the surface of contact between the reactants, but the product grows coherently within one of the reacting structures. A converse process is also possible and is exemplified by several reactions of the block structures in which two solid products are formed. If one of these has a structure that is topologically incapable of coherent intergrowth with the parent compound, it must be formed heterogeneously, as its components diffuse to and across some interface. If the second product could, in principle, intergrow coherently with the parent structure, it might be formed in two ways. The first is by a purely topotactic process, in which a reaction interface between the new structure and the completely unreacted structure moves progressively through the reacting crystals. A second way is by an internal, quasi-homogeneous route, in which the new structure is formed *within* the parent as Wadsley defects and expanding domains of reorganized structure. The locus of reaction, in that case, would be at the periphery of each sheet of product structure as it grows through the reacting crystal. The experimental evidence is that the second mechanism undoubtedly operates, although it cannot be positively stated that it provides the only reaction pathway.

Several examples of this behavior are now known. When the titanium–niobium oxides $TiNb_2O_7$ [with the $(3 \times 3)_\infty$ block structure] and $Ti_2Nb_{19}O_{29}$ [a $(4 \times 3)_\infty$ structure] are reduced, some titanium is rejected as Nb^{5+} is progressively reduced to Nb^{4+}. One product is a rutile phase $(Ti,Nb)O_2$; the other is a block structure, relatively enriched in niobium, which is formed in coherent intergrowth with the starting material (*18*). More detailed evidence may be gained from certain reversible phase reactions in the Nb_2O_5–WO_3 system. As was shown in a thorough equilibrium study by Roth and Waring (*19*), the double oxides $8Nb_2O_5 \cdot 5WO_3$ [$W_5Nb_{16}O_{55}$, with $(5 \times 4)_1$ blocks] and $9Nb_2O_5 \cdot 8WO_3$ [$W_8Nb_{18}O_{69}$, with $(5 \times 5)_1$ blocks] are actually high-temperature phases, contrary to the commonly held assumption that CS planes are formed only by low-temperature ordering processes. At lower temperatures they undergo a reversible disproportionation, giving $W_3Nb_{14}O_{44}$ [with $(4 \times 4)_1$ blocks] and WNb_2O_8 below 1090°C, and $W_5Nb_{16}O_{55}$ and $9WO_3 \cdot 4Nb_2O_5$ [$(NbO)_4Nb_{12}W_{18}O_{90}$, a tetragonal tungsten bronze] below 1265°C, respectively. WNb_2O_8 and $4WO_3 \cdot 9Nb_2O_5$ have structures based on five-sided tunnels in a network of vertex-linked octahedra, itself (like the block structures) derived from the ReO_3 type. In these structures, all the cations are coplanar, and no possible interface can be found for coherent intergrowth with the block structures, in which the cations are necessarily at two levels to furnish the edge sharing between octahedra in the CS planes. In both disproportionations, therefore, one product must be formed incoherently, whereas the other can and does form coherent intergrowths with the parent structure (*20*).

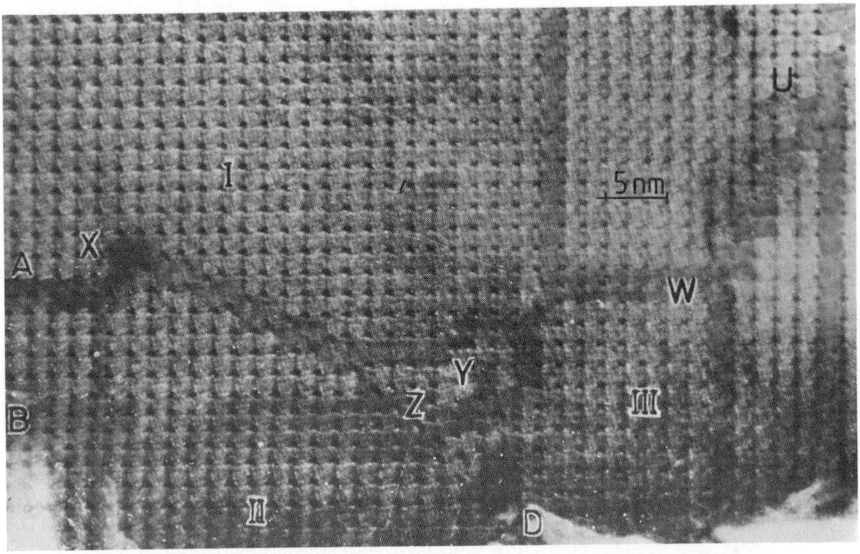

Fig. 21. Disproportionation of $8WO_3 \cdot 9Nb_2O_5$. Complete rows of the product $(5 \times 4)_1$ blocks traverse the micrograph horizontally at A, B, W, etc. One row of completely transformed blocks terminates at a dislocation X. Running vertically are rows of WO_3-enriched, larger $(6 \times 5)_1$ blocks.

Figures 21 and 22 show two micrographs of partially disproportionated crystals of $8WO_3 \cdot 9Nb_2O_5$; it can be seen that the product structure forms lines (i.e., sheets in three dimensions) of blocks running across the crystal as individuals or small groups, with no evidence for ordering or a systematic spreading into domains of converted material. The reaction has consistently operated so as to transform one line of blocks at a time, block for block, into the reaction product. These lines of reacted structure may terminate where they intersect anomalous, faulted structure, and there is evidence for their lateral migration (by the processes discussed earlier) so as to aggregate into domains. The nature of the meandering faults and also of lines of blocks with larger cross section than those of the starting material is considered further below.

The overall reaction in these disproportionations clearly requires an outward migration of WO_3, which might, at first sight, be thought to form the pentagonal tunnel structure simply by attack on the material of the surface layer:

$$3WO_3 + 8Nb_2O_5 \cdot 5WO_3 \rightarrow 8(WNb_2O_8)$$

However, the evidence of the block by block conversion makes it possible to analyze the local chemistry and shows that the real internal processes involve

Fig. 22. Disproportionation of $8WO_3 \cdot 9Nb_2O_5$. Block by block conversion along linear paths of propagation. Reacted rows are displaced across an oblique, meandering fault, much enriched in WO_3.

the transport of both niobium and tungsten, with oxygen, from the locus of the reaction:

Disproportionation of $W_5Nb_{46}O_{55}$: $(5 \times 4)_1$ block $\rightarrow$ $(4 \times 4)_1$ block

$$8Nb_2O_5 \cdot 5WO_3 \rightarrow 7Nb_2O_5 \cdot 3WO_3 + (Nb_2O_5 + 2WO_3)$$

Disproportionation of $W_8Nb_{18}O_{69}$: $(5 \times 5)_1$ block $\rightarrow$ $(5 \times 4)_1$ block

$$9Nb_2O_5 \cdot 8WO_3 \rightarrow 8Nb_2O_5 \cdot 5WO_3 + (Nb_2O_5 + 3WO_3)$$

Thus, there must be a composite diffusion flux, either through the crystal lattice or along some high-conductivity path arising from the nature of the reaction process, which has much in common with the introduction of CS planes into a crystal, considered earlier. Both cations and oxygen must be transported, and the nature of the diffusing species remains uncertain. The reaction itself strips a complete row of octahedra out of the structure. In the process, the abstraction of cations is selective and may require some reshuffling of atoms within each block, since the crystallographic evidence suggests that niobium and tungsten are randomly distributed over the cation sites. The linear propagation of reaction without disturbing, although displacing, the "good crystal" on either side suggests that a dislocation mechanism may apply, the actual locus of reaction being a climbing dislocation loop which provides the requisite path for facile diffusion. Dislocations with edge character have been found in the partially disproportionated crystals at terminating lines of product blocks, but there is no direct evidence that dislocations mediate the reaction.

Insofar as the diffusion of (in effect) WO_3 and Nb_2O_5 takes place by normal lattice processes, it produces local fluctuations of composition, enriched in WO_3. The blocks of enlarged cross section found in meandering faults and as Wadsley defects are relevant in this connection. In Fig. 21, rows of blocks $(6 \times 5)_1$ in cross section can be seen running normal to the rows of $(5 \times 4)_1$ product blocks. These are strips of a structure with the composition $10Nb_2O_5 \cdot 11WO_3$; no stable phase of this composition appears in the equilibrium diagram, and the ratio WO_3/Nb_2O_5 is, in fact, higher than that of the first of the pentagonal tunnel structure series, WNb_2O_8. Similarly, the meandering faults in Figs. 21 and 22 involve fluctuations of composition that are accommodated by the formation of abnormally large blocks, including successive $(7 \times 5)_1$ blocks with the composition $11Nb_2O_5 \cdot 14WO_3$. These are undoubtedly metastable structures, and they serve to emphasize that reactions in these block structures are subject to constraints that have no counterpart in fully reconstructive solid-state reactions.

In the CS structures (and in the tetragonal tungsten bronze type of structures) derived from the ReO_3 type, the architectural principles make it possible to build up ordered crystal structures that could accommodate both wide and subtle variations in composition. Which of those possible structures correspond to phases that can be prepared, and that appear in equilibrium diagrams, is determined by thermodynamic factors. In the nonreconstructive or semireconstructive reactions, however, preservation of the coherence of the crystalline fabric imposes a new requirement. Out of a local fluctuation of composition, structures may be locally formed and stabilized which conform to the requirement of coherence and space filling. They may be transient, formed as a stage in reaction and ultimately eliminated by annealing; their existence and their constitution can be revealed only by those experimental methods that enable the local structure of crystals to be determined.

It can be seen that high-resolution electron microscopy has become a powerful technique for studying both the phenomena and the mechanism of reactions in CS structures. There are other families of crystal structures that also provide great flexibility of adjustment to changes in composition, together with possibilities for coherent intergrowth between different members. It is to be expected that comparable studies of local structures and ordering in these will show features that parallel those that have been reviewed here.

REFERENCES

1. B. G. Hyde, A. N. Bagshaw, S. Andersson, and M. O'Keeffe, *Ann. Rev. Mater. Sci.* **4,** 43 (1974).
2. E. Iguchi and R. J. D. Tilley, *Philos. Trans. R. Soc., Ser. A* **286,** 55 (1977); *J. Solid State Chem.* **21,** 49 (1977); **24,** 121, 131 (1978); **29,** 435 (1979).
3. R. R. Merrit and B. G. Hyde, *Philos. Trans. R. Soc. London, Ser. A* **274,** 627 (1973).
4. D. K. Philp and L. A. Bursill, *J. Solid State Chem.* **8,** 86 (1973).
5. L. A. Bursill and B. G. Hyde, *Philos. Mag.* **23,** 3 (1971).
6. R. J. D. Tilley, *Mater. Res. Bull.* **5,** 813 (1970); J. G. Allpress, R. J. D. Tilley, and M. J. Sienko, *J. Solid State Chem.* **3,** 440 (1971).
7. E. Yagi, A. Koyama, H. Sakairi, and R. R. Hasiguti, *Conf. Ser.—Inst. Phys.* No. 37, p. 485 (1977).
8. J. F. Baumard, D. Panis, and A. M. Anthony, *J. Solid State Chem.* **20,** 43 (1977).
9. P. L. Gai and P. B. Hirsch, personal communication (1980).
10. J. S. Anderson and B. G. Hyde, *J. Phys. Chem. Solids* **28,** 1393 (1967).
11. L. A. Bursill and B. G. Hyde, *Prog. Solid State Chem.* **6,** 177 (1972).
12. L. A. Bursill, I. E. Grey, and D. J. Lloyd, *J. Solid State Chem.* **16,** 331 (1976).
13. L. A. Bursill and G. G. Stone, *Philos. Mag.* **32,** 1151 (1975); R. M. Gibb and J. S. Anderson, *J. Solid State Chem.* **5,** 212 (1972).
14. G. G. Stone and L. A. Bursill, *Philos. Mag.* **35,** 1397 (1977).
15. S. Anderson and A. D. Wadsley, *Nature (London)* **211,** 581 (1966).
16. J. M. Browne and J. S. Anderson, *Proc. R. Soc. London, Ser. A* **339,** 463 (1974).
17. J. S. Anderson and E. S. Crawford, *Philos. Trans. R. Soc., Ser. A* (in press, 1981).
18. J. S. Anderson and S. K. E. Forghany, *J. Solid State Chem.* **40** (in press); S. K. E. Forghany, D.Phil. Thesis, Oxford University, 1978.
19. R. S. Roth and J. L. Waring, *J. Res. Natl. Bur. Stand., Sect. A* **70,** 281 (1966).
20. H. Obayashi and J. S. Anderson, *J. Solid State Chem.* **19,** 331 (1976).
21. S. Iijima, S. Kimura, and M. Goto, *Acta Crystallogr.* **A29,** 632 (1973).

16
Oxide Insertion Compounds

P. G. DICKENS

and

M. F. PYE

I. INTRODUCTION

Recently, novel ambient temperature insertion compounds of the type A_xMO_n, where A is an electropositive group or element, most commonly H

INTERCALATION CHEMISTRY

ISBN 0-12-747380-7

or Li, and MO_n is a transition metal oxide, have been extensively investigated. These materials are of interest as candidate cathode materials in high-energy-density secondary batteries of the type

$$Li(s)|Li^+(\text{nonaqueous electrolyte})|Li_xMO_n(s)$$

where the overall cell reaction is

$$Li + \frac{1}{x}MO_n \rightarrow \frac{1}{x}Li_xMO_n \tag{1}$$

(*1–4*). The compounds are intensely colored, often in marked contrast to the parent oxides, and this has led to the study of their application in electrochromic display systems (*5–8*). The formation reaction of the hydrogen tungsten bronze H_xWO_3 is a well-known example.

$$\underset{}{H^+ + e +} \underset{\text{Colorless}}{\frac{1}{x}WO_3} \rightarrow \underset{\text{Dark blue}}{\frac{1}{x}H_xWO_3} \qquad (x \sim 0.3) \tag{2}$$

In view of the current interest in these materials, this chapter is confined to those phases, A_xMO_n, that are formed at ambient temperatures and in which the parent oxide matrix is essentially unaltered. Excluded from this discussion, therefore, are the more familiar, essentially high-temperature oxide bronze phases of the same general formula in the formation of which considerable rearrangement of the host lattice occurs. Structural, electronic, and thermodynamic properties of these materials have been extensively reviewed elsewhere (*9–11*).

The materials discussed here fall into two broad categories: those with layered host lattices and those formed from oxides containing tunnel networks. The former class is exemplified by MoO_3, in which the interlayer spacing can be increased to accommodate a particular guest species. An example of the latter class is WO_3, which has an interpenetrating three-dimensional network of tunnels. In this case there is a limit to the size of the insertion atom that can be incorporated without breaking bonds in the host oxide matrix, and only elements with a small atomic radius, such as H and Li, can be inserted at ambient temperatures.

The emphasis in this chapter is on selected systems that have been well studied and illustrate the main features of their class.

II. PREPARATIVE METHODS

Synthetic routes to ambient temperature oxide insertion compounds A_xMO_n (A = H, Li) make use of techniques used in other areas of intercalation chemistry and are described in greater detail elsewhere (*12*). Some typical examples illustrating the main lines of approach, applied to oxides, are given below.

A. Use of *n*-Butyllithium

Under an inert atmosphere *n*-butyllithium in hexane solution reacts smoothly with many transition metal oxides (*13*) to give products Li_xMO_n according to the main reaction

$$C_4H_9Li + \frac{1}{x}MO_n \rightarrow \frac{1}{x}Li_xMO_n + \frac{1}{2}C_8H_{18} \quad (3)$$

Homogeneity of the product is facilitated by subsequent equilibration of the solid with a nonaqueous electrolyte solution containing lithium ions.

B. Use of Lithium Iodide

Lithium iodide, either as a solid (*14*) or in acetonitrile solution (*15*), reacts with those transition metal oxides that are sufficiently strong oxidizing agents to incorporate lithium. For example, the reaction

$$x\mathrm{LiI} + V_2O_5 \rightarrow Li_xV_2O_5 + \frac{x}{2}I_2 \quad (4)$$

takes place quantitatively at ambient temperature.

C. Electrochemical Incorporation

Lithium can be incorporated into metal oxides electrochemically by the use of cells, with nonaqueous electrolytes, of the type $Li|Li^+|MO_n$. Typical electrolytes consist of lithium salts, particularly $LiClO_4$ and LiI, dissolved in organic solvents such as dimethylsulfoxide (DMSO) or propylene carbonate. Both $Li_xV_2O_5$ (*14*) and Li_xMoO_3 (*16*) have been prepared by this method, and lithium can similarly be incorporated into a thin film of WO_3 to form Li_xWO_3 in an electrochromic device (*17*). The extent of reaction can be continuously monitored by coulometry. Hydrogen can be incorporated into oxides of molybdenum (*16*) or tungsten (*6*) by cathodic reduction in aqueous solution. Thus, H_xMoO_3 and H_xWO_3 are readily formed electrochemically from aqueous solutions of mineral acids. More generally, cathodic reduction from aqueous solution in the presence of alkali metal cations can lead to the incorporation of hydrated ionic species in oxide cathodes. For example, with MoO_3 in neutral electrolyte solutions, products of the type $A_x(H_2O)_yMoO_3$ are formed (*16*).

D. Incorporation by Chemical Reduction

Many chemical reducing couples have been used to incorporate hydrogen into metal oxides (*18*, *19*). For example, H_xMoO_3 with $x \sim 2.0$ can be formed by vigorous reduction of MoO_3 by Zn/HCl (*20*). From this starting material, H_xMoO_3 intermediate phases in the range $0 < x < 2.0$ can be synthesized by reaction of H_2MoO_3 with MoO_3 in sealed tubes at $<100°C$. In neutral aqueous electrolyte solution, the use of chemical reducing agents can lead again to the incorporation of hydrated species. Thus, MoO_3 is reduced by $Na_2S_2O_4$ solution to give $Na_x(H_2O)_yMoO_3$ (*16*), and many other examples have been discussed by Schöllhorn (*20a*) and Glemser (*18*, *19*).

E. Hydrogen "Spillover"

Small admixtures of platinum or palladium "blacks" to transition metal oxides, particularly MoO_3 or WO_3, lead to facile reaction with hydrogen gas at atmospheric pressure and near ambient temperature ($\lesssim 50°C$) (*21*). The role of the noble metal is presumably that of substrate for the dissociation of the molecular gas. This is a convenient "dry" method of preparation which can be associated with a continuous analysis of composition by measurement of the volumetric uptake. The extent of reaction is usually less than obtained by the more vigorous wet chemical methods. For example, the maximum hydrogen content obtained with MoO_3 is $H_{1.7}MoO_3$ (*21*), whereas Zn/HCl reduction leads to H_2MoO_3 (*20*).

F. Ion Exchange

In the case of layered oxides such as MoO_3, hydrated alkali metal and hydrogen species can be replaced and interchanged by ion exchange from aqueous electrolyte solutions. Typical examples (*16*, *22*) are the rapid exchange of Na ions in $Na_x(H_2O)_yMoO_3$ by other alkali and alkaline earth ions from 1 *M* aqueous solution according to

$$Na_x(H_2O)_yMoO_3 + \frac{x}{n}B^{n+} = B_{x/n}(H_2O)_yMoO_3 + xNa^+ \qquad (5)$$

and the replacement of H by hydrated metal cations in the series $H_{1+x}(H_2O)_yV_3O_8$:

$$H_{1+x}(H_2O)_yV_3O_8 + (1+x)A^+ = A_{1+x}(H_2O)_yV_3O_8 + (1+x)H^+ \qquad (6)$$

Many of the products described above are readily oxidized in air, and preparation and all subsequent manipulation must be carried out in an inert atmosphere.

III. MoO_3-DERIVED INSERTION COMPOUNDS

Molybdenum trioxide possesses a unique layered structure (*23*). It may be thought of as being formed from ReO_3 chains, vertex-sharing chains of MO_6 octahedra (Fig. 1), which share edges with two similar chains to form layers of stoichiometry MoO_3. The layers, which are stacked in a staggered arrangement (Fig. 2) and held together only by weak van der Waals forces, can be intercalated by species such as H, Li, and larger atoms or groups.

A. Hydrogen Insertion Compounds

Early work by Glemser and co-workers showed that reduction of MoO_3 in aqueous acidic media (*18*, *24*) or by hydrogen atoms (*25*) leads to the formation of a series of phases $MoO_{3-x}(OH)_x$ or H_xMoO_3 with $0.5 < x < 2.0$. The similarity of the X-ray patterns of the reduced phases to that of MoO_3 itself suggested that hydrogen was inserted into the parent oxide matrix topotactically. A single-crystal X-ray study of the phase with low hydrogen content showed that the MoO_3 framework was essentially unaltered (*26*). More recently, a phase study of the H/MoO_3 system has confirmed the existence of four phases and established their approximate limits (*20*). The phases are blue orthorhombic, $0.23 < x < 0.4$; blue monoclinic, $0.85 < x < 1.04$; red monoclinic, $1.55 < x < 1.72$; and green monoclinic, $x = 2.0$. The lattice parameters of these compounds are very similar to those of MoO_3 itself.

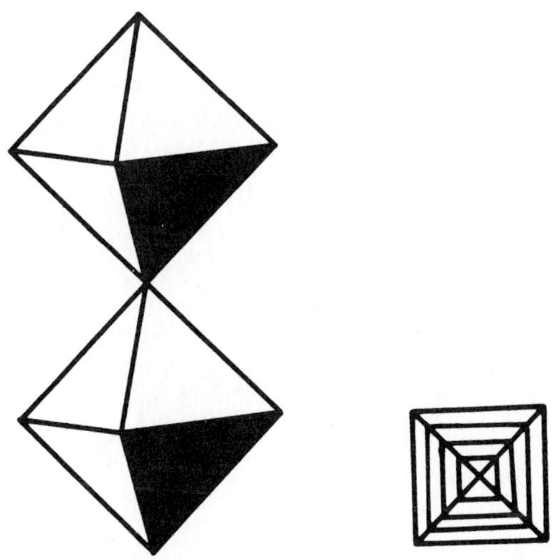

Fig. 1. The ReO_3 chain.

Fig. 2. The MoO_3 structure.

The detailed structures of these materials have been investigated using both elastic and inelastic neutron techniques (*27*, *28*). Neutron diffraction studies of the orthorhombic phase with low hydrogen content show that the hydrogen is incorporated as hydroxyl groups without any major structural change in the MoO_3 lattice (Fig. 3). These groups are not involved in hydrogen bonding between the layers but are located in intralayer sites. The presence of hydroxyl groups has been confirmed by inelastic neutron scattering experiments. Similar studies of the more highly reduced phases suggest that the hydrogen is present as $—OH_2$ groups. It is likely that the hydrogens are in interlayer sites in at least some of these phases.

Nuclear magnetic resonance studies of $H_{0.36}MoO_3$ and $H_{1.7}MoO_3$ (*29*, *30*) indicate rapid hydrogen motion in the latter phase, with $D(300\ K) \sim 10^{-8}\ cm^2/sec$, although hydrogen mobility in the former is low [$D(300\ K) = 10^{-11}\ cm^2/sec$]. Rigid lattice spectra are consistent with the presence of hydroxyl groups in $H_{0.36}MoO_3$ and show that the majority of hydrogens must be present as $—OH_2$ groups in $H_{1.7}MoO_3$. The asymmetry of the absorption spectrum observed for $H_{1.7}MoO_3$ has been interpreted in terms of an asymmetric chemical shielding tensor (*29*).

Enthalpies of formation of all four phases H_xMoO_3, in the range $0 < x < 2$, have been determined by solution calorimetry (*31*). For all, the formation reaction from $H_2(g)$ at 298 K is exothermic. However, the stepwise insertion reaction

$$\frac{1}{2}H_2(g) + \left(\frac{1}{x - y}\right)H_yMoO_3(s) = \left(\frac{1}{x - y}\right)H_xMoO_3(s) \qquad (7)$$

where x and y represent the compositions of successive phases, becomes increasingly unfavorable for higher x values and, for the formation of

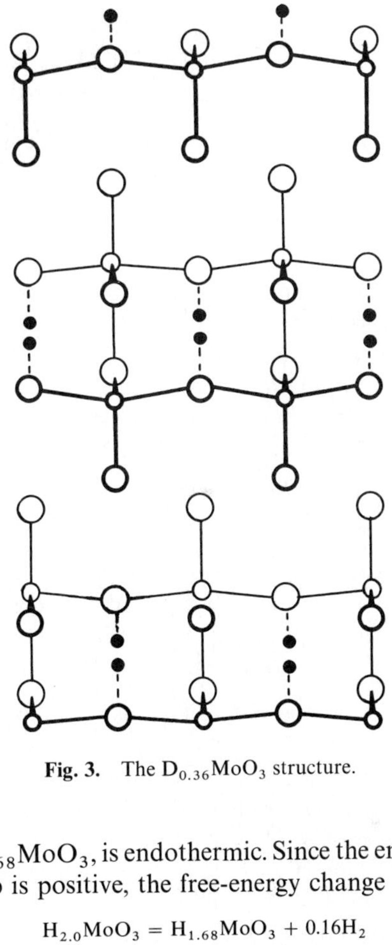

Fig. 3. The $D_{0.36}MoO_3$ structure.

H_2MoO_3 from $H_{1.68}MoO_3$, is endothermic. Since the entropy change for the decomposition step is positive, the free-energy change for the reaction

$$H_{2.0}MoO_3 = H_{1.68}MoO_3 + 0.16H_2 \tag{8}$$

is negative ($\Delta G^0 \sim -10$ kJ/mole), and $H_{2.0}MoO_3$ is actually thermodynamically unstable toward dissociation. At room temperature, H_2MoO_3 slowly evolves hydrogen, and complete decomposition to $H_{1.68}MoO_3$ occurs at 110°C. The remaining phases are stable toward disproportionation to neighboring phases H_xMoO_3, but all except the lowest phase, blue orthorhombic $H_{0.36}MoO_3$, are metastable with respect to decomposition to MoO_2 and H_2O. That is, only kinetic constraints allow the compounds H_xMoO_3 to exist indefinitely at room temperature in the absence of air. All phases of H_xMoO_3 are thermodynamically unstable with respect to reaction with oxygen at 298 K. In practice the phases with higher x values oxidize rapidly in air, but for the blue orthorhombic phase oxidation to MoO_3 is slow both chemically and electrochemically (*16*). Calculated and measured standard electrode potentials for the blue orthorhombic phase in standard acid solution are in good agreement:

$$H^+(aq) + e + \frac{1}{x}MoO_3 = \frac{1}{x}H_xMoO_3 \qquad E^0 = +0.47\ V \tag{9}$$

B. Lithium Insertion Compounds

In contrast to the H/MoO_3 system, to date no detailed systematic study of the structure and thermochemistry of the Li/MoO_3 system has been published, and our knowledge of these materials comes mainly from electrochemical studies. Early work (*2–4*) showed MoO_3 to be an active cathode material in cells of the type $Li(s)|Li^+$ (nonaqueous solvent)$|MoO_3(s)$. It was recognized that the open-circuit voltages produced by such cells (~2.8 V) were higher than could be explained on the basis of simple cell reactions of the type

$$MoO_3 + 2Li = Li_2O + MoO_2 \tag{10}$$

where MoO_3 is reduced to a lower oxide of molybdenum. An emf of ~2 V is predicted for such reactions using the known free energies of formation of the oxide phases (*4*). This discrepancy was explained by Whittingham (*1*) in terms of ternary-phase formation according to

$$Li + \frac{1}{x}MoO_3 = \frac{1}{x}Li_xMoO_3 \tag{11}$$

The Li_xMoO_3 was prepared both by reaction of *n*-butyllithium with MoO_3 and electrochemically; two well-defined discharge products were found. The maximum x value obtained by chemical lithiation is 1.55 (*13*). A study of the charge and discharge behavior of MoO_3 electrodes in nonaqueous Li^+ electrolytes (*32*) has shown the reaction to be a reversible topotactic redox process with lithium inserting between the MoO_3 layers to form the ternary phase Li_xMoO_3. An accompanying increase in the interlayer spacing of ~1 Å is observed, consistent with intercalation of lithium. It was found that reduction beyond $x = 1.5$ caused some irreversible change in the structure. It is worth noting that the high-temperature lithium molybdenum bronzes Li_xMoO_3 ($0.31 < x < 0.39$) and $Li_xMo_6O_{17}$ ($x = 0.9$) have structures very different from that of MoO_3 itself (*33*).

C. Miscellaneous

An extensive investigation by Schöllhorn *et al.* (*16*) has shown that a wide range of guest species can be intercalated into the MoO_3 lattice with an accompanying increase in the interlayer spacing. Hydrated bronzes of the

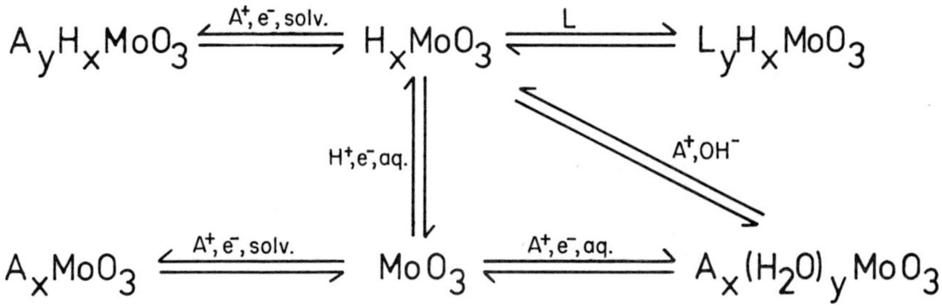

Fig. 4. Summary of the major topochemical reactions of MoO_3. [After Schöllhorn *et al.* (*16*).]

type $A_x(H_2O)_yMoO_3$ are formed by reduction of MoO_3 in neutral aqueous electrolyte solutions either chemically, using a reagent such as sodium dithionite ($Na_2S_2O_4$), or electrochemically. Thus, the dark blue compound $Na_{0.5}(H_2O)_yMoO_3$ is formed on reaction of MoO_3 with 1 *M* $Na_2S_2O_4$ with an increase in the interlayer spacing of 4.5 Å. Another route to these phases involves reaction of "$H_{0.5}MoO_3$" with aqueous alkali. Alkali metal cations are highly mobile in such phases and exchange rapidly with positive ions of electrolyte solutions, including protons in acids, according to Eq. (5). The mixed insertion compound $Na_{0.94}H_{0.85}MoO_3$ has been prepared by boiling H_2MoO_3 with 2 *M* NaOH (*20*).

Many bronzes solvated with species other than water have been prepared (*16*). The hydrated insertion compounds react with polar organic solvents, for example, DMSO, to form solvated phases such as $Na_{0.5}(DMSO)_yMoO_3$ with an accompanying increase in the interlayer spacing. Similar compounds, $H_xL_yMoO_3$, are prepared by the reaction of the hydrogen molybdenum bronzes with Lewis bases such as ammonia and aliphatic and aromatic amines, also with an increase in the interlayer separation. A summary of the major topochemical reactions of MoO_3 is given in Fig. 4.

IV. V_2O_5-DERIVED INSERTION COMPOUNDS

The V_2O_5 structure may be considered as being constructed of "V_2O_5 ribbons" formed by shearing two ReO_3 chains (Fig. 1) together, as in Fig. 5. These join as shown in Fig. 6 to give V_2O_5 itself. However, this is an oversimplification, and the true structure is considerably distorted from it, the coordination being closer to trigonal bipyramidal than octahedral, making V_2O_5 a layer compound, as shown in Fig. 7 (*34*). As with MoO_3 this arrangement allows facile insertion of electropositive species into the oxide matrix.

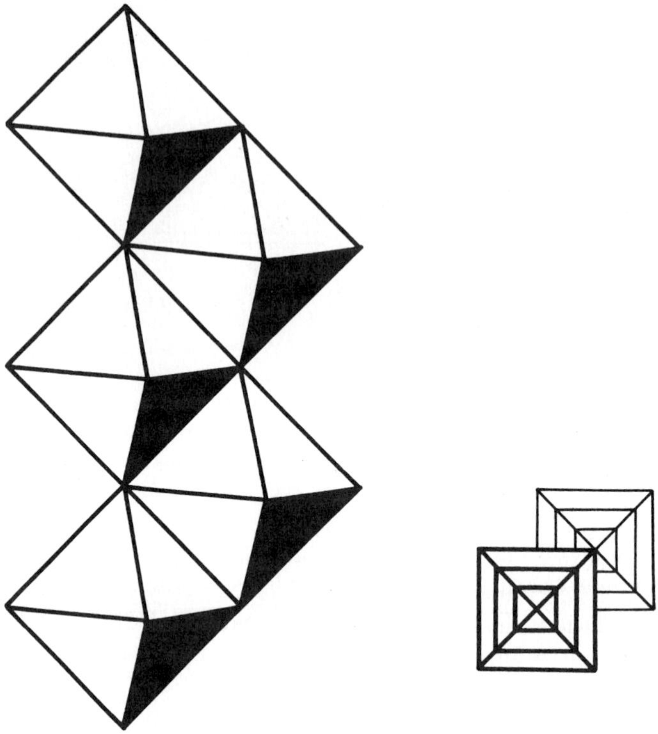

Fig. 5. The V_2O_5 ribbon.

A. Hydrogen Insertion Compounds

In contrast to the MoO_3 and WO_3 systems, no thorough study of the ambient temperature insert on of hydrogen into the V_2O_5 matrix has been made, probably because of the high solubility of the parent oxide in more traditional reducing media such as Zn/HCl. Many compounds of the type H_xVO_n, both naturally occurring and synthetic, have been investigated

Fig. 6. The idealized V_2O_5 structure.

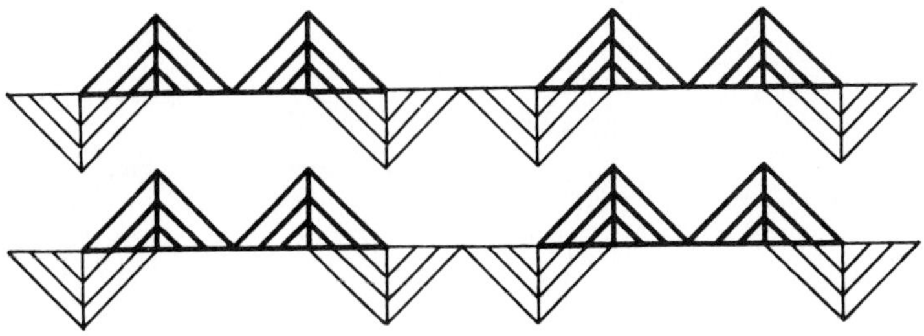

Fig. 7. The V_2O_5 structure—trigonal bipyramidal representation.

(*35–40*), but these materials are not formed by topochemical insertion reactions. However, solid solution of hydrogen in the V_2O_5 matrix appears to occur on electrolytic reduction of the oxide in aqueous media (*41*, *42*).

B. Lithium Insertion Compounds

Lithium may be intercalated chemically or electrochemically into the V_2O_5 lattice to give the ternary phases $Li_xV_2O_5$ ($0 < x < 2.0$) (*13*, *1*). It has been shown that V_2O_5 can be used reversibly in nonaqueous high-power lithium cells (*43*) and a Li/V_2O_5 reserve cell has been described (*44*). The phases in the ambient temperature $Li_xV_2O_5$ system have been investigated in the range $0 < x < 1.0$ (*14*, *15*). At low lithium content ($x < 0.1$) phase I, with lattice parameters very close to those of V_2O_5 and isostructural with the known high-temperature α-$Li_xV_2O_5$ (*45*), is formed by insertion of lithium between the V_2O_5 layers. There are two other distinct phases occurring between $0.1 < x < 1.0$, both with ranges of homogeneity. These also have lattice constants closely related to those of the parent oxide, indicating topochemical insertion of lithium. It has been suggested that a puckering of the oxide layers may occur (*1*). The open-circuit emf versus x curve for the cell $Li(s)|Li^+$(nonaqueous solvent)$|Li_xV_2O_5(s)$ (Fig. 8) is useful in the identification of two-phase and solid solution regions. The plateaus represent regions of constant chemical potential of lithium, that is, regions where two phases coexist in equilibrium. The sloping parts denote regions of varying μ_{Li} and indicate ranges of homogeneity of a single phase. The latter two phases, phase II at $x \sim 0.45$ and phase III at $x \sim 1.0$, have structures that differ considerably from the corresponding high-temperature phases. High-temperature β'-$Li_xV_2O_5$ found at $x \sim 0.45$ has a tunnel structure (*46*) with lithium atoms in the tunnels (Fig. 9), and γ-LiV_2O_5 (Fig. 10) has a layer structure, but the layers differ somewhat from those of V_2O_5 (*47*).

Calorimetric determinations of the enthalpics of formation of representative members of phases I, II, and III for the $Li_xV_2O_5$ system have been

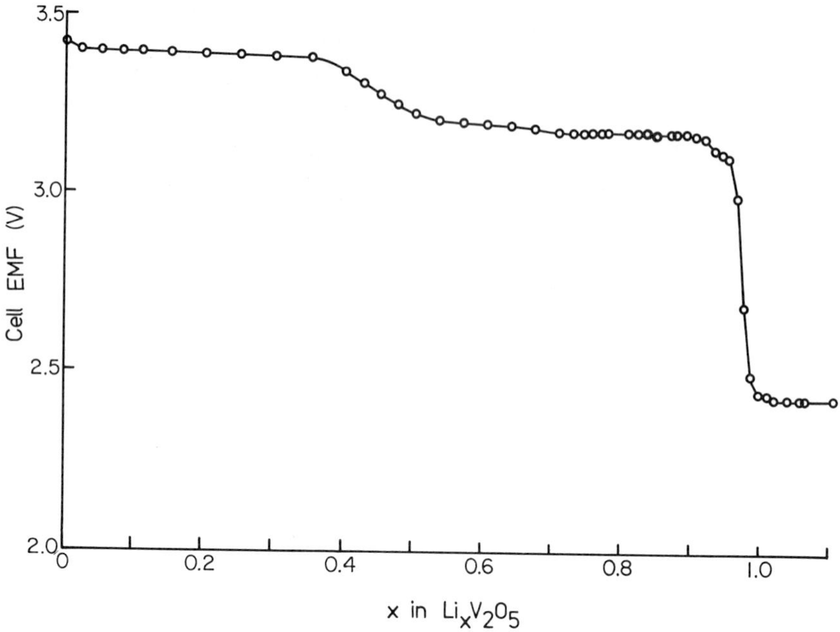

Fig. 8. Electromotive force versus x curve for $Li_xV_2O_5$.

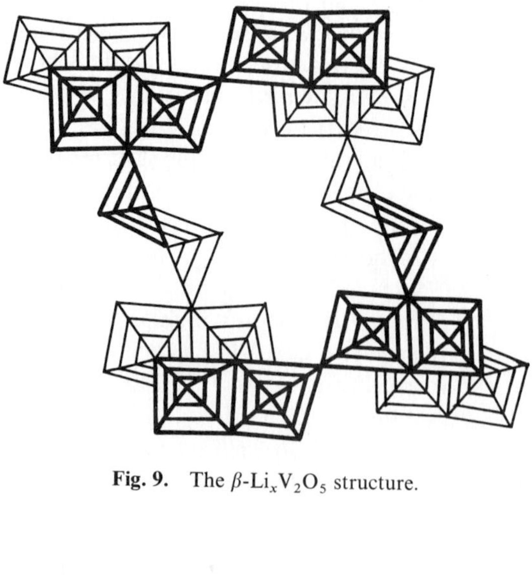

Fig. 9. The β-$Li_xV_2O_5$ structure.

Fig. 10. The γ-$Li_xV_2O_5$ structure.

made (*48*). The exothermicity of the reaction $Li(s) + (1/x)V_2O_5(s) = (1/x)Li_xV_2O_5(s)$ falls with increasing x from $\Delta H = -334$ kJ/mole ($x = 0.10$) to $\Delta H = -320$ kJ/mole (at $x = 1.03$). The fall matches closely the free-energy changes determined by open-circuit emf measurements for the cell $Li(s)|Li^+$(nonaqueous solvent)$|Li_xV_2O_5(s)$ over the same range of composition. This emphasizes how closely the cell voltage of the Li/V_2O_5 battery system is controlled by the enthalpy of formation of the corresponding ambient temperature ternary phase $Li_xV_2O_5$. The low-temperature phases $Li_xV_2O_5$ II and III are metastable with respect to the high-temperature phases β' and γ-$Li_xV_2O_5$ with the same chemical compositions. The enthalpy change for $LiV_2O_5(III) \rightarrow LiV_2O_5(\gamma)$ is approximately -15 to 20 kJ/mole (*15*, *48*). In addition, the low-temperature phases are thermodynamically unstable (or only marginally stable) to the disproportionation reaction

$$\frac{1}{x}Li_xV_2O_5 = LiVO_3 + VO_2 + \left(\frac{1}{x} - 1\right)V_2O_5 \qquad (x < 1) \tag{12}$$

or

$$= LiVO_3 + \left(1 - \frac{1}{x}\right)V_2O_3 + \left(\frac{4}{x} - 3\right)VO_2 \qquad (x > 1) \tag{13}$$

All phases are thermodynamically unstable with respect to oxidation at 298 K.

$$\frac{1}{x}Li_xV_2O_5 + \frac{1}{4}O_2(g) = LiVO_3(s) + \left(\frac{1}{x} - \frac{1}{2}\right)V_2O_5 \tag{14}$$

The apparent chemical inertness of the low-temperature phases is again attributable to kinetic rather than true thermodynamic stability.

V. WO_3-DERIVED INSERTION COMPOUNDS

The WO_3 structure consists of vertex-sharing ReO_3 chains (Fig. 1) linked to give an infinite array of four-sided tunnels. This is the ReO_3 structure (Fig. 11); WO_3 has a similar although somewhat distorted monoclinic form at room temperature (*49*). This, unlike MoO_3 and V_2O_5, has a framework structure with an interpenetrating three-dimensional array of square tunnels, and the cavities formed may be filled only by ions of a compatible size. Occupation of all tunnel sites gives a stoichiometry AMO_3; this is the well-known perovskite structure and is adopted by a vast range of ABO_3 mixed-metal oxides (*50*). Deformed perovskite structures are found for many tungsten bronzes at low x values, the exceptions being bronzes with the larger insertion ions, and, as x is increased, the symmetry of the lattice rises until in some cases a cubic phase is reached. Many tungsten bronzes adopt structures other than those based on the simple ReO_3 structure, especially with the larger insertion ions, for example, hexagonal K_xWO_3 (*51*) and tetragonal I K_xWO_3 (*52*). These, however, are formed, as are most tungsten bronzes, by high-temperature reactions, and although there are many examples of perovskite-type tungsten bronzes with insertion ions as diverse

Fig. 11. The ReO_3 structure.

as Na (*53*), Ge (*54*), and Eu (*55*) only Li and H have sufficient mobility in the WO_3 lattice to be formed in ambient temperature preparations.

A. Hydrogen Insertion Compounds

Traditionally, the hydrogen tungsten bronzes have been prepared by chemically reducing WO_3 with reducing agents such as Zn/HCl (*56, 57*). More recently the techniques of hydrogen "spillover" (*21*) and electrochemical reduction (*58, 59*) have been employed. The ease and reversibility of reduction together with the dramatic color change to dark blue has resulted in the study of the uses of WO_3 films in electrochromic devices (*5, 6, 8, 60*). Three well-defined phases have been characterized in this system; they are tetragonal B, $0.15 < x < 0.23$; tetragonal A, $0.33 < x < 0.50$ (*61*), and cubic $0.50 < x < 0.6$ (*62*). In all cases the framework of the parent oxide is retained. A neutron diffraction study of the cubic phase (*62*) has shown that the perovskite framework is distorted, with hydrogens randomly attached to the oxygen atoms as hydroxyl groups; the crystal structure is closely related to that of $In(OH)_3$ (*63*).

Proton mobility in the hydrogen tungsten bronze $H_{0.46}WO_3$ has been studied by pulsed NMR techniques (*64*). These data indicate slow hydrogen diffusion, $D(300\ K) \sim 10^{-12}$ cm^2/sec; this is similar to that observed in $H_{0.36}MoO_3$, which has the hydrogens in a similar environment. Rigid lattice absorption spectra confirm the structure determined by neutron diffraction. Other studies (*65, 66*) give seemingly inconsistent results, but these used powders of high surface area, which are susceptible to contamination by water.

Enthalpies of formation of the tetragonal B and A phases $H_xWO_3 (x = 0.18$ and $x = 0.35$, respectively) have been determined (*67*). Both $H_{0.35}WO_3$ and $H_{0.18}WO_3$ are stable toward decomposition to $WO_3(s)$ and $H_2(g)$ at 298 K. They are also marginally stable toward disproportionation to $H_2O(l)$ and lower oxides of tungsten. However, H_xWO_3 phases are very unstable toward oxidation, as is observed experimentally. Vacuum decomposition of crystalline samples of H_xWO_3 shows that decomposition occurs over the range 470–670 K (*67*).

B. Lithium Insertion Compounds

Traditionally, the lithium tungsten bronzes have been prepared by high-temperature routes, for example,

$$\frac{x}{2}Li_2WO_4 + \left(\frac{3 - 2x}{3}\right)WO_3 + \frac{x}{6}W \rightarrow Li_xWO_3 \tag{15}$$

Four distinct phases have been found at high temperature (*68*): monoclinic, $0 < x \leq 0.008$; orthorhombic, $0.028 \leq x \leq 0.032$; tetragonal II, $0.09 \leq x \leq 0.11$; and cubic $0.26 \leq x \leq 0.50$. All phases possess the basic WO_3 framework, and the symmetry of the lattice increases with increasing x, as is typically observed in such systems. A neutron diffraction study of the cubic $Li_{0.36}WO_3$ (*53*) showed that the lithiums are located in distorted perovskite-type cavities. Lithium ion mobility in Li_xWO_3 is quite low, making WO_3 unsuitable for use as a battery cathode material, but in spite of this Li_xWO_3 can be prepared by the reaction of WO_3 with *n*-butyllithium with $x < 1.5$ (*69*). Although it has not been verified experimentally, it seems likely that the high- and low-temperature phases will be isostructural. Electrochemical formation of the bronze Li_xWO_3 in thin amorphous films of WO_3 and at ambient temperature has been studied in connection with electrochromic display systems (*7*, *17*). As anticipated, the mobility of Li in WO_3 films is much lower than that of H, reflecting the relative sizes of the insertion species.

C. Related Compounds

Although ReO_3 has the same basic structure as WO_3, its insertion compounds are far less well investigated. However, a bronze Na_xReO_3 ($x \sim 0.6$) with a perovskite-type structure has been reported (*70*). Recently, two insertion compounds have been prepared at temperatures close to ambient. H_xReO_3 was prepared by grinding ReO_3 under H_2O and then boiling the resulting mixture for several hours. The golden-colored product analyzed to $H_{0.15}ReO_3$ and was electrically conducting. X-ray diffraction showed the bronze to have an orthorhombically distorted ReO_3 structure (*71*).

The reaction of *n*-butyllithium with ReO_3 gives a bronze Li_2ReO_3 which has a hexagonal unit cell again related to that of the parent oxide (*69*).

VI. MISCELLANEOUS

Recently, ambient temperature insertion reactions have been reported for more unusual oxide phases such as $Li_{1+x}V_3O_8$, the reduced oxide V_6O_{13}, and several rutile-type oxides. Some of these are described below.

A. $Li_{1+x}V_3O_8$-Related Insertion Compounds

The structure of the high-temperature bronze phase $Li_{1+x}V_3O_8$, shown in Fig. 12, consists of layers composed of octahedral and trigonal bipyramidal ribbons with the lithium atoms occupying positions of sixfold coordination

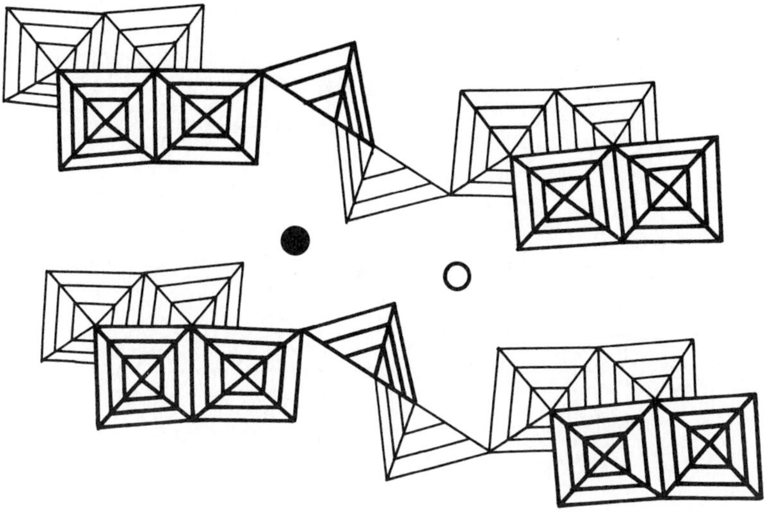

Fig. 12. The $Li_{1+x}V_3O_8$ structure.

(*72*). Crystallographic considerations limit the maximum value of x to 0.5. Besenhard and Schöllhorn showed that this phase is able to take up Li in organic lithium electrolytes reversibly (*32*). More recently a comprehensive investigation of the system (*22*) showed that $Li_{1+x}V_3O_8$ reacts with aqueous acids topotactically to give a layered, hydrated bronze $H_{1+x}(H_2O)_yV_3O_8$. The protons in this phase rapidly exchange with alkali or alkaline earth ions under neutral or alkaline conditions according to Eq. (6). This provides preparative routes to a new series of bronze phases with various A atoms and different interplanar spacings. The hydration behavior found in these systems is complex. The hydrogen bronzes $H_{1+x}(H_2O)_yV_3O_8$ react readily with Lewis bases, such as pyridine, according to

$$nL + H_{1+x}(H_2O)_yV_3O_8 \rightarrow L_nH_{1+x}V_3O_8 + yH_2O \tag{16}$$

The behavior of these phases is closely related to that observed in the MoO_3 system.

B. V_6O_{13} Insertion Compounds

The V_6O_{13} structure, shown in Fig. 13, consists of alternate double and single layers of V_2O_5 ribbons (Fig. 5). The layers are connected by vertex sharing of octahedra, and this leads to a relatively open framework structure (*73, 74*). This, together with its metallic conductivity, has led to the investigation of V_6O_{13} as a potential cathode material for nonaqueous Li cells.

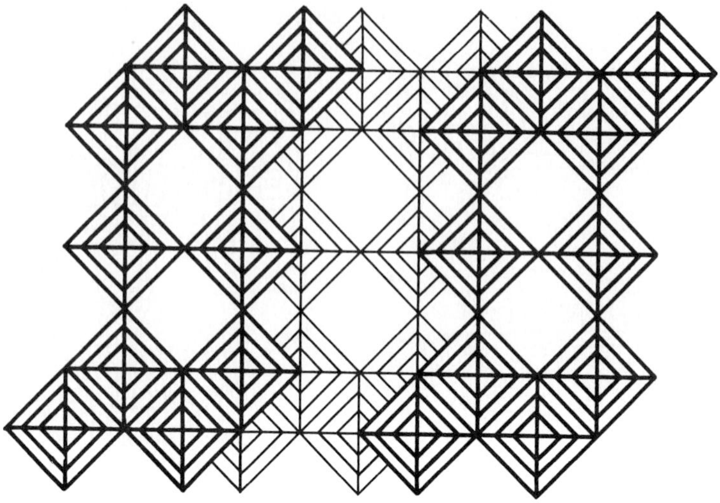

Fig. 13. The V_6O_{13} structure.

Murphy *et al.* (*69*, *75*) prepared the bronze phases $Li_xV_6O_{13}$ at ambient temperatures both electrochemically and using *n*-butyllithium. X-ray powder diffraction patterns show multiphase behavior for $x < 3$ and a single-phase region for $x > 3$. The maximum value of x obtained is 7.6. Insertion of lithium into the oxide matrix causes small changes in the lattice parameters, but the basic framework is preserved. In phases with high lithium content there must be more than one Li atom in each perovskite-type cavity, these probably being located in tunnel sites of distorted square pyramidal coordination. The emf of the cell $Li(s)|Li^+$ (nonaqueous solvent)$|Li_xV_6O_{13}(s)$ varies between 2.8 and 2.2 V, giving a high theoretical energy density of ~800 Wh/kg. The system is readily reversible.

C. Rutile-Type Insertion Compounds

The rutile structure may be represented by joining chains of edge-shared octahedra by vertices, as shown in Fig. 14. This structure is found for a large number of oxides such as RuO_2, TiO_2, MoO_2 (*76*), and WO_2 (*77*). The tunnels are rather small and can accommodate only very small atoms such as H and in some cases Li. No hydrogen bronzes of this type are known, although many oxide hydroxides of general formual MOOH adopt this structure (*78–81*), especially at high pressure.

Some rutile-type oxides incorporate Li topotactically on reaction with *n*-butyllithium (*82*). The oxides of Mo, Ru, Os, and Ir form Li_xMO_2 with

Fig. 14. The rutile structure.

$x > 1.0$. Limiting compositions of $Li_{0.03}TiO_2$, $Li_{0.2}MnO_2$, and $Li_{0.8}CrO_2$ are found, whereas VO_2, NbO_2, and ReO_2 do not react at all. In all cases there is an increase in volume of the parent oxide lattice. Lithium insertion, which can also be accomplished electrochemically, occurs most easily when the parent oxides are metallic. It is interesting that lithiation of WO_2, which is metallic, could be accomplished only electrochemically.

VII. SUMMARY OF THERMODYNAMIC DATA

Quantitative data for the energetics of formation of the oxide insertion compounds discussed here have been obtained by two means. In the first, direct calorimetric determinations of the enthalpy changes $\Delta H_x'$ for the reaction $x\text{A} + \text{MO}_n = \text{A}_x\text{MO}_n$ have been made using synthetic materials. In the second, open-circuit voltages for cells $\text{A}|\text{A}^+|\text{A}_x\text{MO}_n$ have been measured. The corresponding thermodynamic function obtained from the measured voltage is either $\overline{\Delta G}_\text{A}(x)$, the partial molar free-energy change, for the solid solution process $\text{A} = \text{A(soln in A}_x\text{MO}_n)$ or, when a two-phase region occurs, ΔG_A for the reaction $\text{A} + [1/(x-y)]\text{A}_y\text{MO}_n = [1/(x-y)]$-$\text{A}_x\text{MO}_n$, where x and y refer to successive phases. The relationship between voltage and the appropriate free-energy function is simply $EF = -\overline{\Delta G}_\text{A}$ (or ΔG_A). In cases where E has been measured as a function of x, the corresponding integral free-energy change $\Delta G_n'$ for the reaction $x\text{A} + \text{MO}_n = \text{A}_x\text{MO}_n$ can be derived via the Gibbs–Duhem relationship (*18*)

$$\Delta G_x' = -xFE + F\int_{E(0)}^{E(x)} x\,dE$$

For reactions involving only solid phases, entropy changes are small, and $\Delta G_x' \sim \Delta H_x'$. Where gaseous species are involved as, for example, in the

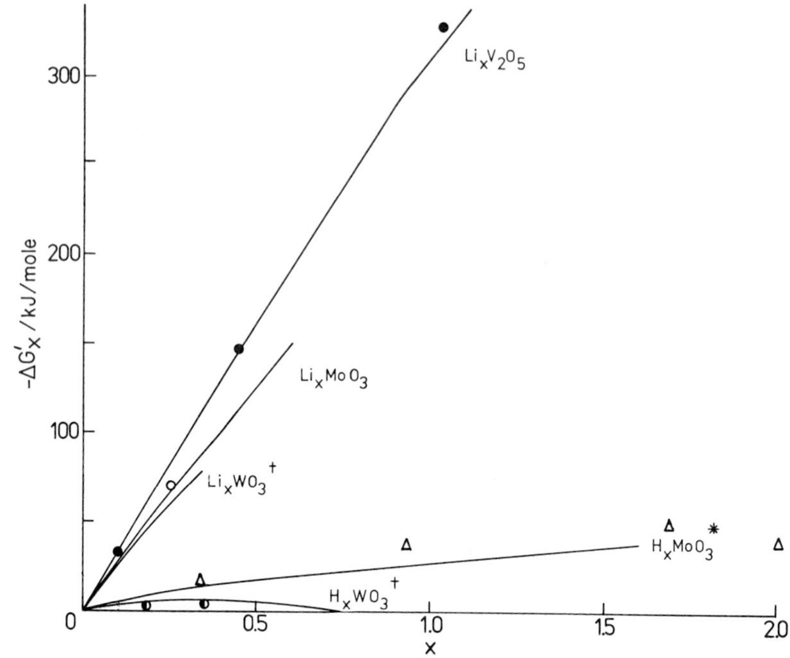

Fig. 15. Plot of $\Delta G_x'$ versus x for ambient temperature A_xMO_n phases. Solid lines from electrochemical data (*8, 16, 17, 48, 83*); circles from solution calorimetry (*31, 48, 67, 83*). Dagger indicates data from measurements on amorphous films; asterisk, from nonequilibrium electrochemical measurements.

reaction $(x/2)H_2 + MO_n = H_xMO_n$ (*19*), the corresponding entropy contributions can be estimated accurately using standard data, and again $\Delta G_x'$ can be obtained from measured values of $\Delta H_x'$. Thus electrochemical and calorimetric determinations of $\Delta G_x'$ can be compared directly. In Fig. 15 are shown values of $\Delta G_x'$ for the ambient temperature phases $Li_xV_2O_5$, Li_xMoO_3, Li_xWO_3, H_xMoO_3, and H_xWO_3 determined by the two approaches.

The calorimetric values refer to well-characterized synthetic samples, whereas the electrochemical values refer to *in situ* cathode materials that in general have not been subjected to subsequent structural or chemical analysis. The agreement between the two sets of data is satisfactory and provides additional confirmation of the identity of the products formed by *in situ* electrochemical preparation and direct chemical synthesis.

The variation of the enthalpy of formation of the phases Li_xMO_n follows the changes in oxidizing character of the oxide matrix $V_2O_5 > MoO_3 > WO_3$ and is compatible with the formulation of insertion compounds as $Li_x^+(MO_n)(xe)$. Similar trends have been noted between the analogous high-temperature oxide bronze phases. The much lower free energies of formation of the hydrogen compounds from H_2(g) compared with the corresponding lithium compounds arise principally from the high dissociation energy of H_2 and not from an intrinsically weak binding of hydrogen in the oxide lattice (either as —OH or —OH_2).

A special feature of the thermochemistry of the low-temperature oxide insertion phases considered in this chapter is that most or all appear to be thermodynamically unstable with respect to disproportionation to lower oxide and corresponding oxy salt, for example, Eqs. (12)–(13). Their apparent stability arises from the unfavorable kinetics of the major lattice rearrangement involved in such a decomposition path.

VIII. CONCLUSIONS

The present discussion has been confined to ambient temperature insertion compounds. Their common feature is that the insertion of electropositive species causes no chemical bonds to be broken. There are two distinct categories. The first involves insertion of guest species into a layer host lattice, such as MoO_3, in which the layers are held together only by van der Waals forces. Electropositive species can be intercalated easily, and their size is irrelevant since the interlayer spacing can increase to accommodate the guest ion. The second category, involving framework structures, is typified by WO_3. Ambient temperature insertion of atoms into WO_3 requires facile passage of the atoms through the narrow square "window" of oxygen atoms, and this essentially restricts guest species to H and Li. The A metal cavity can accommodate larger atoms, such as sodium (in Na_xWO_3), only at high temperatures where the energy required for bond breaking is more readily available.

ACKNOWLEDGMENTS

The authors would like to thank Dr. D. W. Murphy for providing preprints of his work on the lithium vanadium pentoxide system. One of us (M.F.P.) thanks the U. S. Air Force for a research fellowship provided under Grant AFOSR-77-3402.

REFERENCES

1. M. S. Whittingham, *J. Electrochem. Soc.* **123,** 315 (1976).
2. F. W. Dampier, *J. Electrochem. Soc.* **121,** 656 (1974).

3. N. Margalit, *J. Electrochem. Soc.* **121,** 1460 (1974).
4. L. Campanella and G. Pistoia, *J. Electrochem. Soc.* **118,** 1905 (1971).
5. M. L. Hitchman, *J. Electroanal. Chem.* **85,** 135 (1977).
6. B. W. Faughnan, R. S. Crandall, and P. M. Heyman, *RCA Rev.* **36,** 177 (1975).
7. M. Green, W. C. Smith, and J. A. Weiner, *Thin Solid Films* **38,** 89 (1976).
8. R. S. Crandall, P. J. Wojtowicz, and B. W. Faughnan, *Solid State Commun.* **18,** 1409 (1976).
9. P. G. Dickens and P. J. Wiseman, *MTP Int. Rev. Sci.: Inorg. Chem., Ser. Two* **10,** 211 (1975).
10. P. Hagenmuller, *Prog. Solid State Chem.* **5,** 71 (1971).
11. J. B. Goodenough, *Prog. Solid State Chem.* **5,** 145 (1971).
12. M. S. Whittingham, *Prog. Solid State Chem.* **12,** 1 (1978).
13. M. S. Whittingham and M. B. Dines, *J. Electrochem. Soc.* **124,** 1387 (1977).
14. P. G. Dickens, S. J. French, A. T. Hight, and M. F. Pye, *Mater. Res. Bull.* **14,** 1295 (1979).
15. D. W. Murphy, P. A. Christian, F. J. DiSalvo, and J. V. Waszczak, *Inorg. Chem.* **18,** 2800 (1979).
16. R. Schöllhorn, R. Kuhlmann, and J. O. Besenhard, *Mater. Res. Bull.* **11,** 83 (1976).
17. S. K. Mohapatra and S. Wagner, *J. Electrochem. Soc.* **125,** 1603 (1978).
18. O. Glemser and G. Lutz, *Z. Anorg. Allg. Chem.* **264,** 17 (1951).
19. O. Glemser, U. Hauschild, and O. Bimmermann, *Angew. Chem.* **64,** 457 (1952).
20. J. J. Birtill and P. G. Dickens, *Mater. Res. Bull.* **13,** 311 (1978).
20a. R. Schöllhorn, *Angew. Chem. Int. Ed.* **19,** 983 (1980).
21. P. A. Sermon and G. C. Bond, *J.C.S. Faraday I* **72,** 730 (1976).
22. R. Schöllhorn, F. Klein-Reesink, and R. Reimold, *J. C. S. Chem. Commun.* p. 399 (1979).
23. L. Kihlborg, *Ark. Kemi* **21,** 357 (1963).
24. O. Glemser, G. Lutz, and G. Meyer, *Z. Anorg. Allg. Chem.* **285,** 173 (1956).
25. O. Glemser and G. Lutz, *Z. Anorg. Allg. Chem.* **269,** 93 (1952).
26. K.-A. Wilhelmi, *Acta Chem. Scand.* **23,** 419 (1969).
27. P. G. Dickens, J. J. Birtill, and C. J. Wright, *J. Solid State Chem.* **28,** 185 (1979).
28. F. Schröder and H. Weitzel, *Z. Anorg. Allg. Chem.* **435,** 247 (1977).
29. R. C. T. Slade, T. K. Halstead, and P. G. Dickens, *J. Solid State Chem.* **34,** 183 (1980).
30. A. Cirillo and J. J. Fripiat, *J. Phys. (Paris)* **39,** 247 (1978).
31. J. J. Birtill and P. G. Dickens, *J. Solid State Chem.* **29,** 367 (1979).
32. J. O. Besenhard and R. Schöllhorn, *J. Power Sources* **1,** 267 (1976).
33. J. M. Reau, C. Fouassier, and P. Hagenmuller, *J. Solid State Chem.* **1,** 326 (1970).
34. H. G. Bachmann, F. R. Ahmed, and W. H. Barnes, *Z. Kristallogr.* **115,** 110 (1961).
35. H. T. Evans and S. Block, *Am. Mineral.* **38,** 1242 (1953).
36. H. T. Evans and M. E. Mrose, *Acta Crystallogr.* **11,** 56 (1958).
37. H. T. Evans and M. E. Mrose, *Am. Mineral.* **45,** 1144 (1960).
38. F. Theobald, *Bull. Soc. Chim. Fr.* p. 803 (1977).
39. E. Schwarzmann, *Z. Naturforsch., Teil B* **25,** 1485 (1970).
40. O. Glemser and E. Schwarzmann, *Z. Anorg. Allg. Chem.* **278,** 249 (1955).
41. N. V. Watson and A. B. Scott, *J. Electrochem. Soc.* **112,** 883 (1965).
42. K. R. Newby and A. B. Scott, *J. Electrochem. Soc.* **117,** 152 (1970).
43. J. S. Gore and C. R. Walk, U.S. Patent 3,929,504 (1975).
44. M. R. Merriman and G. C. Vezzoli, *Mater. Res. Bull.* **14,** 77 (1979).
45. P. Hagenmuller, J. Galy, M. Pouchard, and A. Casalot, *Mater. Res. Bull.* **1,** 45 (1966).
46. J. Galy, J. Darriet, A. Casalot, and J. B. Goodenough, *J. Solid State Chem.* **1,** 339 (1970).
47. J. Galy and A. Hardy, *Acta Crystallogr.* **19,** 432 (1964).
48. P. G. Dickens, S. J. French, A. T. Hight, M. F. Pye, and G. J. Reynolds, *Solid State Ionics* **2,** 27 (1981).

49. S. Tanisaki, *J. Phys. Soc. Jpn.* **15,** 4 (1960).
50. J. B. Goodenough and J. M. Longo, *in* "Landolt–Börnstein Tabellen," New Series III/4a (K.-H. Hellwege and A. M. Hellwege, eds.), p. 126. Springer-Verlag, Berlin and New York, 1970.
51. A. Magnéli, *Acta Chem. Scand.* **7,** 315 (1953).
52. A. Magnéli, *Ark. Kemi* **1,** 269 (1949).
53. P. J. Wiseman and P. G. Dickens, *J. Solid State Chem.* **17,** 91 (1976).
54. M. Parmentier and C. Gleitzer, *J. Solid State Chem.* **17,** 255 (1976).
55. C. S. Dimbylow, I. J. McColm, C. M. P. Barton, N. N. Greenwood, and G. E. Turner, *J. Solid State Chem.* **10,** 128 (1974).
56. O. Glemser and C. Naumann, *Z. Anorg. Allg. Chem.* **265,** 289 (1951).
57. P. G. Dickens and R. J. Hurditch, *Nature (London)* **215,** 1266 (1967).
58. G. Siclet, J. Chevrier, J. Lenair, and C. Eyraud, *C. R. Hebd. Seances Acad. Sci., Ser. C* **277,** 227 (1973).
59. J. Chevrier and G. Siclet, *Bull. Soc. Chim. Fr.* p. 1038 (1976).
60. R. S. Crandall and B. W. Faughnan, *Appl. Phys. Lett.* **28,** 95 (1976).
61. P. G. Dickens and R. J. Hurditch, *in* "The Chemistry of Extended Defects in Non-Metallic Solids" (L. Eyring and M. O'Keefe, eds.), p. 555. North-Holland Publ., Amsterdam, 1970.
62. P. J. Wiseman and P. G. Dickens, *J. Solid State Chem.* **6,** 374 (1973).
63. A. N. Christensen, N. C. Broch, O. von Heidenstam, and A. Nilsson, *Acta Chem. Scand.* **21,** 1046 (1967).
64. P. G. Dickens, D. J. Murphy, and T. K. Halstead, *J. Solid State Chem.* **6,** 370 (1973).
65. M. A. Vannice, M. Boudart, and J. J. Fripiat, *J. Catal.* **17,** 359 (1970).
66. K. Nishimura, *Solid State Commun.* **20,** 523 (1976).
67. P. G. Dickens, J. H. Moore, and D. J. Neild, *J. Solid State Chem.* **7,** 241 (1973).
68. J. M. Reau, C. Fouassier, G. Le Flem, J. Y. Barraud, J. P. Doumerc, and P. Hagenmuller, *Rev. Chim. Miner.* **7,** 975 (1970).
69. R. J. Cava and D. W. Murphy, International Conference on Fast Ionic Transport in Solids, Gatlinburg, Tennessee, 1981.
70. A. W. Sleight and G. L. Gillson, *Solid State Commun.* **4,** 601 (1966).
71. N. Kimyuka, T. Akahani, S. Matsumoto, and K. Yukimo, *Inorg. Chem.* **15,** 3178 (1976).
72. A. D. Wadsley, *Acta Crystallogr.* **10,** 261 (1957).
73. F. Aeibi, *Helv. Chim. Acta* **31,** 8 (1948).
74. K.-A. Wilhelmi, K. Waltersson, and L. Kihlborg, *Acta Chem. Scand.* **25,** 2675 (1971).
75. D. W. Murphy, P. A. Christian, F. J. DiSalvo, and J. N. Carides, *J. Electrochem. Soc.* **126,** 497 (1979).
76. D. B. Rodgers, R. D. Shannon, A. W. Sleight, and J. L. Gillson, *Inorg. Chem.* **8,** 841 (1968).
77. D. J. Palmer and P. G. Dickens, *Acta Crystallogr., Sect. B* **35,** 2199 (1979).
78. J. Chenevas, J. C. Joubert, J. J. Caponi, and M. Marezio, *J. Solid State Chem.* **6,** 1 (1973).
79. A. N. Christensen, R. Gronback, and S. E. Rosmussen, *Acta Chem. Scand.* **18,** 1261 (1964).
80. A. N. Christensen, *Inorg. Chem.* **5,** 1452 (1966).
81. A. N. Christensen, *Mater. Res. Bull.* **6,** 691 (1971).
82. D. W. Murphy, F. J. DiSalvo, J. N. Carides, and J. V. Waszczak, *Mater. Res. Bull.* **13,** 1395 (1978).
83. P. G. Dickens and G. J. Reynolds, International Conference on Fast Ionic Transport in Solids, Gatlinburg, Tennessee, 1981.

17
Lithium Intercalation Compounds of Vanadium Chalcogenides

D. W. MURPHY

I. INTRODUCTION

The capacity of transition metal compounds of moderate to high oxidation states to form intercalation compounds with alkali metals involves two dominant aspects: oxidation–reduction chemistry (usually of the metal cations) and diffusion of the alkali metal ion into the structure of the host. In order to obtain well-defined intercalation compounds, the amount of alkali metal incorporated cannot exceed the ability of the host cations to change their formal oxidation state while maintaining a similar coordination geometry. Vanadium is a particularly attractive host cation in chalcogenides since compounds in all oxidation states between V^{5+} and V^{2+} are known which have octahedral or distorted octahedral coordination (see Table I). In addition, energy differences among compounds with different oxidation states are not large. In fact, many mixed-valence compounds exist, and some

INTERCALATION CHEMISTRY

ISBN 0-12-747380-7

TABLE I

Oxidation States and Coordination Geometries of Some Vanadium Chalcogenides

Average oxidation state	Coordination number	Geometry	Examples
5	5 or 6	Square pyramidal, distorted octahedral	V_2O_5
5[a]	?	Amorphous	V_2S_5
4.67	5, 6	Trigonal bipyramidal, octahedral	V_3O_7
4.33	6	Octahedral	V_6O_{13}
4	6	Octahedral	VO_2, VS_2
3.2	6	Octahedral	V_5S_8
3	6	Octahedral	V_2O_3
2	6	Octahedral	VO, VS

[a] The possibility exists here for covalent S—S bonding, which would result in a lower formal vanadium oxidation state.

are metallic. Thus, the vanadium chalcogenides appear to be likely candidates for intercalation reactions and offer the prospect of oxidation state changes greater than 1.

Several other points of interest are related to the delicate balance between localization and delocalization of the d electrons. This balance is reflected by metal–insulator transitions or metal–metal bonds in several chalcogenides.

In this chapter a brief account of the descriptive chemistry of the lithium intercalation compounds of vanadium chalcogenides is presented. For the oxides it is instructive to contrast intercalation compounds with the more stable bronzes.

II. OXIDES

A. V_2O_5

It has long been recognized that the vanadium oxide framework of ternary alkali metal vanadium oxides (bronzes) bears close structural similarities to binary vanadium oxides with the same V/O ratio (*1*). The prime example is the family of bronzes $M_xV_2O_5$ (M = Li, Na, K, Ca, Ag, Cu, etc.), for which a variety of framework structures similar to V_2O_5, but involving a somewhat rearranged skeleton, are obtained. It has been recognized for several years that some other $Li_xV_2O_5$ compounds are obtained in reactions at ambient temperature. Recently, two other polymorphs of LiV_2O_5, which can be classified as intercalation compounds, have been prepared and characterized. In the system $Li_xV_2O_5$ ($0 < x < 1$) single-phase regions were observed for $x < 0.15$ and $0.3 < x < 0.7$ on samples prepared at room temperature by treatment of V_2O_5 with LiI in acetonitrile (Fig. 1) (*2*). This method gives well-defined $Li_xV_2O_5$ with sharp X-ray diffraction peaks. These compositions have been further characterized by magnetic susceptibility, differential scanning calorimetry (DSC) data, and infrared spectroscopy. Compositions with $0.7 < x < 1.0$ have two different coexisting forms (δ and ϵ). The δ form transforms reversibly to the ϵ phase with $\Delta H \approx 1$ kcal/mole. Above 300°C these $Li_xV_2O_5$ phases are irreversibly converted to the bronzes of the same composition with $\Delta H \approx 4.0$ kcal/mole. X-ray powder patterns of $Li_xV_2O_5$ obtained from *n*-BuLi reactions or from milling LiI and V_2O_5 in CCl_4 are described as "rather broad" (*3*), but the same room temperature $Li_{1.0}V_2O_5$ phase is obtained as with LiI in acetonitrile, whereas a somewhat smaller single-phase region ($0.35 < x < 0.5$) is observed. Although the detailed crystal structure of these new LiV_2O_5 compounds have not been determined, it is clear that they possess a V_2O_5 skeleton very closely related to V_2O_5 itself. This is evident from the gradual evolution of lattice parameters (especially for the ϵ phase) and the easy reconversion of the LiV_2O_5 to V_2O_5

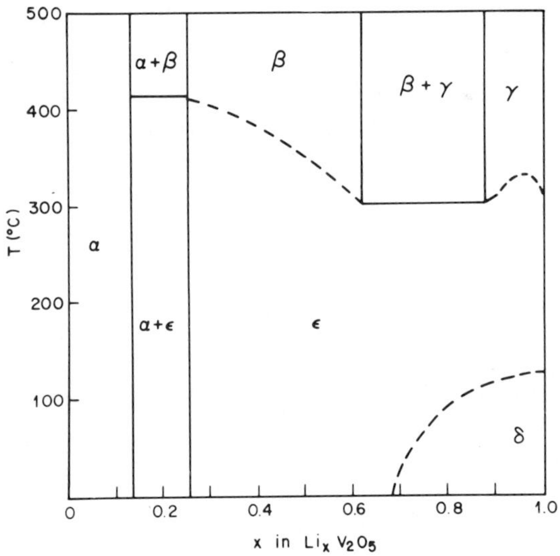

Fig. 1. Relationship between the bronze $Li_xV_2O_5$ phases (α, β, and γ) and intercalation phases (α, ϵ, and δ) (*2, 3*).

with dicyanodichloroquinone (DDQ). The bronze phases are unaffected by this reagent. Electronically, the low-temperature LiV_2O_5 appears to be similar to the bronze. Both show a broad magnetic susceptibility maximum indicative of antiferromagnetic interactions in less than three dimensions.

An important distinction can be made at this point between V_2O_5 as an intercalation host and the van der Waals-bonded hosts such as MoO_3 and MS_2, MPS_3, and MOCl compounds. Whereas V_2O_5 is layerlike, the interaction between layers is sufficient to preclude the intercalation of cations larger than Li^+ or polar organic solvents. An example of a vanadium oxide with true layer separation is $Li_{1+x}V_3O_8$. It is not possible to prepare the parent host since this would require some V^{6+}, but the bronze itself undergoes ion-exchange and solvation reactions in a manner similar to the van der Waals intercalation compounds (*4*).

Strongly reducing sources of lithium such as *n*-BuLi give extensive reaction with V_2O_5. However, products with more than one Li per V_2O_5 are poorly characterized and should not be regarded as intercalation compounds.

B. V_6O_{13}

The highest oxidation state vanadium oxide for which all the vanadium is in reasonably undistorted octahedra is V_6O_{13}, the idealized crystal lattice of which may be thought of as arising from a (2,3) double shear of ReO_3 (Fig. 2) (*5*). The three-dimensional structure consists of alternating planes of double and single chains. Channels made of tricapped cubeoctahedra extend

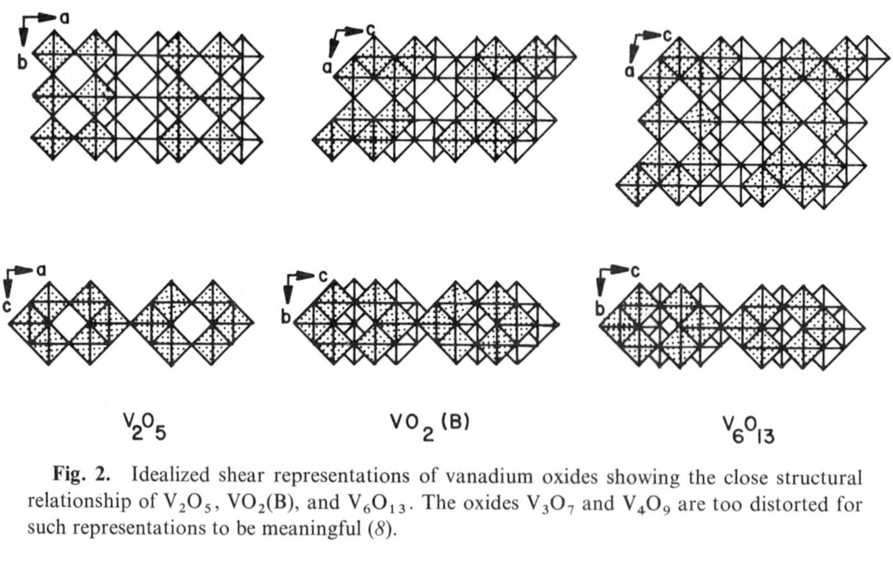

Fig. 2. Idealized shear representations of vanadium oxides showing the close structural relationship of V_2O_5, $VO_2(B)$, and V_6O_{13}. The oxides V_3O_7 and V_4O_9 are too distorted for such representations to be meaningful (*8*).

through the structure. Treatment of V_6O_{13} with *n*-BuLi affords intercalation compounds $Li_xV_6O_{13}$ (*6–8*). The extent of lithiation depends on the conditions of the reaction and the stoichiometry of the V_6O_{13}. Stoichiometric V_6O_{13} gives limiting values for x of 4 at room temperature and 6 at 50°C. The oxygen-rich composition ($V_6O_{13.2}$) gives a limiting value of 8 at room temperature. The characterization of intermediate $Li_xV_6O_{13}$ is complicated by a number of phase changes as a function of x. Hysteresis in these phase changes when $Li_xV_6O_{13}$ is prepared from compositions with opposing limits of x makes homogeneous sample preparation difficult. It is clear, however, that the lithiation should be regarded as an intercalation, since V_6O_{13} is recovered unchanged following oxidative delithiation with DDQ. This is demonstrated by the reappearance of the characteristic metal–insulator transition at 147 K in the magnetic susceptibility. This transition is not observed when only a few percent lithium is present (*9*).

Cations larger than Li do not intercalate V_6O_{13} at any appreciable rate. This is consistent with insertion into a rigid lattice.

Detailed structural studies to determine the location of Li in $Li_xV_6O_{13}$ are lacking. If all the Li were located within the tricapped cubeoctahedral cavity (this is the remnant of the vacant perovskite cavity of ReO_3), $Li_4V_6O_{13}$ would require two Li per cavity and $Li_8V_6O_{13}$ would require four Li per cavity. The presence of two or even three Li per cavity seems reasonable based on the structures of Li_2ReO_3 and $Li_2FeV_3O_8$ (*7*). Since $Li_8V_6O_{13}$ is attainable only for nonstoichiometric V_6O_{13} and the exact nature of the nonstoichiometry is unknown, it seems reasonable that additional sites might be created as a result of it. It is also possible that some of the lithiums occupy smaller octahedral or tetrahedral sites within the double chains of the structure. These sites are similar to those available in rutile VO_2 as described in the following section.

C. VO_2

Many transition metal dioxides have rutile-related structures. These structures may be viewed as being composed of MO_6 octahedra sharing two opposite edges to form linear chains. The chains are cross-linked by sharing corners with adjacent chains such that vacant channels containing one octahedral site per metal cation are also present. Several of these transition metal dioxides (MO_2) form intercalation compounds Li_xMO_2 with $z \geq 1$ (*10*). There are differing reports on the ability of rutile VO_2 to intercalate lithium. Steele (*11*) has observed reversible electrochemical reaction to the extent of ~0.1 Li per V, whereas DiSalvo *et al.* (*12*) observed some slight reaction with *n*-BuLi but concluded that bulk intercalation had not occurred since the characteristic metal–insulator transition observed by magnetic susceptibility was unaffected. These workers also studied intercalation in the alloy system $V_{1-y}Mo_yO_2$ and correlated increased rate and extent of lithiation as y increased with the increase in the volume of the unit cell. Although long-range diffusion of lithium through the channels in VO_2 appears to be difficult, it does not seem unreasonable that diffusion over only a few angstroms into similar sites present in V_6O_{13} could occur, with the bulk of the long-range diffusion occurring in the open channels.

Intercalation occurs readily in a metastable form of VO_2, designated $VO_2(B)$ (*13*). The structure may be regarded as a (2,2) shear of ReO_3 and contains one-dimensional channels formed by interconnected tetracapped cubeoctahedral cavities. It is essentially the V_6O_{13} structure without the single chains (Fig. 2). This structure contains channels of tetracapped cubeoctahedra. The limiting stoichiometry with Li is reported to be $\sim Li_{0.5}VO_2$ (*6*). The compound FeV_3O_8 also has the $VO_2(B)$ structure. Both FeV_3O_8 and $Li_2FeV_3O_8$ are more crystalline than $VO_2(B)$. The structure of $Li_2FeV_3O_8$ has been determined by neutron powder profile analysis (*14*). The lithium ions occupy two of a possible four square-pyramidal sites within the tetracapped cubeoctahedral cavity. The similar sites (three per cavity) in V_6O_{13} are the most likely positions for lithium in that structure as well. It is interesting to note that the sodium ions in $Na_{0.2}TiO_2(B)$ are located in the center of the cubeoctahedral cavity.

D. Other Vanadium Oxides

The remaining oxides with oxidation states greater than 4 are V_4O_9 and V_3O_7. Electrochemical reaction with lithium (largely irreversible) has been noted (*6*). It is not likely that these reactions can be classified as intercalation reactions. There is electrochemical evidence that lithiation of V_3O_7 leads to conversion of a substantial amount of the vanadium oxide framework to V_6O_{13} (*15*).

III. SULFIDES AND SELENIDES

A. VS_4

This compound formally contains only V^{4+}, all the sulfur being present as disulfide ions (S_2^{2-}). No well-characterized lithium intercalation compounds have been reported. There has been a report that lithiation leads to the formation of the discrete anion $(VS_4)^{3-}$ (*16*).

B. V_2S_5

This amorphous compound, which is obtained by thermal decomposition of $(NH_4)_3VS_4$, reacts with 6 equivalents of *n*-BuLi (*16*). Further studies are needed to determine whether this reaction should be regarded as an intercalation. Electrochemical cycling results in gradually lower capacities for lithium until the results resemble those of VS_2.

C. VS_2 and VSe_2

The case of VS_2 presents an interesting contrast to the situation of bronzes in oxides and to that of TiS_2 in sulfides. The synthesis of VS_2 has resulted from oxidative delithiation of $LiVS_2$ to yield a crystalline product isomorphous with TiS_2 (*18*). Ambient temperature reaction of VCl_4 with Li_2S gives amorphous VS_2 (*19*). Above 300°C VS_2 loses sulfur. Thus, whereas for the oxides, lithium intercalation must be carried out at low temperatures in order for the structure of the intercalated product to resemble the host, the reverse is true for VS_2: The host is unstable at elevated temperatures.

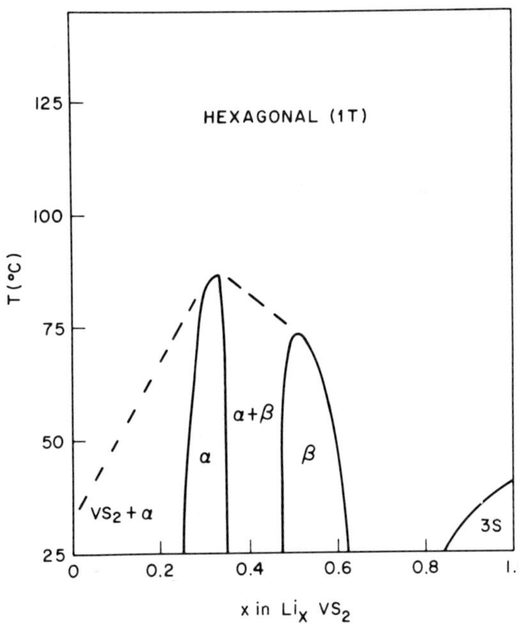

Fig. 3. The Li_xVS_2 phase diagram (*16*).

Jahn–Teller effects or possibly charge density wave instabilities (*20*) apparently prevent the Li_xVS_2 $(0 < x < 1)$ series from being a single nonstoichiometric phase as observed for Li_xTiS_2. The phase diagram for Li_xVS_2 (Fig. 3) shows two distinct distorted (monoclinic) phases for compositions near $x = 0.3$ and 0.5. Detailed structural studies on the intermediate Li_xVS_2 phases have not been reported. Above ~85°C the series becomes single phase. Even at this temperature the hexagonal c/a ratio is abnormally high (>1.85 for $Li_{0.5}VS_2$), indicative of electronic instability. Expansion of the unit cell going from VS_2 to $LiVS_2$ is nearly isotropic. In fact, the maximum interlayer spacing is observed for $Li_{0.5}VS_2$ (Fig. 4), suggesting that electronic effects on V are more important than the size of the Li in determining the lattice parameters.

Compositions $Li_xM_yV_{1-y}S_2$ (M = Fe or Cr) show reduced c/a ratios compared to the pure V compositions and less evidence of lattice distortions (*21*, *22*). Suppression of the lattice distortions has been correlated with

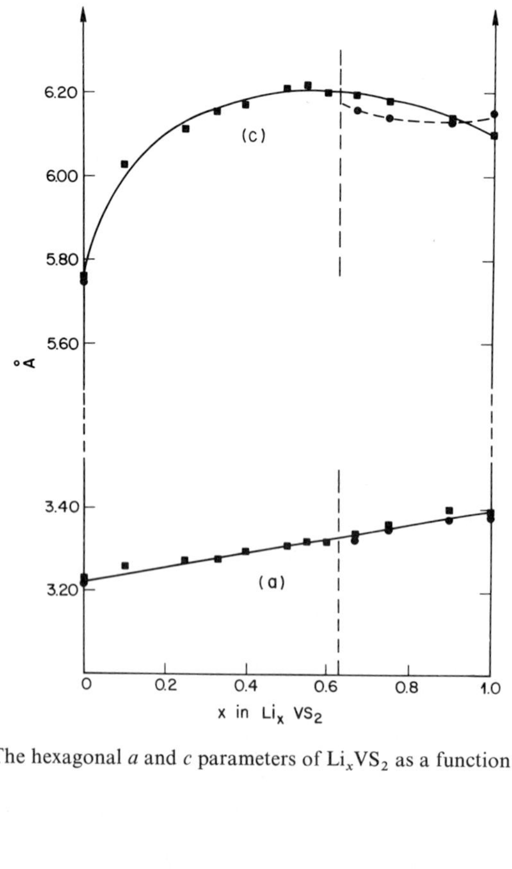

Fig. 4. The hexagonal a and c parameters of Li_xVS_2 as a function of x (*16*).

improved capacities for the electrochemical lithiation–delithiation cycle when these materials are used as cathodes in lithium batteries.

The self-diffusion hopping time from ^{7}Li NMR for $LiVS_2$ is $\gg 100$ μsec at 25°C (*23*). The distorted compositions $Li_{0.5}VS_2$ and $Li_{0.3}VS_2$ show faster hopping times, of the order of 10 μsec (*24*).

The Li_xVSe_2 ($0 < x < 1$) system adopts the V_5S_8 ($V_{0.25}VS_2$) structure for $x = 0.25$ and is two phases between $x = 0.25$ and $x = 1$ (*25*).

Several $LiMS_2$ compounds are well-known and may be prepared either at elevated temperatures by conventional solid-state techniques or by intercalation from the parent MS_2. However, VSe_2 and VS_2 also yield Li_2VSe_2 and Li_2VS_2, respectively, by intercalation at ambient temperature (*26–28*). The selenide may be prepared by the *n*-BuLi technique. The sulfide has been prepared only electrochemically, but the proper choice of lithiating reagent should make chemical preparation possible. Both compounds show a greatly expanded hexagonal axis, reflecting the increased ionic character of the V—X bonds. The magnetic susceptibilities confirm the presence of V^{2+} and show antiferromagnetic interactions. There are one octahedral site and two tetrahedral sites available for Li per VX_2 formula unit within the van der Waals gap, so that at least one Li (and probably both) in Li_2VX_2 must be tetrahedrally coordinated. In $LiVS_2$ the octahedral sites are fully occupied (*29*).

IV. SUMMARY

One of the goals mentioned in Section I, namely, the formation of intercalation compounds with vanadium oxidation state changes greater than 1, has been achieved. The VX_2(X = S, Se)–Li_2VX_2 systems go from V^{4+} to V^{2+}, albeit with a large energy difference ($\approx$20 kcal/mole) between the V^{4+}/V^{3+} and V^{3+}/V^{2+} states. The V_6O_{13}–$Li_8V_6O_{13}$ system indicates average oxidation state changes from $V^{4.33+}$ to V^{3+}.

Several areas in the study of lithium vanadium chalcogenide intercalation compounds are in need of further investigation. The full utilization of many techniques has been hindered by the lack of large single crystals. (This is also true of intercalation compounds in general.) Some structural studies from neutron powder data are now appearing, however. There are still no single-crystal structure determinations for any of the mentioned intercalation compounds. Detailed characterization of the electronic properties would also be aided by single crystals. The detailed mechanism and kinetics, both of the intercalation reaction itself and of the self-diffusion of lithium within the host, are emerging areas of interest.

ACKNOWLEDGMENTS

The author is indebted to R. J. Cava, A. J. Jacobson, A. H. Thompson, and P. G. Dickens for the preliminary communication of results included in this chapter.

REFERENCES

1. See, for example, P. Hagenmuller, *Prog. Solid State Chem.* **5,** 71 (1971); P. G. Dickens, Chap. 16 of this volume.
2. D. W. Murphy, P. A. Christian, F. J. DiSalvo, and J. V. Waszczak, *Inorg. Chem.* **18,** 2800 (1979).
3. P. G. Dickens, S. J. French, A. T. Hight, and M. F. Pye, *Mater. Res. Bull.* **14,** 1295 (1979).
4. R. Schöllhorn, F. Klein-Reesink, and R. Reinhold, *J.C.S. Chem. Commun.* p. 398 (1979).
5. For a general discussion of shear structures, see J. S. Anderson, Chap. 15 of this volume.
6. D. W. Murphy, P. A. Christian, F. J. DiSalvo, and J. N. Carides, *J. Electrochem. Soc.* **126,** 497 (1979).
7. D. W. Murphy, P. A. Christian, J. N. Carides, and F. J. DiSalvo, *in* "Fast Ion Transport in Solids" (P. Vashishta, J. N. Mundy, and G. K. Shenoy, eds.), North-Holland, New York, 1979, pp. 137–140.
8. D. W. Murphy and P. A. Christian, *Science* **205,** 651 (1979).
9. F. J. DiSalvo, D. W. Murphy, P. A. Christian, and J. V. Waszczak, *Bull. Am. Phys. Soc.* **24,** 398 (1979).
10. D. W. Murphy, F. J. DiSalvo, J. N. Carides, and J. V. Waszczak, *Mater. Res. Bull.* **13,** 1395 (1978).
11. B. C. H. Steele, *J. Phys., Colloq.* (*Orsay, Fr.*) **C2,** 198 (1978).
12. F. J. DiSalvo, D. W. Murphy, and J. V. Waszczak, *Synth. Met.* **1,** 29 (1979).
13. F. Theobald, R. Cabala, and J. Bernard, *J. Solid State Chem.* **17,** 431 (1976).
14. R. J. Cava, A. J. Santoro, D. W. Murphy, S. Zahurak, and R. S. Roth, *Solid State Ionics* (in press).
15. D. W. Murphy, P. A. Christian, and J. N. Carides, *J. Electrochem. Soc.* (in press).
16. D. W. Murphy, F. A. Trumbore, and J. N. Carides, *J. Electrochem. Soc.* **124,** 325 (1977).
17. A. J. Jacobson and S. M. Rich, *J. Electrochem. Soc.* **127,** 779 (1980).
18. D. W. Murphy, C. Cros, F. J. DiSalvo, and J. V. Waszczak, *Inorg. Chem.* **16,** 3027 (1977).
19. R. R. Chianelli and M. B. Dines, *Inorg. Chem.* **17,** 2758 (1978).
20. J. A. Wilson, F. J. DiSalvo, and S. Mahajan, *Adv. Phys.* **24,** 117 (1975).
21. F. J. DiSalvo, M. Eibschutz, C. Cros, D. W. Murphy, and J. V. Waszczak, *Phys. Rev. B* **19,** 3441 (1979).
22. D. W. Murphy, F. J. DiSalvo, and J. N. Carides, *J. Solid State Chem.* **29,** 339 (1979).
23. B. G. Silbernagel and M. S. Whittingham, *Mater. Res. Bull.* **12,** 853 (1977).
24. W. W. Warren, Jr., D. W. Murphy, and F. J. DiSalvo, *Bull. Am. Phys. Soc.* **24,** 435 (1979).
25. A. H. Thompson, *Solid State Ionics* **1,** 47 (1980).
26. D. W. Murphy, F. J. DiSalvo, G. W. Hull, Jr., and J. V. Waszczak, *Inorg. Chem.* **15,** 17 (1976).
27. M. S. Whittingham, *Mater. Res. Bull.* **13,** 959 (1978).
28. D. W. Murphy and J. N. Carides, *J. Electrochem. Soc.* **126,** 349 (1979).
29. B. Van Laar and D. J. W. Ijdo, *J. Solid State Chem.* **3,** 590 (1971).

18

Physical Properties of Intercalation Compounds of Transition Metal Dichalcogenides

A. H. THOMPSON

and

F. J. Di SALVO

I. INTRODUCTION

The layered dichalcogenides that are formed from Group IVb, Vb, and VIb metals have received considerable attention recently because they have highly anisotropic physical properties, have electronically driven crystal distortions, and form numerous unusual intercalation compounds. The intercalation compounds are formed when the crystalline layers are separated and a molecular or atomic species is inserted between layers of the host. A number of reviews of intercalation compounds already exist [e.g., (*1–3*)], and we will attempt to build on these and specifically deal with the physical properties of the intercalation compounds. We shall first discuss the properties of the intercalates of metal atom guests in chalcogenide hosts and then review the work on molecular intercalation complexes.

INTERCALATION CHEMISTRY

ISBN 0-12-747380-7

II. METAL INTERCALATES IN CHALCOGENIDE HOSTS

A. Introduction

Metal intercalation compounds of the Group IVb, Vb, or VIb transition metal dichalcogenides have been prepared with a majority of the metals in the periodic table. The chemistry of these intercalates has already been reviewed (*3*).

The Group Vb compounds 2H $NbSe_2$ and 2H TaS_2 have been most widely studied as the host species, and the largest variety of metals have been reported to intercalate in these compounds. The Group VIb intercalation compounds are formed only with the most electropositive metals, since the electron affinity of the Group VIb hosts is lower than that of the Group Vb hosts by about 1.5–2.0 eV. The Group VIb compounds may be unstable with respect to disproportionation, as is $LiMoS_2 \rightarrow MS_{1.5} + \frac{1}{2}Li_2S$ (*4*), unless some organic solvent is also intercalated, such as NH_3, which coordinates strongly to and reduces the activity of the Li. Most intercalation studies have focused on the determination of the structure of the resulting compound versus intercalant concentration. However, in some cases other factors have motivated the studies. (a) In the case of Li and Na, and to a lesser extent Cu and Ag intercalants, the possible utility of such compounds in high-energy-density ambient temperature batteries has resulted in an avid interest in their properties (*3–7*). (b) A number of the intercalation compounds are superconductors, but the maximum T_c values obtained are only about 5–7 K (*1*, *8*). (c) Ferromagnetic (*2*, *9*, *10*), antiferromagnetic (*2*, *10*), and spin glass behavior (*11*) are observed when 3d transition metals are intercalated. (d) The effects of intercalation on the charge density waves that occur in the Group Vb layered compounds (*12–15*) have been the object of some studies (*11*, *13*, *16*). Such instabilities may occur in some host materials only after intercalation.

B. Composition and Structure

Which intercalation compounds of a particular host can be prepared depends on the preparation conditions. For example, the maximum concentration of intercalant, x_{max}, in A_xMX_2 depends on the preparation method and temperature. As an example, consider Li_xTaS_2. When this compound is prepared at 700°C in sealed quartz tubes, the value of x_{max} is 2/3 (*17*), but, when prepared by reaction with a solution of *n*-butyllithium at room temperature, x_{max} is 1.0 (*4*, *5*). A second example is the system Li_xVS_2. When it is prepared by high-temperature techniques, x_{max} is 1.0 (*18*), but, in an electrochemical cell, x_{max} is 2.0 if the potential versus lithium is allowed to decrease to 0.9 V (*19*). However, in the few cases in which a second lithium has been observed to be included, the compounds are metastable; decomposition to lithium chalcogenide occurs at elevated temperatures or after a sufficiently long time at room temperature. In some cases preliminary work indicates that intercalation takes place, but in our opinion the low values of x_{max} observed may be due to the preparation method. In Fig. 1 we show the periodic table with x_{max} listed under each element. We chose the preparation conditions and host that produce the largest x and have cited the publication that reports the largest value of x. If further studies might be expected to alter this value, the x_{max} value is underlined.

The structures of some intercalation compounds deserve further comment. In particular, we consider Sn_xTaS_2, H_xTaS_2, 3d metal, and rare earth metal intercalates. Homogeneous compounds in the Sn_xTaS_2 system are produced for $0 \leq x \leq 1/3$ and at $x = 1/3$ (*8*). In the lower stoichiometry range the structure is "normal"; i.e., the Sn is divalent and in octahedral holes. However, the structure of $SnTaS_2$ is quite unusual (*20*). The Sn is intercalated between trigonal prismatic TaS_2 layers, but the Sn is linearly coordinated by two sulfur atoms above and below the Sn layer and by six near-neighbor Sn atoms in the Sn layer. The structure is consistent with the quadrupolar-split Mössbauer ^{119}Sn spectrum (*8*, *21*) and the single Sn NMR resonance (*22*). This unusual structure and Sn coordination are probably due to an unusual valence configuration for Sn; i.e., it is probably monovalent. (This would be the only monovalent Sn compound of which we are aware, in which obvious Sn clusters are *not* observed.) This possibility seems likely since only one electron is necessary to fill the lower Ta d band, and the upper bands are separated by a gap of about 1.3 eV from this band. The remaining three Sn electrons broaden into a two-dimensional band, as suggested by the NMR Knight shift and its anisotropy (*22*). Furthermore, $InTaS_2$ adopts a similar structure, with almost identical lattice parameters, and In is most certainly monovalent in this case. The unusual state of the Sn would then be a consequence of the geometry imposed by the layers and by their "electron affinity gap" at one electron per Ta.

The H_xTaS_2 compound is unusual. Individual layers of TaS_2 can easily be dispersed in water solutions when $x = 0.25$ and the pH is adusted to near 4 (*23*). Reactions between these individual layers and large organic dyes have been observed (*24*). Furthermore, recent neutron scattering results show that the protons do *not* reside between sulfur layers as for other intercalants, but the protons are in the plane of the Ta atoms (*25*). This bonding is reminiscent of the layered compounds HZrCl, in which the proton resides between the double plane of Zr atoms (*26*), forming a more traditional transition metal hydride compound.

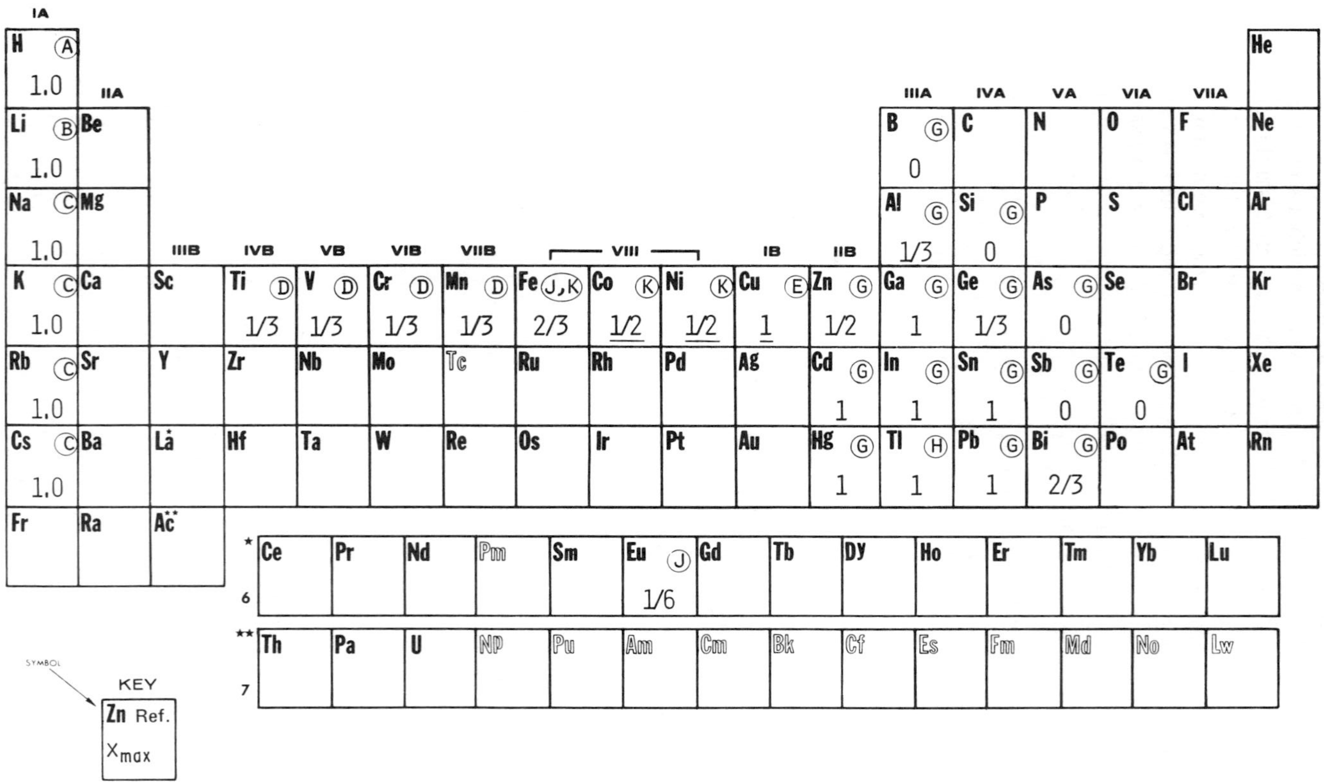

Fig. 1. Periodic table showing metals known to form intercalation products. Each element is labeled with the maximum known composition in a layered host and with the appropriate reference (see circled letter). (A) Murphy *et al.* (*23*). (B) Murphy *et al.* (*5*); Dines (*76*). (C) Leblanc-Soreau *et al.* (*42*). (D) Vandenberg-Voorhoeve (*77*). (E) LeNagard *et al.* (*78*). (F) Engelsman *et al.* (*79*). (G) DiSalvo *et al.* (*8*). (H) Sabrowsky and Platte (*80*). (I) Jellinek (*81*). (J) Boswell *et al.* (*52*). (K) Arnaud *et al.* (*82*).

Frequently, 3d intercalation compounds with higher stoichiometry than shown in Fig. 1 ($x = 1/3$) can be prepared, but these have the defect NiAs structure (*2*) and thus are not included as intercalation compounds (even though NiAs is a closely related structure). Also, as noted previously, lithium concentrations of $x = 2.0$ can be obtained in several layered compounds (VS_2, VSe_2, and TiS_2) in electrochemical cells, but we do not include that stoichiometry in Fig. 1 because these are newly discovered compounds whose properties are just beginning to be explored and they are probably metastable. Finally, we note that $x_{max} = 1/6$ for Eu is probably low and that preparation by other methods, perhaps in electrochemical cells, may lead to larger values. Other rare earth metals ought to intercalate in some layered compounds, particularly those that are frequently found in the divalent state (Sm, Tm, Yb). Rare earth metals that are usually found only in the trivalent state may also intercalate, perhaps in an unusual coordination and valence, as suggested for $SnTaS_2$.

While the usual approach to intercalation is to prepare the layered host and subsequently prepare the intercalation compound, deintercalation of compounds with a layered structure produces the layered host material, even in cases in which synthesis of the host from the elements has proved elusive. For example, the compound $LiVS_2$ has been known for some time (*18*) and has the same structure as $LiTiS_2$. Attempts to synthesize crystalline VS_2 from the elements have so far failed. However, Li can easily be extracted from $LiVS_2$ at room temperature to produce VS_2 (*27*). Attempts to produce CrS_2 from $LiCrS_2$ by a similar method have met with difficulties, but alloys of the type $Cr_xV_{1-x}S_2$ have been prepared for $x \leq 0.75$ (*27*).

C. Phase Transitions

Phase transitions are observed in all the polymorphs of the Group Vb disulfides and diselenides (with the exception of 2H and 3R NbS_2). These transitions are believed to be due to a charge density wave (CDW) instability (*12–15*, *28*, *29*). In most cases an incommensurate lattice distortion is stable in some temperature interval, below which a commensurate distortion (superlattice) is observed. While the upper state is believed to be due to an electron–phonon-driven Fermi surface instability (*12*), the transition to the commensurate state has been empirically related to the crystallographic c/a ratio and the electronegativity difference of the elements involved (*29*). The lattice instability is associated with a marginally stable metal–metal and metal–chalcogen bond geometry. It is interesting that no evidence for CDW instabilities in these materials exists in their intercalation compounds with metal guests when x is near x_{max}. Indeed, studies of 2H H_xTaS_2 show that the anomaly in the magnetic susceptibility due to the CDW is lost for

$x > 0.11$ (*23*). In 2H TaS_2 and 2H $TaSe_2$ the suppression of the CDW susceptibility anomaly produces an increase in the superconducting T_c (*16*), except when the intercalant is a 3d metal with a magnetic moment. An interesting example is the difference between 2H Fe_xTaS_2 and 2H Fe_xTaSe_2. In the sulfide, Fe has no magnetic moment *at low concentrations* ($x < 0.1$) and T_c increases when the CDW anomaly is suppressed (*11*). However, in the selenide, Fe has a magnetic moment at low x and the superconductivity is suppressed by the Fe moment (*11*). As a final example of the interplay between intercalation and structural instabilities, consider Li_xTiSe_2. At 200 K, $TiSe_2$ has a second-order structural transistion that results in a superlattice with doubled a and c axes below the transition (*29*, *30*). $LiTiSe_2$ shows a weak anomaly in the magnetic susceptibility below 40 K, which suggests a phase transition of the CDW type (*5*). (Note, however, that no diffraction results have been obtained to support this speculation.) The only other metal guest intercalation compound that suggests a similar CDW instability is the metallic polymorph of $NaVS_2$ (*31*). There are, of course, phase transitions in intercalated layered compounds that are magnetic in nature, the moment residing on the intercalant (*2*, *10*) or in some cases on the transition metal of the host compounds (*18*, *31*).

D. Electrochemistry and Intercalation

The intercalation of layered compounds is one example of a topochemical reaction (*32*), i.e., a reaction in which the overall morphology of the starting material is left intact (with some possible volume change). As we have seen, the number of species that may intercalate in layered compounds is large. With nonlayered hosts for topochemical reactions, the number of reacting species is frequently much smaller. However, these are also of interest, since they may be of use as battery cathodes. These nonlayered hosts include other transition metal chalcogenides and a number of transition metal oxides. The transition metal trichalcogenides MX_3 react with three lithium atoms per M, but the reaction is not always electrochemically reversible (*3*, *6*). The Cheverel phases based on Mo_3S_4 and Mo_3Se_4 (*33*) are also topochemically reactive. In this section, we first discuss the properties of the oxides and those of the layered dichalcogenides.

Transition metal oxides with a number of different structures react with lithium reversibly and topochemically. The reaction with V_2O_5 and MoO_3 has been known for a number of years (although the topotactic nature of the reaction was only recently recognized). It has been shown that a number of the rutile related metal dioxides (MO_2) react topochemically to give $LiMO_2$ (*34*). Within the rutile family an empirical correlation between rapid reactivity and the unit-cell volume as well as the electrical conductivity has been observed (*35*). Other studies have shown that the oxides with stoichiometries between VO_2 and V_2O_5 (*36*), CrO_2 and CrO_3 (*37*), MoO_2 and MoO_3 (*38*), and WO_2 and WO_3 (*34, 39*) also react reversibly with lithium. It seems clear that most oxides with an anion-to-metal ratio greater than 2 will react with lithium [ReO_3 (*35*), Nb_2O_5 (*40*), etc.]. A particularly interesting example is afforded by V_6O_{13}. Reversible electrochemical behavior versus lithium is observed for up to one lithium per transition metal (*41*). Reaction with *n*-butyllithium suggests that a final stoichiometry of $Li_8V_6O_{13}$ may be attainable electrochemically if the particle size is small enough. For all the above-mentioned oxides, the reaction of the oxide host with lithium at room temperature results in a compound with a metastable structure. Upon heating, irreversible structure changes occur, and frequently the resulting compound(s) are no longer electrochemically reversible with respect to lithium. The decomposition temperature as determined by differential scanning calorimetry (at 10°–20°C/min heating rate) ranges from 70° to 400°C. The former temperatures are too low for the materials to be considered of interest as battery cathode materials in "room temperature" batteries. Of all these oxides, V_6O_{13} presently appears to be the most promising for use as a battery cathode. Its average potential versus lithium is 2.2 V and has a theoretical energy density of 800 Wh/kg based on eight Li per V_6O_{13}.

E. Structural Effects in Metal Intercalates

Structural ordering of the intercalate is a common feature of metal intercalates such as Fe_xTaS_2. The intercalated metal atom tends to form a superlattice apparently associated with the interaction between intercalated atoms. Four classes of structures are distinguished here. (a) The typical transition metal or post-transition metal intercalate forms static ordered structures at room temperature with immobile, electrochemically inactive intercalated metal atoms (for example, Fe_xTaS_2). (b) The alkali metals often distort the host lattice and/or change the layer stacking sequence, resulting in well-defined two-phase regions bounded by first-order transitions to structures of different composition (for example, Na_xTiS_2) (*42*). (c) Some intercalates of metallic hosts appear to interact with layer distortions (for example, CDWs) to form subtle superlattices that may be bounded by either first- or second-order transitions to the undistorted state (for example, 1T Li_xTaS_2) (*43*) or 1T Li_xVS_2 (*27*). (d) Alkali metal intercalates may form a single-phase product at room temperature for $x < x_{max}$, but structure in the electrochemical potential versus x suggests ordering effects that may be only weakly formed, incipient features (for example, Li_xTiS_2) (*43–46*). Although in many cases experimental data are not available, one expects that both ordering effects and CDW distortions will disappear above a first- or second-order transition temperature. Li_xTiS_2 appears to be unique in that such transition temperatures have not been detected, even below room temperature (class d), suggesting that the Li–Li and Li–layer distortion interactions are the weakest of the layer compound–intercalate systems. In Li_xVS_2 these interactions appear stronger and distortions of the VS_2 layers as well as two-phase regions are observed at room temperature. However, at 80°C the Li_xVS_2 system is completely homogeneous, with a structure similar to Li_xTiS_2 (*27*). In the following paragraphs we consider in more detail the behavior of the alkali intercalates.

Since the electrochemistry and structural properties of the dichalcogenides have been extensively reviewed, we concentrate here on recent advances and physical properties. At present, the alkali metal intercalates of layered compounds have been surveyed, and at least the obvious choices for battery cathodes have been studied. Some recent work on these materials is aimed at a more detailed understanding of the intercalation process, including both the kinetics and thermodynamics (*46*). The motivation for these studies is that future electrode developments may depend on our ability to predict electrochemical performance from first-principle arguments or systematic trends. We first describe the structural ordering of intercalated alkali metals and then discuss the modification of the host lattice by an intercalation.

It has been known for a number of years that when the larger alkali metals such as Na and K are intercalated into layered compounds such as TiS_2, there are structural changes in the stacking of the host layers near room temperature. For example, when TiS_2 is intercalated with Na, the Na occupies an octahedral site at low concentrations, moves to a trigonal prismatic site at intermediate compositions, and then at the highest concentration moves into yet another, not fully understood structure (*42*, *47*). These structural changes require a substantial expansion of the crystallographic *c* axis and the shifting of adjacent layers, i.e., the nucleation of a new phase. Electrochemically, substantial changes in cell potential are observed at these phase limits, and the phase transitions apparently inhibit reversible cycling to full alkali metal capacity (*3*). In contrast to the larger alkali metals, Li intercalates in TiS_2 with only small changes in lattice parameter. The interlayer sites (both octahedral and tetrahedral) are large enough to accommodate the Li^+ ion, and there is a continuous range of compositions possible between TiS_2 and Li_1TiS_2. But there are at least two types of structural modifications that can occur in the Li intercalates in addition to the shifting of layers. These are ordering effects on the Li sublattice and distortion of the host layer. Neither of these structural changes necessarily involves nucleation of a new phase at a first-order phase transition.

Recent electrochemical studies on Li_xTiS_2 have supplied evidence suggestive of ordering on the Li sublattice (*43*, *46*). These data are reproduced in

Figs. 2 and 3, in which the voltage–composition relation and the incremental capacity (dx/dV) are presented. These are quasi-equilibrium data obtained by new techniques described elsewhere (*45*). Figures 2 and 3 show that there are subtle features on the voltage-composition relation and that these features appear as sharp peaks in the derivative, dx/dV, near $x = 1/4$ and $1/9$. X-ray examination of Li_xTiS_2 materials at room temperature does not provide any evidence for structural ordering. Direct X-ray observation of the Li ordering would not be expected to be easy because of the small X-ray scattering cross section of Li compared to TiS_2, but the absence of confirming X-ray data does suggest that a large distortion of the TiS_2 layers [e.g., of the type seen in Li_xVS_2 at room temperature (*27*)] is not responsible for the electrochemical effects. Nor have neutron and electron diffraction

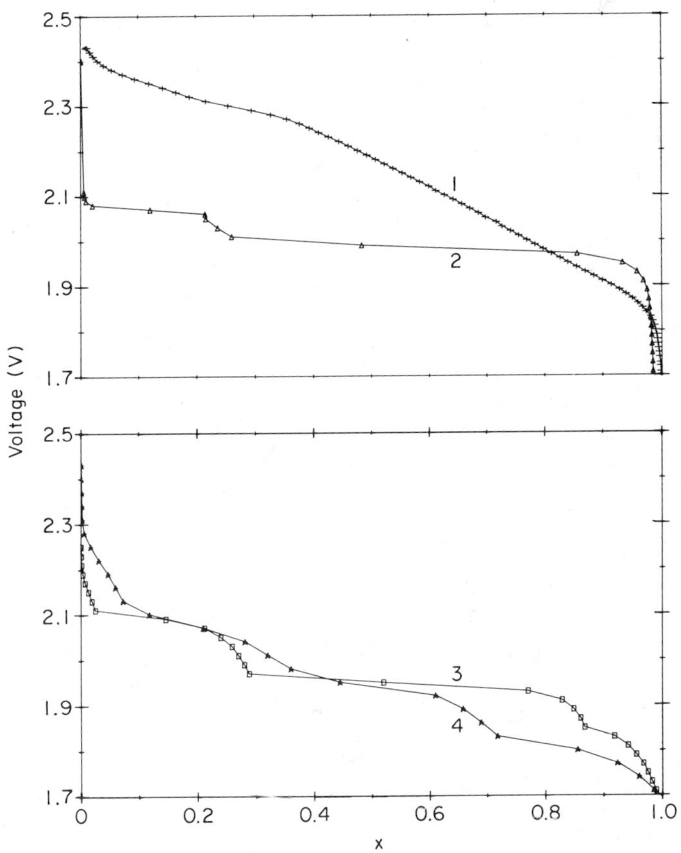

Fig. 2. Voltage–composition relations for intercalation into layered compounds. Curve 1, Li_xTiS_2 (*44*); curve 2, Li_xVSe_2 (*50*); curve 3 $Li_x(1T\ TaS_2)$; curve 4, $Li_xTa_{0.9}Ti_{0.1}S_2$ (*43*).

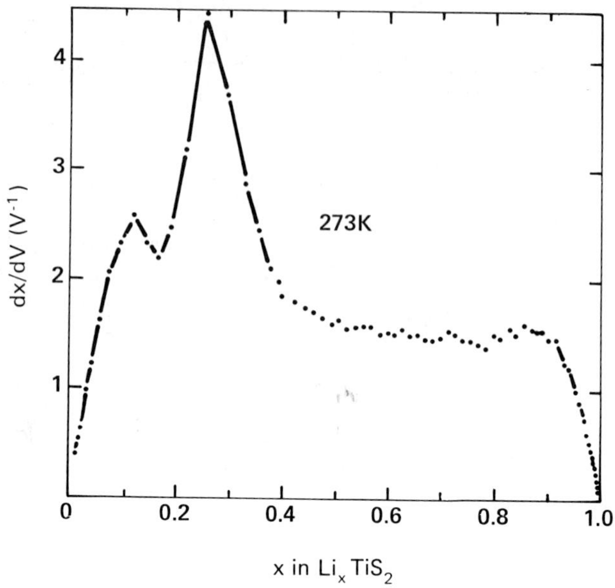

Fig. 3. The negative, reciprocal derivation of the voltage–composition relation of Li_xTiS_2. The peaks correspond to plateaus on Fig. 2, curve 1.

studies yielded supporting structural confirmation. In the absence of structural data and rigorous theoretical analysis, the phase diagram remains uncertain, but the existence of Li ordering seems certain at low enough temperatures. Some discussion concerning the possible lithium ordering arrangement is appropriate, however, since some publications and ongoing experiments have addressed or are addressing this point.

The compound Li_xTiS_2 can be modeled by a system in which the intercalated Li ions are constrained to reside on a triangular lattice supplied by the TiS_2 with interaction between the ions due to coulomb repulsion and strain. The Li^+–Li^+ repulsion would be expected to be screened by the host metallic layers, but in a two-dimension limit (i.e., electrical conductivity perpendicular to layers is zero) this screening will be a multipolar repulsion, so that long-range repulsive forces may be possible. (Note that here a multipolar repulsion is considered to be "long range" compared to an exponentially damped, screened repulsion.) It is also expected that interlayer interactions could be important to ordering effects [especially those due to strain (*48*)]. The fact that the host is undistorted at room temperature implies that there is only a weak intercalant–host interaction with respect to ordering and that the classical lattice gas may be an adequate description of the two-dimensional lattice. A recent calculation of a simple lattice gas model including a few near-neighbor interactions (assumed independent of stoichiometry) does not reproduce the structure in the dx/dV curve for Li_xTiS_2 (*49*), indicating that the observed effects are not well represented by this model.

When the host–guest interaction is very strong or when the host lattice is metastable, CDWs, then the host lattice may distort on intercalation, and these distortions may be associated with the ordering of the intercalates. Effects of these types have been seen in the lithium intercalates of VSe_2 and 1T TaS_2 (*45, 50*). The voltage–composition relations for Li_xVSe_2, Li_xTaS_2, and $Li_xTa_{0.9}Ti_{0.1}S_2$ are shown in Fig. 2. In the case of Li_xVSe_2, there is a step on the voltage–composition curve at $x = 1/4$. This feature is associated with a distortion of the host VSe_2 layers to form a structure analogous to V_5S_8 (*50, 51*). In the V_5S_8 structure the vanadium atoms in partially filled layers form an ordered array with the a axis superlattice of $2a_0$. This superlattice is essentially the same as that described for $Li_{1/4}TiS_2$ (*45*) except that in the vanadium case the two-dimensional ordering is slightly distorted so that the hexagonal c axis is inclined to the hexagonal plane. In comparison to the V_5S_8 structure, the $Li_{1/4}VSe_2$ structure involves a distortion of the host layers and suggests a distorted $2a_0$ superlattice on the lithium lattice. The case of Li/1T TaS_2 is similar to that of Li/VSe_2. Figure 2 shows that there are phase limits in Li_xTaS_2 at $x = 1/4$ and $6/7$. At $x = 1/4$, the structure is again the V_5S_8 structure. In $Li_xTa_{0.9}Ti_{0.01}S_2$, there are three phase boundaries, at $x = 1/13$, $1/3$, and $2/3$. The $x = 1/3$ structure is the V_2S_3 structure.

The $Li_xTa_yTi_{1-y}S_2$ series demonstrates that the electronic properties of the host play an important, sensitive role in determining the distortions on intercalation. Changing the composition from $Ta_{0.9}Ti_{0.1}S_2$ to TaS_2 causes the ordered compositions to change from $x = 1/3$, $2/3$ to $x = 1/4$, $6/7$. These dramatic effects deserve further study as a means of describing lattice distortions associated with electronic effects.

The materials VSe_2, 1T TaS_2, and $Ta_{0.9}Ti_{0.1}S_2$ are treated together because they share the same CdI_2 structure and because they each exhibit CDW distortions in the host below 110, 600, and ~600 K, respectively. We surmise that the presence of the CDW's means that the hosts are susceptible to distortions and that intercalation with lithium readily causes distortion or changes in the distortion of the host. These cases are intermediate between Li_xTiS_2, in which there is a single-phase product, and Na_xTiS_2, in which the intercalation is limited by two-phase boundaries between which the discharge curves are flat.

The ordering of Li in various intercalation complexes provides a useful model system because of the ease with which the compounds can be made and studied by electrochemical means. However, similar ordering effects have been observed in other intercalate systems such as Fe_xNbS_2 (*52*), $Ti_{1+x}S_2$ (*53*), Cu_xNbS_2 (*54, 55*), Mn_xTaS_2 (*56*), and Ni_xTaS_2 (*56*).

In Fe_xNbS_2, electron diffraction studies (*52*, *56*) have found a $2a_0$ superlattice at $x = 1/4$, a $\sqrt{3}a_0$ structure at $x = 1/3$, and a $\sqrt{3}a_0$ structure at $x = 2/3$. These structures are just combinations of the expected close-packed arrangements at each composition with the near neighbors all equidistant (*57*). Similar close-packed structures are observed in $Mn_{1/4}NbS_2$ (*56*), $Ni_{1/3}NbS_2$ (*56*), $Fe_{1/3}TaS_2$ (*56*), and $Fe_{1/4}NbSe_2$ (*58*). In each of these intercalation complexes, the intercalated atoms reside in octahedral interlayer sites. The stacking of the intercalated atoms along the *c* axis is in chains either with intercalated atoms lying directly over each other or with the atoms displaced (*52*). It might be expected that the coulomb repulsion and strain forces between intercalated ions would favor a stacking in which each layer is shifted relative to the next. The observation of chain formation along the *c* axis (*54*) of $Fe_{1/4}NbS_2$ is surprising, and more detailed studies of the $FeNbS_2$ complex are needed. Boswell *et al.* (*54*) also cite electron diffraction evidence for the presence of a $x = 1/4$ structure with a $2\sqrt{3}a_0$ superlattice. This structure was not predicted by two-dimensional lattice gas models (*57*) and so may be stabilized by three-dimensional interlayer effects.

An example of a nonstoichiometric compound having ordering effects in partially filled layers is $Ti_{1+x}S_2$ (*46*, *59*). In these cases, Moret *et al.* (*53*) studied the short-range order in materials prepared at high temperature (900°C). They found that only short-range order in the partially filled Ti layers was detectable by electron diffraction. Earlier studies (*59*) on $Ti_{1+x}S_2$ near TiS_2 showed that preparation at such high temperatures led to the formation of disordered, nonstoichiometric compounds having not only interlayer titanium atoms but also vacancies in the otherwise filled Ti layers. It is perhaps not surprising, then, that only short-range order is found in materials prepared at 900°C. High-temperature treatments of other 3d metal intercalates (*54*) such as $Fe_{1/4}NbS_2$ also show transitions to disordered states (600 K for $Fe_{1/4}NbS_2$), whereas ordered intercalates $Fe_{1/3}TaS_2$ and $Bi_{1/3}TaS_2$ remain ordered below 500°C (*8*).

III. ORGANIC INTERCALATES AND SUPERCONDUCTIVITY

The discovery of superconductivity in the molecular intercalation complexes of metallic layered dichalcogenides (*60–62*) offered the first opportunity to study superconductivity in a crystalline material that is in large part organic. The two-dimensional structure makes it possible to study the interaction between the delocalized electrons of the host and the molecular electrons of the intercalate. Superconductivity in this structure also raised questions regarding the nature of a superconducting state in two dimensions (*63*, *64*).

The early studies of organic intercalation showed that only Lewis base (electron-donating) organic compounds could be intercalated and surveyed the possible candidates to ascertain whether the host–intercalate interaction could lead to a higher superconducting transition temperature. Various authors have proposed that molecular conduction band electron interactions could lead to an excitonic mechanism for superconductivity (*65–67*). The extensive intercalation work on the tantalum and niobium chalcogenides did not result in T_c values above that of unintercalated $NbSe_2$ (7.2 K). The failure to observe an enhancement of T_c (assuming that such an excitonic mechanism exists) may be intrinsic to the intercalation compounds, or it may be that an ideal model system has not yet been synthesized. Little has argued that the position of the intercalated molecule is too far removed from the conduction band electrons to be effectively coupled (*68*). However, it would not appear that an ideal model system for the excitonic mechanism has been made. In such a model compound the intercalated molecule would have low-lying electron excitations that would couple effectively to the conduction electrons. Ideally, a dye molecule is needed, but dye molecules have been difficult to intercalate. One approach to the problem was to disperse TaS_2 in aqueous solution and then to adsorb dye molecules on the monolayers of TaS_2 (*24*). The complex was not a superconductor. A more recent approach (*69*) involved the intercalation of a derivative of a phthalocyanine dye having added amine groups to enhance the likelihood of intercalation. The dye, copper tetra-4-dimethylaminophthalocyanine (TAPcCu), has a first excited state energy of 1.7 eV, significantly lower than that of previous organic intercalates. An intercalation complex of composition $TaS_2(TAPcCu)_{1/2}$ was produced with a superconducting transition temperature of 3.1 K. Although Preobrazhensky *et al.* (*69*) suggest that this T_c is above that expected for an amine intercalate of comparable composition, it does not appear to us that a convincing case for an excitonic interaction can be made.

In summary, the intercalation studies have been discouraging as far as enhanced superconducting transition temperatures are concerned, but the ideal model system is not yet in hand. Such a compound may be difficult to prepare, particularly in light of recent studies that suggest that an ionic model for amine intercalation would be appropriate [see below and Schöllhorn and Zayefka (*70*)].

Despite the lack of success in producing technologically interesting superconductors, the intercalated superconductors have provided a considerable body of new data of interest in studying the superconducting transition in two dimensions or in anisotropic systems. Is the superconductivity of an intercalated layer compound confined to the two dimensional metallic sheets, or does it require three-dimensional coupling between sheets for its existence? Two intercalation experiments have provided data suggesting that interlayer coupling by Josephson coupling (tunneling of superconducting pairs) between layers is not necessary. In the first experiment (*61*), TaS_2 was intercalated with long-chain amine molecules. It was found that the superconducting transition temperature depends only on the stoichiometry of the intercalation complex and not on the interlayer spacing with the interlayer spacing as large as 60 Å. Such a large interlayer spacing would be expected to produce exponentially weak tunneling of superconducting pairs between layers, and T_c would be expected to drop.

In the second experiment (*71*), the isostructural organometallic molecules cobaltocene and chromocene were intercalated in TaS_2, as first reported by Dines (*72*). It was shown (*71*) that these organometallic molecules produce identical lattice expansions and are fully ionized when intercalated, resulting in a diamagnetic complex for cobaltocene and a paramagnetic complex for chromocene. Despite the presence of a local magnetic moment on the chromocene the two complexes were found to have nearly identical superconducting properties. The presence of a localized magnetic moment is known to have a strong pair-breaking influence on the superconducting pairs, and particularly for the tunneling of superconducting pairs, so that the Josephson coupling between the layers of the chromocene complex should be blocked. The similar superconducting properties of the two complexes then suggests that the interlayer Josephson coupling is not an essential ingredient. These studies of intercalated superconductors have not unambiguously answered the question regarding the dimensionality of the superconducting state in layered compounds, particularly regarding other coupling mechanisms such as phonon coupling, but further work on the magnetic intercalation complexes is warranted in order to explore the magnetic superconductivity interactions and possible coexistence of a magnetically ordered intercalate layer in the presence of adjacent superconducting layers.

Another approach to the dimensionality of the layered superconductors has been to study the temperature dependence of the fluctuation-induced diamagnetism in, for example, $(\text{pyridine})_{1/2}TaS_2$ (*73*). These studies and theoretical work (*64*) indicate that the detailed superconducting properties *near* T_c are those characteristic of a three-dimensional, anisotropic superconductor but that at low enough temperatures a crossover from 3d to 2d behavior can occur if the superconducting vortex cores become smaller than the interlayer separation.

The nature of the bond between an intercalated molecule and the host layer has recently been the subject of several studies. Two bonding models are currently under discussion. In the older model (*61*, *62*) the intercalated molecule is bound to the dichalcogenide layer by a weak covalent bond involving little charge transfer, and intermolecular bonding probably contributes to the stabilization of the complex. In more recent studies (*70*) it has been suggested that the molecule–host bond is substantially ionic. In this model part of the intercalated molecules are protonated and ionized with donation of an electron to the layer, whereas the remainder of the molecules are neutral and solvate the ionized species. For example, $(NH_3)TaS_2$ is approximately $(NH_4{}^+)_{0.1}(NH_3)_{0.9}TaS_2$. In this model it is necessary to provide a source for the protons, which may come from molecular decomposition and/or impurities such as water (*74*). It remains to be seen whether the ionic model is appropriate for all situations or whether both types of bond will be observed. If the ionic model is appropriate for previously measured superconducting systems, then some reinterpretation of the data may be required. In particular, the observation by Bray and Sauer (*75*) that the superconducting T_c scales with electron lone pair density seems difficult to reconcile on an ionic model.

NOTE ADDED IN PROOF

Recent experimental and theoretical work has improved our understanding of ordering in Li_xTiS_2. Hibma (*83*) has observed both $2a_0$ and $3a_0$ superlattices in diffuse X-ray analyses. Osorio and Falicov (*84*) have performed cluster-variational calculations on the triangular lattice gas. Their results reproduce the major features of the electrochemical data assuming only a $3a_0$ superlattice corresponding to the $x = 1/3$ composition. Their calculations show that the ordered compositions should be at minima in dx/dV. Direct measurements of the entropy of intercalation (*85*) are in qualitative agreement with the cluster calculations but also provide additional evidence for ordered structures at $x = 1/12$, 1/7, 1/4, and 1/3.

REFERENCES

1. F. R. Gamble and T. H. Geballe, *in* "Treatise on Solid State Chemistry" (N. B. Hannay, ed.), Vol. 3, p. 89. Plenum, New York, 1976.
2. J. M. Vandenberg, *in* "Optical and Electrical Properties" (P. A. Lee, ed.), pp. 423–457. Reidel Publ., Dordrecht, Netherlands, 1976.
3. M. S. Whittingham, *Prog. Solid State Chem.* **12,** 1 (1978).
4. M. S. Whittingham and F. R. Gamble, *Mater. Res. Bull.* **10,** 363 (1975).
5. D. W. Murphy, F. J. DiSalvo, G. W. Hull, Jr., and J. W. Waszczak, *Inorg. Chem.* **15,** 17 (1976).
6. D. W. Murphy and F. A. Trumbore, *J. Cryst. Growth* **39,** 185 (1977).
7. M. S. Whittingham, *J.C.S. Chem. Commun.* p. 328 (1974).
8. F. J. DiSalvo, G. W. Hull, Jr., L. H. Schwartz, J. M. Voorhoeve, and J. V. Waszczak, *J. Chem. Phys.* **59,** 1922 (1973).
9. M. Eibschultz, F. J. DiSalvo, G. W. Hall, and S. Mohajan, *Appl. Phys. Lett.* **27,** 464 (1975).
10. R. H. Friend, A. R. Beal, and A. D. Yoffe, *Philos. Mag.* **35,** 1269 (1977).

11. S. J. Hillenuis, R. V. Coleman, E. R. Domb, and D. J. Sellmyer, *Phys. Rev. B* **19,** 4711 (1979).
12. J. A. Wilson, F. J. DiSalvo, and S. Mahajan, *Adv. Phys.* **24,** 117 (1975).
13. P. M. Williams, *in* "Crystallography and Crystal Chemistry of Materials With Layered Structures" (F. Levy, ed.), Vol. 2, pp. 51–92. Reidel Publ., Dordrecht, Netherlands, 1976.
14. F. J. DiSalvo, *in* "Electron Phonon Interactions and Phase Transitions" (T. Riste, ed.), p. 107. Plenum, New York, 1977.
15. A. H. Thompson, *Comments Solid State Phys.* **7,** 125 (1976).
16. F. J. DiSalvo, *Ferroelectrics* **17,** 361 (1977).
17. W. P. F. A. M. Omloo and F. Jellinek, *J. Less-Common Met.* **20,** 121 (1970).
18. B. VanLaar and D. J. W. Idjo, *J. Solid State Chem.* **3,** 590 (1971).
19. D. W. Murphy and J. N. Carides, *J. Electrochem. Soc.* **126,** 349 (1979).
20. R. Eppinga and G. A. Wiegers, *J. Chem. Phys.* **65,** 37 (1976).
21. R. H. Herber and R. F. Davis, *J. Chem. Phys.* **65,** 3773 (1976).
22. A. C. Gossard, F. J. DiSalvo, and H. Yasuoka, *Phys. Rev. B* **9,** 3965 (1974).
23. D. W. Murphy, F. J. DiSalvo, G. W. Hull, Jr., J. V. Waszczak, S. F. Mayer, G. R. Stewart, S. Early, J. V. Acrivos, and T. H. Geballe, *J. Chem. Phys.* **62,** 967 (1975).
24. D. W. Murphy and G. W. Hall, Jr., *J. Chem. Phys.* **62,** 973 (1975).
25. C. Riekel, H. G. Reznik, R. Schöllhorn, and C. J. Wright, *J. Chem. Phys.* **70,** 5203 (1979).
26. T. Y. Hwang, D. R. Torgeson, and R. G. Barnes, *Phys. Lett. A* **66,** 137 (1978).
27. D. W. Murphy, J. N. Carides, F. J. DiSalvo, C. Cros, and J. V. Waszczak, *Mater. Res. Bull.* **12,** 825 (1977).
28. A. W. Overhauser, *Phys. Rev.* **167,** 691 (1965).
29. A. H. Thompson, *Phys. Rev. Lett.* **34,** 520 (1975).
30. F. J. DiSalvo, D. E. Moncton, and J. V. Waszczak, *Phys. Rev. B* **14,** 4321 (1976).
31. C. F. vanBruggen, J. R. Bloemberge, A. J. A. BosAlerbink, and G. A. Weigers, *J. Less-Common Met.* **60,** 259 (1978).
32. R. R. Chianelli, *J. Cryst. Growth* **34,** 239 (1976).
33. M. Tovar, L. E. DeLong, D. C. Johnston, and M. B. Maple, *Solid State Commun.* **30,** 551 (1979).
34. D. W. Murphy, F. J. DiSalvo, J. N. Carides, and J. V. Waszczak, *Mater. Res. Bull.* **13,** 1395 (1978).
35. D. W. Murphy, P. A. Christian, J. N. Carides, and F. J. DiSalvo, *Proc. Int. Conf. Fast Ion Transp., Lake Geneva, Wis.*, p. 137, 1979.
36. D. W. Murphy and P. A. Christian, *Science* **205,** 651 (1979).
37. J. O. Besenhard and R. Schöllhorn, *J. Electrochem. Soc.* **124,** 968 (1977).
38. P. A. Christian, J. N. Carides, F. J. DiSalvo, and J. V. Waszczak, *J. Electrochem. Soc.* **127,** 2315 (1980).
39. K. H. Cheng and M. S. Whittingham, *J. Solid State Ionics* **1,** 151 (1980).
40. P. A. Christian and D. W. Murphy, personal communication, 1979.
41. D. W. Murphy, P. A. Christian, F. J. DiSalvo, and J. N. Carides, *J. Electrochem. Soc.* **126,** 497 (1979).
42. A. Leblanc-Soreau, M. Danot, L. Trichlet, and J. Rouxel, *Mater. Res. Bull.* **9,** 191 (1974).
43. A. H. Thompson, *Physica B+C (Amsterdam)* **99,** 100 (1980).
44. A. H. Thompson, *Phys. Rev. Lett.* **40,** 1511 (1978).
45. A. H. Thompson, *J. Electrochem. Soc.* **126,** 608 (1979).
46. A. H. Thompson, *Proc. Conf. Fast Ion Conduct. Solids, Lake Geneva, Wis.*, p. 47 (1979).
47. B. G. Silbernagel and M. S. Whittingham, *Mater. Res. Bull.* **11,** 29 (1976).
48. S. A. Safran and D. R. Hamann, *Phys. Rev. Lett.* **42,** 1410 (1979).

49. A. J. Berlinsky W. G. Unruh, W. R. McKinnon, and R. R. Haering, *Solid State Commun.* **31,** 35 (1979).
50. A. H. Thompson, J. S. Scanlon, and C. R. Symon, *J. Solid State Ionics* **1,** 47 (1980).
51. I. Kawada, M. Nakavo-Onoda, M. Ishii, M. Saeki, and M. Nakohira, *J. Solid State Chem.* **15,** 246 (1975).
52. F. W. Boswell, A. Prodan, W. R. Vaughn, and J. M. Corbett, *Phys. Status Solidi* **45,** 469 (1978).
53. R. Moret, M. Huber, and R. Comes, *Phys. Status Solidi* **38,** 695 (1976).
54. F. W. Boswell, A. Prodan, and J. M. Corbett, *Phys. Status Solidi* **35,** 591 (1976).
55. R. deRidder, G. vonTendeloo, J. vanLanduyt, D. vanDyck, and S. Amelinckx, *Phys. Status Solidi* **37,** 591 (1976).
56. G. vonTendeloo, R. deRidder, L. vanGoethen, D. vanDyck, J. vanLanduyt, and S. Amelinckx, *Phys. Status Solidi* **42,** 319 (1977).
57. M. Kaburagi and J. Kanamori, *J. Phys. Soc. Jpn.* **44,** 718 (1978).
58. F. Hulliger and E. Pobitschka, *J. Solid State Chem.* **1,** 117 (1970).
59. A. H. Thompson, F. R. Gamble, and C. R. Symon, *Mater. Res. Bull.* **10,** 915 (1975).
60. F. R. Gamble, F. J. DiSalvo, R. A. Klemm, and T. H. Geballe, *Science* **168,** 568 (1970).
61. F. R. Gamble, J. H. Osiecki, M. Cias, R. Pisharody, F. J. DiSalvo, and T. H. Geballe, *Science* **174,** 493 (1971).
62. F. R. Gamble, J. H. Osiecki, and F. J. DiSalvo, *J. Chem. Phys.* **55,** 3525 (1971).
63. W. E. Lawrence and S. Doniach, *Proc. Int. Conf. Low Temp. Phys., 12th* p. 361 (1971).
64. R. A. Klemm, A. Luther, and M. R. Beasley, *Phys. Rev. B* **12,** 877 (1975).
65. W. A. Little, *Phys. Rev. A* **134,** 1416 (1964).
66. V. L. Ginsburg, *Contemp. Phys.* **9,** 355 (1968).
67. H. M. McConnell, F. R. Gamble, and B. M. Hoffman, *Proc. Natl. Acad. Sci. U.S.A.* **57,** 1131 (1967).
68. W. A. Little, *J. Low Temp. Phys.* **13,** 365 (1973).
69. V. B. Preobrazhensky, Z. J. Zhexnova, A. P. Babichev, I. K. Kobrin, E. A. Lukyanez, and O. L. Kabija, *Solid State Commun.* **27,** 127 (1978).
70. R. Schöllhorn and H. D. Zayefka, *Angew. Chem., Int. Ed. Engl.* **16,** 199 (1977).
71. F. R. Gamble and A. H. Thompson, *Solid State Commun.* **27,** 379 (1978).
72. M. B. Dines, *Science* **188,** 1210 (1975).
73. D. E. Prober, M. R. Beasley, and R. E. Schwall, *Phys. Rev. B* **15,** 5245 (1977).
74. J. W. Johnson, *Physica B+C (Amsterdam)* **99,** 141 (1980).
75. P. J. Bray and E. G. Sauer, *Solid State Commun.* **11,** 1239 (1972).
76. M. B. Dines, *Mater. Res. Bull.* **10,** 287 (1975).
77. J. M. Vandenberg-Voorhoeve, *in* "Optical and Electrical Properties" (P. A. Lee, ed.), p. 423. Reidel Publ., Dordrecht, Netherlands, 1976.
78. N. LeNagard, G. Collin, and O. Gorochov, *Mater. Res. Bull.* **14,** 1411 (1979).
79. F. M. R. Engelsman, G. A. Wiegers, F. Jellinek, and B. van Laar, *J. Solid State Chem.* **6,** 574 (1973).
80. H. Sabrowsky and C. Platte, *Int. Conf. Solid Comp. Trans. Met., Uppsala, Ext. Abstr.* p. 170 (1976).
81. F. Jellinek, *Mater. Res. Bull.* **6,** 169 (1971).
82. Y. Arnaud, M. Chevreton, A. Ahouandjinou, M. Danot, and J. Rouxel, *J. Solid State Chem.* **18,** 9 (1976).
83. T. Hibma, *J. Solid State Chem.* **34,** 97 (1980).
84. R. Osorio and L. M. Falicov, *Phys. Rev. B*, in press.
85. A. H. Thompson, *Physica B+C (Amsterdam)* **105B,** 461 (1981).

Index

MATERIALS SCIENCE AND TECHNOLOGY

A. S. Nowick and B. S. Berry, ANELASTIC RELAXATION IN CRYSTALLINE SOLIDS, 1972

E. A. Nesbitt and J. H. Wernick, RARE EARTH PERMANENT MAGNETS, 1973

W. E. Wallace, RARE EARTH INTERMETALLICS, 1973

J. C. Phillips, BONDS AND BANDS IN SEMICONDUCTORS, 1973

J. H. Richardson and R. V. Peterson (editors), SYSTEMATIC MATERIALS ANALYSIS, VOLUMES I, II, AND III, 1974; IV, 1978

A. J. Freeman and J. B. Darby, Jr. (editors), THE ACTINIDES: ELECTRONIC STRUCTURE AND RELATED PROPERTIES, VOLUMES I AND II, 1974

A. S. Nowick and J. J. Burton (editors), DIFFUSION IN SOLIDS: RECENT DEVELOPMENTS, 1975

J. W. Matthews (editor), EPITAXIAL GROWTH, PARTS A AND B, 1975

J. M. Blakely (editor), SURFACE PHYSICS OF MATERIALS, VOLUMES I AND II, 1975

G. A. Chadwick and D. A. Smith (editors), GRAIN BOUNDARY STRUCTURE AND PROPERTIES, 1975

John W. Hastie, HIGH TEMPERATURE VAPORS: SCIENCE AND TECHNOLOGY, 1975

John K. Tien and George S. Ansell (editors), ALLOY AND MICROSTRUCTURAL DESIGN, 1976

M. T. Sprackling, THE PLASTIC DEFORMATION OF SIMPLE IONIC CRYSTALS, 1976

James J. Burton and Robert L. Garten (editors), ADVANCED MATERIALS IN CATALYSIS, 1977

Gerald Burns, INTRODUCTION TO GROUP THEORY WITH APPLICATIONS, 1977

L. H. Schwartz and J. B. Cohen, DIFFRACTION FROM MATERIALS, 1977

Zenji Nishiyama, MARTENSITIC TRANSFORMATION, 1978

Paul Hagenmuller and W. van Gool (editors), SOLID ELECTROLYTES: GENERAL PRINCIPLES, CHARACTERIZATION, MATERIALS, APPLICATIONS, 1978

G. G. Libowitz and M. S. Whittingham, MATERIALS SCIENCE IN ENERGY TECHNOLOGY, 1978

Otto Buck, John K. Tien, and Harris L. Marcus (editors), ELECTRON AND POSITRON SPECTROSCOPIES IN MATERIALS SCIENCE AND ENGINEERING, 1979

Lawrence L. Kazmerski (editor), POLYCRYSTALLINE AND AMORPHOUS THIN FILMS AND DEVICES, 1980

Manfred von Heimendahl, ELECTRON MICROSCOPY OF MATERIALS: AN INTRODUCTION, 1980

O. Toft Sørensen (editor), NONSTOICHIOMETRIC OXIDES, 1981

M. Stanley Whittingham and Allan J. Jacobson (editors), INTERCALATION CHEMISTRY, 1982

In Preparation

A. Ciferri, W. R. Krigbaum, and R. B. Meyer (editors), POLYMER LIQUID CRYSTALS, 1982